Techniques for Finding the Inverse (If It Exists) of an $n \times n$ Matrix A

- **2×2 case**: The inverse of $\begin{bmatrix} a & b \\ c & d \end{bmatrix}$ exists if and only if $ad - bc \neq 0$. In that case, the inverse is given by $\left(\dfrac{1}{ad - bc} \right) \begin{bmatrix} d & -b \\ -c & a \end{bmatrix}$. (Section 2.4)

- **Row reduction**: Row reduce $[\mathbf{A} \mid \mathbf{I}_n]$ to $[\mathbf{I}_n \mid \mathbf{A}^{-1}]$ (where $\mathbf{A}^{-1}$ does not exist if the process stops prematurely). Advantages: easily computerized; relatively efficient. (Section 2.4)

- **Adjoint matrix**: $\mathbf{A}^{-1} = \left(\frac{1}{|\mathbf{A}|} \right) \mathcal{A}$, where $\mathcal{A}$ is the adjoint matrix of $\mathbf{A}$. Advantage: gives an algebraic formula for $\mathbf{A}^{-1}$. Disadvantage: not very efficient, because $|\mathbf{A}|$ and all n^2 cofactors of $\mathbf{A}$ must be calculated first. (Section 3.3)

Techniques for Finding the Determinant of an $n \times n$ Matrix A

- **2×2 case**: $|\mathbf{A}| = ad - bc$ if $\mathbf{A} = \begin{bmatrix} a & b \\ c & d \end{bmatrix}$. (Sections 2.4 and 3.1)

- **3×3 case**: Basketweaving. (Section 3.1)

- **Row reduction**: Row reduce $\mathbf{A}$ to an upper triangular form matrix $\mathbf{B}$, keeping track of the effect of each row operation on the determinant using a variable P. Then $|\mathbf{A}| = P|\mathbf{B}|$, using the final value of P. Advantages: easily computerized; relatively efficient. (Section 3.2)

- **Cofactor expansion**: Multiply each element along any row or column of $\mathbf{A}$ by its cofactor and sum the results. Advantage: useful for matrices with many zero entries. Disadvantage: not as fast as row reduction. (Sections 3.1 and 3.3)

Also remember: $|\mathbf{A}| = 0$ if $\mathbf{A}$ is row equivalent to a matrix with a row or column of zeroes, or with two identical rows, or with two identical columns.

Techniques for Finding the Eigenvalues of an $n \times n$ Matrix A

- **Characteristic polynomial**: Find the roots of $p_{\mathbf{A}}(x) = |x\mathbf{I}_n - \mathbf{A}|$. We only consider the real roots of $p_{\mathbf{A}}(x)$ in Chapters 1 through 6. Disadvantages: tedious to calculate $p_{\mathbf{A}}(x)$; polynomial becomes more difficult to factor as degree of $p_{\mathbf{A}}(x)$ increases. (Section 3.4)

- **Power Method**: Use for dominant eigenvalues. Choose an initial approximation to a unit eigenvector for the eigenvalue, and multiply $\mathbf{A}$ by this approximation. Normalize the answer and take this result as the next approximation, repeating the process until convergence occurs. The desired eigenvalue is either $\pm$ the length of the final resultant vector. Advantage: iterative method, easy to computerize. Disadvantage: may fail to converge. (Section 9.3)

Technique for Finding a Basis of Eigenvectors for an Eigenvalue λ of an $n \times n$ Matrix A

- **Row reduction**: Row reduce $[(\lambda\mathbf{I}_n - \mathbf{A}) \mid \mathbf{0}]$ to obtain the solution set. To find a basis of eigenvectors for λ, let each independent variable in the solution set in turn equal 1 with all other independent variables equal to 0. (Section 5.5)

Elementary Linear Algebra

Third Edition

Elementary Linear Algebra

Third Edition

Stephen Andrilli
La Salle University

David Hecker
Saint Joseph's University

ELSEVIER
ACADEMIC
PRESS

Amsterdam Boston Heidelberg London New York Oxford Paris
San Diego San Francisco Singapore Sydney Tokyo

Acquisition Editor: Barbara A. Holland
Associate Editor: Tom Singer
Project Manager: Julio Esperas/Troy Lilly
Marketing Manager: Linda Beattie
Marketing Manager: Clare Fleming
Full-Service Provider: Kolam USA
Cover Design: Eric DeCicco
Composition: Integra
Printer: Quebecor Fairfield

Elsevier Academic Press
200 Wheeler Road, Burlington, MA 01803, USA
525 B Street, Suite 1900, San Diego, California 92101-4495, USA
84 Theobald's Road, London WCIX 8RR, UK

This book is printed on acid-free paper. ⊚

British Library Cataloguing-in-Publication Data
A catalogue record for this book is available from the British Library

Library of Congress Cataloging-in-Publication Data
Andrilli, Stephen Francis, 1952-
    Elementary linear algebra/Stephen Andrilli, David Hecker.—3rd ed.
        p. cm.
Includes bibliographical references and index.
    ISBN-13: 978–0–12–058621–9   ISBN-10: 0–12–058621–5 (acid-free paper)
1. Algebras, Linear. I. Hecker, David. II. Title.

QA184.2.A53 2003
512′.5–dc22

                                                              2003021341

For all information on all Academic Press publications
visit our website at www. academicpressbooks.com

ISBN-13: 978-0–12–058621–9
ISBN-10: 0–12–058621–5

Printed in China.
08 09 9 8 7 6 5 4 3 2

To our wives, Ene and Lyn, for all their help and encouragement

Contents

Preface

This textbook is intended for a sophomore- or junior-level introductory course in linear algebra. We assume that the students have had at least one course in calculus.

Philosophy of the Text

In teaching elementary linear algebra, we encountered three major problems.

1. Students had difficulty reading linear algebra textbooks. Frequently, they were too terse, especially where proofs of important results were concerned.
2. Students invariably ran into trouble as the largely computational first half of the course gave way to the more theoretical second half. Students were suddenly asked to work on a much higher level of abstraction and had difficulty with such nontrivial concepts as span, linear independence, one-to-one, onto, etc.
3. Most textbooks contained few, if any, guidelines about reading and writing simple mathematical proofs. However, many instructors have traditionally used a first course in linear algebra as a vehicle for familiarizing students with proof techniques or building upon the introductory material on proofs found in an earlier course in discrete mathematics.

This text addresses these problems by taking particular care, as follows:

Clarity

Above all, we have striven for clarity and used straightforward language throughout the book, occasionally sacrificing brevity for clear and convincing explanations. We strongly encourage students to take advantage of the book's presentation by reading it deeply and thoroughly.

Smooth Transition to Abstraction

To make the transition to the second, more theoretical, half of the course easier, we have students working on proofs as quickly as possible. After a discussion of the basic properties of vectors, there is a special section (Section 1.3) on general proof techniques, with concrete examples using the material on vectors from Sections 1.1 and 1.2. The early placement of Section 1.3 helps to give students a strong foundation and build their confidence in the reading and writing of proofs.

Revisiting Topics

We frequently introduce difficult concepts by revisiting them frequently throughout the text. Students are initially exposed to abstract concepts through concrete examples, and then again in increasingly abstract settings as they progress through the book. Here are several examples:

- Students are introduced to the concept of linear combinations beginning in Section 1.1, long before linear combinations are defined for real vector spaces in Chapter 4.

- The row space of a matrix is introduced in Chapter 2, thereby preparing students for the more general concepts of subspace and span in Sections 4.2 and 4.3.

- The technique behind the first two methods in Section 4.6 for computing bases are introduced earlier in Sections 4.3 and 4.4 in the Simplified Span Method and the Independence Test Method, respectively. In this way, students will become comfortable with these methods in the context of span and linear independence before employing them to find appropriate bases for vector spaces.

- The concepts of eigenvalues and eigenvectors are initially studied in Chapter 3 in the context of matrices. Further properties of eigenvectors are included throughout Chapters 4 and 5 as the underlying vector space concepts are introduced. A full treatment of the subject for linear transformations appears at the end of Chapter 5. The more advanced topics of orthogonal and unitary diagonalization are covered in Chapters 6 and 7.

Major Changes for the Third Edition

Student Solutions Manual

A new supplement has been written that contains full solutions for each exercise in the text bearing a ⋆ (those whose answers appear in the back of the book). It also contains the proofs of most of the theorems whose proofs were left to the exercises. These exercises are marked in the text with a ▶.

Because we have compiled this manual ourselves, it utilizes the same styles of proof-writing and solution techniques that appear in the text itself.

True/False Exercises

True/False exercises have been added at the end of each section in Chapters 1 through 10, as well as in Appendices B and C. These help students to check their understanding of the fundamental concepts introduced in each section. Full explanations of the answers to the true/false exercises appear in the Student Solutions Manual.

Gaussian Elimination

Gaussian elimination is now the first method introduced for solving systems of linear equations (Section 2.1). This is more intuitive and computationally more efficient than row reducing completely to reduced row echelon form. However, we also discuss the Gauss-Jordan method for its theoretical applications (Section 2.2).

Equivalent Linear Systems, Rank, and Row Space

The theoretical concepts of equivalent systems, rank, and row space have been streamlined and placed together in Section 2.3.

Linear Independence

A more straightforward definition and exposition of the concepts of linear independence/dependence is presented at the beginning of Section 4.4.

Complex Linear Algebra

For those who want to include a treatment of complex matrices, eigenvectors, and vector spaces, the corresponding material on these topics (formerly in Section 7.1) has been rearranged and segmented into four shorter sections, Sections 7.1 through 7.4. This allows complex linear algebra topics to be covered more easily in parallel with their corresponding real linear algebra topics. For example, Section 7.1 on complex n-vectors and matrices can be covered directly after completing Chapter 1, thus allowing the integration of these complex structures throughout the course. Formal prerequisites for each section of Chapter 7 are given in a chart following this Preface.

New Applications

Two new application sections are included in this edition: Section 8.6, "Change of Variables and the Jacobian," and Section 8.11, "Max-Min Problems in $\mathbb{R}^n$ and the Hessian Matrix." Also, the application "Computer Graphics" (Section 8.8) has been extensively revised. The material on "Function Spaces" and "Quadratic Forms" has been moved to Sections 10.2 and 10.3, respectively.

Elementary Matrices

By popular demand, the material on elementary matrices from the first edition has been restored in Section 10.1.

Revised Computational Methods

As mentioned earlier, two computational methods that are useful for finding bases, which were formerly in Section 4.6, have been moved earlier in the text to Sections 4.3 and 4.4, respectively, and given new names (Simplified Span Method and Independence Test Method). Also, the methods for diagonalizing and orthogonally diagonalizing a linear operator in Sections 5.5 and 6.3 have been streamlined to make their connection to the Diagonalization Method of Section 3.4 more apparent.

Computational Aid

The material in Appendix D ("Computers and Calculators") has been updated to reflect the current state of various software packages.

Help on the Web

Our web site, http://www.sju.edu/~dhecker/linalg.html, contains appropriate updates on the textbook as well as a way to communicate with the authors. It also contains information on earlier versions of some of the software packages mentioned in Appendix D, in case you are working with an older version. We also expect the web site to have updates for future versions of the software as they are released.

Features

Numerous Examples and Exercises

There are more than 310 numbered examples in the text, at least one for each new concept or application, to ensure that students fully understand the material before proceeding. Almost every theorem has a corresponding example to illustrate its meaning and/or usefulness.

The text also contains an unusually large number of exercises. There are more than 830 numbered exercises, and many of these have multiple parts, for a total of more than 2135 questions. Some are purely computational. Many others ask the students to write short proofs or explore further consequences of the material. The exercises within each section are generally ordered by increasing difficulty, beginning with basic computational problems and moving on to more theoretical problems and proofs. Answers are provided at the end of the book for approximately half the computational exercises; these problems are marked with a star (⋆). Full solutions to the ⋆ exercises appear in the new Student Solutions Manual.

Careful Coverage of Vector Space Topics

Many students have difficulties when abstract vector space topics are introduced. These concepts represent a sharp transition from concrete problem solving to theoretical conceptualizing. The material in Sections 4.1 through 5.4 (vector spaces and subspaces, span, linear independence, basis and dimension, coordinatization, linear transformations and their matrices, kernel and range, and isomorphism) constitutes the "heart" of this linear algebra text, and we have taken great care to help students focus attention on these important concepts. As previously mentioned, we revisit several difficult topics in new settings in order to gently introduce them to students.

Early Introduction of Eigenvalues/Eigenvectors

Eigenvalues and eigenvectors are introduced early in the text (Section 3.4), just before the introduction of abstract vector spaces. Students traditionally find eigenvalues to be a difficult topic. By introducing eigenvalues and eigenvectors early on, we have the opportunity to present these concepts initially on an elementary level and then reinforce them throughout the course with appropriate examples in subsequent sections. This approach enables students to gain confidence with eigenvalues and eigenvectors before encountering a more thorough, detailed treatment in Section 5.5.

Emphasis on Proofs

We have written the proofs of the theorems in the text in a careful, concise manner to give students a model for writing their own proofs. We have also included a special section early in the text (Section 1.3), that reviews the most important proof techniques in the context of linear algebra.

We have left the proofs of some elementary theorems to the student. However, for almost every *nontrivial* theorem in Chapters 1 through 6, we have either included a proof or given detailed hints that should be sufficient to enable students to provide a proof on their own.[1] Most of the proofs that are left as exercises can be found in the Student Solutions Manual. The exercises corresponding to these proofs are marked with the symbol ▶.

We have avoided "clever" or "sneaky" proofs, in which the last line suddenly produces "a rabbit out of a hat" because such proofs invariably frustrate students. They are given no insight into the strategy of the proof or how the deductive process was used. In fact, such proofs tend to reinforce the students'

---

[1] The only exception is Theorem 2.4 (uniqueness of reduced row echelon form).

mistaken belief that they will never become competent in the art of writing proofs.

In this text, proofs longer than one paragraph are often written in a "top-down" manner, a concept borrowed from structured programming. A complex theorem is broken down into a secondary series of results, which together are sufficient to prove the original theorem. In this way, the student has a clear outline of the logical argument and can more easily reproduce the proof if called on to do so.

Applications

Linear algebra is a subject with a multitude of practical applications, and we have included many standard ones so that instructors can choose their favorites. There is a chart following the Preface that lists the most important linear algebra applications in this text. Chapter 8 is devoted entirely to applications of linear algebra, but there are also several shorter applications in Chapters 1 to 6. Another chart following the Preface lists the prerequisites required for each of the application sections in Chapter 8. Instructors may choose to assign some of these applications as reading assignments outside of class.

Help with Technology

Almost all students now have access to appropriate computer software or graphing calculators that reduce the amount of computational drudgery involved in a typical elementary linear algebra course. We believe that once a student has mastered the concepts of matrix multiplication and row reduction, there is no need to waste precious time in or out of class with rote computations. This frees both the instructor and the student to concentrate on the theoretical ideas of the subject without becoming unduly bogged down with calculations.

While the exposition of this text does not depend on the use of any particular technology, Appendix D provides a short introduction to the following prominent computer packages: *Maple 8, Derive 5, Mathematica 4.2, MATLAB 6.5*, the following graphing calculators from Texas Instruments: the *TI-86, TI-89, TI-92, TI-92 Plus*, and *Voyage 200*. While this appendix does not give an in-depth treatment of these packages and calculators, it does illustrate how to perform several fundamental types of vector and matrix computations in each environment. Our web site, http://www.sju.edu/~dhecker/linalg.html, will contain updates for new versions of these software packages, as well as details on other calculators and older versions of the software packages.

Formal Computational Methods

There are 17 computational methods presented in step-by-step form, each illustrating a fundamental process in linear algebra. These have been placed in boxes for easier reference by the students. Several additional numerical methods are presented, especially in Sections 9.1 and 9.2. A chart listing all of these methods appears after this Preface.

Subsections, Summary Charts, and Symbol Table

Almost every section of the text is divided into several manageable subsections to enhance clarity and readability. These subsections are individually titled to highlight the main themes of the section. Condensed versions of some useful charts are printed on the inside front and back covers for easy reference. Finally, for convenience, there is a comprehensive Symbol Table listing all of the major symbols employed in this text related to linear algebra together with their meanings.

Supplements

We have written a **Student Solutions Manual** that contains the full worked-out solution for each exercise marked with a ⋆ in the textbook (whose answer also appears at the back of the book). This manual also contains proofs for most of the theorems whose proofs were left as exercises. The exercises corresponding to these theorems are marked with the symbol ▶. There is also an **Instructor's Manual** that contains the answers to all computational exercises and complete solutions to the theoretical and proof exercises. This manual also includes three versions of a sample test for each of Chapters 1 through 7. These can be used without change or as a test question bank for making tests. Answer keys for the sample tests are also included.

Additional information, as well as appropriate updates that become available after the book is printed will be posted on our web site,

$$\text{http://www.sju.edu/˜dhecker/linalg.html.}$$

Chapter-by-Chapter Summary

The first six chapters constitute the fundamental material covered in most elementary linear algebra courses.

- **Chapter 1 (Vectors and Matrices)** introduces vectors and matrices and their fundamental operations and properties. This chapter includes a special section (Section 1.3) on proof techniques, illustrating some of the most important methods of proof and pointing out some of the pitfalls.

- **Chapter 2 (Systems of Linear Equations)** begins with the solution of systems of linear equations using the Gaussian elimination and Gauss-Jordan row reduction methods. This is followed by a discussion of the uniqueness of reduced row echelon form, equivalent systems, rank, row space, and inverses of matrices.

- **Chapter 3 (Determinants and Eigenvalues)** introduces the determinant (using a cofactor approach) and shows its usefulness in working with systems of linear equations. The chapter ends with an introductory treatment of eigenvalues and eigenvectors for matrices.

- **Chapter 4 (Finite Dimensional Vector Spaces)** begins a treatment of the abstract concepts of vector spaces and subspaces. Span, linear independence, basis and dimension, and coordinatization are covered. Several useful methods for finding bases are illustrated.

- **Chapter 5 (Linear Transformations)** introduces linear transformations. The matrix, kernel, and range of a linear transformation are covered. One-to-one and onto linear transformations are treated in depth, and an isomorphism of any n-dimensional real vector space with $\mathbb{R}^n$ is established. The chapter ends with a more formal treatment of the concepts of eigenvalues and diagonalization in the context of linear transformations.

- **Chapter 6 (Orthogonality)** begins with a study of orthogonal and orthonormal bases, and the Gram-Schmidt Process. Orthogonal matrices, orthogonal complements, and orthogonal projections are treated. The chapter ends with orthogonal diagonalization, a fitting culmination of the material in the first six chapters.

The remaining four chapters contain additional material. The sections in these chapters can be covered at any time after their stated prerequisites have been met.

- **Chapter 7 (Complex Vector Spaces and General Inner Products)** generalizes the material of earlier chapters to complex vector spaces and general inner product spaces. The various sections of Chapter 7 are written so that they may be covered in tandem with the corresponding sections on real vectors and matrices. For this reason, there is no need to wait until the end of the course to discuss these "complex" topics.

- **Chapter 8 (Additional Applications)** is devoted to applications of linear algebra, including elementary graph theory, Ohm's Law, least-squares polynomials, Markov chains, Hill substitution, change of variables and the Jacobian matrix in two and three dimensions, rotation of axes, computer graphics, differential equations, least-squares solutions for inconsistent systems, and max-min problems in $\mathbb{R}^n$ involving the Hessian matrix.

- **Chapter 9 (Numerical Methods)** discusses important considerations when using a computer or calculator to perform computations in linear algebra. Numerical methods such as partial pivoting, the Jacobi and Gauss-Seidel iterative methods, **LDU** decomposition, and the Power Method for calculating dominant eigenvalues are covered.

- **Chapter 10 (Further Horizons)** covers the following three supplemental topics: elementary matrices, function spaces, and quadratic forms.

There are five appendices, the fifth of which is **Answers to Selected Exercises**.

- **Appendix A (Miscellaneous Proofs)** contains proofs of five results that were omitted in the main part of the text because of length or complexity.

- **Appendix B (Functions)** includes a review of basic function terminology and properties, as well as a treatment of one-to-one, onto, inverse, and composite functions.

- **Appendix C (Complex Numbers)** contains a review of the basic properties of complex numbers.

- **Appendix D (Computers and Calculators)** includes a brief introduction to the use of several software packages and graphing calculators in performing basic vector and matrix operations.

Guide for the Instructor

Chapters 1 through 6 have been written in a sequential fashion. Each section is generally needed as a prerequisite for what follows. Therefore, we recommend that these sections be covered in order. However, there are three exceptions.

- **Section 1.3 (An Introduction to Proofs)** can be covered, in whole, or in part, at any time after Section 1.2.

- **Section 3.3 (Further Properties of the Determinant)** contains some material that can be omitted without affecting most of the remaining development. The topics of general cofactor expansion, (classical) adjoint matrix, and Cramer's Rule are used very sparingly in the rest of the text.

- **Section 6.1 (Orthogonal Bases and the Gram-Schmidt Process)** can be covered any time after Chapter 4, as can much of the material in **Section 6.2 (Orthogonal Complements)**.

Prerequisites for the material in Chapters 7 through 10 are listed in a chart following this Preface. Each section of Chapter 7 needs its stated prerequisite as well as all earlier sections of Chapter 7.[2] However, the sections of Chapters 8 through 10 are completely independent of each other, and any of these sections can be covered after its prerequisite has been met.

---

[2] *Most* of Section 7.5 can be covered without having covered Sections 7.1 through 7.4 by concentrating only on real inner products.

Two suggested timetables for covering the material in this text are presented below — one for a 3-credit course, and the other for a 4-credit course. While all the material of Chapters 1 through 6, and some of Chapter 7, would be covered in the 4-credit course, the 3-credit course could deemphasize portions of Sections 1.3, 2.3, 3.3, 5.5, 6.2, and 6.3, and would not include Chapter 7.

	3-Credit Course	4-Credit Course
Chapter 1	5 classes	6 classes
Chapter 2	4 classes	5 classes
Chapter 3	3 classes	6 classes
Chapter 4	12 classes	12 classes
Chapter 5	8 classes	9 classes
Chapter 6	2 classes	5 classes
Chapter 7		3 classes
Chapters 8/9/10 (selections)	2 classes	4 classes
Review	3 classes	3 classes
Tests	3 classes	3 classes
Total	42 classes	56 classes

Acknowledgments

We gratefully thank all those who have helped in the publication of this book. We especially thank Barbara Holland, our senior editor at Elsevier/Academic Press, Tom Singer, our editorial coordinator, Christine Brandt, our project manager and copyeditor, and Julio Esperas, our production designer.

We also want to thank those who have supported our textbook at various stages. In particular, we thank Agnes Rash, chair of the Mathematics and Computer Science Department at Saint Joseph's University, for her continual support of this project. We also thank Paul Klingsberg and Richard Cavaliere of Saint Joseph's University, both of whom gave us many suggestions for improvements to the second edition.

We thank those students who have classroom-tested versions of the earlier editions of the manuscript. Their comments and suggestions have been extremely useful and have guided us in shaping the text in many ways.

We acknowledge those reviewers who have supplied many worthwhile suggestions. For reviewing the first edition, we thank the following people:

C. S. Ballantine, Oregon State University
Yuh-ching Chen, Fordham University
Susan Jane Colley, Oberlin College
Roland di Franco, University of the Pacific
Colin Graham, Northwestern University
K. G. Jinadasa, Illinois State University
Ralph Kelsey, Denison University
Masood Otarod, University of Scranton
J. Bryan Sperry, Pittsburg State University
Robert Tyler, Susquehanna University

For reviewing the second edition, we thank the following people:

Ruth Favro, Lawrence Technological University
Howard Hamilton, California State University
Ray Heitmann, University of Texas, Austin
Richard Hodel, Duke University
James Hurley, University of Connecticut
Jack Lawlor, University of Vermont
Peter Nylen, Auburn University
Ed Shea, California State University, Sacramento

For reviewing the third edtion, we thank the following people:

John Lawlor, University of Vermont
Susan Jane Colley, Oberlin College
Joel Robbin, University of Wisconsin
Ian Morrison, Fordham University
Ali Miri, University of Ottawa
Vania Mascioni, Ball State University
Sergei Bezrukov, University of Wisconsin Superior
Don Passman, University of Wisconsin

Last, but most important of all, we want to thank our wives, Ene and Lyn, for bearing extra hardships so that we could work on this text. Their love and support has been an inspiration. We also thank Ene, who conveniently works at Saint Joseph's University, for ferrying various revisions and files of the manuscript between us.

Coming to Terms with Linear Algebra

As students vector through the space of this text from its initial point to its terminal point, we hope that on a one-to-one basis, they will undergo a real transformation from the norm. Their induction into the domain of linear algebra should be sufficient to produce a pivotal change in their abilities.

One characteristic that we expect students to manifest is a greater linear independence in problem-solving. After much reflection on the kernel of ideas presented in this book, the range of new methods available to them should be graphically augmented in a multiplicity of ways. An associative feature of this transition is that all of the new techniques they learn should become a consistent and normalized part of their identity in the future. In addition, students will gain a singular new appreciation of their mathematical skills. Consequently, the resultant change in their self-image should be one of no minor magnitude.

One obvious implication is that the level of the students' success is an isomorphic reflection of the amount of homogeneous energy they expend on this complex material. That is, we can often trace the rank of their achievement to the depth of their resolve to be a scalar of new distances. Similarly, we make this symmetric claim — the students' positive, definite growth is clearly a function of their overall coordinatization of effort. Naturally, the matrix of thought behind this parallel assertion is that students should avoid the negative consequences of sparse learning. Instead, it is the inverse approach of systematic and iterative study that will ultimately lead them to less error and not rotate them into useless dead-ends and diagonal tangents of zero worth.

Of course some nontrivial length of time is necessary to transpose a student with an empty set of knowledge on this subject into higher echelons of understanding. But our projection is that the unique dimensions of this text will be a determinant factor in enriching the span of students' lives, and translate them onto new orthogonal paths of wisdom.

Stephen Andrilli
David Hecker
August, 2003

Prerequisite Chart for Chapters 7 through 10

Section	Prerequisite
Section 7.1 (Complex n-Vectors and Matrices)	Section 1.5 (Matrix Multiplication)
Section 7.2 (Complex Eigenvalues and Complex Eigenvectors)[1]	Section 3.4 (Eigenvalues and Diagonalization)
Section 7.3 (Complex Vector Spaces)[1]	Section 5.2 (The Matrix of a Linear Transformation)
Section 7.4 (Orthogonality in $\mathbb{C}^n$)[1]	Section 6.3 (Orthogonal Diagonalization)
Section 7.5 (Inner Product Spaces)[1]	Section 6.3 (Orthogonal Diagonalization)
Section 8.1 (Graph Theory)	Section 1.5 (Matrix Multiplication)
Section 8.2 (Ohm's Law)	Section 2.2 (Gauss-Jordan Row Reduction and Reduced Row Echelon Form)
Section 8.3 (Least-Squares Polynomials)	Section 2.2 (Gauss-Jordan Row Reduction and Reduced Row Echelon Form)
Section 8.4 (Markov Chains)	Section 2.2 (Gauss-Jordan Row Reduction and Reduced Row Echelon Form)
Section 8.5 (Hill Substitution: An Introduction to Coding Theory)	Section 2.4 (Inverses of Matrices)
Section 8.6 (Change of Variables and the Jacobian)[2]	Section 3.1 (Introduction to Determinants)
Section 8.7 (Rotation of Axes)	Section 4.7 (Coordinatization)
Section 8.8 (Computer Graphics)	Section 5.2 (The Matrix of a Linear Transformation)
Section 8.9 (Differential Equations)[3]	Section 5.5 (Diagonalization of Linear Operators)
Section 8.10 (Least-Squares Solutions for Inconsistent Systems)	Section 6.2 (Orthogonal Complements)
Section 8.11 (Max-Min Problems in $\mathbb{R}^3$ and the Hessian Matrix)	Section 6.3 (Orthogonal Diagonalization)

Section	Prerequisite
Section 9.1 (Numerical Methods for Solving Systems)	Section 2.3 (Equivalent Systems, Rank, and Row Space)
Section 9.2 (**LDU** Decomposition)	Section 2.4 (Inverses of Matrices)
Section 9.3 (The Power Method for Finding Eigenvalues)[4]	Section 5.5 (Diagonalization of Linear Operators)
Section 10.1 (Elementary Matrices)	Section 2.4 (Inverses of Matrices)
Section 10.2 (Function Spaces)[5]	Section 4.7 (Coordinatization)
Section 10.3 (Quadratic Forms)	Section 6.3 (Orthogonal Diagonalization)

[1] In addition to the prerequisites listed, each section in Chapter 7 requires the sections of Chapter 7 that precede it, although *most* of Section 7.5 can be covered without having covered Sections 7.1 through 7.4.

[2] Section 8.6 uses the fact that $|\mathbf{A}| = |\mathbf{A}^T|$ from Section 3.3, but we believe that it is more appropriate to cover this section directly after Section 3.1 to provide a deeper geometric understanding of the determinant.

[3] The techniques presented for solving differential equations in Section 8.9 require only Section 3.4 as a prerequisite. However, terminology from Chapters 4 and 5 is used throughout Section 8.9.

[4] The Power Method in Section 9.3 requires only material from Section 3.4 for its implementation. However, topics from Chapters 4 and 5 are needed for the justification of the Power Method and are used throughout Section 9.3.

[5] The material in Section 10.2 requires only a knowledge of Section 4.4 (Linear Independence). However, several exercises in Section 10.2 involve material from Sections 4.5, 4.6, and 4.7.

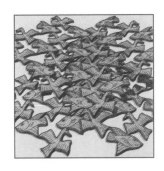

Applications

The following chart gives a list of the major applications of linear algebra presented throughout the text (see the Prerequisite Chart for the applications in Chapters 8 and 10):

Application	Section
Resultant Velocity	Section 1.1
Newton's Second Law	Section 1.1
Work	Section 1.2
Shipping Cost and Profit	Section 1.5
Curve Fitting	Section 2.1
Balancing Chemical Equations	Section 2.2
Areas and Volumes	Section 3.1
Large Powers of a Matrix	Section 3.4
Orthogonal Projections and Reflections in $\mathbb{R}^n$	Section 6.2
Distance from a Point to a Subspace	Section 6.2
Graph Theory	Section 8.1
Ohm's Law	Section 8.2
Least-Squares Polynomials	Section 8.3
Markov Chains	Section 8.4
Hill Substitution (Coding Theory)	Section 8.5
Change of Variables and the Jacobian	Section 8.6
Rotation of Axes	Section 8.7
Computer Graphics	Section 8.8
Differential Equations	Section 8.9
Least-Squares Solutions for Inconsistent Systems	Section 8.10
Max-Min Problems in $\mathbb{R}^n$ and the Hessian Matrix	Section 8.11
Quadratic Forms	Section 10.3

Formal Methods

The following is a list of the formal computational methods presented throughout the text:

Section	Formal Method
Section 2.4	Inverse Method (finding the inverse of a matrix)
Section 3.4	Diagonalization Method (diagonalizing a square matrix)
Section 4.3	Simplified Span Method (determining span using row reduction)
Section 4.4	Independence Test Method (determining linear independence using row reduction)
Section 4.6	Inspection Method (finding a basis by inspection)
Section 4.6	Enlarging Method (enlarging a linearly independent set to a basis)
Section 4.7	Coordinatization Method (coordinatizing a vector with respect to an ordered basis)
Section 4.7	Transition Matrix Method (calculating a transition matrix using row reduction)
Section 5.3	Kernel Method (finding a basis for a kernel of a linear transformation)
Section 5.3	Range Method (finding a basis for the range of a linear transformation)
Section 5.5	Generalized Diagonalization Method (diagonalizing a linear operator)
Section 6.1	Gram-Schmidt Process (creating an orthogonal set from a linearly independent set)
Section 6.3	Orthogonal Diagonalization Method (orthogonally diagonalizing a symmetric operator)
Section 7.2	Generalized Gram-Schmidt Process (creating an orthogonal set in an inner product space)
Section 8.8	Similarity Method (in computer graphics, finding a matrix for a transformation centered at a point other than the origin)
Section 9.3	Power Method (finding the dominant eigenvalue of a square matrix)
Section 10.3	Quadratic Form Method (diagonalizing a quadratic form)

Numerical Methods

The following 15 methods in numerical linear algebra are discussed throughout the text:

Section	Numerical Method
Section 2.1	Gaussian elimination
Section 2.1	Back substitution
Section 2.2	Gauss-Jordan row reduction
Section 2.4	Solving a system using the inverse of the coefficient matrix
Section 3.1	Basketweaving
Section 3.2	Finding a determinant by row reduction
Section 3.3	Cofactor expansion (general)
Section 3.3	Cramer's Rule
Section 8.3	Linear regression (line of best fit)
Section 8.10	Approximate solutions for inconsistent systems (least-squares)
Section 9.1	Partial pivoting
Section 9.1	Jacobi method
Section 9.1	Gauss-Seidel method
Section 9.2	**LDU** decomposition
Section 9.3	Power method for finding eigenvalues

Symbol Table

$\oplus$	addition on a vector space (unusual)
$\mathcal{A}$	adjoint (classical) of a matrix $\mathbf{A}$
I	ampere (unit of current)
$\approx$	approximately equal to
$[\mathbf{A} \mid \mathbf{B}]$	augmented matrix formed from matrices $\mathbf{A}$ and $\mathbf{B}$
$\mathbf{p_A}(x)$	characteristic polynomial of a matrix $\mathbf{A}$
$\mathbf{p}_L(x)$	characteristic polynomial of a linear operator L
$\mathcal{A}_{ij}$	cofactor, (i, j), of a matrix $\mathbf{A}$
$\overline{z}$	complex conjugate of a complex number z
$\overline{\mathbf{z}}$	complex conjugate of $\mathbf{z} \in \mathbb{C}^n$
$\overline{\mathbf{Z}}$	complex conjugate of $\mathbf{Z} \in \mathcal{M}_{mn}^{\mathbb{C}}$
$\mathbb{C}$	complex numbers, set of
$\mathbb{C}^n$	complex n-vectors, set of (ordered n-tuples of complex numbers)
$g \circ f$	composition of functions f and g
$L_2 \circ L_1$	composition of linear transformations L_1 and L_2
$\mathbf{Z}^*$	conjugate transpose of $\mathbf{Z} \in \mathcal{M}_{mn}^{\mathbb{C}}$
$C^0(\mathbb{R})$	continuous real-valued functions with domain $\mathbb{R}$, set of
$C^1(\mathbb{R})$	continuously differentiable functions with domain $\mathbb{R}$, set of
$[\mathbf{w}]_B$	coordinatization of a vector $\mathbf{w}$ with respect to a basis B
$\mathbf{x} \times \mathbf{y}$	cross product of vectors $\mathbf{x}$ and $\mathbf{y}$
$f^{(n)}$	derivative, nth, of a function f
$|\mathbf{A}|$	determinant of a matrix $\mathbf{A}$
δ	determinant of a 2×2 matrix, $ad - bc$
$\mathcal{D}_n$	diagonal $n \times n$ matrices, set of
$\dim(\mathcal{V})$	dimension of a vector space $\mathcal{V}$
$\mathbf{x} \cdot \mathbf{y}$	dot product of vectors $\mathbf{x}$ and $\mathbf{y}$; complex dot product of $\mathbf{x}$ and $\mathbf{y}$
λ	eigenvalue of a matrix
E_λ	eigenspace corresponding to eigenvalue λ

$\{\,\}$	empty set
a_{ij}	entry, (i, j), of a matrix $\mathbf{A}$
$f\colon X \to Y$	function f from a set X (domain) to a set Y (codomain)
∇f	gradient of f
$\mathbf{H}$	Hessian matrix
$\mathbf{I}, \mathbf{I}_n$	identity matrix; $n \times n$ identity matrix
$\Longleftrightarrow$, iff	if and only if
$f(x)$	image of an element x under a function f
$f(S)$	image of a set S under a function f
i	imaginary number whose square $= -1$
$\Rightarrow$	implies; if . . . then
$\langle \mathbf{x}, \mathbf{y} \rangle$	inner product of $\mathbf{x}$ and $\mathbf{y}$
$\mathbb{Z}$	integers, set of
f^{-1}	inverse of a function f
$\mathbf{A}^{-1}$	inverse of a matrix $\mathbf{A}$
L^{-1}	inverse of a linear transformation L
$\cong$	isomorphic
$\mathbf{J}; |\mathbf{J}|$	Jacobian matrix; Jacobian determinant
$\ker(L)$	kernel of a linear transformation L
δ_{ij}	Kronecker delta
$\|\mathbf{a}\|$	length, or norm, of a vector $\mathbf{a}$
$\mathbf{M}_f$	limit matrix of a Markov chain
$\mathbf{p}_f$	limit vector of a Markov chain
$\mathcal{L}_n$	lower triangular $n \times n$ matrices, set of
$|z|$	magnitude (absolute value) of a complex number z
$\mathbf{A}|_{\mathbf{x}_0}$	matrix $\mathbf{A}$ evaluated at point (vector) $\mathbf{x}_0$
$\mathbf{A}_{BC}$	matrix for a linear transformation with respect to ordered bases B and C
$\mathcal{M}_{mn}$	matrices of size $m \times n$, set of
$\mathcal{M}_{mn}^{\mathbb{C}}$	matrices of size $m \times n$ with complex entries, set of
$|\mathbf{A}_{ij}|$	minor, (i, j), of a matrix $\mathbf{A}$
$\mathbb{N}$	natural numbers, set of
not A	negation of statement A
$|S|$	number of elements in a set S
Ω	ohm (unit of resistance)
$(\mathbf{v}_1, \mathbf{v}_2, \ldots, \mathbf{v}_n)$	ordered basis containing vectors $\mathbf{v}_1, \mathbf{v}_2, \ldots, \mathbf{v}_n$
$\mathcal{W}^{\perp}$	orthogonal complement of a subspace $\mathcal{W}$
$\perp$	perpendicular to
$\mathcal{P}$	polynomials, set of all
$\mathcal{P}_n$	polynomials of degree $\leq n$, set of
$\mathcal{P}_n^{\mathbb{C}}$	polynomials of degree $\leq n$ with complex coefficients, set of
$\mathbb{R}^{+}$	positive real numbers, set of
$\mathbf{A}^k$	power, kth, of a matrix $\mathbf{A}$
$f^{-1}(x)$	pre-image of an element x under a function f
$f^{-1}(S)$	pre-image of a set S under a function f

$\mathbf{proj_a b}$	projection of $\mathbf{b}$ onto $\mathbf{a}$
$\mathbf{proj}_{\mathcal{W}} \mathbf{v}$	projection of $\mathbf{v}$ onto a subspace $\mathcal{W}$
range(L)	range of a linear transformation L
rank($\mathbf{A}$)	rank of a matrix $\mathbf{A}$
$\mathbb{R}$	real numbers, set of
$\mathbb{R}^n$	real n-vectors, set of (ordered n-tuples of real numbers)
$\langle i \rangle$	row, ith, of a matrix
$\langle i \rangle \leftarrow c \langle i \rangle$	row operation of type (I)
$\langle i \rangle \leftarrow c \langle j \rangle + \langle i \rangle$	row operation of type (II)
$\langle i \rangle \leftrightarrow \langle j \rangle$	row operation of type (III)
$R(\mathbf{A})$	row operation R applied to matrix $\mathbf{A}$
$\odot$	scalar multiplication on a vector space (unusual)
$m \times n$	size of a matrix with m rows and n columns
span(S)	span of a set S
$\boldsymbol{\Psi}_{ij}$	standard basis vector (matrix) in $\mathcal{M}_{nn}$
$\mathbf{i}, \mathbf{j}, \mathbf{k}$	standard basis vectors in $\mathbb{R}^3$
$\mathbf{e}_1, \mathbf{e}_2, \ldots, \mathbf{e}_n$	standard basis vectors in $\mathbb{R}^n$; standard basis vectors in $\mathbb{C}^n$
$\mathbf{p}_n$	state vector, nth, of a Markov chain
$\mathbf{A}_{ij}$	submatrix, (i, j), of a matrix $\mathbf{A}$
$\sum$	sum of
trace($\mathbf{A}$)	trace of a matrix $\mathbf{A}$
$\mathbf{A}^T$	transpose of a matrix $\mathbf{A}$
$C^2(\mathbb{R})$	twice continuously differentiable functions with domain $\mathbb{R}$, set of
$\mathcal{U}_n$	upper triangular $n \times n$ matrices, set of
$\mathbf{V}_n$	Vandermonde $n \times n$ matrix
V	volt (unit of voltage)
$\mathbf{O}; \mathbf{O}_n; \mathbf{O}_{mn}$	zero matrix; $n \times n$ zero matrix; $m \times n$ zero matrix
$\mathbf{0}$	zero vector in a vector space

Chapter 1

Vectors and Matrices

Proof Positive

The concept of proof is central to higher mathematics. Mathematicians claim no statement as a "fact" until it is proven true using logical deduction. Therefore, no one can succeed in higher mathematics without mastering the techniques required to supply such a proof.

Linear algebra, in addition to having a multitude of practical applications in science and engineering, also can be used to introduce proof-writing skills. Section 1.3 gives an introductory overview of the basic proof-writing tools that a mathematician uses on a daily basis. Other proofs given throughout the text should be taken as models for constructing proofs of your own when completing the exercises. With these tools and models, you can begin to develop the proof-writing skills crucial to your future success in mathematics.

Our study of linear algebra begins with vectors and matrices—two of the most practical concepts in mathematics. You are probably already familiar with the use of vectors to describe positions, movements, and forces. And as we will see later, matrices are the key to representing motions that are "linear" in nature, such as the rigid motion of an object in space or the movement of an image on a computer screen.

"YOU WANT PROOF? I'LL GIVE YOU PROOF!"

Ⅰn linear algebra, the most fundamental object is the vector. We define vectors in Sections 1.1 and 1.2 and describe their algebraic and geometric properties. The link between algebraic manipulation and geometric intuition is a recurring theme in linear algebra, which we use to establish many important results.

In Section 1.3, we examine techniques that are useful for reading and writing proofs. In Sections 1.4 and 1.5, we introduce the matrix, another fundamental object, whose basic properties parallel those of the vector. However, we will eventually find many differences between the more advanced properties of vectors and matrices, especially regarding matrix multiplication.

1.1 Fundamental Operations with Vectors

In this section, we introduce vectors and consider the following two operations on vectors: scalar multiplication and addition. Let $\mathbb{R}$ denote the set of all **real numbers** (that is, all coordinate values on the real number line).

Definition of a Vector

DEFINITION

A **real n-vector** is an ordered sequence of n real numbers (sometimes referred to as an **ordered n-tuple** of real numbers). The set of all n-vectors is denoted $\mathbb{R}^n$.

For example, $\mathbb{R}^2$ is the set of all 2-vectors (ordered 2-tuples = ordered pairs) of real numbers; it includes $[2, -4]$ and $[-6.2, 3.14]$. $\mathbb{R}^3$ is the set of all 3-vectors (ordered 3-tuples = ordered triples) of real numbers; it includes $[2, -3, 0]$ and $[-\sqrt{2}, 42.7, \pi]$.[1]

The vector in $\mathbb{R}^n$ that has all n entries equal to zero is called the **zero n-vector**. In $\mathbb{R}^2$ and $\mathbb{R}^3$, the zero vectors are $[0, 0]$ and $[0, 0, 0]$, respectively.

Two vectors in $\mathbb{R}^n$ are **equal** if and only if all corresponding entries (called **coordinates**) in their n-tuples agree. That is, $[x_1, x_2, \ldots, x_n] = [y_1, y_2, \ldots, y_n]$ if and only if $x_1 = y_1, x_2 = y_2, \ldots,$ and $x_n = y_n$.

A single number (such as -10 or 2.6) is often called a **scalar** to distinguish it from a vector.

Geometric Interpretation of Vectors

Vectors in $\mathbb{R}^2$ frequently represent movement from one point to another in a coordinate plane. From initial point $(3, 2)$ to terminal point $(1, 5)$, there is a

---
[1] Many texts distinguish between *row* vectors, such as $[2, -3]$, and *column* vectors, such as $\begin{bmatrix} 2 \\ -3 \end{bmatrix}$. However, in this text, we express vectors as row or column vectors as the situation warrants.

net decrease of 2 units along the x-axis and a net increase of 3 units along the y-axis. A vector representing this change would thus be $[-2, 3]$, as indicated by the arrow in Figure 1.1.

Figure 1.1

Movement represented
by the vector

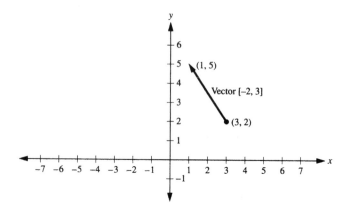

Vectors can be positioned at any desired starting point. For example, $[-2, 3]$ could also represent a movement from initial point $(9, -6)$ to terminal point $(7, -3)$.[2]

Vectors in $\mathbb{R}^3$ have a similar geometric interpretation: a 3-vector is used to represent movement between points in three-dimensional space. For example, $[2, -2, 6]$ can represent movement from initial point $(2, 3, -1)$ to terminal point $(4, 1, 5)$, as shown in Figure 1.2.

Figure 1.2

The vector $[2, -2, 6]$
with initial point
$(2, 3, -1)$

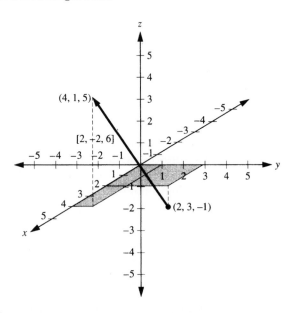

---

[2] We use italicized capital letters and parentheses for the points of a coordinate system, such as $A = (3, 2)$, and boldface lowercase letters and brackets for vectors, such as $\mathbf{x} = [3, 2]$.

Three-dimensional movements are usually graphed on a two-dimensional page by slanting the x-axis at an angle to create the optical illusion of three mutually perpendicular axes. Movements are determined on such a graph by breaking them down into components parallel to each of the coordinate axes.

Visualizing vectors in $\mathbb{R}^4$ and higher dimensions is difficult. However, the same algebraic principles are involved. For example, the vector $\mathbf{x} = [2, 7, -3, 10]$ can represent a movement between points $(5, -6, 2, -1)$ and $(7, 1, -1, 9)$ in a four-dimensional coordinate system.

Length of a Vector

Recall the **distance formula** in the plane; the distance between two points (x_1, y_1) and (x_2, y_2) is $d = \sqrt{(x_2 - x_1)^2 + (y_2 - y_1)^2}$ (see Figure 1.3). This formula arises from the Pythagorean Theorem for right triangles. The 2-vector between the points is $[a_1, a_2]$, where $a_1 = x_2 - x_1$ and $a_2 = y_2 - y_1$, so $d = \sqrt{a_1^2 + a_2^2}$. This formula motivates the following definition:

Figure 1.3

The line segment (and vector) connecting points A and B, with length

$$\sqrt{(x_2 - x_1)^2 + (y_2 - y_1)^2}$$
$$= \sqrt{a_1^2 + a_2^2}$$

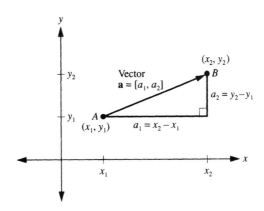

DEFINITION

The **length** (also known as the **norm** or **magnitude**) of a vector $\mathbf{a} = [a_1, a_2, \ldots, a_n]$ in $\mathbb{R}^n$ is $\|\mathbf{a}\| = \sqrt{a_1^2 + a_2^2 + \cdots + a_n^2}$.

EXAMPLE 1

The length of the vector $\mathbf{a} = [4, -3, 0, 2]$ is given by

$$\|\mathbf{a}\| = \sqrt{4^2 + (-3)^2 + 0^2 + 2^2} = \sqrt{16 + 9 + 4} = \sqrt{29}.$$

∎

Note that the length of any vector in $\mathbb{R}^n$ is always nonnegative (that is, ≥ 0). (Do you know why this statement is true?) Also, the only vector with length 0 in $\mathbb{R}^n$ is the zero vector $[0, 0, \ldots, 0]$ (why?).

Vectors of length 1 play an important role in linear algebra.

DEFINITION

Any vector of length 1 is called a **unit vector**.

In $\mathbb{R}^2$, the vector $\left[\frac{3}{5}, -\frac{4}{5}\right]$ is a unit vector, because $\sqrt{\left(\frac{3}{5}\right)^2 + \left(-\frac{4}{5}\right)^2} = 1$. Similarly, $\left[0, \frac{3}{5}, 0, -\frac{4}{5}\right]$ is a unit vector in $\mathbb{R}^4$. Certain unit vectors are particularly useful: those with a single coordinate equal to 1 and all other coordinates equal to 0. In $\mathbb{R}^2$ these vectors are denoted $\mathbf{i} = [1, 0]$ and $\mathbf{j} = [0, 1]$; in $\mathbb{R}^3$ they are denoted $\mathbf{i} = [1, 0, 0]$, $\mathbf{j} = [0, 1, 0]$, and $\mathbf{k} = [0, 0, 1]$. In $\mathbb{R}^n$, these vectors, the **standard unit vectors**, are denoted $\mathbf{e}_1 = [1, 0, 0, \ldots, 0]$, $\mathbf{e}_2 = [0, 1, 0, \ldots, 0]$, $\ldots$, $\mathbf{e}_n = [0, 0, 0, \ldots, 1]$.

Scalar Multiplication and Parallel Vectors

DEFINITION

Let $\mathbf{x} = [x_1, x_2, \ldots, x_n]$ be a vector in $\mathbb{R}^n$, and let c be any scalar (real number). Then $c\mathbf{x}$, the **scalar multiple of x by** c, is the vector $[cx_1, cx_2, \ldots, cx_n]$.

For example, if $\mathbf{x} = [4, -5]$, then $2\mathbf{x} = [8, -10]$, $-3\mathbf{x} = [-12, 15]$, and $-\frac{1}{2}\mathbf{x} = \left[-2, \frac{5}{2}\right]$. These vectors are graphed in Figure 1.4. From the graph, you can see that the vector $2\mathbf{x}$ points in the same direction as $\mathbf{x}$ but is twice as long. The vectors $-3\mathbf{x}$ and $-\frac{1}{2}\mathbf{x}$ indicate movements in the direction opposite to $\mathbf{x}$, with $-3\mathbf{x}$ being three times as long as $\mathbf{x}$ and $-\frac{1}{2}\mathbf{x}$ being half as long.

Figure 1.4

Scalar multiples of $\mathbf{x} = [4, -5]$ (all vectors drawn with initial point at origin)

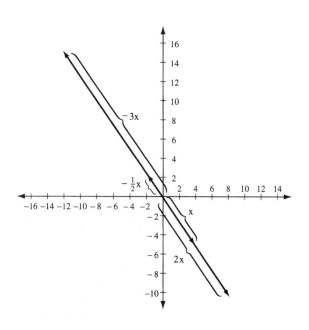

In general, in $\mathbb{R}^n$, multiplication by c **dilates** (expands) the length of the vector when $|c| > 1$ and **contracts** (shrinks) the length when $|c| < 1$. Scalar multiplication by 1 or -1 does not affect the length. Scalar multiplication by 0 always yields the zero vector. These properties are all special cases of the following theorem:

THEOREM 1.1

Let $\mathbf{x} \in \mathbb{R}^n$, and let c be any real number (scalar). Then $\|c\mathbf{x}\| = |c| \, \|\mathbf{x}\|$. That is, the length of $c\mathbf{x}$ is the absolute value of c times the length of $\mathbf{x}$.

The proof of Theorem 1.1 is left as Exercise 22 at the end of this section.

We have noted that in $\mathbb{R}^2$, the vector $c\mathbf{x}$ is in the same direction as $\mathbf{x}$ when c is positive and in the direction opposite to $\mathbf{x}$ when c is negative, but have not yet discussed "direction" in higher-dimensional coordinate systems. We use scalar multiplication to give a precise definition for vectors having the same or opposite directions.

DEFINITION

Two nonzero vectors $\mathbf{x}$ and $\mathbf{y}$ in $\mathbb{R}^n$ are **in the same direction** if and only if there is a positive real number c such that $\mathbf{y} = c\mathbf{x}$. Two nonzero vectors $\mathbf{x}$ and $\mathbf{y}$ are **in opposite directions** if and only if there is a negative real number c such that $\mathbf{y} = c\mathbf{x}$. Two nonzero vectors are **parallel** if and only if they are either in the same direction or in the opposite direction.

Hence, vectors $[1, -3, 2]$ and $[3, -9, 6]$ are in the same direction, because $[3, -9, 6] = 3[1, -3, 2]$ (or because $[1, -3, 2] = \frac{1}{3}[3, -9, 6]$), as shown in Figure 1.5. Similarly, vectors $[-3, 6, 0, 15]$ and $[4, -8, 0, -20]$ are in opposite directions, because $[4, -8, 0, -20] = -\frac{4}{3}[-3, 6, 0, 15]$.

Figure 1.5

The parallel vectors $[1, -3, 2]$ and $[3, -9, 6]$

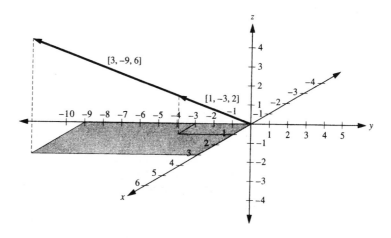

The next result follows from Theorem 1.1:

COROLLARY 1.2

If $\mathbf{x}$ is a nonzero vector in $\mathbb{R}^n$, then $\mathbf{u} = (1/\|\mathbf{x}\|)\,\mathbf{x}$ is a unit vector in the same direction as $\mathbf{x}$.

PROOF OF COROLLARY 1.2

The vector $\mathbf{u}$ in Corollary 1.2 is certainly in the same direction as $\mathbf{x}$ because $\mathbf{u}$ is a positive scalar multiple of $\mathbf{x}$ (the scalar is $1/\|\mathbf{x}\|$). Also by Theorem 1.1, $\|\mathbf{u}\| = \|(1/\|\mathbf{x}\|)\,\mathbf{x}\| = (1/\|\mathbf{x}\|)\,\|\mathbf{x}\| = 1$, so $\mathbf{u}$ is a unit vector. ∎

This process of "dividing" a vector by its length to obtain a unit vector in the same direction is called **normalizing** the vector (see Figure 1.6).

Figure 1.6

Normalizing a vector $\mathbf{x}$ to obtain a unit vector $\mathbf{u}$ in the same direction (with $\|\mathbf{x}\| > 1$)

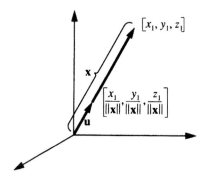

EXAMPLE 2

Consider the vector $[2, 3, -1, 1]$ in $\mathbb{R}^4$. Because $\|[2, 3, -1, 1]\| = \sqrt{15}$, normalizing $[2, 3, -1, 1]$ gives a unit vector $\mathbf{u}$ in the same direction as $[2, 3, -1, 1]$, which is

$$\mathbf{u} = \left(\frac{1}{\sqrt{15}}\right)[2, 3, -1, 1] = \left[\frac{2}{\sqrt{15}}, \frac{3}{\sqrt{15}}, \frac{-1}{\sqrt{15}}, \frac{1}{\sqrt{15}}\right].$$ ∎

Addition and Subtraction with Vectors

DEFINITION

Let $\mathbf{x} = [x_1, x_2, \ldots, x_n]$ and $\mathbf{y} = [y_1, y_2, \ldots, y_n]$ be vectors in $\mathbb{R}^n$. Then $\mathbf{x} + \mathbf{y}$, the **sum** of $\mathbf{x}$ and $\mathbf{y}$, is the vector $[x_1 + y_1, x_2 + y_2, \ldots, x_n + y_n]$ in $\mathbb{R}^n$.

Vectors are added by summing their respective coordinates. For example, if $\mathbf{x} = [2, -3, 5]$ and $\mathbf{y} = [-6, 4, -2]$, then $\mathbf{x} + \mathbf{y} = [2 - 6, -3 + 4, 5 - 2] = [-4, 1, 3]$. Vectors cannot be added unless they have the same number of coordinates.

There is a natural geometric interpretation for the sum of vectors in a plane or in space. Draw a vector $\mathbf{x}$. Then draw a vector $\mathbf{y}$ from the terminal point of $\mathbf{x}$. The sum of $\mathbf{x}$ and $\mathbf{y}$ is the vector whose *initial* point is the same as that of $\mathbf{x}$ and whose *terminal* point is the same as that of $\mathbf{y}$. The total movement $(\mathbf{x}+\mathbf{y})$ is equivalent to first moving along $\mathbf{x}$ and then along $\mathbf{y}$. Figure 1.7 illustrates this in $\mathbb{R}^2$.

Figure 1.7

Addition of vectors in $\mathbb{R}^2$

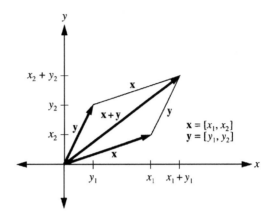

Let $-\mathbf{y}$ denote the scalar multiple $-1\mathbf{y}$. We can now define **subtraction** of vectors in a natural way: if $\mathbf{x}$ and $\mathbf{y}$ are both vectors in $\mathbb{R}^n$, let $\mathbf{x} - \mathbf{y}$ be the vector $\mathbf{x} + (-\mathbf{y})$. A geometric interpretation of this is in Figure 1.8 (movement $\mathbf{x}$ followed by movement $-\mathbf{y}$). An alternative interpretation is described in Exercise 11.

Figure 1.8

Subtraction of vectors in $\mathbb{R}^2 : \mathbf{x} - \mathbf{y} = \mathbf{x} + (-\mathbf{y})$

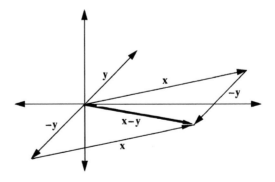

Fundamental Properties of Addition and Scalar Multiplication

Theorem 1.3 contains the basic properties of addition and scalar multiplication of vectors. The **commutative**, **associative**, and **distributive** laws are so named because they resemble the corresponding laws for real numbers.

THEOREM 1.3

Let $\mathbf{x} = [x_1, x_2, \ldots, x_n]$, $\mathbf{y} = [y_1, y_2, \ldots, y_n]$ and $\mathbf{z} = [z_1, z_2, \ldots, z_n]$ be any vectors in $\mathbb{R}^n$, and let c and d be any real numbers (scalars). Let $\mathbf{0}$ represent the zero vector in $\mathbb{R}^n$. Then

(1)	$\mathbf{x} + \mathbf{y} = \mathbf{y} + \mathbf{x}$	Commutative Law of Addition
(2)	$\mathbf{x} + (\mathbf{y} + \mathbf{z}) = (\mathbf{x} + \mathbf{y}) + \mathbf{z}$	Associative Law of Addition
(3)	$\mathbf{0} + \mathbf{x} = \mathbf{x} + \mathbf{0} = \mathbf{x}$	Existence of Identity Element for Addition
(4)	$\mathbf{x} + (-\mathbf{x}) = (-\mathbf{x}) + \mathbf{x} = \mathbf{0}$	Existence of Inverse Elements for Addition
(5)	$c(\mathbf{x} + \mathbf{y}) = c\mathbf{x} + c\mathbf{y}$	Distributive Laws of Scalar
(6)	$(c + d)\mathbf{x} = c\mathbf{x} + d\mathbf{x}$	Multiplication over Addition
(7)	$(cd)\mathbf{x} = c(d\mathbf{x})$	Associativity of Scalar Multiplication
(8)	$1\mathbf{x} = \mathbf{x}$	Identity Property for Scalar Multiplication

In part (3), the vector $\mathbf{0}$ is called an **identity element** for addition because $\mathbf{0}$ does not change the identity of any vector to which it is added. A similar statement is true in part (8) for the scalar 1 with scalar multiplication. In part (4), the vector $-\mathbf{x}$ is called the **additive inverse element of x** because it "cancels out $\mathbf{x}$" to produce the zero vector.

Each part of the theorem is proved by calculating the entries in each coordinate of the vectors and applying a corresponding law for real-number arithmetic. We illustrate this *coordinate-wise* technique by proving part (6). You are asked to prove other parts of the theorem in Exercise 23.

PROOF OF THEOREM 1.3, PART (6):

$$
\begin{aligned}
(c + d)\mathbf{x} &= (c + d)[x_1, x_2, \ldots, x_n] \\
&= [(c + d)x_1, (c + d)x_2, \ldots, (c + d)x_n] && \text{definition of scalar multiplication} \\
&= [cx_1 + dx_1, cx_2 + dx_2, \ldots, cx_n + dx_n] && \text{coordinate-wise use of distributive law in } \mathbb{R} \\
&= [cx_1, cx_2, \ldots, cx_n] + [dx_1, dx_2, \ldots, dx_n] && \text{definition of vector addition} \\
&= c[x_1, x_2, \ldots, x_n] + d[x_1, x_2, \ldots, x_n] && \text{definition of scalar multiplication} \\
&= c\mathbf{x} + d\mathbf{x}. && \blacksquare
\end{aligned}
$$

The following theorem is very useful (the proof is left as Exercise 24):

THEOREM 1.4
Let $\mathbf{x}$ be a vector in $\mathbb{R}^n$, and let c be a scalar. If $c\mathbf{x} = \mathbf{0}$, then either $c = 0$ or $\mathbf{x} = \mathbf{0}$.

Linear Combinations of Vectors

DEFINITION
Let $\mathbf{v}_1, \mathbf{v}_2, \ldots, \mathbf{v}_k$ be vectors in $\mathbb{R}^n$. Then the vector $\mathbf{v}$ is a **linear combination** of $\mathbf{v}_1, \mathbf{v}_2, \ldots, \mathbf{v}_k$ if and only if there are scalars $c_1, c_2, \ldots, c_k$ such that $\mathbf{v} = c_1\mathbf{v}_1 + c_2\mathbf{v}_2 + \cdots + c_k\mathbf{v}_k$.

Thus, a linear combination of vectors is a sum of scalar multiples of those vectors. For example, the vector $[-2, 8, 5, 0]$ is a linear combination of $[3, 1, -2, 2]$, $[1, 0, 3, -1]$, and $[4, -2, 1, 0]$ because $2[3, 1, -2, 2] + 4[1, 0, 3, -1] - 3[4, -2, 1, 0] = [-2, 8, 5, 0]$.

Note that any vector in $\mathbb{R}^3$ can be expressed in a unique way as a linear combination of $\mathbf{i}$, $\mathbf{j}$, and $\mathbf{k}$. For example, $[3, -2, 5] = 3[1, 0, 0] - 2[0, 1, 0] + 5[0, 0, 1] = 3\mathbf{i} - 2\mathbf{j} + 5\mathbf{k}$. In general, $[a, b, c] = a\mathbf{i} + b\mathbf{j} + c\mathbf{k}$. Also, every vector in $\mathbb{R}^n$ can be expressed as a linear combination of the standard unit vectors $\mathbf{e}_1 = [1, 0, 0, \ldots, 0]$, $\mathbf{e}_2 = [0, 1, 0, \ldots, 0]$, $\ldots$, $\mathbf{e}_n = [0, 0, \ldots, 0, 1]$ (why?).

One helpful way to picture linear combinations of the vectors $\mathbf{v}_1, \mathbf{v}_2, \ldots, \mathbf{v}_k$ is to remember that each vector represents a certain amount of movement in a particular direction. When we combine these vectors using addition and scalar multiplication, the endpoint of each linear combination vector represents a "destination" that can be reached using these operations. For example, the linear combination $\mathbf{w} = 2[1, 3] - \frac{1}{2}[4, -5] + 3[2, -1] = [6, \frac{11}{2}]$ is the destination reached by traveling in the direction of $[1, 3]$, but traveling twice its length, then traveling in the direction opposite to $[4, -5]$, but half its length, and finally traveling in the direction $[2, -1]$, but three times its length (see Figure 1.9(a)).

We can also consider the set of all possible destinations that can be reached using linear combinations of a certain set of vectors. For example, the set of all linear combinations in $\mathbb{R}^3$ of $\mathbf{v}_1 = [2, 0, 1]$ and $\mathbf{v}_2 = [0, 1, -2]$ is the set of all vectors (beginning at the origin) with endpoints lying in the plane through the origin containing $\mathbf{v}_1$ and $\mathbf{v}_2$ (see Figure 1.9(b)).

Physical Applications of Addition and Scalar Multiplication

Addition and scalar multiplication of vectors are often used to solve problems in elementary physics. Recall the trigonometric fact that if $\mathbf{v}$ is a vector in $\mathbb{R}^2$ forming an angle of θ with the positive x-axis then $\mathbf{v} = [\|\mathbf{v}\| \cos\theta, \|\mathbf{v}\| \sin\theta]$, as in Figure 1.10.

Figure 1.9

(a) The destination
$\mathbf{w} = 2[1, 3] - \frac{1}{2}[4, -5] + 3[2, -1] = [6, \frac{11}{2}]$;
(b) The plane in
$\mathbb{R}^3$ containing all
linear combinations of
$[2, 0, 1]$ and $[0, 1, -2]$

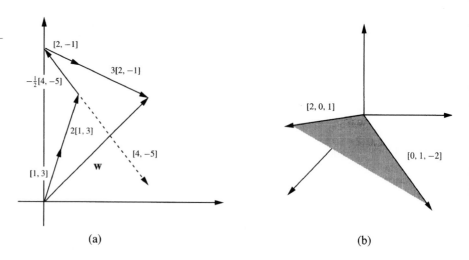

(a)   (b)

Figure 1.10

The vector $\mathbf{v} = [\|\mathbf{v}\| \cos\theta, \|\mathbf{v}\| \sin\theta]$
forming an angle of θ
with the positive x-axis

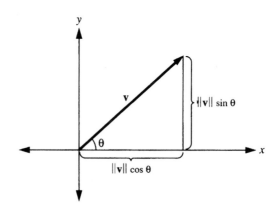

EXAMPLE 3   **Resultant Velocity:** Suppose a man swims 5 km/hr in calm water. If he is swimming toward the east in a wide stream with a northwest current of 3 km/hr, what is his **resultant velocity** (net speed and direction)?

   The velocities of the swimmer and current are shown as vectors in Figure 1.11, where we have, for convenience, placed the swimmer at the origin. Now, $\mathbf{v}_1 = [5, 0]$ and $\mathbf{v}_2 = [3\cos 135°, 3\sin 135°] = \left[-3\sqrt{2}/2, 3\sqrt{2}/2\right]$. Thus, the total (resultant) velocity of the swimmer is the sum of these velocities, $\mathbf{v}_1 + \mathbf{v}_2$, which is $\left[5 - 3\sqrt{2}/2, 3\sqrt{2}/2\right] \approx [2.88, 2.12]$. Hence, each hour the swimmer is traveling about 2.9 km east and 2.1 km north. The resultant speed of the swimmer is $\left\|\left[5 - 3\sqrt{2}/2, 3\sqrt{2}/2\right]\right\| \approx 3.58$ km/hr. ■

Figure 1.11

Velocity $\mathbf{v}_1$ of swimmer, velocity $\mathbf{v}_2$ of current, and resultant velocity $\mathbf{v}_1 + \mathbf{v}_2$

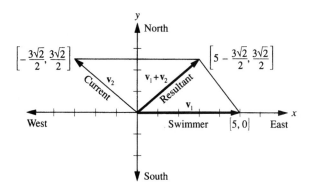

| EXAMPLE 4 |

Newton's Second Law: Newton's famous **Second Law of Motion** asserts that the sum, $\mathbf{f}$, of the vector forces on an object is equal to the scalar multiple of the mass m of the object times the vector acceleration $\mathbf{a}$ of the object; that is, $\mathbf{f} = m\mathbf{a}$. For example, suppose a mass of 5 kg (kilograms) in a three-dimensional coordinate system has two forces acting on it: a force $\mathbf{f}_1$ of 10 newtons[3] in the direction of the vector $[-2, 1, 2]$ and a force $\mathbf{f}_2$ of 20 newtons in the direction of the vector $[6, 3, -2]$. What is the acceleration of the object?

We must first normalize the direction vectors $[-2, 1, 2]$ and $[6, 3, -2]$ so that their lengths do not contribute to the magnitude of the forces $\mathbf{f}_1$ and $\mathbf{f}_2$. Therefore, $\mathbf{f}_1 = 10\,([-2, 1, 2]/\,\|[-2, 1, 2]\|)$, and $\mathbf{f}_2 = 20([6, 3, -2]/\,\|[6, 3, -2]\|)$. The net force on the object is $\mathbf{f} = \mathbf{f}_1 + \mathbf{f}_2$. Thus, the net acceleration on the object is

$$\mathbf{a} = \frac{1}{m}\mathbf{f} = \frac{1}{m}(\mathbf{f}_1 + \mathbf{f}_2) = \frac{1}{5}\left(10\left(\frac{[-2, 1, 2]}{\|[-2, 1, 2]\|}\right) + 20\left(\frac{[6, 3, -2]}{\|[6, 3, -2]\|}\right)\right),$$

which equals $\frac{2}{3}[-2, 1, 2] + \frac{4}{7}[6, 3, -2] = \left[\frac{44}{21}, \frac{50}{21}, \frac{4}{21}\right]$. The length of $\mathbf{a}$ is approximately 3.18, so pulling out a factor of 3.18 from each coordinate, we can approximate $\mathbf{a}$ as $3.18[0.66, 0.75, 0.06]$, where $[0.66, 0.75, 0.06]$ is a *unit* vector. Hence, the acceleration is about 3.18 m/sec^2 in the direction $[0.66, 0.75, 0.06]$. ∎

If the sum of the forces on an object is $\mathbf{0}$, then the object is in **equilibrium**; there is no acceleration in any direction (Exercise 20).

Exercises for Section 1.1

Note: A star ($\star$) next to an exercise indicates that the answer for that exercise appears in the back of the book, and the full solution appears in the Student

---

[3] 1 newton = 1kg-m/ sec^2 (kilogram-meter/second^2), or the force needed to push 1 kg at a speed 1 m/sec (meter per second) faster every second.

Solutions Manual. A triangle ($\blacktriangleright$) next to an exercise that requires a proof indicates that the proof is supplied in the Student Solution Manual.

1. In each of the following cases, find a vector that represents a movement from the first (initial) point to the second (terminal) point. Then use this vector to find the distance between the given points.

   $\star$(a) $(-4, 3)$, $(5, -1)$
   (b) $(2, -1, 4)$, $(-3, 0, 2)$
   $\star$(c) $(1, -2, 0, 2, 3)$, $(0, -3, 2, -1, -1)$

2. In each of the following cases, draw a directed line segment in space that represents the movement associated with each of the vectors if the initial point is $(1, 1, 1)$. What is the terminal point in each case?

   $\star$(a) $[2, 3, 1]$ (b) $[-1, 4, 2]$
   $\star$(c) $[0, -3, -1]$ (d) $[2, -1, -1]$

3. In each of the following cases, find the initial point, given the vector and the terminal point.

   $\star$(a) $[-1, 4]$, $(6, -9)$
   (b) $[2, -2, 5]$, $(-4, 1, 7)$
   $\star$(c) $[3, -4, 0, 1, -2]$, $(2, -1, -1, 5, 4)$

4. In each of the following cases, find a point that is two-thirds of the distance from the first (initial) point to the second (terminal) point.

   $\star$(a) $(-4, 7, 2)$, $(10, -10, 11)$
   (b) $(2, -1, 0, -7)$, $(-11, -1, -9, 2)$

5. In each of the following cases, find a unit vector in the same direction as the given vector. Is the resulting (normalized) vector longer or shorter than the original? Why?

   $\star$(a) $[3, -5, 6]$ (b) $[4, 1, 0, -2]$
   $\star$(c) $[0.6, -0.8]$ (d) $\left[\frac{1}{5}, -\frac{2}{5}, -\frac{1}{5}, \frac{1}{5}, \frac{2}{5}\right]$

6. Which of the following pairs of vectors are parallel?

   $\star$(a) $[12, -16]$, $[9, -12]$ (b) $[4, -14]$, $[-2, 7]$
   $\star$(c) $[-2, 3, 1]$, $[6, -4, -3]$ (d) $[10, -8, 3, 0, 27]$, $\left[\frac{5}{6}, -\frac{2}{3}, \frac{3}{4}, 0, -\frac{5}{2}\right]$

7. If $\mathbf{x} = [-2, 4, 5]$, $\mathbf{y} = [-1, 0, 3]$, and $\mathbf{z} = [4, -1, 2]$, find the following:

   $\star$(a) $3\mathbf{x}$ (b) $-2\mathbf{y}$
   $\star$(c) $\mathbf{x} + \mathbf{y}$ (d) $\mathbf{y} - \mathbf{z}$
   $\star$(e) $4\mathbf{y} - 5\mathbf{x}$ (f) $2\mathbf{x} + 3\mathbf{y} - 4\mathbf{z}$

8. Given $\mathbf{x}$ and $\mathbf{y}$ as follows, calculate $\mathbf{x} + \mathbf{y}$, $\mathbf{x} - \mathbf{y}$, and $\mathbf{y} - \mathbf{x}$, and sketch $\mathbf{x}$, $\mathbf{y}$, $\mathbf{x} + \mathbf{y}$, $\mathbf{x} - \mathbf{y}$, and $\mathbf{y} - \mathbf{x}$ in the same coordinate system.

   $\star$(a) $\mathbf{x} = [-1, 5]$, $\mathbf{y} = [2, -4]$
   (b) $\mathbf{x} = [10, -2]$, $\mathbf{y} = [-7, -3]$
   $\star$(c) $\mathbf{x} = [2, 5, -3]$, $\mathbf{y} = [-1, 3, -2]$
   (d) $\mathbf{x} = [1, -2, 5]$, $\mathbf{y} = [-3, -2, -1]$

9. Show that the points $(7, -3, 6)$, $(11, -5, 3)$, and $(10, -7, 8)$ are the vertices of an isosceles triangle. Is this an equilateral triangle?

10. A certain clock has a minute hand that is 10 cm long. Find the vector representing the displacement of the tip of the minute hand of the clock.

⋆(a) From 12 PM to 12:15 PM

⋆(b) From 12 PM to 12:40 PM (Hint: use trigonometry)

(c) From 12 PM to 1 PM

11. Show that if **x** and **y** are vectors in $\mathbb{R}^2$, then **x** + **y** and **x** − **y** are the two diagonals of the parallelogram whose sides are **x** and **y**.

12. Consider the vectors in $\mathbb{R}^3$ in Figure 1.12. Verify that **x** + (**y** + **z**) is a diagonal of the parallelepiped with sides **x**, **y**, **z**. Does (**x**+**y**)+**z** represent the same diagonal vector? Why or why not?

Figure 1.12

Parallelepiped with sides **x**, **y**, **z**

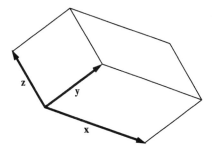

⋆**13.** At a certain green on a golf course, a golfer takes three putts to sink the ball. If the first putt moved the ball 1 m (meter) southwest, the second putt moved the ball 0.5 m east, and the third putt moved the ball 0.2 m northwest, what single putt (expressed as a vector) would have had the same final result?

14. (a) Show that every unit vector in $\mathbb{R}^2$ is of the form $[\cos(\theta_1), \cos(\theta_2)]$, where θ_1 is the angle the vector makes with the positive x-axis and θ_2 is the angle the vector makes with the positive y-axis.

(b) Show that every unit vector in $\mathbb{R}^3$ is of the form $[\cos(\alpha_1), \cos(\alpha_2), \cos(\alpha_3)]$, where α_1, α_2, and α_3 are the angles the vector makes with the positive x-, y-, and z-axes, respectively. (Note: The coordinates of this unit vector are often called the **direction cosines** of the vector.)

⋆**15.** A rower can propel a boat 4 km/hr on a calm river. If the rower rows northwestward against a current of 3 km/hr southward, what is the net velocity of the boat? What is its resultant speed?

16. A singer is walking 3 km/hr southwestward on a moving parade float that is being pulled northward at 4 km/hr. What is the net velocity of the singer? What is the singer's resultant speed?

⋆**17.** A woman rowing on a wide river wants the resultant (net) velocity of her boat to be 8 km/hr westward. If the current is moving 2 km/hr northeastward, what velocity vector should she maintain?

*18. Using Newton's Second Law of Motion, find the acceleration vector on a 20-kg object in a three-dimensional coordinate system when the following three forces are simultaneously applied:

   $\mathbf{f}_1$: A force of 4 newtons in the direction of the vector $[3, -12, 4]$
   $\mathbf{f}_2$: A force of 2 newtons in the direction of the vector $[0, -4, -3]$
   $\mathbf{f}_3$: A force of 6 newtons in the direction of the unit vector $\mathbf{k}$

19. Using Newton's Second Law of Motion, find the resultant sum of the forces on a 30-kg object in a three-dimensional coordinate system undergoing an acceleration of 6 m/sec^2 in the direction of the vector $[-2, 3, 1]$.

*20. Two forces, $\mathbf{a}$ and $\mathbf{b}$, are simultaneously applied along cables attached to a weight, as in Figure 1.13, to keep the weight in equilibrium by balancing the force of gravity (which is $m\mathbf{g}$, where m is the mass of the weight and $\mathbf{g} = [0, -g]$ is the downward acceleration due to gravity). Solve for the coordinates of forces $\mathbf{a}$ and $\mathbf{b}$ in terms of m and g.

Figure 1.13

Forces in equilibrium

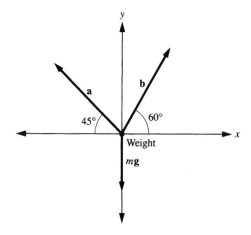

21. (a) Prove that the length of each vector in $\mathbb{R}^n$ is nonnegative (that is, ≥ 0).
    (b) Prove that the only vector in $\mathbb{R}^n$ of length 0 is the zero vector.

▶22. Prove Theorem 1.1.

23. (a) Prove part (2) of Theorem 1.3.
    ▶(b) Prove part (4) of Theorem 1.3.
    ▶(c) Prove part (5) of Theorem 1.3.
    (d) Prove part (7) of Theorem 1.3.

▶24. Prove Theorem 1.4.

25. If $\mathbf{x}$ is a vector in $\mathbb{R}^n$ and $c_1 \neq c_2$, show that $c_1\mathbf{x} = c_2\mathbf{x}$ implies that $\mathbf{x} = \mathbf{0}$ (zero vector).

*26. True or False:
    (a) The length of $\mathbf{a} = [a_1, a_2, a_3]$ is $a_1^2 + a_2^2 + a_3^2$.
    (b) For any vectors $\mathbf{x}, \mathbf{y}, \mathbf{z}$ in $\mathbb{R}^n$, $(\mathbf{x} + \mathbf{y}) + \mathbf{z} = \mathbf{z} + (\mathbf{y} + \mathbf{x})$.

 (c) $[2, 0, -3]$ is a linear combination of $[1, 0, 0]$ and $[0, 0, 1]$.

 (d) The vectors $[3, -5, 2]$ and $[6, -10, 5]$ are parallel.

 (e) Let $\mathbf{x} \in \mathbb{R}^n$, and let d be a scalar. If $d\mathbf{x} = \mathbf{0}$, and $d \neq 0$, then $\mathbf{x} = \mathbf{0}$.

 (f) If two nonzero vectors in $\mathbb{R}^n$ are parallel, then they are in the same direction.

 (g) The properties in Theorem 1.3 are only true if all the vectors have their initial points at the origin.

1.2 The Dot Product

We now discuss another important vector operation: the dot product. After explaining several properties of the dot product, we show how to calculate the angle between vectors and to "project" one vector onto another.

Definition and Properties of the Dot Product

DEFINITION

Let $\mathbf{x} = [x_1, x_2, \ldots, x_n]$ and $\mathbf{y} = [y_1, y_2, \ldots, y_n]$ be two vectors in $\mathbb{R}^n$. The **dot (inner) product** of $\mathbf{x}$ and $\mathbf{y}$ is given by

$$\mathbf{x} \cdot \mathbf{y} = x_1 y_1 + x_2 y_2 + \cdots + x_n y_n = \sum_{k=1}^{n} x_k y_k.$$

For example, if $\mathbf{x} = [2, -4, 3]$ and $\mathbf{y} = [1, 5, -2]$, then $\mathbf{x} \cdot \mathbf{y} = (2)(1) + (-4)(5) + (3)(-2) = -24$. Notice that the dot product involves two vectors and the result is a *scalar*, whereas scalar multiplication involves a scalar and a vector and the result is a *vector*. Also, the dot product is not defined for vectors having different numbers of coordinates. The next theorem states some elementary results involving the dot product.

THEOREM 1.5

If $\mathbf{x}$, $\mathbf{y}$, and $\mathbf{z}$ are any vectors in $\mathbb{R}^n$, and if c is any scalar, then

(1) $\mathbf{x} \cdot \mathbf{y} = \mathbf{y} \cdot \mathbf{x}$ Commutativity of Dot Product

(2) $\mathbf{x} \cdot \mathbf{x} = \|\mathbf{x}\|^2 \geq 0$ Relationship between Dot Product and Length

(3) $\mathbf{x} \cdot \mathbf{x} = 0$ if and only if $\mathbf{x} = \mathbf{0}$

(4) $c(\mathbf{x} \cdot \mathbf{y}) = (c\mathbf{x}) \cdot \mathbf{y} = \mathbf{x} \cdot (c\mathbf{y})$ Relationship between Scalar Multiplication and Dot Product

(5) $\mathbf{x} \cdot (\mathbf{y} + \mathbf{z}) = (\mathbf{x} \cdot \mathbf{y}) + (\mathbf{x} \cdot \mathbf{z})$ Distributive Laws of Dot Product

(6) $(\mathbf{x} + \mathbf{y}) \cdot \mathbf{z} = (\mathbf{x} \cdot \mathbf{z}) + (\mathbf{y} \cdot \mathbf{z})$ over Addition

---

 The proofs of parts (1), (2), (4), (5), and (6) are done by expanding the expressions on each side of the equation and then showing they are equal.

We illustrate this with the proof of part (5). The remaining proofs are left as Exercise 6.

PROOF OF THEOREM 1.5, PART (5)
Let $\mathbf{x} = [x_1, x_2, \ldots, x_n]$, $\mathbf{y} = [y_1, y_2, \ldots, y_n]$, and $\mathbf{z} = [z_1, z_2, \ldots, z_n]$. Then,

$$
\begin{aligned}
\mathbf{x} \cdot (\mathbf{y} + \mathbf{z}) &= [x_1, x_2, \ldots, x_n] \cdot ([y_1, y_2, \ldots, y_n] + [z_1, z_2, \ldots, z_n]) \\
&= [x_1, x_2, \ldots, x_n] \cdot [y_1 + z_1, y_2 + z_2, \ldots, y_n + z_n] \\
&= x_1(y_1 + z_1) + x_2(y_2 + z_2) + \cdots + x_n(y_n + z_n) \\
&= (x_1 y_1 + x_2 y_2 + \cdots + x_n y_n) + (x_1 z_1 + x_2 z_2 + \cdots + x_n z_n).
\end{aligned}
$$

Also,

$$
\begin{aligned}
(\mathbf{x} \cdot \mathbf{y}) + (\mathbf{x} \cdot \mathbf{z}) &= ([x_1, x_2, \ldots, x_n] \cdot [y_1, y_2, \ldots, y_n]) \\
&\quad + ([x_1, x_2, \ldots, x_n] \cdot [z_1, z_2, \ldots, z_n]) \\
&= (x_1 y_1 + x_2 y_2 + \cdots + x_n y_n) + (x_1 z_1 + x_2 z_2 + \cdots + x_n z_n).
\end{aligned}
$$

Hence, $\mathbf{x} \cdot (\mathbf{y} + \mathbf{z}) = (\mathbf{x} \cdot \mathbf{y}) + (\mathbf{x} \cdot \mathbf{z})$. ∎

The properties in Theorem 1.5 allow us to simplify dot product expressions just as in elementary algebra. For example,

$$
\begin{aligned}
(5\mathbf{x} - 4\mathbf{y}) \cdot (-2\mathbf{x} + 3\mathbf{y}) &= [(5\mathbf{x} - 4\mathbf{y}) \cdot (-2\mathbf{x})] + [(5\mathbf{x} - 4\mathbf{y}) \cdot (3\mathbf{y})] \\
&= [(5\mathbf{x}) \cdot (-2\mathbf{x})] + [(-4\mathbf{y}) \cdot (-2\mathbf{x})] + [(5\mathbf{x}) \cdot (3\mathbf{y})] \\
&\quad + [(-4\mathbf{y}) \cdot (3\mathbf{y})] \\
&= -10(\mathbf{x} \cdot \mathbf{x}) + 8(\mathbf{y} \cdot \mathbf{x}) + 15(\mathbf{x} \cdot \mathbf{y}) - 12(\mathbf{y} \cdot \mathbf{y}) \\
&= -10 \|\mathbf{x}\|^2 + 23(\mathbf{x} \cdot \mathbf{y}) - 12 \|\mathbf{y}\|^2.
\end{aligned}
$$

Inequalities Involving the Dot Product

The next theorem gives an upper and lower bound on the dot product.

THEOREM 1.6 (Cauchy-Schwarz Inequality)

If $\mathbf{x}$ and $\mathbf{y}$ are vectors in $\mathbb{R}^n$, then $|\mathbf{x} \cdot \mathbf{y}| \leq (\|\mathbf{x}\|)(\|\mathbf{y}\|)$.

PROOF OF THEOREM 1.6
If either $\mathbf{x} = \mathbf{0}$ or $\mathbf{y} = \mathbf{0}$, the theorem is certainly true. Hence, we need only examine the case when both $\|\mathbf{x}\|$ and $\|\mathbf{y}\|$ are nonzero. We need to prove $-(\|\mathbf{x}\|)(\|\mathbf{y}\|) \leq \mathbf{x} \cdot \mathbf{y} \leq (\|\mathbf{x}\|)(\|\mathbf{y}\|)$. This statement is true if and only if

$$
-1 \leq \frac{\mathbf{x} \cdot \mathbf{y}}{(\|\mathbf{x}\|)(\|\mathbf{y}\|)} \leq 1.
$$

Now, $(\mathbf{x} \cdot \mathbf{y}) / (\|\mathbf{x}\|)(\|\mathbf{y}\|)$ is equal to $(\mathbf{x}/\|\mathbf{x}\|) \cdot (\mathbf{y}/\|\mathbf{y}\|)$. Note that $\mathbf{x}/\|\mathbf{x}\|$ and $\mathbf{y}/\|\mathbf{y}\|$ are both *unit* vectors. Thus, it is enough to show that $-1 \leq \mathbf{a} \cdot \mathbf{b} \leq 1$ for any unit vectors $\mathbf{a}$ and $\mathbf{b}$.

The term $\mathbf{a} \cdot \mathbf{b}$ occurs as part of the expansion of $(\mathbf{a} + \mathbf{b}) \cdot (\mathbf{a} + \mathbf{b})$, as well as part of $(\mathbf{a} - \mathbf{b}) \cdot (\mathbf{a} - \mathbf{b})$. The first expansion gives

$$
\begin{aligned}
(\mathbf{a} + \mathbf{b}) \cdot (\mathbf{a} + \mathbf{b}) &= \|\mathbf{a} + \mathbf{b}\|^2 \geq 0 && \text{using part (2) of Theorem 1.5} \\
\Rightarrow (\mathbf{a} \cdot \mathbf{a}) + (\mathbf{b} \cdot \mathbf{a}) + (\mathbf{a} \cdot \mathbf{b}) + (\mathbf{b} \cdot \mathbf{b}) &\geq 0 && \\
\Rightarrow \|\mathbf{a}\|^2 + 2(\mathbf{a} \cdot \mathbf{b}) + \|\mathbf{b}\|^2 &\geq 0 && \text{by parts (1) and (2) of Theorem 1.5} \\
\Rightarrow 1 + 2(\mathbf{a} \cdot \mathbf{b}) + 1 &\geq 0 && \text{because } \mathbf{a} \text{ and } \mathbf{b} \text{ are unit vectors} \\
\Rightarrow \mathbf{a} \cdot \mathbf{b} &\geq -1. &&
\end{aligned}
$$

A similar argument beginning with $(\mathbf{a} - \mathbf{b}) \cdot (\mathbf{a} - \mathbf{b}) = \|\mathbf{a} - \mathbf{b}\|^2 \geq 0$ shows $\mathbf{a} \cdot \mathbf{b} \leq 1$ (see Exercise 8). Hence, $-1 \leq \mathbf{a} \cdot \mathbf{b} \leq 1$. ∎

EXAMPLE 1

Let $\mathbf{x} = [-1, 4, 2, 0, -3]$ and let $\mathbf{y} = [2, 1, -4, -1, 0]$. We verify the Cauchy-Schwarz Inequality in this specific case. Now, $\mathbf{x} \cdot \mathbf{y} = -2 + 4 - 8 + 0 + 0 = -6$. Also, $\|\mathbf{x}\| = \sqrt{1 + 16 + 4 + 0 + 9} = \sqrt{30}$, and $\|\mathbf{y}\| = \sqrt{4 + 1 + 16 + 1 + 0} = \sqrt{22}$. Then, $|\mathbf{x} \cdot \mathbf{y}| \leq (\|\mathbf{x}\|)(\|\mathbf{y}\|)$, because $|-6| = 6 \leq \sqrt{(30)(22)} = 2\sqrt{165} \approx 25.7$. ∎

Another useful result, sometimes known as **Minkowski's Inequality**, is

---

THEOREM 1.7 (Triangle Inequality)

If $\mathbf{x}$ and $\mathbf{y}$ are vectors in $\mathbb{R}^n$, then $\|\mathbf{x} + \mathbf{y}\| \leq \|\mathbf{x}\| + \|\mathbf{y}\|$.

We can prove this theorem geometrically in $\mathbb{R}^2$ and $\mathbb{R}^3$ by noting that the length of $\mathbf{x} + \mathbf{y}$, one side of the triangles in Figure 1.14, is never larger than the sum of the lengths of the other two sides, $\mathbf{x}$ and $\mathbf{y}$. The following algebraic proof extends this result to $\mathbb{R}^n$ for $n > 3$.

PROOF OF THEOREM 1.7

It is enough to show that $\|\mathbf{x} + \mathbf{y}\|^2 \leq (\|\mathbf{x}\| + \|\mathbf{y}\|)^2$ (why?). But

$$
\begin{aligned}
\|\mathbf{x} + \mathbf{y}\|^2 &= (\mathbf{x} + \mathbf{y}) \cdot (\mathbf{x} + \mathbf{y}) \\
&= (\mathbf{x} \cdot \mathbf{x}) + 2(\mathbf{x} \cdot \mathbf{y}) + (\mathbf{y} \cdot \mathbf{y}) \\
&= \|\mathbf{x}\|^2 + 2(\mathbf{x} \cdot \mathbf{y}) + \|\mathbf{y}\|^2 \\
&\leq \|\mathbf{x}\|^2 + 2|\mathbf{x} \cdot \mathbf{y}| + \|\mathbf{y}\|^2 \\
&\leq \|\mathbf{x}\|^2 + 2(\|\mathbf{x}\|)(\|\mathbf{y}\|) + \|\mathbf{y}\|^2 \\
&\quad \text{by the Cauchy-Schwarz Inequality} \\
&= (\|\mathbf{x}\| + \|\mathbf{y}\|)^2. \quad ∎
\end{aligned}
$$

The Angle between Two Vectors

The dot product enables us to find the angle θ between two nonzero vectors $\mathbf{x}$ and $\mathbf{y}$ in $\mathbb{R}^2$ or $\mathbb{R}^3$ that begin at the same initial point. There are actually

Figure 1.14

Triangle inequality in
$\mathbb{R}^2$:

$\|\mathbf{x} + \mathbf{y}\| \leq \|\mathbf{x}\| + \|\mathbf{y}\|$

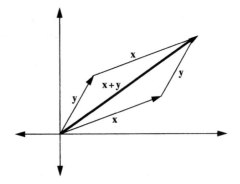

two angles formed by the vectors $\mathbf{x}$ and $\mathbf{y}$, but we always choose the angle θ between two vectors to be the one measuring between 0 and π radians.

Consider the vector $\mathbf{x} - \mathbf{y}$ in Figure 1.15, which begins at the terminal point of $\mathbf{y}$ and ends at the terminal point of $\mathbf{x}$. Because $0 \leq \theta \leq \pi$, it follows from the Law of Cosines that $\|\mathbf{x} - \mathbf{y}\|^2 = \|\mathbf{x}\|^2 + \|\mathbf{y}\|^2 - 2\,(\|\mathbf{x}\|)\,(\|\mathbf{y}\|)\cos\theta$. But,

$$\begin{aligned}\|\mathbf{x} - \mathbf{y}\|^2 &= (\mathbf{x} - \mathbf{y}) \cdot (\mathbf{x} - \mathbf{y}) \\ &= (\mathbf{x} \cdot \mathbf{x}) - 2\,(\mathbf{x} \cdot \mathbf{y}) + (\mathbf{y} \cdot \mathbf{y}) \\ &= \|\mathbf{x}\|^2 - 2\,(\mathbf{x} \cdot \mathbf{y}) + \|\mathbf{y}\|^2\,.\end{aligned}$$

Hence, $-2\,\|\mathbf{x}\|\,\|\mathbf{y}\|\cos\theta = -2(\mathbf{x} \cdot \mathbf{y})$, which implies $\|\mathbf{x}\|\,\|\mathbf{y}\|\cos\theta = \mathbf{x} \cdot \mathbf{y}$, and so

$$\cos\theta = \frac{\mathbf{x} \cdot \mathbf{y}}{(\|\mathbf{x}\|)\,(\|\mathbf{y}\|)}\,.$$

Figure 1.15

The angle θ between two nonzero vectors $\mathbf{x}$ and $\mathbf{y}$ in $\mathbb{R}^2$

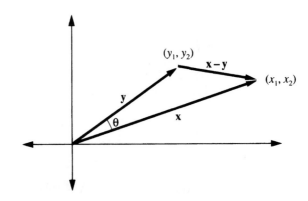

EXAMPLE 2   Suppose $\mathbf{x} = [6, -4]$ and $\mathbf{y} = [-2, 3]$ and θ is the angle between $\mathbf{x}$ and $\mathbf{y}$. Then,

$$\cos\theta = \frac{\mathbf{x} \cdot \mathbf{y}}{(\|\mathbf{x}\|)\,(\|\mathbf{y}\|)} = \frac{(6)(-2) + (-4)(3)}{\sqrt{52}\sqrt{13}} = -\frac{12}{13} \approx -0.9231.$$

Using a calculator, we find that $\theta \approx 2.74$ radians, or $157.4°$. (Remember that $0 \leq \theta \leq \pi$.) ∎

In higher-dimensional spaces, we are outside the geometry of everyday experience, and in such cases, we have not yet defined the angle between two vectors. However, by the Cauchy-Schwarz Inequality, $(\mathbf{x} \cdot \mathbf{y}) / \|\mathbf{x}\| \|\mathbf{y}\|$ always has a value between -1 and 1 for any nonzero vectors $\mathbf{x}$ and $\mathbf{y}$ in $\mathbb{R}^n$. Thus, this value equals $\cos \theta$ for a unique θ between 0 and π radians. Hence, we can define the angle between two vectors in $\mathbb{R}^n$ so it is consistent with the situation in $\mathbb{R}^2$ and $\mathbb{R}^3$.

DEFINITION

Let $\mathbf{x}$ and $\mathbf{y}$ be two nonzero vectors in $\mathbb{R}^n$, for $n \geq 2$. Then the **angle between x and y** is the unique angle between 0 and π radians whose cosine is $(\mathbf{x} \cdot \mathbf{y}) / (\|\mathbf{x}\|) (\|\mathbf{y}\|)$.

EXAMPLE 3

For $\mathbf{x} = [-1, 4, 2, 0, -3]$ and $\mathbf{y} = [2, 1, -4, -1, 0]$, we have $(\mathbf{x} \cdot \mathbf{y}) / (\|\mathbf{x}\|) (\|\mathbf{y}\|) = -6/\left(2\sqrt{165}\right) \approx -0.234$. Using a calculator, we find the angle θ between $\mathbf{x}$ and $\mathbf{y}$ is approximately 1.8 radians, or $103.5°$. ∎

The following theorem is an immediate consequence of the last definition:

THEOREM 1.8

Let $\mathbf{x}$ and $\mathbf{y}$ be nonzero vectors in $\mathbb{R}^n$, and let θ be the angle between $\mathbf{x}$ and $\mathbf{y}$. Then,

(1) $\mathbf{x} \cdot \mathbf{y} > 0$ if and only if $0 \leq \theta < \dfrac{\pi}{2}$ radians ($0°$ or *acute*).

(2) $\mathbf{x} \cdot \mathbf{y} = 0$ if and only if $\theta = \dfrac{\pi}{2}$ radians ($90°$).

(3) $\mathbf{x} \cdot \mathbf{y} < 0$ if and only if $\dfrac{\pi}{2} < \theta \leq \pi$ radians ($180°$ or *obtuse*).

Special Cases: Orthogonal and Parallel Vectors

DEFINITION

Two vectors $\mathbf{x}$ and $\mathbf{y}$ in $\mathbb{R}^n$ are **orthogonal (perpendicular)** if and only if $\mathbf{x} \cdot \mathbf{y} = 0$.

EXAMPLE 4

The vectors $\mathbf{x} = [2, -5]$ and $\mathbf{y} = [-10, -4]$ are orthogonal in $\mathbb{R}^2$ because $\mathbf{x} \cdot \mathbf{y} = 0$. By Theorem 1.8, $\mathbf{x}$ and $\mathbf{y}$ form a right angle, as shown in Figure 1.16. ∎

In $\mathbb{R}^3$, the vectors $\mathbf{i}$, $\mathbf{j}$, and $\mathbf{k}$ are **mutually orthogonal**; that is, the dot product of any pair of these vectors equals zero. In general, in $\mathbb{R}^n$ the standard

Figure 1.16

The orthogonal vectors
$\mathbf{x} = [2, -5]$ and
$\mathbf{y} = [-10, -4]$

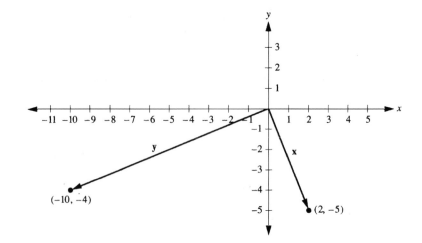

unit vectors $\mathbf{e}_1 = [1, 0, 0, \ldots, 0]$, $\mathbf{e}_2 = [0, 1, 0, \ldots, 0]$, ..., $\mathbf{e}_n = [0, 0, 0, \ldots, 1]$ form a mutually orthogonal set of vectors.

The next theorem gives an alternative way of describing parallel vectors in terms of the angle between them. A proof for the case $\mathbf{x} \cdot \mathbf{y} = + \|\mathbf{x}\| \|\mathbf{y}\|$ appears in Section 1.3 (see Result 4), and the proof of the other case is similar.

THEOREM 1.9

Let $\mathbf{x}$ and $\mathbf{y}$ be nonzero vectors in $\mathbb{R}^n$. Then $\mathbf{x}$ and $\mathbf{y}$ are parallel if and only if $\mathbf{x} \cdot \mathbf{y} = \pm \|\mathbf{x}\| \|\mathbf{y}\|$ (that is, $\cos\theta = \pm 1$, where θ is the angle between $\mathbf{x}$ and $\mathbf{y}$).

EXAMPLE 5   Let $\mathbf{x} = [8, -20, 4]$ and $\mathbf{y} = [6, -15, 3]$. Then, if θ is the angle between $\mathbf{x}$ and $\mathbf{y}$,

$$\cos\theta = \frac{\mathbf{x} \cdot \mathbf{y}}{(\|\mathbf{x}\|)\,(\|\mathbf{y}\|)} = \frac{48 + 300 + 12}{\sqrt{480}\sqrt{270}} = \frac{360}{\sqrt{129600}} = 1.$$

Thus, by Theorem 1.9, $\mathbf{x}$ and $\mathbf{y}$ are parallel. (Notice also that $\mathbf{x}$ and $\mathbf{y}$ are parallel by the definition of parallel vectors in Section 1.1 because $[8, -20, 4] = \frac{4}{3}[6, -15, 3]$.) ■

Projection Vectors

The projection of one vector onto another is useful in physics, engineering, computer graphics, and statistics. Suppose $\mathbf{a}$ and $\mathbf{b}$ are nonzero vectors, both in $\mathbb{R}^2$ or both in $\mathbb{R}^3$, drawn at the same initial point. Let θ represent the angle between $\mathbf{a}$ and $\mathbf{b}$. Drop a perpendicular line segment from the terminal point of $\mathbf{b}$ to the straight line ℓ containing the vector $\mathbf{a}$, as in Figure 1.17.

By the projection $\mathbf{p}$ of $\mathbf{b}$ onto $\mathbf{a}$, we mean the vector from the initial point of $\mathbf{a}$ to the point where the dropped perpendicular meets the line ℓ. Note that $\mathbf{p}$ is in the same direction as $\mathbf{a}$ when $0 \le \theta < \frac{\pi}{2}$ radians (see Figure 1.17) and in the opposite direction to $\mathbf{a}$ when $\frac{\pi}{2} < \theta \le \pi$ radians, as in Figure 1.18.

Figure 1.17

The projection **p** of the vector **b** onto **a** (when θ is acute)

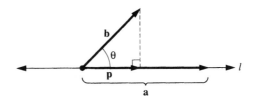

Figure 1.18

The projection **p** of **b** onto **a** (when θ is obtuse)

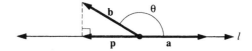

Using trigonometry, we see that when $0 \leq \theta \leq \frac{\pi}{2}$, the vector **p** has length $\|\mathbf{b}\| \cos\theta$ and is in the direction of the unit vector $\mathbf{a}/\|\mathbf{a}\|$. Also, when $\frac{\pi}{2} < \theta \leq \pi$, **p** has length $-\|\mathbf{b}\| \cos\theta$ and is in the direction of the unit vector $-\mathbf{a}/\|\mathbf{a}\|$. Therefore, we can express **p** in all cases as

$$\mathbf{p} = (\|\mathbf{b}\| \cos\theta) \left(\frac{\mathbf{a}}{\|\mathbf{a}\|} \right).$$

But we know that $\cos\theta = (\mathbf{a} \cdot \mathbf{b}) / (\|\mathbf{a}\| \|\mathbf{b}\|)$, and hence

$$\mathbf{p} = \left(\frac{\mathbf{a} \cdot \mathbf{b}}{\|\mathbf{a}\|^2} \right) \mathbf{a}.$$

The projection **p** of vector **b** onto **a** is often denoted by $\mathbf{proj_a b}$.

EXAMPLE 6 Let $\mathbf{a} = [4, 0, -3]$ and $\mathbf{b} = [3, 1, -7]$. Then

$$\mathbf{proj_a b} = \mathbf{p} = \left(\frac{\mathbf{a} \cdot \mathbf{b}}{\|\mathbf{a}\|^2} \right) \mathbf{a} = \frac{(4)(3) + (0)(1) + (-3)(-7)}{\left(\sqrt{16 + 0 + 9} \right)^2} \mathbf{a} = \frac{33}{25} \mathbf{a}$$

$$= \frac{33}{25} [4, 0, -3] = \left[\frac{132}{25}, 0, -\frac{99}{25} \right]. \qquad \blacksquare$$

Next, we algebraically define projection vectors in $\mathbb{R}^n$ to be consistent with the geometric definition in $\mathbb{R}^2$ and $\mathbb{R}^3$.

DEFINITION

If **a** and **b** are vectors in $\mathbb{R}^n$, with $\mathbf{a} \neq \mathbf{0}$, then the **projection vector of b onto a** is

$$\mathbf{proj_a b} = \left(\frac{\mathbf{a} \cdot \mathbf{b}}{\|\mathbf{a}\|^2} \right) \mathbf{a}.$$

The projection vector can be used to decompose a given vector **b** into the sum of two **component vectors**. Suppose $\mathbf{a} \neq \mathbf{0}$. Notice that if $\mathbf{proj_a b} \neq \mathbf{0}$, then it is parallel to **a** by definition because it is a scalar multiple of **a** (see Figure 1.19). Also, $\mathbf{b} - \mathbf{proj_a b}$ is orthogonal to **a** because

$$(\mathbf{b} - \mathbf{proj_a b}) \cdot \mathbf{a} = \mathbf{b} \cdot \mathbf{a} - (\mathbf{proj_a b}) \cdot \mathbf{a}$$
$$= \mathbf{b} \cdot \mathbf{a} - \left(\frac{\mathbf{a} \cdot \mathbf{b}}{\|\mathbf{a}\|^2} \right) (\mathbf{a} \cdot \mathbf{a})$$
$$= \mathbf{b} \cdot \mathbf{a} - \left(\frac{\mathbf{a} \cdot \mathbf{b}}{\|\mathbf{a}\|^2} \right) \|\mathbf{a}\|^2$$
$$= 0.$$

Figure 1.19

Decomposition of a vector **b** into two components, one parallel to **a** and the other orthogonal to **a**

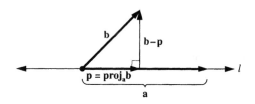

Because $\mathbf{proj_a b} + (\mathbf{b} - \mathbf{proj_a b}) = \mathbf{b}$, we have proved Theorem 1.10.

THEOREM 1.10

Let **a** be a nonzero vector in $\mathbb{R}^n$, and let **b** be any vector in $\mathbb{R}^n$. Then **b** can be decomposed as the sum of two component vectors, $\mathbf{proj_a b}$ and $\mathbf{b} - \mathbf{proj_a b}$, where the first (if nonzero) is parallel to **a** and the second is orthogonal to **a**.

EXAMPLE 7   Consider $\mathbf{a} = [4, 0, -3]$ and $\mathbf{b} = [3, 1, -7]$ from Example 6, where we found the component of **b** in the direction of the vector **a** is $\mathbf{proj_a b} = [132/25, 0, -99/25]$. Then the component of **b** orthogonal to **a** (and **p** as well) is $\mathbf{b} - \mathbf{proj_a b} = [-57/25, 1, -76/25]$. We can easily check that $\mathbf{b} - \mathbf{p}$ is orthogonal to **a** as follows:

$$(\mathbf{b} - \mathbf{p}) \cdot \mathbf{a} = \left(-\frac{57}{25} \right) (4) + (1)(0) + \left(-\frac{76}{25} \right)(-3) = -\frac{228}{25} + \frac{228}{25} = 0.$$ ∎

Application: Work

Suppose that a vector force **f** is exerted on an object and causes the object to undergo a vector displacement **d**. Let θ be the angle between these vectors. In physics, when measuring the work done on the object, only the component of the force that acts in the direction of movement is important. But the component of **f** in the direction of **d** is $\|\mathbf{f}\| \cos \theta$, as shown in Figure 1.20. Thus, the **work** accomplished by the force is defined to be the product of this force

Figure 1.20

Projection $\|\mathbf{f}\| \cos\theta$ of a vector force $\mathbf{f}$ onto a vector displacement $\mathbf{d}$, with angle θ between $\mathbf{f}$ and $\mathbf{d}$

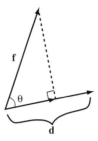

component, $\|\mathbf{f}\| \cos\theta$, times the length $\|\mathbf{d}\|$ of the displacement, which equals $(\|\mathbf{f}\| \cos\theta) \|\mathbf{d}\| = \mathbf{f} \cdot \mathbf{d}$. That is, we can calculate the work simply by finding the dot product of $\mathbf{f}$ and $\mathbf{d}$.

Work is measured in *joules*, where 1 joule is the work done when a force of 1 newton (nt) moves an object 1 meter.

EXAMPLE 8

Suppose that a force of 8 nt is exerted on an object in the direction of the vector $[1, -2, 1]$ and that the object travels 5 m in the direction of the vector $[2, -1, 0]$. Then, $\mathbf{f}$ is 8 times a unit vector in the direction of $[1, -2, 1]$ and $\mathbf{d}$ is 5 times a unit vector in the direction of $[2, -1, 0]$. Therefore, the total work performed is

$$\mathbf{f} \cdot \mathbf{d} = 8 \left(\frac{[1, -2, 1]}{\|[1, -2, 1]\|} \right) \cdot 5 \left(\frac{[2, -1, 0]}{\|[2, -1, 0]\|} \right) = \frac{40 \, (2 + 2 + 0)}{\sqrt{6}\sqrt{5}} \approx 29.2 \text{ joules.}$$

Exercises for Section 1.2

Note: Some exercises ask for proofs. If you have difficulty with these, try them again after working through Section 1.3 in which proof techniques are discussed.

1. Use a calculator to find the angle θ (to the nearest degree) between the following given vectors $\mathbf{x}$ and $\mathbf{y}$:
   ⋆(a) $\mathbf{x} = [-4, 3], \mathbf{y} = [6, -1]$
   (b) $\mathbf{x} = [0, -3, 2], \mathbf{y} = [1, -7, -4]$
   ⋆(c) $\mathbf{x} = [7, -4, 2], \mathbf{y} = [-6, -10, 1]$
   (d) $\mathbf{x} = [-18, -4, -10, 2, -6], \mathbf{y} = [9, 2, 5, -1, 3]$

2. Show that points $A_1(9, 19, 16), A_2(11, 12, 13)$, and $A_3(14, 23, 10)$ are the vertices of a right triangle. (Hint: Construct vectors between the points and check for an orthogonal pair.)

3. (a) Show that $[a, b]$ and $[-b, a]$ are orthogonal. Show that $[a, -b]$ and $[b, a]$ are orthogonal.
   (b) Show that the lines given by the equations $ax + by + c = 0$ and $bx - ay + d = 0$ (where $a, b, c, d \in \mathbb{R}$) are perpendicular by finding a vector in the direction of each line and showing that these vectors are orthogonal. (Hint: Watch out for the cases in which a or b equals zero.)

4. (a) Calculate (in joules) the total work performed by a force $\mathbf{f} = 3\mathbf{i} + 2\mathbf{j} - \mathbf{k}$ (nt) on an object which causes a displacement $\mathbf{d} = -\mathbf{i} + 6\mathbf{j} - 3\mathbf{k}$ (m).

   ⋆(b) Calculate (in joules) the total work performed by a force of 26 nt acting in the direction of the vector $-2\mathbf{i} + 4\mathbf{j} + 5\mathbf{k}$ on an object displaced a total of 10 m in the direction of the vector $-\mathbf{i} + 2\mathbf{j} + 2\mathbf{k}$.

   (c) Calculate (in joules) the total work performed by a force of 6 nt acting in the direction of the vector $3\mathbf{i} - 2\mathbf{j} + 6\mathbf{k}$ on an object displaced a total of 21 m in the direction of the vector $-4\mathbf{i} + 4\mathbf{j} - 7\mathbf{k}$.

5. Why isn't it true that if $\mathbf{x}, \mathbf{y}, \mathbf{z} \in \mathbb{R}^n$, then $\mathbf{x} \cdot (\mathbf{y} \cdot \mathbf{z}) = (\mathbf{x} \cdot \mathbf{y}) \cdot \mathbf{z}$?

▶6. Prove parts (1), (2), (3), (4), and (6) of Theorem 1.5.

⋆7. Does the Cancellation Law of algebra hold for the dot product; that is, assuming that $\mathbf{z} \neq \mathbf{0}$, does $\mathbf{x} \cdot \mathbf{z} = \mathbf{y} \cdot \mathbf{z}$ always imply that $\mathbf{x} = \mathbf{y}$?

8. Finish the proof of Theorem 1.6 by showing that for unit vectors $\mathbf{a}$ and $\mathbf{b}$, $(\mathbf{a} - \mathbf{b}) \cdot (\mathbf{a} - \mathbf{b}) \geq 0$ implies $\mathbf{a} \cdot \mathbf{b} \leq 1$.

9. Prove that if $(\mathbf{x} + \mathbf{y}) \cdot (\mathbf{x} - \mathbf{y}) = 0$, then $\|\mathbf{x}\| = \|\mathbf{y}\|$. (Hence, if the diagonals of a parallelogram are perpendicular, then the parallelogram is a rhombus.)

10. Prove that $\frac{1}{2}\left(\|\mathbf{x} + \mathbf{y}\|^2 + \|\mathbf{x} - \mathbf{y}\|^2\right) = \|\mathbf{x}\|^2 + \|\mathbf{y}\|^2$ for any vectors $\mathbf{x}, \mathbf{y}$ in $\mathbb{R}^n$. (This equation is known as the **Parallelogram Identity** because it asserts that the sum of the squares of the lengths of all four sides of a parallelogram equals the sum of the squares of the diagonals.)

11. (a) Prove that for vectors $\mathbf{x}, \mathbf{y}$ in $\mathbb{R}^n$, $\|\mathbf{x} + \mathbf{y}\|^2 = \|\mathbf{x}\|^2 + \|\mathbf{y}\|^2$ if and only if $\mathbf{x} \cdot \mathbf{y} = 0$.

    (b) Prove that if $\mathbf{x}, \mathbf{y}, \mathbf{z}$ are mutually orthogonal vectors in $\mathbb{R}^n$, then $\|\mathbf{x} + \mathbf{y} + \mathbf{z}\|^2 = \|\mathbf{x}\|^2 + \|\mathbf{y}\|^2 + \|\mathbf{z}\|^2$.

    (c) Prove that $\mathbf{x} \cdot \mathbf{y} = \frac{1}{4}\left(\|\mathbf{x} + \mathbf{y}\|^2 - \|\mathbf{x} - \mathbf{y}\|^2\right)$, if $\mathbf{x}$ and $\mathbf{y}$ are vectors in $\mathbb{R}^n$. (This result, a form of the **Polarization Identity**, gives a way of defining the dot product using the norms of vectors.)

12. Given $\mathbf{x}, \mathbf{y}, \mathbf{z}$ in $\mathbb{R}^n$, with $\mathbf{x}$ orthogonal to both $\mathbf{y}$ and $\mathbf{z}$, prove that $\mathbf{x}$ is orthogonal to $c_1\mathbf{y} + c_2\mathbf{z}$, where $c_1, c_2 \in \mathbb{R}$.

⋆13. Let $\mathbf{x} = [a, b, c]$ be a vector in $\mathbb{R}^3$. If θ_1, θ_2, and θ_3 are the angles that $\mathbf{x}$ forms with the x-, y-, and z-axes, respectively, find formulas for $\cos\theta_1$, $\cos\theta_2$, and $\cos\theta_3$ in terms of a, b, c, and show that $\cos^2\theta_1 + \cos^2\theta_2 + \cos^2\theta_3 = 1$. (Note: $\cos\theta_1$, $\cos\theta_2$, and $\cos\theta_3$ are commonly known as the **direction cosines** of the vector $\mathbf{x}$. See Exercise 14(b) in Section 1.1.)

⋆14. (a) If the side of a cube has length s, what is the length of the cube's diagonal?

    (b) Using vectors, find the angle that the diagonal makes with one of the sides of the cube.

15. Calculate $\mathbf{proj_a b}$ in each case, and verify $\mathbf{b} - \mathbf{proj_a b}$ is orthogonal to $\mathbf{a}$.

    ⋆(a) $\mathbf{a} = [2, 1, 5]$, $\mathbf{b} = [1, 4, -3]$

    (b) $\mathbf{a} = [-5, 3, 0]$, $\mathbf{b} = [3, -7, 1]$

    ⋆(c) $\mathbf{a} = [1, 0, -1, 2]$, $\mathbf{b} = [3, -1, 0, -1]$

16. (a) Suppose that $\mathbf{a}$ is orthogonal to $\mathbf{b}$ in $\mathbb{R}^n$. What is $\mathbf{proj_a b}$? Why? Give a geometric interpretation in $\mathbb{R}^2$ or $\mathbb{R}^3$.

    (b) Suppose $\mathbf{a}$ and $\mathbf{b}$ are parallel vectors in $\mathbb{R}^n$. What is $\mathbf{proj_a b}$? Why? Give a geometric interpretation in $\mathbb{R}^2$ or $\mathbb{R}^3$.

★17. What are the projections of the general vector $[a, b, c]$ onto each of the vectors $\mathbf{i}, \mathbf{j}$, and $\mathbf{k}$ in turn?

18. Let $\mathbf{x} = [-6, 2, 7]$ represent the force on an object in a three-dimensional coordinate system. Decompose $\mathbf{x}$ into two component forces in directions parallel and orthogonal to each vector given.

    ★(a) $[2, -3, 4]$

    (b) $[-1, 2, -1]$

    ★(c) $[3, -2, 6]$

19. Show that if ℓ is any line through the origin in $\mathbb{R}^3$ and $\mathbf{x}$ is any vector with its initial point at the origin, then the **reflection** of $\mathbf{x}$ through the line ℓ (acting as a mirror) is equal to $2(\mathbf{proj_r x}) - \mathbf{x}$, where $\mathbf{r}$ is any nonzero vector parallel to the line ℓ (see Figure 1.21).

Figure 1.21

Reflection of $\mathbf{x}$ through the line ℓ

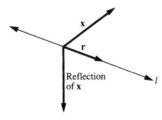

20. Prove the **Reverse Triangle Inequality;** that is, for any vectors $\mathbf{x}$ and $\mathbf{y}$ in $\mathbb{R}^n$, $\left| \|\mathbf{x}\| - \|\mathbf{y}\| \right| \le \|\mathbf{x} + \mathbf{y}\|$. (Hint: Consider the cases $\|\mathbf{x}\| \le \|\mathbf{y}\|$ and $\|\mathbf{x}\| \ge \|\mathbf{y}\|$ separately.)

21. Let $\mathbf{x}$ and $\mathbf{y}$ be nonzero vectors in $\mathbb{R}^n$.

    (a) Prove that $\mathbf{y} = c\mathbf{x} + \mathbf{w}$ for some scalar c and some vector $\mathbf{w}$ such that $\mathbf{w}$ is orthogonal to $\mathbf{x}$.

    (b) Show that the vector $\mathbf{w}$ and the scalar c in part (a) are unique; that is, show that if $\mathbf{y} = c\mathbf{x} + \mathbf{w}$ and $\mathbf{y} = d\mathbf{x} + \mathbf{v}$, where $\mathbf{w}$ and $\mathbf{v}$ are both orthogonal to $\mathbf{x}$, then $c = d$ and $\mathbf{w} = \mathbf{v}$. (Hint: Compute $\mathbf{x} \cdot \mathbf{y}$.)

22. If $\mathbf{x}, \mathbf{y} \in \mathbb{R}^n$ such that $\mathbf{x} \cdot \mathbf{y} \ne 0$, prove that the angle between $\mathbf{x}$ and $\mathbf{y}$ equals the angle between $\mathbf{proj_x y}$ and $\mathbf{proj_y x}$.

★23. True or False:

    (a) For any vectors $\mathbf{x}, \mathbf{y}$ in $\mathbb{R}^n$, and any scalar d, $\mathbf{x} \cdot (d\mathbf{y}) = (d\mathbf{x}) \cdot \mathbf{y}$.

    (b) For all $\mathbf{x}, \mathbf{y}$ in $\mathbb{R}^n$ with $\mathbf{x} \ne \mathbf{0}$, $(\mathbf{x} \cdot \mathbf{y})/\|\mathbf{x}\| \le \|\mathbf{y}\|$.

    (c) For all $\mathbf{x}, \mathbf{y}$ in $\mathbb{R}^n$, $\|\mathbf{x} - \mathbf{y}\| \le \|\mathbf{x}\| - \|\mathbf{y}\|$.

    (d) If θ is the angle between vectors $\mathbf{x}$ and $\mathbf{y}$ in $\mathbb{R}^n$, and $\theta > \frac{\pi}{2}$, then $\mathbf{x} \cdot \mathbf{y} > 0$.

    (e) The standard unit vectors in $\mathbb{R}^n$ are mutually orthogonal.

    (f) If $\mathbf{proj_a b} = \mathbf{b}$, then $\mathbf{a}$ is perpendicular to $\mathbf{b}$.

1.3 An Introduction to Proof Techniques

In reading this book, you will spend much time studying the proofs of theorems, and for the exercises, you will often write proofs. Hence, in this section we discuss several methods of proving theorems in order to sharpen your skills in reading and writing proofs.

The "results" (not all new) proved in this section are intended only to illustrate various proof techniques. Therefore, they are not labeled as "theorems."

Proof Techniques: Direct Proof

The most straightforward proof method is **direct proof**, a logical step-by-step argument concluding with the statement to be proved. The following is a direct proof for a familiar result from Theorem 1.5:

RESULT 1

Let $\mathbf{x}$ be a vector in $\mathbb{R}^n$. Then $\mathbf{x} \cdot \mathbf{x} = \|\mathbf{x}\|^2$.

PROOF OF RESULT 1

Step 1:	Let $\mathbf{x} = [x_1, \ldots, x_n]$	because $\mathbf{x} \in \mathbb{R}^n$
Step 2:	$\mathbf{x} \cdot \mathbf{x} = x_1^2 + \cdots + x_n^2$	definition of dot product
Step 3:	$\|\mathbf{x}\| = \sqrt{x_1^2 + \cdots + x_n^2}$	definition of $\|\mathbf{x}\|$
Step 4:	$\|\mathbf{x}\|^2 = x_1^2 + \cdots + x_n^2$	squaring both sides of Step 3
Step 5:	$\mathbf{x} \cdot \mathbf{x} = \|\mathbf{x}\|^2$	from Steps 2 and 4 ∎

Each step in a direct proof should follow immediately from a definition, a previous step, or a known fact. The reasons for each step should be clearly stated when necessary for the intended reader. However, the above type of presentation is infrequently used. A more typical paragraph version of the same argument is

PROOF OF RESULT 1

If $\mathbf{x}$ is a vector in $\mathbb{R}^n$, then we can express $\mathbf{x}$ as $[x_1, x_2, \ldots, x_n]$ for some real numbers $x_1, \ldots, x_n$. Now, $\mathbf{x} \cdot \mathbf{x} = x_1^2 + \cdots + x_n^2$, by definition of the dot product. However, $\|\mathbf{x}\| = \sqrt{x_1^2 + \cdots + x_n^2}$, by definition of the length of a vector. Therefore, $\|\mathbf{x}\|^2 = \mathbf{x} \cdot \mathbf{x}$, because both sides are equal to $x_1^2 + \cdots + x_n^2$. ∎

The paragraph form should contain the same information as the step-by-step form and be presented in such a way that a corresponding step-by-step proof occurs naturally to the reader. We present most proofs in this book in paragraph style. But you may want to begin writing proofs in the step-by-step format and then change to paragraph style once you have more experience with proofs.

Stating the definitions of the relevant terms is usually a good beginning when tackling a proof because it helps to clarify what you must prove. For example, the first four of the five steps in the step-by-step proof of Result 1 merely involve writing what each side of $\mathbf{x} \cdot \mathbf{x} = \|\mathbf{x}\|^2$ means. The final result then follows naturally.

Working "Backward" to Discover a Proof

A method often used when there is no obvious direct proof is to work "backward" — that is, to start with the desired conclusion and work in reverse toward the given facts. Although these "reversed" steps do not constitute a proof, they may provide sufficient insight to make construction of a "forward" proof easier, as we now illustrate.

RESULT 2

Let $\mathbf{x}$ and $\mathbf{y}$ be nonzero vectors in $\mathbb{R}^n$. If $\mathbf{x} \cdot \mathbf{y} \geq 0$, then $\|\mathbf{x} + \mathbf{y}\| > \|\mathbf{y}\|$.

We begin with the desired conclusion $\|\mathbf{x} + \mathbf{y}\| > \|\mathbf{y}\|$ and try to work "backward" toward the given fact $\mathbf{x} \cdot \mathbf{y} \geq 0$, as follows:

$$\|\mathbf{x} + \mathbf{y}\| > \|\mathbf{y}\|$$

$$\|\mathbf{x} + \mathbf{y}\|^2 > \|\mathbf{y}\|^2$$

$$(\mathbf{x} + \mathbf{y}) \cdot (\mathbf{x} + \mathbf{y}) > \|\mathbf{y}\|^2$$

$$\mathbf{x} \cdot \mathbf{x} + 2\mathbf{x} \cdot \mathbf{y} + \mathbf{y} \cdot \mathbf{y} > \|\mathbf{y}\|^2$$

$$\|\mathbf{x}\|^2 + 2\mathbf{x} \cdot \mathbf{y} + \|\mathbf{y}\|^2 > \|\mathbf{y}\|^2$$

$$\|\mathbf{x}\|^2 + 2\mathbf{x} \cdot \mathbf{y} > 0.$$

At this point, we cannot easily continue going "backward." However, the last inequality is true if $\mathbf{x} \cdot \mathbf{y} \geq 0$. Therefore, we *reverse* the above steps to create the following "forward" proof of Result 2:

PROOF OF RESULT 2

Step 1:	$\|\mathbf{x}\|^2 > 0$	$\mathbf{x}$ is nonzero
Step 2:	$2(\mathbf{x} \cdot \mathbf{y}) \geq 0$	because $\mathbf{x} \cdot \mathbf{y} \geq 0$
Step 3:	$\|\mathbf{x}\|^2 + 2\mathbf{x} \cdot \mathbf{y} > 0$	from Steps 1 and 2
Step 4:	$\|\mathbf{x}\|^2 + 2\mathbf{x} \cdot \mathbf{y} + \|\mathbf{y}\|^2 > \|\mathbf{y}\|^2$	
Step 5:	$\mathbf{x} \cdot \mathbf{x} + 2\mathbf{x} \cdot \mathbf{y} + \mathbf{y} \cdot \mathbf{y} > \|\mathbf{y}\|^2$	from Theorem 1.5, part (2)
Step 6:	$(\mathbf{x} + \mathbf{y}) \cdot (\mathbf{x} + \mathbf{y}) > \|\mathbf{y}\|^2$	from Theorem 1.5, parts (5) and (6)
Step 7:	$\|\mathbf{x} + \mathbf{y}\|^2 > \|\mathbf{y}\|^2$	from Theorem 1.5, part (2)
Step 8:	$\|\mathbf{x} + \mathbf{y}\| > \|\mathbf{y}\|$	take square root of both sides; length is always nonnegative ∎

When "working backward," your steps must be reversed for the final proof. Therefore, each step must be carefully examined to determine if it is "reversible." For example, if t is a real number, then $t > 5 \Rightarrow t^2 > 25$ is a valid step, but reversing this yields $t^2 > 25 \Rightarrow t > 5$, which is certainly an invalid step if $t < -5$. Notice that we were very careful in Step 8 of the proof when we took the square root of both sides to ensure the step was indeed valid.

"If A Then B" Proofs

Frequently, a theorem is given in the form "If A then B," where A and B represent statements. An example is "If $\|\mathbf{x}\| = 0$, then $\mathbf{x} = \mathbf{0}$" for vectors $\mathbf{x}$ in $\mathbb{R}^n$, where A is "$\|\mathbf{x}\| = 0$" and B is "$\mathbf{x} = \mathbf{0}$." The entire "If A then B" statement is called an **implication**; A alone is the **premise**, and B is the **conclusion**. The meaning of "If A then B" is that, whenever A is true, B is true as well. Thus, the implication "If $\|\mathbf{x}\| = 0$, then $\mathbf{x} = \mathbf{0}$" means that, if we know $\|\mathbf{x}\| = 0$ for some particular vector $\mathbf{x}$ in $\mathbb{R}^n$, then we can conclude that $\mathbf{x}$ is the zero vector.

Note that the implication "If A then B" asserts nothing about the truth or falsity of B unless A is true.[4] Therefore, to prove "If A then B," we assume A is true and try to prove B is also true. This is illustrated in the proof of the next result, a part of Theorem 1.8.

RESULT 3

If $\mathbf{x}$ and $\mathbf{y}$ are nonzero vectors in $\mathbb{R}^n$ such that $\mathbf{x} \cdot \mathbf{y} > 0$, then the angle between $\mathbf{x}$ and $\mathbf{y}$ is acute.

PROOF OF RESULT 3

The premise in this result is "$\mathbf{x}$ and $\mathbf{y}$ are nonzero vectors and $\mathbf{x} \cdot \mathbf{y} > 0$." The conclusion is "the angle between $\mathbf{x}$ and $\mathbf{y}$ is acute." We begin by assuming that both parts of the premise are true.

Step 1:	$\mathbf{x}$ and $\mathbf{y}$ are nonzero	first part of premise
Step 2:	$\|\mathbf{x}\| > 0$ and $\|\mathbf{y}\| > 0$	Theorem 1.5, parts (2) and (3)
Step 3:	$\mathbf{x} \cdot \mathbf{y} > 0$	second part of premise
Step 4:	$\cos\theta = \frac{\mathbf{x}\cdot\mathbf{y}}{\|\mathbf{x}\|\|\mathbf{y}\|}$, where θ is the angle between $\mathbf{x}$ and $\mathbf{y}$, and $0 \le \theta \le \pi$	definition of the angle between two vectors
Step 5:	$\cos\theta > 0$	quotient of positive reals is positive
Step 6:	θ is acute	since $0 \le \theta \le \pi$, $\cos\theta > 0$ only if $0 < \theta < \frac{\pi}{2}$ ∎

---

[4] In formal logic, when A is false, the implication "If A then B" is considered true but worthless because it tells us absolutely nothing about B. For example, the implication "If every vector in $\mathbb{R}^3$ is a unit vector, then the inflation rate will be 8% next year" is considered true because the premise "every vector in $\mathbb{R}^3$ is a unit vector" is clearly false. However, the implication is useless. It tells us nothing about next year's inflation rate, which is free to take any value, such as 6%.

Beware! An implication is not always written in the form "If A then B."

Some Equivalent Forms for "If A Then B"	
A implies B	B if A
$A \Rightarrow B$	A is a sufficient condition for B
A only if B	B is a necessary condition for A

Another common practice is to place some of the conditions of the premise before the "If . . . then." For example, Result 3 might be rewritten as

Let $\mathbf{x}$ and $\mathbf{y}$ be nonzero vectors in $\mathbb{R}^n$. If $\mathbf{x} \cdot \mathbf{y} > 0$, then the angle between $\mathbf{x}$ and $\mathbf{y}$ is acute.

The condition "$\mathbf{x}$ and $\mathbf{y}$ are nonzero vectors in $\mathbb{R}^n$" sets the stage for the implication to come. Such conditions are treated as given information along with the premise in the actual proof.

"A If and Only If B" Proofs

Some theorems have the form "A if and only if B." This is really a combination of two statements: "If A then B" and "If B then A." Both of these statements must be shown true to fully complete the proof of the original statement. In essence, we must show A and B are logically equivalent: the "If A then B" half means that whenever A is true, B must follow; the "If B then A" half means that whenever B is true, A must follow. Therefore, A is true exactly when B is true. For an example of an "if and only if" argument, we prove the following special case of Theorem 1.9.

RESULT 4
Let $\mathbf{x}$ and $\mathbf{y}$ be nonzero vectors in $\mathbb{R}^n$. Then $\mathbf{x} \cdot \mathbf{y} = \|\mathbf{x}\| \, \|\mathbf{y}\|$ if and only if $\mathbf{y}$ is a positive scalar multiple of $\mathbf{x}$.

In an "if and only if" proof, it is usually good to begin by stating the two halves of the "if and only if" statement. This gives a clearer picture of what is given and what must be proved in each half. In Result 4, the two halves are

1. Suppose that $\mathbf{y} = c\mathbf{x}$ for some positive $c \in \mathbb{R}$. Prove that $\mathbf{x} \cdot \mathbf{y} = \|\mathbf{x}\| \, \|\mathbf{y}\|$.
2. Suppose that $\mathbf{x} \cdot \mathbf{y} = \|\mathbf{x}\| \, \|\mathbf{y}\|$. Prove that there is some $c > 0$ such that $\mathbf{y} = c\mathbf{x}$.

The assumption "Let $\mathbf{x}$ and $\mathbf{y}$ be nonzero vectors in $\mathbb{R}^n$" is considered given information for both halves.

PROOF OF RESULT 4

Part 1: We suppose that $\mathbf{y} = c\mathbf{x}$ for some $c > 0$. Then,

$$
\begin{aligned}
\mathbf{x} \cdot \mathbf{y} \quad &= \mathbf{x} \cdot (c\mathbf{x}) && \text{because } \mathbf{y} = c\mathbf{x} \\
&= c\,(\mathbf{x} \cdot \mathbf{x}) && \text{Theorem 1.5, part (4)} \\
&= c\,\|\mathbf{x}\|^2 && \text{Theorem 1.5, part (2)} \\
&= \|\mathbf{x}\|\,(c\,\|\mathbf{x}\|) && \text{associative law of multiplication} \\
& && \qquad \text{for real numbers} \\
&= \|\mathbf{x}\|\,(|c|\,\|\mathbf{x}\|) && \text{because } c > 0 \\
&= \|\mathbf{x}\|\,\|c\mathbf{x}\| && \text{Theorem 1.1} \\
&= \|\mathbf{x}\|\,\|\mathbf{y}\| && \text{because } \mathbf{y} = c\mathbf{x}.
\end{aligned}
$$

Part 2: We assume that $\mathbf{x} \cdot \mathbf{y} = \|\mathbf{x}\|\,\|\mathbf{y}\|$ and show that there is some $c > 0$ such that $\mathbf{y} = c\mathbf{x}$. By Theorem 1.10, $\mathbf{y}$ can be expressed as $\mathbf{proj_x y} + \mathbf{w}$, where $\mathbf{w}$ is orthogonal to $\mathbf{x}$. Our strategy is first to show that $\mathbf{proj_x y}$ is a positive scalar multiple of $\mathbf{x}$ and then to show that $\mathbf{w} = \mathbf{0}$. Therefore, $\mathbf{y} = c\mathbf{x}$ with $c > 0$, and the proof is done.

First, note that

$$
\begin{aligned}
\mathbf{proj_x y} \quad &= \left(\frac{\mathbf{x} \cdot \mathbf{y}}{\|\mathbf{x}\|^2}\right)\mathbf{x} && \text{formula for } \mathbf{proj_x y} \\
&= \left(\frac{\|\mathbf{x}\|\|\mathbf{y}\|}{\|\mathbf{x}\|^2}\right)\mathbf{x} && \text{because } \mathbf{x} \cdot \mathbf{y} = \|\mathbf{x}\|\,\|\mathbf{y}\| \\
&= \left(\frac{\|\mathbf{y}\|}{\|\mathbf{x}\|}\right)\mathbf{x}.
\end{aligned}
$$

Let $c = \|\mathbf{y}\| \,/\, \|\mathbf{x}\|$. Note that c is positive.

Finally, we conclude by showing $\mathbf{w} = \mathbf{0}$. Now,

$$
\begin{aligned}
\|\mathbf{w}\|^2 &= \mathbf{w} \cdot \mathbf{w} && \text{Theorem 1.5, part (2)} \\
&= (\mathbf{y} - c\mathbf{x}) \cdot (\mathbf{y} - c\mathbf{x}) && \text{because } \mathbf{y} = c\mathbf{x} + \mathbf{w} \\
&= (\mathbf{y} \cdot \mathbf{y}) - 2c\,(\mathbf{x} \cdot \mathbf{y}) + c^2\,(\mathbf{x} \cdot \mathbf{x}) && \text{distributive law of dot product} \\
& && \qquad \text{over addition} \\
&= \|\mathbf{y}\|^2 - 2c\,\|\mathbf{x}\|\,\|\mathbf{y}\| + c^2\,\|\mathbf{x}\|^2 && \text{Theorem 1.5, part (2), and} \\
& && \qquad \mathbf{x} \cdot \mathbf{y} = \|\mathbf{x}\|\,\|\mathbf{y}\| \\
&= \|\mathbf{y}\|^2 - 2\,\|\mathbf{y}\|^2 + \|\mathbf{y}\|^2 && \text{because } c = \frac{\|\mathbf{y}\|}{\|\mathbf{x}\|},
\end{aligned}
$$

which equals zero, and so $\mathbf{w} = \mathbf{0}$. The proof is complete. ∎

Note that two proofs are required to prove an "if and only if" type of statement — one for each of the implications involved. Also, each half is not necessarily just a reversal of the steps in the other half. Sometimes the two halves must be proved very differently, as for Result 4.

Other common alternate forms for "if and only if" are

Some Equivalent Forms for "A If and Only If B"
A iff B
$A \Leftrightarrow B$
A is a necessary and sufficient condition for B

"If A Then (B or C)" Proofs

Sometimes we must prove a statement of the form "If A then (B or C)." [5] This is an implication whose conclusion has two parts. Note that B is either true or false. Now, if B is true, there is no need for a proof, because we only need to establish that *either* B or C holds. For this reason, "If A then (B or C)" is equivalent to "If A is true and B is *false*, then C is true." That is, we are allowed to assume that B is false, and then use this extra information to prove C is true. This strategy often makes the proof easier. As an example, consider the following result:

> **RESULT 5**
> If $\mathbf{x}$ is a nonzero vector in $\mathbb{R}^2$, then $\mathbf{x} \cdot [1, 0] \neq 0$ or $\mathbf{x} \cdot [0, 1] \neq 0$.

In this case, $A =$ "$\mathbf{x}$ is a nonzero vector in $\mathbb{R}^2$," $B =$ "$\mathbf{x} \cdot [1, 0] \neq 0$," and $C =$ "$\mathbf{x} \cdot [0, 1] \neq 0$." Assuming B is false, we obtain the following statement equivalent to Result 5:

If $\mathbf{x}$ is a nonzero vector in $\mathbb{R}^n$ and $\mathbf{x} \cdot [1, 0] = 0$, then $\mathbf{x} \cdot [0, 1] \neq 0$.

Proving this (which can be done with a direct proof – try it!) has the effect of proving the original statement in Result 5.

Of course, an alternate way of proving "If A then (B or C)" is to assume instead that C is false and use this assumption to prove B is true.

Proof Techniques: Proof by Contrapositive

Related to the implication "If A then B" is its **contrapositive**: "If not B, then not A." For example, for an integer n, the statement "If n^2 is even, then n is even" has the contrapositive "If n is odd (that is, not even), then n^2 is odd." A statement and its contrapositive are always logically equivalent; that is, they are either both true or both false together. Therefore, proving the contrapositive of any statement (known as **proof by contrapositive**) has the effect of proving the

---

[5] In this text, *or* is used in the **inclusive** sense. That is, "A or B" always means "A or B or both." For example, "n is even or prime" means that n could be even or n could be prime or n could be both. Therefore, "n is even or prime" is true for $n = 2$, which is both even and prime, as well as for $n = 6$ (even but not prime) and $n = 7$ (prime but not even). However, in English, the word *or* is frequently used in the **exclusive** sense, as in "You may have the prize behind the curtain *or* the cash in my hand," where you are not meant to have both prizes. The "exclusive or" is rarely used in mathematics.

original statement as well. In many cases, the contrapositive is easier to prove. The following result illustrates this method:

RESULT 6

Let $\mathbf{x}$ be a vector in $\mathbb{R}^n$. If $\|\mathbf{x}\| = 0$, then $\mathbf{x} = \mathbf{0}$.

PROOF OF RESULT 6

To prove this result, we give a direct proof of its contrapositive: if $\mathbf{x} \neq \mathbf{0}$, then $\|\mathbf{x}\| \neq 0$.

Step 1:	Let $\mathbf{x} = [x_1, \ldots, x_n] \neq \mathbf{0}$	premise of contrapositive
Step 2:	For some i, $1 \leq i \leq n$, we have $x_i \neq 0$	
Step 3:	$\|\mathbf{x}\| = \sqrt{x_1^2 + \cdots + x_i^2 + \cdots + x_n^2}$	
Step 4:	$\|\mathbf{x}\|^2 = x_1^2 + \cdots + x_i^2 + \cdots + x_n^2$	
Step 5:	$\|\mathbf{x}\|^2 \geq x_i^2 > 0$	
Step 6:	$\|\mathbf{x}\|^2 > 0$	
Step 7:	$\|\mathbf{x}\| \neq 0$	

You should fill in the missing reasons for Steps 2 through 7 to complete the proof of the contrapositive and hence the proof of the result itself. ∎

Converse and Inverse

Along with the contrapositive, there are two other related statements of interest — the **converse** and **inverse**:

Original Statement	If A then B
Contrapositive	If not B then not A
Converse	If B then A
Inverse	If not A then not B

Notice that, when "If A then B" and its converse "If B then A" are combined together, they form the familiar "A if and only if B" statement.

Although the converse and inverse may resemble the contrapositive, take care: neither the converse nor the inverse is logically equivalent to the original statement. However, the converse and inverse of a statement are equivalent to each other, and are both true or both false together. For example, consider "If $\mathbf{x} = \mathbf{y}$, then $\mathbf{x} \cdot \mathbf{y} = \|\mathbf{x}\|^2$," for vectors in $\mathbb{R}^n$.

Original Statement	If $\mathbf{x} = \mathbf{y}$, then $\mathbf{x} \cdot \mathbf{y} = \|\mathbf{x}\|^2$	equivalent to each other
Contrapositive	If $\mathbf{x} \cdot \mathbf{y} \neq \|\mathbf{x}\|^2$, then $\mathbf{x} \neq \mathbf{y}$	
Converse	If $\mathbf{x} \cdot \mathbf{y} = \|\mathbf{x}\|^2$, then $\mathbf{x} = \mathbf{y}$	equivalent to each other
Inverse	If $\mathbf{x} \neq \mathbf{y}$, then $\mathbf{x} \cdot \mathbf{y} \neq \|\mathbf{x}\|^2$	

Notice that in this case the original statement and its contrapositive are both true; the converse and the inverse are both false (see Exercise 5).

Beware! It is possible for a statement and its converse to have the same truth value. For example, the converse of Result 6 is "If $\mathbf{x} = \mathbf{0}$, then $\|\mathbf{x}\| = 0$," and this is also a true statement. The moral here is that a statement and its converse are logically independent, and thus, proving the converse (or inverse) is never acceptable as a valid proof of the original statement.

Finally, when constructing the contrapositive, converse, or inverse of an "If A then B" statement, you should not change the accompanying conditions. For instance, consider the condition "Let $\mathbf{x}$ and $\mathbf{y}$ be nonzero vectors in $\mathbb{R}^n$" of Result 2. The contrapositive, converse, and inverse should all begin with this condition. For example, the contrapositive of Result 2 is "Let $\mathbf{x}$ and $\mathbf{y}$ be nonzero vectors in $\mathbb{R}^n$. If $\|\mathbf{x} + \mathbf{y}\| \leq \|\mathbf{y}\|$, then $\mathbf{x} \cdot \mathbf{y} < 0$."

Proof Technique: Proof by Contradiction

Another common proof method is **proof by contradiction,** in which we assume the statement to be proved is false and use this assumption to contradict a known fact. In effect, we prove a result by showing that if it were false it would be inconsistent with some other true statement, as in the proof of the following result:

RESULT 7

Let $S = \{\mathbf{x}_1, \ldots, \mathbf{x}_k\}$ be a set of mutually orthogonal nonzero vectors in $\mathbb{R}^n$. Then no vector in S can be expressed as a linear combination of the other vectors in S.

Recall that a set $\{\mathbf{x}_1, \ldots, \mathbf{x}_k\}$ of nonzero vectors is mutually orthogonal if and only if $\mathbf{x}_i \cdot \mathbf{x}_j = 0$ whenever $i \neq j$.

PROOF OF RESULT 7

To prove this by contradiction, we assume it is false; that is, *some* vector in S *can* be expressed as a linear combination of the other vectors in S. That is, some $\mathbf{x}_i = a_1\mathbf{x}_1 + \cdots + a_{i-1}\mathbf{x}_{i-1} + a_{i+1}\mathbf{x}_{i+1} + \cdots + a_k\mathbf{x}_k$, for some $a_1, \ldots, a_{i-1}, a_{i+1}, \ldots, a_k \in \mathbb{R}$. We then show this assumption leads to a contradiction:

$$\begin{aligned}
\mathbf{x}_i \cdot \mathbf{x}_i &= \mathbf{x}_i \cdot (a_1\mathbf{x}_1 + \cdots + a_{i-1}\mathbf{x}_{i-1} + a_{i+1}\mathbf{x}_{i+1} + \cdots + a_k\mathbf{x}_k) \\
&= a_1(\mathbf{x}_i \cdot \mathbf{x}_1) + \cdots + a_{i-1}(\mathbf{x}_i \cdot \mathbf{x}_{i-1}) + a_{i+1}(\mathbf{x}_i \cdot \mathbf{x}_{i+1}) + \cdots + a_k(\mathbf{x}_i \cdot \mathbf{x}_k) \\
&= a_1(0) + \cdots + a_{i-1}(0) + a_{i+1}(0) + \cdots + a_k(0) = 0.
\end{aligned}$$

Hence, $\mathbf{x}_i = \mathbf{0}$, by part (3) of Theorem 1.5. This equation contradicts the given fact that $\mathbf{x}_1, \ldots, \mathbf{x}_k$ are all nonzero vectors, thus completing the proof. ■

A mathematician generally constructs a proof by contradiction by assuming that the given statement is false and then investigates where this assumption leads until some absurdity appears. Of course, any "blind alleys" encountered in the investigation should not appear in the final proof.

In the preceding proof we assumed that some chosen vector $\mathbf{x}_i$ could be expressed as a linear combination of the other vectors. However, we could easily have renumbered the vectors so that $\mathbf{x}_i$ becomes $\mathbf{x}_1$, and the other vectors are $\mathbf{x}_2$ through $\mathbf{x}_k$. A mathematician would express this by writing: "We assume some vector in S can be expressed as a linear combination of the others. Without loss of generality, choose $\mathbf{x}_1$ to be this vector." This phrase **"without loss of generality"** implies here that the vectors have been suitably rearranged if necessary, so that $\mathbf{x}_1$ now has the desired property. Then our assumption in the proof of Result 7 would be: $\mathbf{x}_1 = a_2\mathbf{x}_2 + \cdots + a_k\mathbf{x}_k$. The proof is now simpler to express, since we do not have to skip over subscript "i."

$$\begin{aligned} \mathbf{x}_1 \cdot \mathbf{x}_1 &= \mathbf{x}_1 \cdot (a_2\mathbf{x}_2 + \cdots + a_k\mathbf{x}_k) \\ &= a_2(\mathbf{x}_1 \cdot \mathbf{x}_2) + \cdots + a_k(\mathbf{x}_1 \cdot \mathbf{x}_k) \\ &= a_2(0) + \cdots + a_k(0) = 0. \end{aligned}$$

Proof Technique: Proof by Induction

The method of **proof by induction** is used to show that a statement is true for all values of an integer variable greater than or equal to some initial value i. For example, $A =$ "For every integer $n \geq 1$, $1^2 + 2^2 + \cdots + n^2 = n(n+1)(2n+1)/6$" can be proved by induction for all integers n greater than or equal to the initial value $i = 1$. You may have seen such a proof in your calculus course.

There are two steps in any induction proof, the **Base Step** and the **Inductive Step**.

> **(1) Base Step:** Prove that the desired statement is true for the initial value i of the (integer) variable.
> **(2) Inductive Step:** Prove that if the statement is true for an integer value k of the variable (with $k \geq i$), then the statement is true for the next integer value $k + 1$ as well.

These two steps together show that the statement is true for every integer greater than or equal to the initial value i because the Inductive Step sets up a "chain of implications," as in Figure 1.22. First, the Base Step implies that the initial statement, A_i, is true. But A_i is the premise for the first implication in the chain. Hence, the Inductive Step tells us that the conclusion of this implication, A_{i+1}, must also be true. However, A_{i+1} is the premise of the second implication; hence, the Inductive Step tells us that the conclusion A_{i+2} must be true. In this way, the statement is true for each integer value $\geq i$.

The process of induction can be likened to knocking down a line of dominoes—one domino for each integer greater than or equal to the initial value. Keep in mind that the Base Step is needed to knock over the first domino

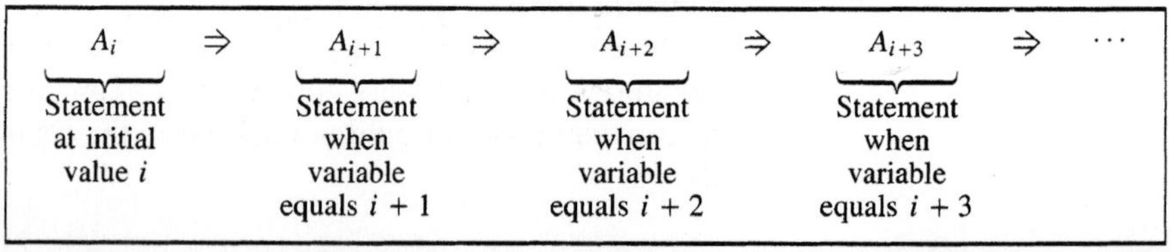

Figure 1.22

Chain of implications set up by the Inductive Step

A_i	$\Rightarrow$	A_{i+1}	$\Rightarrow$	A_{i+2}	$\Rightarrow$	A_{i+3}	$\Rightarrow$	$\cdots$
Statement at initial value i		Statement when variable equals $i + 1$		Statement when variable equals $i + 2$		Statement when variable equals $i + 3$		

and thus start the entire process. Without the Base Step, we cannot be sure that the given statement is true for any integer value at all. The next proof illustrates the induction technique:

RESULT 8

Let $\mathbf{z}$, and $\mathbf{x}_1, \mathbf{x}_2, \ldots, \mathbf{x}_n$ (for $n \geq 1$) be vectors in $\mathbb{R}^m$, and let $c_1, c_2, \ldots, c_n \in \mathbb{R}$. Then,

$$(c_1\mathbf{x}_1 + c_2\mathbf{x}_2 + \cdots + c_n\mathbf{x}_n) \cdot \mathbf{z} = c_1(\mathbf{x}_1 \cdot \mathbf{z}) + c_2(\mathbf{x}_2 \cdot \mathbf{z}) + \cdots + c_n(\mathbf{x}_n \cdot \mathbf{z}).$$

This is a generalization of part (6) of Theorem 1.5, where a linear combination replaces a single addition of vectors.

PROOF OF RESULT 8

The integer induction variable is n, with initial value $i = 1$.

Base Step: The Base Step is typically proved by plugging in the initial value and verifying the result is true in that case. When $n = 1$, the left-hand side of the equation in Result 8 has only one term: $(c_1\mathbf{x}_1) \cdot \mathbf{z}$, while the right-hand side yields $c_1(\mathbf{x}_1 \cdot \mathbf{z})$. But $(c_1\mathbf{x}_1) \cdot \mathbf{z} = c_1(\mathbf{x}_1 \cdot \mathbf{z})$ by part (4) of Theorem 1.5, and so we have completed the Base Step.

Inductive Step: Assume in what follows that $c_1, c_2, \ldots, c_k, c_{k+1} \in \mathbb{R}$, $\mathbf{z}, \mathbf{x}_1, \mathbf{x}_2, \ldots, \mathbf{x}_k, \mathbf{x}_{k+1} \in \mathbb{R}^m$, and $k \geq 1$. The Inductive Step requires us to prove the following:

If

$$(c_1\mathbf{x}_1 + c_2\mathbf{x}_2 + \cdots + c_k\mathbf{x}_k) \cdot \mathbf{z} = c_1(\mathbf{x}_1 \cdot \mathbf{z}) + c_2(\mathbf{x}_2 \cdot \mathbf{z}) + \cdots + c_k(\mathbf{x}_k \cdot \mathbf{z}),$$

then

$$(c_1\mathbf{x}_1 + c_2\mathbf{x}_2 + \cdots + c_k\mathbf{x}_k + c_{k+1}\mathbf{x}_{k+1}) \cdot \mathbf{z}$$
$$= c_1(\mathbf{x}_1 \cdot \mathbf{z}) + c_2(\mathbf{x}_2 \cdot \mathbf{z}) + \cdots + c_k(\mathbf{x}_k \cdot \mathbf{z}) + c_{k+1}(\mathbf{x}_{k+1} \cdot \mathbf{z}).$$

We assume that the premise of this implication is true, and use it to prove its conclusion.

$$(c_1\mathbf{x}_1 + c_2\mathbf{x}_2 + \cdots + c_k\mathbf{x}_k + c_{k+1}\mathbf{x}_{k+1}) \cdot \mathbf{z}$$
$$= ((c_1\mathbf{x}_1 + c_2\mathbf{x}_2 + \cdots + c_k\mathbf{x}_k) + (c_{k+1}\mathbf{x}_{k+1})) \cdot \mathbf{z}$$
$$= (c_1\mathbf{x}_1 + c_2\mathbf{x}_2 + \cdots + c_k\mathbf{x}_k) \cdot \mathbf{z} + (c_{k+1}\mathbf{x}_{k+1}) \cdot \mathbf{z}$$

by part (6) of Theorem 1.5, where $\mathbf{x} = c_1\mathbf{x}_1 + c_2\mathbf{x}_2 + \cdots + c_k\mathbf{x}_k$, and $\mathbf{y} = c_{k+1}\mathbf{x}_{k+1}$

$$= (c_1\mathbf{x}_1 + c_2\mathbf{x}_2 + \cdots + c_k\mathbf{x}_k) \cdot \mathbf{z} + c_{k+1}(\mathbf{x}_{k+1} \cdot \mathbf{z})$$

by part (4) of Theorem 1.5

$$= c_1(\mathbf{x}_1 \cdot \mathbf{z}) + c_2(\mathbf{x}_2 \cdot \mathbf{z}) + \cdots + c_k(\mathbf{x}_k \cdot \mathbf{z}) + c_{k+1}(\mathbf{x}_{k+1} \cdot \mathbf{z})$$

by the induction premise.

Thus, we have proven the conclusion and completed the Inductive Step. Because we have completed both parts of the induction proof, the proof is finished. ∎

Note that in the Inductive Step we are proving an implication, and so we get the powerful advantage of assuming the premise of that implication. This premise is called the **inductive hypothesis**. In Result 8, the induction hypothesis is

$$(c_1\mathbf{x}_1 + c_2\mathbf{x}_2 + \cdots + c_k\mathbf{x}_k) \cdot \mathbf{z} = c_1(\mathbf{x}_1 \cdot \mathbf{z}) + c_2(\mathbf{x}_2 \cdot \mathbf{z}) + \cdots + c_k(\mathbf{x}_k \cdot \mathbf{z}).$$

It allows us to make the crucial substitution for $(c_1\mathbf{x}_1 + c_2\mathbf{x}_2 + \cdots + c_k\mathbf{x}_k) \cdot \mathbf{z}$ in the Inductive Step. A successful proof by induction ultimately depends on using the inductive hypothesis to reach the final conclusion.

Negating Statements with Quantifiers and Connectives

When considering some statement A, we are frequently interested in its **negation**, "not A." For example, negation is used in constructing a contrapositive, as well as in proof by contradiction. Of course, "not A" is true precisely when A is false, and "not A" is false precisely when A is true. That is, A and "not A" always have opposite truth values. Negating a simple statement is usually easy. However, when a statement involves **quantifiers** (such as *all*, *some*, or *none*) or involves **connectives** (such as *and* or *or*), the negation process can be tricky.

We first discuss negating statements with quantifiers. As an example, suppose S represents some set of vectors in $\mathbb{R}^3$ and $A =$ "All vectors in S are unit vectors." The correct negation of A is "not A" = "Some vector in S is not a unit vector." These statements have opposite truth values in all cases. Students frequently err in giving $B =$ "No vector in S is a unit vector" as the negation of A. This is incorrect, because if S contained unit and non-unit vectors, then both A and B would be false. Hence, A and B do not have opposite truth values in all cases.

Next consider $C =$ "There is a real number c such that $\mathbf{y} = c\mathbf{x}$," referring to specific vectors $\mathbf{x}$ and $\mathbf{y}$. Then "not C" = "No real number c exists such that $\mathbf{y} = c\mathbf{x}$." Alternately, "not C" = "For every real number c, $\mathbf{y} \neq c\mathbf{x}$."

There are two types of quantifiers. **Universal quantifiers** (such as *every*, *all*, *no*, and *none*) say that a statement is true or false in every instance, and **existential quantifiers** (such as *some* and *there exists*) claim that there is at least one instance in which the statement is satisfied. The statements A and "not C" in the preceding examples involve universal quantifiers; "not A" and C use existential quantifiers. These examples follow a general pattern.

Rules for Negating Statements with Quantifiers

The negation of a statement involving a universal quantifier uses an existential quantifier.

The negation of a statement involving an existential quantifier uses a universal quantifier.

Hence, negating a statement changes the type of quantifier used.

Next, consider negating with the connectives *and* or *or*. The formal rules for negating such statements are known as **DeMorgan's Laws**.

Rules for Negating Statements with Connectives (DeMorgan's Laws)

The negation of "*A* or *B*" is "(not *A*) and (not *B*)."
The negation of "*A* and *B*" is "(not *A*) or (not *B*)."

Note that when negating, *or* is converted to *and* and vice versa.

Table 1.1 illustrates the rules for negating quantifiers and connectives. In the table, *S* refers to a set of vectors in $\mathbb{R}^3$, and *n* represents a positive integer. Only some of the statements are true. Regardless, each statement has the opposite truth value of its negation.

Table 1.1

Several statements and their negations

Original Statement	Negation of the Statement
n is an even number or a prime.	*n* is odd and not prime.
$\mathbf{x}$ is a unit vector and $\mathbf{x} \in S$.	$\|\mathbf{x}\| \neq 1$ or $\mathbf{x} \notin S$.
Some prime numbers are odd.	Every prime number is even.
There is a unit vector in *S*.	No elements of *S* are unit vectors.
There is a vector $\mathbf{x}$ in *S* with $\mathbf{x} \cdot [1, 1, -1] = 0$.	For every vector $\mathbf{x}$ in *S*, $\mathbf{x} \cdot [1, 1, -1] \neq 0$.
All numbers divisible by 4 are even.	Some number divisible by 4 is odd.
Every vector in *S* is either a unit vector or is parallel to $[1, -2, 1]$.	There is a non-unit vector in *S* that is not parallel to $[1, -2, 1]$.
For every nonzero vector $\mathbf{x}$ in $\mathbb{R}^3$, there is a vector in *S* that is parallel to $\mathbf{x}$.	There is a nonzero vector $\mathbf{x}$ in $\mathbb{R}^3$ that is not parallel to any vector in *S*.
There is a real number *K* such that for every $\mathbf{x} \in S$, $\|\mathbf{x}\| \leq K$.	For every real number *K* there is a vector $\mathbf{x} \in S$ such that $\|\mathbf{x}\| > K$.

Disproving Statements

Frequently we must prove that a given statement is false rather than true. To disprove a statement *A*, we must instead prove "not *A*." There are two cases.

Case 1: Statements involving universal quantifiers: A statement *A* with a universal quantifier is disproved by finding a single **counterexample** that makes *A* false. For example, consider *B* = "For all $\mathbf{x}$ and $\mathbf{y}$ in $\mathbb{R}^3$, $\|\mathbf{x} + \mathbf{y}\| = \|\mathbf{x}\| + \|\mathbf{y}\|$." We disprove *B* by finding a counterexample — that is, a specific case where *B* is false. Letting $\mathbf{x} = [3, 0, 0]$ and $\mathbf{y} = [0, 0, 4]$, we get $\|\mathbf{x} + \mathbf{y}\| = \|[3, 0, 4]\| = 5$. However, $\|\mathbf{x}\| = 3$ and $\|\mathbf{y}\| = 4$, so $\|\mathbf{x} + \mathbf{y}\| \neq \|\mathbf{x}\| + \|\mathbf{y}\|$, and *B* is disproved.

Sometimes we want to disprove an implication "If *A* then *B*." This implication involves a universal quantifier because it asserts "In all cases in which *A* is true, *B* is also true." Therefore,

Disproving "If A then B" entails finding a specific counterexample for which A is true but B is false.

To illustrate, consider $C =$ "If $\mathbf{x}$ and $\mathbf{y}$ are unit vectors in $\mathbb{R}^4$, then $\mathbf{x} \cdot \mathbf{y} = 1$." To disprove C, we must find a counterexample in which the premise "$\mathbf{x}$ and $\mathbf{y}$ are unit vectors in $\mathbb{R}^4$" is true and the conclusion "$\mathbf{x} \cdot \mathbf{y} = 1$" is false. Consider $\mathbf{x} = [1, 0, 0, 0]$ and $\mathbf{y} = [0, 1, 0, 0]$, which are unit vectors in $\mathbb{R}^4$. Then $\mathbf{x} \cdot \mathbf{y} = 0 \neq 1$. This counterexample disproves C.

Case 2: Statements involving existential quantifiers: Recall that an existential quantifier changes to a universal quantifier under negation. For example, consider $D =$ "There is a nonzero vector $\mathbf{x}$ in $\mathbb{R}^2$ such that $\mathbf{x} \cdot [1, 0] = 0$ and $\mathbf{x} \cdot [0, 1] = 0$." To disprove D, we must prove "not D" $=$ "For every nonzero vector $\mathbf{x}$ in $\mathbb{R}^2$, either $\mathbf{x} \cdot [1, 0] \neq 0$ or $\mathbf{x} \cdot [0, 1] \neq 0$."[6] We cannot prove this statement by giving a single example. Instead, we must show "not D" is true for *every* nonzero vector in $\mathbb{R}^2$. This can be done with a direct proof. (You were asked to supply its proof earlier, since "not D" is actually Result 5.)

The moral here is we cannot disprove a statement having an existential quantifier with a counterexample. Instead, a proof of the negation must be given.

Exercises for Section 1.3

1. (a) Give a direct proof that, if $\mathbf{x}$ and $\mathbf{y}$ are vectors in $\mathbb{R}^n$, then $\|4\mathbf{x} + 7\mathbf{y}\| \leq 7 (\|\mathbf{x}\| + \|\mathbf{y}\|)$.
   - ★(b) Can you generalize your proof in part (a) to draw any conclusions about $\|c\mathbf{x} + d\mathbf{y}\|$, where $c, d \in \mathbb{R}$? What about $\|c\mathbf{x} - d\mathbf{y}\|$?

2. (a) Give a direct proof that if an integer has the form $6j - 5$, then it also has the form $3k + 1$, where j and k are integers.
   - ★(b) Find a counterexample to show that the converse of part (a) is not true.

3. Let $\mathbf{x}$ and $\mathbf{y}$ be nonzero vectors in $\mathbb{R}^n$. Prove $\mathbf{proj_x y} = \mathbf{0}$ if and only if $\mathbf{proj_y x} = \mathbf{0}$.

4. Let $\mathbf{x}$ and $\mathbf{y}$ be nonzero vectors in $\mathbb{R}^n$. Prove $\|\mathbf{x} + \mathbf{y}\| = \|\mathbf{x}\| + \|\mathbf{y}\|$ if and only if $\mathbf{y} = c\mathbf{x}$ for some $c > 0$. (Hint: Be sure to prove both halves of this statement. Result 4 may make one half of the proof easier.)

★5. Consider the statement $A =$ "If $\mathbf{x} \cdot \mathbf{y} = \|\mathbf{x}\|^2$, then $\mathbf{x} = \mathbf{y}$."
   - (a) Show that A is false by exhibiting a counterexample.
   - (b) State the contrapositive of A.
   - (c) Does your counterexample from part (a) also show that the contrapositive from part (b) is false?

6. Prove the following statements of the form "If A, then B or C."
   - (a) If $\|\mathbf{x} + \mathbf{y}\| = \|\mathbf{x}\|$, then $\mathbf{y} = \mathbf{0}$ or $\mathbf{x}$ is not orthogonal to $\mathbf{y}$.
   - (b) If $\mathbf{proj_x y} = \mathbf{x}$, then either $\mathbf{x}$ is a unit vector or $\mathbf{x} \cdot \mathbf{y} \neq 1$.

---

[6] Notice that along with the change in the quantifier, the *and* connective changes to *or*.

7. Prove the following by contrapositive: Assume that $\mathbf{x}$ and $\mathbf{y}$ are vectors in $\mathbb{R}^n$. If $\mathbf{x} \cdot \mathbf{y} \neq 0$, then $\|\mathbf{x} + \mathbf{y}\|^2 \neq \|\mathbf{x}\|^2 + \|\mathbf{y}\|^2$.

8. State the contrapositive, converse, and inverse of each of the following statements for vectors in $\mathbb{R}^n$:

   ★(a) If $\mathbf{x}$ is a unit vector, then $\mathbf{x}$ is nonzero.

   (b) Let $\mathbf{x}$ and $\mathbf{y}$ be nonzero vectors. If $\mathbf{x}$ is parallel to $\mathbf{y}$, then $\mathbf{y} = \mathbf{proj}_{\mathbf{x}}\mathbf{y}$.

   ★(c) Let $\mathbf{x}$ and $\mathbf{y}$ be nonzero vectors. If $\mathbf{proj}_{\mathbf{x}}\mathbf{y} = \mathbf{0}$, then $\mathbf{proj}_{\mathbf{y}}\mathbf{x} = \mathbf{0}$.

9. (a) State the converse of Result 2.

   (b) Show that this converse is false by finding a counterexample.

10. Each of the following statements has the opposite truth value as its converse; that is, one of them is true, and the other is false. In each case,

   (i) State the converse of the given statement.

   (ii) Which is true — the statement or its converse?

   (iii) Prove the one from part (ii) that is true.

   (iv) Disprove the other one by finding a counterexample.

   (a) Let $\mathbf{x}$, $\mathbf{y}$, and $\mathbf{z}$ be vectors in $\mathbb{R}^n$. If $\mathbf{x} \cdot \mathbf{y} = \mathbf{x} \cdot \mathbf{z}$, then $\mathbf{y} = \mathbf{z}$.

   ★(b) Let $\mathbf{x}$ and $\mathbf{y}$ be vectors in $\mathbb{R}^n$. If $\mathbf{x} \cdot \mathbf{y} = 0$, then $\|\mathbf{x} + \mathbf{y}\| \geq \|\mathbf{y}\|$.

   (c) Assume that $\mathbf{x}$ and $\mathbf{y}$ are vectors in $\mathbb{R}^n$ with $n > 1$. If $\mathbf{x} \cdot \mathbf{y} = 0$, then $\mathbf{x} = \mathbf{0}$ or $\mathbf{y} = \mathbf{0}$.

11. Let $\mathbf{x}$ and $\mathbf{y}$ be vectors in $\mathbb{R}^n$ such that each coordinate of both $\mathbf{x}$ and $\mathbf{y}$ is equal to either 1 or -1. Prove by contradiction that if $\mathbf{x}$ is orthogonal to $\mathbf{y}$, then n is even.

12. Prove the following by contradiction: three mutually orthogonal nonzero vectors do not exist in $\mathbb{R}^2$. (Hint: Assume three such vectors $[x_1, x_2]$, $[y_1, y_2]$, and $[z_1, z_2]$ exist. First, show that at least one of x_1, y_1, or z_1 is nonzero. Without loss of generality, you may assume $x_1 \neq 0$. Next, show you may also assume that $y_1 \neq 0$. Let $a = x_2/x_1$ and $b = y_2/y_1$. Then, prove that $[1, a]$,$[1, b]$, and $[z_1, z_2]$ are also mutually orthogonal. Finally, show that $z_1 + az_2 = z_1 + bz_2$, and obtain a contradiction.)

13. Prove by induction: If $\mathbf{x}_1, \mathbf{x}_2, \ldots, \mathbf{x}_{n-1}, \mathbf{x}_n$ (for $n \geq 1$) are vectors in $\mathbb{R}^m$, then $\mathbf{x}_1 + \mathbf{x}_2 + \cdots + \mathbf{x}_{n-1} + \mathbf{x}_n = \mathbf{x}_n + \mathbf{x}_{n-1} + \cdots + \mathbf{x}_2 + \mathbf{x}_1$.

14. Prove by induction: For each integer $m \geq 1$, let $\mathbf{x}_1, \ldots, \mathbf{x}_m$ be vectors in $\mathbb{R}^n$. Then, $\|\mathbf{x}_1 + \mathbf{x}_2 + \cdots + \mathbf{x}_m\| \leq \|\mathbf{x}_1\| + \|\mathbf{x}_2\| + \cdots + \|\mathbf{x}_m\|$.

15. Let $\mathbf{x}_1, \ldots, \mathbf{x}_k$ be a mutually orthogonal set of nonzero vectors in $\mathbb{R}^n$. Use induction to show that

$$\left\| \sum_{i=1}^{k} \mathbf{x}_i \right\|^2 = \sum_{i=1}^{k} \|\mathbf{x}_i\|^2.$$

16. Prove by induction: Let $\mathbf{x}_1, \ldots, \mathbf{x}_k$ be unit vectors in $\mathbb{R}^n$, and let $a_1, \ldots, a_k$ be real numbers. Then, for every $\mathbf{y}$ in $\mathbb{R}^n$,

$$\left(\sum_{i=1}^{k} a_i \mathbf{x}_i \right) \cdot \mathbf{y} \leq \left(\sum_{i=1}^{k} |a_i| \right) \|\mathbf{y}\|.$$

17. Let $\mathbf{x} = [x_1, \ldots, x_n]$ be a vector in $\mathbb{R}^n$. Prove that $\|\mathbf{x}\| \leq \Sigma_{i=1}^n |\mathbf{x}_i|$. (Hint: Use a proof by induction on n to prove that $\sqrt{\Sigma_{i=1}^n x_i^2} \leq \Sigma_{i=1}^n |x_i|$. For the Inductive Step, let $\mathbf{y} = [x_1, \ldots, x_k, x_{k+1}]$, $\mathbf{z} = [x_1, \ldots, x_k, 0]$, and $\mathbf{w} = [0, 0, \ldots, 0, x_{k+1}]$. Note that $\mathbf{y} = \mathbf{z} + \mathbf{w}$. Then apply the Triangle Inequality.)

⋆18. Which steps in the following argument cannot be "reversed?" Why? Assume that $y = f(x)$ is nonzero everywhere and that d^2y/dx^2 exists for all x.

Step 1: $y = x^2 + 2$ $\Rightarrow$ $y^2 = x^4 + 4x^2 + 4$

Step 2: $y^2 = x^4 + 4x^2 + 4$ $\Rightarrow$ $2y\dfrac{dy}{dx} = 4x^3 + 8x$

Step 3: $2y\dfrac{dy}{dx} = 4x^3 + 8x$ $\Rightarrow$ $\dfrac{dy}{dx} = \dfrac{4x^3 + 8x}{2y}$

Step 4: $\dfrac{dy}{dx} = \dfrac{4x^3 + 8x}{2y}$ $\Rightarrow$ $\dfrac{dy}{dx} = \dfrac{4x^3 + 8x}{2\left(x^2 + 2\right)}$

Step 5: $\dfrac{dy}{dx} = \dfrac{4x^3 + 8x}{2\left(x^2 + 2\right)}$ $\Rightarrow$ $\dfrac{dy}{dx} = 2x$

Step 6: $\dfrac{dy}{dx} = 2x$ $\Rightarrow$ $\dfrac{d^2y}{dx^2} = 2$

19. State the negation of each of the following statements involving quantifiers and connectives. (The statements are not necessarily true.)

⋆(a) There is a unit vector in $\mathbb{R}^3$ perpendicular to $[1, -2, 3]$.

(b) $\mathbf{x} = \mathbf{0}$ or $\mathbf{x} \cdot \mathbf{y} > 0$, for all vectors $\mathbf{x}$ and $\mathbf{y}$ in $\mathbb{R}^n$.

⋆(c) $\mathbf{x} \neq \mathbf{0}$ and $\|\mathbf{x} + \mathbf{y}\| = \|\mathbf{y}\|$, for some vectors $\mathbf{x}$ and $\mathbf{y}$ in $\mathbb{R}^n$.

(d) For every vector $\mathbf{x}$ in $\mathbb{R}^n$, $\mathbf{x} \cdot \mathbf{x} > 0$.

⋆(e) For every $\mathbf{x} \in \mathbb{R}^3$, there is a nonzero $\mathbf{y} \in \mathbb{R}^3$ such that $\mathbf{x} \cdot \mathbf{y} = 0$.

(f) There is an $\mathbf{x} \in \mathbb{R}^4$ such that for every $\mathbf{y} \in \mathbb{R}^4$, $\mathbf{x} \cdot \mathbf{y} = 0$.

20. State the contrapositive, converse, and inverse of the following statements involving connectives. (The statements are not necessarily true.)

⋆(a) If $\mathbf{x} \cdot \mathbf{y} = 0$, then either $\mathbf{x} = \mathbf{0}$ or $\|\mathbf{x} - \mathbf{y}\| > \|\mathbf{y}\|$.

(b) If $\mathbf{x} \neq \mathbf{0}$ and $\mathbf{x} \cdot \mathbf{y} = 0$, then $\|\mathbf{x} - \mathbf{y}\| > \|\mathbf{y}\|$.

21. Prove the following by contrapositive: Let $\mathbf{x}$ be a vector in $\mathbb{R}^n$. If $\mathbf{x} \cdot \mathbf{y} = 0$ for every vector $\mathbf{y}$ in $\mathbb{R}^n$, then $\mathbf{x} = \mathbf{0}$.

22. Prove the following by contrapositive: Let $\mathbf{u}$ and $\mathbf{v}$ be nonzero vectors in $\mathbb{R}^n$. If for all $\mathbf{x}$ in $\mathbb{R}^n$, either $\mathbf{u} \cdot \mathbf{x} \leq 0$ or $\mathbf{v} \cdot \mathbf{x} \leq 0$, then $\mathbf{u}$ and $\mathbf{v}$ are in opposite directions. (Hint: Consider a vector that bisects the angle between $\mathbf{u}$ and $\mathbf{v}$.)

23. Disprove the following: If $\mathbf{x}$ and $\mathbf{y}$ are vectors in $\mathbb{R}^n$, then $\|\mathbf{x} - \mathbf{y}\| \leq \|\mathbf{x}\| - \|\mathbf{y}\|$.

24. Use Result 2 to disprove the following: there is a vector $\mathbf{x}$ in $\mathbb{R}^3$ such that $\mathbf{x} \cdot [1, -2, 2] = 0$ and $\|\mathbf{x} + [1, -2, 2]\| < 3$.

⋆25. True or False:

(a) After "working backward" to complete a proof, it is enough to reverse your steps to give a valid "forward" proof.

(b) "If A then B" has the same truth value as "If not B then not A."
(c) The converse of "A only if B" is "If B then A."
(d) "A if and only if B" is logically equivalent to "A is a necessary condition for B."
(e) "A if and only if B" is logically equivalent to "A is a necessary condition for B" together with "B is a sufficient condition for A."
(f) The converse and inverse of a statement must have opposite truth values.
(g) A proof of a given statement by induction is valid if, whenever the statement is true for any integer k, it is also true for the next integer $k+1$.
(h) When negating a statement, universal quantifiers change to existential quantifiers, and vice versa.
(i) The negation of "A and B" is "not A and not B."

1.4 Fundamental Operations with Matrices

We now introduce a new algebraic structure — the matrix. Matrices are two-dimensional arrays created by arranging vectors into rows and columns. We examine several fundamental types of matrices, as well as three basic operations on matrices and their properties.

Definition of a Matrix

DEFINITION
An $m \times n$ **matrix** is a rectangular array of real numbers, arranged in m rows and n columns. The elements of a matrix are called the **entries**. The expression $m \times n$ denotes the **size** of the matrix.

For example, each of the following is a matrix, listed with its correct size:

$$\mathbf{A} = \underbrace{\begin{bmatrix} 2 & 3 & -1 \\ 4 & 0 & -5 \end{bmatrix}}_{2 \times 3 \text{ matrix}} \qquad \mathbf{B} = \underbrace{\begin{bmatrix} 4 & -2 \\ 1 & 7 \\ -5 & 3 \end{bmatrix}}_{3 \times 2 \text{ matrix}}$$

$$\mathbf{C} = \underbrace{\begin{bmatrix} 1 & 2 & 3 \\ 4 & 5 & 6 \\ 7 & 8 & 9 \end{bmatrix}}_{3 \times 3 \text{ matrix}} \qquad \mathbf{D} = \underbrace{\begin{bmatrix} 7 \\ 1 \\ -2 \end{bmatrix}}_{3 \times 1 \text{ matrix}}$$

$$\mathbf{E} = \underbrace{\begin{bmatrix} 4 & -3 & 0 \end{bmatrix}}_{1 \times 3 \text{ matrix}} \qquad \mathbf{F} = \underbrace{\begin{bmatrix} 4 \end{bmatrix}}_{1 \times 1 \text{ matrix}}$$

Here are some conventions to remember regarding matrices.

- We use a single (or subscripted) bold capital letter to denote a matrix (such as $\mathbf{A}, \mathbf{B}, \mathbf{C}_1, \mathbf{C}_2$) in contrast to the lowercase bold letters used to represent vectors. The capital letters $\mathbf{I}$ and $\mathbf{O}$ are usually reserved for special types of matrices discussed later.

- The size of a matrix is always specified by stating the number of rows first. For example, a 3×4 matrix always has three rows and four columns, never four rows and three columns.

- An $m \times n$ matrix can be thought of either as a collection of m row vectors, each having n coordinates, or as a collection of n column vectors, each having m coordinates. A matrix with just one row (or column) is essentially equivalent to a vector with coordinates in row (or column) form.

- We often write a_{ij} to represent the entry in the ith row and jth column of a matrix $\mathbf{A}$. For example, in the previous matrix $\mathbf{A}$, a_{23} is the entry -5 in the second row and third column. A typical 3×4 matrix $\mathbf{C}$ has entries symbolized by

$$\mathbf{C} = \begin{bmatrix} c_{11} & c_{12} & c_{13} & c_{14} \\ c_{21} & c_{22} & c_{23} & c_{24} \\ c_{31} & c_{32} & c_{33} & c_{34} \end{bmatrix}.$$

- $\mathcal{M}_{mn}$ represents the set of all matrices with real-number entries having m rows and n columns. For example, $\mathcal{M}_{34}$ is the set of all matrices having three rows and four columns. A typical matrix in $\mathcal{M}_{34}$ has the form of the preceding matrix $\mathbf{C}$.

- The **main diagonal** entries of a matrix $\mathbf{A}$ are $a_{11}, a_{22}, a_{33}, \ldots$, those that lie on a diagonal line drawn down to the right, beginning from the upper-left corner of the matrix.

Matrices occur naturally in many contexts. For example, two-dimensional tables (having rows and columns) of real numbers are matrices. The following table represents a 50×3 matrix with integer entries:

U.S. State	Population (2000)	Area (sq. mi.)	Year Admitted to Union
Alabama	4447100	51609	1819
Alaska	626932	589757	1959
Arizona	5130632	113909	1912
⋮	⋮	⋮	⋮
Wyoming	493782	97914	1890

Two $m \times n$ matrices $\mathbf{A}$ and $\mathbf{B}$ are **equal** if and only if all of their corresponding entries are equal. That is, $\mathbf{A} = \mathbf{B}$ if $a_{ij} = b_{ij}$ for all $i, 1 \leq i \leq m$, and for all $j, 1 \leq j \leq n$.

Note that the following may be considered equal as vectors but not as matrices:

$$[3, -2, \ 4] \quad \text{and} \quad \begin{bmatrix} 3 \\ -2 \\ 4 \end{bmatrix},$$

since the former is a 1×3 matrix, but the latter is a 3×1 matrix.

Special Types of Matrices

We now describe a few important types of matrices.

A **square matrix** is an $n \times n$ matrix; that is, a matrix having the same number of rows as columns. For example, the following matrices are square:

$$\mathbf{A} = \begin{bmatrix} 5 & 0 \\ 9 & -2 \end{bmatrix} \quad \text{and} \quad \mathbf{B} = \begin{bmatrix} 1 & 2 & 3 \\ 4 & 5 & 6 \\ 7 & 8 & 9 \end{bmatrix}.$$

A **diagonal matrix** is a square matrix in which all entries that are not on the main diagonal are zero. That is, $\mathbf{A}$ is diagonal if and only if $a_{ij} = 0$ for $i \neq j$. For example, the following are diagonal matrices:

$$\mathbf{E} = \begin{bmatrix} 6 & 0 & 0 \\ 0 & 7 & 0 \\ 0 & 0 & -2 \end{bmatrix}, \quad \mathbf{F} = \begin{bmatrix} 4 & 0 & 0 & 0 \\ 0 & 0 & 0 & 0 \\ 0 & 0 & -2 & 0 \\ 0 & 0 & 0 & 0 \end{bmatrix}, \quad \text{and} \quad \mathbf{G} = \begin{bmatrix} -4 & 0 \\ 0 & 5 \end{bmatrix}.$$

However, the matrices are *not* diagonal. (The main diagonal elements have been shaded in each case.) We use $\mathcal{D}_n$ to represent the **set of all** $n \times n$ **diagonal matrices**.

$$\mathbf{H} = \begin{bmatrix} 4 & 3 \\ 0 & 1 \end{bmatrix} \quad \text{and} \quad \mathbf{J} = \begin{bmatrix} 0 & 4 & 3 \\ -7 & 0 & 6 \\ 5 & -2 & 0 \end{bmatrix}.$$

An **identity matrix** is a diagonal matrix with all main diagonal entries equal to 1. That is, an $n \times n$ matrix $\mathbf{A}$ is an identity matrix if and only if $a_{ij} = 0$ for $i \neq j$ and $a_{ii} = 1$ for $1 \leq i \leq n$. The $n \times n$ identity matrix is denoted by $\mathbf{I}_n$. For example, the following are identity matrices:

$$\mathbf{I}_2 = \begin{bmatrix} 1 & 0 \\ 0 & 1 \end{bmatrix} \quad \text{and} \quad \mathbf{I}_4 = \begin{bmatrix} 1 & 0 & 0 & 0 \\ 0 & 1 & 0 & 0 \\ 0 & 0 & 1 & 0 \\ 0 & 0 & 0 & 1 \end{bmatrix}.$$

If the size of the identity matrix is clear from the context, $\mathbf{I}$ alone may be used.

An **upper triangular matrix** is a square matrix with all entries *below* the main diagonal equal to zero. That is, an $n \times n$ matrix $\mathbf{A}$ is upper triangular if and only if $a_{ij} = 0$ for $i > j$. For example, the following are upper triangular:

$$\mathbf{P} = \begin{bmatrix} 6 & 9 & 11 \\ 0 & -2 & 3 \\ 0 & 0 & 5 \end{bmatrix} \quad \text{and} \quad \mathbf{Q} = \begin{bmatrix} 7 & -2 & 2 & 0 \\ 0 & -4 & 9 & 5 \\ 0 & 0 & 0 & 8 \\ 0 & 0 & 0 & 3 \end{bmatrix}$$

$$\mathbf{R} = \begin{bmatrix} 3 & 0 & 0 \\ 9 & -2 & 0 \\ 14 & -6 & 1 \end{bmatrix}$$

Similarly, a **lower triangular matrix** is one in which all entries *above* the main diagonal equal zero; for example,

We use $\mathcal{U}_n$ to represent the **set of all** $n \times n$ **upper triangular matrices** and $\mathcal{L}_n$ to represent the **set of all** $n \times n$ **lower triangular matrices**.

A **zero matrix** is any matrix all of whose entries are zero. $\mathbf{O}_{mn}$ denotes the $m \times n$ zero matrix, and $\mathbf{O}_n$ denotes the $n \times n$ zero matrix. For example,

$$\mathbf{O}_{23} = \begin{bmatrix} 0 & 0 & 0 \\ 0 & 0 & 0 \end{bmatrix} \quad \text{and} \quad \mathbf{O}_2 = \begin{bmatrix} 0 & 0 \\ 0 & 0 \end{bmatrix}.$$

If the size of the zero matrix is clear from the context, $\mathbf{O}$ alone may be used.

Addition and Scalar Multiplication with Matrices

DEFINITION

Let $\mathbf{A}$ and $\mathbf{B}$ both be $m \times n$ matrices. The **sum** of $\mathbf{A}$ and $\mathbf{B}$ is the $m \times n$ matrix $(\mathbf{A} + \mathbf{B})$ whose (i, j) entry is equal to $a_{ij} + b_{ij}$.

As with vectors, matrices are summed simply by adding their corresponding entries together. For example,

$$\begin{bmatrix} 6 & -3 & 2 \\ -7 & 0 & 4 \end{bmatrix} + \begin{bmatrix} 5 & -6 & -3 \\ -4 & -2 & -4 \end{bmatrix} = \begin{bmatrix} 11 & -9 & -1 \\ -11 & -2 & 0 \end{bmatrix}.$$

Notice that the definition does not allow addition of matrices with different sizes. For example, the following matrices cannot be added:

$$\mathbf{A} = \begin{bmatrix} -2 & 3 & 0 \\ 1 & 4 & -5 \end{bmatrix} \quad \text{and} \quad \mathbf{B} = \begin{bmatrix} 6 & 7 \\ -2 & 5 \\ 4 & -1 \end{bmatrix},$$

since $\mathbf{A}$ is a 2×3 matrix, and $\mathbf{B}$ is a 3×2 matrix.

DEFINITION

Let $\mathbf{A}$ be an $m \times n$ matrix, and let c be a scalar. Then the matrix $c\mathbf{A}$, the **scalar multiplication** of c and $\mathbf{A}$, is the $m \times n$ matrix whose (i, j) entry is equal to ca_{ij}.

As with vectors, scalar multiplication with matrices is done by multiplying every entry by the given scalar. For example, if $c = -2$ and

$$\mathbf{A} = \begin{bmatrix} 4 & -1 & 6 & 7 \\ 2 & 4 & 9 & -5 \end{bmatrix}, \text{ then } -2\mathbf{A} = \begin{bmatrix} -8 & 2 & -12 & -14 \\ -4 & -8 & -18 & 10 \end{bmatrix}.$$

Note that if $\mathbf{A}$ is any $m \times n$ matrix, then $0\mathbf{A} = \mathbf{O}_{mn}$.

Let $-\mathbf{A}$ denote the matrix $-1\mathbf{A}$, the scalar multiple of $\mathbf{A}$ by (-1). For example, if

$$\mathbf{A} = \begin{bmatrix} 3 & -2 \\ 10 & 6 \end{bmatrix}, \text{ then } -1\mathbf{A} = -\mathbf{A} = \begin{bmatrix} -3 & 2 \\ -10 & -6 \end{bmatrix}.$$

Also, we define **subtraction** of matrices as $\mathbf{A} - \mathbf{B} = \mathbf{A} + (-\mathbf{B})$.

As with vectors, sums of scalar multiples of matrices are called **linear combinations.** For example, $-2\mathbf{A} + 6\mathbf{B} - 3\mathbf{C}$ is a linear combination of $\mathbf{A}$, $\mathbf{B}$, and $\mathbf{C}$.

Fundamental Properties of Addition and Scalar Multiplication

The properties in the next theorem are similar to the vector properties of Theorem 1.3.

THEOREM 1.11

Let $\mathbf{A}$, $\mathbf{B}$, and $\mathbf{C}$ be $m \times n$ matrices (elements of $\mathcal{M}_{mn}$), and let c and d be scalars. Then

(1)	$\mathbf{A} + \mathbf{B} = \mathbf{B} + \mathbf{A}$	Commutative Law of Addition
(2)	$\mathbf{A} + (\mathbf{B} + \mathbf{C}) = (\mathbf{A} + \mathbf{B}) + \mathbf{C}$	Associative Law of Addition
(3)	$\mathbf{O}_{mn} + \mathbf{A} = \mathbf{A} + \mathbf{O}_{mn} = \mathbf{A}$	Existence of Identity Element for Addition
(4)	$\mathbf{A} + (-\mathbf{A}) = (-\mathbf{A}) + \mathbf{A} = \mathbf{O}_{mn}$	Existence of Inverse Elements for Addition
(5)	$c(\mathbf{A} + \mathbf{B}) = c\mathbf{A} + c\mathbf{B}$	Distributive Laws of Scalar
(6)	$(c + d)\mathbf{A} = c\mathbf{A} + d\mathbf{A}$	Multiplication over Addition
(7)	$(cd)\mathbf{A} = c(d\mathbf{A})$	Associativity of Scalar Multiplication
(8)	$1(\mathbf{A}) = \mathbf{A}$	Identity Property for Scalar Multiplication

---

To prove each property, calculate corresponding entries on both sides and show they agree by applying an appropriate law of real numbers. We prove part (1) as an example and leave some of the remaining proofs as Exercise 10.

PROOF OF THEOREM 1.11, PART (1):

For any i, j, where $1 \leq i \leq m$ and $1 \leq j \leq n$, the (i, j) entry of $(\mathbf{A} + \mathbf{B})$ is the sum of the entries a_{ij} and b_{ij} from $\mathbf{A}$ and $\mathbf{B}$, respectively. Similarly, the

(i, j) entry of $\mathbf{B} + \mathbf{A}$ is the sum of b_{ij} and a_{ij}. But $a_{ij} + b_{ij} = b_{ij} + a_{ij}$, by the commutative property of addition for real numbers. Hence, $\mathbf{A} + \mathbf{B} = \mathbf{B} + \mathbf{A}$, because their corresponding entries agree. ∎

The Transpose of a Matrix and Its Properties

DEFINITION

If $\mathbf{A}$ is an $m \times n$ matrix, then the **transpose**, $\mathbf{A}^T$, is the $n \times m$ matrix whose (i, j) entry is the same as the (j, i) entry of $\mathbf{A}$.

Thus, transposing $\mathbf{A}$ moves the (i, j) entry of $\mathbf{A}$ to the (j, i) entry of $\mathbf{A}^T$. Notice that the entries on the main diagonal do not move as we convert $\mathbf{A}$ to $\mathbf{A}^T$. However, all entries above the main diagonal are moved below it, and vice versa. For example,

$$\text{if } \mathbf{A} = \begin{bmatrix} 6 & 10 \\ -2 & 4 \\ 3 & 0 \\ 1 & 8 \end{bmatrix} \quad \text{and} \quad \mathbf{B} = \begin{bmatrix} 1 & 5 & -3 \\ 0 & -4 & 6 \\ 0 & 0 & -5 \end{bmatrix},$$

$$\text{then } \mathbf{A}^T = \begin{bmatrix} 6 & -2 & 3 & 1 \\ 10 & 4 & 0 & 8 \end{bmatrix} \quad \text{and} \quad \mathbf{B}^T = \begin{bmatrix} 1 & 0 & 0 \\ 5 & -4 & 0 \\ -3 & 6 & -5 \end{bmatrix}.$$

Notice that the transpose changes the rows of $\mathbf{A}$ into the columns of $\mathbf{A}^T$. Similarly, the columns of $\mathbf{A}$ become the rows of $\mathbf{A}^T$. Also note that the transpose of an upper triangular matrix (such as $\mathbf{B}$) is lower triangular, and vice versa.

Three useful properties of the transpose are given in the next theorem. We prove one and leave the others as Exercise 11.

THEOREM 1.12

Let $\mathbf{A}$ and $\mathbf{B}$ both be $m \times n$ matrices, and let c be a scalar. Then

$$(1) \quad (\mathbf{A}^T)^T = \mathbf{A}$$
$$(2) \quad (\mathbf{A} + \mathbf{B})^T = \mathbf{A}^T + \mathbf{B}^T$$
$$(3) \quad (c\mathbf{A})^T = c(\mathbf{A}^T)$$

PROOF OF THEOREM 1.12, PART (2):
Notice that both $(\mathbf{A} + \mathbf{B})^T$ and $(\mathbf{A}^T) + (\mathbf{B}^T)$ are $n \times m$ matrices (why?). We need to show that the (i, j) entries of both are equal, for $1 \le i \le n$ and $1 \le j \le m$. Now, the (i, j) entry of $(\mathbf{A} + \mathbf{B})^T$ equals the (j, i) entry of $\mathbf{A} + \mathbf{B}$, which is $a_{ji} + b_{ji}$. But the (i, j) entry of $\mathbf{A}^T + \mathbf{B}^T$ equals the (i, j) entry of $\mathbf{A}^T$ plus the (i, j) entry of $\mathbf{B}^T$, which is also $a_{ji} + b_{ji}$. ∎

Symmetric and Skew-Symmetric Matrices

DEFINITION

A matrix $\mathbf{A}$ is **symmetric** if and only if $\mathbf{A} = \mathbf{A}^T$. A matrix $\mathbf{A}$ is **skew-symmetric** if and only if $\mathbf{A} = -\mathbf{A}^T$.

In Exercise 5, you are asked to show that any symmetric or skew-symmetric matrix is a square matrix.

EXAMPLE 1 Consider the following matrices:

$$\mathbf{A} = \begin{bmatrix} 2 & 6 & 4 \\ 6 & -1 & 0 \\ 4 & 0 & -3 \end{bmatrix} \quad \text{and} \quad \mathbf{B} = \begin{bmatrix} 0 & -1 & 3 & 6 \\ 1 & 0 & 2 & -5 \\ -3 & -2 & 0 & 4 \\ -6 & 5 & -4 & 0 \end{bmatrix}.$$

$\mathbf{A}$ is symmetric and $\mathbf{B}$ is skew-symmetric, because their respective transposes are

$$\mathbf{A}^T = \begin{bmatrix} 2 & 6 & 4 \\ 6 & -1 & 0 \\ 4 & 0 & -3 \end{bmatrix} \quad \text{and} \quad \mathbf{B}^T = \begin{bmatrix} 0 & 1 & -3 & -6 \\ -1 & 0 & -2 & 5 \\ 3 & 2 & 0 & -4 \\ 6 & -5 & 4 & 0 \end{bmatrix},$$

which equal $\mathbf{A}$ and $-\mathbf{B}$, respectively. However, neither of the following is symmetric or skew-symmetric (why?):

$$\mathbf{C} = \begin{bmatrix} 3 & -2 & 1 \\ 2 & 4 & 0 \\ -1 & 0 & -2 \end{bmatrix} \quad \text{and} \quad \mathbf{D} = \begin{bmatrix} 1 & -2 \\ 3 & 4 \\ 5 & -6 \end{bmatrix}. \qquad \blacksquare$$

Notice that an $n \times n$ matrix $\mathbf{A}$ is symmetric [skew-symmetric] if and only if $a_{ij} = a_{ji}$ [$a_{ij} = -a_{ji}$] for all i, j such that $1 \le i, j \le n$. In other words, the entries above the main diagonal are reflected into equal (for symmetric) or opposite (for skew-symmetric) entries below the diagonal. Since the main diagonal elements are reflected into themselves, *all of the main diagonal elements of a skew-symmetric matrix must be zeroes* ($a_{ii} = -a_{ii}$ only if $a_{ii} = 0$).

Notice that any diagonal matrix is equal to its transpose, and so such matrices are automatically symmetric. Another useful result is the following:

THEOREM 1.13

Every square matrix $\mathbf{A}$ can be decomposed uniquely as the sum of two matrices $\mathbf{S}$ and $\mathbf{V}$, where $\mathbf{S}$ is symmetric and $\mathbf{V}$ is skew-symmetric.

An outline of the proof of Theorem 1.13 is given in Exercise 13, which also states that $\mathbf{S} = \frac{1}{2}\left(\mathbf{A} + \mathbf{A}^T\right)$ and $\mathbf{V} = \frac{1}{2}\left(\mathbf{A} - \mathbf{A}^T\right)$.

EXAMPLE 2   We can decompose the matrix

$$A = \begin{bmatrix} -4 & 2 & 5 \\ 6 & 3 & 7 \\ -1 & 0 & 2 \end{bmatrix}$$

as the sum of a symmetric matrix **S** and a skew-symmetric matrix **V**, where

$$S = \frac{1}{2}\left(A + A^T\right) = \frac{1}{2}\left(\begin{bmatrix} -4 & 2 & 5 \\ 6 & 3 & 7 \\ -1 & 0 & 2 \end{bmatrix} + \begin{bmatrix} -4 & 6 & -1 \\ 2 & 3 & 0 \\ 5 & 7 & 2 \end{bmatrix} \right) = \begin{bmatrix} -4 & 4 & 2 \\ 4 & 3 & \frac{7}{2} \\ 2 & \frac{7}{2} & 2 \end{bmatrix}$$

and

$$V = \frac{1}{2}\left(A - A^T\right) = \frac{1}{2}\left(\begin{bmatrix} -4 & 2 & 5 \\ 6 & 3 & 7 \\ -1 & 0 & 2 \end{bmatrix} - \begin{bmatrix} -4 & 6 & -1 \\ 2 & 3 & 0 \\ 5 & 7 & 2 \end{bmatrix} \right) = \begin{bmatrix} 0 & -2 & 3 \\ 2 & 0 & \frac{7}{2} \\ -3 & -\frac{7}{2} & 0 \end{bmatrix}.$$

Notice that **S** and **V** really are, respectively, symmetric and skew-symmetric and that **S** + **V** really does equal **A**. ∎

Exercises for Section 1.4

1. Compute the following, if possible, for the matrices

$$A = \begin{bmatrix} -4 & 2 & 3 \\ 0 & 5 & -1 \\ 6 & 1 & -2 \end{bmatrix} \quad B = \begin{bmatrix} 6 & -1 & 0 \\ 2 & 2 & -4 \\ 3 & -1 & 1 \end{bmatrix} \quad C = \begin{bmatrix} 5 & -1 \\ -3 & 4 \end{bmatrix}$$

$$D = \begin{bmatrix} -7 & 1 & -4 \\ 3 & -2 & 8 \end{bmatrix} \quad E = \begin{bmatrix} 3 & -3 & 5 \\ 1 & 0 & -2 \\ 6 & 7 & -2 \end{bmatrix} \quad F = \begin{bmatrix} 8 & -1 \\ 2 & 0 \\ 5 & -3 \end{bmatrix}.$$

⋆(a) **A** + **B**

(b) **C** + **D**

⋆(c) 4**A**

(d) 2**A** − 4**B**

⋆(e) **C** + 3**F** − **E**

(f) **A** − **B** + **E**

⋆(g) 2**A** − 3**E** − **B**

(h) 2**D** − 3**F**

⋆(i) **A**^{T} + **E**^{T}

(j) (**A** + **E**)^T

(k) 4**D** + 2**F**^{T}

⋆(l) 2**C**^{T} − 3**F**

(m) 5$\left(\mathbf{F}^T - \mathbf{D}^T\right)$

⋆(n) $\left((\mathbf{B} - \mathbf{A})^T + \mathbf{E}^T\right)^T$

⋆**2.** Indicate which of the following matrices are square, diagonal, upper or lower triangular, symmetric, or skew-symmetric. Calculate the transpose for each matrix.

$$\mathbf{A} = \begin{bmatrix} -1 & 4 \\ 0 & 1 \\ 6 & 0 \end{bmatrix} \quad \mathbf{B} = \begin{bmatrix} 2 & 0 \\ 0 & -1 \end{bmatrix} \quad \mathbf{C} = \begin{bmatrix} -1 & 1 \\ -1 & 1 \end{bmatrix} \quad \mathbf{D} = \begin{bmatrix} -1 \\ 4 \\ 2 \end{bmatrix}$$

$$\mathbf{E} = \begin{bmatrix} 0 & 0 & 6 \\ 0 & -6 & 0 \\ -6 & 0 & 0 \end{bmatrix} \quad \mathbf{F} = \begin{bmatrix} 1 & 0 & 0 & 1 \\ 0 & 0 & 1 & 1 \\ 0 & 1 & 0 & 0 \\ 1 & 1 & 0 & 1 \end{bmatrix} \quad \mathbf{G} = \begin{bmatrix} 6 & 0 & 0 \\ 0 & 6 & 0 \\ 0 & 0 & 6 \end{bmatrix}$$

$$\mathbf{H} = \begin{bmatrix} 0 & -1 & 6 & 2 \\ 1 & 0 & -7 & 1 \\ -6 & 7 & 0 & -4 \\ -2 & -1 & 4 & 0 \end{bmatrix} \quad \mathbf{J} = \begin{bmatrix} 0 & 1 & 0 & 0 \\ 1 & 0 & 1 & 1 \\ 0 & 1 & 1 & 1 \\ 0 & 1 & 1 & 0 \end{bmatrix} \quad \mathbf{K} = \begin{bmatrix} 1 & 2 & 3 & 4 \\ -2 & 1 & 5 & 6 \\ -3 & -5 & 1 & 7 \\ -4 & -6 & -7 & 1 \end{bmatrix}$$

$$\mathbf{L} = \begin{bmatrix} 1 & 1 & 1 \\ 0 & 1 & 1 \\ 0 & 0 & 1 \end{bmatrix} \quad \mathbf{M} = \begin{bmatrix} 0 & 0 & 0 \\ 1 & 0 & 0 \\ 1 & 1 & 0 \end{bmatrix} \quad \mathbf{N} = \begin{bmatrix} 1 & 0 & 0 \\ 0 & 1 & 0 \\ 0 & 0 & 1 \end{bmatrix}$$

$$\mathbf{P} = \begin{bmatrix} 0 & 1 \\ 1 & 0 \end{bmatrix} \quad \mathbf{Q} = \begin{bmatrix} -2 & 0 & 0 \\ 4 & 0 & 0 \\ -1 & 2 & 3 \end{bmatrix} \quad \mathbf{R} = \begin{bmatrix} 6 & 2 \\ 3 & -2 \\ -1 & 0 \end{bmatrix}$$

3. Decompose each of the following as the sum of a symmetric and a skew-symmetric matrix:

⋆(a) $\begin{bmatrix} 3 & -1 & 4 \\ 0 & 2 & 5 \\ 1 & -3 & 2 \end{bmatrix}$
(b) $\begin{bmatrix} 1 & 0 & -4 \\ 3 & 3 & -1 \\ 4 & -1 & 0 \end{bmatrix}$

(c) $\begin{bmatrix} 2 & 3 & 4 & -1 \\ -3 & 5 & -1 & 2 \\ -4 & 1 & -2 & 0 \\ 1 & -2 & 0 & 5 \end{bmatrix}$

4. Prove that if $\mathbf{A}^T = \mathbf{B}^T$, then $\mathbf{A} = \mathbf{B}$.

5. (a) Prove that any symmetric or skew-symmetric matrix is square.
(b) Prove that every diagonal matrix is symmetric.
(c) Show that $(\mathbf{I}_n)^T = \mathbf{I}_n$. (Hint: Use part (b).)
⋆(d) Describe completely every matrix that is both diagonal and skew-symmetric.

6. Assume that $\mathbf{A}$ and $\mathbf{B}$ are square matrices of the same size.

(a) If $\mathbf{A}$ and $\mathbf{B}$ are diagonal, prove that $\mathbf{A} + \mathbf{B}$ is diagonal.
(b) If $\mathbf{A}$ and $\mathbf{B}$ are symmetric, prove that $\mathbf{A} + \mathbf{B}$ is symmetric.

7. Use induction to prove that, if $\mathbf{A}_1, \ldots, \mathbf{A}_n$ are upper triangular matrices of the same size, then $\Sigma_{i=1}^{n} \mathbf{A}_i$ is upper triangular.

8. (a) If $\mathbf{A}$ is a symmetric matrix, show that $\mathbf{A}^T$ and $c\mathbf{A}$ are also symmetric.

(b) If $\mathbf{A}$ is a skew-symmetric matrix, show that $\mathbf{A}^T$ and $c\mathbf{A}$ are also skew-symmetric.

9. The **Kronecker Delta** δ_{ij} is defined as follows: $\delta_{ij} = \begin{cases} 1 \text{ if } i = j \\ 0 \text{ if } i \neq j \end{cases}$. If $\mathbf{A} = \mathbf{I}_n$, explain why $a_{ij} = \delta_{ij}$.

10. Prove parts (4), (5), and (7) of Theorem 1.11.

▶**11.** Prove parts (1) and (3) of Theorem 1.12.

12. Let $\mathbf{A}$ be an $m \times n$ matrix. Prove that if $c\mathbf{A} = \mathbf{O}_{mn}$, the $m \times n$ zero matrix, then $c = 0$ or $\mathbf{A} = \mathbf{O}_{mn}$.

13. This exercise provides an outline for the proof of Theorem 1.13. Let $\mathbf{A}$ be an $n \times n$ matrix.

(a) Prove that $\frac{1}{2}(\mathbf{A} + \mathbf{A}^T)$ is a symmetric matrix.

(b) Prove that $\frac{1}{2}(\mathbf{A} - \mathbf{A}^T)$ is a skew-symmetric matrix.

(c) Show that $\mathbf{A} = \frac{1}{2}(\mathbf{A} + \mathbf{A}^T) + \frac{1}{2}(\mathbf{A} - \mathbf{A}^T)$.

(d) Suppose that $\mathbf{S}_1$ and $\mathbf{S}_2$ are symmetric matrices and that $\mathbf{V}_1$ and $\mathbf{V}_2$ are skew-symmetric matrices such that $\mathbf{S}_1 + \mathbf{V}_1 = \mathbf{S}_2 + \mathbf{V}_2$. Derive a second equation involving $\mathbf{S}_1, \mathbf{S}_2, \mathbf{V}_1$, and $\mathbf{V}_2$ by taking the transpose of both sides of the equation and simplifying.

(e) Prove that $\mathbf{S}_1 = \mathbf{S}_2$ by adding the two equations from part (d) together.

(f) Use parts (d) and (e) to prove that $\mathbf{V}_1 = \mathbf{V}_2$.

(g) Explain how parts (a) through (f) together prove Theorem 1.13.

14. The **trace** of a square matrix $\mathbf{A}$ is the sum of the elements along the main diagonal.

⋆(a) Find the trace of each square matrix in Exercise 2.

(b) If $\mathbf{A}$ and $\mathbf{B}$ are both $n \times n$ matrices, prove that:

   (i) $\text{trace}(\mathbf{A} + \mathbf{B}) = \text{trace}(\mathbf{A}) + \text{trace}(\mathbf{B})$

   (ii) $\text{trace}(c\mathbf{A}) = c(\text{trace}(\mathbf{A}))$

   (iii) $\text{trace}(\mathbf{A}) = \text{trace}(\mathbf{A}^T)$

⋆(c) Suppose that $\text{trace}(\mathbf{A}) = \text{trace}(\mathbf{B})$ for two $n \times n$ matrices $\mathbf{A}$ and $\mathbf{B}$. Does $\mathbf{A} = \mathbf{B}$? Prove your answer.

⋆**15.** True or False:

(a) A 5×6 matrix has exactly 6 entries on its main diagonal.

(b) The transpose of a lower triangular matrix is upper triangular.

(c) No skew-symmetric matrix is diagonal.

(d) If $\mathbf{V}$ is a skew-symmetric matrix, then $-\mathbf{V}^T = \mathbf{V}$.

(e) For all scalars c, and $n \times n$ matrices $\mathbf{A}$ and $\mathbf{B}$, $(c(\mathbf{A}^T + \mathbf{B}))^T = c\mathbf{B}^T + c\mathbf{A}$.

1.5 Matrix Multiplication

Another useful operation is matrix multiplication, which is a generalization of the dot product of vectors.

Definition of Matrix Multiplication

Two matrices $\mathbf{A}$ and $\mathbf{B}$ can be multiplied (in that order) only if the number of columns of $\mathbf{A}$ is equal to the number of rows of $\mathbf{B}$. In that case,

Size of product $\mathbf{AB}$ = (number of rows of $\mathbf{A}$) × (number of columns of $\mathbf{B}$).

That is, if $\mathbf{A}$ is an $m \times n$ matrix, then $\mathbf{AB}$ is defined only when the number of rows of $\mathbf{B}$ is n — that is, when $\mathbf{B}$ is an $n \times p$ matrix, for some integer p. In this case, $\mathbf{AB}$ is an $m \times p$ matrix, because $\mathbf{A}$ has m rows and $\mathbf{B}$ has p columns. The actual entries of $\mathbf{AB}$ are given by the following definition:

DEFINITION

If $\mathbf{A}$ is an $m \times n$ matrix and $\mathbf{B}$ is an $n \times p$ matrix, their matrix product $\mathbf{C} = \mathbf{AB}$ is the $m \times p$ matrix whose (i, j) entry is the dot product of the ith row of $\mathbf{A}$ with the jth column of $\mathbf{B}$. That is,

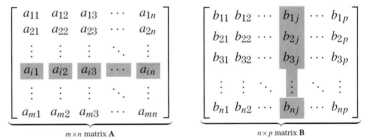

where $c_{ij} = a_{i1}b_{1j} + a_{i2}b_{2j} + a_{i3}b_{3j} + \cdots + a_{in}b_{nj} = \sum_{k=1}^{n} a_{ik}b_{kj}$.

Since the number of columns in **A** equals the number of rows in **B** in this definition, each row of **A** contains the same number of entries as each column of **B**. Thus, it is possible to perform the dot products needed to calculate **C** = **AB**.

Consider

$$
\mathbf{A} = \begin{bmatrix} 5 & -1 & 4 \\ -3 & 6 & 0 \end{bmatrix} \quad \text{and} \quad \mathbf{B} = \begin{bmatrix} 9 & 4 & -8 & 2 \\ 7 & 6 & -1 & 0 \\ -2 & 5 & 3 & -4 \end{bmatrix}.
$$

Since **A** is a 2×3 matrix and **B** is a 3×4 matrix, the number of columns of **A** equals the number of rows of **B** (three in each case). Therefore, **A** and **B** can be multiplied, and the product matrix **C** = **AB** is a 2×4 matrix, because **A** has two rows and **B** has four columns. To calculate each entry of **C**, we take the dot product of the appropriate row of **A** with the appropriate column of **B**. For example, to find c_{11}, we take the dot product of the 1st row of **A** with the 1st column of **B**:

$$
c_{11} = [5, -1, 4] \cdot \begin{bmatrix} 9 \\ 7 \\ -2 \end{bmatrix} = (5)(9) + (-1)(7) + (4)(-2) = 45 - 7 - 8 = 30.
$$

To find c_{23}, we take the dot product of the 2nd row of **A** with the 3rd column of **B**:

$$
c_{23} = [-3, 6, 0] \cdot \begin{bmatrix} -8 \\ -1 \\ 3 \end{bmatrix} = (-3)(-8) + (6)(-1) + (0)(3) = 24 - 6 + 0 = 18.
$$

The other entries are computed similarly, yielding

$$
\mathbf{C} = \mathbf{AB} = \begin{bmatrix} 30 & 34 & -27 & -6 \\ 15 & 24 & 18 & -6 \end{bmatrix}.
$$

∎

Consider the following matrices:

$$
\mathbf{D} = \underbrace{\begin{bmatrix} -2 & 1 \\ 0 & 5 \\ 4 & -3 \end{bmatrix}}_{3 \times 2 \text{ matrix}}, \quad \mathbf{E} = \underbrace{\begin{bmatrix} 1 & -6 \\ 0 & 2 \end{bmatrix}}_{2 \times 2 \text{ matrix}}, \quad \mathbf{F} = \underbrace{[-4\ 2\ 1]}_{1 \times 3 \text{ matrix}},
$$

$$
\mathbf{G} = \underbrace{\begin{bmatrix} 7 \\ -1 \\ 5 \end{bmatrix}}_{3 \times 1 \text{ matrix}}, \quad \text{and } \mathbf{H} = \underbrace{\begin{bmatrix} 5 & 0 \\ 1 & -3 \end{bmatrix}}_{2 \times 2 \text{ matrix}}.
$$

The only possible products of two of these matrices that are defined are

$$\mathbf{DE} = \begin{bmatrix} -2 & 14 \\ 0 & 10 \\ 4 & -30 \end{bmatrix}, \qquad \mathbf{DH} = \begin{bmatrix} -9 & -3 \\ 5 & -15 \\ 17 & 9 \end{bmatrix}, \qquad \mathbf{GF} = \begin{bmatrix} -28 & 14 & 7 \\ 4 & -2 & -1 \\ -20 & 10 & 5 \end{bmatrix},$$

$$\mathbf{EE} = \begin{bmatrix} 1 & -18 \\ 0 & 4 \end{bmatrix}, \quad \mathbf{EH} = \begin{bmatrix} -1 & 18 \\ 2 & -6 \end{bmatrix}, \quad \mathbf{HE} = \begin{bmatrix} 5 & -30 \\ 1 & -12 \end{bmatrix}, \quad \mathbf{HH} = \begin{bmatrix} 25 & 0 \\ 2 & 9 \end{bmatrix},$$

$\mathbf{FG} = [-25]$ $(1 \times 1$ matrix), and $\mathbf{FD} = [12, \ 3]$ $(1 \times 2$ matrix). (Verify!) ∎

Example 2 points out that the order in which matrix multiplication is performed is extremely important. In fact, for two given matrices, we have seen the following:

- Neither product may be defined (for example, **DG** or **GD**).

- One product may be defined but not the other. (**DE** is defined, but not **ED**.)

- Both products may be defined, but the resulting sizes may not agree. (**FG** is 1×1, but **GF** is 3×3.)

- Both products may be defined, and the resulting sizes may agree, but the entries may differ. (**EH** and **HE** are both 2×2, but have different entries.)

In unusual cases, where $\mathbf{AB} = \mathbf{BA}$, we say that **A** and **B commute**, or that "**A commutes** with **B**." But as we have seen, there is no general commutative law for matrix multiplication, although there is a commutative law for addition.

If **A** is any 2×2 matrix, then $\mathbf{AI_2} = \mathbf{I_2A}$ $(= \mathbf{A})$, where $\mathbf{I_2}$ is the identity matrix $\begin{bmatrix} 1 & 0 \\ 0 & 1 \end{bmatrix}$. For example, if $\mathbf{A} = \begin{bmatrix} -4 & 2 \\ 5 & 6 \end{bmatrix}$, then $\begin{bmatrix} -4 & 2 \\ 5 & 6 \end{bmatrix} \begin{bmatrix} 1 & 0 \\ 0 & 1 \end{bmatrix} = \mathbf{A} = \begin{bmatrix} 1 & 0 \\ 0 & 1 \end{bmatrix} \begin{bmatrix} -4 & 2 \\ 5 & 6 \end{bmatrix}$. In Exercise 13, we generalize this to show that if **A** is any $m \times n$ matrix, then $\mathbf{AI_n} = \mathbf{I_mA} = \mathbf{A}$. This is why **I** is called the (**multiplicative**) **identity** matrix — because it preserves the "identity" of any matrices multiplied by it. In particular, for an $n \times n$ matrix **A**, $\mathbf{AI_n} = \mathbf{I_nA} = \mathbf{A}$, and so **A** commutes with $\mathbf{I_n}$.

Application: Shipping Cost and Profit

Matrix products are vital in modeling certain geometric transformations (as we will see in Sections 5.1 and 8.8). They are also widely used in graph theory, coding theory, physics, and chemistry. Here is a simple application in business.

EXAMPLE 3   Suppose four popular DVDs — say, W, X, Y, and Z — are being sold online by a video company that operates three different warehouses. After purchase, the shipping cost is added to the price of the DVDs when they are mailed to the customer. The number of each type of DVD shipped from each warehouse during the past week is shown in matrix **A** below. The shipping cost and profit collected for each DVD sold is shown in matrix **B**.

$$
\mathbf{A} = \begin{array}{c} \\ \text{Warehouse 1} \\ \text{Warehouse 2} \\ \text{Warehouse 3} \end{array} \begin{array}{cccc} \text{DVD W} & \text{DVD X} & \text{DVD Y} & \text{DVD Z} \\ \left[\begin{array}{cccc} 130 & 160 & 240 & 190 \\ 210 & 180 & 320 & 240 \\ 170 & 200 & 340 & 220 \end{array}\right] \end{array}
$$

$$
\mathbf{B} = \begin{array}{c} \\ \text{DVD W} \\ \text{DVD X} \\ \text{DVD Y} \\ \text{DVD Z} \end{array} \begin{array}{cc} \text{Shipping Cost} & \text{Profit} \\ \left[\begin{array}{cc} \$3 & \$3 \\ \$4 & \$2 \\ \$3 & \$4 \\ \$2 & \$2 \end{array}\right] \end{array}
$$

The product **AB** represents the combined total shipping costs and profits last week.

$$
\mathbf{AB} = \begin{array}{c} \\ \text{Warehouse 1} \\ \text{Warehouse 2} \\ \text{Warehouse 3} \end{array} \begin{array}{cc} \text{Total Shipping Cost} & \text{Total Profit} \\ \left[\begin{array}{cc} \$2130 & \$2050 \\ \$2790 & \$2750 \\ \$2770 & \$2710 \end{array}\right] \end{array}
$$

In particular, the entry in the 2nd row and 2nd column of **AB** is calculated by taking the dot product of the 2nd row of **A** with the 2nd column of **B**; that is,

$$(210)(\$3) + (180)(\$2) + (320)(\$4) + (240)(\$2) = \$2750.$$

In this case, we are multiplying the number of each type of DVD sold from Warehouse 2 times the profit per DVD, which equals the total profit for Warehouse 2. ■

Often we need to find only a particular row or column of a matrix product:

> If the product **AB** is defined, then the kth row of **AB** is the product (kth row of **A**)**B**. Also, the lth column of **AB** is the product **A**(lth column of **B**).

Thus, in Example 3, if we only want the results for Warehouse 3, we only need to compute the 3rd row of **AB**. This is

$$
\underbrace{\left[\begin{array}{cccc} 170 & 200 & 340 & 220 \end{array}\right]}_{\text{3rd row of } \mathbf{A}} \underbrace{\left[\begin{array}{cc} \$3 & \$3 \\ \$4 & \$2 \\ \$3 & \$4 \\ \$2 & \$2 \end{array}\right]}_{\mathbf{B}} = \underbrace{\left[\begin{array}{cc} \$2770 & \$2710 \end{array}\right]}_{\text{3rd row of } \mathbf{AB}}.
$$

Fundamental Properties of Matrix Multiplication

If the zero matrix **O** is multiplied times any matrix **A**, or if **A** is multiplied times **O**, the result is **O** (see Exercise 12). The following theorem lists some other important properties of matrix multiplication:

> ### THEOREM 1.14
> Suppose that **A**, **B**, and **C** are matrices for which the following sums and products are defined. Let c be a scalar. Then
>
> (1) $\mathbf{A}(\mathbf{BC}) = (\mathbf{AB})\mathbf{C}$ Associative Law of Multiplication
> (2) $\mathbf{A}(\mathbf{B} + \mathbf{C}) = \mathbf{AB} + \mathbf{AC}$ Distributive Laws of Matrix
> (3) $(\mathbf{A} + \mathbf{B})\mathbf{C} = \mathbf{AC} + \mathbf{BC}$ Multiplication over Addition
> (4) $c(\mathbf{AB}) = (c\mathbf{A})\mathbf{B} = \mathbf{A}(c\mathbf{B})$ Associative Law of Scalar and Matrix Multiplication

The proof of part (1) of Theorem 1.14 is more difficult than the others, and so it is included in Appendix A for the interested reader. You are asked to provide the proofs of parts (2), (3), and (4) in Exercise 11.

Other expected properties do not hold for matrix multiplication (such as the commutative law). For example, the **cancellation laws** of algebra do not hold in general. That is, if $\mathbf{AB} = \mathbf{AC}$, with $\mathbf{A} \neq \mathbf{O}$, it does not necessarily follow that $\mathbf{B} = \mathbf{C}$. For example, if

$$\mathbf{A} = \begin{bmatrix} 2 & 1 \\ 6 & 3 \end{bmatrix}, \quad \mathbf{B} = \begin{bmatrix} -1 & 0 \\ 5 & 2 \end{bmatrix}, \quad \text{and} \quad \mathbf{C} = \begin{bmatrix} 3 & 1 \\ -3 & 0 \end{bmatrix},$$

then

$$\mathbf{AB} = \begin{bmatrix} 2 & 1 \\ 6 & 3 \end{bmatrix} \begin{bmatrix} -1 & 0 \\ 5 & 2 \end{bmatrix} = \begin{bmatrix} 3 & 2 \\ 9 & 6 \end{bmatrix}$$

and

$$\mathbf{AC} = \begin{bmatrix} 2 & 1 \\ 6 & 3 \end{bmatrix} \begin{bmatrix} 3 & 1 \\ -3 & 0 \end{bmatrix} = \begin{bmatrix} 3 & 2 \\ 9 & 6 \end{bmatrix}.$$

Here, $\mathbf{AB} = \mathbf{AC}$, but $\mathbf{B} \neq \mathbf{C}$. Similarly, if $\mathbf{AB} = \mathbf{CB}$, it does not necessarily follow that $\mathbf{A} = \mathbf{C}$.

Also, if $\mathbf{AB} = \mathbf{O}$, it is not necessarily true that $\mathbf{A} = \mathbf{O}$ or $\mathbf{B} = \mathbf{O}$. For example, if

$$\mathbf{A} = \begin{bmatrix} 2 & 1 \\ 6 & 3 \end{bmatrix} \quad \text{and} \quad \mathbf{B} = \begin{bmatrix} -1 & 2 \\ 2 & -4 \end{bmatrix},$$

then

$$\mathbf{AB} = \begin{bmatrix} 2 & 1 \\ 6 & 3 \end{bmatrix} \begin{bmatrix} -1 & 2 \\ 2 & -4 \end{bmatrix} = \begin{bmatrix} 0 & 0 \\ 0 & 0 \end{bmatrix}.$$

Here, $\mathbf{AB} = \mathbf{O}_2$, but neither **A** nor **B** equals $\mathbf{O}_2$.

Powers of Square Matrices

Any square matrix can be multiplied by itself because the number of rows is the same as the number of columns. In fact, square matrices are the only matrices that can be multiplied by themselves (why?). The various nonnegative powers of a square matrix are defined in a natural way.

DEFINITION

Let $\mathbf{A}$ be any $n \times n$ matrix. Then the (nonnegative) powers of $\mathbf{A}$ are given by $\mathbf{A}^0 = \mathbf{I}_n$, $\mathbf{A}^1 = \mathbf{A}$, and for $k \geq 2$, $\mathbf{A}^k = (\mathbf{A}^{k-1})(\mathbf{A})$.

EXAMPLE 4
Suppose that $\mathbf{A} = \begin{bmatrix} 2 & 1 \\ -4 & 3 \end{bmatrix}$. Then

$$\mathbf{A}^2 = (\mathbf{A})(\mathbf{A}) = \begin{bmatrix} 2 & 1 \\ -4 & 3 \end{bmatrix} \begin{bmatrix} 2 & 1 \\ -4 & 3 \end{bmatrix} = \begin{bmatrix} 0 & 5 \\ -20 & 5 \end{bmatrix}, \text{ and}$$

$$\mathbf{A}^3 = (\mathbf{A}^2)(\mathbf{A}) = \begin{bmatrix} 0 & 5 \\ -20 & 5 \end{bmatrix} \begin{bmatrix} 2 & 1 \\ -4 & 3 \end{bmatrix} = \begin{bmatrix} -20 & 15 \\ -60 & -5 \end{bmatrix}.$$
∎

EXAMPLE 5
The identity matrix $\mathbf{I}_n$ is square, and so $\mathbf{I}_n^k$ exists, for all $k \geq 0$. However, since $\mathbf{I}_n \mathbf{A} = \mathbf{A}$, for any $n \times n$ matrix $\mathbf{A}$, we have $\mathbf{I}_n \mathbf{I}_n = \mathbf{I}_n$. Thus, $\mathbf{I}_n^k = \mathbf{I}_n$, for all $k \geq 0$.
∎

The next theorem asserts that two familiar laws of exponents in algebra are still valid for powers of a square matrix. The proof is left as Exercise 16.

THEOREM 1.15

If $\mathbf{A}$ is a square matrix, and if s and t are nonnegative integers, then

(1) $\mathbf{A}^{s+t} = (\mathbf{A}^s)(\mathbf{A}^t)$
(2) $(\mathbf{A}^s)^t = \mathbf{A}^{st} = (\mathbf{A}^t)^s$.

As an example of part (1) of this theorem, we have $\mathbf{A}^{4+6} = (\mathbf{A}^4)(\mathbf{A}^6) = \mathbf{A}^{10}$. As an example of part (2), we have $(\mathbf{A}^3)^2 = \mathbf{A}^{(3)(2)} = (\mathbf{A}^2)^3 = \mathbf{A}^6$.

One law of exponents in elementary algebra that does not carry over to matrix algebra is $(xy)^q = x^q y^q$. In fact, if $\mathbf{A}$ and $\mathbf{B}$ are square matrices of the same size, usually $(\mathbf{AB})^q \neq \mathbf{A}^q \mathbf{B}^q$, if q is an integer ≥ 2. Even in the simplest case $q = 2$, usually $(\mathbf{AB})(\mathbf{AB}) \neq (\mathbf{AA})(\mathbf{BB})$ because the *order* of matrix multiplication is important.

EXAMPLE 6
Let

$$\mathbf{A} = \begin{bmatrix} 2 & -4 \\ 1 & 3 \end{bmatrix} \quad \text{and} \quad \mathbf{B} = \begin{bmatrix} 3 & 2 \\ -1 & 5 \end{bmatrix}.$$

Then

$$(\mathbf{AB})^2 = \begin{bmatrix} 10 & -16 \\ 0 & 17 \end{bmatrix}^2 = \begin{bmatrix} 100 & -432 \\ 0 & 289 \end{bmatrix}.$$

However,

$$\mathbf{A}^2 \mathbf{B}^2 = \begin{bmatrix} 0 & -20 \\ 5 & 5 \end{bmatrix} \begin{bmatrix} 7 & 16 \\ -8 & 23 \end{bmatrix} = \begin{bmatrix} 160 & -460 \\ -5 & 195 \end{bmatrix}.$$

Hence, in this particular case, $(\mathbf{AB})^2 \neq \mathbf{A}^2 \mathbf{B}^2$.
∎

The Transpose of a Matrix Product

THEOREM 1.16
If $\mathbf{A}$ is an $m \times n$ matrix and $\mathbf{B}$ is an $n \times p$ matrix, then $(\mathbf{AB})^T = \mathbf{B}^T\mathbf{A}^T$.

This result may seem unusual at first because you might expect $(\mathbf{AB})^T$ to equal $\mathbf{A}^T\mathbf{B}^T$. But notice that $\mathbf{A}^T\mathbf{B}^T$ may not be defined, because $\mathbf{A}^T$ is an $n \times m$ matrix and $\mathbf{B}^T$ is a $p \times n$ matrix. Instead, the transpose of the product of two matrices is the product of their transposes *in reverse order*.

PROOF OF THEOREM 1.16
Because $\mathbf{AB}$ is an $m \times p$ matrix and $\mathbf{B}^T$ is a $p \times n$ matrix and $\mathbf{A}^T$ is an $n \times m$ matrix, it follows that $(\mathbf{AB})^T$ and $\mathbf{B}^T\mathbf{A}^T$ are both $p \times m$ matrices. Hence, we only need to show the (i, j) entries of $(\mathbf{AB})^T$ and $\mathbf{B}^T\mathbf{A}^T$ are equal, for $1 \le i \le p$ and $1 \le j \le m$. Now, the (i, j) entry of $(\mathbf{AB})^T$ is the (j, i) entry of $\mathbf{AB}$, which is [jth row of $\mathbf{A}$]·[ith column of $\mathbf{B}$]. However, the (i, j) entry of $\mathbf{B}^T\mathbf{A}^T$ is [ith row of $\mathbf{B}^T$]·[jth column of $\mathbf{A}^T$], which equals [ith column of $\mathbf{B}$]·[jth row of $\mathbf{A}$]. Thus, the (i, j) entries of $(\mathbf{AB})^T$ and $\mathbf{B}^T\mathbf{A}^T$ agree. ■

EXAMPLE 7   For the matrices $\mathbf{A}$ and $\mathbf{B}$ of Example 6, we have

$$\mathbf{AB} = \begin{bmatrix} 10 & -16 \\ 0 & 17 \end{bmatrix}, \quad \mathbf{B}^T = \begin{bmatrix} 3 & -1 \\ 2 & 5 \end{bmatrix}, \quad \text{and} \quad \mathbf{A}^T = \begin{bmatrix} 2 & 1 \\ -4 & 3 \end{bmatrix}.$$

Hence

$$\mathbf{B}^T\mathbf{A}^T = \begin{bmatrix} 3 & -1 \\ 2 & 5 \end{bmatrix}\begin{bmatrix} 2 & 1 \\ -4 & 3 \end{bmatrix} = \begin{bmatrix} 10 & 0 \\ -16 & 17 \end{bmatrix} = (\mathbf{AB})^T.$$

Notice, however, that

$$\mathbf{A}^T\mathbf{B}^T = \begin{bmatrix} 2 & 1 \\ -4 & 3 \end{bmatrix}\begin{bmatrix} 3 & -1 \\ 2 & 5 \end{bmatrix} = \begin{bmatrix} 8 & 3 \\ -6 & 19 \end{bmatrix} \ne (\mathbf{AB})^T.$$

■

♦ **Supplemental Material:** You have now covered the prerequisites for Section 7.1, "Complex n-Vectors and Matrices."
   ♦ **Application:** You have now covered the prerequisites for Section 8.1, "Graph Theory."

Exercises for Section 1.5

Note: Exercises 1 through 3 refer to the following matrices:

$$A = \begin{bmatrix} -2 & 3 \\ 6 & 5 \\ 1 & -4 \end{bmatrix} \quad B = \begin{bmatrix} -5 & 3 & 6 \\ 3 & 8 & 0 \\ -2 & 0 & 4 \end{bmatrix} \quad C = \begin{bmatrix} 11 & -2 \\ -4 & -2 \\ 3 & -1 \end{bmatrix}$$

$$D = \begin{bmatrix} -1 & 4 & 3 & 7 \\ 2 & 1 & 7 & 5 \\ 0 & 5 & 5 & -2 \end{bmatrix} \quad E = \begin{bmatrix} 1 & 1 & 0 & 1 \\ 1 & 0 & 1 & 0 \\ 0 & 0 & 0 & 1 \\ 1 & 0 & 1 & 0 \end{bmatrix} \quad F = \begin{bmatrix} 9 & -3 \\ 5 & -4 \\ 2 & 0 \\ 8 & -3 \end{bmatrix}$$

$$G = \begin{bmatrix} 5 & 1 & 0 \\ 0 & -2 & -1 \\ 1 & 0 & 3 \end{bmatrix} \quad H = \begin{bmatrix} 6 & 3 & 1 \\ 1 & -15 & -5 \\ -2 & -1 & 10 \end{bmatrix} \quad J = \begin{bmatrix} 8 \\ -1 \\ 4 \end{bmatrix}$$

$$K = \begin{bmatrix} 2 & 1 & -5 \\ 0 & 2 & 7 \end{bmatrix} \quad L = \begin{bmatrix} 10 & 9 \\ 8 & 7 \end{bmatrix} \quad M = \begin{bmatrix} 7 & -1 \\ 11 & 3 \end{bmatrix}$$

$$N = \begin{bmatrix} 0 & 0 \\ 0 & 0 \end{bmatrix} \quad P = \begin{bmatrix} 3 & -1 \\ 4 & 7 \end{bmatrix} \quad Q = \begin{bmatrix} 1 & 4 & -1 & 6 \\ 8 & 7 & -3 & 3 \end{bmatrix}$$

$$R = \begin{bmatrix} -3 & 6 & -2 \end{bmatrix} \quad S = \begin{bmatrix} 6 & -4 & 3 & 2 \end{bmatrix} \quad T = \begin{bmatrix} 4 & -1 & 7 \end{bmatrix}$$

1. Which of these products are possible? If possible, then calculate the product.

   (a) **AB**   ⋆(b) **BA**

   ⋆(c) **JM**   (d) **DF**

   ⋆(e) **RJ**   ⋆(f) **JR**

   ⋆(g) **RT**   (h) **SF**

   (i) **KN**   ⋆(j) F^2

   (k) B^2 ⋆(l) E^3

   (m) $(TJ)^3$ ⋆(n) **D (FK)**

   (o) **(CL) G**

2. Determine whether these pairs of matrices commute.

   ⋆(a) **L** and **M**   (b) **G** and **H**

   ⋆(c) **A** and **K**   ⋆(d) **N** and **P**

   (e) **F** and **Q**

3. Find only the indicated row or column of each given matrix product.

   ⋆(a) The 2nd row of **BG**   (b) The 3rd column of **DE**

   ⋆(c) The 1st column of **SE**   (d) The 3rd row of **FQ**

⋆4. Assuming that all of the following products exist, which of these equations are always valid? If valid, specify which theorems (and parts, if appropriate) apply.

   (a) $(RG)H = R(GH)$ (b) $LP = PL$

   (c) $E(FK) = (EF)K$ (d) $K(A + C) = KA + KC$

   (e) $(QF)^T = F^T Q^T$ (f) $L(ML) = L^2 M$

   (g) $GC + HC = (G + H)C$ (h) $R(J + T^T) = RJ + RT^T$

   (i) $(AK)^T = A^T K^T$ (j) $(Q + F^T)E^T = QE^T + (EF)^T$

⋆**5.** The following matrices detail the number of employees at four different retail outlets and their wages and benefits (per year). Calculate the total salaries and fringe benefits paid by each outlet per year to its employees.

	Executives	Salespersons	Others
Outlet 1	3	7	8
Outlet 2	2	4	5
Outlet 3	6	14	18
Outlet 4	3	6	9

	Salary	Fringe Benefits
Executives	$30000	$7500
Salespersons	$22500	$4500
Others	$15000	$3000

⋆**6.** Matrix **A** gives the percentage of nitrogen, phosphates, and potash in three fertilizers. Matrix **B** gives the amount (in tons) of each type of fertilizer spread on three different fields. Use matrix operations to find the total amount of nitrogen, phosphates, and potash on each field.

		Nitrogen	Phosphates	Potash
	Fertilizer 1	10%	10%	5%
$A =$	Fertilizer 2	25%	5%	5%
	Fertilizer 3	0%	10%	20%

		Field 1	Field 2	Field 3
	Fertilizer 1	5	2	4
$B =$	Fertilizer 2	2	1	1
	Fertilizer 3	3	1	3

⋆**7.** (a) Find a nondiagonal matrix **A** such that $A^2 = I_2$.
   (b) Find a nondiagonal matrix **A** such that $A^2 = I_3$. (Hint: Modify your answer to part (a).)
   (c) Find a nonidentity matrix **A** such that $A^3 = I_3$.

8. Let **A** be an $m \times n$ matrix, and let **B** be an $n \times m$ matrix, with $m, n \geq 5$. Each of the following sums represents an entry of either **AB** or **BA**. Determine which product is involved and which entry of that product is represented.

⋆(a) $\sum_{k=1}^{n} a_{3k} b_{k4}$ (b) $\sum_{q=1}^{n} a_{4q} b_{q1}$

⋆(c) $\sum_{k=1}^{m} a_{k2} b_{3k}$ (d) $\sum_{q=1}^{m} b_{2q} a_{q5}$

9. Let **A** be an $m \times n$ matrix, and let **B** be an $n \times m$ matrix, where $m, n \geq 4$. Use sigma (Σ) notation to express the following entries symbolically:

⋆(a) The entry in the 3rd row and 2nd column of **AB**
   (b) The entry in the 4th row and 1st column of **BA**

10. (a) Consider the unit vectors **i**, **j**, and **k** in $\mathbb{R}^3$. Show that, if **A** is an $m \times 3$ matrix, then $\mathbf{Ai} =$ first column of **A**, $\mathbf{Aj} =$ second column of **A**, and $\mathbf{Ak} =$ third column of **A**.

    (b) Generalize part (a) to a similar result involving an $m \times n$ matrix $\mathbf{A}$ and the standard unit vectors $\mathbf{e}_1, \ldots, \mathbf{e}_n$ in $\mathbb{R}^n$.

    (c) Let $\mathbf{A}$ be an $m \times n$ matrix. Use part (b) to show that, if $\mathbf{A}\mathbf{x} = \mathbf{0}$ for all vectors $\mathbf{x} \in \mathbb{R}^n$, then $\mathbf{A} = \mathbf{O}_{mn}$.

▶**11.** Prove parts (2), (3), and (4) of Theorem 1.14.

12. Let $\mathbf{A}$ be an $m \times n$ matrix. Prove $\mathbf{A}\mathbf{O}_{np} = \mathbf{O}_{mp}$.

13. Let $\mathbf{A}$ be an $m \times n$ matrix. Prove $\mathbf{A}\mathbf{I}_n = \mathbf{I}_m\mathbf{A} = \mathbf{A}$.

14. (a) Prove that the product of two diagonal matrices is diagonal. (Hint: If $\mathbf{C} = \mathbf{AB}$ where $\mathbf{A}$ and $\mathbf{B}$ are diagonal, show that $c_{ij} = 0$ when $i \neq j$.)

    (b) Prove that the product of two upper triangular matrices is upper triangular. (Hint: Let $\mathbf{A}$ and $\mathbf{B}$ be upper triangular and $\mathbf{C} = \mathbf{AB}$. Show $c_{ij} = 0$ when $i > j$ by checking that all terms $a_{ik}b_{kj}$ in the formula for c_{ij} have at least one zero factor. Consider the following two cases: $i > k$ and $i \leq k$.)

    (c) Prove that the product of two lower triangular matrices is lower triangular. (Hint: Use Theorem 1.16 and part (b) of this exercise.)

15. Show that if $c \in \mathbb{R}$ and $\mathbf{A}$ is a square matrix, then $(c\mathbf{A})^n = c^n\mathbf{A}^n$ for any integer $n \geq 1$. (Hint: Use a proof by induction.)

▶**16.** Prove each part of Theorem 1.15 using the method of induction. (Hint: Use induction on t for both parts. Part (1) will be useful in proving part (2).)

17. (a) Show $\mathbf{AB} = \mathbf{BA}$ only if $\mathbf{A}$ and $\mathbf{B}$ are square matrices of the same size.

    (b) Prove two square matrices $\mathbf{A}$ and $\mathbf{B}$ of the same size commute if and only if $(\mathbf{A} + \mathbf{B})^2 = \mathbf{A}^2 + 2\mathbf{AB} + \mathbf{B}^2$.

18. If $\mathbf{A}$, $\mathbf{B}$, and $\mathbf{C}$ are all square matrices of the same size, show that $\mathbf{AB}$ commutes with $\mathbf{C}$ if $\mathbf{A}$ and $\mathbf{B}$ both commute with $\mathbf{C}$.

19. Show that $\mathbf{A}$ and $\mathbf{B}$ commute if and only if $\mathbf{A}^T$ and $\mathbf{B}^T$ commute.

20. Let $\mathbf{A}$ be any matrix. Show that $\mathbf{A}\mathbf{A}^T$ and $\mathbf{A}^T\mathbf{A}$ are both symmetric.

21. Let $\mathbf{A}$ and $\mathbf{B}$ both be $n \times n$ matrices.

    (a) Show that $(\mathbf{AB})^T = \mathbf{BA}$ if $\mathbf{A}$ and $\mathbf{B}$ are both symmetric or both skew-symmetric.

    (b) If $\mathbf{A}$ and $\mathbf{B}$ are both symmetric, show that $\mathbf{AB}$ is symmetric if and only if $\mathbf{A}$ and $\mathbf{B}$ commute.

22. Recall the definition of the **trace** of a matrix given in Exercise 14 of Section 1.4. If $\mathbf{A}$ and $\mathbf{B}$ are both $n \times n$ matrices, show the following:

    (a) Trace$(\mathbf{A}\mathbf{A}^T)$ is the sum of the squares of all entries of $\mathbf{A}$.

    (b) If trace$(\mathbf{A}\mathbf{A}^T) = 0$, then $\mathbf{A} = \mathbf{O}_n$. (Hint: Use part (a) of this exercise.)

    (c) Trace$(\mathbf{AB})$ = trace$(\mathbf{BA})$. (Hint: Calculate trace$(\mathbf{AB})$ and trace$(\mathbf{BA})$ in the 3×3 case to discover how to prove the general $n \times n$ case.)

23. An **idempotent** matrix is a square matrix $\mathbf{A}$ for which $\mathbf{A}^2 = \mathbf{A}$. (Note that if $\mathbf{A}$ is idempotent, then $\mathbf{A}^n = \mathbf{A}$ for every integer $n \geq 1$.)

★(a) Find a 2×2 idempotent matrix (besides $\mathbf{I}_n$ and $\mathbf{O}_n$).

(b) Show $\begin{bmatrix} -1 & 1 & 1 \\ -1 & 1 & 1 \\ -1 & 1 & 1 \end{bmatrix}$ is idempotent.

(c) If $\mathbf{A}$ is an $n \times n$ idempotent matrix, show $\mathbf{I}_n - \mathbf{A}$ is also idempotent.

(d) Use parts (b) and (c) to get another example of an idempotent matrix.

(e) Let $\mathbf{A}$ and $\mathbf{B}$ be $n \times n$ matrices. Show that $\mathbf{A}$ is idempotent if both $\mathbf{AB} = \mathbf{A}$ and $\mathbf{BA} = \mathbf{B}$.

24. (a) Let $\mathbf{A}$ be an $m \times n$ matrix, and let $\mathbf{B}$ be an $n \times p$ matrix. Prove that $\mathbf{AB} = \mathbf{O}_{mp}$ if and only if every (vector) row of $\mathbf{A}$ is orthogonal to each column of $\mathbf{B}$.

★(b) Find a 2×3 matrix $\mathbf{A} \neq \mathbf{O}$ and a 3×2 matrix $\mathbf{B} \neq \mathbf{O}$ such that $\mathbf{AB} = \mathbf{O}_2$.

(c) Using your answers from part (b), find a matrix $\mathbf{C} \neq \mathbf{B}$ such that $\mathbf{AB} = \mathbf{AC}$.

★25. What form does a 2×2 matrix have if it commutes with every other 2×2 matrix? Prove that your answer is correct.

26. Let $\mathbf{A}$ be an $n \times n$ matrix. Consider the $n \times n$ matrix $\mathbf{\Psi}_{ij}$, which has all entries zero except for an entry of 1 in the (i, j) position.

(a) Show that the jth column of $\mathbf{A}\mathbf{\Psi}_{ij}$ equals the ith column of $\mathbf{A}$ and all other columns of $\mathbf{A}\mathbf{\Psi}_{ij}$ have only zero entries.

(b) Show that the ith row of $\mathbf{\Psi}_{ij}\mathbf{A}$ equals the jth row of $\mathbf{A}$ and all other rows of $\mathbf{\Psi}_{ij}\mathbf{A}$ have only zero entries.

(c) Use parts (a) and (b) to prove that an $n \times n$ matrix $\mathbf{A}$ commutes with all other $n \times n$ matrices if and only if $\mathbf{A} = c\mathbf{I}_n$, for some $c \in \mathbb{R}$. (Hint: Use $\mathbf{A}\mathbf{\Psi}_{kk} = \mathbf{\Psi}_{kk}\mathbf{A}$, for $1 \leq k \leq n$, to prove $a_{ij} = 0$ for $i \neq j$. Then use $\mathbf{A}\mathbf{\Psi}_{ij} = \mathbf{\Psi}_{ij}\mathbf{A}$ to show $a_{ii} = a_{jj}$.)

★27. True or False:

(a) If $\mathbf{AB}$ is defined, the jth column of $\mathbf{AB} = \mathbf{A}(j$th column of $\mathbf{B})$.

(b) If $\mathbf{A}, \mathbf{B}, \mathbf{D}$ are $n \times n$ matrices, then $\mathbf{D}(\mathbf{A} + \mathbf{B}) = \mathbf{DB} + \mathbf{DA}$.

(c) If t is a scalar, and $\mathbf{D}$ and $\mathbf{E}$ are $n \times n$ matrices, then $(t\mathbf{D})\mathbf{E} = \mathbf{D}(t\mathbf{E})$.

(d) If $\mathbf{D}, \mathbf{E}$ are $n \times n$ matrices, then $(\mathbf{DE})^2 = \mathbf{D}^2\mathbf{E}^2$.

(e) If $\mathbf{D}, \mathbf{E}$ are $n \times n$ matrices, then $(\mathbf{DE})^T = \mathbf{D}^T\mathbf{E}^T$.

(f) If $\mathbf{DE} = \mathbf{O}$, then $\mathbf{D} = \mathbf{O}$ or $\mathbf{E} = \mathbf{O}$.

Chapter 2

Systems of Linear Equations

A Systematic Approach

One important mathematical problem that arises frequently is the need to unscramble data that has been mixed together by an apparently irreversible process. A common problem of this type is the calculation of the exact ratios of chemical elements that were combined to produce a certain compound. To solve this problem, we must unscramble the given mix of given elements to determine the original ratios involved. An analogous type of problem involves the deciphering of a coded message, where in order to find the answer we must recover the original message before it was scrambled into code.

We will see that whenever information is scrambled in a "linear" fashion, a matrix multiplication is involved. And a system of linear equations corresponding to that matrix can be constructed. Unscrambling the data is then accomplished by solving that system of linear equations. In this chapter, we develop a systematic method for solving such systems and then study some of the theoretical consequences of that technique.

A ttempts to solve systems of linear equations inspired much of the development of linear algebra. In Sections 2.1 and 2.2, we present Gaussian elimination and Gauss-Jordan row reduction, which are important techniques for solving linear systems. The study of linear systems leads to the examination of further properties of matrices, including row equivalence, rank, and the row space of a matrix in Section 2.3, and inverses of matrices in Section 2.4.

2.1 Solving Linear Systems Using Gaussian Elimination

In this section, we introduce systems of linear equations and the Gaussian elimination method for solving such systems.

Systems of Linear Equations

A **linear equation** is an equation involving one or more variables in which only the operations of multiplication by real numbers and summing of terms are allowed. For example, $6x - 3y = 4$ and $8x_1 + 3x_2 - 4x_3 = -20$ are linear equations in two and three variables, respectively.

When several linear equations involving the same variables are considered together, we have a **system of linear equations**. For example, the following system has four equations and three variables:

$$\begin{cases} 3x_1 & - & 2x_2 & - & 5x_3 & = & 4 \\ 2x_1 & + & 4x_2 & - & x_3 & = & 2 \\ 6x_1 & - & 4x_2 & - & 10x_3 & = & 8 \\ -4x_1 & + & 8x_2 & + & 9x_3 & = & -6 \end{cases}.$$

We often need to find the solutions to a given system. The ordered triple, or 3-tuple, $(x_1, x_2, x_3) = (4, -1, 2)$ is a solution to the preceding system because each equation in the system is satisfied for these values of x_1, x_2, and x_3. Notice that $\left(-\frac{3}{2}, \frac{3}{4}, -2\right)$ is another solution for that same system. These two particular solutions are part of the complete set of all solutions for that system.

We now formally define linear systems and their solutions.

DEFINITION

A **system of** m (simultaneous) **linear equations** in n variables

$$\begin{cases} a_{11}x_1 + a_{12}x_2 + a_{13}x_3 + \cdots + a_{1n}x_n = b_1 \\ a_{21}x_1 + a_{22}x_2 + a_{23}x_3 + \cdots + a_{2n}x_n = b_2 \\ \vdots \qquad \vdots \qquad \vdots \qquad \ddots \qquad \vdots \qquad \vdots \\ a_{m1}x_1 + a_{m2}x_2 + a_{m3}x_3 + \cdots + a_{mn}x_n = b_m \end{cases}$$

is a collection of m equations, each containing a linear combination of the same n variables summing to a scalar. A **particular solution** to a system of linear equations in the variables $x_1, x_2, \ldots, x_n$ is an n-tuple

$(s_1, s_2, \ldots, s_n)$ that satisfies each equation in the system when s_1 is substituted for x_1, s_2 for x_2, and so on. The **(complete) solution set** for a system of linear equations in n variables is the collection of all n-tuples that form solutions to the system.

In this definition, the coefficients of $x_1, x_2, \ldots, x_n$ can be collected together in an $m \times n$ **coefficient matrix**

$$\mathbf{A} = \begin{bmatrix} a_{11} & a_{12} & \cdots & a_{1n} \\ a_{21} & a_{22} & \cdots & a_{2n} \\ \vdots & \vdots & \ddots & \vdots \\ a_{m1} & a_{m2} & \cdots & a_{mn} \end{bmatrix}.$$

If we also let

$$\mathbf{X} = \begin{bmatrix} x_1 \\ x_2 \\ \vdots \\ x_n \end{bmatrix} \quad \text{and} \quad \mathbf{B} = \begin{bmatrix} b_1 \\ b_2 \\ \vdots \\ b_m \end{bmatrix},$$

then the linear system is equivalent to the matrix equation $\mathbf{AX} = \mathbf{B}$ (verify!).

An alternate way to express this system is to form the **augmented matrix**

$$[\mathbf{A}\,|\,\mathbf{B}] = \left[\begin{array}{cccc|c} a_{11} & a_{12} & \cdots & a_{1n} & b_1 \\ a_{21} & a_{22} & \cdots & a_{2n} & b_2 \\ \vdots & \vdots & \ddots & \vdots & \vdots \\ a_{m1} & a_{m2} & \cdots & a_{mn} & b_m \end{array} \right].$$

Each row of $[\mathbf{A}\,|\,\mathbf{B}]$ represents one equation in the original system, and each column to the left of the vertical bar represents one of the variables in the system. Hence, this augmented matrix contains all the vital information from the original system.

EXAMPLE 1 Consider the linear system

$$\begin{cases} 4w & - & 2x & + & y & - & 3z & = & 5 \\ 3w & + & x & & & + & 5z & = & 12 \end{cases}.$$

Letting

$$\mathbf{A} = \begin{bmatrix} 4 & -2 & 1 & -3 \\ 3 & 1 & 0 & 5 \end{bmatrix}, \quad \mathbf{X} = \begin{bmatrix} w \\ x \\ y \\ z \end{bmatrix}, \quad \text{and} \quad \mathbf{B} = \begin{bmatrix} 5 \\ 12 \end{bmatrix},$$

we see that the system is equivalent to $\mathbf{AX} = \mathbf{B}$, or,

$$\begin{bmatrix} 4 & -2 & 1 & -3 \\ 3 & 1 & 0 & 5 \end{bmatrix} \begin{bmatrix} w \\ x \\ y \\ z \end{bmatrix} = \begin{bmatrix} 4w - 2x + y - 3z \\ 3w + x + 5z \end{bmatrix} = \begin{bmatrix} 5 \\ 12 \end{bmatrix}.$$

This system can also be represented by the augmented matrix

$$[\mathbf{A}\,|\,\mathbf{B}] = \begin{bmatrix} 4 & -2 & 1 & -3 & \Big| & 5 \\ 3 & 1 & 0 & 5 & \Big| & 12 \end{bmatrix}. \qquad \blacksquare$$

Number of Solutions to a System

There are only three possibilities for the size of the solution set of a linear system: a single solution, an infinite number of solutions, or no solutions. There are no other possibilities because if at least two solutions exist, we can show that an infinite number of solutions must exist (Exercise 10). For instance, in a system of two equations and two variables — say, x and y — the solution set for each equation forms a line in the xy-plane. The solution to the system is the intersection of the lines corresponding to each equation. But any two given lines in the plane either intersect in exactly one point (unique solution), are equal (infinite number of solutions, all points on the common line), or are parallel (no solutions).

For example, the system

$$\begin{cases} 4x_1 - 3x_2 = 0 \\ 2x_1 + 3x_2 = 18 \end{cases}$$

(where x_1 and x_2 are used instead of x and y) has the unique solution $(3, 4)$ because that is the only intersection point of the two lines. On the other hand, the system

$$\begin{cases} 4x - 6y = 10 \\ 6x - 9y = 15 \end{cases}$$

has an infinite number of solutions because the two given lines are really the same, and so every point on one line is also on the other. Finally, the system

$$\begin{cases} 2x_1 + x_2 = 3 \\ 2x_1 + x_2 = 1 \end{cases}$$

has no solutions at all because the two lines are parallel but not equal. (Both of their slopes are -2.) The solution set for this system is the empty set $\{\} = \emptyset$. All three systems are pictured in Figure 2.1.

Any system that has at least one solution (either unique or infinitely many) is said to be **consistent**. A system whose solution set is empty is called **inconsistent**. The first two systems in Figure 2.1 are consistent, and the last one is inconsistent.

Gaussian Elimination

Many methods are available for finding the complete solution set for a given linear system. The first one we present, **Gaussian elimination**, involves systematically replacing most of the coefficients in the system with simpler numbers (1's and 0's) to make the solution apparent.

Figure 2.1

Three systems: unique
solution, infinite
number of solutions, no
solution

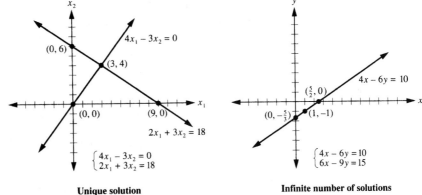

Unique solution   **Infinite number of solutions**

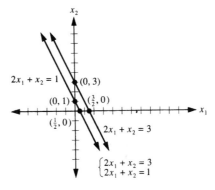

No solution

In Gaussian elimination, we begin with the augmented matrix for the given
system, then examine each column in turn from left to right. In each column,
if possible, we choose a special entry, called a **pivot entry**, convert that pivot
entry to "1," and then perform further operations to zero out the entries below
the pivot. The pivots will be "staggered" so that as we proceed from column
to column, each new pivot occurs in a lower row.

Row Operations and their Notation

There are three operations that we are allowed to use on the augmented matrix
in the Gaussian elimination method. These are as follows:

Row Operations
(I)   Multiplying a row by a *nonzero* scalar
(II) Adding a scalar multiple of one row to another row
(III) Switching the positions of two rows in the matrix

To save space, we will use a shorthand notation for these row operations. For example, a row operation of type (I) in which each entry of row 3 is multiplied by $\frac{1}{2}$ times that entry is represented by (I): $\langle 3 \rangle \leftarrow \frac{1}{2} \langle 3 \rangle$. That is, each entry of row 3 is multiplied by $\frac{1}{2}$, and the result replaces the previous row 3. A type (II) row operation in which $(-3) \times$ (row 4) is added to row 2 is represented by (II): $\langle 2 \rangle \leftarrow -3 \langle 4 \rangle + \langle 2 \rangle$. That is, a multiple ($-3$, in this case) of one row (in this case, row 4) is added to row 2, and the result replaces the previous row 2. Finally, a type (III) row operation in which the second and third rows are exchanged is represented by (III): $\langle 2 \rangle \leftrightarrow \langle 3 \rangle$. (Note that a double arrow is used for type (III) operations.)

We now illustrate the use of the first two operations with the following example:

EXAMPLE 2   Let us solve the following system of linear equations:

$$\begin{cases} 5x - 5y - 15z = 40 \\ 4x - 2y - 6z = 19 \\ 3x - 6y - 17z = 41 \end{cases}.$$

The augmented matrix associated with this system is

$$\left[\begin{array}{ccc|c} 5 & -5 & -15 & 40 \\ 4 & -2 & -6 & 19 \\ 3 & -6 & -17 & 41 \end{array} \right].$$

We now perform row operations on this matrix to give it a simpler form, proceeding through the columns from left to right. Starting with the first column, we choose the $(1, 1)$ position as our first pivot entry. We want to place a 1 in this position. The row containing the current pivot is often referred to as the **pivot row**, and so row 1 is currently our pivot row. Now, when placing 1 in the matrix, we generally use a type (I) operation to multiply the pivot row by the reciprocal of the pivot entry. In this case, we multiply each entry of the first row by $\frac{1}{5}$.

$$\text{Type (I) operation:} \quad \langle 1 \rangle \leftarrow \frac{1}{5} \langle 1 \rangle$$

$$\left[\begin{array}{ccc|c} ① & -1 & -3 & 8 \\ 4 & -2 & -6 & 19 \\ 3 & -6 & -17 & 41 \end{array} \right].$$

For reference, we circle all pivot entries as we proceed.

Next we want to convert all entries below this pivot to 0. We will refer to this as "targeting" these entries. As each entry is changed to 0 it is called the **target**, and its row is called the **target row**. To change a target entry to 0, we always use the following type (II) row operation:

$$\text{(II):} \quad \langle \text{target row} \rangle \leftarrow (-\text{target value}) \times \langle \text{pivot row} \rangle + \langle \text{target row} \rangle$$

For example, to zero out (target) the $(1, 2)$ entry, we use the type (II) operation $\langle 2 \rangle \leftarrow (-4) \times \langle 1 \rangle + \langle 2 \rangle$. (That is, we add (-4) times the pivot row to the target row.) To perform this operation, we first do the following side calculation:

$$
\begin{array}{rrrr|r}
(-4)\times \text{(row 1)} & -4 & 4 & 12 & -32 \\
\text{(row 2)} & 4 & -2 & -6 & 19 \\
\hline
\text{(sum)} & 0 & 2 & 6 & -13
\end{array}
$$

The resulting sum is now substituted in place of the old row 2, producing

Type (II) operation: $\langle 2 \rangle \leftarrow (-4) \times \langle 1 \rangle + \langle 2 \rangle$

$$
\left[
\begin{array}{rrr|r}
① & -1 & -3 & 8 \\
0 & 2 & 6 & -13 \\
3 & -6 & -17 & 41
\end{array}
\right].
$$

Note that even though we multiplied row 1 by -4 in the side calculation, row 1 itself was not changed in the matrix. Only row 2, the target row, was altered by this type (II) row operation.

Similarly, to target the $(1, 3)$ position (that is, convert the $(1, 3)$ entry to 0), row 3 becomes the target row, and we use another type (II) row operation. We replace row 3 with $(-3) \times \text{(row 1)} + \text{(row 3)}$. This gives

Type (II) operation: $\langle 3 \rangle \leftarrow (-3) \times \langle 1 \rangle + \langle 3 \rangle$

$$
\begin{array}{c|c}
\textbf{Side Calculation} & \textbf{Resultant Matrix} \\
\begin{array}{rrrr|r}
(-3)\times \text{(row 1)} & -3 & 3 & 9 & -24 \\
\text{(row 3)} & 3 & -6 & -17 & 41 \\
\hline
\text{(sum)} & 0 & -3 & -8 & 17
\end{array}
&
\left[
\begin{array}{rrr|r}
① & -1 & -3 & 8 \\
0 & 2 & 6 & -13 \\
0 & -3 & -8 & 17
\end{array}
\right]
\end{array}
$$

Now, the last resultant matrix is associated with the linear system

$$
\begin{cases}
x & - & y & - & 3z & = & 8 \\
& & 2y & + & 6z & = & -13 \\
& & -3y & - & 8z & = & 17
\end{cases}.
$$

Note that x has been eliminated from the second and third equations, which makes this system simpler than the original. However, as we will prove later, this new system has the same solution set.

Our work on the first column is finished, and we proceed to the second column. The pivot entry for this column must be in a lower row than the previous pivot, so we choose the $(2, 2)$ position as our next pivot entry. Thus, row 2 is now the pivot row. We first perform a type (I) operation on the pivot row to convert the pivot entry to 1. Multiplying each entry of row 2 by $\frac{1}{2}$ (the reciprocal of the pivot entry), we obtain

Type (I) operation: $\langle 2 \rangle \leftarrow \dfrac{1}{2} \langle 2 \rangle$

$$\text{Resultant matrix } = \begin{bmatrix} 1 & -1 & -3 & | & 8 \\ 0 & ① & 3 & | & -\frac{13}{2} \\ 0 & -3 & -8 & | & 17 \end{bmatrix}.$$

We now use a type (II) operation to target the $(3, 2)$ entry. The target row is now row 3.

$$\text{Type (II) operation: } \langle 3 \rangle \leftarrow 3 \times \langle 2 \rangle + \langle 3 \rangle$$

Side Calculation					**Resultant Matrix**

$$\begin{array}{l|cccr} (3)\times(\text{row } 2) & 0 & 3 & 9 & -\frac{39}{2} \\ (\text{row } 3) & 0 & -3 & -8 & 17 \\ \hline (\text{sum}) & 0 & 0 & 1 & -\frac{5}{2} \end{array} \qquad \begin{bmatrix} 1 & -1 & -3 & | & 8 \\ 0 & ① & 3 & | & -\frac{13}{2} \\ 0 & 0 & 1 & | & -\frac{5}{2} \end{bmatrix}$$

The last matrix corresponds to

$$\begin{cases} x & - & y & - & 3z & = & 8 \\ & & y & + & 3z & = & -\frac{13}{2} \\ & & & & z & = & -\frac{5}{2} \end{cases}.$$

Notice that y has been eliminated from the third equation. Again, this new system has exactly the same solution set as the original system.

At this point, we know from the third equation that $z = -\frac{5}{2}$. Substituting this result into the second equation and solving for y, we obtain $y + 3(-\frac{5}{2}) = -\frac{13}{2}$, and hence, $y = 1$. Finally, substituting these values for y and z into the first equation, we obtain $x - 1 - 3(-\frac{5}{2}) = 8$, and hence $x = \frac{3}{2}$. This process of working backwards through the set of equations to solve for each variable in turn is called **back substitution**.

Thus, the final system has a unique solution—the ordered triple $\left(\frac{3}{2}, 1, -\frac{5}{2}\right)$.

However, you can check by substitution that $\left(\frac{3}{2}, 1, -\frac{5}{2}\right)$ is also a solution to the original system. In fact, Gaussian elimination always produces the complete solution set, and so $\left(\frac{3}{2}, 1, -\frac{5}{2}\right)$ is the unique solution to the original linear system.   ∎

The Strategy in the Simplest Case

In Gaussian elimination, we work on one column of the augmented matrix at a time. Beginning with the first column, we choose row 1 as our initial pivot row, convert the $(1, 1)$ pivot entry to 1, and target (zero out) the entries below that pivot. After each column is simplified, we proceed to the next column to the right. In each column, if possible, we choose a **pivot entry** that is in the next row lower than the previous pivot, and this entry is converted to 1. The row containing the current pivot is referred to as the **pivot row.** The entries below each pivot are targeted (converted to 0) before proceeding to the next column. The process advances to additional columns until we reach the augmentation bar or run out of rows to use as the pivot row.

We generally convert pivot entries to 1 by multiplying the pivot row by the reciprocal of the current pivot entry. Then we use type (II) operations of the form

$$\langle\text{target row}\rangle \leftarrow (-\text{target value}) \times \langle\text{pivot row}\rangle + \langle\text{target row}\rangle$$

to target (zero out) each entry below the pivot entry. This eliminates the variable corresponding to that column from each equation in the system below the pivot row. Note that in type (II) operations, we add an appropriate multiple of the *pivot* row to the target row. (Any other type (II) operation could destroy work done in previous columns.)

Using Type (III) Operations

So far, we have used only type (I) and type (II) operations. However, when we begin work on a new column, if the logical choice for a pivot entry in that column is 0, it is impossible to convert the pivot to 1 using a type (I) operation. Frequently, this dilemma can be resolved by first using a type (III) operation to switch the pivot row with another row *below* it. (We never switch the pivot row with a row *above* it, because such a type (III) operation could destroy work done in previous columns.)

EXAMPLE 3 Let us solve the following system using Gaussian elimination:

$$\begin{cases} 3x & + & y & = & -5 \\ -6x & - & 2y & = & 10 \\ 4x & + & 5y & = & 8 \end{cases}, \quad \text{with augmented matrix} \quad \left[\begin{array}{cc|c} 3 & 1 & -5 \\ -6 & -2 & 10 \\ 4 & 5 & 8 \end{array}\right].$$

We start with the first column, and establish row 1 as the pivot row. We convert the pivot entry in the $(1, 1)$ position to 1 by multiplying the pivot row by the reciprocal of the pivot entry.

$$\text{Type (I) operation: } \langle 1\rangle \leftarrow \frac{1}{3}\langle 1\rangle$$

$$\text{Resultant matrix} = \left[\begin{array}{cc|c} ① & \frac{1}{3} & -\frac{5}{3} \\ -6 & -2 & 10 \\ 4 & 5 & 8 \end{array}\right]$$

Next, we use type (II) operations to target the rest of the first column by adding appropriate multiples of the pivot row (the first row) to the target rows.

$$\text{Type (II) operation: } \langle 2\rangle \leftarrow 6 \times \langle 1\rangle + \langle 2\rangle$$

$$\text{Type (II) operation: } \langle 3\rangle \leftarrow (-4) \times \langle 1\rangle + \langle 3\rangle$$

$$\text{Resultant matrix} = \left[\begin{array}{cc|c} ① & \frac{1}{3} & -\frac{5}{3} \\ 0 & 0 & 0 \\ 0 & \frac{11}{3} & \frac{44}{3} \end{array}\right]$$

We now advance to the second column, and designate row 2 as the pivot row. We want to convert the pivot entry $(2, 2)$ to 1, but because the pivot is 0, a type (I) operation will not work. Instead, we first perform a type (III) operation, switching the pivot row with the row below it, in order to change the pivot to a nonzero number.

$$\text{Type (III) operation: } \langle 2 \rangle \leftrightarrow \langle 3 \rangle$$

$$\text{Resultant matrix} = \begin{bmatrix} 1 & \frac{1}{3} & -\frac{5}{3} \\ 0 & \frac{11}{3} & \frac{44}{3} \\ 0 & 0 & 0 \end{bmatrix}$$

Now, using a type (I) operation, we can convert the $(2, 2)$ pivot entry to 1.

$$\text{Type (I) operation: } \langle 2 \rangle \leftarrow \frac{3}{11} \langle 2 \rangle$$

$$\text{Resultant matrix} = \begin{bmatrix} 1 & \frac{1}{3} & -\frac{5}{3} \\ 0 & ① & 4 \\ 0 & 0 & 0 \end{bmatrix}$$

Since the entry below the current pivot is already 0, the second column is now simplified. Because there are no more columns to the left of the augmentation bar, we stop. The final matrix corresponds to the following system:

$$\begin{cases} x + \frac{1}{3}y = -\frac{5}{3} \\ \qquad y = 4 \\ \qquad 0 = 0 \end{cases}.$$

The third equation is always satisfied, no matter what values x and y have, and provides us with no information. The second equation gives $y = 4$. Back substituting into the first equation, we obtain $x + \frac{1}{3}(4) = -\frac{5}{3}$, and so $x = -3$. Thus, the unique solution for our original system is $(-3, 4)$. ∎

The general rule for using type (III) operations is

> When starting a new column, if the pivot entry is 0, look for a nonzero number in the current column *below* the pivot row. If you find one, use a type (III) operation to switch the pivot row with the row containing this nonzero number.

Skipping a Column

Occasionally when we progress to a new column, the pivot entry as well as all lower entries in that column are zero. Here, a type (III) operation cannot help. In such cases, we skip over the current column and advance to the next column to the right. Hence, the new pivot entry is located horizontally to the right from where we would normally expect it. We illustrate the use of this rule in the next few examples. Example 4 involves an inconsistent system, and Examples 5, 6, and 7 involve infinitely many solutions.

Inconsistent Systems

EXAMPLE 4 Let us solve the following system using Gaussian elimination:

$$\begin{cases} 3x_1 & - & 6x_2 & & & + & 3x_4 & = & 9 \\ -2x_1 & + & 4x_2 & + & 2x_3 & - & x_4 & = & -11 \\ 4x_1 & - & 8x_2 & + & 6x_3 & + & 7x_4 & = & -5 \end{cases}.$$

First, we set up the augmented matrix

$$\left[\begin{array}{cccc|c} 3 & -6 & 0 & 3 & 9 \\ -2 & 4 & 2 & -1 & -11 \\ 4 & -8 & 6 & 7 & -5 \end{array}\right].$$

We begin with the first column and establish row 1 as the pivot row. We use a type (I) operation to convert the current pivot entry, the $(1, 1)$ entry, to 1.

$$(I): \langle 1 \rangle \leftarrow \frac{1}{3} \langle 1 \rangle$$

$$\text{Resultant matrix} = \left[\begin{array}{cccc|c} ① & -2 & 0 & 1 & 3 \\ -2 & 4 & 2 & -1 & -11 \\ 4 & -8 & 6 & 7 & -5 \end{array}\right]$$

Next, we target the entries below the pivot using type (II) row operations.

$$(II): \langle 2 \rangle \leftarrow 2 \langle 1 \rangle + \langle 2 \rangle$$

$$(II): \langle 3 \rangle \leftarrow -4 \langle 1 \rangle + \langle 3 \rangle$$

$$\text{Resultant matrix} = \left[\begin{array}{cccc|c} ① & -2 & 0 & 1 & 3 \\ 0 & 0 & 2 & 1 & -5 \\ 0 & 0 & 6 & 3 & -17 \end{array}\right]$$

We are finished with the first column, so we advance to the second column. The pivot row now advances to row 2, and so the pivot is now the $(2, 2)$ entry, which unfortunately is 0. We search for a nonzero entry below the pivot but do not find one. Hence, we skip over this column and advance horizontally to the third column, still maintaining row 2 as the pivot row.

We now change the current pivot entry (the $(2, 3)$ entry) into 1.

$$(I): \langle 2 \rangle \leftarrow \frac{1}{2} \langle 2 \rangle$$

$$\text{Resultant matrix} = \left[\begin{array}{cccc|c} 1 & -2 & 0 & 1 & 3 \\ 0 & 0 & ① & \frac{1}{2} & -\frac{5}{2} \\ 0 & 0 & 6 & 3 & -17 \end{array}\right]$$

Targeting the entry below this pivot, we obtain

$$(\text{II}) : \langle 3 \rangle \leftarrow -6 \langle 2 \rangle + \langle 3 \rangle$$

$$\text{Resultant matrix} = \begin{bmatrix} 1 & -2 & 0 & 1 & 3 \\ 0 & 0 & ① & \frac{1}{2} & -\frac{5}{2} \\ 0 & 0 & 0 & 0 & -2 \end{bmatrix}$$

We proceed to the fourth column, and the pivot row advances to row 3. However, the pivot entry, the $(3, 4)$ entry, is also 0. Since there is no row below the pivot row (row 3) to switch with, the fourth column is finished. We attempt to move the pivot horizontally to the right, but we have reached the augmentation bar, so we stop. The resultant system is

$$\begin{cases} x_1 - 2x_2 \quad + \quad x_4 = \quad 3 \\ \qquad\qquad x_3 + \frac{1}{2}x_4 = -\frac{5}{2} \\ \qquad\qquad\qquad\quad 0 = -2 \end{cases}.$$

Regardless of the values of x_1, x_2, x_3, and x_4, the last equation, $0 = -2$, is *never* satisfied. This equation has no solutions. But any solution to the system must satisfy every equation in the system. Therefore, this system is inconsistent, as is the original system with which we started. ∎

For inconsistent systems, the final augmented matrix always contains at least one row of the form

$$\begin{bmatrix} 0 & 0 & \cdots & 0 & c \end{bmatrix},$$

with all zeroes on the left of the augmentation bar and a nonzero number c on the right. Such a row corresponds to the equation $0 = c$, for some $c \neq 0$, which certainly has no solutions. In fact, if you encounter such a row at *any* stage of the Gaussian elimination process, the original system is inconsistent.

Beware! An entire row of zeroes, with *zero* on the right of the augmentation bar, does not imply the system is inconsistent. Such a row is simply ignored, as in Example 3.

Infinite Solution Sets

EXAMPLE 5   Let us solve the following system using Gaussian elimination:

$$\begin{cases} 3x_1 + \quad x_2 + \quad 7x_3 + 2x_4 = \quad 13 \\ 2x_1 - \quad 4x_2 + 14x_3 - \quad x_4 = -10 \\ 5x_1 + 11x_2 - \quad 7x_3 + 8x_4 = \quad 59 \\ 2x_1 + \quad 5x_2 - \quad 4x_3 - 3x_4 = \quad 39 \end{cases}.$$

The augmented matrix for this system is

$$\begin{bmatrix} 3 & 1 & 7 & 2 & 13 \\ 2 & -4 & 14 & -1 & -10 \\ 5 & 11 & -7 & 8 & 59 \\ 2 & 5 & -4 & -3 & 39 \end{bmatrix}.$$

After simplifying the first two columns as in earlier examples, we obtain

$$\begin{bmatrix} ① & \frac{1}{3} & \frac{7}{3} & \frac{2}{3} & \frac{13}{3} \\ 0 & ① & -2 & \frac{1}{2} & 4 \\ 0 & 0 & 0 & 0 & 0 \\ 0 & 0 & 0 & -\frac{13}{2} & 13 \end{bmatrix}.$$

There is no nonzero pivot in the third column, so we advance to the fourth column and use row operation (III): $\langle 3 \rangle \leftrightarrow \langle 4 \rangle$ to put a nonzero number into the $(3, 4)$ pivot position, obtaining

$$\begin{bmatrix} 1 & \frac{1}{3} & \frac{7}{3} & \frac{2}{3} & \frac{13}{3} \\ 0 & 1 & -2 & \frac{1}{2} & 4 \\ 0 & 0 & 0 & -\frac{13}{2} & 13 \\ 0 & 0 & 0 & 0 & 0 \end{bmatrix}.$$

Converting the pivot entry in the fourth column to 1 leads to the final augmented matrix

$$\begin{bmatrix} 1 & \frac{1}{3} & \frac{7}{3} & \frac{2}{3} & \frac{13}{3} \\ 0 & 1 & -2 & \frac{1}{2} & 4 \\ 0 & 0 & 0 & ① & -2 \\ 0 & 0 & 0 & 0 & 0 \end{bmatrix}.$$

This matrix corresponds to

$$\begin{cases} x_1 + \frac{1}{3}x_2 + \frac{7}{3}x_3 + \frac{2}{3}x_4 = \frac{13}{3} \\ \quad\quad x_2 - 2x_3 + \frac{1}{2}x_4 = 4 \\ \quad\quad\quad\quad\quad\quad x_4 = -2 \\ \quad\quad\quad\quad\quad\quad\quad 0 = 0 \end{cases}.$$

We discard the last equation, which gives no information about the solution set. The third equation gives $x_4 = -2$, but values for the other three variables are not uniquely determined — there are infinitely many solutions. We can let x_3 take on any value whatsoever, which then determines the values for x_1 and x_2. For example, if we let $x_3 = 5$, then back substituting into the second equation for x_2 yields $x_2 - 2(5) + \frac{1}{2}(-2) = 4$, which gives $x_2 = 15$. Back substituting into the first equation gives $x_1 + \frac{1}{3}(15) + \frac{7}{3}(5) + \frac{2}{3}(-2) = \frac{13}{3}$, which reduces to $x_1 = -11$. Thus, *one* solution is $(-11, 15, 5, -2)$. However, different solutions can be found by choosing alternate values for x_3. For example, letting $x_3 = -4$ gives the solution $x_1 = 16$, $x_2 = -3$, $x_3 = -4$, $x_4 = -2$. All such solutions satisfy the original system.

How can we express the complete solution set? Of course, $x_4 = -2$. If we use a variable, say c, to represent x_3, then from the second equation, we obtain $x_2 - 2c + \frac{1}{2}(-2) = 4$, which gives $x_2 = 5 + 2c$. Then from the first equation, we obtain $x_1 + \frac{1}{3}(5 + 2c) + \frac{7}{3}(c) + \frac{2}{3}(-2) = \frac{13}{3}$, which leads to $x_1 = 4 - 3c$. Thus, the infinite solution set can be expressed as

$$\{ (4 - 3c, \ 5 + 2c, \ c, \ -2) \,|\, c \in \mathbb{R} \}. \qquad \blacksquare$$

After Gaussian elimination, the columns having no pivot entries are often referred to as **nonpivot columns**, while those with pivots are called **pivot columns**. Recall that the columns to the left of the augmentation bar correspond to the variables x_1, x_2, and so on, in the system. The variables for nonpivot columns are called **independent variables**, while those for pivot columns are **dependent variables**. If a given system is consistent, solutions are found by letting each independent variable take on any real value whatsoever. The values of the dependent variables are then calculated from these choices. Thus, in Example 5, the third column is the only nonpivot column. Hence, x_3 is an independent variable, while x_1, x_2, and x_4 are dependent variables. We found a general solution by letting x_3 take on any value, and we determined the remaining variables from that choice.

EXAMPLE 6   Suppose that the final matrix after Gaussian elimination is

$$\begin{bmatrix} ① & -2 & 0 & 3 & 5 & -1 & | & 1 \\ 0 & 0 & ① & 4 & 23 & 0 & | & -9 \\ 0 & 0 & 0 & 0 & 0 & ① & | & 16 \\ 0 & 0 & 0 & 0 & 0 & 0 & | & 0 \end{bmatrix},$$

which corresponds to the system

$$\begin{cases} x_1 - 2x_2 & + 3x_4 + 5x_5 - x_6 = 1 \\ & x_3 + 4x_4 + 23x_5 = -9 \\ & x_6 = 16 \end{cases}.$$

Note that we have ignored the row of zeroes. Because the nonpivot columns are columns 2, 4, and 5, x_2, x_4, and x_5 are the independent variables. Therefore, we can let x_2, x_4, and x_5 take on any real values — say, $x_2 = b$, $x_4 = d$, and $x_5 = e$. We know $x_6 = 16$. We now use back substitution to solve the remaining equations in the system for the dependent variables x_1 and x_3, yielding $x_3 = -9 - 4d - 23e$, $x_1 = 17 + 2b - 3d - 5e$. Hence, the complete solution set is

$$\{(17 + 2b - 3d - 5e,\ b,\ -9 - 4d - 23e,\ d,\ e,\ 16)|\ b, d, e \in \mathbb{R}\}.$$

Particular solutions can be found by choosing values for b, d, and e. For example, choosing $b = 1$, $d = -1$, and $e = 0$ yields $(22, 1, -5, -1, 0, 16)$.   ∎

EXAMPLE 7   Suppose that the final matrix after Gaussian elimination is

$$\begin{bmatrix} ① & 4 & -1 & 2 & 1 & | & 8 \\ 0 & ① & 3 & -2 & 6 & | & -11 \\ 0 & 0 & 0 & ① & -3 & | & 9 \\ 0 & 0 & 0 & 0 & 0 & | & 0 \end{bmatrix}.$$

Because the third and fifth columns are nonpivot columns, x_3 and x_5 are the independent variables. Therefore, we can let x_3 and x_5 take on any real values — say, $x_3 = c$ and $x_5 = e$. We now use back substitution to solve the remaining equations in the system for the dependent variables x_1, x_2, and x_4, yielding $x_4 = 9 + 3e$, $x_2 = -11 - 3c + 2(9 + 3e) - 6e = 7 - 3c + 6e$, and $x_1 = 8 - 4(7 - 3c + 6e) + c - 2(9 + 3e) - e = -38 + 13c - 31e$. Hence, the complete solution set is

$$\{(-38 + 13c - 31e, \ 7 - 3c + 6e, \ c, \ 9 + 3e, \ e) | \ c, e \in \mathbb{R}\}.$$

Particular solutions can be found by choosing values for c and e. For example, choosing $c = -1$, and $e = 2$ yields $(-113, 22, -1, 15, 2)$. ∎

Application: Curve Fitting

 Let us find the unique quadratic equation of the form $y = ax^2 + bx + c$ whose graph passes through the points $(-2, 20)$, $(1, 5)$, and $(3, 25)$ in the xy-plane. By substituting each of the (x, y) pairs in turn into the equation, we get

$$\begin{cases} a(-2)^2 + b(-2) + c = 20 \\ a(1)^2 + b(1) + c = 5 \\ a(3)^2 + b(3) + c = 25 \end{cases},$$

which leads to the system

$$\begin{cases} 4a - 2b + c = 20 \\ a + b + c = 5 \\ 9a + 3b + c = 25 \end{cases}.$$

Using Gaussian elimination on this system leads to the final augmented matrix

$$\left[\begin{array}{ccc|c} 1 & -\frac{1}{2} & \frac{1}{4} & 5 \\ 0 & 1 & \frac{1}{2} & 0 \\ 0 & 0 & 1 & 4 \end{array} \right].$$

Thus, $c = 4$, and after back substituting, we find $b = -2$, and $a = 3$, and so the desired quadratic equation is $y = 3x^2 - 2x + 4$. ∎

The Effect of Row Operations on Matrix Multiplication

We conclude this section with a property involving row operations and matrix multiplication that will be useful later. The following notation is helpful: if a row operation R is performed on a matrix $\mathbf{A}$, we represent the resultant matrix by $R(\mathbf{A})$.

THEOREM 2.1

Let **A** and **B** be matrices for which the product **AB** is defined.
  (1) If R is any row operation, then $R(\mathbf{AB}) = (R(\mathbf{A}))\mathbf{B}$.
  (2) If $R_1, \ldots, R_n$ are row operations, then
$$R_n(\cdots(R_2(R_1(\mathbf{AB})))\cdots) = (R_n(\cdots(R_2(R_1(\mathbf{A})))\cdots))\mathbf{B}.$$

Part (1) of this theorem asserts that whenever a row operation is performed on the product of two matrices, the same answer is obtained by performing the row operation on the first matrix alone before multiplying. Part (1) is proved by considering each type of row operation in turn. Part (2) generalizes this result to any finite number of row operations, and is proved by using part (1) and induction. We leave the proof of Theorem 2.1 for you to do in Exercise 8.

EXAMPLE 9   Let

$$\mathbf{A} = \begin{bmatrix} 1 & -2 & 1 \\ 3 & 4 & 2 \end{bmatrix}, \quad \mathbf{B} = \begin{bmatrix} 3 & 7 \\ 0 & -1 \\ 5 & 2 \end{bmatrix},$$

and let R be the row operation $\langle 2 \rangle \leftarrow -2\langle 1 \rangle + \langle 2 \rangle$. Then

$$R(\mathbf{AB}) = R\left(\begin{bmatrix} 8 & 11 \\ 19 & 21 \end{bmatrix}\right) = \begin{bmatrix} 8 & 11 \\ 3 & -1 \end{bmatrix}, \quad \text{and}$$

$$(R(\mathbf{A}))\mathbf{B} = \left(R\left(\begin{bmatrix} 1 & -2 & 1 \\ 3 & 4 & 2 \end{bmatrix}\right)\right)\begin{bmatrix} 3 & 7 \\ 0 & -1 \\ 5 & 2 \end{bmatrix}$$

$$= \begin{bmatrix} 1 & -2 & 1 \\ 1 & 8 & 0 \end{bmatrix}\begin{bmatrix} 3 & 7 \\ 0 & -1 \\ 5 & 2 \end{bmatrix} = \begin{bmatrix} 8 & 11 \\ 3 & -1 \end{bmatrix}.$$

Similarly, with R_1: $\langle 1 \rangle \leftrightarrow \langle 2 \rangle$, R_2: $\langle 1 \rangle \leftarrow -3\langle 2 \rangle + \langle 1 \rangle$, and R_3: $\langle 1 \rangle \leftarrow 4\langle 1 \rangle$, you can verify that

$$R_3(R_2(R_1(\mathbf{AB}))) = R_3\left(R_2\left(R_1\left(\begin{bmatrix} 8 & 11 \\ 19 & 21 \end{bmatrix}\right)\right)\right) = \begin{bmatrix} -20 & -48 \\ 8 & 11 \end{bmatrix}, \quad \text{and}$$

$$(R_3(R_2(R_1(\mathbf{A}))))\mathbf{B} = \left(R_3\left(R_2\left(R_1\left(\begin{bmatrix} 1 & -2 & 1 \\ 3 & 4 & 2 \end{bmatrix}\right)\right)\right)\right)\begin{bmatrix} 3 & 7 \\ 0 & -1 \\ 5 & 2 \end{bmatrix}$$

$$= \begin{bmatrix} 0 & 40 & -4 \\ 1 & -2 & 1 \end{bmatrix}\begin{bmatrix} 3 & 7 \\ 0 & -1 \\ 5 & 2 \end{bmatrix} = \begin{bmatrix} -20 & -48 \\ 8 & 11 \end{bmatrix} \quad \text{also.}$$

∎

Exercises for Section 2.1

1. Use the Gaussian elimination method to solve each of the following systems of linear equations. In each case, indicate whether the system is consistent or inconsistent. Give the complete solution set, and if the solution set is infinite, specify three particular solutions.

★(a) $\begin{cases} -5x_1 - 2x_2 + 2x_3 = 14 \\ 3x_1 + x_2 - x_3 = -8 \\ 2x_1 + 2x_2 - x_3 = -3 \end{cases}$

(b) $\begin{cases} 3x_1 - 3x_2 - 2x_3 = 23 \\ -6x_1 + 4x_2 + 3x_3 = -38 \\ -2x_1 + x_2 + x_3 = -11 \end{cases}$

★(c) $\begin{cases} 3x_1 - 2x_2 + 4x_3 = -54 \\ -x_1 + x_2 - 2x_3 = 20 \\ 5x_1 - 4x_2 + 8x_3 = -83 \end{cases}$

(d) $\begin{cases} -2x_1 + 3x_2 - 4x_3 + x_4 = -17 \\ 8x_1 - 5x_2 + 2x_3 - 4x_4 = 47 \\ -5x_1 + 9x_2 - 13x_3 + 3x_4 = -44 \\ -4x_1 + 3x_2 - 2x_3 + 2x_4 = -25 \end{cases}$

★(e) $\begin{cases} 6x_1 - 12x_2 - 5x_3 + 16x_4 - 2x_5 = -53 \\ -3x_1 + 6x_2 + 3x_3 - 9x_4 + x_5 = 29 \\ -4x_1 + 8x_2 + 3x_3 - 10x_4 + x_5 = 33 \end{cases}$

(f) $\begin{cases} 5x_1 - 5x_2 - 15x_3 - 3x_4 = -34 \\ -2x_1 + 2x_2 + 6x_3 + x_4 = 12 \end{cases}$

★(g) $\begin{cases} 4x_1 - 2x_2 - 7x_3 = 5 \\ -6x_1 + 5x_2 + 10x_3 = -11 \\ -2x_1 + 3x_2 + 4x_3 = -3 \\ -3x_1 + 2x_2 + 5x_3 = -5 \end{cases}$

(h) $\begin{cases} 5x_1 - x_2 - 9x_3 - 2x_4 = 26 \\ 4x_1 - x_2 - 7x_3 - 2x_4 = 21 \\ -2x_1 + 4x_3 + x_4 = -12 \\ -3x_1 + 2x_2 + 4x_3 + 2x_4 = -11 \end{cases}$

2. Suppose that each of the following is the final augmented matrix obtained after Gaussian elimination. In each case, give the complete solution set for the corresponding system of linear equations.

★(a) $\left[\begin{array}{ccccc|c} 1 & -5 & 2 & 3 & -2 & -4 \\ 0 & 1 & -1 & -3 & -7 & -2 \\ 0 & 0 & 0 & 1 & 2 & 5 \\ 0 & 0 & 0 & 0 & 0 & 0 \end{array} \right]$

(b) $\left[\begin{array}{cccccc|c} 1 & -3 & 6 & 0 & -2 & 4 & -3 \\ 0 & 0 & 1 & -2 & 8 & -1 & 5 \\ 0 & 0 & 0 & 0 & 0 & 1 & -2 \\ 0 & 0 & 0 & 0 & 0 & 0 & 0 \end{array} \right]$

$$\star\text{(c)} \quad \begin{bmatrix} 1 & 4 & -8 & -1 & 2 & -3 & | & -4 \\ 0 & 1 & -7 & 2 & -9 & -1 & | & -3 \\ 0 & 0 & 0 & 0 & 1 & -4 & | & 2 \\ 0 & 0 & 0 & 0 & 0 & 0 & | & 0 \end{bmatrix}$$

$$\text{(d)} \quad \begin{bmatrix} 1 & -7 & -3 & -2 & -1 & | & -5 \\ 0 & 0 & 1 & 2 & 3 & | & 1 \\ 0 & 0 & 0 & 1 & -1 & | & 4 \\ 0 & 0 & 0 & 0 & 0 & | & -2 \end{bmatrix}$$

$\star$**3.** Solve the following problem by using a linear system: A certain number of nickels, dimes, and quarters totals \$16.50. There are twice as many dimes as quarters, and the total number of nickels and quarters is twenty more than the number of dimes. Find the correct number of each type of coin.

$\star$**4.** Find the quadratic equation $y = ax^2 + bx + c$ that goes through the points $(3, 18)$, $(2, 9)$, and $(-2, 13)$.

5. Find the cubic equation $y = ax^3 + bx^2 + cx + d$ that goes through the points $(1, 1)$, $(2, -18)$, $(-2, 46)$, and $(3, -69)$.

$\star$**6.** The general equation of a circle is $x^2 + y^2 + ax + by = c$. Find the equation of the circle that goes through the points $(6, 8)$, $(8, 4)$, and $(3, 9)$.

7. Let $\mathbf{A} = \begin{bmatrix} 2 & 3 & 4 \\ 0 & 1 & 1 \\ -2 & 1 & 5 \\ 3 & 0 & 1 \end{bmatrix}$, $\mathbf{B} = \begin{bmatrix} 2 & 1 & -5 \\ 2 & 3 & 0 \\ 4 & 1 & 1 \end{bmatrix}$. Compute $R(\mathbf{AB})$ and $(R(\mathbf{A}))\mathbf{B}$ to verify that they are equal, if

$\star$(a) R: $\langle 3 \rangle \leftarrow -3\langle 2 \rangle + \langle 3 \rangle$.

(b) R: $\langle 2 \rangle \leftrightarrow \langle 4 \rangle$.

8. ▶(a) Prove part (1) of Theorem 2.1 by showing that $R(\mathbf{AB}) = (R(\mathbf{A}))\mathbf{B}$ for each type of row operation ((I), (II), (III)) in turn. (Hint: Use the fact from Section 1.5 that the kth row of $(\mathbf{AB})$ = (kth row of $\mathbf{A}$)$\mathbf{B}$.)

(b) Use part (a) and induction to prove part (2) of Theorem 2.1.

9. Explain why the scalar used in a type (I) row operation must be nonzero.

10. Prove that if more than one solution to a system of linear equations exists, then an infinite number of solutions exists. (Hint: Show that if $\mathbf{X}_1$ and $\mathbf{X}_2$ are different solutions to $\mathbf{AX} = \mathbf{B}$, then $\mathbf{X}_1 + c(\mathbf{X}_2 - \mathbf{X}_1)$ is also a solution, for every real number c. Also, show that all these solutions are different.)

$\star$**11.** True or False:

(a) The augmented matrix for a linear system contains all the essential information from the system.

(b) It is possible for a linear system of equations to have exactly three solutions.

(c) A consistent system must have exactly one solution.

(d) Type (II) row operations are used to convert nonzero pivot entries to 1.

(e) A type (III) row operation is used to replace a zero pivot entry with a nonzero entry below it.

(f) Multiplying matrices and then performing a row operation on the product has the same effect as performing the row operation on the first matrix and then calculating the product.

2.2 Gauss-Jordan Row Reduction and Reduced Row Echelon Form

In this section, we introduce the Gauss-Jordan row reduction method, an extension of the Gaussian elimination method. We also examine homogeneous linear systems and their solutions.

Introduction to Gauss-Jordan Row Reduction

In the Gaussian elimination method, we created the augmented matrix for a given linear system and systematically proceeded through the columns from left to right, creating pivots and targeting (zeroing out) entries below the pivots. Although we occasionally skipped over a column, we placed pivots into successive rows, and so the overall effect was to create a **staircase pattern of pivots**, as in

$$
\left[\begin{array}{ccc|c}
① & 3 & -7 & 3 \\
0 & ① & 5 & 2 \\
0 & 0 & ① & 8
\end{array}\right],
\qquad
\left[\begin{array}{cccc|c}
① & 2 & 17 & 9 & 6 \\
0 & 0 & ① & 3 & 9 \\
0 & 0 & 0 & ① & -2 \\
0 & 0 & 0 & 0 & 0
\end{array}\right],
$$

and

$$
\left[\begin{array}{cccccc|c}
① & -3 & 6 & -2 & 4 & -5 & -3 \\
0 & 0 & ① & -5 & 2 & -3 & -1 \\
0 & 0 & 0 & 0 & 0 & ① & 2 \\
0 & 0 & 0 & 0 & 0 & 0 & 0
\end{array}\right].
$$

Such matrices are said to be in **row echelon form**. However, we can extend the Gaussian elimination method further to target (zero out) the entries *above* each pivot as well, as we proceed from column to column. This extension is called the **Gauss-Jordan row reduction** method, sometimes simply referred to as "row reduction."

EXAMPLE 1

We will solve the following system of equations using the Gauss-Jordan method:

$$
\begin{cases}
2x_1 + x_2 + 3x_3 &= 16 \\
3x_1 + 2x_2 \quad\;\; + x_4 &= 16 \\
2x_1 \quad\quad + 12x_3 - 5x_4 &= 5
\end{cases}.
$$

This system has the corresponding augmented matrix

$$\left[\begin{array}{cccc|c} 2 & 1 & 3 & 0 & 16 \\ 3 & 2 & 0 & 1 & 16 \\ 2 & 0 & 12 & -5 & 5 \end{array}\right].$$

As in Gaussian elimination, we begin with the first column and set row 1 as the pivot row. The following operation places 1 in the $(1, 1)$ pivot position:

Row Operation	Resulting Matrix
(I): $\langle 1 \rangle \leftarrow \frac{1}{2} \langle 1 \rangle$	$\left[\begin{array}{cccc|c} ① & \frac{1}{2} & \frac{3}{2} & 0 & 8 \\ 3 & 2 & 0 & 1 & 16 \\ 2 & 0 & 12 & -5 & 5 \end{array}\right].$

The next operations target (zero out) the entries below the $(1, 1)$ pivot.

Row Operations	Resulting Matrix
(II): $\langle 2 \rangle \leftarrow (-3) \langle 1 \rangle + \langle 2 \rangle$ (II): $\langle 3 \rangle \leftarrow (-2) \langle 1 \rangle + \langle 3 \rangle$	$\left[\begin{array}{cccc|c} ① & \frac{1}{2} & \frac{3}{2} & 0 & 8 \\ 0 & \frac{1}{2} & -\frac{9}{2} & 1 & -8 \\ 0 & -1 & 9 & -5 & -11 \end{array}\right]$

Proceeding to the second column, we set row 2 as the pivot row. The following operation places a 1 in the $(2, 2)$ pivot position.

Row Operation	Resulting Matrix
(I): $\langle 2 \rangle \leftarrow 2 \langle 2 \rangle$	$\left[\begin{array}{cccc|c} ① & \frac{1}{2} & \frac{3}{2} & 0 & 8 \\ 0 & ① & -9 & 2 & -16 \\ 0 & -1 & 9 & -5 & -11 \end{array}\right]$

The next operations target the entries *above* and *below* the $(2, 2)$ pivot.

Row Operations	Resulting Matrix
(II): $\langle 1 \rangle \leftarrow -\frac{1}{2} \langle 2 \rangle + \langle 1 \rangle$ (II): $\langle 3 \rangle \leftarrow 1 \langle 2 \rangle + \langle 3 \rangle$	$\left[\begin{array}{cccc|c} ① & 0 & 6 & -1 & 16 \\ 0 & ① & -9 & 2 & -16 \\ 0 & 0 & 0 & -3 & -27 \end{array}\right]$

We cannot place a nonzero pivot in the third column, so we proceed to the fourth column and set row 3 as the pivot row. The following operation places 1 in the $(3, 4)$ pivot position.

Row Operation	Resulting Matrix
(I): $\langle 3 \rangle \leftarrow -\frac{1}{3} \langle 3 \rangle$	$\left[\begin{array}{cccc|c} ① & 0 & 6 & -1 & 16 \\ 0 & ① & -9 & 2 & -16 \\ 0 & 0 & 0 & ① & 9 \end{array}\right]$

The next operations target the entries *above* the $(3, 4)$ pivot.

Row Operations	**Resulting Matrix**

(II): $\langle 1 \rangle \leftarrow 1 \langle 3 \rangle + \langle 1 \rangle$
(II): $\langle 2 \rangle \leftarrow -2 \langle 3 \rangle + \langle 2 \rangle$

$$\left[\begin{array}{cccc|c} ① & 0 & 6 & 0 & 25 \\ 0 & ① & -9 & 0 & -34 \\ 0 & 0 & 0 & ① & 9 \end{array}\right]$$

Since we have reached the augmentation bar, we stop. (Notice the staircase pattern of pivots in the final augmented matrix.) The corresponding system for this final matrix is

$$\begin{cases} x_1 & + 6x_3 & = & 25 \\ & x_2 - 9x_3 & = & -34 \\ & x_4 & = & 9 \end{cases}.$$

The third equation gives $x_4 = 9$. Since the third column is not a pivot column, the independent variable x_3 can take on any real value, say c. The other variables x_1 and x_2 are now determined to be $x_1 = 25 - 6c$ and $x_2 = -34 + 9c$. Then the complete solution set is $\{ (25 - 6c,\ 9c - 34,\ c,\ 9) \mid c \in \mathbb{R} \}$.   ■

One disadvantage of the Gauss-Jordan method is that more type (II) operations generally need to be performed on the augmented matrix in order to target the entries *above* the pivots. Hence, Gaussian elimination is faster. It is also more accurate when using a calculator or computer because there is less opportunity for the compounding of roundoff errors during the process. On the other hand, with the Gauss-Jordan method there are fewer nonzero numbers in the final augmented matrix, which makes the solution set more apparent.

Reduced Row Echelon Form

In the final augmented matrix in Example 1, each step on the staircase begins with a nonzero pivot, although the steps are not uniform in width. As in row echelon form, all entries below the staircase are 0, but now all entries above a nonzero pivot are 0 as well. When a matrix satisfies these conditions, it is said to be in **reduced row echelon form**. The following definition states these conditions more formally:

DEFINITION

A matrix is in **[reduced] row echelon form** if and only if all the following conditions hold:
(1) The first nonzero entry in each row is 1.
(2) Each successive row has its first nonzero entry in a later column.
(3) All entries [above and] below the first nonzero entry of each row are zero.
(4) All full rows of zeroes are the final rows of the matrix.

Condition (3) asserts that if the entries above each pivot are zero in a row echelon form matrix, then the matrix is in reduced row echelon form as well.

Technically speaking, to put an augmented matrix into reduced row echelon form, this definition requires us to row reduce *all* columns. Therefore, putting an augmented matrix into reduced row echelon form may require proceeding beyond the augmentation bar. However, we have seen that the solution set of a linear system can actually be determined without simplifying the column to the right of the augmentation bar.

EXAMPLE 2 The following augmented matrices are all in reduced row echelon form:

$$
\mathbf{A} = \left[\begin{array}{ccc|c} ① & 0 & 0 & 6 \\ 0 & ① & 0 & -2 \\ 0 & 0 & ① & 3 \end{array} \right], \quad
\mathbf{B} = \left[\begin{array}{cccc|c} ① & 0 & 2 & 0 & -1 \\ 0 & ① & 3 & 0 & 4 \\ 0 & 0 & 0 & ① & 2 \\ 0 & 0 & 0 & 0 & 0 \end{array} \right],
$$

$$
\text{and} \quad \mathbf{C} = \left[\begin{array}{cccc|c} ① & 4 & 0 & -3 & 0 \\ 0 & 0 & ① & 2 & 0 \\ 0 & 0 & 0 & 0 & ① \end{array} \right].
$$

Notice the staircase pattern of pivots in each matrix, with 1 as the first nonzero entry in each row. The linear system corresponding to $\mathbf{A}$ has a unique solution $(6, -2, 3)$. The system corresponding to $\mathbf{B}$ has an infinite number of solutions since the third column has no pivot entry, and its corresponding variable can take on any real value. (The complete solution set for this system is $\{(-1 - 2c,\ 4 - 3c,\ c, 2) \mid c \in \mathbb{R}\}$.) However, the system corresponding to $\mathbf{C}$ has no solutions since the third row is equivalent to the equation $0 = 1$. ∎

Number of Solutions

The Gauss-Jordan row reduction method also implies the following:

NUMBER OF SOLUTIONS OF A LINEAR SYSTEM

Let $\mathbf{AX} = \mathbf{B}$ be a system of linear equations. Let $\mathbf{C}$ be the reduced row echelon form augmented matrix obtained by row reducing $[\mathbf{A} \,|\, \mathbf{B}]$.

▶If there is a row of $\mathbf{C}$ having all zeroes to the left of the augmentation bar but with its last entry nonzero, then $\mathbf{AX} = \mathbf{B}$ has no solution.

▶If not, but if one of the columns of $\mathbf{C}$ to the left of the augmentation bar has no nonzero pivot entry, then $\mathbf{AX} = \mathbf{B}$ has an infinite number of solutions. The nonpivot columns correspond to (independent) variables that can take on any value, and the values of the remaining (dependent) variables are determined from those.

▶Otherwise, $\mathbf{AX} = \mathbf{B}$ has a unique solution.

Homogeneous Systems

DEFINITION

A system of linear equations having matrix form $\mathbf{AX} = \mathbf{O}$, where $\mathbf{O}$ represents a zero column matrix, is called a **homogeneous system**.

For example, the following are homogeneous systems:

$$\begin{cases} 2x - 3y = 0 \\ -4x + 6y = 0 \end{cases} \quad \text{and} \quad \begin{cases} 5x_1 - 2x_2 + 3x_3 = 0 \\ 6x_1 + x_2 - 7x_3 = 0 \\ -x_1 + 3x_2 + x_3 = 0 \end{cases}.$$

Notice that homogeneous systems are always consistent. This is because all of the variables can be set equal to zero to satisfy all of the equations. This special solution, $(0, 0, \ldots, 0)$, is called the **trivial solution**. Any other solution of a homogeneous system is called a **nontrivial solution**. For example, $(0, 0)$ is the trivial solution to the first homogeneous system shown, but $(9, 6)$ is a nontrivial solution. Whenever a homogeneous system has a nontrivial solution, it actually has infinitely many solutions (why?).

An important result about homogeneous systems is the following:

If the reduced row echelon form augmented matrix for a homogeneous system in n variables has fewer than n nonzero pivot entries, then the system has a nontrivial solution.

EXAMPLE 3   Consider the following 3×3 homogeneous systems:

$$\begin{cases} 2x_1 + x_2 + 4x_3 = 0 \\ 3x_1 + 2x_2 + 5x_3 = 0 \\ -x_2 + x_3 = 0 \end{cases} \quad \text{and} \quad \begin{cases} 4x_1 - 8x_2 - 2x_3 = 0 \\ 3x_1 - 5x_2 - 2x_3 = 0 \\ 2x_1 - 8x_2 + x_3 = 0 \end{cases}.$$

After Gauss-Jordan row reduction, the final augmented matrices for these systems are, respectively,

$$\left[\begin{array}{ccc|c} ① & 0 & 0 & 0 \\ 0 & ① & 0 & 0 \\ 0 & 0 & ① & 0 \end{array}\right] \quad \text{and} \quad \left[\begin{array}{ccc|c} ① & 0 & -\frac{3}{2} & 0 \\ 0 & ① & -\frac{1}{2} & 0 \\ 0 & 0 & 0 & 0 \end{array}\right].$$

The first system has only the trivial solution because all 3 columns are pivot columns. However, the second system has a nontrivial solution because only 2 of its 3 variable columns are pivot columns, (that is, there is at least one nonpivot column). The complete solution set for the second system is

$$\left\{ \left(\frac{3}{2}c, \frac{1}{2}c, c \right) \Big| c \in \mathbb{R} \right\} = \left\{ c \left(\frac{3}{2}, \frac{1}{2}, 1 \right) \Big| c \in \mathbb{R} \right\}.$$

■

Notice that if there are fewer equations than variables in a homogeneous system, we are bound to get at least one nonpivot column. Therefore, such a homogeneous system always has nontrivial solutions.

EXAMPLE 4   Consider the following homogeneous system:

$$\begin{cases} x_1 - 3x_2 + 2x_3 - 4x_4 + 8x_5 + 17x_6 = 0 \\ 3x_1 - 9x_2 + 6x_3 - 12x_4 + 24x_5 + 49x_6 = 0 \\ -2x_1 + 6x_2 - 5x_3 + 11x_4 - 18x_5 - 40x_6 = 0 \end{cases}.$$

Because this homogeneous system has fewer equations than variables, it has a nontrivial solution. To find all the solutions, we row reduce to obtain the final augmented matrix

$$\begin{bmatrix} ① & -3 & 0 & 2 & 4 & 0 & | & 0 \\ 0 & 0 & ① & -3 & 2 & 0 & | & 0 \\ 0 & 0 & 0 & 0 & 0 & ① & | & 0 \end{bmatrix}.$$

The second, fourth, and fifth columns are nonpivot columns, so we can let $x_2, x_4,$ and x_5 take on any real values — say, $b, d,$ and $e,$ respectively. The values of the remaining variables are then determined by solving the equations $x_1 - 3b + 2d + 4e = 0, x_3 - 3d + 2e = 0,$ and $x_6 = 0.$ The complete solution set is

$$\{(3b - 2d - 4e, \ b, \ 3d - 2e, \ d, \ e, \ 0)|\ b, d, e \in \mathbb{R}\}.$$

■

The solutions for the homogeneous system in Example 4 can be expressed as linear combinations of three particular solutions as follows:

$$(3b - 2d - 4e, \ b, \ 3d - 2e, \ d, \ e, \ 0)$$
$$= b(3, 1, 0, 0, 0, 0) + d(-2, 0, 3, 1, 0, 0) + e(-4, 0, -2, 0, 1, 0).$$

Each particular solution was found by setting one independent variable equal to 1 and the others equal to 0. We will frequently find it useful to express solutions in this way.

Application: Balancing Chemical Equations

Homogeneous systems frequently occur when balancing chemical equations. In chemical reactions, we often know the **reactants** (initial substances) and **products** (results of the reaction). For example, it is known that the reactants phosphoric acid and calcium hydroxide produce calcium phosphate and water. This reaction can be symbolized as

$$\underset{\text{Phosphoric acid}}{H_3PO_4} \ + \ \underset{\text{Calcium hydroxide}}{Ca(OH)_2} \ \rightarrow \ \underset{\text{Calcium phosphate}}{Ca_3(PO_4)_2} \ + \underset{\text{Water}}{H_2O}.$$

An **empirical formula** for this reaction is an equation containing the minimal integer multiples of the reactants and products so that the number of atoms of each element agrees on both sides. (Finding the empirical formula is called **balancing** the equation.) In the preceding example, we are looking for minimal positive integer values of a, b, c, and d such that

$$aH_3PO_4 + bCa(OH)_2 \rightarrow cCa_3(PO_4)_2 + dH_2O$$

balances the number of hydrogen (H), phosphorus (P), oxygen (O), and calcium (Ca) atoms on both sides.[1] Considering each element in turn, we get

$$\begin{cases} 3a + 2b = \quad\quad\ 2d & (\text{H}) \\ a \quad\quad\ = 2c & (\text{P}) \\ 4a + 2b = 8c + \ d & (\text{O}) \\ b = 3c & (\text{Ca}) \end{cases}.$$

Bringing the c and d terms to the left side of each equation, we get the following augmented matrix for this system:

$$\left[\begin{array}{cccc|c} 3 & 2 & 0 & -2 & 0 \\ 1 & 0 & -2 & 0 & 0 \\ 4 & 2 & -8 & -1 & 0 \\ 0 & 1 & -3 & 0 & 0 \end{array}\right], \quad \text{which row reduces to} \quad \left[\begin{array}{cccc|c} ① & 0 & 0 & -\frac{1}{3} & 0 \\ 0 & ① & 0 & -\frac{1}{2} & 0 \\ 0 & 0 & ① & -\frac{1}{6} & 0 \\ 0 & 0 & 0 & 0 & 0 \end{array}\right].$$

The only variable having a nonpivot column is d. We choose $d = 6$ because this is the minimum positive integer value we can assign to d so that a, b, and c are also integers (why?). We then have $a = 2$, $b = 3$, and $c = 1$. Thus, the empirical formula for this reaction is

$$2H_3PO_4 + 3Ca(OH)_2 \rightarrow Ca_3(PO_4)_2 + 6H_2O.$$

Solving Several Systems Simultaneously

In many cases, we need to solve two or more systems having the same coefficient matrix. Suppose we wanted to solve both of the systems

$$\begin{cases} 3x_1 + \ x_2 - 2x_3 = \quad 1 \\ 4x_1 \quad\quad\ - \ x_3 = \quad 7 \\ 2x_1 - 3x_2 + 5x_3 = 18 \end{cases} \quad \text{and} \quad \begin{cases} 3x_1 + \ x_2 - 2x_3 = \quad\ 8 \\ 4x_1 \quad\quad\ - \ x_3 = \ -1 \\ 2x_1 - 3x_2 + 5x_3 = -32 \end{cases}.$$

It is wasteful to do two almost identical row reductions on the augmented matrices

$$\left[\begin{array}{ccc|c} 3 & 1 & -2 & 1 \\ 4 & 0 & -1 & 7 \\ 2 & -3 & 5 & 18 \end{array}\right] \quad \text{and} \quad \left[\begin{array}{ccc|c} 3 & 1 & -2 & 8 \\ 4 & 0 & -1 & -1 \\ 2 & -3 & 5 & -32 \end{array}\right].$$

---

[1] In expressions like $(OH)_2$ and $(PO_4)_2$, the number immediately following the parentheses indicates that every term in the unit should be considered to appear that many times. Hence, $(PO_4)_2$ is equivalent to PO_4PO_4 for our purposes.

Instead, we can create the following "simultaneous" matrix containing the information from both systems:

$$\begin{bmatrix} 3 & 1 & -2 & \bigm| & 1 & 8 \\ 4 & 0 & -1 & \bigm| & 7 & -1 \\ 2 & -3 & 5 & \bigm| & 18 & -32 \end{bmatrix}.$$

Row reducing this matrix completely yields

$$\begin{bmatrix} ① & 0 & 0 & \bigm| & 2 & -1 \\ 0 & ① & 0 & \bigm| & -3 & 5 \\ 0 & 0 & ① & \bigm| & 1 & -3 \end{bmatrix}.$$

By considering each of the right-hand columns separately, we discover that the unique solution of the first system is $x_1 = 2$, $x_2 = -3$, and $x_3 = 1$ and that the unique solution of the second system is $x_1 = -1$, $x_2 = 5$, and $x_3 = -3$.

Any number of systems with the same coefficient matrix can be handled similarly, with one column on the right side of the augmented matrix for each system.

♦ **Applications**: You now have covered the prerequisites for Section 8.2, "Ohm's Law," Section 8.3, "Least-Squares Polynomials," and Section 8.4, "Markov Chains."

Exercises for Section 2.2

★**1.** Which of these matrices are not in reduced row echelon form? Why?

(a) $\begin{bmatrix} 1 & 0 & 0 & 0 \\ 0 & 0 & 1 & 0 \\ 0 & 1 & 0 & 0 \end{bmatrix}$

(b) $\begin{bmatrix} 1 & -2 & 0 & 0 \\ 0 & 0 & 0 & 0 \\ 0 & 0 & 1 & 0 \\ 0 & 0 & 0 & 1 \end{bmatrix}$

(c) $\begin{bmatrix} 1 & 0 & 0 & 3 \\ 0 & 2 & 0 & -2 \\ 0 & 0 & 3 & 0 \end{bmatrix}$

(d) $\begin{bmatrix} 1 & -4 & 0 & 0 \\ 0 & 0 & 1 & 0 \\ 0 & 0 & 2 & 0 \end{bmatrix}$

(e) $\begin{bmatrix} 1 & 0 & 4 \\ 0 & 1 & -2 \\ 0 & 0 & 0 \end{bmatrix}$

(f) $\begin{bmatrix} 1 & -2 & 0 & -2 & 3 \\ 0 & 0 & 1 & 5 & 4 \\ 0 & 0 & 0 & 0 & 1 \end{bmatrix}$

2. Use the Gauss-Jordan method to convert these matrices to reduced row echelon form, and draw in the correct staircase pattern.

★(a) $\begin{bmatrix} 5 & 20 & -18 & \bigm| & -11 \\ 3 & 12 & -14 & \bigm| & 3 \\ -4 & -16 & 13 & \bigm| & 13 \end{bmatrix}$

★(b) $\begin{bmatrix} -2 & 1 & 1 & 15 \\ 6 & -1 & -2 & -36 \\ 1 & -1 & -1 & -11 \\ -5 & -5 & -5 & -14 \end{bmatrix}$

★(c) $\begin{bmatrix} -5 & 10 & -19 & -17 & | & 20 \\ -3 & 6 & -11 & -11 & | & 14 \\ -7 & 14 & -26 & -25 & | & 31 \\ 9 & -18 & 34 & 31 & | & -37 \end{bmatrix}$

(d) $\begin{bmatrix} 2 & -5 & -20 \\ 0 & 2 & 7 \\ 1 & -5 & -19 \\ -5 & 16 & 64 \\ 3 & -9 & -36 \end{bmatrix}$

★(e) $\begin{bmatrix} -3 & 6 & -1 & -5 & 0 & | & -5 \\ -1 & 2 & 3 & -5 & 10 & | & 5 \end{bmatrix}$

(f) $\begin{bmatrix} -2 & 1 & -1 & -1 & 3 \\ 3 & 1 & -4 & -2 & -4 \\ 7 & 1 & -6 & -2 & -3 \\ -8 & -1 & 6 & 2 & 3 \\ -3 & 0 & 2 & 1 & 2 \end{bmatrix}$

★**3.** In parts (a), (e), and (g) of Exercise 1 in Section 2.1, take the final row echelon form matrix that you obtained from Gaussian elimination and convert it to reduced row echelon form. Then check that the reduced row echelon form leads to the same solution set that you obtained using Gaussian elimination.

4. Each of the following homogeneous systems has a nontrivial solution since the number of variables is greater than the number of equations. Use the Gauss-Jordan method to determine the complete solution set for each system, and give one particular nontrivial solution.

★(a) $\begin{cases} -2x_1 - 3x_2 + 2x_3 - 13x_4 = 0 \\ -4x_1 - 7x_2 + 4x_3 - 29x_4 = 0 \\ x_1 + 2x_2 - x_3 + 8x_4 = 0 \end{cases}$

(b) $\begin{cases} 2x_1 + 4x_2 - x_3 + 5x_4 + 2x_5 = 0 \\ 3x_1 + 3x_2 - x_3 + 3x_4 = 0 \\ -5x_1 - 6x_2 + 2x_3 - 6x_4 - x_5 = 0 \end{cases}$

★(c) $\begin{cases} 7x_1 + 28x_2 + 4x_3 - 2x_4 + 10x_5 + 19x_6 = 0 \\ -9x_1 - 36x_2 - 5x_3 + 3x_4 - 15x_5 - 29x_6 = 0 \\ 3x_1 + 12x_2 + 2x_3 + 6x_5 + 11x_6 = 0 \\ 6x_1 + 24x_2 + 3x_3 - 3x_4 + 10x_5 + 20x_6 = 0 \end{cases}$

5. Use the Gauss-Jordan method to find the complete solution set for each of the following homogeneous systems, and express each solution set as linear combinations of particular solutions, as shown after Example 4.

★(a) $\begin{cases} -2x_1 + x_2 + 8x_3 = 0 \\ 7x_1 - 2x_2 - 22x_3 = 0 \\ 3x_1 - x_2 - 10x_3 = 0 \end{cases}$

(b) $\begin{cases} 5x_1 - 2x_3 = 0 \\ -15x_1 - 16x_2 - 9x_3 = 0 \\ 10x_1 + 12x_2 + 7x_3 = 0 \end{cases}$

$$\star(c) \begin{cases} 2x_1 + 6x_2 + 13x_3 + x_4 = 0 \\ x_1 + 4x_2 + 10x_3 + x_4 = 0 \\ 2x_1 + 8x_2 + 20x_3 + x_4 = 0 \\ 3x_1 + 10x_2 + 21x_3 + 2x_4 = 0 \end{cases}$$

$$(d) \begin{cases} 2x_1 - 6x_2 + 3x_3 - 21x_4 = 0 \\ 4x_1 - 5x_2 + 2x_3 - 24x_4 = 0 \\ -x_1 + 3x_2 - x_3 + 10x_4 = 0 \\ -2x_1 + 3x_2 - x_3 + 13x_4 = 0 \end{cases}$$

6. Use the Gauss-Jordan method to find the minimal integer values for the variables that will balance each of the following chemical equations:[2]

$\star$(a) $aC_6H_6 + bO_2 \rightarrow cCO_2 + dH_2O$

(b) $aC_8H_{18} + bO_2 \rightarrow cCO_2 + dH_2O$

$\star$(c) $aAgNO_3 + bH_2O \rightarrow cAg + dO_2 + eHNO_3$

(d) $aHNO_3 + bHCl + cAu \rightarrow dNOCl + eHAuCl_4 + fH_2O$

7. Use the Gauss-Jordan method to find the values of A, B, C (and D in part (b)) in the following partial fractions problems:

$\star$(a) $\dfrac{5x^2 + 23x - 58}{(x-1)(x-3)(x+4)} = \dfrac{A}{x-1} + \dfrac{B}{x-3} + \dfrac{C}{x+4}$

(b) $\dfrac{-3x^3 + 29x^2 - 91x + 94}{(x-2)^2(x-3)^2} = \dfrac{A}{(x-2)^2} + \dfrac{B}{x-2} + \dfrac{C}{(x-3)^2} + \dfrac{D}{x-3}$

$\star$**8.** Solve the systems $\mathbf{AX} = \mathbf{B}_1$ and $\mathbf{AX} = \mathbf{B}_2$ simultaneously, as illustrated in this section, where

$$\mathbf{A} = \begin{bmatrix} 9 & 2 & 2 \\ 3 & 2 & 4 \\ 27 & 12 & 22 \end{bmatrix}, \quad \mathbf{B}_1 = \begin{bmatrix} -6 \\ 0 \\ 12 \end{bmatrix}, \quad \text{and} \quad \mathbf{B}_2 = \begin{bmatrix} -12 \\ -3 \\ 8 \end{bmatrix}.$$

9. Solve the systems $\mathbf{AX} = \mathbf{B}_1$ and $\mathbf{AX} = \mathbf{B}_2$ simultaneously, as illustrated in this section, where

$$\mathbf{A} = \begin{bmatrix} 12 & 2 & 0 & 3 \\ -24 & -4 & 1 & -6 \\ -4 & -1 & -1 & 0 \\ -30 & -5 & 0 & -6 \end{bmatrix}, \quad \mathbf{B}_1 = \begin{bmatrix} 3 \\ 8 \\ -4 \\ 6 \end{bmatrix}, \quad \text{and} \quad \mathbf{B}_2 = \begin{bmatrix} 2 \\ 4 \\ -24 \\ 0 \end{bmatrix}.$$

10. Let $\mathbf{A} = \begin{bmatrix} 0 & 3 & -12 \\ 2 & 4 & -10 \end{bmatrix}$ and $\mathbf{B} = \begin{bmatrix} 2 & 1 \\ 3 & -3 \\ -4 & 1 \end{bmatrix}$.

(a) Find row operations $R_1, \ldots, R_n$ such that $R_n(R_{n-1}(\cdots(R_2(R_1(\mathbf{A})))\cdots))$ is in reduced row echelon form.

(b) Verify part (2) of Theorem 2.1 using $\mathbf{A}$, $\mathbf{B}$, and the row operations from part (a).

---

[2] The chemical elements used in these equations are silver (Ag), gold (Au), carbon (C), chlorine (Cl), hydrogen (H), nitrogen (N), and oxygen (O). The compounds are water (H_2O), carbon dioxide (CO_2), benzene (C_6H_6), octane (C_8H_{18}), silver nitrate ($AgNO_3$), nitric acid (HNO_3), hydrochloric acid (HCl), nitrous chloride (NOCl), and hydrogen tetrachloroaurate (III) ($HAuCl_4$).

11. Consider the homogeneous system $\mathbf{AX} = \mathbf{O}$ having m equations and n variables.

    (a) Prove that, if $\mathbf{X}_1$ and $\mathbf{X}_2$ are both solutions to this system, then $\mathbf{X}_1 + \mathbf{X}_2$ and any scalar multiple $c\mathbf{X}_1$ are also solutions.

    ★(b) Give a counterexample to show that the results of part (a) do not necessarily hold if the system is nonhomogeneous.

    (c) Consider a nonhomogeneous system $\mathbf{AX} = \mathbf{B}$ having the same coefficient matrix as the homogeneous system $\mathbf{AX} = \mathbf{O}$. Prove that, if $\mathbf{X}_1$ is a solution of $\mathbf{AX} = \mathbf{B}$ and if $\mathbf{X}_2$ is a solution of $\mathbf{AX} = \mathbf{O}$, then $\mathbf{X}_1 + \mathbf{X}_2$ is also a solution of $\mathbf{AX} = \mathbf{B}$.

    (d) Show that if $\mathbf{AX} = \mathbf{B}$ has a unique solution, with $\mathbf{B} \neq \mathbf{O}$, then the corresponding homogeneous system $\mathbf{AX} = \mathbf{O}$ can have only the trivial solution. (Hint: Use part (c).)

12. Prove that the following homogeneous system has a nontrivial solution if and only if $ad - bc = 0$:

$$\begin{cases} ax_1 + bx_2 = 0 \\ cx_1 + dx_2 = 0 \end{cases}.$$

(Hint: First, suppose that $a \neq 0$, and show that under the Gauss-Jordan method, the second column has a nonzero pivot entry if and only if $ad - bc \neq 0$. Then consider the case $a = 0$.)

13. Suppose that $\mathbf{AX} = \mathbf{O}$ is a homogeneous system of n equations in n variables.

    (a) If the system $\mathbf{A}^2\mathbf{X} = \mathbf{O}$ has a nontrivial solution, show that $\mathbf{AX} = \mathbf{O}$ also has a nontrivial solution. (Hint: Prove the contrapositive.)

    (b) Generalize the result of part (a) to show that, if the system $\mathbf{A}^n\mathbf{X} = \mathbf{O}$ has a nontrivial solution for some positive integer n, then $\mathbf{AX} = \mathbf{O}$ has a nontrivial solution. (Hint: Use a proof by induction.)

★**14.** True or False:

    (a) In Gaussian elimination, a descending "staircase" pattern of pivots is created, in which each step starts with 1 and the entries below the staircase are all 0.

    (b) Gauss-Jordan row reduction differs from Gaussian elimination by targeting (zeroing out) entries above each nonzero pivot as well as those below the pivot.

    (c) In a reduced row echelon form matrix, the nonzero pivot entries are always located in successive rows and columns.

    (d) No homogeneous system is inconsistent.

    (e) Nontrivial solutions to a homogeneous system are found by setting the dependent (pivot column) variables equal to any real number and then determining the independent (nonpivot column) variables from those choices.

    (f) If a homogeneous system has more equations than variables, then the system has a nontrivial solution.

2.3 Equivalent Systems, Rank, and Row Space

In this section, we continue discussing the solution sets of linear systems. First we introduce row equivalence of matrices, and use this to prove our assertion in the last two sections that the Gaussian elimination and Gauss-Jordan row reduction methods always produce the complete solution set for a given linear system. We also note that every matrix has a unique corresponding matrix in reduced row echelon form and use this fact to define the rank of the matrix. We then introduce an important set of linear combinations of vectors associated with a matrix, called the row space of the matrix, and show it is invariant under row operations.

Equivalent Systems and Row Equivalence of Matrices

The first two definitions below involve related concepts. The connection between them will be shown in Theorem 2.3.

DEFINITION

Two systems of m linear equations in n variables are **equivalent** if and only if they have exactly the same solution set.

For example, the systems

$$\begin{cases} 2x - y = 1 \\ 3x + y = 9 \end{cases} \quad \text{and} \quad \begin{cases} x + 4y = 14 \\ 5x - 2y = 4 \end{cases}$$

are equivalent, because the solution set of both is exactly $\{(2, 3)\}$.

DEFINITION

An (augmented) matrix $\mathbf{D}$ is **row equivalent** to a matrix $\mathbf{C}$ if and only if $\mathbf{D}$ is obtained from $\mathbf{C}$ by a finite number of row operations of types (I), (II), and (III).

For example, given any matrix, either Gaussian elimination or the Gauss-Jordan row reduction method produces a matrix that is row equivalent to the original.

Now, if $\mathbf{D}$ is row equivalent to $\mathbf{C}$, then $\mathbf{C}$ is also row equivalent to $\mathbf{D}$. The reason is that each row operation is reversible; that is, the effect of any row operation can be undone by performing another row operation. These **reverse**, or **inverse**, row operations are shown in Table 2.1. Notice a row operation of type (I) is reversed by using the reciprocal $1/c$ and an operation of type (II) is reversed by using the additive inverse $-c$. (Do you see why?)

Thus, if $\mathbf{D}$ is obtained from $\mathbf{C}$ by the sequence

$$\mathbf{C} \overset{R_1}{\to} \mathbf{A}_1 \overset{R_2}{\to} \mathbf{A}_2 \overset{R_3}{\to} \cdots \overset{R_n}{\to} \mathbf{A}_n \overset{R_{n+1}}{\to} \mathbf{D},$$

Table 2.1

Row operations and
their inverses

Type of Operation	Operation	Reverse Operation
(I)	$\langle i \rangle \leftarrow c \langle i \rangle$	$\langle i \rangle \leftarrow \frac{1}{c} \langle i \rangle$
(II)	$\langle j \rangle \leftarrow c \langle i \rangle + \langle j \rangle$	$\langle j \rangle \leftarrow -c \langle i \rangle + \langle j \rangle$
(III)	$\langle i \rangle \leftrightarrow \langle j \rangle$	$\langle i \rangle \leftrightarrow \langle j \rangle$

then $\mathbf{C}$ can be obtained from $\mathbf{D}$ using the reverse operations in reverse order

$$\mathbf{D} \overset{R_{n+1}^{-1}}{\to} \mathbf{A}_n \overset{R_n^{-1}}{\to} \mathbf{A}_{n-1} \overset{R_{n-1}^{-1}}{\to} \cdots \overset{R_2^{-1}}{\to} \mathbf{A}_1 \overset{R_1^{-1}}{\to} \mathbf{C}.$$

(R_i^{-1} represents the inverse operation of R_i, as indicated in Table 2.1.) These comments provide a sketch for the proof of the following theorem. You are asked to fill in the details of the proof in Exercise 13(a).

THEOREM 2.2
If a matrix $\mathbf{D}$ is row equivalent to a matrix $\mathbf{C}$, then $\mathbf{C}$ is row equivalent to $\mathbf{D}$.

The next theorem asserts that if two augmented matrices are obtained from each other using only row operations, then their corresponding systems have the same solution set. This result guarantees that the Gaussian elimination and Gauss-Jordan methods provided in Sections 2.1 and 2.2 are correct because the only steps allowed in those procedures were row operations. Therefore, a final augmented matrix produced by either method represents a system equivalent to the original — that is, a system with precisely the same solution set.

THEOREM 2.3
Let $\mathbf{AX} = \mathbf{B}$ be a system of linear equations. If $[\mathbf{C} | \mathbf{D}]$ is row equivalent to $[\mathbf{A} | \mathbf{B}]$, then the system $\mathbf{CX} = \mathbf{D}$ is equivalent to $\mathbf{AX} = \mathbf{B}$.

PROOF OF THEOREM 2.3
(abridged) Let S_A represent the complete solution set of the system $\mathbf{AX} = \mathbf{B}$, and let S_C be the solution set of $\mathbf{CX} = \mathbf{D}$. Our goal is to prove that if $[\mathbf{C} | \mathbf{D}]$ is row equivalent to $[\mathbf{A} | \mathbf{B}]$, then $S_A = S_C$. It will be enough to show that $[\mathbf{C} | \mathbf{D}]$ row equivalent to $[\mathbf{A} | \mathbf{B}]$ implies $S_A \subseteq S_C$. This fact, together with Theorem 2.2, implies the reverse inclusion, $S_C \subseteq S_A$ (why?).

Also, it is enough to assume that $[\mathbf{C} | \mathbf{D}] = R([\mathbf{A} | \mathbf{B}])$ for a single row operation R because an induction argument extends the result to the case where any (finite) number of row operations are required to produce $[\mathbf{C} | \mathbf{D}]$ from $[\mathbf{A} | \mathbf{B}]$. Therefore, we need only consider the effect of each type of row operation in turn. We present the proof for a type (II) operation and leave the proofs for the other types as Exercise 13(b).

Type (II) Operation: Suppose that the original system has the form

$$\begin{cases} a_{11}x_1 + a_{12}x_2 + a_{13}x_3 + \cdots + a_{1n}x_n = b_1 \\ a_{21}x_1 + a_{22}x_2 + a_{23}x_3 + \cdots + a_{2n}x_n = b_2 \\ \vdots \qquad \vdots \qquad \vdots \qquad \ddots \qquad \vdots \qquad \vdots \\ a_{m1}x_1 + a_{m2}x_2 + a_{m3}x_3 + \cdots + a_{mn}x_n = b_m \end{cases}$$

and that the row operation used is $\langle j \rangle \leftarrow q \langle i \rangle + \langle j \rangle$ (where $i \neq j$). When this row operation is applied to the corresponding augmented matrix, all rows except the jth row remain unchanged. The new jth equation then has the form

$$(qa_{i1} + a_{j1})x_1 + (qa_{i2} + a_{j2})x_2 + \cdots + (qa_{in} + a_{jn})x_n = qb_i + b_j.$$

We must show that any solution $(s_1, s_2, \ldots, s_n)$ of the original system is a solution of the new one. Now, since $(s_1, s_2, \ldots, s_n)$ is a solution of both the ith and jth equations in the original system, we have

$$a_{i1}s_1 + a_{i2}s_2 + \cdots + a_{in}s_n = b_i \quad \text{and} \quad a_{j1}s_1 + a_{j2}s_2 + \cdots + a_{jn}s_n = b_j.$$

Multiplying the first equation by q and then adding equations yields

$$(qa_{i1} + a_{j1})s_1 + (qa_{i2} + a_{j2})s_2 + \cdots + (qa_{in} + a_{jn})s_n = qb_i + b_j.$$

Hence, $(s_1, s_2, \ldots, s_n)$ is also a solution of the new jth equation. And $(s_1, s_2, \ldots, s_n)$ is certainly a solution of every other equation in the new system as well, since none of those have changed. ∎

Rank of a Matrix

When the Gauss-Jordan method is performed on a matrix, only one final augmented matrix can result. This fact is stated in the following theorem, which we present without proof:

THEOREM 2.4
Every matrix is row equivalent to a unique matrix in reduced row echelon form.

While each matrix is row equivalent to exactly one matrix in *reduced* row echelon form, there may be many matrices in row echelon form to which it is row equivalent. This is one of the advantages of Gauss-Jordan row reduction over Gaussian elimination.

Because each matrix has a *unique* corresponding reduced row echelon form matrix, we can make the following definition:

DEFINITION
Let **A** be a matrix. Then the **rank** of **A** is the number of nonzero rows (that is, rows with nonzero pivot entries) in the unique reduced row echelon form matrix that is row equivalent to **A**.

EXAMPLE 1   Consider the following matrices:

$$\mathbf{A} = \begin{bmatrix} 2 & 1 & 4 \\ 3 & 2 & 5 \\ 0 & -1 & 1 \end{bmatrix} \quad \text{and} \quad \mathbf{B} = \begin{bmatrix} 3 & 1 & 0 & 1 & -9 \\ 0 & -2 & 12 & -8 & -6 \\ 2 & -3 & 22 & -14 & -17 \end{bmatrix}.$$

The unique reduced row echelon form matrices for $\mathbf{A}$ and $\mathbf{B}$ are, respectively:

$$\begin{bmatrix} 1 & 0 & 0 \\ 0 & 1 & 0 \\ 0 & 0 & 1 \end{bmatrix} (= \mathbf{I}_3) \quad \text{and} \quad \begin{bmatrix} 1 & 0 & 2 & -1 & -4 \\ 0 & 1 & -6 & 4 & 3 \\ 0 & 0 & 0 & 0 & 0 \end{bmatrix}.$$

Therefore, the rank of $\mathbf{A}$ is 3 since the reduced row echelon form of $\mathbf{A}$ has 3 nonzero rows (and hence 3 nonzero pivot entries). On the other hand, the rank of $\mathbf{B}$ is 2, since the reduced row echelon form of $\mathbf{B}$ has 2 nonzero rows (and hence 2 nonzero pivot entries).   ■

Homogeneous Systems and Rank

We can now restate our observations about homogeneous systems from Section 2.2 in terms of rank.

THEOREM 2.5

Let $\mathbf{AX} = \mathbf{O}$ be a homogeneous system in n variables.
(1) If rank($\mathbf{A}$) $< n$, then the system has a nontrivial solution.
(2) If rank($\mathbf{A}$) $= n$, then the system has only the trivial solution.

Note that the presence of a nontrivial solution when rank($\mathbf{A}$) $< n$ means that the homogeneous system has an infinite number of solutions.

PROOF OF THEOREM 2.5
After the Gauss-Jordan method is applied to the augmented matrix $[\mathbf{A}|\mathbf{O}]$, the number of nonzero pivots equals rank($\mathbf{A}$). Suppose rank($\mathbf{A}$) $< n$. Then at least one of the columns is a nonpivot column, and so at least one of the n variables on the left side of $[\mathbf{A}|\mathbf{O}]$ is independent. Now, because this system is homogeneous, it is consistent. Therefore, the solution set is infinite, with particular solutions found by choosing arbitrary values for all independent variables and then solving for the dependent variables. Choosing a nonzero value for at least one independent variable yields a nontrivial solution.

On the other hand, suppose rank($\mathbf{A}$) $= n$. Then, because $\mathbf{A}$ has n columns, every column on the left side of $[\mathbf{A}|\mathbf{O}]$ is a pivot column, and each variable must equal zero. Hence, in this case, $\mathbf{AX} = \mathbf{O}$ has only the trivial solution.   ■

The following corollary (illustrated by Example 4 in Section 2.2) follows immediately from Theorem 2.5:

COROLLARY 2.6

Let $\mathbf{AX} = \mathbf{O}$ be a homogeneous system of m linear equations in n variables. If $m < n$, then the system has a nontrivial solution.

Linear Combinations of Vectors

In Section 1.1, we introduced linear combinations of vectors. Recall that a linear combination of vectors is a sum of scalar multiples of the vectors.

EXAMPLE 2

Let $\mathbf{a}_1 = [-4, 1, 2]$, $\mathbf{a}_2 = [2, 1, 0]$, and $\mathbf{a}_3 = [6, -3, -4]$ in $\mathbb{R}^3$. Consider the vector $[-18, 15, 16]$. Because

$$[-18, 15, 16] = 2[-4, 1, 2] + 4[2, 1, 0] - 3[6, -3, -4],$$

the vector $[-18, 15, 16]$ is a linear combination of the vectors $\mathbf{a}_1$, $\mathbf{a}_2$, and $\mathbf{a}_3$. This combination shows us how to reach the "destination" $[-18, 15, 16]$ by traveling in directions parallel to the vectors $\mathbf{a}_1$, $\mathbf{a}_2$, and $\mathbf{a}_3$.

Now consider the vector $[16, -3, 8]$. This vector is not a linear combination of $\mathbf{a}_1$, $\mathbf{a}_2$, and $\mathbf{a}_3$. For if it were, the equation

$$[16, -3, 8] = c_1[-4, 1, 2] + c_2[2, 1, 0] + c_3[6, -3, -4]$$

would have a solution. But, equating coordinates, we get the following system:

$$\begin{cases} -4c_1 + 2c_2 + 6c_3 = 16 & \text{first coordinates} \\ c_1 + c_2 - 3c_3 = -3 & \text{second coordinates} \\ 2c_1 - 4c_3 = 8 & \text{third coordinates.} \end{cases}$$

We solve this system by row reducing

$$\begin{bmatrix} -4 & 2 & 6 & \bigm| & 16 \\ 1 & 1 & -3 & \bigm| & -3 \\ 2 & 0 & -4 & \bigm| & 8 \end{bmatrix} \quad \text{to obtain} \quad \begin{bmatrix} 1 & 0 & -2 & \bigm| & -\frac{11}{3} \\ 0 & 1 & -1 & \bigm| & \frac{2}{3} \\ 0 & 0 & 0 & \bigm| & \frac{46}{3} \end{bmatrix}.$$

The third row of this final matrix indicates that the system has no solutions, and hence, there are no values of c_1, c_2, and c_3 that together satisfy the equation

$$[16, -3, 8] = c_1[-4, 1, 2] + c_2[2, 1, 0] + c_3[6, -3, -4].$$

Therefore, $[16, -3, 8]$ is not a linear combination of the vectors $[-4, 1, 2]$, $[2, 1, 0]$, and $[6, -3, -4]$. This means that it is impossible to reach the "destination" $[16, -3, 8]$ by traveling in directions parallel to the vectors $\mathbf{a}_1$, $\mathbf{a}_2$, and $\mathbf{a}_3$. ∎

The next example shows that a vector $\mathbf{x}$ can sometimes be expressed as a linear combination of vectors $\mathbf{a}_1$, $\mathbf{a}_2$, ..., $\mathbf{a}_k$ in more than one way.

EXAMPLE 3

To determine whether $[14, -21, 7]$ is a linear combination of $[2, -3, 1]$ and $[-4, 6, -2]$, we need to find scalars c_1 and c_2 such that

$$[14, -21, 7] = c_1[2, -3, 1] + c_2[-4, 6, -2].$$

This is equivalent to solving the system

$$\begin{cases} 2c_1 - 4c_2 = 14 \\ -3c_1 + 6c_2 = -21 \\ c_1 - 2c_2 = 7 \end{cases}.$$

We solve this system by row reducing

$$\begin{bmatrix} 2 & -4 & | & 14 \\ -3 & 6 & | & -21 \\ 1 & -2 & | & 7 \end{bmatrix} \quad \text{to obtain} \quad \begin{bmatrix} 1 & -2 & | & 7 \\ 0 & 0 & | & 0 \\ 0 & 0 & | & 0 \end{bmatrix}.$$

Because c_2 is an independent variable, we may take c_2 to be any real value. Then $c_1 = 2c_2 + 7$. Hence, there are an infinite number of solutions to the system.

For example, we could let $c_2 = 1$, which forces $c_1 = 2(1) + 7 = 9$, yielding

$$[14, -21, 7] = 9[2, -3, 1] + 1[-4, 6, -2].$$

On the other hand, we could let $c_2 = 0$, which forces $c_1 = 7$, yielding

$$[14, -21, 7] = 7[2, -3, 1] + 0[-4, 6, -2].$$

Thus, we have expressed $[14, -21, 7]$ as a linear combination of $[2, -3, 1]$ and $[-4, 6, -2]$ in more than one way. ∎

In Examples 2 and 3 we saw that to find the coefficients to express a given vector **a** as a linear combination of other vectors, we row reduce an augmented matrix whose rightmost column is **a**, and whose remaining *columns* are the other vectors.

It is possible to have a linear combination of a single vector: any scalar multiple of **a** is considered a linear combination of **a**. For example, if $\mathbf{a} = [3, -1, 5]$, then $-2\mathbf{a} = [-6, 2, -10]$ is a linear combination of **a**.

The Row Space of a Matrix

Suppose **A** is an $m \times n$ matrix. Recall that each of the m rows of **A** is a vector with n entries — that is, a vector in $\mathbb{R}^n$.

DEFINITION

Let **A** be an $m \times n$ matrix. The subset of $\mathbb{R}^n$ consisting of all vectors that are linear combinations of the rows of **A** is called the **row space** of **A**.

Recall that we consider a linear combination of vectors to be a "possible destination" obtained by traveling in the directions of those vectors. Hence, the row space of a matrix is the set of "*all* possible destinations" using the rows of **A** as our fundamental directions.

EXAMPLE 4   Consider the matrix

$$\mathbf{A} = \begin{bmatrix} 3 & 1 & -2 \\ 4 & 0 & 1 \\ -2 & 4 & -3 \end{bmatrix}.$$

We want to determine whether $[5, 17, -20]$ is in the row space of $\mathbf{A}$. If so, $[5, 17, -20]$ can be expressed as a linear combination of the rows of $\mathbf{A}$, as follows:

$$[5, 17, -20] = c_1[3, 1, -2] + c_2[4, 0, 1] + c_3[-2, 4, -3].$$

Equating the coordinates on each side leads to the following system:

$$\begin{cases} 3c_1 + 4c_2 - 2c_3 = 5 \\ c_1 \qquad + 4c_3 = 17 \\ -2c_1 + c_2 - 3c_3 = -20 \end{cases}, \text{ whose matrix row reduces to } \begin{bmatrix} 1 & 0 & 0 & | & 5 \\ 0 & 1 & 0 & | & -1 \\ 0 & 0 & 1 & | & 3 \end{bmatrix}.$$

Hence, $c_1 = 5$, $c_2 = -1$, and $c_3 = 3$, and

$$[5, 17, -20] = 5[3, 1, -2] - 1[4, 0, 1] + 3[-2, 4, -3].$$

Therefore, $[5, 17, -20]$ is in the row space of $\mathbf{A}$.   ∎

Example 4 shows that to check whether a vector $\mathbf{X}$ is in the row space of $\mathbf{A}$, we row reduce the augmented matrix $\left[\mathbf{A}^T \,\middle|\, \mathbf{X} \right]$ to determine whether its corresponding system has a solution.

EXAMPLE 5   The vector $\mathbf{X} = [3, 5]$ is not in the row space of $\mathbf{B} = \begin{bmatrix} 2 & -4 \\ -1 & 2 \end{bmatrix}$ because there is no way to express $[3, 5]$ as a linear combination of the rows $[2, -4]$ and $[-1, 2]$ of $\mathbf{B}$. That is, row reducing

$$\left[\mathbf{B}^T \,\middle|\, \mathbf{X} \right] = \begin{bmatrix} 2 & -1 & | & 3 \\ -4 & 2 & | & 5 \end{bmatrix} \text{ yields } \begin{bmatrix} 1 & -\frac{1}{2} & | & \frac{3}{2} \\ 0 & 0 & | & 11 \end{bmatrix},$$

thus showing that the corresponding linear system is inconsistent.   ∎

If $\mathbf{A}$ is any $m \times n$ matrix, then $[0, 0, \ldots, 0]$ in $\mathbb{R}^n$ is always in the row space of $\mathbf{A}$. This is because the zero vector can always be expressed as a linear combination of the rows of $\mathbf{A}$ simply by multiplying each row by zero and adding the results. Similarly, each individual row of $\mathbf{A}$ is in the row space of $\mathbf{A}$, because any particular row of $\mathbf{A}$ can be expressed as a linear combination of all the rows of $\mathbf{A}$ simply by multiplying that row by 1, multiplying all other rows by zero, and summing.

Row Equivalence Determines the Row Space

The following lemma is used in the proof of Theorem 2.8:

LEMMA 2.7

Suppose that $\mathbf{x}$ is a linear combination of $\mathbf{q}_1, \ldots, \mathbf{q}_k$, and suppose also that each of $\mathbf{q}_1, \ldots, \mathbf{q}_k$ is itself a linear combination of $\mathbf{r}_1, \ldots, \mathbf{r}_l$. Then $\mathbf{x}$ is a linear combination of $\mathbf{r}_1, \ldots, \mathbf{r}_l$.

If we create a matrix $\mathbf{Q}$ whose rows are the vectors $\mathbf{q}_1, \ldots, \mathbf{q}_k$ and a matrix $\mathbf{R}$ whose rows are the vectors $\mathbf{r}_1, \ldots, \mathbf{r}_l$, then Lemma 2.7 can be rephrased as:

> If $\mathbf{x}$ is in the row space of $\mathbf{Q}$ and each row of $\mathbf{Q}$ is in the row space of $\mathbf{R}$, then $\mathbf{x}$ is in the row space of $\mathbf{R}$.

PROOF OF LEMMA 2.7

Because $\mathbf{x}$ is a linear combination of $\mathbf{q}_1, \ldots, \mathbf{q}_k$, we can write $\mathbf{x} = c_1\mathbf{q}_1 + c_2\mathbf{q}_2 + \cdots + c_k\mathbf{q}_k$ for some scalars $c_1, c_2, \ldots, c_k$. But, since each of $\mathbf{q}_1, \ldots, \mathbf{q}_k$ can be expressed as a linear combination of $\mathbf{r}_1, \ldots, \mathbf{r}_l$, there are scalars $d_{11}, \ldots, d_{kl}$ such that

$$\begin{cases} \mathbf{q}_1 = d_{11}\mathbf{r}_1 + d_{12}\mathbf{r}_2 + \cdots + d_{1l}\mathbf{r}_l \\ \mathbf{q}_2 = d_{21}\mathbf{r}_1 + d_{22}\mathbf{r}_2 + \cdots + d_{2l}\mathbf{r}_l \\ \vdots \qquad \vdots \qquad \vdots \qquad \ddots \qquad \vdots \\ \mathbf{q}_k = d_{k1}\mathbf{r}_1 + d_{k2}\mathbf{r}_2 + \cdots + d_{kl}\mathbf{r}_l \end{cases}$$

Substituting these equations into the equation for $\mathbf{x}$, we obtain

$$\begin{aligned} \mathbf{x} = c_1(d_{11}\mathbf{r}_1 &+ d_{12}\mathbf{r}_2 + \cdots + d_{1l}\mathbf{r}_l) \\ + c_2(d_{21}\mathbf{r}_1 &+ d_{22}\mathbf{r}_2 + \cdots + d_{2l}\mathbf{r}_l) \\ \vdots \qquad &\vdots \qquad \ddots \qquad \vdots \\ + c_k(d_{k1}\mathbf{r}_1 &+ d_{k2}\mathbf{r}_2 + \cdots + d_{kl}\mathbf{r}_l) \end{aligned}$$

Collecting all $\mathbf{r}_1$ terms, all $\mathbf{r}_2$ terms, and so on, we get

$$\begin{aligned} \mathbf{x} = (c_1d_{11} &+ c_2d_{21} + \cdots + c_kd_{k1})\mathbf{r}_l \\ + (c_1d_{12} &+ c_2d_{22} + \cdots + c_kd_{k2})\mathbf{r}_2 \\ \vdots \qquad &\vdots \qquad \ddots \qquad \vdots \\ + (c_1d_{1l} &+ c_2d_{2l} + \cdots + c_kd_{kl})\mathbf{r}_l \end{aligned}$$

Thus, $\mathbf{x}$ can be expressed as a linear combination of $\mathbf{r}_1, \mathbf{r}_2, \ldots, \mathbf{r}_l$. ∎

The next theorem illustrates an important connection between row equivalence and row space.

THEOREM 2.8

Suppose that $\mathbf{A}$ and $\mathbf{B}$ are row equivalent matrices. Then the row space of $\mathbf{A}$ equals the row space of $\mathbf{B}$.

In other words, if $\mathbf{A}$ and $\mathbf{B}$ are row equivalent, then any vector that is a linear combination of the rows of $\mathbf{A}$ must be a linear combination of the rows of **B**, and vice versa. Theorem 2.8 assures us that we do not gain or lose any linear combinations of the rows when we perform row operations. That is, the same set of "destination vectors" is obtained from the rows of row equivalent matrices.

PROOF OF THEOREM 2.8
(abridged) Let **A** and **B** be row equivalent $m \times n$ matrices. We will show that if **x** is a vector in the row space of **B**, then **x** is in the row space of **A**. (A similar argument can then be used to show that if **x** is in the row space of **A**, then **x** is in the row space of **B**.)

First consider the case in which **B** is obtained from **A** by performing a single row operation. In this case, the definition for each type of row operation implies that each row of **B** is a linear combination of the rows of **A** (see Exercise 19(a)). Now, suppose **x** is in the row space of **B**. Then **x** is a linear combination of the rows of **B**. But since each of the rows of **B** is a linear combination of the rows of **A**, Lemma 2.7 indicates that **x** is in the row space of **A**. By induction, this argument can be extended to the case where **B** is obtained from **A** by any (finite) sequence of row operations (see Exercise 20).   ■

EXAMPLE 6   Consider the matrix

$$\mathbf{A} = \begin{bmatrix} 5 & 10 & 12 & 33 & 19 \\ 3 & 6 & -4 & -25 & -11 \\ 1 & 2 & -2 & -11 & -5 \\ 2 & 4 & -1 & -10 & -4 \end{bmatrix}.$$

The reduced row echelon form matrix for **A** is

$$\mathbf{B} = \begin{bmatrix} 1 & 2 & 0 & -3 & -1 \\ 0 & 0 & 1 & 4 & 2 \\ 0 & 0 & 0 & 0 & 0 \\ 0 & 0 & 0 & 0 & 0 \end{bmatrix}.$$

Theorem 2.8 asserts that the row spaces of **A** and **B** are equal. Hence the linear combinations that can be created from the rows of **A** are identical to those that can be created from **B**. For example, the vector $\mathbf{x} = [4, 8, -30, -132, -64]$ is in both row spaces:

$$\mathbf{x} = -1[5, 10, 12, 33, 19] + 3[3, 6, -4, -25, -11]$$
$$+4[1, 2, -2, -11, -5] - 2[2, 4, -1, -10, -4],$$

which shows **x** is in the row space of **A**. But **x** is in the row space of **B** since

$$\mathbf{x} = 4[1, 2, 0, -3, -1] - 30[0, 0, 1, 4, 2].   ■$$

The matrix **A** in Example 6 essentially has two unneeded, or "redundant" rows. Thus, from the reduced row echelon form matrix **B** of **A**, we obtain a smaller number of rows (those that are nonzero in **B**) producing the same row space. In other words, we can reach the same "destinations" using just the two vector directions of the nonzero rows of **B** as we could using all four of the vector directions of the rows of **A**. In fact, we will prove in Chapter 4 that the rank of **A** gives precisely the minimal number of rows of **A** needed to produce the same set of linear combinations.

♦ **Numerical Methods:** You have now covered the prerequisites for Section 9.1, "Numerical Methods for Solving Systems."

Exercises for Section 2.3

Note: To save time, you should use a calculator or an appropriate software package to perform nontrivial row reductions. See Appendix D for instructions on using several popular calculator and software packages.

1. For each of the following pairs of matrices **A** and **B**, give a reason why **A** and **B** are row equivalent:

    ⋆(a) $\mathbf{A} = \begin{bmatrix} 1 & 0 & 0 \\ 0 & 1 & 0 \\ 0 & 0 & 1 \end{bmatrix}$, $\mathbf{B} = \begin{bmatrix} 1 & 0 & 0 \\ 0 & -5 & 0 \\ 0 & 0 & 1 \end{bmatrix}$

    (b) $\mathbf{A} = \begin{bmatrix} 12 & 9 & -5 \\ 4 & 6 & -2 \\ 0 & 1 & 3 \end{bmatrix}$, $\mathbf{B} = \begin{bmatrix} 0 & 1 & 3 \\ 4 & 6 & -2 \\ 12 & 9 & -5 \end{bmatrix}$

    ⋆(c) $\mathbf{A} = \begin{bmatrix} 3 & 2 & 7 \\ -4 & 1 & 6 \\ 2 & 5 & 4 \end{bmatrix}$, $\mathbf{B} = \begin{bmatrix} 3 & 2 & 7 \\ -2 & 6 & 10 \\ 2 & 5 & 4 \end{bmatrix}$

2. (a) Find the reduced row echelon form **B** of the following matrix **A**, keeping track of the row operations used:

$$\mathbf{A} = \begin{bmatrix} 4 & 0 & -20 \\ -2 & 0 & 11 \\ 3 & 1 & -15 \end{bmatrix}.$$

    ⋆(b) Use your answer to part (a) to give a sequence of row operations that converts **B** back to **A**. Check your answer. (Hint: Use the inverses of the row operations from part (a), but in reverse order.)

⋆3. (a) Verify that the following matrices are row equivalent by showing they have the same reduced row echelon form:

$$\mathbf{A} = \begin{bmatrix} 1 & 0 & 9 \\ 0 & 1 & -3 \\ 0 & -2 & 5 \end{bmatrix} \quad \text{and} \quad \mathbf{B} = \begin{bmatrix} -5 & 3 & 0 \\ -2 & 1 & 0 \\ -3 & 0 & 1 \end{bmatrix}.$$

    (b) Find a sequence of row operations that converts **A** into **B**. (Hint: Let **C** be the common matrix in reduced row echelon form corresponding to **A** and **B**. In part (a) you found a sequence of row operations that converts **A** to **C** and another sequence that converts **B** to **C**. Reverse the operations in the second sequence to obtain a sequence

that converts **C** to **B**. Finally, combine the first sequence with these "reversed" operations to create a sequence from **A** to **B**.)

4. Verify that the following matrices are not row equivalent by showing that their corresponding matrices in reduced row echelon form are different:

$$\mathbf{A} = \begin{bmatrix} 1 & -2 & 0 & 0 & 3 \\ 2 & -5 & -3 & -2 & 6 \\ 0 & 5 & 15 & 10 & 0 \\ 2 & 6 & 18 & 8 & 6 \end{bmatrix} \quad \text{and} \quad \mathbf{B} = \begin{bmatrix} 0 & 0 & 1 & 1 & 0 \\ 0 & 0 & 0 & 0 & 1 \\ 0 & 1 & 3 & 2 & 0 \\ -1 & 2 & 0 & 0 & -3 \end{bmatrix}.$$

5. Find the rank of each of the following matrices:

⋆(a) $\begin{bmatrix} 1 & -1 & 3 \\ 2 & 0 & 4 \\ -1 & -3 & 1 \end{bmatrix}$

(b) $\begin{bmatrix} -1 & 3 & 2 \\ 2 & -6 & -4 \end{bmatrix}$

⋆(c) $\begin{bmatrix} 4 & 0 & 0 \\ 0 & 0 & 0 \\ 0 & 0 & 5 \end{bmatrix}$

(d) $\begin{bmatrix} 3 & 5 & 2 \\ 4 & 2 & 3 \\ -1 & 2 & 4 \end{bmatrix}$

⋆(e) $\begin{bmatrix} -1 & -1 & 0 & 0 \\ 0 & 0 & 2 & 3 \\ 4 & 0 & -2 & 1 \\ 3 & -1 & 0 & 4 \end{bmatrix}$

(f) $\begin{bmatrix} 1 & 1 & -1 & 0 & 1 \\ 2 & -4 & 3 & 1 & 0 \\ 3 & 15 & -13 & -2 & 7 \end{bmatrix}$

6. Does Corollary 2.6 apply to the following homogeneous systems? Why or why not? Find the rank of the augmented matrix for each system. From the rank, what does Theorem 2.5 predict about the solution set? Find the complete solution set to verify this prediction.

⋆(a) $\begin{cases} -2x_1 + 6x_2 + 3x_3 = 0 \\ 5x_1 - 9x_2 - 4x_3 = 0 \\ 4x_1 - 8x_2 - 3x_3 = 0 \\ 6x_1 - 11x_2 - 5x_3 = 0 \end{cases}$

(b) $\begin{cases} -x_1 + 4x_2 + 19x_3 = 0 \\ 5x_1 + x_2 - 11x_3 = 0 \\ 4x_1 - 5x_2 - 32x_3 = 0 \\ 2x_1 + x_2 - 2x_3 = 0 \\ x_1 - 2x_2 - 11x_3 = 0 \end{cases}$

7. Assume that for each type of system below there is at least one variable with a nonzero coefficient. Find the smallest and largest rank possible for the corresponding augmented matrix in each case.

⋆(a) Four equations, three variables, nonhomogeneous

(b) Three equations, four variables

⋆(c) Three equations, four variables, inconsistent

(d) Five equations, three variables, nonhomogeneous, consistent

8. In each of the following cases, express the vector **x** as a linear combination of the other vectors, if possible:

⋆(a) $\mathbf{x} = [-3, -6]$, $\mathbf{a}_1 = [1, 4]$, $\mathbf{a}_2 = [-2, 3]$

(b) $\mathbf{x} = [5, 9, 5]$, $\mathbf{a}_1 = [2, 1, 4]$, $\mathbf{a}_2 = [1, -1, 3]$, $\mathbf{a}_3 = [3, 2, 5]$

$\star$(c) $\mathbf{x} = [2, -1, 4]$, $\mathbf{a}_1 = [3, 6, 2]$, $\mathbf{a}_2 = [2, 10, -4]$

(d) $\mathbf{x} = [2, 2, 3]$, $\mathbf{a}_1 = [6, -2, 3]$, $\mathbf{a}_2 = [0, -5, -1]$, $\mathbf{a}_3 = [-2, 1, 2]$

$\star$(e) $\mathbf{x} = [7, 2, 3]$, $\mathbf{a}_1 = [1, -2, 3]$, $\mathbf{a}_2 = [5, -2, 6]$, $\mathbf{a}_3 = [4, 0, 3]$

(f) $\mathbf{x} = [1, 1, 1, 1]$, $\mathbf{a}_1 = [2, 1, 0, 3]$, $\mathbf{a}_2 = [3, -1, 5, 2]$, $\mathbf{a}_3 = [-1, 0, 2, 1]$

$\star$(g) $\mathbf{x} = [2, 3, -7, 3]$, $\mathbf{a}_1 = [3, 2, -2, 4]$, $\mathbf{a}_2 = [-2, 0, 1, -3]$, $\mathbf{a}_3 = [6, 1, 2, 8]$

(h) $\mathbf{x} = [-3, 1, 2, 0, 1]$, $\mathbf{a}_1 = [-6, 2, 4, -1, 7]$

9. In each of the following cases, determine whether the given vector is in the row space of the given matrix:

$\star$(a) $[7, 1, 18]$, with $\begin{bmatrix} 3 & 6 & 2 \\ 2 & 10 & -4 \\ 2 & -1 & 4 \end{bmatrix}$

(b) $[4, 0, -3]$, with $\begin{bmatrix} 3 & 1 & 1 \\ 2 & -1 & 5 \\ -4 & -3 & 3 \end{bmatrix}$

$\star$(c) $[2, 2, -3]$, with $\begin{bmatrix} 4 & -1 & 2 \\ -2 & 3 & 5 \\ 6 & 1 & 9 \end{bmatrix}$

(d) $[1, 2, 5, -1]$, with $\begin{bmatrix} 2 & -1 & 0 & 3 \\ 7 & -1 & 5 & 8 \end{bmatrix}$

$\star$(e) $[1, 11, -4, 11]$, with $\begin{bmatrix} 2 & -4 & 1 & -3 \\ 7 & -1 & -1 & 2 \\ 3 & 7 & -3 & 8 \end{bmatrix}$

$\star$**10.** (a) Express the vector $[13, -23, 60]$ as a linear combination of the vectors

$$\mathbf{q}_1 = [-1, -5, 11], \quad \mathbf{q}_2 = [-10, 3, -8], \quad \text{and} \quad \mathbf{q}_3 = [7, -12, 30].$$

(b) Express each of the vectors $\mathbf{q}_1$, $\mathbf{q}_2$, and $\mathbf{q}_3$ in turn as a linear combination of the vectors $\mathbf{r}_1 = [3, -2, 4]$, $\mathbf{r}_2 = [2, 1, -3]$, and $\mathbf{r}_3 = [4, -1, 2]$.

(c) Use the results of parts (a) and (b) to express the vector $[13, -23, 60]$ as a linear combination of the vectors $\mathbf{r}_1$, $\mathbf{r}_2$, and $\mathbf{r}_3$. (Hint: Use the method given in the proof of Lemma 2.7.)

11. For each given matrix $\mathbf{A}$, perform the following steps:

(i) Find $\mathbf{B}$, the reduced row echelon form of $\mathbf{A}$.

(ii) Show that every nonzero row of $\mathbf{B}$ is in the row space of $\mathbf{A}$ by solving for the appropriate linear combination.

(iii) Show that every row of $\mathbf{A}$ is in the row space of $\mathbf{B}$ by solving for the appropriate linear combination.

$\star$(a) $\begin{bmatrix} 0 & 4 & 12 & 8 \\ 2 & 7 & 19 & 18 \\ 1 & 2 & 5 & 6 \end{bmatrix}$

(b) $\begin{bmatrix} 1 & 2 & 3 & -4 & -21 \\ -2 & -4 & -6 & 5 & 27 \\ 13 & 26 & 39 & 5 & 12 \\ 2 & 4 & 6 & -1 & -7 \end{bmatrix}$

12. Let $\mathbf{A}$ be a diagonal $n \times n$ matrix. Prove that $\mathbf{A}$ is row equivalent to $\mathbf{I}_n$ if and only if $a_{ii} \neq 0$, for all i, $1 \leq i \leq n$.

▶**13.** (a) Finish the proof of Theorem 2.2 by showing that the three inverse row operations given in Table 2.1 correctly reverse their corresponding type (I), (II), and (III) row operations.

(b) Finish the proof of Theorem 2.3 by showing that when a single row operation of type (I) or type (III) is applied to the augmented matrix $[\mathbf{A}\,|\,\mathbf{B}]$, every solution of the original system is also a solution of the new system.

★**14.** Let $\mathbf{A}$ be an $m \times n$ matrix. If $\mathbf{B}$ is a nonzero m-vector, explain why the systems $\mathbf{AX} = \mathbf{B}$ and $\mathbf{AX} = \mathbf{O}$ are not equivalent.

★**15.** Show that the converse to Theorem 2.3 is not true by exhibiting two inconsistent systems (with the same number of equations and variables) whose corresponding augmented matrices are not row equivalent.

16. (a) Show that, if five distinct points in the plane are given, then they must lie on a conic section: an equation of the form $ax^2 + bxy + cy^2 + dx + ey + f = 0$. (Hint: Create the corresponding homogeneous system of five equations and use Corollary 2.6.)

(b) Is this result also true when fewer than five points are given? Why or why not?

17. Explain why the proof of Theorem 2.5 does not necessarily work for a nonhomogeneous system.

18. Let $\mathbf{A}$ and $\mathbf{B}$ be $m \times n$ and $n \times p$ matrices, respectively, and let R be a row operation.

(a) Prove that $\mathrm{rank}(R(\mathbf{A})) = \mathrm{rank}(\mathbf{A})$.

(b) Show that if $\mathbf{A}$ has k rows of all zeroes, then $\mathrm{rank}(\mathbf{A}) \leq m - k$.

(c) Show that if $\mathbf{A}$ is in reduced row echelon form, then $\mathrm{rank}(\mathbf{AB}) \leq \mathrm{rank}(\mathbf{A})$. (Hint: Use part (b).)

(d) Use parts (a) and (c) to prove that for a general matrix $\mathbf{A}$, $\mathrm{rank}(\mathbf{AB}) \leq \mathrm{rank}(\mathbf{A})$.

▶**19.** Suppose a matrix $\mathbf{B}$ is created from a matrix $\mathbf{A}$ by a single row operation (of type (I), (II), or (III)).

(a) Verify the assertion in the proof of Theorem 2.8 that each row of $\mathbf{B}$ is a linear combination of the rows of $\mathbf{A}$.

(b) Prove that the row space of $\mathbf{B}$ is contained in the row space of $\mathbf{A}$. (Hint: The argument needed here is contained in the proof of Theorem 2.8.)

▶**20.** Complete the proof of Theorem 2.8 by showing that if a matrix $\mathbf{B}$ is obtained from a matrix $\mathbf{A}$ by any finite sequence of row operations, then the row space of $\mathbf{B}$ is contained in the row space of $\mathbf{A}$. (Hint: The case for a single row operation follows from Exercise 19. Use induction and Lemma 2.7 to extend this result to the case of more than one row operation.)

21. Let $\mathbf{x}_1, \ldots, \mathbf{x}_{n+1}$ be vectors in $\mathbb{R}^n$.

(a) Show that there exist real numbers $a_1, \ldots, a_{n+1}$, not all zero, such that the linear combination $a_1\mathbf{x}_1 + \cdots + a_{n+1}\mathbf{x}_{n+1}$ equals $\mathbf{0}$. (Hint: Solve an appropriate homogeneous system.)

(b) Using part (a), show that

$$\mathbf{x}_i = b_1\mathbf{x}_1 + \cdots + b_{i-1}\mathbf{x}_{i-1} + b_{i+1}\mathbf{x}_{i+1} + \cdots + b_{n+1}\mathbf{x}_{n+1},$$

for some i, $1 \le i \le n+1$, and some $b_1, \ldots, b_{i-1}, b_{i+1}, \ldots, b_{n+1} \in \mathbb{R}$.

★**22.** True or False:

(a) Two linear systems are equivalent if their corresponding augmented matrices are row equivalent.

(b) If $\mathbf{A}$ is row equivalent to $\mathbf{B}$, and $\mathbf{B}$ has rank 3, then $\mathbf{A}$ has rank 3.

(c) The inverse of a type (I) row operation is a type (II) row operation.

(d) If the matrix for a linear system with n variables has rank $< n$, then the system must have a nontrivial solution.

(e) If the matrix for a homogeneous system with n variables has rank n, then the system has a nontrivial solution.

(f) If $\mathbf{x}$ is a linear combination of the rows of $\mathbf{A}$, and $\mathbf{B}$ is row equivalent to $\mathbf{A}$, then $\mathbf{x}$ is in the row space of $\mathbf{B}$.

2.4 Inverses of Matrices

In this section, we consider whether a given $n \times n$ (square) matrix $\mathbf{A}$ has a *multiplicative* inverse matrix (that is, a matrix $\mathbf{A}^{-1}$ such that $\mathbf{A}\mathbf{A}^{-1} = \mathbf{I}_n$). Interestingly, not all square matrices have multiplicative inverses, but most do. We examine some properties of multiplicative inverses and illustrate methods for finding these inverses when they exist.

Multiplicative Inverse of a Matrix

When the word "inverse" is used with matrices, it usually refers to the *multiplicative* inverse in the next definition, rather than the additive inverse of Theorem 1.11, part (4).

DEFINITION

Let $\mathbf{A}$ be an $n \times n$ matrix. Then an $n \times n$ matrix $\mathbf{B}$ is a **(multiplicative) inverse** of $\mathbf{A}$ if and only if $\mathbf{AB} = \mathbf{BA} = \mathbf{I}_n$.

Note that if $\mathbf{B}$ is an inverse of $\mathbf{A}$, then $\mathbf{A}$ is also an inverse of $\mathbf{B}$, as is seen by switching the roles of $\mathbf{A}$ and $\mathbf{B}$ in the definition.

EXAMPLE 1   The matrices

$$\mathbf{A} = \begin{bmatrix} 1 & -4 & 1 \\ 1 & 1 & -2 \\ -1 & 1 & 1 \end{bmatrix} \quad \text{and} \quad \mathbf{B} = \begin{bmatrix} 3 & 5 & 7 \\ 1 & 2 & 3 \\ 2 & 3 & 5 \end{bmatrix}$$

are inverses of each other because

$$\underbrace{\begin{bmatrix} 1 & -4 & 1 \\ 1 & 1 & -2 \\ -1 & 1 & 1 \end{bmatrix}}_{\mathbf{A}} \underbrace{\begin{bmatrix} 3 & 5 & 7 \\ 1 & 2 & 3 \\ 2 & 3 & 5 \end{bmatrix}}_{\mathbf{B}} = \underbrace{\begin{bmatrix} 1 & 0 & 0 \\ 0 & 1 & 0 \\ 0 & 0 & 1 \end{bmatrix}}_{\mathbf{I}_3} = \underbrace{\begin{bmatrix} 3 & 5 & 7 \\ 1 & 2 & 3 \\ 2 & 3 & 5 \end{bmatrix}}_{\mathbf{B}} \underbrace{\begin{bmatrix} 1 & -4 & 1 \\ 1 & 1 & -2 \\ -1 & 1 & 1 \end{bmatrix}}_{\mathbf{A}}.$$

However, $\mathbf{C} = \begin{bmatrix} 2 & 1 \\ 6 & 3 \end{bmatrix}$ has no inverse because there is no $\begin{bmatrix} a & b \\ c & d \end{bmatrix}$ such that

$$\begin{bmatrix} 2 & 1 \\ 6 & 3 \end{bmatrix}\begin{bmatrix} a & b \\ c & d \end{bmatrix} = \begin{bmatrix} 1 & 0 \\ 0 & 1 \end{bmatrix}.$$

For, if so, then multiplying out the left side of this equation would give

$$\begin{bmatrix} 2a+c & 2b+d \\ 6a+3c & 6b+3d \end{bmatrix} = \begin{bmatrix} 1 & 0 \\ 0 & 1 \end{bmatrix}.$$

This would force $2a + c = 1$ and $6a + 3c = 0$, but these are contradictory equations, since $6a + 3c = 3(2a + c)$. ∎

When checking whether two given square matrices $\mathbf{A}$ and $\mathbf{B}$ are inverses, we do not need to multiply both products $\mathbf{AB}$ and $\mathbf{BA}$, as the next theorem asserts.

THEOREM 2.9

Let $\mathbf{A}$ and $\mathbf{B}$ be $n \times n$ matrices. If either product $\mathbf{AB}$ or $\mathbf{BA}$ equals $\mathbf{I}_n$, then the other product also equals $\mathbf{I}_n$, and $\mathbf{A}$ and $\mathbf{B}$ are inverses of each other.

The proof is tedious and is in Appendix A for the interested reader.

DEFINITION

A square matrix is **singular** if and only if it does not have an inverse.
A square matrix is **nonsingular** if and only if it has an inverse.

For example, the 2×2 matrix $\mathbf{C}$ from Example 1 is a singular matrix since we proved that it does not have an inverse. Another example of a singular matrix is the $n \times n$ zero matrix $\mathbf{O}_n$ (why?). On the other hand, the 3×3 matrix $\mathbf{A}$ from Example 1 is nonsingular, because we found an inverse $\mathbf{B}$ for $\mathbf{A}$.

Properties of the Matrix Inverse

The next theorem shows that the inverse of a matrix must be unique (when it exists).

THEOREM 2.10 (Uniqueness of Inverse Matrix)
If **B** and **C** are both inverses of an $n \times n$ matrix **A**, then **B** = **C**.

PROOF OF THEOREM 2.10

$$\mathbf{B} = \mathbf{B}\mathbf{I}_n = \mathbf{B}(\mathbf{A}\mathbf{C}) = (\mathbf{B}\mathbf{A})\mathbf{C} = \mathbf{I}_n\mathbf{C} = \mathbf{C}.$$

∎

Because Theorem 2.10 asserts that a nonsingular matrix **A** can have exactly one inverse, we denote the unique inverse of **A** by $\mathbf{A}^{-1}$.

For a nonsingular matrix **A**, we can use the inverse to define negative integral powers of **A**.

DEFINITION
Let **A** be a nonsingular $n \times n$ matrix. Then the negative powers of **A** are given as follows: $\mathbf{A}^{-1}$ is the (unique) inverse of **A**, and for $k \geq 2$,
$\mathbf{A}^{-k} = \left(\mathbf{A}^{-1}\right)^{k}$.

EXAMPLE 2 We know from Example 1 that

$$\mathbf{A} = \begin{bmatrix} 1 & -4 & 1 \\ 1 & 1 & -2 \\ -1 & 1 & 1 \end{bmatrix} \quad \text{has} \quad \mathbf{A}^{-1} = \begin{bmatrix} 3 & 5 & 7 \\ 1 & 2 & 3 \\ 2 & 3 & 5 \end{bmatrix}$$

as its unique inverse. Since $\mathbf{A}^{-3} = \left(\mathbf{A}^{-1}\right)^{3}$, we have

$$\mathbf{A}^{-3} = \begin{bmatrix} 3 & 5 & 7 \\ 1 & 2 & 3 \\ 2 & 3 & 5 \end{bmatrix}^{3} = \begin{bmatrix} 272 & 445 & 689 \\ 107 & 175 & 271 \\ 184 & 301 & 466 \end{bmatrix}.$$

∎

THEOREM 2.11
Let **A** and **B** be nonsingular $n \times n$ matrices. Then
(1) $\mathbf{A}^{-1}$ is nonsingular, and $(\mathbf{A}^{-1})^{-1} = \mathbf{A}$
(2) $\mathbf{A}^{k}$ is nonsingular, and $(\mathbf{A}^{k})^{-1} = (\mathbf{A}^{-1})^{k} = \mathbf{A}^{-k}$, for any integer k
(3) $\mathbf{A}\mathbf{B}$ is nonsingular, and $(\mathbf{A}\mathbf{B})^{-1} = \mathbf{B}^{-1}\mathbf{A}^{-1}$
(4) $\mathbf{A}^{T}$ is nonsingular, and $\left(\mathbf{A}^{T}\right)^{-1} = (\mathbf{A}^{-1})^{T}$

Part (3) says that the inverse of a product equals the product of the inverses in *reverse* order. To prove each part of this theorem, show that the right side of each equation is the inverse of the term in parentheses on the left side. This is done by simply multiplying them together and observing that their product is $\mathbf{I}_n$. We prove parts (3) and (4) here and leave the others as Exercise 15(a).

PROOF OF THEOREM 2.11

(abridged)

Part (3): We must show that $\mathbf{B}^{-1}\mathbf{A}^{-1}$ (right side) is the inverse of $\mathbf{AB}$ (in parentheses on the left side). Multiplying them together gives $(\mathbf{AB})(\mathbf{B}^{-1}\mathbf{A}^{-1}) = \mathbf{A}\left(\mathbf{BB}^{-1}\right)\mathbf{A}^{-1} = \mathbf{AI}_n\mathbf{A}^{-1} = \mathbf{AA}^{-1} = \mathbf{I}_n$.

Part (4): We must show that $\left(\mathbf{A}^{-1}\right)^T$ (right side) is the inverse of $\mathbf{A}^T$ (in parentheses on the left side). Multiplying them together gives $\mathbf{A}^T\left(\mathbf{A}^{-1}\right)^T = \left(\mathbf{A}^{-1}\mathbf{A}\right)^T$ (by Theorem 1.16) $= (\mathbf{I}_n)^T = \mathbf{I}_n$, since $\mathbf{I}_n$ is symmetric. ∎

Using a proof by induction, part (3) of Theorem 2.11 generalizes as follows: if $\mathbf{A}_1, \mathbf{A}_2, \ldots, \mathbf{A}_k$ are nonsingular matrices of the same size, then

$$(\mathbf{A}_1\mathbf{A}_2\cdots\mathbf{A}_k)^{-1} = \mathbf{A}_k^{-1}\cdots\mathbf{A}_2^{-1}\mathbf{A}_1^{-1}$$

(Exercise 15(b)). Notice that the order of the matrices on the right side is reversed. Theorem 1.15 can also be generalized to show that the laws of exponents hold for negative integer powers, as follows:

THEOREM 2.12 (Expanded Version of Theorem 1.15)

If $\mathbf{A}$ is a nonsingular matrix and if s and t are integers, then

(1) $\mathbf{A}^{s+t} = (\mathbf{A}^s)(\mathbf{A}^t)$

(2) $(\mathbf{A}^s)^t = \mathbf{A}^{st} = \left(\mathbf{A}^t\right)^s$

The proof of this theorem is a bit tedious. Some special cases are considered in Exercise 17.

Recall that in Section 1.5 we observed that if $\mathbf{AB} = \mathbf{AC}$ for three matrices $\mathbf{A}$, $\mathbf{B}$, and $\mathbf{C}$, it does not necessarily follow that $\mathbf{B} = \mathbf{C}$. However, if $\mathbf{A}$ is a nonsingular matrix, then $\mathbf{B} = \mathbf{C}$ because you can multiply both sides of $\mathbf{AB} = \mathbf{AC}$ by $\mathbf{A}^{-1}$ on the left to effectively cancel out the $\mathbf{A}$'s.

Inverses for 2×2 Matrices

So far, we have studied many properties of the matrix inverse, but we have not discussed methods for finding inverses. In fact, there is an immediate way to find the inverse (if it exists) of a 2×2 matrix. Note that if we let $\delta = ad - bc$, then

$$\begin{bmatrix} a & b \\ c & d \end{bmatrix}\begin{bmatrix} d & -b \\ -c & a \end{bmatrix} = \begin{bmatrix} \delta & 0 \\ 0 & \delta \end{bmatrix} = \delta\mathbf{I}_n.$$

Hence, if $\delta \neq 0$, we can divide this equation by δ to prove one half of the following theorem:

THEOREM 2.13

The matrix $\mathbf{A} = \begin{bmatrix} a & b \\ c & d \end{bmatrix}$ has an inverse if and only if $\delta = ad - bc \neq 0$. In that case,

$$\mathbf{A}^{-1} = \begin{bmatrix} a & b \\ c & d \end{bmatrix}^{-1} = \frac{1}{\delta} \begin{bmatrix} d & -b \\ -c & a \end{bmatrix}.$$

For the other half of the proof, note that if $\delta = ad - bc = 0$, then $\begin{bmatrix} a & b \\ c & d \end{bmatrix}$ $\begin{bmatrix} d & -b \\ -c & a \end{bmatrix} = \mathbf{O}_2$, and it can then be shown that $\mathbf{A}^{-1}$ does not exist (see Exercise 10). Hence, the condition $\delta = ad - bc \neq 0$ is both a *necessary* and a *sufficient* condition for the inverse to exist. The quantity $\delta = ad - bc$ is called the **determinant** of $\mathbf{A}$. We will discuss determinants in more detail in Chapter 3.

EXAMPLE 3

There is no inverse for $\begin{bmatrix} 12 & -4 \\ 9 & -3 \end{bmatrix}$, since $\delta = (12)(-3) - (-4)(9) = 0$. On the other hand, $\mathbf{M} = \begin{bmatrix} -5 & 2 \\ 9 & -4 \end{bmatrix}$ does have an inverse because $\delta = (-5)(-4) - (2)(9) = 2 \neq 0$. This inverse is

$$\mathbf{M}^{-1} = \frac{1}{2} \begin{bmatrix} -4 & -2 \\ -9 & -5 \end{bmatrix} = \begin{bmatrix} -2 & -1 \\ -\frac{9}{2} & -\frac{5}{2} \end{bmatrix}.$$

Verify this by checking that $\mathbf{M}\mathbf{M}^{-1} = \mathbf{I}_2$. ∎

Inverses of Larger Matrices

Let $\mathbf{A}$ be an $n \times n$ matrix. We now describe a process for calculating $\mathbf{A}^{-1}$, if it exists.

METHOD FOR FINDING THE INVERSE OF A MATRIX (IF IT EXISTS) (INVERSE METHOD)

Suppose that $\mathbf{A}$ is a given $n \times n$ matrix.
Step 1: Augment $\mathbf{A}$ to a $n \times 2n$ matrix, whose first n columns constitute $\mathbf{A}$ itself and whose last n columns constitute $\mathbf{I}_n$.
Step 2: Convert $[\mathbf{A}|\mathbf{I}_n]$ into reduced row echelon form.
Step 3: If the first n columns of $[\mathbf{A}|\mathbf{I}_n]$ cannot be converted into $\mathbf{I}_n$, then $\mathbf{A}$ is singular. Stop.
Step 4: Otherwise, $\mathbf{A}$ is nonsingular, and the last n columns of the augmented matrix in reduced row echelon form constitute $\mathbf{A}^{-1}$. That is, $[\mathbf{A}|\mathbf{I}_n]$ row reduces to $[\mathbf{I}_n|\mathbf{A}^{-1}]$.

Before proving that this procedure is valid, we consider some examples.

To find the inverse of the matrix

$$\mathbf{A} = \begin{bmatrix} 2 & -6 & 5 \\ -4 & 12 & -9 \\ 2 & -9 & 8 \end{bmatrix},$$

we first enlarge this to a 3×6 matrix by adjoining the identity matrix $\mathbf{I}_3$:

$$\begin{bmatrix} 2 & -6 & 5 & 1 & 0 & 0 \\ -4 & 12 & -9 & 0 & 1 & 0 \\ 2 & -9 & 8 & 0 & 0 & 1 \end{bmatrix}.$$

Row reduction yields

$$\begin{bmatrix} 1 & 0 & 0 & \frac{5}{2} & \frac{1}{2} & -1 \\ 0 & 1 & 0 & \frac{7}{3} & 1 & -\frac{1}{3} \\ 0 & 0 & 1 & 2 & 1 & 0 \end{bmatrix}.$$

The last three columns give the inverse of the original matrix $\mathbf{A}$. This is

$$\mathbf{A}^{-1} = \begin{bmatrix} \frac{5}{2} & \frac{1}{2} & -1 \\ \frac{7}{3} & 1 & -\frac{1}{3} \\ 2 & 1 & 0 \end{bmatrix}.$$

You should check that this matrix really is the inverse of $\mathbf{A}$ by showing that its product with $\mathbf{A}$ is equal to $\mathbf{I}_3$. ∎

Using Row Reduction to Show That a Matrix Is Singular

As we have seen, not every square matrix has an inverse. For a singular matrix $\mathbf{A}$, row reduction of $[\mathbf{A} \mid \mathbf{I}_n]$ does not produce $\mathbf{I}_n$ to the left of the augmentation bar. Now, the only way this can happen is if, during row reduction, we reach a column whose main diagonal entry and all entries below it are zero. In that case, there is no way to use a type (I) or type (III) operation to place a nonzero entry in the main diagonal position for that column. Hence, we cannot transform the leftmost columns into the identity matrix. This situation is illustrated in the following example:

  We attempt to find an inverse for the singular matrix

$$\mathbf{A} = \begin{bmatrix} 4 & 2 & 8 & 1 \\ -2 & 0 & -4 & 1 \\ 1 & 4 & 2 & 0 \\ 3 & -1 & 6 & -2 \end{bmatrix}.$$

Beginning with $[\mathbf{A}\,|\,\mathbf{I}_4]$ and simplifying the first two columns, we obtain

$$\left[\begin{array}{cccc|cccc} 1 & 0 & 2 & -\frac{1}{2} & 0 & -\frac{1}{2} & 0 & 0 \\ 0 & 1 & 0 & \frac{3}{2} & \frac{1}{2} & 1 & 0 & 0 \\ 0 & 0 & 0 & -\frac{11}{2} & -2 & -\frac{7}{2} & 1 & 0 \\ 0 & 0 & 0 & 1 & \frac{1}{2} & \frac{5}{2} & 0 & 1 \end{array}\right].$$

Continuing on to the third column, we see that the $(3, 3)$ entry is zero. Thus, a type (I) operation cannot be used to make the pivot 1. Because the $(4, 3)$ entry is also zero, no type (III) operation (switching the pivot row with a row below it) can make the pivot nonzero. We conclude that there is no way to transform the first four columns into the identity matrix $\mathbf{I}_4$ using the row reduction process, and so the original matrix $\mathbf{A}$ has no inverse. ∎

Justification of the Inverse Method

To verify that the Inverse Method is valid, we must prove that for a given square matrix $\mathbf{A}$, the algorithm correctly predicts whether $\mathbf{A}$ has an inverse and, if it does, calculates its (unique) inverse.

Now, from the technique of solving simultaneous systems in Section 2.2, we know that row reduction of

$$[\mathbf{A}\,|\,\mathbf{I}_n] = \left[\begin{array}{c|cccc} \mathbf{A} & \begin{matrix}\text{1st}\\\text{column}\\\text{of }\mathbf{I}_n\end{matrix} & \begin{matrix}\text{2nd}\\\text{column}\\\text{of }\mathbf{I}_n\end{matrix} & \begin{matrix}\text{3rd}\\\text{column}\\\text{of }\mathbf{I}_n\end{matrix} \cdots & \begin{matrix}n\text{th}\\\text{column}\\\text{of }\mathbf{I}_n\end{matrix} \end{array}\right]$$

is equivalent to separately using row reduction to solve each of the n linear systems whose augmented matrices are

$$\left[\begin{array}{c|c} \mathbf{A} & \begin{matrix}\text{1st}\\\text{column}\\\text{of }\mathbf{I}_n\end{matrix} \end{array}\right], \left[\begin{array}{c|c} \mathbf{A} & \begin{matrix}\text{2nd}\\\text{column}\\\text{of }\mathbf{I}_n\end{matrix} \end{array}\right], \ldots, \left[\begin{array}{c|c} \mathbf{A} & \begin{matrix}n\text{th}\\\text{column}\\\text{of }\mathbf{I}_n\end{matrix} \end{array}\right].$$

First, suppose $\mathbf{A}$ is a nonsingular $n \times n$ matrix (that is, $\mathbf{A}^{-1}$ exists). Now, because $\mathbf{A}\mathbf{A}^{-1} = \mathbf{I}_n$, we know $\mathbf{A} \begin{bmatrix} i\text{th}\\\text{column}\\\text{of }\mathbf{A}^{-1}\end{bmatrix} = \begin{bmatrix} i\text{th}\\\text{column}\\\text{of }\mathbf{I}_n\end{bmatrix}$. Therefore, the columns of $\mathbf{A}^{-1}$ are respective solutions of the n systems above. Thus, these systems are all consistent. Now, if any one of these systems has more than one solution, then a second solution for that system can be used to replace the corresponding column in $\mathbf{A}^{-1}$ to give a second inverse for $\mathbf{A}$. But by Theorem 2.10, the inverse of $\mathbf{A}$ is unique, and so each of these systems must have a unique solution. Therefore, each column to the left of the augmentation bar must be a pivot column, or else there would be independent variables, giving an infinite number of solutions. Thus, $[\mathbf{A}\,|\,\mathbf{I}_n]$ must row reduce to $\left[\mathbf{I}_n\,|\,\mathbf{A}^{-1}\right]$, since the columns of $\mathbf{A}^{-1}$ are the unique solutions for these simultaneous systems.

Now consider the case where $\mathbf{A}$ is singular. Because an inverse for $\mathbf{A}$ cannot be found, at least one of the original n systems, such as

$$\left[\begin{array}{c|c} \mathbf{A} & \begin{array}{c} k\text{th} \\ \text{column} \\ \text{of } \mathbf{I}_n \end{array} \end{array}\right],$$

has no solutions. But this occurs only if the final augmented matrix after row reduction contains a row of the form

$$\left[\, 0 \ 0 \ 0 \ \cdots \ 0 \,\middle|\, r \,\right],$$

where $r \neq 0$. Hence, there is a row that contains no pivot entry in the first n columns, and so we cannot obtain $\mathbf{I}_n$ to the left of the augmentation bar. Step 3 of the formal algorithm correctly concludes that $\mathbf{A}$ is singular.

Recall that $\mathbf{A}$ row reduces to $\mathbf{I}_n$ if and only if rank($\mathbf{A}$)= n. Now, since the inverse algorithm is valid, we have the following:

THEOREM 2.14

An $n \times n$ matrix $\mathbf{A}$ is nonsingular if and only if rank($\mathbf{A}$) = n.

Solving a System Using the Inverse of the Coefficient Matrix

The following result gives us another method for solving certain linear systems:

THEOREM 2.15

Let $\mathbf{AX} = \mathbf{B}$ represent a system where the coefficient matrix $\mathbf{A}$ is square.

(1) If $\mathbf{A}$ is nonsingular, then the system has a unique solution $\left(\mathbf{X} = \mathbf{A}^{-1}\mathbf{B}\right)$.

(2) If $\mathbf{A}$ is singular, then the system either has no solutions or an infinite number of solutions.

Hence, $\mathbf{AX} = \mathbf{B}$ has a unique solution *if and only if* $\mathbf{A}$ is nonsingular.

PROOF OF THEOREM 2.15

If $\mathbf{A}$ is nonsingular, then $\mathbf{A}^{-1}\mathbf{B}$ is a solution for the system $\mathbf{AX} = \mathbf{B}$ because $\mathbf{A}(\mathbf{A}^{-1}\mathbf{B}) = (\mathbf{A}\mathbf{A}^{-1})\mathbf{B} = \mathbf{I}_n\mathbf{B} = \mathbf{B}$. To show that this solution is unique, suppose $\mathbf{Y}$ is any solution to the system; that is, suppose that $\mathbf{AY} = \mathbf{B}$. Then we can multiply both sides of $\mathbf{AY} = \mathbf{B}$ on the left by $\mathbf{A}^{-1}$ to get

$$\begin{aligned} \mathbf{A}^{-1}(\mathbf{AY}) = \mathbf{A}^{-1}\mathbf{B} &\implies \left(\mathbf{A}^{-1}\mathbf{A}\right)\mathbf{Y} = \mathbf{A}^{-1}\mathbf{B} \\ &\implies \mathbf{I}_n\mathbf{Y} = \mathbf{A}^{-1}\mathbf{B} \\ &\implies \mathbf{Y} = \mathbf{A}^{-1}\mathbf{B}. \end{aligned}$$

Therefore, $\mathbf{A}^{-1}\mathbf{B}$ is the only solution of $\mathbf{AX} = \mathbf{B}$.

On the other hand, if **A** is singular, then by Theorem 2.14, rank(**A**) $< n$, and so not every column of **A** becomes a pivot column in the row reduction of the augmented matrix $[\mathbf{A}|\mathbf{B}]$. Now, suppose $\mathbf{AX} = \mathbf{B}$ has at least one solution. Then this system has at least one independent variable (which can take on any real value), and hence, the system has an infinite number of solutions. ∎

Theorem 2.15 indicates that when $\mathbf{A}^{-1}$ is known, the matrix **X** of variables can be found by a simple matrix multiplication of $\mathbf{A}^{-1}$ and **B**.

EXAMPLE 6   We will solve the 3×3 system

$$\begin{cases} -7x_1 + 5x_2 + 3x_3 = 6 \\ 3x_1 - 2x_2 - 2x_3 = -3 \\ 3x_1 - 2x_2 - x_3 = 2 \end{cases}$$

using the inverse of the coefficient matrix. This system has matrix form

$$\underbrace{\begin{bmatrix} -7 & 5 & 3 \\ 3 & -2 & -2 \\ 3 & -2 & -1 \end{bmatrix}}_{\mathbf{A}} \underbrace{\begin{bmatrix} x_1 \\ x_2 \\ x_3 \end{bmatrix}}_{\mathbf{X}} = \underbrace{\begin{bmatrix} 6 \\ -3 \\ 2 \end{bmatrix}}_{\mathbf{B}}.$$

Calculating the inverse matrix of **A**, we find

$$\mathbf{A}^{-1} = \begin{bmatrix} 2 & 1 & 4 \\ 3 & 2 & 5 \\ 0 & -1 & 1 \end{bmatrix}.$$

Now by Theorem 2.15, $\mathbf{X} = \mathbf{A}^{-1}\mathbf{B}$, and so

$$\begin{bmatrix} x_1 \\ x_2 \\ x_3 \end{bmatrix} = \begin{bmatrix} 2 & 1 & 4 \\ 3 & 2 & 5 \\ 0 & -1 & 1 \end{bmatrix} \begin{bmatrix} 6 \\ -3 \\ 2 \end{bmatrix} = \begin{bmatrix} 17 \\ 22 \\ 5 \end{bmatrix}.$$ ∎

This method for solving an $n \times n$ system is not as efficient as the Gauss-Jordan method because it involves finding an inverse as well as performing a matrix multiplication. It is sometimes used when many systems, all having the same nonsingular coefficient matrix, must be solved. In that case, the inverse of the coefficient matrix can be calculated first, and then each system can be solved with a single matrix multiplication.

♦ **Application**: You have now covered the prerequisites for Section 8.5, "Hill Substitution: An Introduction to Coding Theory."

♦ **Numerical Method:** You have now covered the prerequisites for Section 9.2, "**LDU** Decomposition."

♦ **Supplemental Material:** You have now covered the prerequisites for Section 10.1, "Elementary Matrices."

Exercises for Section 2.4

Note: You should be using a calculator or appropriate computer software to perform nontrivial row reductions. See Appendix D for instructions on using several popular calculator and software packages.

1. Verify that the following pairs of matrices are inverses:

    (a) $\begin{bmatrix} 10 & 41 & -5 \\ -1 & -12 & 1 \\ 3 & 20 & -2 \end{bmatrix}$, $\begin{bmatrix} 4 & -18 & -19 \\ 1 & -5 & -5 \\ 16 & -77 & -79 \end{bmatrix}$

    (b) $\begin{bmatrix} 1 & 0 & -1 & 5 \\ -1 & 1 & 0 & -3 \\ 0 & 2 & -3 & 7 \\ 2 & -1 & -2 & 12 \end{bmatrix}$, $\begin{bmatrix} 1 & -4 & 1 & -2 \\ 4 & 6 & -2 & 1 \\ 5 & 11 & -4 & 3 \\ 1 & 3 & -1 & 1 \end{bmatrix}$

2. Determine whether each of the following matrices is nonsingular by calculating its rank:

    ★(a) $\begin{bmatrix} 4 & -9 \\ -2 & 3 \end{bmatrix}$

    (b) $\begin{bmatrix} 3 & -1 & 4 \\ 2 & -2 & 1 \\ -1 & 3 & 2 \end{bmatrix}$

    ★(c) $\begin{bmatrix} -6 & -6 & 1 \\ 2 & 3 & -1 \\ 8 & 6 & -1 \end{bmatrix}$

    (d) $\begin{bmatrix} -10 & -3 & 1 & -3 \\ 18 & 5 & -2 & 6 \\ 6 & 2 & -1 & 6 \\ 12 & 3 & -1 & 3 \end{bmatrix}$

    ★(e) $\begin{bmatrix} 2 & 1 & -7 & 14 \\ -6 & -3 & 19 & -38 \\ 1 & 0 & -3 & 6 \\ 2 & 1 & -6 & 12 \end{bmatrix}$

3. Find the inverse, if it exists, for each of the following 2×2 matrices:

    ★(a) $\begin{bmatrix} 4 & 2 \\ 9 & -3 \end{bmatrix}$

    (b) $\begin{bmatrix} 10 & -5 \\ -4 & 2 \end{bmatrix}$

    ★(c) $\begin{bmatrix} -3 & 5 \\ -12 & -8 \end{bmatrix}$

    (d) $\begin{bmatrix} 1 & 2 \\ 4 & -3 \end{bmatrix}$

    ★(e) $\begin{bmatrix} -6 & 12 \\ 4 & -8 \end{bmatrix}$

    (f) $\begin{bmatrix} -\frac{1}{2} & \frac{3}{4} \\ \frac{1}{3} & -\frac{1}{2} \end{bmatrix}$

4. Use row reduction to find the inverse, if it exists, for each of the following:

    ★(a) $\begin{bmatrix} -4 & 7 & 6 \\ 3 & -5 & -4 \\ -2 & 4 & 3 \end{bmatrix}$

    (b) $\begin{bmatrix} 5 & 7 & -6 \\ 3 & 1 & -2 \\ 1 & -5 & 2 \end{bmatrix}$

    ★(c) $\begin{bmatrix} 2 & -2 & 3 \\ 8 & -4 & 9 \\ -4 & 6 & -9 \end{bmatrix}$

    (d) $\begin{bmatrix} 0 & 0 & -2 & -1 \\ -2 & 0 & -1 & 0 \\ -1 & -2 & -1 & -5 \\ 0 & 1 & 1 & 3 \end{bmatrix}$

$$
\star\text{(e)} \quad
\begin{bmatrix}
2 & 0 & -1 & 3 \\
1 & -2 & 3 & 1 \\
4 & 1 & 0 & -1 \\
1 & 3 & -2 & -5
\end{bmatrix}
\qquad
\text{(f)} \quad
\begin{bmatrix}
3 & 3 & 0 & -2 \\
14 & 15 & 0 & -11 \\
-3 & 1 & 2 & -5 \\
-2 & 0 & 1 & -2
\end{bmatrix}
$$

5. Assuming that all main diagonal entries are nonzero, find the inverse of each of the following:

(a) $\begin{bmatrix} a_{11} & 0 \\ 0 & a_{22} \end{bmatrix}$

(b) $\begin{bmatrix} a_{11} & 0 & 0 \\ 0 & a_{22} & 0 \\ 0 & 0 & a_{33} \end{bmatrix}$

$\star$(c) $\begin{bmatrix} a_{11} & 0 & 0 & \cdots & 0 \\ 0 & a_{22} & 0 & \cdots & 0 \\ \vdots & \vdots & \vdots & \ddots & \vdots \\ 0 & 0 & 0 & \cdots & a_{nn} \end{bmatrix}$

$\star$**6.** The following matrices are useful in computer graphics for rotating vectors (see Section 5.1). Find the inverse of each matrix, and then state what the matrix and its inverse are when $\theta = \frac{\pi}{6}, \frac{\pi}{4}$, and $\frac{\pi}{2}$.

(a) $\begin{bmatrix} \cos\theta & -\sin\theta \\ \sin\theta & \cos\theta \end{bmatrix}$

(b) $\begin{bmatrix} \cos\theta & -\sin\theta & 0 \\ \sin\theta & \cos\theta & 0 \\ 0 & 0 & 1 \end{bmatrix}$ (Hint: Modify your answer from part (a).)

7. In each case, find the inverse of the coefficient matrix and use it to solve the system by matrix multiplication.

$\star$(a) $\begin{cases} 5x_1 - x_2 = 20 \\ -7x_1 + 2x_2 = -31 \end{cases}$

(b) $\begin{cases} -5x_1 + 3x_2 + 6x_3 = 4 \\ 3x_1 - x_2 - 7x_3 = 11 \\ -2x_1 + x_2 + 2x_3 = 2 \end{cases}$

$\star$(c) $\begin{cases} -2x_2 + 5x_3 + x_4 = 25 \\ -7x_1 - 4x_2 + 5x_3 + 22x_4 = -15 \\ 5x_1 + 3x_2 - 4x_3 - 16x_4 = 9 \\ -3x_1 - x_2 + 9x_4 = -16 \end{cases}$

$\star$**8.** A matrix with the property $\mathbf{A}^2 = \mathbf{I}_n$ is called an **involutory** matrix.

(a) Find an example of a 2×2 involutory matrix other than $\mathbf{I}_2$.

(b) Find an example of a 3×3 involutory matrix other than $\mathbf{I}_3$.

(c) What is $\mathbf{A}^{-1}$ if $\mathbf{A}$ is involutory?

9. (a) Give an example to show that $\mathbf{A} + \mathbf{B}$ can be singular if $\mathbf{A}$ and $\mathbf{B}$ are both nonsingular.

(b) Give an example to show that $\mathbf{A} + \mathbf{B}$ can be nonsingular if $\mathbf{A}$ and $\mathbf{B}$ are both singular.

(c) Give an example to show that even when $\mathbf{A}$, $\mathbf{B}$, and $\mathbf{A} + \mathbf{B}$ are all nonsingular, $(\mathbf{A} + \mathbf{B})^{-1}$ is not necessarily equal to $\mathbf{A}^{-1} + \mathbf{B}^{-1}$.

$\star$**10.** Let $\mathbf{A}$, $\mathbf{B}$, and $\mathbf{C}$ be $n \times n$ matrices.

(a) Suppose that $\mathbf{AB} = \mathbf{O}_n$, and $\mathbf{A}$ is nonsingular. What must $\mathbf{B}$ be?

(b) If $\mathbf{AB} = \mathbf{I}_n$, is it possible for $\mathbf{AC}$ to equal $\mathbf{O}_n$ without $\mathbf{C} = \mathbf{O}_n$? Why or why not?

$\star$**11.** If $\mathbf{A}^4 = \mathbf{I}_n$, but $\mathbf{A} \neq \mathbf{I}_n$, $\mathbf{A}^2 \neq \mathbf{I}_n$, and $\mathbf{A}^3 \neq \mathbf{I}_n$, which powers of $\mathbf{A}$ are equal to $\mathbf{A}^{-1}$?

$\star$**12.** If the matrix product $\mathbf{A}^{-1}\mathbf{B}$ is known, how could you calculate $\mathbf{B}^{-1}\mathbf{A}$ without necessarily knowing what $\mathbf{A}$ and $\mathbf{B}$ are?

13. Let $\mathbf{A}$ be a symmetric nonsingular matrix. Prove that $\mathbf{A}^{-1}$ is symmetric.

14. $\star$(a) You have already seen in this section that every square matrix containing a row of zeroes must be singular. Why must every square matrix containing a column of zeroes be singular?

(b) Why must every diagonal matrix with at least one zero main diagonal entry be singular?

(c) Why must every upper triangular matrix with no zero entries on the main diagonal be nonsingular?

(d) Use part (c) and the transpose to show that every lower triangular matrix with no zero entries on the main diagonal must be nonsingular.

15. ▶(a) Prove parts (1) and (2) of Theorem 2.11. (Hint: In proving part (2), consider the cases $k \geq 0$ and $k < 0$ separately.)

(b) Use the method of induction to prove the following generalization of part (3) of Theorem 2.11: if $\mathbf{A}_1, \mathbf{A}_2, \ldots, \mathbf{A}_m$ are nonsingular matrices of the same size, then $(\mathbf{A}_1\mathbf{A}_2 \cdots \mathbf{A}_m)^{-1} = \mathbf{A}_m^{-1} \cdots \mathbf{A}_2^{-1}\mathbf{A}_1^{-1}$.

16. If $\mathbf{A}$ is a nonsingular matrix and $c \in \mathbb{R}$ with $c \neq 0$, prove that $(c\mathbf{A})^{-1} = \left(\frac{1}{c}\right)\mathbf{A}^{-1}$.

17. ▶(a) Prove part (1) of Theorem 2.12 if $s < 0$ and $t < 0$.

(b) Prove part (2) of Theorem 2.12 if $s \geq 0$ and $t < 0$.

18. Assume that $\mathbf{A}$ and $\mathbf{B}$ are nonsingular $n \times n$ matrices. Prove that $\mathbf{A}$ and $\mathbf{B}$ commute (that is, $\mathbf{AB} = \mathbf{BA}$) if and only if $(\mathbf{AB})^2 = \mathbf{A}^2\mathbf{B}^2$.

19. Prove that if $\mathbf{A}$ and $\mathbf{B}$ are nonsingular matrices of the same size, then $\mathbf{AB} = \mathbf{BA}$ if and only if $(\mathbf{AB})^q = \mathbf{A}^q\mathbf{B}^q$ for every positive integer $q \geq 2$. (Hint: To prove the "if" part, let $q = 2$. For the "only if" part, first show by induction that if $\mathbf{AB} = \mathbf{BA}$, then $\mathbf{AB}^q = \mathbf{B}^q\mathbf{A}$, for any positive integer $q \geq 2$. Finish the proof with a second induction argument to show $(\mathbf{AB})^q = \mathbf{A}^q\mathbf{B}^q$.)

20. Prove that, if $\mathbf{A}$ is an $n \times n$ matrix and $\mathbf{A} - \mathbf{I}_n$ is nonsingular, then for every integer $k \geq 0$,

$$\mathbf{I}_n + \mathbf{A} + \mathbf{A}^2 + \mathbf{A}^3 + \cdots + \mathbf{A}^k = \left(\mathbf{A}^{k+1} - \mathbf{I}_n\right)(\mathbf{A} - \mathbf{I}_n)^{-1}.$$

$\star$**21.** True or False:

(a) Every $n \times n$ matrix $\mathbf{A}$ has a unique inverse.

(b) If $\mathbf{A}$, $\mathbf{B}$ are $n \times n$ matrices, and $\mathbf{BA} = \mathbf{I}_n$, then $\mathbf{A}$ and $\mathbf{B}$ are inverses.

(c) If $\mathbf{A}$, $\mathbf{B}$ are nonsingular $n \times n$ matrices, then $((\mathbf{AB})^T)^{-1} = (\mathbf{A}^{-1})^T(\mathbf{B}^{-1})^T$

(d) $\mathbf{A} = \begin{bmatrix} a & b \\ c & d \end{bmatrix}$ is singular if and only if $ad - bc \neq 0$.

(e) If $\mathbf{A}$ is an $n \times n$ matrix, then $\mathbf{A}$ is nonsingular if and only if $[\mathbf{A} | \mathbf{I}_n]$ has less than n nonzero pivots before the augmentation bar after row reduction.

(f) If $\mathbf{A}$ is an $n \times n$ matrix, then rank($\mathbf{A}$) = n if and only if any system of the form $\mathbf{AX} = \mathbf{B}$ has a unique solution for $\mathbf{X}$.

Chapter 3

Determinants and Eigenvalues

The Determining Factor

Amazingly, many important geometric and algebraic properties of a square matrix are revealed by a single real number associated with the matrix, known as its determinant. For example, the areas and volumes of certain figures can be found by creating a matrix based on the figure's edges and then calculating the determinant of that matrix. The determinant also provides a quick method for discovering whether or not certain linear systems have a unique solution.

In this chapter, we also use determinants to introduce the concept of eigenvectors. An eigenvector of a square matrix is a special vector that, when multiplied by the matrix, produces a parallel vector. Such vectors provide a new way to look at matrix multiplication and help to solve many intractable problems. Eigenvectors are practical tools in linear algebra with applications in differential equations, probability, statistics, and related disciplines, such as economics, physics, chemistry, and computer graphics.

$\mathbf{I}$n this chapter, we introduce the determinant, a particular real number associated with each square matrix. In Section 3.1, we define the determinant using cofactor expansion, and illustrate a geometric application. In Section 3.2, we examine a technique for finding the determinant of a given square matrix using row reduction. In Section 3.3, we present several useful properties of the determinant. In Section 3.4, we introduce eigenvalues and eigenvectors. This leads to diagonalization of matrices, which will be revisited in greater detail in Section 5.5.

3.1 Introduction to Determinants

Determinants of 1×1, 2×2, and 3×3 Matrices

For a 1×1 matrix $\mathbf{A} = [a_{11}]$, the **determinant** $|\mathbf{A}|$ is defined to be a_{11}, its only entry. For example, the determinant of $\mathbf{A} = [-4]$ is simply $|\mathbf{A}| = -4$. We will represent a determinant by placing absolute value signs around the matrix, even though the determinant could be negative.

For a 2×2 matrix $\mathbf{A} = \begin{bmatrix} a_{11} & a_{12} \\ a_{21} & a_{22} \end{bmatrix}$, the **determinant** $|\mathbf{A}|$ is defined to be $a_{11}a_{22} - a_{12}a_{21}$. For example, the determinant of $\mathbf{A} = \begin{bmatrix} 4 & -3 \\ 2 & 5 \end{bmatrix}$ is $|\mathbf{A}| = \begin{vmatrix} 4 & -3 \\ 2 & 5 \end{vmatrix} = (4)(5) - (-3)(2) = 26$. Recall that in Section 2.4 we proved $\begin{bmatrix} a_{11} & a_{12} \\ a_{21} & a_{22} \end{bmatrix}$ has an inverse if and only if $|\mathbf{A}| = a_{11}a_{22} - a_{12}a_{21} \neq 0$.

For the 3×3 matrix

$$\mathbf{A} = \begin{bmatrix} a_{11} & a_{12} & a_{13} \\ a_{21} & a_{22} & a_{23} \\ a_{31} & a_{32} & a_{33} \end{bmatrix},$$

we define the **determinant** $|\mathbf{A}|$ to be the following expression, which has six terms:

$$|\mathbf{A}| = a_{11}a_{22}a_{33} + a_{12}a_{23}a_{31} + a_{13}a_{21}a_{32} - a_{13}a_{22}a_{31} - a_{11}a_{23}a_{32} - a_{12}a_{21}a_{33}.$$

This expression may look complicated, but its terms can be obtained by multiplying the following entries linked by arrows. Notice that the first two columns of the original 3×3 matrix have been repeated. Also, the arrows pointing right indicate terms with a positive sign, while those pointing left indicate terms with a negative sign.

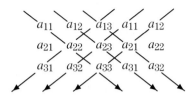

This technique is sometimes referred to as the **basketweaving** method for calculating the determinant of a 3×3 matrix.

EXAMPLE 1   Find the determinant of

$$\mathbf{A} = \begin{bmatrix} 4 & -2 & 3 \\ -1 & 5 & 0 \\ 6 & -1 & -2 \end{bmatrix}.$$

Repeating the first two columns and forming terms using the basketweaving method, we have

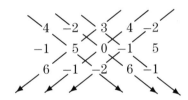

which gives

$$(4)(5)(-2)+(-2)(0)(6)+(3)(-1)(-1)-(3)(5)(6)-(4)(0)(-1)-(-2)(-1)(-2).$$

This reduces to $-40 + 0 + 3 - 90 - 0 - (-4) = -123$. Thus,

$$|\mathbf{A}| = \begin{vmatrix} 4 & -2 & 3 \\ -1 & 5 & 0 \\ 6 & -1 & -2 \end{vmatrix} = -123.$$ ∎

Application: Areas and Volumes

The next theorem illustrates why 2×2 and 3×3 determinants are sometimes interpreted as areas and volumes, respectively.

THEOREM 3.1

(1) Let $\mathbf{x} = [x_1, x_2]$ and $\mathbf{y} = [y_1, y_2]$ be two nonparallel vectors in $\mathbb{R}^2$ beginning at a common point (see Figure 3.1(a)). Then the area of the parallelogram determined by $\mathbf{x}$ and $\mathbf{y}$ is the absolute value of the determinant

$$\begin{vmatrix} x_1 & x_2 \\ y_1 & y_2 \end{vmatrix}.$$

(2) Let $\mathbf{x} = [x_1, x_2, x_3]$, $\mathbf{y} = [y_1, y_2, y_3]$, and $\mathbf{z} = [z_1, z_2, z_3]$ be three vectors not all in the same plane beginning at a common initial point (see Figure 3.1(b)). Then the volume of the parallelepiped determined by $\mathbf{x}$, $\mathbf{y}$, and $\mathbf{z}$ is the absolute value of the determinant

$$\begin{vmatrix} x_1 & x_2 & x_3 \\ y_1 & y_2 & y_3 \\ z_1 & z_2 & z_3 \end{vmatrix}.$$

Figure 3.1

(a) The parallelogram determined by $\mathbf{x}$ and $\mathbf{y}$ (Theorem 3.1);
(b) the parallelepiped determined by $\mathbf{x}$, $\mathbf{y}$, and $\mathbf{z}$ (Theorem 3.1)

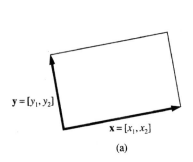

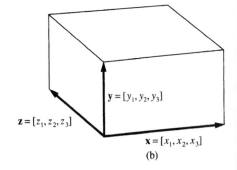

(a)   (b)

The proof of this theorem is straightforward (see Exercises 10 and 12).

The volume of the parallelepiped whose sides are $\mathbf{x} = [-2, 1, 3]$, $\mathbf{y} = [3, 0, -2]$, and $\mathbf{z} = [-1, 3, 7]$ is given by the absolute value of the determinant

$$\begin{vmatrix} -2 & 1 & 3 \\ 3 & 0 & -2 \\ -1 & 3 & 7 \end{vmatrix}.$$

Calculating this determinant, we obtain -4, so the volume is $|-4| = 4$. ∎

Cofactors

Before defining determinants for square matrices larger than 3×3, we first introduce a few new terms.

DEFINITION

Let $\mathbf{A}$ be an $n \times n$ matrix, with $n \geq 2$. The (i, j) **submatrix**, $\mathbf{A}_{ij}$, of $\mathbf{A}$, is the $(n-1) \times (n-1)$ matrix obtained by deleting all entries of the ith row and all entries of the jth column of $\mathbf{A}$. The (i, j) **minor**, $\left|\mathbf{A}_{ij}\right|$, of $\mathbf{A}$, is the determinant of the submatrix $\mathbf{A}_{ij}$ of $\mathbf{A}$.

EXAMPLE 3 Consider the following matrices:

$$\mathbf{A} = \begin{bmatrix} 5 & -2 & 1 \\ 0 & 4 & -3 \\ 2 & -7 & 6 \end{bmatrix} \quad \text{and} \quad \mathbf{B} = \begin{bmatrix} 9 & -1 & 4 & 7 \\ -3 & 2 & 6 & -2 \\ -8 & 0 & 1 & 3 \\ 4 & 7 & -5 & -1 \end{bmatrix}.$$

The $(1, 3)$ submatrix of $\mathbf{A}$ obtained by deleting all entries in the 1st row and all entries in the 3rd column is $\mathbf{A}_{13} = \begin{bmatrix} 0 & 4 \\ 2 & -7 \end{bmatrix}$, and the $(3, 4)$ submatrix of $\mathbf{B}$ obtained by deleting all entries in the 3rd row and all entries in the 4th column is

$$\mathbf{B}_{34} = \begin{bmatrix} 9 & -1 & 4 \\ -3 & 2 & 6 \\ 4 & 7 & -5 \end{bmatrix}.$$

The corresponding minors associated with these submatrices are

$$\left|\mathbf{A}_{13}\right| = \begin{vmatrix} 0 & 4 \\ 2 & -7 \end{vmatrix} = -8 \quad \text{and} \quad \left|\mathbf{B}_{34}\right| = \begin{vmatrix} 9 & -1 & 4 \\ -3 & 2 & 6 \\ 4 & 7 & -5 \end{vmatrix} = -593. \quad \blacksquare$$

An $n \times n$ matrix has a total of n^2 minors — one for each entry of the matrix. In particular, a 3×3 matrix has nine minors. For the matrix $\mathbf{A}$ in Example 3, the minors are

$$\left|\mathbf{A}_{11}\right| = \begin{vmatrix} 4 & -3 \\ -7 & 6 \end{vmatrix} = 3 \quad \left|\mathbf{A}_{12}\right| = \begin{vmatrix} 0 & -3 \\ 2 & 6 \end{vmatrix} = 6 \quad \left|\mathbf{A}_{13}\right| = \begin{vmatrix} 0 & 4 \\ 2 & -7 \end{vmatrix} = -8$$

$$\left|\mathbf{A}_{21}\right| = \begin{vmatrix} -2 & 1 \\ -7 & 6 \end{vmatrix} = -5 \quad \left|\mathbf{A}_{22}\right| = \begin{vmatrix} 5 & 1 \\ 2 & 6 \end{vmatrix} = 28 \quad \left|\mathbf{A}_{23}\right| = \begin{vmatrix} 5 & -2 \\ 2 & -7 \end{vmatrix} = -31$$

$$\left|\mathbf{A}_{31}\right| = \begin{vmatrix} -2 & 1 \\ 4 & -3 \end{vmatrix} = 2 \quad \left|\mathbf{A}_{32}\right| = \begin{vmatrix} 5 & 1 \\ 0 & -3 \end{vmatrix} = -15 \quad \left|\mathbf{A}_{33}\right| = \begin{vmatrix} 5 & -2 \\ 0 & 4 \end{vmatrix} = 20.$$

We now define a "cofactor" for each entry based on its minor.

DEFINITION
Let $\mathbf{A}$ be an $n \times n$ matrix, with $n \geq 2$. The (i, j) **cofactor** of $\mathbf{A}$, $\mathcal{A}_{ij}$, is $(-1)^{i+j}$ times the (i, j) minor of $\mathbf{A}$ — that is, $\mathcal{A}_{ij} = (-1)^{i+j} \left|\mathbf{A}_{ij}\right|$.

EXAMPLE 4 For the matrices $\mathbf{A}$ and $\mathbf{B}$ in Example 3, the cofactor $\mathcal{A}_{13}$ of $\mathbf{A}$ is $(-1)^{1+3} \left|\mathbf{A}_{13}\right| = (-1)^4(-8) = -8$, and the cofactor $\mathcal{B}_{34}$ of $\mathbf{B}$ is $(-1)^{3+4} \left|\mathbf{B}_{34}\right| = (-1)^7(-593) = 593$. $\blacksquare$

An $n \times n$ matrix has n^2 cofactors, one for each matrix entry. In particular, a 3×3 matrix has nine cofactors. For the matrix $\mathbf{A}$ from Example 3, these cofactors are

$$
\begin{aligned}
\mathcal{A}_{11} &= (-1)^{1+1} |\mathbf{A}_{11}| = (-1)^2 (3) &= 3 \\
\mathcal{A}_{12} &= (-1)^{1+2} |\mathbf{A}_{12}| = (-1)^3 (6) &= -6 \\
\mathcal{A}_{13} &= (-1)^{1+3} |\mathbf{A}_{13}| = (-1)^4 (-8) &= -8 \\
\mathcal{A}_{21} &= (-1)^{2+1} |\mathbf{A}_{21}| = (-1)^3 (-5) &= 5 \\
\mathcal{A}_{22} &= (-1)^{2+2} |\mathbf{A}_{22}| = (-1)^4 (28) &= 28 \\
\mathcal{A}_{23} &= (-1)^{2+3} |\mathbf{A}_{23}| = (-1)^5 (-31) &= 31 \\
\mathcal{A}_{31} &= (-1)^{3+1} |\mathbf{A}_{31}| = (-1)^4 (2) &= 2 \\
\mathcal{A}_{32} &= (-1)^{3+2} |\mathbf{A}_{32}| = (-1)^5 (-15) &= 15 \\
\mathcal{A}_{33} &= (-1)^{3+3} |\mathbf{A}_{33}| = (-1)^6 (20) &= 20 \ .
\end{aligned}
$$

Formal Definition of the Determinant

We are now ready to define the determinant of a general $n \times n$ matrix. We will see shortly that the following definition agrees with our earlier formulas for determinants of size 1×1, 2×2, and 3×3.

DEFINITION
Let $\mathbf{A}$ be an $n \times n$ (square) matrix. The **determinant** of $\mathbf{A}$, denoted $|\mathbf{A}|$, is defined as follows:

If $n = 1$ (so that $\mathbf{A} = [a_{11}]$), then $|\mathbf{A}| = a_{11}$.

If $n > 1$, then $|\mathbf{A}| = a_{n1}\mathcal{A}_{n1} + a_{n2}\mathcal{A}_{n2} + \cdots + a_{nn}\mathcal{A}_{nn}$.

For $n > 1$, this defines the determinant as a sum of products. Each entry a_{ni} of the last row of the matrix $\mathbf{A}$ is multiplied by its corresponding cofactor $\mathcal{A}_{ni}$, and we sum the results. This process is often referred to as **cofactor expansion** (or, **Laplace expansion**) **along the last row** of the matrix. Since the cofactors of an $n \times n$ matrix are calculated by finding determinants of appropriate $(n-1) \times (n-1)$ submatrices, we see that this definition is actually recursive. That is, we can find the determinant of any matrix once we know how to find the determinant of any smaller size matrix!

EXAMPLE 5   Consider again the matrix from Example 3

$$
\mathbf{A} = \begin{bmatrix} 5 & -2 & 1 \\ 0 & 4 & -3 \\ 2 & -7 & 6 \end{bmatrix}.
$$

Multiplying every entry of the last row by its cofactor, and summing, we have

$$
|\mathbf{A}| = a_{31}\mathcal{A}_{31} + a_{32}\mathcal{A}_{32} + a_{33}\mathcal{A}_{33} = 2(2) + (-7)(15) + 6(20) = 19.
$$

You can verify that using "basketweaving" also produces $|\mathbf{A}| = 19$. ∎

Note that this new definition for the determinant agrees with the previous definitions for 2×2 and 3×3 matrices. For, if **B** is a 2×2 matrix, then cofactor expansion on **B** yields

$$
\begin{aligned}
|\mathbf{B}| &= b_{21}\mathcal{B}_{21} + b_{22}\mathcal{B}_{22} \\
&= b_{21}(-1)^{2+1}|\mathbf{B}_{21}| + b_{22}(-1)^{2+2}|\mathbf{B}_{22}| \\
&= -b_{21}(b_{12}) + b_{22}(b_{11}) \\
&= b_{11}b_{22} - b_{12}b_{21},
\end{aligned}
$$

which is correct. Similarly, if **C** is a 3×3 matrix, then

$$
\begin{aligned}
|\mathbf{C}| &= c_{31}\mathcal{C}_{31} + c_{32}\mathcal{C}_{32} + c_{33}\mathcal{C}_{33} \\
&= c_{31}(-1)^{3+1}|\mathbf{C}_{31}| + c_{32}(-1)^{3+2}|\mathbf{C}_{32}| + c_{33}(-1)^{3+3}|\mathbf{C}_{33}| \\
&= c_{31}\begin{vmatrix} c_{12} & c_{13} \\ c_{22} & c_{23} \end{vmatrix} - c_{32}\begin{vmatrix} c_{11} & c_{13} \\ c_{21} & c_{23} \end{vmatrix} + c_{33}\begin{vmatrix} c_{11} & c_{12} \\ c_{21} & c_{22} \end{vmatrix} \\
&= c_{31}(c_{12}c_{23} - c_{13}c_{22}) - c_{32}(c_{11}c_{23} - c_{13}c_{21}) + c_{33}(c_{11}c_{22} - c_{12}c_{21}) \\
&= c_{11}c_{22}c_{33} + c_{12}c_{23}c_{31} + c_{13}c_{21}c_{32} - c_{13}c_{22}c_{31} - c_{11}c_{23}c_{32} - c_{12}c_{21}c_{33},
\end{aligned}
$$

which agrees with the formula for a 3×3 determinant.

We now compute the determinant of a 4×4 matrix.

EXAMPLE 6   Consider the matrix

$$
\mathbf{A} = \begin{bmatrix} 3 & 2 & 0 & 5 \\ 4 & 1 & 3 & -1 \\ 2 & -1 & 3 & 6 \\ 5 & 0 & 2 & -1 \end{bmatrix}.
$$

Then, using cofactor expansion along the last row, we have

$$
\begin{aligned}
|\mathbf{A}| &= a_{41}\mathcal{A}_{41} + a_{42}\mathcal{A}_{42} + a_{43}\mathcal{A}_{43} + a_{44}\mathcal{A}_{44} \\
&= 5(-1)^{4+1}|\mathbf{A}_{41}| + 0(-1)^{4+2}|\mathbf{A}_{42}| + 2(-1)^{4+3}|\mathbf{A}_{43}| + (-1)(-1)^{4+4}|\mathbf{A}_{44}| \\
&= -5\begin{vmatrix} 2 & 0 & 5 \\ 1 & 3 & -1 \\ -1 & 3 & 6 \end{vmatrix} + 0 - 2\begin{vmatrix} 3 & 2 & 5 \\ 4 & 1 & -1 \\ 2 & -1 & 6 \end{vmatrix} - 1\begin{vmatrix} 3 & 2 & 0 \\ 4 & 1 & 3 \\ 2 & -1 & 3 \end{vmatrix}.
\end{aligned}
$$

At this point, we could use basketweaving to finish the calculation. Instead, we evaluate each of the remaining determinants using cofactor expansion along the last row to illustrate the recursive nature of the method. Now,

$$
\begin{aligned}
\begin{vmatrix} 2 & 0 & 5 \\ 1 & 3 & -1 \\ -1 & 3 & 6 \end{vmatrix} &= (-1)(-1)^{3+1}\begin{vmatrix} 0 & 5 \\ 3 & -1 \end{vmatrix} + 3(-1)^{3+2}\begin{vmatrix} 2 & 5 \\ 1 & -1 \end{vmatrix} + 6(-1)^{3+3}\begin{vmatrix} 2 & 0 \\ 1 & 3 \end{vmatrix} \\
&= (-1)(0 - 15) + (-3)(-2 - 5) + (6)(6 - 0) \\
&= 15 + 21 + 36 = 72,
\end{aligned}
$$

$$\begin{vmatrix} 3 & 2 & 5 \\ 4 & 1 & -1 \\ 2 & -1 & 6 \end{vmatrix} = (2)(-1)^{3+1}\begin{vmatrix} 2 & 5 \\ 1 & -1 \end{vmatrix} + (-1)(-1)^{3+2}\begin{vmatrix} 3 & 5 \\ 4 & -1 \end{vmatrix}$$

$$+6(-1)^{3+3}\begin{vmatrix} 3 & 2 \\ 4 & 1 \end{vmatrix}$$

$$= (2)(-2-5) + (1)(-3-20) + (6)(3-8)$$

$$= -14 - 23 - 30 = -67, \quad \text{and}$$

$$\begin{vmatrix} 3 & 2 & 0 \\ 4 & 1 & 3 \\ 2 & -1 & 3 \end{vmatrix} = (2)(-1)^{3+1}\begin{vmatrix} 2 & 0 \\ 1 & 3 \end{vmatrix} + (-1)(-1)^{3+2}\begin{vmatrix} 3 & 0 \\ 4 & 3 \end{vmatrix} + 3(-1)^{3+3}\begin{vmatrix} 3 & 2 \\ 4 & 1 \end{vmatrix}$$

$$= (2)(6-0) + (1)(9-0) + (3)(3-8)$$

$$= 12 + 9 - 15 = 6.$$

Hence, $|\mathbf{A}| = (-5)(72) - 2(-67) - 1(6) = -360 + 134 - 6 = -232.$ ∎

The computation of the 4×4 determinant in Example 6 is quite cumbersome. Finding the determinant of a 5×5 matrix would involve the computation of five 4×4 determinants! As the size of the matrix increases, the calculation of the determinant can become tedious. In Section 3.2, we present an alternative method for calculating determinants that is computationally more efficient for larger matrices. After that, we will generally use methods other than cofactor expansion, except in cases in which enough zeroes in the matrix allow us to avoid computing many of the corresponding cofactors. (For instance, in Example 6, we did not need to calculate $\mathcal{A}_{42}$.)

♦ **Application:** You have now covered the prerequisites for Section 8.6, "Change of Variables and the Jacobian."

Exercises for Section 3.1

1. Calculate the determinant of each of the following matrices using the quick formulas given at the beginning of this section:

★(a) $\begin{bmatrix} -2 & 5 \\ 3 & 1 \end{bmatrix}$

(b) $\begin{bmatrix} 5 & -3 \\ 2 & 0 \end{bmatrix}$

★(c) $\begin{bmatrix} 6 & -12 \\ -4 & 8 \end{bmatrix}$

(d) $\begin{bmatrix} \cos\theta & \sin\theta \\ -\sin\theta & \cos\theta \end{bmatrix}$

★(e) $\begin{bmatrix} 2 & 0 & 5 \\ -4 & 1 & 7 \\ 0 & 3 & -3 \end{bmatrix}$

(f) $\begin{bmatrix} 3 & -2 & 4 \\ 5 & 1 & -2 \\ -1 & 3 & 6 \end{bmatrix}$

★(g) $\begin{bmatrix} 5 & 0 & 0 \\ 3 & -2 & 0 \\ -1 & 8 & 4 \end{bmatrix}$

(h) $\begin{bmatrix} -6 & 0 & 0 \\ 0 & 2 & 0 \\ 0 & 0 & 5 \end{bmatrix}$

★(i) $\begin{bmatrix} 3 & 1 & -2 \\ -1 & 4 & 5 \\ 3 & 1 & -2 \end{bmatrix}$

★(j) $[-3]$

2. Calculate the indicated minors for each given matrix.

*(a) $|A_{21}|$, for $\mathbf{A} = \begin{bmatrix} -2 & 4 & 3 \\ 3 & -1 & 6 \\ 5 & -2 & 4 \end{bmatrix}$

(b) $|B_{34}|$, for $\mathbf{B} = \begin{bmatrix} 0 & 2 & -3 & 1 \\ 1 & 4 & 2 & -1 \\ 3 & -2 & 4 & 0 \\ 4 & -1 & 1 & 0 \end{bmatrix}$

*(c) $|C_{42}|$, for $\mathbf{C} = \begin{bmatrix} -3 & 3 & 0 & 5 \\ 2 & 1 & -1 & 4 \\ 6 & -3 & 4 & 0 \\ -1 & 5 & 1 & -2 \end{bmatrix}$

3. Calculate the indicated cofactors for each given matrix.

*(a) $\mathcal{A}_{22}$, for $\mathbf{A} = \begin{bmatrix} 4 & 1 & -3 \\ 0 & 2 & -2 \\ 9 & 14 & -7 \end{bmatrix}$

(b) $\mathcal{B}_{23}$, for $\mathbf{B} = \begin{bmatrix} -9 & 6 & 7 \\ 2 & -1 & 0 \\ 4 & 3 & -8 \end{bmatrix}$

*(c) $\mathcal{C}_{43}$, for $\mathbf{C} = \begin{bmatrix} -5 & 2 & 2 & 13 \\ -8 & 2 & -5 & 22 \\ -6 & -3 & 0 & -16 \\ 4 & -1 & 7 & -8 \end{bmatrix}$

*(d) $\mathcal{D}_{12}$, for $\mathbf{D} = \begin{bmatrix} x+1 & x & x-7 \\ x-4 & x+5 & x-3 \\ x-1 & x & x+2 \end{bmatrix}$, where $x \in \mathbb{R}$

4. Calculate the determinant of each of the matrices in Exercise 1 using the formal definition of the determinant.

5. Calculate the determinant of each of the following matrices:

*(a) $\begin{bmatrix} 5 & 2 & 1 & 0 \\ -1 & 3 & 5 & 2 \\ 4 & 1 & 0 & 2 \\ 0 & 2 & 3 & 0 \end{bmatrix}$

(b) $\begin{bmatrix} 0 & 5 & 4 & 0 \\ 4 & 1 & -2 & 7 \\ -1 & 0 & 3 & 0 \\ 0 & 2 & 1 & 5 \end{bmatrix}$

(c) $\begin{bmatrix} 2 & 1 & 9 & 7 \\ 0 & -1 & 3 & 8 \\ 0 & 0 & 5 & 2 \\ 0 & 0 & 0 & 6 \end{bmatrix}$

*(d) $\begin{bmatrix} 0 & 4 & 1 & 3 & -2 \\ 2 & 2 & 3 & -1 & 0 \\ 3 & 1 & 2 & -5 & 1 \\ 1 & 0 & -4 & 0 & 0 \\ 0 & 3 & 0 & 0 & 2 \end{bmatrix}$

6. For a general 4×4 matrix $\mathbf{A}$, write out the formula for $|\mathbf{A}|$ using cofactor expansion along the last row, and simplify as far as possible. (Your final answer should have 24 terms, each being a product of four entries of $\mathbf{A}$.)

*7. Give a counterexample to show that for square matrices $\mathbf{A}$ and $\mathbf{B}$ of the same size, it is not always true that $|\mathbf{A} + \mathbf{B}| = |\mathbf{A}| + |\mathbf{B}|$.

8. (a) Show that the **cross product** $\mathbf{a} \times \mathbf{b} = [a_2b_3 - a_3b_2,\ a_3b_1 - a_1b_3,\ a_1b_2 - a_2b_1]$ of $\mathbf{a} = [a_1, a_2, a_3]$ and $\mathbf{b} = [b_1, b_2, b_3]$ can be expressed in "determinant notation" as

$$\begin{bmatrix} \mathbf{i} & \mathbf{j} & \mathbf{k} \\ a_1 & a_2 & a_3 \\ b_1 & b_2 & b_3 \end{bmatrix}.$$

(b) Show that $\mathbf{a} \times \mathbf{b}$ is orthogonal to both $\mathbf{a}$ and $\mathbf{b}$.

9. Calculate the area of the parallelogram in $\mathbb{R}^2$ determined by the following:

⋆(a) $\mathbf{x} = [3, 2],\ \mathbf{y} = [4, 5]$　　(b) $\mathbf{x} = [-4, 3],\ \mathbf{y} = [-2, 6]$
⋆(c) $\mathbf{x} = [5, -1],\ \mathbf{y} = [-3, 3]$　　(d) $\mathbf{x} = [-2, 3],\ \mathbf{y} = [6, -9]$

▶10. Prove part (1) of Theorem 3.1. (Hint: See Figure 3.2. The area of the parallelogram is the length of the base $\mathbf{x}$ multiplied by the length of the perpendicular height $\mathbf{h}$. Note that if $\mathbf{p} = \text{proj}_{\mathbf{x}}\mathbf{y}$, then $\mathbf{h} = \mathbf{y} - \mathbf{p}$.)

Figure 3.2

Parallelogram
determined by $\mathbf{x}$ and $\mathbf{y}$

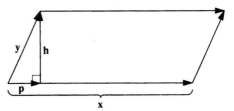

11. Calculate the volume of the parallelepiped in $\mathbb{R}^3$ determined by the following:

⋆(a) $\mathbf{x} = [-2, 3, 1],\ \mathbf{y} = [4, 2, 0],\ \mathbf{z} = [-1, 3, 2]$
(b) $\mathbf{x} = [1, 2, 3],\ \mathbf{y} = [0, -1, 0],\ \mathbf{z} = [4, -1, 5]$
⋆(c) $\mathbf{x} = [-3, 4, 0],\ \mathbf{y} = [6, -2, 1],\ \mathbf{z} = [0, -3, 3]$
(d) $\mathbf{x} = [1, 2, 0],\ \mathbf{y} = [3, 2, -1],\ \mathbf{z} = [5, -2, -1]$

▶12. Prove part (2) of Theorem 3.1. (Hint: See Figure 3.3. Let $\mathbf{h}$ be the perpendicular dropped from $\mathbf{z}$ to the plane of the parallelogram. From Exercise 8, $\mathbf{x} \times \mathbf{y}$ is perpendicular to both $\mathbf{x}$ and $\mathbf{y}$, and so $\mathbf{h}$ is actually the projection of $\mathbf{z}$ onto $\mathbf{x} \times \mathbf{y}$. Hence, the volume of the parallelepiped is the area of the parallelogram determined by $\mathbf{x}$ and $\mathbf{y}$ multiplied by the length of $\mathbf{h}$. A calculation similar to that in Exercise 10 shows that the area of the parallelogram is $\sqrt{(x_2y_3 - x_3y_2)^2 + (x_1y_3 - x_3y_1)^2 + (x_1y_2 - x_2y_1)^2}$.)

Figure 3.3

Parallelepiped
determined by $\mathbf{x}$, $\mathbf{y}$,
and $\mathbf{z}$

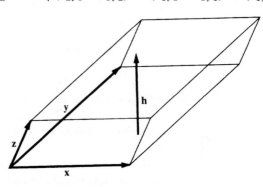

13. (a) If $\mathbf{A}$ is an $n \times n$ matrix, and c is a scalar, prove that $|c\mathbf{A}| = c^n |\mathbf{A}|$. (Hint: Use a proof by induction on n.)

(b) Use part (a) together with part (2) of Theorem 3.1 to explain why, when each side of a parallelepiped is doubled, the volume is multiplied by 8.

14. Show that, for $x \in \mathbb{R}$, $x^4 + a_3 x^3 + a_2 x^2 + a_1 x + a_0$ is the determinant of

$$\begin{bmatrix} x & -1 & 0 & 0 \\ 0 & x & -1 & 0 \\ 0 & 0 & x & -1 \\ a_0 & a_1 & a_2 & a_3 + x \end{bmatrix}.$$

15. Solve the following determinant equations for $x \in \mathbb{R}$:

★(a) $\begin{vmatrix} x & 2 \\ 5 & x+3 \end{vmatrix} = 0$ (b) $\begin{vmatrix} 15 & x-4 \\ x+7 & -2 \end{vmatrix} = 0$

★(c) $\begin{vmatrix} x-3 & 5 & -19 \\ 0 & x-1 & 6 \\ 0 & 0 & x-2 \end{vmatrix} = 0$

16. (a) Show that the determinant of the 3×3 **Vandermonde matrix**

$$\begin{bmatrix} 1 & 1 & 1 \\ a & b & c \\ a^2 & b^2 & c^2 \end{bmatrix}$$

is equal to $(a - b)(b - c)(c - a)$.

★(b) Using part (a), calculate the determinant of

$$\begin{bmatrix} 1 & 1 & 1 \\ 2 & 3 & -2 \\ 4 & 9 & 4 \end{bmatrix}.$$

17. The purpose of this exercise is to show that it is impossible to have an equilateral triangle whose three vertices all lie on **lattice points** in the plane; that is, points whose coordinates are both integers. Suppose T is such an equilateral triangle. Use the following steps to reach a contradiction:

(a) If s is the length of a side of T, use elementary geometry to find a formula for the area of T in terms of s.

(b) Use your answer for part (a) to show that the area of T is an irrational number. (You may assume $\sqrt{3}$ is irrational.)

(c) Suppose the three vertices of a triangle in the plane are given. Use part (1) of Theorem 3.1 to express the area of the triangle using a determinant.

(d) Use your answer for part (c) to show that the area of T is a rational number, thus contradicting part (b).

★**18.** True or False:

(a) The basketweaving technique can be used to find determinants of 3×3 and larger square matrices.

(b) The area of the parallelogram determined by nonparallel vectors $[x_1, x_2]$ and $[y_1, y_2]$ is $|x_1 y_2 - x_2 y_1|$.

(c) An $n \times n$ matrix has $2n$ associated cofactors.

(d) The cofactor $\mathcal{B}_{23}$ for a square matrix $\mathbf{B}$ equals the minor $|\mathbf{B}_{23}|$.

(e) The determinant of a 4×4 matrix $\mathbf{A}$ is $a_{41}\mathcal{A}_{41} + a_{42}\mathcal{A}_{42} + a_{43}\mathcal{A}_{43} + a_{44}\mathcal{A}_{44}$.

3.2 Determinants and Row Reduction

In this section, we provide a method for calculating the determinant of a matrix by using row reduction. For large matrices, this technique is computationally more efficient than cofactor expansion. We will also use the relationship between determinants and row reduction to establish a link between determinants and rank.

Determinants of Upper Triangular Matrices

We begin by proving the following simple formula for the determinant of an upper triangular matrix. Our goal will be to reduce every other determinant computation to this special case using row reduction.

THEOREM 3.2

Let $\mathbf{A}$ be an upper triangular $n \times n$ matrix. Then $|\mathbf{A}| = a_{11}a_{22} \cdots a_{nn}$, the product of the entries of $\mathbf{A}$ along the main diagonal.

Because we have defined the determinant recursively, we prove Theorem 3.2 by induction.

PROOF OF THEOREM 3.2

We use induction on n.

Base Step: $n = 1$. In this case, $\mathbf{A} = [a_{11}]$, and $|\mathbf{A}| = a_{11}$, which verifies the formula in the theorem.

Inductive Step: Let $n > 1$. Assume that for any upper triangular $(n-1) \times (n-1)$ matrix $\mathbf{B}$, $|\mathbf{B}| = b_{11}b_{22} \cdots b_{(n-1)(n-1)}$. We must prove that the formula given in the theorem holds for any $n \times n$ matrix $\mathbf{A}$.

Now, $|\mathbf{A}| = a_{n1}\mathcal{A}_{n1} + a_{n2}\mathcal{A}_{n2} + \cdots + a_{nn}\mathcal{A}_{nn} = 0\mathcal{A}_{n1} + 0\mathcal{A}_{n2} + \cdots + 0\mathcal{A}_{(n-1)(n-1)} + a_{nn}\mathcal{A}_{nn}$, because $a_{ni} = 0$ for $i < n$ since $\mathbf{A}$ is upper triangular. Thus, $|\mathbf{A}| = a_{nn}\mathcal{A}_{nn} = a_{nn}(-1)^{n+n}|\mathbf{A}_{nn}| = a_{nn}|\mathbf{A}_{nn}|$ (since $n + n$ is even). However, the $(n-1) \times (n-1)$ submatrix $\mathbf{A}_{nn}$ is itself an upper triangular matrix, since $\mathbf{A}$ is upper triangular. Thus, by the inductive hypothesis, $|\mathbf{A}_{nn}| = a_{11}a_{22} \cdots a_{(n-1)(n-1)}$. Hence, $|\mathbf{A}| = a_{nn}(a_{11}a_{22} \cdots a_{(n-1)(n-1)}) = a_{11}a_{22} \cdots a_{nn}$, completing the proof. ∎

EXAMPLE 1 By Theorem 3.2,

$$\begin{vmatrix} 4 & 2 & 0 & 1 \\ 0 & 3 & 9 & 6 \\ 0 & 0 & -1 & 5 \\ 0 & 0 & 0 & 7 \end{vmatrix} = (4)(3)(-1)(7) = -84.$$ ∎

As a special case of Theorem 3.2, notice that for all $n \geq 1$, we have $|\mathbf{I}_n| = 1$, since $\mathbf{I}_n$ is upper triangular with all its main diagonal entries equal to 1.

Effect of Row Operations on the Determinant

The following theorem describes explicitly how each type of row operation affects the determinant:

THEOREM 3.3

Let $\mathbf{A}$ be an $n \times n$ matrix, with determinant $|\mathbf{A}|$, and let c be a scalar.
(1) If R_1 is the row operation $\langle i \rangle \leftarrow c \langle i \rangle$ of type (I), then $|R_1(\mathbf{A})| = c|\mathbf{A}|$.
(2) If R_2 is the row operation $\langle j \rangle \leftarrow c \langle i \rangle + \langle j \rangle$ of type (II), then $|R_2(\mathbf{A})| = |\mathbf{A}|$.
(3) If R_3 is the row operation $\langle i \rangle \leftrightarrow \langle j \rangle$ of type (III), then $|R_3(\mathbf{A})| = -|\mathbf{A}|$.

All three parts of Theorem 3.3 are proved by induction. The proof of part (1) is easiest and is outlined in Exercise 8. Part (2) is easier to prove after part (3) is proven, and we outline the proof of part (2) in Exercises 9 and 10. The proof of part (3) is done by induction. Most of the proof of part (3) is given after the next example, except for one tedious case which has been placed in Appendix A.

EXAMPLE 2 Let

$$\mathbf{A} = \begin{bmatrix} 5 & -2 & 1 \\ 4 & 3 & -1 \\ 2 & 1 & 0 \end{bmatrix}.$$

You can quickly verify by the basketweaving method that $|\mathbf{A}| = 7$. Consider the following matrices:

$$\mathbf{B}_1 = \begin{bmatrix} 5 & -2 & 1 \\ 4 & 3 & -1 \\ -6 & -3 & 0 \end{bmatrix}, \quad \mathbf{B}_2 = \begin{bmatrix} 5 & -2 & 1 \\ 4 & 3 & -1 \\ 12 & -3 & 2 \end{bmatrix} \quad \text{and} \quad \mathbf{B}_3 = \begin{bmatrix} 4 & 3 & -1 \\ 5 & -2 & 1 \\ 2 & 1 & 0 \end{bmatrix}.$$

Now, $\mathbf{B}_1$ is obtained from $\mathbf{A}$ by the operation $\langle 3 \rangle \leftarrow -3 \langle 3 \rangle$ of type (I). Hence, part (1) of Theorem 3.3 asserts that $|\mathbf{B}_1| = -3|\mathbf{A}| = (-3)(7) = -21$.

Next, $\mathbf{B}_2$ is obtained from $\mathbf{A}$ by the operation $\langle 3 \rangle \leftarrow 2 \langle 1 \rangle + \langle 3 \rangle$ of type (II). By part (2) of Theorem 3.3, $|\mathbf{B}_2| = |\mathbf{A}| = 7$.

Finally, $\mathbf{B}_3$ is obtained from $\mathbf{A}$ by the operation $\langle 1 \rangle \leftrightarrow \langle 2 \rangle$ of type (III). Then, by part (3) of Theorem 3.3, $|\mathbf{B}_3| = -|\mathbf{A}| = -7$.

You can use basketweaving on $\mathbf{B}_1$, $\mathbf{B}_2$, and $\mathbf{B}_3$ to verify that the values given for their determinants are indeed correct. ∎

PROOF OF PART (3) OF THEOREM 3.3:

We proceed by induction on n. Notice that for $n = 1$, we cannot have a type (III) row operation, so $n = 2$ for the Base Step.

Base Step: $n = 2$. Then R must be the row operation $\langle 1 \rangle \leftrightarrow \langle 2 \rangle$, and

$$|R(\mathbf{A})| = \left| R \left(\begin{bmatrix} a_{11} & a_{12} \\ a_{21} & a_{22} \end{bmatrix} \right) \right| = \begin{vmatrix} a_{21} & a_{22} \\ a_{11} & a_{12} \end{vmatrix} = a_{21}a_{12} - a_{22}a_{11}$$
$$= -(a_{11}a_{22} - a_{12}a_{21}) = -|\mathbf{A}|.$$

Inductive Step: Assume $n \geq 3$, and that switching two rows of an $(n-1) \times (n-1)$ matrix results in a matrix whose determinant has the opposite sign. We consider three separate cases.

Case 1: Suppose R is the row operation $\langle i \rangle \leftrightarrow \langle j \rangle$, where $i \neq n$ and $j \neq n$. Let $\mathbf{B} = R(\mathbf{A})$. Then, since the last row of $\mathbf{A}$ is not changed, $b_{nk} = a_{nk}$, for $1 \leq k \leq n$. Also, $\mathbf{B}_{nk}$, the (n, k) submatrix of $\mathbf{B}$, equals $R(\mathbf{A}_{nk})$ (why?). Therefore, by the inductive hypothesis, $|\mathbf{B}_{nk}| = -|\mathbf{A}_{nk}|$, implying $\mathcal{B}_{nk} = (-1)^{n+k}|\mathbf{B}_{nk}| = (-1)^{n+k}(-1)|\mathbf{A}_{nk}| = -\mathcal{A}_{nk}$, for $1 \leq k \leq n$. Hence, $|\mathbf{B}| = b_{n1}\mathcal{B}_{n1} + \cdots + b_{nn}\mathcal{B}_{nn} = a_{n1}(-\mathcal{A}_{n1}) + \cdots + a_{nn}(-\mathcal{A}_{nn}) = -(a_{n1}\mathcal{A}_{n1} + \cdots + a_{nn}\mathcal{A}_{nn}) = -|\mathbf{A}|$.

Case 2: Suppose R is the row operation $\langle n-1 \rangle \leftrightarrow \langle n \rangle$, switching the last two rows. This case is proved by brute-force calculation, the details of which appear in Appendix A.

Case 3: Suppose R is the row operation $\langle i \rangle \leftrightarrow \langle n \rangle$, with $i \leq n - 2$. In this case, our strategy is to express R as a sequence of row swaps from the two previous cases. Let R_1 be the row operation $\langle i \rangle \leftrightarrow \langle n-1 \rangle$ and R_2 be the row operation $\langle n-1 \rangle \leftrightarrow \langle n \rangle$. Then $\mathbf{B} = R(\mathbf{A}) = R_1(R_2(R_1(\mathbf{A})))$ (why?). Using the previous two cases, we have $|\mathbf{B}| = |R(\mathbf{A})| = |R_1(R_2(R_1(\mathbf{A})))| = -|R_2(R_1(\mathbf{A}))| = (-1)^2|R_1(\mathbf{A})| = (-1)^3|\mathbf{A}| = -|\mathbf{A}|$.

This completes the proof. ∎

Theorem 3.3 can be used to prove that if a matrix $\mathbf{A}$ has a row with all entries zero, or has two identical rows, then $|\mathbf{A}| = 0$ (see Exercises 11 and 12).

Part (1) of Theorem 3.3 can be used to multiply each of the n rows of a matrix $\mathbf{A}$ by c in turn, thus proving the following corollary[1]:

COROLLARY 3.4

If $\mathbf{A}$ is an $n \times n$ matrix, and c is any scalar, then $|c\mathbf{A}| = c^n|\mathbf{A}|$.

| EXAMPLE 3 | A quick calculation shows that

$$\begin{vmatrix} 0 & 2 & 1 \\ 3 & -3 & -2 \\ 16 & 7 & 1 \end{vmatrix} = -1.$$

---

[1] You were also asked to prove this result in Exercise 13 of Section 3.1 directly from the definition of the determinant using induction.

Therefore,

$$\begin{vmatrix} 0 & -4 & -2 \\ -6 & 6 & 4 \\ -32 & -14 & -2 \end{vmatrix} = \begin{vmatrix} -2 \begin{bmatrix} 0 & 2 & 1 \\ 3 & -3 & -2 \\ 16 & 7 & 1 \end{bmatrix} \end{vmatrix} = (-2)^3 \begin{vmatrix} 0 & 2 & 1 \\ 3 & -3 & -2 \\ 16 & 7 & 1 \end{vmatrix}$$

$$= (-8)(-1) = 8. \qquad\blacksquare$$

Calculating the Determinant by Row Reduction

We will now illustrate how to use row operations to calculate the determinant of a given matrix $\mathbf{A}$ by finding an upper triangular matrix $\mathbf{B}$ that is row equivalent to $\mathbf{A}$.

EXAMPLE 4   Let

$$\mathbf{A} = \begin{bmatrix} 0 & -14 & -8 \\ 1 & 3 & 2 \\ -2 & 0 & 6 \end{bmatrix}.$$

We find $|\mathbf{A}|$ by row reducing $\mathbf{A}$ to produce an upper triangular form, as follows:

$$\mathbf{A} = \begin{bmatrix} 0 & -14 & -8 \\ 1 & 3 & 2 \\ -2 & 0 & 6 \end{bmatrix}$$

$$(\text{III}): \langle 1 \rangle \leftrightarrow \langle 2 \rangle \quad \Rightarrow \quad \mathbf{B}_1 = \begin{bmatrix} 1 & 3 & 2 \\ 0 & -14 & -8 \\ -2 & 0 & 6 \end{bmatrix}$$

$$(\text{II}): \langle 3 \rangle \leftarrow 2\langle 1 \rangle + \langle 3 \rangle \quad \Rightarrow \quad \mathbf{B}_2 = \begin{bmatrix} 1 & 3 & 2 \\ 0 & -14 & -8 \\ 0 & 6 & 10 \end{bmatrix}$$

$$(\text{I}): \langle 2 \rangle \leftarrow -\tfrac{1}{14}\langle 2 \rangle \quad \Rightarrow \quad \mathbf{B}_3 = \begin{bmatrix} 1 & 3 & 2 \\ 0 & 1 & \frac{4}{7} \\ 0 & 6 & 10 \end{bmatrix}$$

$$(\text{II}): \langle 3 \rangle \leftarrow -6\langle 2 \rangle + \langle 3 \rangle \quad \Rightarrow \quad \mathbf{B} = \begin{bmatrix} 1 & 3 & 2 \\ 0 & 1 & \frac{4}{7} \\ 0 & 0 & \frac{46}{7} \end{bmatrix}.$$

Because the last matrix $\mathbf{B}$ is in upper triangular form, we stop. (Notice that we do not target the entries above the main diagonal, as in reduced row echelon form.) From Theorem 3.2, $|\mathbf{B}| = (1)(1)\left(\frac{46}{7}\right) = \frac{46}{7}$. We now work "backward" through the preceding row operations to find $|\mathbf{A}|$. By Theorem 3.3, we have

$$\begin{aligned} |\mathbf{B}| &= |\mathbf{B}_3| & \text{type (II) row operation} \\ &= -\tfrac{1}{14}|\mathbf{B}_2| & \text{type (I) row operation} \end{aligned}$$

$$= -\tfrac{1}{14}\,|\mathbf{B}_1| \quad \text{type (II) row operation}$$

$$= \tfrac{1}{14}\,|\mathbf{A}| \quad \text{type (III) row operation.}$$

Thus, $|\mathbf{A}| = 14\,|\mathbf{B}| = 14\left(\tfrac{46}{7}\right) = 92.$ ∎

A more convenient method of calculating $|\mathbf{A}|$ is to create a variable P (for "product") with initial value 1, and update P appropriately as each row operation is performed. That is, we replace the current value of P by

$$\begin{cases} P \times (1/c) & \text{for type (I) row operations} \\ P \times (-1) & \text{for type (III) row operations} \end{cases}$$

Of course, row operations of type (II) do not affect the determinant.[2] Then, using the final value of P, we have $|\mathbf{A}| = P\,|\mathbf{B}|$, where $\mathbf{B}$ is the upper triangular result of the row reduction process. This method is illustrated in the next example.

EXAMPLE 5

Let us redo the calculation for $|\mathbf{A}|$ in Example 4. We create a variable P and initialize P to 1. Listed below are the row operations used in that example to convert $\mathbf{A}$ into upper triangular form $\mathbf{B}$, with $|\mathbf{B}| = \tfrac{46}{7}$. After each operation, we update the value of P accordingly.

Row Operation	Effect	P
(III): $\langle 1 \rangle \leftrightarrow \langle 2 \rangle$	Multiply P by -1	-1
(II): $\langle 3 \rangle \leftarrow 2\langle 1 \rangle + \langle 3 \rangle$	No change	-1
(I): $\langle 2 \rangle \leftarrow -\tfrac{1}{14}\langle 2 \rangle$	Multiply P by -14	14
(II): $\langle 3 \rangle \leftarrow -6\langle 2 \rangle + \langle 3 \rangle$	No change	14

Then $|\mathbf{A}|$ equals the final value of P times $|\mathbf{B}|$, so $|\mathbf{A}| = 14 \times \tfrac{46}{7} = 92.$ ∎

Determinant Criterion for Matrix Singularity

The next theorem gives an alternative way of determining whether the inverse of a given square matrix exists.

THEOREM 3.5

An $n \times n$ matrix $\mathbf{A}$ is nonsingular if and only if $|\mathbf{A}| \neq 0$.

PROOF OF THEOREM 3.5

Let $\mathbf{D}$ be the unique matrix in reduced row echelon form for $\mathbf{A}$. Now, using Theorem 3.3, we see that a single row operation of type (I), (II), or (III) cannot

---

[2] We multiply by $\tfrac{1}{c}$ (rather than c) when a row operation of type (I) is performed because as the row reduction goes forward from $\mathbf{A}$ to $\mathbf{B}$, the determinant calculation goes backward from $\mathbf{B}$ to $\mathbf{A}$.

convert a matrix having a nonzero determinant to a matrix having a zero determinant (why?). Because $\mathbf{A}$ is converted to $\mathbf{D}$ using a finite number of such row operations, Theorem 3.3 assures us that $|\mathbf{A}|$ and $|\mathbf{D}|$ are either both zero or both nonzero.

Now, if $\mathbf{A}$ is nonsingular (which implies $\mathbf{D} = \mathbf{I}_n$), we know that $|\mathbf{D}| = 1 \neq 0$ and therefore $|\mathbf{A}| \neq 0$, and we have completed half of the proof.

For the other half, assume that $|\mathbf{A}| \neq 0$. Then $|\mathbf{D}| \neq 0$. Because $\mathbf{D}$ is a square matrix with a staircase pattern of pivots, it is upper triangular. Because $|\mathbf{D}| \neq 0$, Theorem 3.2 asserts that all main diagonal entries of $\mathbf{D}$ are nonzero. Hence, they are all pivots, and $\mathbf{D} = \mathbf{I}_n$. Therefore, row reduction transforms $\mathbf{A}$ to $\mathbf{I}_n$, so $\mathbf{A}$ is nonsingular. ∎

Notice that Theorem 3.5 agrees with Theorem 2.13 in asserting that an inverse for $\begin{bmatrix} a & b \\ c & d \end{bmatrix}$ exists if and only if $\begin{vmatrix} a & b \\ c & d \end{vmatrix} = ad - bc \neq 0$.

Theorem 2.14 and Theorem 3.5 together imply the following:

COROLLARY 3.6

Let $\mathbf{A}$ be an $n \times n$ matrix. Then $\text{rank}(\mathbf{A}) = n$ if and only if $|\mathbf{A}| \neq 0$.

EXAMPLE 6

Consider the matrix $\mathbf{A} = \begin{bmatrix} 1 & 6 \\ -3 & 5 \end{bmatrix}$. Now, $|\mathbf{A}| = 23 \neq 0$. Hence, $\text{rank}(\mathbf{A}) = 2$ by Corollary 3.6. Also, because $\mathbf{A}$ is the coefficient matrix of the system

$$\begin{cases} x + 6y = 20 \\ -3x + 5y = 9 \end{cases}$$

and $|\mathbf{A}| \neq 0$, this system has a unique solution by Theorem 2.15. In fact, the solution is $(2, 3)$.

On the other hand, the matrix

$$\mathbf{B} = \begin{bmatrix} 1 & 5 & 1 \\ 2 & 1 & -7 \\ -1 & 2 & 6 \end{bmatrix}$$

has determinant zero. Thus, $\text{rank}(\mathbf{B}) < 3$. Also, because $\mathbf{B}$ is the coefficient matrix for the homogeneous system

$$\begin{cases} x_1 + 5x_2 + x_3 = 0 \\ 2x_1 + x_2 - 7x_3 = 0 \\ -x_1 + 2x_2 + 6x_3 = 0 \end{cases},$$

this system has nontrivial solutions by Theorem 2.5. You can verify that its solution set is $\{c(4, -1, 1) \mid c \in \mathbb{R}\}$. ∎

For reference, we summarize many of the results obtained in Chapters 2 and 3 in Table 3.1. You should be able to justify each equivalence in Table 3.1 by citing a relevant definition or result.

Table 3.1

Equivalent conditions for singular and nonsingular matrices

Assume that A is an $n \times n$ matrix. Then the following are all equivalent:	Assume that A is an $n \times n$ matrix. Then the following are all equivalent:
A is singular ($\mathbf{A}^{-1}$ does not exist).	A is nonsingular ($\mathbf{A}^{-1}$ exists).
Rank($\mathbf{A}$) $\neq n$.	Rank($\mathbf{A}$) $= n$.
$|\mathbf{A}| = 0$.	$|\mathbf{A}| \neq 0$.
A is not row equivalent to $\mathbf{I}_n$.	A is row equivalent to $\mathbf{I}_n$.
$\mathbf{AX} = \mathbf{O}$ has a nontrivial solution for **X**.	$\mathbf{AX} = \mathbf{O}$ has only the trivial solution for **X**.
$\mathbf{AX} = \mathbf{B}$ does not have a unique solution (no solutions or infinitely many solutions).	$\mathbf{AX} = \mathbf{B}$ has a unique solution for **X** (namely, $\mathbf{X} = \mathbf{A}^{-1}\mathbf{B}$).

Exercises for Section 3.2

1. Each of the following matrices is obtained from $\mathbf{I}_3$ by performing a single row operation of type (I), (II), or (III). Identify the operation, and use Theorem 3.3 to give the determinant of each matrix.

★(a) $\begin{bmatrix} 1 & -3 & 0 \\ 0 & 1 & 0 \\ 0 & 0 & 1 \end{bmatrix}$ (b) $\begin{bmatrix} 1 & 0 & 0 \\ 0 & 0 & 1 \\ 0 & 1 & 0 \end{bmatrix}$

★(c) $\begin{bmatrix} 1 & 0 & 0 \\ 0 & 1 & 0 \\ 0 & 0 & -4 \end{bmatrix}$ (d) $\begin{bmatrix} 1 & 0 & 0 \\ 2 & 1 & 0 \\ 0 & 0 & 1 \end{bmatrix}$

(e) $\begin{bmatrix} \frac{1}{2} & 0 & 0 \\ 0 & 1 & 0 \\ 0 & 0 & 1 \end{bmatrix}$ ★(f) $\begin{bmatrix} 0 & 1 & 0 \\ 1 & 0 & 0 \\ 0 & 0 & 1 \end{bmatrix}$

2. Calculate the determinant of each of the following matrices by using row reduction to produce an upper triangular form:

★(a) $\begin{bmatrix} 10 & 4 & 21 \\ 0 & -4 & 3 \\ -5 & -1 & -12 \end{bmatrix}$ (b) $\begin{bmatrix} 18 & -9 & -14 \\ 6 & -3 & -5 \\ -3 & 1 & 2 \end{bmatrix}$

★(c) $\begin{bmatrix} 1 & -1 & 5 & 1 \\ -2 & 1 & -7 & 1 \\ -3 & 2 & -12 & -2 \\ 2 & -1 & 9 & 1 \end{bmatrix}$ (d) $\begin{bmatrix} -8 & 4 & -3 & 2 \\ 2 & 1 & -1 & -1 \\ -3 & -5 & 4 & 0 \\ 2 & -4 & 3 & -1 \end{bmatrix}$

★(e) $\begin{bmatrix} 5 & 3 & -8 & 4 \\ \frac{15}{2} & \frac{1}{2} & -1 & -7 \\ -\frac{5}{2} & \frac{3}{2} & -4 & 1 \\ 10 & -3 & 8 & -8 \end{bmatrix}$ (f) $\begin{bmatrix} 1 & 2 & -1 & 3 & 0 \\ 2 & 4 & -3 & 1 & -4 \\ 2 & 6 & 4 & 8 & -4 \\ -3 & -8 & -1 & 1 & 0 \\ 1 & 3 & 3 & 10 & 1 \end{bmatrix}$

3. By calculating the determinant of each matrix, decide whether it is non-singular.

★(a) $\begin{bmatrix} 5 & 6 \\ -3 & -4 \end{bmatrix}$
(b) $\begin{bmatrix} \cos\theta & -\sin\theta \\ \sin\theta & \cos\theta \end{bmatrix}$

★(c) $\begin{bmatrix} -12 & 7 & -27 \\ 4 & -1 & 2 \\ 3 & 2 & -8 \end{bmatrix}$
(d) $\begin{bmatrix} 31 & -20 & 106 \\ -11 & 7 & -37 \\ -9 & 6 & -32 \end{bmatrix}$

4. By calculating the determinant of the coefficient matrix, decide whether each of the following homogeneous systems has a nontrivial solution. (You do not need to find the actual solutions.)

★(a) $\begin{cases} -6x + 3y - 22z = 0 \\ -7x + 4y - 31z = 0 \\ 11x - 6y + 46z = 0 \end{cases}$

(b) $\begin{cases} 4x_1 - x_2 + x_3 = 0 \\ -x_1 + x_2 - 2x_3 = 0 \\ -6x_1 + 9x_2 - 19x_3 = 0 \end{cases}$

(c) $\begin{cases} 2x_1 - 2x_2 + x_3 + 4x_4 = 0 \\ 4x_1 + 2x_2 + x_3 = 0 \\ -x_1 - x_2 - x_4 = 0 \\ -12x_1 - 7x_2 - 5x_3 + 2x_4 = 0 \end{cases}$

5. Let $\mathbf{A}$ be an upper triangular matrix. Prove that $|\mathbf{A}| \neq 0$ if and only if all the main diagonal elements of $\mathbf{A}$ are nonzero.

★6. Find the determinant of the following matrix:

$$\mathbf{A} = \begin{bmatrix} 0 & 0 & 0 & 0 & 0 & a_{16} \\ 0 & 0 & 0 & 0 & a_{25} & a_{26} \\ 0 & 0 & 0 & a_{34} & a_{35} & a_{36} \\ 0 & 0 & a_{43} & a_{44} & a_{45} & a_{46} \\ 0 & a_{52} & a_{53} & a_{54} & a_{55} & a_{56} \\ a_{61} & a_{62} & a_{63} & a_{64} & a_{65} & a_{66} \end{bmatrix}.$$

(Hint: Use part (3) of Theorem 3.3 and then Theorem 3.2.)

7. Suppose that $\mathbf{AB} = \mathbf{AC}$ and $|\mathbf{A}| \neq 0$. Show that $\mathbf{B} = \mathbf{C}$.

8. The purpose of this exercise is to outline a proof by induction of part (1) of Theorem 3.3. Let $\mathbf{A}$ be an $n \times n$ matrix, let R be the row operation $\langle i \rangle \leftarrow c \langle i \rangle$, and let $\mathbf{B} = R(\mathbf{A})$.

   (a) Prove $|\mathbf{B}| = c|\mathbf{A}|$ when $n = 1$. (This is the Base Step.)
   (b) State the inductive hypothesis for the Inductive Step.
   (c) Complete the Inductive Step for the case in which R is not performed on the last row of $\mathbf{A}$.
   (d) Complete the Inductive Step for the case in which R is performed on the last row of $\mathbf{A}$.

9. The purpose of this exercise and the next is to outline a proof by induction of part (2) of Theorem 3.3. This exercise completes the Base Step.

(a) Explain why $n \neq 1$ in this problem.

(b) Prove that applying the row operation $\langle 1 \rangle \leftarrow c \langle 2 \rangle + \langle 1 \rangle$ to a 2×2 matrix does not change the determinant.

(c) Repeat part (b) for the row operation $\langle 2 \rangle \leftarrow c \langle 1 \rangle + \langle 2 \rangle$.

10. The purpose of this exercise is to outline the Inductive Step in the proof of part (2) of Theorem 3.3. You may assume that part (3) of Theorem 3.3 has already been proved. Let **A** be an $n \times n$ matrix, for $n \geq 3$, and let R be the row operation $\langle i \rangle \leftarrow c \langle j \rangle + \langle i \rangle$.

   (a) State the inductive hypothesis and the statement to be proved for the Inductive Step. (Assume for size $n - 1$, and prove for size n.)

   (b) Prove the Inductive Step in the case where $i \neq n$ and $j \neq n$. (Your proof should be similar to that for Case 1 in the proof of part (3) of Theorem 3.3.)

   (c) Consider the case $i = n$. Suppose $k \neq j$ and $k \neq n$. Let R_1 be the row operation $\langle k \rangle \leftrightarrow \langle n \rangle$ and R_2 be the row operation $\langle k \rangle \leftarrow c \langle j \rangle + \langle k \rangle$. Prove that $R(\mathbf{A}) = R_1(R_2(R_1(\mathbf{A})))$.

   (d) Finish the proof of the Inductive Step for the case $i = n$. (Your proof should be similar to that for Case 3 in the proof of part (3) of Theorem 3.3.)

   (e) Finally, consider the case $j = n$. Suppose $k \neq i$ and $k \neq n$. Let R_1 be the row operation $\langle k \rangle \leftrightarrow \langle n \rangle$ and R_3 be the row operation $\langle i \rangle \leftarrow c \langle k \rangle + \langle i \rangle$. Prove that $R(\mathbf{A}) = R_1(R_3(R_1(\mathbf{A})))$.

   (f) Finish the proof of the Inductive Step for the case $j = n$.

11. Let **A** be an $n \times n$ matrix having an entire row of zeroes.

   (a) Use part (1) of Theorem 3.3 to prove that $|\mathbf{A}| = 0$.

   (b) Use Corollary 3.6 to provide an alternate proof that $|\mathbf{A}| = 0$.

12. Let **A** be an $n \times n$ matrix having two identical rows.

   (a) Use part (3) of Theorem 3.3 to prove that $|\mathbf{A}| = 0$.

   (b) Use Corollary 3.6 to provide an alternate proof that $|\mathbf{A}| = 0$.

13. Let **A** be an $n \times n$ matrix.

   (a) Show that if the entries of some row of **A** are proportional to those in another row, then $|\mathbf{A}| = 0$.

   (b) Show that if the entries in every row of **A** add up to zero, then $|\mathbf{A}| = 0$. (Hint: Consider the system $\mathbf{AX} = \mathbf{O}$, and note that the $n \times 1$ vector **X** having every entry equal to 1 is a nontrivial solution.)

14. (a) Use row reduction to show that the determinant of the $n \times n$ matrix symbolically represented by $\begin{bmatrix} \mathbf{A} & \mathbf{C} \\ \mathbf{O} & \mathbf{B} \end{bmatrix}$ is $|\mathbf{A}|\,|\mathbf{B}|$, where

   A is an $m \times m$ submatrix,
   B is an $(n - m) \times (n - m)$ submatrix,
   C is an $m \times (n - m)$ submatrix, and
   O is an $(n - m) \times m$ zero submatrix.

(b) Use part (a) to compute

$$\begin{vmatrix} -2 & 6 & 7 & -1 \\ 3 & -9 & 2 & -2 \\ 0 & 0 & 4 & -3 \\ 0 & 0 & -1 & 5 \end{vmatrix}.$$

15. Suppose that $f \colon \mathcal{M}_{nn} \to \mathbb{R}$ such that $f(\mathbf{I}_n) = 1$, and that whenever a single row operation is performed on $\mathbf{A} \in \mathcal{M}_{nn}$ to create $\mathbf{B}$,

$$f(\mathbf{B}) = \begin{cases} cf(\mathbf{A}) & \text{for a type (I) row operation with } c \neq 0 \\ f(\mathbf{A}) & \text{for a type (II) row operation} \\ -f(\mathbf{A}) & \text{for a type (III) row operation} \end{cases}.$$

Prove that $f(\mathbf{A}) = |\mathbf{A}|$, for all $\mathbf{A} \in \mathcal{M}_{nn}$. (Hint: If $\mathbf{A}$ is row equivalent to $\mathbf{I}_n$, then the given properties of f guarantee that $f(\mathbf{A}) = |\mathbf{A}|$ (why?). Otherwise, $\mathbf{A}$ is row equivalent to a matrix with a row of zeroes, and $|\mathbf{A}| = 0$. In this case, apply a type (I) operation with $c = -1$ to obtain $f(\mathbf{A}) = 0$.)

★16. True or False:

(a) The determinant of a square matrix is the product of its main diagonal entries.

(b) Two row operations of type (III) performed in succession have no overall effect on the determinant.

(c) If every row of a 4×4 matrix is multiplied by 3, the determinant is multiplied by 3 also.

(d) If two rows of a matrix $\mathbf{A}$ are identical, then $|\mathbf{A}| = 1$.

(e) A square matrix $\mathbf{A}$ is nonsingular if and only if $|\mathbf{A}| = 0$.

(f) An $n \times n$ matrix $\mathbf{A}$ has determinant zero if and only if rank$(\mathbf{A}) < n$.

3.3 Further Properties of the Determinant

In this section, we investigate the determinant of a product and the determinant of a transpose. We also introduce the classical adjoint of a matrix. Finally, we present Cramer's Rule, an alternative technique for solving certain linear systems using determinants.

Theorems 3.9, 3.10, 3.11 and 3.13 are not proven in this section. An interrelated progressive development of these proofs is left as Exercises 23 through 36.

Determinant of a Matrix Product

We begin by proving that the determinant of a product of two matrices $\mathbf{A}$ and $\mathbf{B}$ is equal to the product of their determinants $|\mathbf{A}|$ and $|\mathbf{B}|$.

THEOREM 3.7

If $\mathbf{A}$ and $\mathbf{B}$ are both $n \times n$ matrices, then $|\mathbf{AB}| = |\mathbf{A}| \, |\mathbf{B}|$.

PROOF OF THEOREM 3.7

First, suppose $\mathbf{A}$ is singular. Then $|\mathbf{A}| = 0$ by Theorem 3.5. If $|\mathbf{AB}| = 0$, then $|\mathbf{AB}| = |\mathbf{A}|\,|\mathbf{B}|$ and we will be done. We assume $|\mathbf{AB}| \neq 0$ and get a contradiction. If $|\mathbf{AB}| \neq 0$, $(\mathbf{AB})^{-1}$ exists, and $\mathbf{I}_n = \mathbf{AB}(\mathbf{AB})^{-1}$. Hence, $\mathbf{B}(\mathbf{AB})^{-1}$ is a right inverse for $\mathbf{A}$. But then by Theorem 2.9, $\mathbf{A}^{-1}$ exists, contradicting the fact that $\mathbf{A}$ is singular.

Now suppose $\mathbf{A}$ is nonsingular. In the special case where $\mathbf{A} = \mathbf{I}_n$, we have $|\mathbf{A}| = 1$ (why?), and so $|\mathbf{AB}| = |\mathbf{I}_n\mathbf{B}| = |\mathbf{B}| = 1|\mathbf{B}| = |\mathbf{A}|\,|\mathbf{B}|$. Finally, if $\mathbf{A}$ is any other nonsingular matrix, then $\mathbf{A}$ is row equivalent to $\mathbf{I}_n$, so there is a sequence $R_1, R_2, \ldots, R_k$ of row operations such that $R_k(\cdots(R_2(R_1(\mathbf{I}_n)))\cdots) = \mathbf{A}$. (These are the inverses of the row operations that row reduce $\mathbf{A}$ to $\mathbf{I}_n$.) Now, each row operation R_i has an associated real number r_i, so that applying R_i to a matrix multiplies its determinant by r_i (as in Theorem 3.3). Hence,

$$
\begin{aligned}
|\mathbf{AB}| &= |R_k(\cdots(R_2(R_1(\mathbf{I}_n)))\cdots)\mathbf{B}| \\
&= |R_k(\cdots(R_2(R_1(\mathbf{I}_n\mathbf{B})))\cdots)| && \text{by Theorem 2.1, part (2)} \\
&= r_k\cdots r_2 r_1 |\mathbf{I}_n\mathbf{B}| && \text{by Theorem 3.3} \\
&= r_k\cdots r_2 r_1 |\mathbf{I}_n||\mathbf{B}| && \text{by the } \mathbf{I}_n \text{ special case} \\
&= |R_k(\cdots(R_2(R_1(\mathbf{I}_n)))\cdots)||\mathbf{B}| && \text{by Theorem 3.3} \\
&= |\mathbf{A}|\,|\mathbf{B}| .
\end{aligned}
$$

$\blacksquare$

EXAMPLE 1 Let

$$
\mathbf{A} = \begin{bmatrix} 3 & 2 & 1 \\ 5 & 0 & -2 \\ -3 & 1 & 4 \end{bmatrix} \quad \text{and} \quad \mathbf{B} = \begin{bmatrix} 1 & -1 & 0 \\ 4 & 2 & -1 \\ -2 & 0 & 3 \end{bmatrix} .
$$

Quick calculations show that $|\mathbf{A}| = -17$ and $|\mathbf{B}| = 16$. Therefore, the determinant of

$$
\mathbf{AB} = \begin{bmatrix} 9 & 1 & 1 \\ 9 & -5 & -6 \\ -7 & 5 & 11 \end{bmatrix}
$$

is $|\mathbf{AB}| = |\mathbf{A}|\,|\mathbf{B}| = (-17)(16) = -272$.

$\blacksquare$

One consequence of Theorem 3.7 is that $|\mathbf{AB}| = 0$ if and only if either $|\mathbf{A}| = 0$ or $|\mathbf{B}| = 0$. (See Exercise 6(a).) Therefore, it follows that $\mathbf{AB}$ is singular if and only if either $\mathbf{A}$ or $\mathbf{B}$ is singular. Another important result is

---

COROLLARY 3.8

If $\mathbf{A}$ is nonsingular, then $\left|\mathbf{A}^{-1}\right| = \frac{1}{|\mathbf{A}|}$.

PROOF OF COROLLARY 3.8

If $\mathbf{A}$ is nonsingular, then $\mathbf{AA}^{-1} = \mathbf{I}_n$. By Theorem 3.7, $|\mathbf{A}||\mathbf{A}^{-1}| = |\mathbf{I}_n| = 1$, so $\left|\mathbf{A}^{-1}\right| = 1/|\mathbf{A}|$.

$\blacksquare$

Determinant of the Transpose

THEOREM 3.9

If $\mathbf{A}$ is an $n \times n$ matrix, then $|\mathbf{A}| = \left|\mathbf{A}^T\right|$.

See Exercises 23 through 31 for an outline of the proof of Theorem 3.9.

EXAMPLE 2   A quick calculation shows that if

$$\mathbf{A} = \begin{bmatrix} -1 & 4 & 1 \\ 2 & 0 & 3 \\ -1 & -1 & 2 \end{bmatrix},$$

then $|\mathbf{A}| = -33$. Hence, by Theorem 3.9,

$$|\mathbf{A}^T| = \begin{vmatrix} -1 & 2 & -1 \\ 4 & 0 & -1 \\ 1 & 3 & 2 \end{vmatrix} = -33. \qquad \blacksquare$$

Theorem 3.9 can be used to prove "column versions" of several earlier results involving determinants. For example, the determinant of a *lower* triangular matrix equals the product of its main diagonal entries, just as for an upper triangular matrix. Also, if a square matrix has an entire *column* of zeroes, or if it has two identical *columns*, then its determinant is zero, just as with rows.

Also, column operations analogous to the familiar row operations can be defined. For example, a type (I) column operation multiplies all entries of a given column of a matrix by a nonzero scalar. Theorem 3.9 can be combined with Theorem 3.3 to show that each type of column operation has the same effect on the determinant of a matrix as its corresponding row operation.

EXAMPLE 3   Let

$$\mathbf{A} = \begin{bmatrix} 2 & 5 & 1 \\ 1 & 2 & 3 \\ -3 & 1 & -1 \end{bmatrix}.$$

After the type (II) *column* operation $\langle \text{col. 2} \rangle \leftarrow -3 \langle \text{col. 1} \rangle + \langle \text{col. 2} \rangle$, we have

$$\mathbf{B} = \begin{bmatrix} 2 & -1 & 1 \\ 1 & -1 & 3 \\ -3 & 10 & -1 \end{bmatrix}.$$

A quick calculation checks that $|\mathbf{A}| = -43 = |\mathbf{B}|$. Thus, this column operation of type (II) has no effect on the determinant, as we would expect. $\blacksquare$

A More General Cofactor Expansion

Our definition of the determinant specifies that we multiply the elements a_{ni} of the last row of an $n \times n$ matrix $\mathbf{A}$ by their corresponding cofactors $\mathcal{A}_{ni}$, and sum the results. The next theorem shows the same result is obtained when a cofactor expansion is performed across *any* row or *any* column of the matrix!

THEOREM 3.10

Let $\mathbf{A}$ be an $n \times n$ matrix, with $n \geq 2$. Then,

(1) $a_{i1}\mathcal{A}_{i1} + a_{i2}\mathcal{A}_{i2} + \cdots + a_{in}\mathcal{A}_{in} = |\mathbf{A}|$, for each i, $1 \leq i \leq n$

(2) $a_{1j}\mathcal{A}_{1j} + a_{2j}\mathcal{A}_{2j} + \cdots + a_{nj}\mathcal{A}_{nj} = |\mathbf{A}|$, for each j, $1 \leq j \leq n$.

The formulas for $|\mathbf{A}|$ given in Theorem 3.10 are called the **cofactor expansion** (or, **Laplace expansion**) **along the** ith **row** (part (1)) **and** jth **column** (part (2)). An outline of the proof of this theorem is provided in Exercises 23 through 32. The proof that any row can be used, not simply the last row, is established by considering the effect of certain row swaps of the matrix. Then the $|\mathbf{A}| = |\mathbf{A}^T|$ formula explains why any column expansion is allowable.

EXAMPLE 4   Consider the matrix

$$\mathbf{A} = \begin{bmatrix} 5 & 0 & 1 & -2 \\ 2 & 2 & 3 & 1 \\ -1 & 3 & 2 & 5 \\ 6 & 0 & 1 & 1 \end{bmatrix}.$$

After some calculation, we find that the 16 cofactors of $\mathbf{A}$ are

$\mathcal{A}_{11} = -12$	$\mathcal{A}_{12} = -74$	$\mathcal{A}_{13} = 50$	$\mathcal{A}_{14} = 22$
$\mathcal{A}_{21} = 9$	$\mathcal{A}_{22} = 42$	$\mathcal{A}_{23} = -51$	$\mathcal{A}_{24} = -3$
$\mathcal{A}_{31} = -6$	$\mathcal{A}_{32} = -46$	$\mathcal{A}_{33} = 34$	$\mathcal{A}_{34} = 2$
$\mathcal{A}_{41} = -3$	$\mathcal{A}_{42} = 40$	$\mathcal{A}_{43} = -19$	$\mathcal{A}_{44} = -17$

We will use these values to compute $|\mathbf{A}|$ by a cofactor expansion across several different rows and columns of $\mathbf{A}$. Along the 2nd row, we have

$$\begin{aligned} |\mathbf{A}| &= a_{21}\mathcal{A}_{21} + a_{22}\mathcal{A}_{22} + a_{23}\mathcal{A}_{23} + a_{24}\mathcal{A}_{24} \\ &= 2(9) + 2(42) + 3(-51) + 1(-3) = -54. \end{aligned}$$

Along the 2nd column, we have

$$\begin{aligned} |\mathbf{A}| &= a_{12}\mathcal{A}_{12} + a_{22}\mathcal{A}_{22} + a_{32}\mathcal{A}_{32} + a_{42}\mathcal{A}_{42} \\ &= 0(-74) + 2(42) + 3(-46) + 0(40) = -54. \end{aligned}$$

Along the 4th column, we have

$$\begin{aligned} |\mathbf{A}| &= a_{14}\mathcal{A}_{14} + a_{24}\mathcal{A}_{24} + a_{34}\mathcal{A}_{34} + a_{44}\mathcal{A}_{44} \\ &= -2(22) + 1(-3) + 5(2) + 1(-17) = -54. \end{aligned}$$

■

Note in Example 4 that cofactor expansion is easiest along the second column because that column has two zeroes (entries a_{12} and a_{42}). In this case, only two cofactors, $\mathcal{A}_{22}$ and $\mathcal{A}_{32}$ were really needed to compute $|\mathbf{A}|$. We generally choose the row or column containing the largest number of zero entries for cofactor expansion.

The Adjoint Matrix

DEFINITION

Let $\mathbf{A}$ be an $n \times n$ matrix, with $n \geq 2$. The (**classical**) **adjoint** $\mathcal{A}$ of $\mathbf{A}$ is the $n \times n$ matrix whose (i, j) entry is $\mathcal{A}_{ji}$, the (j, i) cofactor of $\mathbf{A}$.

Notice that the (i, j) entry of the adjoint is not the cofactor $\mathcal{A}_{ij}$ of $\mathbf{A}$ but is $\mathcal{A}_{ji}$ instead. Hence, the general form of the adjoint of an $n \times n$ matrix $\mathbf{A}$ is

$$
\mathcal{A} = \begin{bmatrix}
\mathcal{A}_{11} & \mathcal{A}_{21} & \cdots & \mathcal{A}_{n1} \\
\mathcal{A}_{12} & \mathcal{A}_{22} & \cdots & \mathcal{A}_{n2} \\
\vdots & \vdots & \ddots & \vdots \\
\mathcal{A}_{1n} & \mathcal{A}_{2n} & \cdots & \mathcal{A}_{nn}
\end{bmatrix}.
$$

EXAMPLE 5 Recall the matrix

$$
\mathbf{A} = \begin{bmatrix}
5 & 0 & 1 & -2 \\
2 & 2 & 3 & 1 \\
-1 & 3 & 2 & 5 \\
6 & 0 & 1 & 1
\end{bmatrix}
$$

whose cofactors $\mathcal{A}_{ij}$ were given in Example 4. Grouping these cofactors into a matrix gives the adjoint matrix for $\mathbf{A}$.

$$
\mathcal{A} = \begin{bmatrix}
-12 & 9 & -6 & -3 \\
-74 & 42 & -46 & 40 \\
50 & -51 & 34 & -19 \\
22 & -3 & 2 & -17
\end{bmatrix}
$$

Note that the cofactors are "transposed;" that is, the cofactors for entries in the same *row* of $\mathbf{A}$ are placed in the same *column* of $\mathcal{A}$. ∎

The next theorem shows that the adjoint $\mathcal{A}$ of $\mathbf{A}$ is "almost" an inverse for $\mathbf{A}$.

THEOREM 3.11

If $\mathbf{A}$ is an $n \times n$ matrix with adjoint matrix $\mathcal{A}$, then

$$
\mathbf{A}\mathcal{A} = \mathcal{A}\mathbf{A} = (|\mathbf{A}|)\,\mathbf{I}_n.
$$

The fact that the diagonal entries of $\mathbf{A}\mathcal{A}$ and $\mathcal{A}\mathbf{A}$ equal $|\mathbf{A}|$ follows immediately from Theorem 3.10 (why?). The proof that the other entries of $\mathbf{A}\mathcal{A}$ and $\mathcal{A}\mathbf{A}$ equal zero is outlined in Exercises 23 through 35.

EXAMPLE 6   Using $\mathbf{A}$ and $\mathcal{A}$ from Example 5, we have

$$\mathbf{A}\mathcal{A} = \begin{bmatrix} 5 & 0 & 1 & -2 \\ 2 & 2 & 3 & 1 \\ -1 & 3 & 2 & 5 \\ 6 & 0 & 1 & 1 \end{bmatrix} \begin{bmatrix} -12 & 9 & -6 & -3 \\ -74 & 42 & -46 & 40 \\ 50 & -51 & 34 & -19 \\ 22 & -3 & 2 & -17 \end{bmatrix}$$

$$= \begin{bmatrix} -54 & 0 & 0 & 0 \\ 0 & -54 & 0 & 0 \\ 0 & 0 & -54 & 0 \\ 0 & 0 & 0 & -54 \end{bmatrix} = (-54)\mathbf{I}_4$$

(verify!), as predicted by Theorem 3.10, since $|\mathbf{A}| = -54$ (see Example 4). Similarly, you can check that $\mathcal{A}\mathbf{A} = (-54)\mathbf{I}_4$ as well.   ∎

Calculating Inverses with the Adjoint Matrix

If $|\mathbf{A}| \neq 0$ we can divide the equation in Theorem 3.11 by the scalar $|\mathbf{A}|$ to obtain $(1/|\mathbf{A}|)(\mathbf{A}\mathcal{A}) = \mathbf{I}_n$. But then, $\mathbf{A}((1/|\mathbf{A}|)\mathcal{A}) = \mathbf{I}_n$. Therefore, the scalar multiple $1/|\mathbf{A}|$ of the adjoint $\mathcal{A}$ must be the inverse matrix of $\mathbf{A}$, and we have proved

---

COROLLARY 3.12

If $\mathbf{A}$ is a nonsingular $n \times n$ matrix with adjoint $\mathcal{A}$, then $\mathbf{A}^{-1} = \left(\frac{1}{|\mathbf{A}|}\right)\mathcal{A}$.

This corollary gives an algebraic formula for the inverse of a matrix (when it exists).

EXAMPLE 7   The adjoint matrix for

$$\mathbf{B} = \begin{bmatrix} -2 & 0 & -3 \\ 0 & 1 & 0 \\ 0 & 0 & 4 \end{bmatrix} \quad \text{is} \quad \mathcal{B} = \begin{bmatrix} \mathcal{B}_{11} & \mathcal{B}_{21} & \mathcal{B}_{31} \\ \mathcal{B}_{12} & \mathcal{B}_{22} & \mathcal{B}_{32} \\ \mathcal{B}_{13} & \mathcal{B}_{23} & \mathcal{B}_{33} \end{bmatrix},$$

where each $\mathcal{B}_{ij}$ (for $1 \leq i, j \leq 3$) is the (i, j) cofactor of $\mathbf{B}$. But a quick computation of these cofactors (try it!) gives

$$\mathcal{B} = \begin{bmatrix} 4 & 0 & 3 \\ 0 & -8 & 0 \\ 0 & 0 & -2 \end{bmatrix}.$$

Now, $|\mathbf{B}| = -8$ (because $\mathbf{B}$ is upper triangular), and so

$$\mathbf{B}^{-1} = \frac{1}{|\mathbf{B}|}\mathcal{B} = -\frac{1}{8}\begin{bmatrix} 4 & 0 & 3 \\ 0 & -8 & 0 \\ 0 & 0 & -2 \end{bmatrix} = \begin{bmatrix} -\frac{1}{2} & 0 & -\frac{3}{8} \\ 0 & 1 & 0 \\ 0 & 0 & \frac{1}{4} \end{bmatrix}. \qquad \blacksquare$$

Finding the inverse by row reduction is usually quicker than using the adjoint. However, Corollary 3.12 is often useful for proving other results (see Exercise 19).

Cramer's Rule

We conclude this section by stating an explicit formula, known as **Cramer's Rule**, for the solution to a system of n equations and n variables when it is unique:

THEOREM 3.13 (Cramer's Rule)

Let $\mathbf{AX} = \mathbf{B}$ be a system of n equations in n variables with $|\mathbf{A}| \neq 0$. For $1 \leq i \leq n$, let $\mathbf{A}_i$ be the $n \times n$ matrix obtained by replacing the ith column of $\mathbf{A}$ with $\mathbf{B}$. Then the entries of the unique solution $\mathbf{X}$ are

$$x_1 = \frac{|\mathbf{A}_1|}{|\mathbf{A}|}, \quad x_2 = \frac{|\mathbf{A}_2|}{|\mathbf{A}|}, \quad \ldots, \quad x_n = \frac{|\mathbf{A}_n|}{|\mathbf{A}|}.$$

The proof of this theorem is outlined in Exercise 36. Cramer's Rule cannot be used for a system $\mathbf{AX} = \mathbf{B}$ in which $|\mathbf{A}| = 0$ (why?). It is frequently used on 3×3 systems having a unique solution, because the determinants involved can be calculated quickly by hand.

EXAMPLE 8 We will solve

$$\begin{cases} 5x_1 & - & 3x_2 & - & 10x_3 & = & -9 \\ 2x_1 & + & 2x_2 & - & 3x_3 & = & 4 \\ -3x_1 & - & x_2 & + & 5x_3 & = & -1 \end{cases}$$

using Cramer's Rule. This system is equivalent to $\mathbf{AX} = \mathbf{B}$ where

$$\mathbf{A} = \begin{bmatrix} 5 & -3 & -10 \\ 2 & 2 & -3 \\ -3 & -1 & 5 \end{bmatrix} \quad \text{and} \quad \mathbf{B} = \begin{bmatrix} -9 \\ 4 \\ -1 \end{bmatrix}.$$

A quick calculation shows that $|\mathbf{A}| = -2$. Let

$$\mathbf{A}_1 = \begin{bmatrix} -9 & -3 & -10 \\ 4 & 2 & -3 \\ -1 & -1 & 5 \end{bmatrix}, \; \mathbf{A}_2 = \begin{bmatrix} 5 & -9 & -10 \\ 2 & 4 & -3 \\ -3 & -1 & 5 \end{bmatrix}$$

and $\mathbf{A}_3 = \begin{bmatrix} 5 & -3 & -9 \\ 2 & 2 & 4 \\ -3 & -1 & -1 \end{bmatrix}$.

The matrix $\mathbf{A}_1$ is identical to $\mathbf{A}$, except in the first column, where its entries are taken from $\mathbf{B}$. $\mathbf{A}_2$ and $\mathbf{A}_3$ are created in an analogous manner. A quick computation shows that $|\mathbf{A}_1| = 8$, $|\mathbf{A}_2| = -6$, and $|\mathbf{A}_3| = 4$. Therefore,

$$x_1 = \frac{|\mathbf{A}_1|}{|\mathbf{A}|} = \frac{8}{-2} = -4, \quad x_2 = \frac{|\mathbf{A}_2|}{|\mathbf{A}|} = \frac{-6}{-2} = 3, \quad \text{and} \quad x_3 = \frac{|\mathbf{A}_3|}{|\mathbf{A}|} = \frac{4}{-2} = -2.$$

Hence, the unique solution to the given system is $(x_1, x_2, x_3) = (-4, 3, -2)$. ∎

Notice that solving the system in Example 8 essentially amounts to calculating four determinants: $|\mathbf{A}|$, $|\mathbf{A}_1|$, $|\mathbf{A}_2|$, and $|\mathbf{A}_3|$.

Exercises for Section 3.3

1. For a general 4×4 matrix $\mathbf{A}$, write out the formula for $|\mathbf{A}|$ using a cofactor expansion along the indicated row or column.
   ⋆(a) Third row   (b) First row
   ⋆(c) Fourth column   (d) First column

2. Find the determinant of each of the following matrices by performing a cofactor expansion along the indicated row or column:
   ⋆(a) Second row of $\begin{bmatrix} 2 & -1 & 4 \\ 0 & 3 & -2 \\ 5 & -2 & -3 \end{bmatrix}$

   (b) First row of $\begin{bmatrix} 10 & -2 & 7 \\ 3 & 2 & -8 \\ 6 & 5 & -2 \end{bmatrix}$

   ⋆(c) First column of $\begin{bmatrix} 4 & -2 & 3 \\ 5 & -1 & -2 \\ 3 & 3 & 2 \end{bmatrix}$

   (d) Third column of $\begin{bmatrix} 4 & -2 & 0 & -1 \\ -1 & 3 & -3 & 2 \\ 2 & 4 & -4 & -3 \\ 3 & 6 & 0 & -2 \end{bmatrix}$

3. Calculate the adjoint matrix for each of the following by finding the associated cofactor for each entry. Then use the adjoint to find the inverse of the original matrix (if it exists).

$$\star\text{(a)} \begin{bmatrix} 14 & -1 & -21 \\ 2 & 0 & -3 \\ 20 & -2 & -33 \end{bmatrix} \qquad \text{(b)} \begin{bmatrix} -15 & -6 & -2 \\ 5 & 3 & 2 \\ 5 & 6 & 5 \end{bmatrix}$$

$$\star\text{(c)} \begin{bmatrix} -2 & 1 & 0 & -1 \\ 7 & -4 & 1 & 4 \\ -14 & 11 & -2 & -8 \\ -12 & 10 & -2 & -7 \end{bmatrix} \qquad \text{(d)} \begin{bmatrix} -4 & 0 & 0 \\ -3 & 2 & 0 \\ 0 & 0 & 3 \end{bmatrix}$$

$$\star\text{(e)} \begin{bmatrix} 3 & -1 & 0 \\ 0 & -3 & 2 \\ 0 & 0 & -1 \end{bmatrix} \qquad \text{(f)} \begin{bmatrix} 2 & 1 & 0 & 0 \\ 0 & -1 & 1 & 0 \\ 0 & 0 & 1 & -1 \\ 0 & 0 & 0 & -2 \end{bmatrix}$$

4. Use Cramer's Rule to solve each of the following systems:

$$\star\text{(a)} \begin{cases} 3x_1 - x_2 - x_3 = -8 \\ 2x_1 - x_2 - 2x_3 = 3 \\ -9x_1 + x_2 = 39 \end{cases}$$

$$\text{(b)} \begin{cases} -2x_1 + 5x_2 - 4x_3 = -3 \\ 3x_1 - 3x_2 + 4x_3 = 6 \\ 2x_1 - x_2 + 2x_3 = 5 \end{cases}$$

$$\text{(c)} \begin{cases} -5x_1 + 6x_2 + 2x_3 = -16 \\ 3x_1 - 5x_2 - 3x_3 = 13 \\ -3x_1 + 3x_2 + x_3 = -11 \end{cases}$$

$$\text{*(d)} \begin{cases} -5x_1 + 2x_2 - 2x_3 + x_4 = -10 \\ 2x_1 - x_2 + 2x_3 - 2x_4 = -9 \\ 5x_1 - 2x_2 + 3x_3 - x_4 = 7 \\ -6x_1 + 2x_2 - 2x_3 + x_4 = -14 \end{cases}$$

5. Let $\mathbf{A}$ and $\mathbf{B}$ be $n \times n$ matrices.

(a) Show that $\mathbf{A}$ is nonsingular if and only if $\mathbf{A}^T$ is nonsingular.

(b) Show that $|\mathbf{AB}| = |\mathbf{BA}|$. (Remember that, in general, $\mathbf{AB} \neq \mathbf{BA}$.)

6. Let $\mathbf{A}$ and $\mathbf{B}$ be $n \times n$ matrices.

(a) Show that $|\mathbf{AB}| = 0$ if and only if $|\mathbf{A}| = 0$ or $|\mathbf{B}| = 0$.

(b) Show that if $\mathbf{AB} = -\mathbf{BA}$ and n is odd, then $\mathbf{A}$ or $\mathbf{B}$ is singular.

7. Let $\mathbf{A}$ and $\mathbf{B}$ be $n \times n$ matrices.

(a) Show that $\left|\mathbf{AA}^T\right| \geq 0$.

(b) Show that $\left|\mathbf{AB}^T\right| = \left|\mathbf{A}^T\right| |\mathbf{B}|$.

8. Let $\mathbf{A}$ be an $n \times n$ skew-symmetric matrix.

(a) If n is odd, show that $|\mathbf{A}| = 0$.

$\star$(b) If n is even, give an example where $|\mathbf{A}| \neq 0$.

9. An **orthogonal** matrix is a (square) matrix $\mathbf{A}$ with $\mathbf{A}^T = \mathbf{A}^{-1}$.

(a) Why is $\mathbf{I}_n$ orthogonal?

$\star$(b) Find a 3×3 orthogonal matrix other than $\mathbf{I}_3$.

(c) Show that $|\mathbf{A}| = \pm 1$ if $\mathbf{A}$ is orthogonal.

10. Show that there is no matrix $\mathbf{A}$ such that

$$\mathbf{A}^2 = \begin{bmatrix} 9 & 0 & -3 \\ 3 & 2 & -1 \\ -6 & 0 & 1 \end{bmatrix}.$$

11. Give a proof by induction in each case.

(a) **General form of Theorem 3.7:** Assuming Theorem 3.7, prove $|\mathbf{A}_1\mathbf{A}_2\cdots\mathbf{A}_k| = |\mathbf{A}_1|\,|\mathbf{A}_2|\cdots|\mathbf{A}_k|$ for any $n \times n$ matrices $\mathbf{A}_1, \mathbf{A}_2, \ldots, \mathbf{A}_k$.

(b) Prove $\left|\mathbf{A}^k\right| = |\mathbf{A}|^k$ for any $n \times n$ matrix $\mathbf{A}$ and any integer $k \geq 1$.

(c) Let $\mathbf{A}$ be an $n \times n$ matrix. Show that if $\mathbf{A}^k = \mathbf{O}_n$, for some integer $k \geq 1$, then $|\mathbf{A}| = 0$.

12. Suppose that $|\mathbf{A}|$ is an integer.

(a) Prove that $|\mathbf{A}^n|$ is not prime, for $n \geq 2$. (Recall that a **prime** number is an integer > 1 with no positive integer divisors except itself and 1.)

(b) Prove that if $\mathbf{A}^n = \mathbf{I}$, for some $n \geq 1$, n odd, then $|\mathbf{A}| = 1$.

13. We say that a matrix $\mathbf{B}$ is **similar** to a matrix $\mathbf{A}$ if there exists some (nonsingular) matrix $\mathbf{P}$ such that $\mathbf{P}\mathbf{A}\mathbf{P}^{-1} = \mathbf{B}$.

(a) Show that if $\mathbf{A}$ and $\mathbf{B}$ are similar, then they are both square matrices of the same size.

$\star$(b) Find two different matrices $\mathbf{B}$ similar to $\mathbf{A} = \begin{bmatrix} 1 & 2 \\ 3 & 4 \end{bmatrix}$.

(c) Show that if $\mathbf{B}$ is similar to $\mathbf{A}$, then $\mathbf{A}$ is similar to $\mathbf{B}$.

(d) Show that if $\mathbf{A}$ and $\mathbf{B}$ are similar, then $|\mathbf{A}| = |\mathbf{B}|$.

$\star$**14.** Let $\mathbf{A}$ and $\mathbf{B}$ be nonsingular matrices of the same size, with adjoints $\mathcal{A}$ and $\mathcal{B}$. Express $(\mathbf{A}\mathbf{B})^{-1}$ in terms of $\mathcal{A}$, $\mathcal{B}$, $|\mathbf{A}|$, and $|\mathbf{B}|$.

15. If all entries of a (square) matrix $\mathbf{A}$ are integers and $|\mathbf{A}| = \pm 1$, show that all entries of $\mathbf{A}^{-1}$ are integers.

16. If $\mathbf{A}$ is an $n \times n$ matrix with adjoint $\mathcal{A}$, show that $\mathbf{A}\mathcal{A} = \mathbf{O}_n$ if and only if $\mathbf{A}$ is singular.

17. Let $\mathbf{A}$ be an $n \times n$ matrix with adjoint $\mathcal{A}$.

(a) Show that the adjoint of $\mathbf{A}^T$ is $\mathcal{A}^T$.

(b) Show that the adjoint of $k\mathbf{A}$ is $k^{n-1}\mathcal{A}$, for any scalar k.

18. (a) Prove that if $\mathbf{A}$ is symmetric with adjoint matrix $\mathcal{A}$, then $\mathcal{A}$ is symmetric. (Hint: Show that the cofactors $\mathcal{A}_{ij}$ and $\mathcal{A}_{ji}$ of $\mathbf{A}$ are equal.)

$\star$(b) Give an example to show that part (a) is not necessarily true when "symmetric" is replaced by "skew-symmetric."

19. If $\mathbf{A}$ is nonsingular and upper triangular, show that $\mathbf{A}^{-1}$ is also upper triangular. (Hint: Use Corollary 3.12.)

20. Let $\mathbf{A}$ be an $n \times n$ matrix with adjoint $\mathcal{A}$.

(a) Prove that if $\mathbf{A}$ is singular, then $\mathcal{A}$ is singular. (Hint: Use Exercise 16 and a proof by contradiction.)

(b) Prove that $|\mathcal{A}| = |\mathbf{A}|^{n-1}$. (Hint: Consider the cases $|\mathbf{A}| = 0$ and $|\mathbf{A}| \neq 0$.)

21. Recall the 3×3 Vandermonde matrix from Exercise 16 of Section 3.1. For $n \geq 3$, the **general** $n \times n$ **Vandermonde matrix** is

$$\mathbf{V}_n = \begin{bmatrix} 1 & 1 & 1 & \cdots & 1 \\ x_1 & x_2 & x_3 & \cdots & x_n \\ x_1^2 & x_2^2 & x_3^2 & \cdots & x_n^2 \\ \vdots & \vdots & \vdots & \ddots & \vdots \\ x_1^{n-1} & x_2^{n-1} & x_3^{n-1} & \cdots & x_n^{n-1} \end{bmatrix}.$$

If $x_1, x_2, \ldots, x_n$ are distinct real numbers, show that

$$|\mathbf{V}_n| = (-1)^{n+1}(x_1 - x_n)(x_2 - x_n) \cdots (x_{n-1} - x_n)|\mathbf{V}_{n-1}|.$$

(Hint: Subtract the last column from every other column, and use cofactor expansion along the first row to show that $|\mathbf{V}_n|$ is equal or opposite to the determinant of a matrix $\mathbf{W}$ of size $(n-1) \times (n-1)$. Next, divide each column of $\mathbf{W}$ by the first element of that column, using the "column" version of part (1) of Theorem 3.3 to pull out the factors $x_1 - x_n, x_2 - x_n, \ldots, x_{n-1} - x_n$. (Note that $\left(x_1^k - x_n^k\right)/(x_1 - x_n) = x_1^{k-1} + x_1^{k-2}x_n + x_1^{k-3}x_n^2 + \cdots + x_1 x_n^{k-2} + x_n^{k-1}$.) Finally, create $|\mathbf{V}_{n-1}|$ from the resulting matrix by going through each row from 2 to n in *reverse order* and adding $-x_n$ times the previous row to it.)

★22. True or False:
  (a) If $\mathbf{A}$ is a nonsingular matrix, then $|\mathbf{A}^{-1}| = \frac{1}{|\mathbf{A}^T|}$.
  (b) If $\mathbf{A}$ is a 5×5 matrix, a cofactor expansion along the second row gives the same result as a cofactor expansion along the third column.
  (c) If $\mathbf{B}$ is obtained from a type (III) column operation on a square matrix $\mathbf{A}$, then $|\mathbf{B}| = |\mathbf{A}|$.
  (d) The (i, j) entry of the adjoint of $\mathbf{A}$ is $(-1)^{i+j}|\mathbf{A}_{ji}|$.
  (e) For every nonsingular matrix $\mathbf{A}$, we have $\mathbf{A}\mathcal{A} = \mathbf{I}$.
  (f) For the system $\begin{cases} 4x_1 - 2x_2 - x_3 = -6 \\ -3x_2 + 4x_3 = 5 \\ x_3 = 3 \end{cases}$, $x_2 = -\frac{1}{12} \begin{vmatrix} 4 & -6 & -1 \\ 0 & 5 & 4 \\ 0 & 3 & 1 \end{vmatrix}$.

Taken together, the remaining exercises outline the proofs of Theorems 3.9, 3.10, 3.11, and 3.13 but not in the order in which these theorems were stated. Almost every exercise in this group is dependent on those which precede it.

▶**23.** This exercise will prove part (1) of Theorem 3.10.
  (a) Show that if part (1) of Theorem 3.10 is true for some $i = k$ with $2 \leq k \leq n$, then it is also true for $i = k - 1$. (Hint: Let $\mathbf{B} = R(\mathbf{A})$, where R is the row operation $\langle k \rangle \leftrightarrow \langle k - 1 \rangle$. Show that $|\mathbf{B}_{kj}| = |\mathbf{A}_{(k-1)j}|$ for each j. Then apply part (1) of Theorem 3.10 along the kth row of $\mathbf{B}$.)
  (b) Use part (a) to complete the proof of part (1) of Theorem 3.10.

▶**24.** Let $\mathbf{A}$ be an $n \times n$ matrix. Prove that if $\mathbf{A}$ has two identical rows, then $|\mathbf{A}| = 0$. (This was also proven in Exercise 12 in Section 3.2.)

▶**25.** Let $\mathbf{A}$ be an $n \times n$ matrix. Prove that $a_{i1}\mathcal{A}_{j1} + a_{i2}\mathcal{A}_{j2} + \cdots + a_{in}\mathcal{A}_{jn} = 0$, for $i \neq j$, $1 \leq i, j \leq n$. (Hint: Form a new matrix $\mathbf{B}$, which has all entries equal to $\mathbf{A}$, except that both the ith and jth rows of $\mathbf{B}$ equal the ith row of $\mathbf{A}$. Show that the cofactor expansion along the jth row of $\mathbf{B}$ equals $a_{i1}\mathcal{A}_{j1} + a_{i2}\mathcal{A}_{j2} + \cdots + a_{in}\mathcal{A}_{jn}$. Then apply Exercises 23 and 24.)

▶**26.** Let $\mathbf{A}$ be an $n \times n$ matrix. Prove that $\mathbf{A}\mathcal{A} = (|\mathbf{A}|)\,\mathbf{I}_n$. (Hint: Use Exercises 23 and 25.)

▶**27.** Let $\mathbf{A}$ be a nonsingular $n \times n$ matrix. Prove that $\mathcal{A}\mathbf{A} = (|\mathbf{A}|)\,\mathbf{I}_n$. (Hint: Use Exercise 26 and Theorem 2.9.)

▶**28.** Prove part (2) of Theorem 3.10 if $\mathbf{A}$ is nonsingular. (Hint: Use Exercise 27.)

▶**29.** Let $\mathbf{A}$ be a singular $n \times n$ matrix. Prove that $|\mathbf{A}| = |\mathbf{A}^T|$. (Hint: Use a proof by contradiction to show $\mathbf{A}^T$ is also singular, and then use Theorem 3.5.)

▶**30.** Let $\mathbf{A}$ be an $n \times n$ matrix. Show that $(\mathbf{A}_{jm})^T = (\mathbf{A}^T)_{mj}$, for $1 \leq j, m \leq n$, where $(\mathbf{A}^T)_{mj}$ refers to the (m, j) submatrix of $\mathbf{A}^T$.

▶**31.** Let $\mathbf{A}$ be a nonsingular $n \times n$ matrix. Prove that $|\mathbf{A}| = |\mathbf{A}^T|$. (Hint: Note $\mathbf{A}^T$ is also nonsingular by part (4) of Theorem 2.11. Use induction on n. The Base Step ($n = 1$) is straightforward. For the Inductive Step, show that a cofactor expansion along the last column of $\mathbf{A}$ equals a cofactor expansion along the last row of $\mathbf{A}^T$. (Use Exercise 30 to obtain that each minor $|(\mathbf{A}^T)_{ni}| = |(\mathbf{A}_{in})^T|$, and then use either the inductive hypothesis or Exercise 29 to show $|(\mathbf{A}_{in})^T| = |\mathbf{A}_{in}|$.) Finally, note that a cofactor expansion along the last column of $\mathbf{A}$ equals $|\mathbf{A}|$ by Exercise 28.) (This exercise completes the proof of Theorem 3.9.)

▶**32.** Prove part (2) of Theorem 3.10 if $\mathbf{A}$ is singular. (Hint: Show that a cofactor expansion along the jth column of $\mathbf{A}$ is equal to a cofactor expansion along the jth row of $\mathbf{A}^T$. (Note that each $|\mathbf{A}_{kj}| = |(\mathbf{A}_{kj})^T|$ (from Exercises 29 and 31) $= |(\mathbf{A}^T)_{jk}|$ (by Exercise 30). Next, apply Exercise 23 to $\mathbf{A}^T$. Finally, use Exercise 29.) (This exercise completes the proof of Theorem 3.10.)

▶**33.** Let $\mathbf{A}$ be an $n \times n$ matrix. Prove that if $\mathbf{A}$ has two identical columns, then $|\mathbf{A}| = 0$. (Hint: Use Exercises 29 and 31 together with Exercise 24.)

▶**34.** Let $\mathbf{A}$ be an $n \times n$ matrix. Prove that $a_{1i}\mathcal{A}_{1j} + a_{2i}\mathcal{A}_{2j} + \cdots + a_{ni}\mathcal{A}_{nj} = 0$, for $i \neq j$, $1 \leq i, j \leq n$. (Hint: Use an argument similar to that in Exercise 25, but with columns instead of rows. Use Exercises 28 and 32 together with Exercise 33.)

▶**35.** Let $\mathbf{A}$ be a singular $n \times n$ matrix. Prove that $\mathcal{A}\mathbf{A} = (|\mathbf{A}|)\,\mathbf{I}_n$. (Hint: Use Exercises 32 and 34.) (This exercise completes the proof of Theorem 3.11.)

▶**36.** This exercise outlines the proof that Cramer's Rule (Theorem 3.13) is valid. We want to solve $\mathbf{AX} = \mathbf{B}$, where $\mathbf{A}$ is an $n \times n$ matrix with $|\mathbf{A}| \neq 0$. Assume $n \geq 2$ (since the case $n = 1$ is trivial).

(a) Show that $\mathbf{X} = (1/|\mathbf{A}|)(\mathcal{A}\mathbf{B})$.
(b) Prove that the kth entry of $\mathbf{X}$ is $(1/|\mathbf{A}|)(b_1\mathcal{A}_{1k} + \cdots + b_n\mathcal{A}_{nk})$.
(c) Prove that $|\mathbf{A}_k| = b_1\mathcal{A}_{1k} + \cdots + b_n\mathcal{A}_{nk}$, where $\mathbf{A}_k$ is defined as in the statement of Theorem 3.13. (Hint: Perform a cofactor expansion along the kth column of $\mathbf{A}_k$, and use part (2) of Theorem 3.10.)
(d) Explain how parts (b) and (c) together prove Theorem 3.13.

3.4 Eigenvalues and Diagonalization

In this section, we define eigenvalues and eigenvectors in the context of matrices, in order to find, when possible, a diagonal form for a square matrix. Some of the theoretical details involved cannot be discussed fully until we have introduced vector spaces and linear transformations, which are covered in Chapters 4 and 5. Thus, we will take a more comprehensive look at eigenvalues and eigenvectors at the end of Chapter 5, as well as in Chapters 6 and 7.

Eigenvalues and Eigenvectors

DEFINITION

Let $\mathbf{A}$ be an $n \times n$ matrix. A real number λ is an **eigenvalue** of $\mathbf{A}$ if and only if there is a nonzero n-vector $\mathbf{X}$ such that $\mathbf{A}\mathbf{X} = \lambda\mathbf{X}$. Also, any nonzero vector $\mathbf{X}$ for which $\mathbf{A}\mathbf{X} = \lambda\mathbf{X}$ is an **eigenvector** corresponding to the eigenvalue λ.

In some textbooks, eigenvalues are called **characteristic values** and eigenvectors are called **characteristic vectors**.

Notice that an eigenvalue can be zero. However, by definition, an eigenvector is never the zero vector.

If $\mathbf{X}$ is an eigenvector associated with an eigenvalue λ for an $n \times n$ matrix $\mathbf{A}$, then the matrix product $\mathbf{A}\mathbf{X}$ is equivalent to performing the scalar product $\lambda\mathbf{X}$. Thus, $\mathbf{A}\mathbf{X}$ is parallel to the vector $\mathbf{X}$, **dilating** (or lengthening) $\mathbf{X}$ if $|\lambda| > 1$ and **contracting** (or shortening) $\mathbf{X}$ if $|\lambda| < 1$. Of course, if $\lambda = 0$, then $\mathbf{A}\mathbf{X} = \mathbf{0}$.

EXAMPLE 1   Consider the 3×3 matrix

$$\mathbf{A} = \begin{bmatrix} -4 & 8 & -12 \\ 6 & -6 & 12 \\ 6 & -8 & 14 \end{bmatrix}.$$

Now, $\lambda = 2$ is an eigenvalue for $\mathbf{A}$ because a vector $\mathbf{X}$ exists such that $\mathbf{A}\mathbf{X} = 2\mathbf{X}$. In particular,

$$\mathbf{A}\begin{bmatrix} 4 \\ 3 \\ 0 \end{bmatrix} = \begin{bmatrix} -4 & 8 & -12 \\ 6 & -6 & 12 \\ 6 & -8 & 14 \end{bmatrix}\begin{bmatrix} 4 \\ 3 \\ 0 \end{bmatrix} = \begin{bmatrix} 8 \\ 6 \\ 0 \end{bmatrix} = 2\begin{bmatrix} 4 \\ 3 \\ 0 \end{bmatrix}.$$

Hence, $\mathbf{X} = [4, 3, 0]$ is an eigenvector corresponding to the eigenvalue 2. In fact, any nonzero scalar multiple c of $[4, 3, 0]$ is also an eigenvector corresponding to 2, because $\mathbf{A}(c\mathbf{X}) = c(\mathbf{AX}) = c(2\mathbf{X}) = 2(c\mathbf{X})$. Therefore, there are infinitely many eigenvectors corresponding to the eigenvalue $\lambda = 2$. ∎

DEFINITION

Let $\mathbf{A}$ be an $n \times n$ matrix and λ be an eigenvalue for $\mathbf{A}$. Then the set $E_\lambda = \{\mathbf{X} \mid \mathbf{AX} = \lambda\mathbf{X}\}$ is called the **eigenspace** of λ.

The eigenspace E_λ for a particular eigenvalue λ of $\mathbf{A}$ consists of the set of all eigenvectors for $\mathbf{A}$ associated with λ, together with the zero vector $\mathbf{0}$, since $\mathbf{A0} = \mathbf{0} = \lambda\mathbf{0}$, for any λ. Thus, for the matrix $\mathbf{A}$ in Example 1, E_2 contains (at least) all of the scalar multiples of $[4, 3, 0]$.

The Characteristic Polynomial of a Matrix

Our next goal is to find a method for determining the eigenvalues and eigenvectors of an $n \times n$ matrix $\mathbf{A}$. Now, if $\mathbf{X}$ is an eigenvector for $\mathbf{A}$ corresponding to the eigenvalue λ, then we have

$$\mathbf{AX} = \lambda\mathbf{X} = \lambda\mathbf{I}_n\mathbf{X}, \quad \text{or} \quad (\lambda\mathbf{I}_n - \mathbf{A})\mathbf{X} = \mathbf{0}.$$

Therefore, $\mathbf{X}$ is a nontrivial solution to the homogeneous system whose coefficient matrix is $\lambda\mathbf{I}_n - \mathbf{A}$. Theorem 2.5 and Corollary 3.6 then show that $|\lambda\mathbf{I}_n - \mathbf{A}| = 0$. Since all of the steps in this argument are reversible, we have proved

---

THEOREM 3.14

Let $\mathbf{A}$ be an $n \times n$ matrix and let λ be a real number. Then λ is an eigenvalue of $\mathbf{A}$ if and only if $|\lambda\mathbf{I}_n - \mathbf{A}| = 0$. The eigenvectors corresponding to λ are the nontrivial solutions of the homogeneous system $(\lambda\mathbf{I}_n - \mathbf{A})\mathbf{X} = \mathbf{0}$. The eigenspace E_λ is the complete solution set for this homogeneous system.

EXAMPLE 2   Recall the matrix

$$\mathbf{A} = \begin{bmatrix} -4 & 8 & -12 \\ 6 & -6 & 12 \\ 6 & -8 & 14 \end{bmatrix}$$

from Example 1. We discovered that $\lambda_1 = 2$ is an eigenvalue for $\mathbf{A}$. Notice that

$$|2\mathbf{I}_3 - \mathbf{A}| = \left| \begin{bmatrix} 2 & 0 & 0 \\ 0 & 2 & 0 \\ 0 & 0 & 2 \end{bmatrix} - \begin{bmatrix} -4 & 8 & -12 \\ 6 & -6 & 12 \\ 6 & -8 & 14 \end{bmatrix} \right| = \begin{vmatrix} 6 & -8 & 12 \\ -6 & 8 & -12 \\ -6 & 8 & -12 \end{vmatrix} = 0.$$

We can find all eigenvectors corresponding to $\lambda_1 = 2$ by row reducing

$$[2\mathbf{I}_3 - \mathbf{A} \mid \mathbf{0}] = \begin{bmatrix} 6 & -8 & 12 & 0 \\ -6 & 8 & -12 & 0 \\ -6 & 8 & -12 & 0 \end{bmatrix} \quad \text{to obtain} \quad \begin{bmatrix} 1 & -\frac{4}{3} & 2 & 0 \\ 0 & 0 & 0 & 0 \\ 0 & 0 & 0 & 0 \end{bmatrix}.$$

Using the method of Section 2.2 to express the solution set as a combinations of particular solutions, we obtain $E_2 = \{a[4, 3, 0] + v_1$ ..., $a, b \in \mathbb{R}\}$. Setting $a = 1$, $b = 0$ produces the eigenvector $\mathbf{X} = [4, 3, 0]$ from Example 1. However, with $a = 0$, $b = 1$, we also discover the eigenvector $\mathbf{Y} = [-2, 0, 1]$. You can verify that $\mathbf{AY} = 2\mathbf{Y}$. Also, any nontrivial linear combination of $\mathbf{X}$ and $\mathbf{Y}$ is also an eigenvector for $\mathbf{A}$ corresponding to λ (why?).

To find all the eigenvalues for $\mathbf{A}$ using Theorem 3.14, we set $|\lambda \mathbf{I}_3 - \mathbf{A}| = 0$ and solve for λ:

$$0 = \left| \begin{bmatrix} \lambda & 0 & 0 \\ 0 & \lambda & 0 \\ 0 & 0 & \lambda \end{bmatrix} - \begin{bmatrix} -4 & 8 & -12 \\ 6 & -6 & 12 \\ 6 & -8 & 14 \end{bmatrix} \right| = \begin{vmatrix} \lambda + 4 & -8 & 12 \\ -6 & \lambda + 6 & -12 \\ -6 & 8 & \lambda - 14 \end{vmatrix}.$$

After some simplification, this equation becomes $0 = \lambda^3 - 4\lambda^2 + 4\lambda = \lambda(\lambda - 2)^2$, which yields two solutions: $\lambda_1 = 2$, and $\lambda_2 = 0$. Thus, along with $\lambda_1 = 2$, we have now discovered another eigenvalue for $\mathbf{A}$, namely, $\lambda_2 = 0$.

We can find eigenvectors corresponding to λ_2 by row reducing

$$[0\mathbf{I}_3 - \mathbf{A} \,|\, \mathbf{0}] = \left[\begin{array}{ccc|c} 4 & -8 & 12 & 0 \\ -6 & 6 & -12 & 0 \\ -6 & 8 & -14 & 0 \end{array} \right] \quad \text{to obtain} \quad \left[\begin{array}{ccc|c} 1 & 0 & 1 & 0 \\ 0 & 1 & -1 & 0 \\ 0 & 0 & 0 & 0 \end{array} \right],$$

which has the solution set

$$E_0 = \{c[-1, 1, 1] \mid c \in \mathbb{R}\}.$$

Therefore, the eigenvectors for $\mathbf{A}$ corresponding to $\lambda_2 = 0$ are $\mathbf{Z} = [-1, 1, 1]$ as well as all nonzero scalar multiples of $\mathbf{Z}$. We can check this by noting that

$$\mathbf{AZ} = \begin{bmatrix} -4 & 8 & -12 \\ 6 & -6 & 12 \\ 6 & -8 & 14 \end{bmatrix} \begin{bmatrix} -1 \\ 1 \\ 1 \end{bmatrix} = \begin{bmatrix} 0 \\ 0 \\ 0 \end{bmatrix} = 0 \begin{bmatrix} -1 \\ 1 \\ 1 \end{bmatrix} = 0\mathbf{Z}. \quad \blacksquare$$

Because the determinant $|\lambda \mathbf{I}_n - \mathbf{A}|$ is useful for finding eigenvalues, we make the following definition:

DEFINITION

If $\mathbf{A}$ is an $n \times n$ matrix, then the **characteristic polynomial** of $\mathbf{A}$ is the polynomial $p_\mathbf{A}(x) = |x\mathbf{I}_n - \mathbf{A}|$.

It can be shown that if $\mathbf{A}$ is an $n \times n$ matrix, then $p_\mathbf{A}(x)$ is a polynomial of degree n (see Exercise 23). From calculus, we know that $p_\mathbf{A}(x)$ has at most n real roots. Now, using this terminology, we can rephrase the first assertion of Theorem 3.14 as

The eigenvalues of an $n \times n$ matrix $\mathbf{A}$ are precisely the real roots of the characteristic polynomial $p_\mathbf{A}(x)$.

EXAMPLE 3 The characteristic polynomial of $\mathbf{A} = \begin{bmatrix} 12 & -51 \\ 2 & -11 \end{bmatrix}$ is

$$
\begin{aligned}
p_{\mathbf{A}}(x) &= \left| \begin{bmatrix} x & 0 \\ 0 & x \end{bmatrix} - \begin{bmatrix} 12 & -51 \\ 2 & -11 \end{bmatrix} \right| = \begin{vmatrix} x-12 & 51 \\ -2 & x+11 \end{vmatrix} \\
&= (x-12)(x+11) + 102 \\
&= x^2 - x - 30 = (x-6)(x+5).
\end{aligned}
$$

Therefore, the eigenvalues of $\mathbf{A}$ are the solutions to $p_{\mathbf{A}}(x) = 0$, or, $\lambda_1 = 6$ and $\lambda_2 = -5$.

Similarly, the characteristic polynomial of

$$
\mathbf{B} = \begin{bmatrix} 7 & 1 & -1 \\ -11 & -3 & 2 \\ 18 & 2 & -4 \end{bmatrix} \quad \text{is} \quad p_{\mathbf{B}}(x) = \begin{vmatrix} x-7 & -1 & 1 \\ 11 & x+3 & -2 \\ -18 & -2 & x+4 \end{vmatrix},
$$

which simplifies to $p_{\mathbf{B}}(x) = x^3 - 12x - 16$, or, $p_{\mathbf{B}}(x) = (x+2)^2(x-4)$. Hence, $\alpha_1 = -2$ and $\alpha_2 = 4$ are the eigenvalues for $\mathbf{B}$. ∎

Calculating the characteristic polynomial of a 4×4 or larger matrix can be tedious. Computing the roots of the characteristic polynomial may also be difficult. Thus, in practice, you should use a calculator or computer with appropriate software to compute the eigenvalues of a matrix. Numerical techniques for finding eigenvalues without the characteristic polynomial are discussed in Section 9.3.

Diagonalization

One of the most important uses of eigenvalues and eigenvectors is in the diagonalization of matrices. Because diagonal matrices have such a simple structure, it is relatively easy to compute a matrix product when one of the matrices is diagonal. As we will see later, other important matrix computations are also easier when using diagonal matrices. Hence, if a given square matrix can be replaced by a corresponding diagonal matrix, it could greatly simplify computations involving the original matrix. Therefore, our next goal is to present a formal method for using eigenvalues and eigenvectors to find a diagonal form for a given square matrix, if possible. Before stating the method, we motivate it with an example.

EXAMPLE 4 Consider again the 3×3 matrix

$$
\mathbf{A} = \begin{bmatrix} -4 & 8 & -12 \\ 6 & -6 & 12 \\ 6 & -8 & 14 \end{bmatrix}.
$$

In Example 2, we found the eigenvalues $\lambda_1 = 2$ and $\lambda_2 = 0$ of **A**. We also found eigenvectors $\mathbf{X} = [4, 3, 0]$ and $\mathbf{Y} = [-2, 0, 1]$ for $\lambda_1 = 2$ and an eigenvector $\mathbf{Z} = [-1, 1, 1]$ for $\lambda_2 = 0$. We will use these three vectors as columns for a 3×3 matrix

$$\mathbf{P} = \begin{bmatrix} 4 & -2 & -1 \\ 3 & 0 & 1 \\ 0 & 1 & 1 \end{bmatrix}.$$

Now, $|\mathbf{P}| = -1$ (verify!), and so **P** is nonsingular. A quick calculation yields

$$\mathbf{P}^{-1} = \begin{bmatrix} 1 & -1 & 2 \\ 3 & -4 & 7 \\ -3 & 4 & -6 \end{bmatrix}.$$

We now use **A**, **P**, and $\mathbf{P}^{-1}$ to compute a diagonal matrix **D**.

$$\mathbf{D} = \mathbf{P}^{-1}\mathbf{A}\mathbf{P} = \begin{bmatrix} 1 & -1 & 2 \\ 3 & -4 & 7 \\ -3 & 4 & -6 \end{bmatrix} \begin{bmatrix} -4 & 8 & -12 \\ 6 & -6 & 12 \\ 6 & -8 & 14 \end{bmatrix} \begin{bmatrix} 4 & -2 & -1 \\ 3 & 0 & 1 \\ 0 & 1 & 1 \end{bmatrix} = \begin{bmatrix} 2 & 0 & 0 \\ 0 & 2 & 0 \\ 0 & 0 & 0 \end{bmatrix}.$$

Notice that each main diagonal entry d_{ii} of **D** is an eigenvalue having an associated eigenvector in the corresponding column of **P**. ■

The diagonalization process presented in Example 4 works for many matrices.

THEOREM 3.15

Let **A** and **P** be $n \times n$ matrices such that each column of **P** is an eigenvector for **A**. If **P** is nonsingular, then $\mathbf{D} = \mathbf{P}^{-1}\mathbf{A}\mathbf{P}$ is a diagonal matrix, with its ith main diagonal entry d_{ii} equal to the eigenvalue for the eigenvector which is the ith column of **P**.

The proof of Theorem 3.15 is not difficult, and we leave it, with hints, as Exercise 20. Note also that since $\mathbf{D} = \mathbf{P}^{-1}\mathbf{A}\mathbf{P}$, then $\mathbf{A} = \mathbf{P}\mathbf{D}\mathbf{P}^{-1}$ (why?).

Thus, the following technique can be used to diagonalize a matrix:

> **METHOD FOR DIAGONALIZING AN $n \times n$ MATRIX A (IF POSSIBLE) (DIAGONALIZATION METHOD)**
>
> **Step 1:** Calculate $p_{\mathbf{A}}(x) = |x\mathbf{I}_n - \mathbf{A}|$.
> **Step 2:** Find all real roots of $p_{\mathbf{A}}(x)$ (that is, all real solutions to $p_{\mathbf{A}}(x) = 0$). These are the eigenvalues $\lambda_1, \lambda_2, \lambda_3, \ldots, \lambda_k$ for $\mathbf{A}$.
> **Step 3:** For each eigenvalue λ_m in turn:
>   Row reduce the augmented matrix $[\lambda_m\mathbf{I}_n - \mathbf{A} \,|\, \mathbf{0}]$. Use the result to obtain as many solutions as possible of the homogeneous system $(\lambda_m\mathbf{I}_n - \mathbf{A}) = \mathbf{0}$ by setting each independent variable in turn equal to 1 and all other independent variables equal to 0. (Eliminate fractions, if desired, from these solutions by replacing them with nonzero scalar multiples.)
> **Step 4:** If after repeating Step 3 for each eigenvalue, you have a total of less than n eigenvectors for $\mathbf{A}$, then $\mathbf{A}$ cannot be put into diagonal form. Stop.
> **Step 5:** Otherwise, form a matrix $\mathbf{P}$ whose columns are these n eigenvectors. (This matrix $\mathbf{P}$ is nonsingular.)
> **Step 6:** To check your work, verify that $\mathbf{D} = \mathbf{P}^{-1}\mathbf{AP}$ is a diagonal matrix whose d_{ii} entry is the eigenvalue for the eigenvector which is the ith column of $\mathbf{P}$. Also note that $\mathbf{A} = \mathbf{PDP}^{-1}$.

The assertions in Step 4 that $\mathbf{A}$ cannot be diagonalized, and in Step 5 that $\mathbf{P}$ is nonsingular, will not be proved here, but will follow from results in Section 5.5.

EXAMPLE 5   Consider the 4×4 matrix

$$\mathbf{A} = \begin{bmatrix} -4 & 7 & 1 & 4 \\ 6 & -16 & -3 & -9 \\ 12 & -27 & -4 & -15 \\ -18 & 43 & 7 & 24 \end{bmatrix}.$$

Step 1: A lengthy calculation gives $p_{\mathbf{A}}(x) = x^4 - 3x^2 - 2x = x(x-2)(x+1)^2$.
Step 2: The eigenvalues of $\mathbf{A}$ are the roots of $p_{\mathbf{A}}(x)$, namely, $\lambda_1 = -1$, $\lambda_2 = 2$, and $\lambda_3 = 0$.
Step 3: We first compute eigenvectors for $\lambda_1 = -1$. Row reducing $[(-1)\mathbf{I}_4 - \mathbf{A} \,|\, \mathbf{0}]$ yields

$$\begin{bmatrix} 1 & 0 & 2 & 1 & | & 0 \\ 0 & 1 & 1 & 1 & | & 0 \\ 0 & 0 & 0 & 0 & | & 0 \\ 0 & 0 & 0 & 0 & | & 0 \end{bmatrix}.$$

Setting the first independent variable (corresponding to column 3) equal to 1 and the second independent variable (column 4) equal to 0 gives the solution $\mathbf{X}_1 = [-2, -1, 1, 0]$. Setting the second independent variable equal to 1 and the first independent variable equal to 0 gives the solution $\mathbf{X}_2 = [-1, -1, 0, 1]$.

Similarly, we row reduce $[2\mathbf{I}_4 - \mathbf{A} \,|\, \mathbf{0}]$ to obtain the eigenvector $\left[\frac{1}{6}, -\frac{1}{3}, -\frac{2}{3}, 1\right]$. We multiply this by 6 to avoid fractions, yielding the eigenvector $\mathbf{X}_3 = [1, -2, -4, 6]$. Finally, from $[0\mathbf{I}_4 - \mathbf{A} \,|\, \mathbf{0}]$, we obtain the eigenvector $\mathbf{X}_4 = [1, -3, -3, 7]$.

Step 4: We have produced 4 eigenvectors, so we proceed to Step 5.

Step 5: Let

$$\mathbf{P} = \begin{bmatrix} -2 & -1 & 1 & 1 \\ -1 & -1 & -2 & -3 \\ 1 & 0 & -4 & -3 \\ 0 & 1 & 6 & 7 \end{bmatrix},$$

the matrix whose columns are the eigenvectors $\mathbf{X}_1, \mathbf{X}_2, \mathbf{X}_3, \mathbf{X}_4$.

Step 6: Calculating $\mathbf{D} = \mathbf{P}^{-1}\mathbf{A}\mathbf{P}$, we verify that $\mathbf{D}$ is the diagonal matrix whose corresponding entries on the main diagonal are the eigenvalues $-1, -1, 2$, and 0, respectively. ∎

Notice in Example 5 that, for each eigenvalue λ, the set of eigenvectors found in Step 3 is not the entire eigenspace E_λ. Instead, it is a special set of particular solutions to the homogeneous system having properties that make the method work.

Theorem 3.15 requires a nonsingular matrix $\mathbf{P}$ whose columns are eigenvectors for $\mathbf{A}$, as in Examples 4 and 5. However, such a matrix $\mathbf{P}$ does not always exist in general. Thus, we have the following definition[3]:

DEFINITION

An $n \times n$ matrix $\mathbf{A}$ is **diagonalizable** if and only if there exists a nonsingular $n \times n$ matrix $\mathbf{P}$ such that $\mathbf{D} = \mathbf{P}^{-1}\mathbf{A}\mathbf{P}$ is diagonal.

Nondiagonalizable Matrices

In the next two examples, we illustrate some square matrices that are not diagonalizable.

EXAMPLE 6   Consider the matrix

$$\mathbf{B} = \begin{bmatrix} 7 & 1 & -1 \\ -11 & -3 & 2 \\ 18 & 2 & -4 \end{bmatrix}$$

from Example 3, where we found $p_\mathbf{B}(x) = (x+2)^2(x-4)$, thus giving us the eigenvalues $\alpha_1 = -2$ and $\alpha_2 = 4$. Using Step 3 of the Diagonalization Method produces the eigenvectors $[1, -7, 2]$ for $\alpha_1 = -2$, and $[1, -1, 2]$ for $\alpha_2 = 4$. Since the method yielded only two eigenvectors for this 3×3 matrix, $\mathbf{B}$ cannot be diagonalized. ∎

---

[3] Although not explicitly stated in the definition, it can be shown that if such a matrix $\mathbf{P}$ exists, then the columns of $\mathbf{P}$ are eigenvectors of $\mathbf{A}$ (see Exercise 21).

EXAMPLE 7   Consider the 2×2 matrix

$$\mathbf{A} = \left[\begin{array}{cc} \cos \theta & -\sin \theta \\ \sin \theta & \cos \theta \end{array} \right],$$

for some angle θ (in radians). In Chapter 5, we will see that if a 2-vector $\mathbf{X}$ has its initial point at the origin, then $\mathbf{AX}$ is the vector obtained by rotating $\mathbf{X}$ counterclockwise about the origin through an angle of θ radians. Now,

$$p_{\mathbf{A}}(x) = \left| \begin{array}{cc} (x - \cos \theta) & \sin \theta \\ -\sin \theta & (x - \cos \theta) \end{array} \right| = x^2 - (2 \cos \theta)x + 1.$$

Using the Quadratic Formula to solve for eigenvalues yields

$$\lambda = \frac{2 \cos \theta \pm \sqrt{4 \cos^2 \theta - 4}}{2} = \cos \theta \pm \sqrt{-\sin^2 \theta}.$$

Thus, there are no eigenvalues unless θ is an integral multiple of π. When there are no eigenvalues, there cannot be any eigenvectors, and so in most cases $\mathbf{A}$ cannot be diagonalized.

The lack of eigenvectors for $\mathbf{A}$ makes perfect sense geometrically. If we rotate a vector $\mathbf{X}$ beginning at the origin through an angle which is not a multiple of π radians, then the new vector $\mathbf{AX}$ points in a direction that is not parallel to $\mathbf{X}$. Thus, $\mathbf{AX}$ cannot be a scalar multiple of $\mathbf{X}$, and hence there are no eigenvalues. If θ is an even multiple of π, then $\mathbf{A} = \mathbf{I}_2$, and $\mathbf{X}$ is rotated into itself. Therefore, 1 is an eigenvalue. (Here, $\mathbf{AX} = +1\mathbf{X}$.) If θ is an odd multiple of π, then $\mathbf{AX}$ is in the opposite direction as $\mathbf{X}$, so -1 is an eigenvalue. (Here, $\mathbf{AX} = -1\mathbf{X}$.)   ∎

Algebraic Multiplicity of an Eigenvalue

DEFINITION

Let $\mathbf{A}$ be an $n \times n$ matrix, and let λ be an eigenvalue for $\mathbf{A}$. Suppose that $(x - \lambda)^k$ is the highest power of $(x - \lambda)$ that divides $p_{\mathbf{A}}(x)$. Then k is called the **algebraic multiplicity of** λ.

EXAMPLE 8   Recall the matrix $\mathbf{A}$ in Example 5 whose characteristic polynomial is $p_{\mathbf{A}}(x) = x(x - 2)(x + 1)^2$. The algebraic multiplicity of $\lambda_1 = -1$ is 2 (because the factor $(x+1)$ appears to the second power in $p_{\mathbf{A}}(x)$), while the algebraic multiplicities of $\lambda_2 = 2$ and $\lambda_3 = 0$ are both 1.   ∎

Note that in Example 8, the algebraic multiplicity of each eigenvalue agrees with the number of eigenvectors produced for that eigenvalue in Example 5 by the Diagonalization Method. In Chapter 5, we will prove results that imply that, for any eigenvalue, the number of eigenvectors produced by Step 3 of the Diagonalization Method is always *less than or equal to* its algebraic multiplicity.

EXAMPLE 9 Recall the matrix **B** from Example 3 with $p_\mathbf{B}(x) = (x + 2)^2(x - 4)$. The eigenvalue $\alpha_1 = -2$ for **B** has algebraic multiplicity 2 because the factor $(x + 2)$ appears to the second power in $p_\mathbf{B}(x)$. By the remark just before this example, we know the Diagonalization Method must produce two or fewer eigenvectors for $\alpha_1 = -2$. In fact, in Example 6, we found that Step 3 of the Diagonalization Method applied to this matrix **B** produced only one eigenvector for $\alpha_1 = -2$. ■

EXAMPLE 10 Consider the 3×3 matrix

$$\mathbf{A} = \begin{bmatrix} -3 & -1 & -2 \\ -2 & 16 & -18 \\ 2 & 9 & -7 \end{bmatrix},$$

for which $p_\mathbf{A}(x) = |x\mathbf{I}_3 - \mathbf{A}| = x^3 - 6x^2 + 25x = x(x^2 - 6x + 25)$ (verify!). Since $x^2 - 6x + 25$ has no real solutions (try the Quadratic Formula!), **A** has only 1 eigenvalue, $\lambda = 0$, which has algebraic multiplicity 1. Thus, Step 3 of the Diagonalization Method can produce only one eigenvector for λ, and hence a total of only 1 eigenvector overall. Hence, according to Step 4, **A** cannot be diagonalized. ■

We have already seen that if the Diagonalization Method does not produce n eigenvectors overall for an $n \times n$ matrix **A**, then **A** is not diagonalizable, as in Example 6. However, Example 10 illustrates that if the sum of the algebraic multiplicities of all the eigenvalues for an $n \times n$ matrix **A** is less than n, then there is no need to proceed beyond Step 2 of the Diagonalization Method, since we are assured that Step 3 can not produce a sufficient number of eigenvectors, and so **A** cannot be diagonalized.

Application: Large Powers of a Matrix

If **D** is a diagonal matrix, any positive integer power of **D** can be obtained by merely raising the diagonal entries of **D** to that power (why?). For example,

$$\begin{bmatrix} 3 & 0 \\ 0 & -2 \end{bmatrix}^{12} = \begin{bmatrix} 3^{12} & 0 \\ 0 & (-2)^{12} \end{bmatrix} = \begin{bmatrix} 531441 & 0 \\ 0 & 4096 \end{bmatrix}.$$

Now, suppose that **A** and **P** are $n \times n$ matrices such that $\mathbf{P}^{-1}\mathbf{A}\mathbf{P} = \mathbf{D}$, a diagonal matrix. We know $\mathbf{A} = \mathbf{P}\mathbf{D}\mathbf{P}^{-1}$. But then,

$$\mathbf{A}^2 = \mathbf{A}\mathbf{A} = \left(\mathbf{P}\mathbf{D}\mathbf{P}^{-1}\right)\left(\mathbf{P}\mathbf{D}\mathbf{P}^{-1}\right) = \mathbf{P}\mathbf{D}\left(\mathbf{P}^{-1}\mathbf{P}\right)\mathbf{D}\mathbf{P}^{-1} = \mathbf{P}\mathbf{D}\mathbf{I}_n\mathbf{D}\mathbf{P}^{-1} = \mathbf{P}\mathbf{D}^2\mathbf{P}^{-1}$$

More generally, a straightforward proof by induction can be used to show that for all positive integers k, $\mathbf{A}^k = \mathbf{P}\mathbf{D}^k\mathbf{P}^{-1}$ (see Exercise 15). Hence, calculating positive powers of **A** is relatively easy if the corresponding matrices **P** and **D** are known.

EXAMPLE 11   We will use eigenvalues and eigenvectors to compute $\mathbf{A}^{11}$ for the matrix

$$\mathbf{A} = \begin{bmatrix} -4 & 7 & 1 & 4 \\ 6 & -16 & -3 & -9 \\ 12 & -27 & -4 & -15 \\ -18 & 43 & 7 & 24 \end{bmatrix}$$

in Example 5. Recall that in that example, we found $\mathbf{A} = \mathbf{PDP}^{-1}$ with

$$\mathbf{P} = \begin{bmatrix} -2 & -1 & 1 & 1 \\ -1 & -1 & -2 & -3 \\ 1 & 0 & -4 & -3 \\ 0 & 1 & 6 & 7 \end{bmatrix} \quad \text{and} \quad \mathbf{D} = \begin{bmatrix} -1 & 0 & 0 & 0 \\ 0 & -1 & 0 & 0 \\ 0 & 0 & 2 & 0 \\ 0 & 0 & 0 & 0 \end{bmatrix}.$$

Then,

$$\mathbf{A}^{11} = \mathbf{PD}^{11}\mathbf{P}^{-1}$$

$$= \begin{bmatrix} -2 & -1 & 1 & 1 \\ -1 & -1 & -2 & -3 \\ 1 & 0 & -4 & -3 \\ 0 & 1 & 6 & 7 \end{bmatrix} \begin{bmatrix} -1 & 0 & 0 & 0 \\ 0 & -1 & 0 & 0 \\ 0 & 0 & 2048 & 0 \\ 0 & 0 & 0 & 0 \end{bmatrix} \begin{bmatrix} -4 & 11 & 4 & 7 \\ 6 & -19 & -7 & -12 \\ -1 & 2 & 0 & 1 \\ 0 & 1 & 1 & 1 \end{bmatrix}$$

$$= \begin{bmatrix} -2050 & 4099 & 1 & 2050 \\ 4098 & -8200 & -3 & -4101 \\ 8196 & -16395 & -4 & -8199 \\ -12294 & 24595 & 7 & 12300 \end{bmatrix}.$$

∎

The technique illustrated in Example 11 can also be adapted to calculate square roots and cube roots of matrices (see Exercises 7 and 8).

♦ **Supplemental Material:** You have now covered the prerequisites for Section 7.2, "Complex Eigenvalues and Complex Eigenvectors."

Exercises for Section 3.4

1. Find the characteristic polynomial of each given matrix. (Hint: For part (e), do a cofactor expansion along the third row.)

★(a) $\begin{bmatrix} 3 & 1 \\ -2 & 4 \end{bmatrix}$

(b) $\begin{bmatrix} 2 & 5 & 8 \\ 0 & -1 & 9 \\ 0 & 0 & 5 \end{bmatrix}$

★(c) $\begin{bmatrix} 2 & 1 & -1 \\ -6 & 6 & 0 \\ 3 & 0 & 0 \end{bmatrix}$

(d) $\begin{bmatrix} 5 & 1 & 4 \\ 1 & 2 & 3 \\ 3 & -1 & 1 \end{bmatrix}$

★(e) $\begin{bmatrix} 0 & -1 & 0 & 1 \\ -5 & 2 & -1 & 2 \\ 0 & 1 & 1 & 0 \\ 4 & -1 & 3 & 0 \end{bmatrix}$

2. Solve for the eigenspace E_λ corresponding to the given eigenvalue λ for each of the following matrices. Express E_λ as a set of linear combinations of particular eigenvectors, as in Example 2.

★(a) $\begin{bmatrix} 1 & 1 \\ -2 & 4 \end{bmatrix}$, $\lambda = 2$ (b) $\begin{bmatrix} 1 & -1 & -1 \\ 1 & 3 & 2 \\ -3 & -3 & -2 \end{bmatrix}$, $\lambda = 2$

★(c) $\begin{bmatrix} -5 & 2 & 0 \\ -8 & 3 & 0 \\ 4 & -2 & -1 \end{bmatrix}$, $\lambda = -1$

3. Find all eigenvalues corresponding to each given matrix and their corresponding algebraic multiplicities. Also, express each eigenspace as a set of linear combinations of particular eigenvectors, as in Example 2.

★(a) $\begin{bmatrix} 1 & 3 \\ 0 & 1 \end{bmatrix}$ (b) $\begin{bmatrix} 2 & -1 \\ 0 & 3 \end{bmatrix}$

★(c) $\begin{bmatrix} 1 & 0 & 1 \\ 0 & 2 & -3 \\ 0 & 0 & -5 \end{bmatrix}$ (d) $\begin{bmatrix} 8 & -21 \\ 3 & -8 \end{bmatrix}$

★(e) $\begin{bmatrix} 4 & 0 & -2 \\ 6 & 2 & -6 \\ 4 & 0 & -2 \end{bmatrix}$ (f) $\begin{bmatrix} 3 & 4 & 12 \\ 4 & -12 & 3 \\ 12 & 3 & -4 \end{bmatrix}$

(g) $\begin{bmatrix} 2 & 1 & -2 & -4 \\ -2 & -4 & 4 & 10 \\ 3 & 4 & -5 & -12 \\ -2 & -3 & 4 & 9 \end{bmatrix}$ ★(h) $\begin{bmatrix} 3 & -1 & 4 & -1 \\ 0 & 3 & -3 & 3 \\ -6 & 2 & -8 & 2 \\ -6 & -4 & -2 & -4 \end{bmatrix}$

4. Use the Diagonalization Method to determine whether each of the following matrices is diagonalizable. If so, specify the matrices **D** and **P** and check your work by verifying that $\mathbf{D} = \mathbf{P}^{-1}\mathbf{A}\mathbf{P}$.

★(a) $\mathbf{A} = \begin{bmatrix} 19 & -48 \\ 8 & -21 \end{bmatrix}$ (b) $\mathbf{A} = \begin{bmatrix} -18 & 40 \\ -8 & 18 \end{bmatrix}$

★(c) $\mathbf{A} = \begin{bmatrix} 13 & -34 \\ 5 & -13 \end{bmatrix}$ ★(d) $\mathbf{A} = \begin{bmatrix} -13 & -3 & 18 \\ -20 & -4 & 26 \\ -14 & -3 & 19 \end{bmatrix}$

(e) $\mathbf{A} = \begin{bmatrix} -3 & 3 & -1 \\ 2 & 2 & 4 \\ 6 & -3 & 4 \end{bmatrix}$ ★(f) $\mathbf{A} = \begin{bmatrix} 5 & -8 & -12 \\ -2 & 3 & 4 \\ 4 & -6 & -9 \end{bmatrix}$

★(g) $\mathbf{A} = \begin{bmatrix} 2 & 0 & 0 \\ -3 & 4 & 1 \\ 3 & -2 & 1 \end{bmatrix}$ (h) $\mathbf{A} = \begin{bmatrix} -5 & 18 & 6 \\ -2 & 7 & 2 \\ 1 & -3 & 0 \end{bmatrix}$

★(i) $\mathbf{A} = \begin{bmatrix} 3 & 1 & -6 & -2 \\ 4 & 0 & -6 & -4 \\ 2 & 0 & -3 & -2 \\ 0 & 1 & -2 & 1 \end{bmatrix}$

5. Use diagonalization to calculate the indicated powers of $\mathbf{A}$ in each case.

   ⋆(a) $\mathbf{A}^{15}$, where $\mathbf{A} = \begin{bmatrix} 4 & -6 \\ 3 & -5 \end{bmatrix}$

   (b) $\mathbf{A}^{30}$, where $\mathbf{A} = \begin{bmatrix} 11 & -6 & -12 \\ 13 & -6 & -16 \\ 5 & -3 & -5 \end{bmatrix}$

   ⋆(c) $\mathbf{A}^{49}$, where $\mathbf{A}$ is the matrix of part (b)

   (d) $\mathbf{A}^{11}$, where $\mathbf{A} = \begin{bmatrix} 4 & -4 & 6 \\ -1 & 2 & -1 \\ -1 & 4 & -3 \end{bmatrix}$

   ⋆(e) $\mathbf{A}^{10}$, where $\mathbf{A} = \begin{bmatrix} 7 & 9 & -12 \\ 10 & 16 & -22 \\ 8 & 12 & -16 \end{bmatrix}$

6. Prove that if $\mathbf{D} = \mathbf{P}^{-1}\mathbf{A}\mathbf{P}$, then $\mathbf{A} = \mathbf{P}\mathbf{D}\mathbf{P}^{-1}$ for matrices $\mathbf{A}$, $\mathbf{D}$, and $\mathbf{P}$.

7. Let $\mathbf{A}$ be a diagonalizable $n \times n$ matrix.

   (a) Show that $\mathbf{A}$ has a cube root — that is, that there is a matrix $\mathbf{B}$ such that $\mathbf{B}^3 = \mathbf{A}$.

   ⋆(b) Give a sufficient condition for $\mathbf{A}$ to have a square root. Prove that your condition is valid.

⋆8. Find a matrix $\mathbf{A}$ such that $\mathbf{A}^3 = \begin{bmatrix} 15 & -14 & -14 \\ -13 & 16 & 17 \\ 20 & -22 & -23 \end{bmatrix}$. (Hint: See Exercise 7.)

9. Prove that $\begin{bmatrix} a & b \\ c & d \end{bmatrix}$ has two distinct eigenvalues if $(a-d)^2 + 4bc > 0$, one eigenvalue if $(a-d)^2 + 4bc = 0$, and no eigenvalues if $(a-d)^2 + 4bc < 0$.

10. Let $\mathbf{A}$ be an $n \times n$ matrix, and let k be a positive integer.

    (a) Prove that if λ is an eigenvalue of $\mathbf{A}$, then λ^k is an eigenvalue of $\mathbf{A}^k$.

    ⋆(b) Give a 2×2 matrix $\mathbf{A}$ and an integer k that provide a counterexample to the converse of part (a).

11. Suppose that $\mathbf{A}$ is a nonsingular $n \times n$ matrix. Prove that

$$p_{\mathbf{A}^{-1}}(x) = (-x)^n \left| \mathbf{A}^{-1} \right| p_{\mathbf{A}}\left(\frac{1}{x}\right).$$

(Hint: First express $p_{\mathbf{A}}\left(\frac{1}{x}\right)$ as $\left| \left(\frac{1}{x}\right)\mathbf{I}_n - \mathbf{A} \right|$. Then collect the right-hand side into one determinant.)

12. Let $\mathbf{A}$ be an upper triangular $n \times n$ matrix. (Note: The following assertions are also true if $\mathbf{A}$ is a lower triangular matrix.)

    (a) Prove that λ is an eigenvalue for $\mathbf{A}$ if and only if λ appears on the main diagonal of $\mathbf{A}$.

    (b) Show that the algebraic multiplicity of an eigenvalue λ of $\mathbf{A}$ equals the number of times λ appears on the main diagonal.

13. Let $\mathbf{A}$ be an $n \times n$ matrix. Prove that $\mathbf{A}$ and $\mathbf{A}^T$ have the same characteristic polynomial and hence the same eigenvalues.

14. (Note: You must have covered the material in Section 8.4 in order to do this exercise.) Suppose that $\mathbf{A}$ is a stochastic $n \times n$ matrix. Prove that $\lambda = 1$ is an eigenvalue for $\mathbf{A}$. (Hint: Let $\mathbf{X} = [1, 1, \ldots, 1]$, and consider $\mathbf{A}^T \mathbf{X}$. Then use Exercise 13.) (This exercise implies that every stochastic matrix has a fixed point. However, not all initial conditions reach this fixed point, as demonstrated in Example 3 in Section 8.4.)

15. Let $\mathbf{A}$, $\mathbf{P}$, and $\mathbf{D}$ be $n \times n$ matrices with $\mathbf{P}$ nonsingular and $\mathbf{P}^{-1}\mathbf{A}\mathbf{P} = \mathbf{D}$. Use a proof by induction to show that $\mathbf{A}^k = \mathbf{P}\mathbf{D}^k\mathbf{P}^{-1}$, for every integer $k > 0$.

16. Let $\mathbf{A}$ be an $n \times n$ upper triangular matrix with all main diagonal entries distinct. Prove that $\mathbf{A}$ is diagonalizable.

17. Suppose $\mathbf{A}$ and $\mathbf{C}$ are similar $n \times n$ matrices (see Exercise 13 in Section 3.3). Prove that $p_{\mathbf{A}}(x) = p_{\mathbf{C}}(x)$.

18. Let $\mathbf{A}$ be a diagonalizable matrix. Prove that $\mathbf{A}^T$ is diagonalizable.

19. Let $\mathbf{A}$ be a nonsingular diagonalizable matrix. Prove that $\mathbf{A}^{-1}$ is diagonalizable.

20. This exercise outlines a proof of Theorem 3.15. Let $\mathbf{A}$ and $\mathbf{P}$ be given as stated in the theorem.

    (a) Suppose λ_i is the eigenvalue corresponding to $\mathbf{P}_i = i$th column of $\mathbf{P}$. Prove that the ith column of $\mathbf{A}\mathbf{P}$ equals $\lambda_i \mathbf{P}_i$.

    (b) Use the fact that $\mathbf{P}^{-1}\mathbf{P} = \mathbf{I}_n$ to prove that $\mathbf{P}^{-1}\lambda_i\mathbf{P}_i = \lambda_i\mathbf{e}_i$.

    (c) Use parts (a) and (b) to finish the proof of Theorem 3.15.

21. Prove that if $\mathbf{A}$ and $\mathbf{P}$ are $n \times n$ matrices such that $\mathbf{P}$ is nonsingular and $\mathbf{D} = \mathbf{P}^{-1}\mathbf{A}\mathbf{P}$ is diagonal, then each column of $\mathbf{P}$ is an eigenvector for $\mathbf{A}$. (Hint: Note that $\mathbf{P}\mathbf{D} = \mathbf{A}\mathbf{P}$, and calculate the ith column of both sides to show that $d_{ii}\mathbf{P}_i = \mathbf{A}\mathbf{P}_i$, where $\mathbf{P}_i$ is the ith column of $\mathbf{P}$.)

22. Prove the following: Let $\mathbf{A}$, $\mathbf{B}$, $\mathbf{C}$ be $n \times n$ matrices such that $\mathbf{C} = \mathbf{A}x + \mathbf{B}$. If at most k rows of $\mathbf{A}$ have nonzero entries, then $|\mathbf{C}|$ is a polynomial in x of degree $\leq k$. (Hint: Use induction on n.)

23.   (a) Show that the characteristic polynomial of a 2×2 matrix $\mathbf{A}$ is given by $x^2 - (\text{trace}(\mathbf{A}))x + |\mathbf{A}|$.

    (b) Prove that the characteristic polynomial of an $n \times n$ matrix always has degree n, with the coefficient of x^n equal to 1. (Hint: Use induction and Exercise 22.)

    (c) If $\mathbf{A}$ is an $n \times n$ matrix, show that the constant term of $p_{\mathbf{A}}(x)$ is $(-1)^n |\mathbf{A}|$. (Hint: The constant term of $p_{\mathbf{A}}(x)$ equals $p_{\mathbf{A}}(0)$.)

    (d) If $\mathbf{A}$ is an $n \times n$ matrix, show that the coefficient of x^{n-1} in $p_{\mathbf{A}}(x)$ is $-\text{trace}(\mathbf{A})$. (Hint: Use induction and Exercise 22.)

★24. True or False:

    (a) If $\mathbf{A}$ is a square matrix, then 5 is an eigenvalue of $\mathbf{A}$ if $\mathbf{A}\mathbf{X} = 5\mathbf{X}$ for some nonzero vector $\mathbf{X}$.

    (b) The eigenvalues of an $n \times n$ matrix $\mathbf{A}$ are the solutions of $x\mathbf{I}_n - \mathbf{A} = \mathbf{O}$.

    (c) If λ is an eigenvalue for an $n \times n$ matrix $\mathbf{A}$, then any nontrivial solution of $(\lambda\mathbf{I}_n - \mathbf{A})\mathbf{X} = \mathbf{0}$ is an eigenvector for $\mathbf{A}$ corresponding to λ.

(d) If **D** is the diagonal matrix created from an $n \times n$ matrix **A** in Step 6 of the Diagonalization Method, then the main diagonal entries of **D** are eigenvalues of **A**.

(e) If **A**, **P** are $n \times n$ matrices and each column of **P** is an eigenvector for **A**, then **P** is nonsingular and $\mathbf{P}^{-1}\mathbf{AP}$ is a diagonal matrix.

(f) If **A** is a square matrix and $p_{\mathbf{A}}(x) = (x - 3)^2(x + 1)$, then the Diagonalization Method cannot produce more than one eigenvector for the eigenvalue -1.

(g) If a 3×3 matrix **A** has 3 distinct eigenvalues, then **A** is diagonalizable.

(h) If $\mathbf{A} = \mathbf{PDP}^{-1}$, where **D** is a diagonal matrix, then $\mathbf{A}^n = \mathbf{P}^n\mathbf{D}^n(\mathbf{P}^{-1})^n$.

Summary of Techniques

We summarize here many of the computational techniques developed in Chapters 2 and 3. These computations should be done using calculators or computer software packages if they cannot be done easily by hand.

Techniques for Solving a System AX = B of m Linear Equations in n Unknowns

- **Gaussian elimination:** Use row operations to find a matrix in row echelon form that is row equivalent to $[\mathbf{A}|\mathbf{B}]$. Assign values to the independent variables and use back substitution to determine the values of the dependent variables. Advantages: finds the complete solution set for any linear system; fewer computational roundoff errors than Gauss-Jordan row reduction (Section 2.1).

- **Gauss-Jordan row reduction:** Use row operations to find the matrix in reduced row echelon form for $[\mathbf{A}|\mathbf{B}]$. Assign values to the independent variables and solve for the dependent variables. Advantages: easily computerized; finds the complete solution set for any linear system (Section 2.2).

- **Multiplication by inverse matrix:** Use when $m = n$ and $|\mathbf{A}| \neq 0$. The solution is $\mathbf{X} = \mathbf{A}^{-1}\mathbf{B}$. Disadvantage: $\mathbf{A}^{-1}$ must be known or calculated first, and therefore the method is only useful when there are several systems to be solved with the same coefficient matrix **A** (Section 2.4).

- **Cramer's Rule**: Use when $m = n$ and $|\mathbf{A}| \neq 0$. The solution is $x_1 = |\mathbf{A}_1| / |\mathbf{A}|$, $x_2 = |\mathbf{A}_2| / |\mathbf{A}|, \ldots, x_n = |\mathbf{A}_n| / |\mathbf{A}|$, where $\mathbf{A}_i$ (for $1 \leq i \leq n$) and **A** are identical except that the ith column of $\mathbf{A}_i$ equals **B**. Disadvantage: efficient only for small systems because it involves calculating $n + 1$ determinants of size n (Section 3.3).

Other techniques for solving systems are discussed in Chapter 9. Among these are **LDU** decomposition and iterative methods, such as the Gauss-Seidel and Jacobi techniques.

Also remember that if $m < n$ and $\mathbf{B} = \mathbf{0}$ (homogeneous case), then there is an infinite number of solutions to $\mathbf{AX} = \mathbf{B}$.

Techniques for Finding the Inverse (If It Exists) of an $n \times n$ Matrix A

- **2 × 2 case:** The inverse of $\begin{bmatrix} a & b \\ c & d \end{bmatrix}$ exists if and only if $ad - bc \neq 0$. In that case, the inverse is given by $\left(\frac{1}{ad-bc} \right) \begin{bmatrix} d & -b \\ -c & a \end{bmatrix}$ (Section 2.4).

- **Row reduction:** Row reduce $[\mathbf{A} | \mathbf{I}_n]$ to $[\mathbf{I}_n | \mathbf{A}^{-1}]$ (where $\mathbf{A}^{-1}$ does not exist if the process stops prematurely). Advantages: easily computerized; relatively efficient (Section 2.4).

- **Adjoint matrix:** $\mathbf{A}^{-1} = \left(\frac{1}{|\mathbf{A}|} \right) \mathcal{A}$, where $\mathcal{A}$ is the adjoint matrix of $\mathbf{A}$. Advantage: gives an algebraic formula for $\mathbf{A}^{-1}$. Disadvantage: not very efficient, because $|\mathbf{A}|$ and all n^2 cofactors of $\mathbf{A}$ must be calculated first (Section 3.3).

Techniques for Finding the Determinant of an $n \times n$ Matrix A

- **2 × 2 case:** $|\mathbf{A}| = a_{11}a_{22} - a_{12}a_{21}$ (Sections 2.4 and 3.1).

- **3 × 3 case:** Basketweaving (Section 3.1).

- **Row reduction:** Row reduce $\mathbf{A}$ to an upper triangular form matrix $\mathbf{B}$, keeping track of the effect of each row operation on the determinant using a variable P. Then $|\mathbf{A}| = P|\mathbf{B}|$, using the final value of P. Advantages: easily computerized; relatively efficient (Section 3.2).

- **Cofactor expansion:** Multiply each element along any row or column of $\mathbf{A}$ by its cofactor and sum the results. Advantage: useful for matrices with many zero entries. Disadvantage: not as fast as row reduction (Sections 3.1 and 3.3).

Also remember that $|\mathbf{A}| = 0$ if $\mathbf{A}$ is row equivalent to a matrix with a row or column of zeroes, or with two identical rows, or with two identical columns.

Technique for Finding the Eigenvalues of an $n \times n$ Matrix A

- **Characteristic polynomial:** Find the roots of $p_{\mathbf{A}}(x) = |x\mathbf{I}_n - \mathbf{A}|$. (We only consider the real roots of $p_{\mathbf{A}}(x)$ in Chapters 1 through 6.) Disadvantages: tedious to calculate $p_{\mathbf{A}}(x)$; polynomial becomes more difficult to factor as degree of $p_{\mathbf{A}}(x)$ increases (Section 3.4).

A more computationally efficient technique for finding eigenvalues is the Power Method in Chapter 9.

Technique for Finding the Eigenvectors of an $n \times n$ Matrix A

- **Row reduction:** For a given eigenvalue λ of $\mathbf{A}$, solve $(\lambda \mathbf{I}_n - \mathbf{A})\mathbf{X} = \mathbf{0}$ by row reducing the augmented matrix $[(\lambda \mathbf{I}_n - \mathbf{A})| \mathbf{0}]$ and taking the nontrivial solutions (Section 3.4).

Chapter 4

Finite Dimensional Vector Spaces

Driven to Abstraction

Students frequently wonder why mathematicians often feel the need to work in abstract terms. Could abstract generalizations of common mathematical concepts have any real-world applications? Most often, the answer is "Yes!" The inspiration for such generalizations in linear algebra comes from considering the properties of vectors and matrices.

Generalization is necessary in linear algebra because studying $\mathbb{R}^n$ can take us only so far. But as we will see, many other sets of mathematical objects, such as functions, matrices, infinite series, etc., have properties in common with $\mathbb{R}^n$. This suggests that we should generalize our discussion of vectors to other sets of objects, which we call vector spaces. By studying vector spaces whose objects share many of the same properties of vectors in $\mathbb{R}^n$, we reveal a more abstract theory with a wider range of applications than we would obtain from a study of $\mathbb{R}^n$ alone.

In Chapter 1, we saw that the operations of addition and scalar multiplication on the set $\mathcal{M}_{mn}$ possess many of the same algebraic properties as addition and scalar multiplication on the set $\mathbb{R}^n$. In fact, there are many other sets with comparable operations, and it is profitable to study them together. In this chapter, we define vector spaces to be algebraic structures with operations having properties similar to those of addition and scalar multiplication on $\mathbb{R}^n$. We then establish many important results relating to vector spaces. Because we are studying vector spaces as a class, this chapter is more abstract than previous chapters. But the advantage of working in this more general setting is that we generate theorems that apply to all vector spaces, not just $\mathbb{R}^n$.

4.1 Introduction to Vector Spaces

Definition of a Vector Space

In Theorems 1.3 and 1.11, we proved eight properties of addition and scalar multiplication in $\mathbb{R}^n$ and $\mathcal{M}_{mn}$. These properties are important because all other results involving these operations can be derived from them. We now introduce the general class of sets called **vector spaces**,[1] having operations of addition and scalar multiplication with these same eight properties, as well as two **closure properties**.

DEFINITION

A **vector space** is a set $\mathcal{V}$ together with an operation called **vector addition** (a rule for adding two elements of $\mathcal{V}$ to obtain a third element of $\mathcal{V}$) and another operation called **scalar multiplication** (a rule for multiplying a real number times an element of $\mathcal{V}$ to obtain a second element of $\mathcal{V}$) on which the following ten properties hold:

For every $\mathbf{u}$, $\mathbf{v}$, and $\mathbf{w}$ in $\mathcal{V}$, and for every a and b in $\mathbb{R}$

(A)	$\mathbf{u} + \mathbf{v} \in \mathcal{V}$	Closure Property of Addition
(B)	$a\mathbf{u} \in \mathcal{V}$	Closure Property of Scalar Multiplication
(1)	$\mathbf{u} + \mathbf{v} = \mathbf{v} + \mathbf{u}$	Commutative Law of Addition
(2)	$\mathbf{u} + (\mathbf{v} + \mathbf{w}) = (\mathbf{u} + \mathbf{v}) + \mathbf{w}$	Associative Law of Addition
(3)	There is an element $\mathbf{0}$ of $\mathcal{V}$ so that for every $\mathbf{y}$ in $\mathcal{V}$ we have $\mathbf{0} + \mathbf{y} = \mathbf{y} = \mathbf{y} + \mathbf{0}$.	Existence of Identity Element for Addition

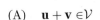

---

[1] We actually define what are called *real vector spaces*, rather than just vector spaces. The word *real* implies that the scalars involved in the scalar multiplication are real numbers. In Chapter 7, we consider *complex vector spaces*, where the scalars are complex numbers. Other types of vector spaces involving more general sets of scalars are not considered in this book.

✳ (4)   There is an element $-\mathbf{u}$ in $\mathcal{V}$ such   Existence of Additive Inverse
that $\mathbf{u} + (-\mathbf{u}) = \mathbf{0} = (-\mathbf{u}) + \mathbf{u}$.

(5) $a(\mathbf{u} + \mathbf{v}) = (a\mathbf{u}) + (a\mathbf{v})$ Distributive Laws for Scalar
(6) $(a + b)\mathbf{u} = (a\mathbf{u}) + (b\mathbf{u})$ Multiplication over Addition
(7) $(ab)\mathbf{u} = a(b\mathbf{u})$ Associativity of Scalar
Multiplication

(8) $1\mathbf{u} = \mathbf{u}$ Identity Property for Scalar
Multiplication

The elements of a vector space $\mathcal{V}$ are called **vectors**.

The two closure properties require that both the operations of vector addition and scalar multiplication always produce an element of the vector space as a result.

The standard plus sign, "+," is used to indicate both vector addition and the sum of real numbers, two different operations. All sums in properties (1), (2), (3), (4), and (5) are vector sums. In property (6), the "+" on the left side of the equation represents addition of real numbers; the "+" on the right side stands for the sum of two vectors. In property (7), the left side of the equation contains one product of real numbers, ab, and one instance of scalar multiplication, (ab) times $\mathbf{u}$. The right side of property (7) involves two scalar multiplications — first, b times $\mathbf{u}$, then, a times the vector $(b\mathbf{u})$. Usually, we can tell from the context which type of operation is being used.

In any vector space, the additive identity element in property (3) is unique, and the additive inverse (property (4)) of each vector is unique (see Exercise 12).

Examples of Vector Spaces

EXAMPLE 1   Let $\mathcal{V} = \mathbb{R}^n$, with addition and scalar multiplication of n-vectors as defined in Section 1.1. Since these operations always produce vectors in $\mathbb{R}^n$, the clos-ure properties certainly hold for $\mathbb{R}^n$. By Theorem 1.3, the remaining eight properties hold as well. Thus, $\mathcal{V} = \mathbb{R}^n$ is a vector space with these operations.

Similarly, consider $\mathcal{M}_{mn}$, the set of $m \times n$ matrices. The usual operations of matrix addition and scalar multiplication on $\mathcal{M}_{mn}$ always produce $m \times n$ matrices, and so the closure properties certainly hold for $\mathcal{M}_{mn}$. By Theorem 1.11, the remaining eight properties hold as well. Hence, $\mathcal{M}_{mn}$ is a vector space with these operations. ∎

$\mathbb{R}^n$ and $\mathcal{M}_{mn}$ (with the usual operations of addition and scalar multiplication) are representative of most of the vector spaces we consider here. Keep $\mathbb{R}^n$ and $\mathcal{M}_{mn}$ in mind as examples later, as we consider theorems involving general vector spaces.

Some vector spaces can have additional operations. For example, $\mathbb{R}^n$ has the dot product, and $\mathcal{M}_{mn}$ has matrix multiplication and the transpose. But these additional structures are not shared by all vector spaces because they are not included in the definition. We cannot assume the existence of any additional operations in a general discussion of vector spaces. In particular, there is no such operation as multiplication or division of one vector by another in general vector spaces.

The only general vector space operation that combines two *vectors* is vector addition.

EXAMPLE 2

The set $V = \{0\}$ is a vector space with the rules for addition and multiplication given by $0 + 0 = 0$ and $a0 = 0$ for every scalar (real number) a. Since 0 is the only possible result of either operation, V must be closed under both addition and scalar multiplication. A quick check verifies that the remaining eight properties also hold for V. This vector space is called the **trivial vector space**, and no smaller vector space is possible (why?).   ■

EXAMPLE 3

Consider $\mathbb{R}^3$ as the set of 3-vectors in three-dimensional space, all with initial points at the origin. Let W be any plane containing the origin. W can also be considered as the set of all 3-vectors whose terminal point lies in this plane (that is, W is the set of all 3-vectors that lie entirely in the plane when drawn on a graph, since both the initial point and terminal point of each vector lie in the plane). For example, in Figure 4.1, W is the plane containing the vectors **u** and **v** (elements of W); **q** is not in W because its terminal point does not lie in the plane. We will prove that W is a vector space.

Figure 4.1

A plane W in $\mathbb{R}^3$
containing the origin

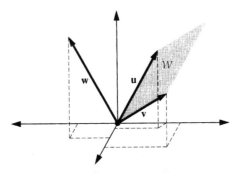

To check the closure properties, we must show that the sum of any two vectors in W is a vector in W and that any scalar multiple of a vector in W also lies in W.

If **x** and **y** are elements of W, then the parallelogram they form lies entirely in the plane, because **x** and **y** do. Hence, the diagonal **x** + **y** of this parallelogram also lies in the plane, so **x** + **y** is in W. This observation verifies that W is closed under vector addition (that is, the closure property holds for vector addition). Notice that it is not enough to know that the sum of two 3-vectors in W produces another 3-vector. We have to show that the sum they produce is actually in the set W.

Next consider scalar multiplication. If **x** is a vector in W, then any scalar multiple of **x**, a**x**, is either parallel to **x** or equal to 0. Therefore, a**x** lies in any plane through the origin that contains **x** (in particular, W). Hence, a**x** is in W, and W is closed under scalar multiplication.

We now check that the remaining eight vector space properties hold. Properties (1), (2), (5), (6), (7), and (8) are true for all vectors in $\mathcal{W}$ by Theorem 1.3, since $\mathcal{W} \subseteq \mathbb{R}^3$. However, properties (3) and (4) must be checked separately for $\mathcal{W}$ because they are *existence* properties. We know that the zero vector and additive inverses exist in $\mathbb{R}^3$, but are they in $\mathcal{W}$? Now, $\mathbf{0} = [0, 0, 0]$ is in $\mathcal{W}$, because the plane $\mathcal{W}$ passes through the origin, thus proving property (3). Also, the opposite (additive inverse) of any vector lying in the plane $\mathcal{W}$ also lies in $\mathcal{W}$, thus proving property (4). Hence, all eight properties and the closure properties are true, so $\mathcal{W}$ is a vector space. ∎

EXAMPLE 4

Let $\mathcal{P}_n$ be the set of polynomials of degree $\leq n$, with real coefficients. The vectors in $\mathcal{P}_n$ have the form $\mathbf{p} = a_n x^n + \cdots + a_1 x + a_0$ for some real numbers $a_0, a_1, \ldots, a_n$. We define addition of polynomials in the usual manner — that is, by adding corresponding coefficients. Then the sum of any two polynomials of degree $\leq n$ also has degree $\leq n$ and so is in $\mathcal{P}_n$. Thus, the closure property of addition holds. Similarly, if b is a real number and $\mathbf{p} = a_n x^n + \cdots + a_1 x + a_0$ is in $\mathcal{P}_n$, we define $b\mathbf{p}$ to be the polynomial $(ba_n)x^n + \cdots + (ba_1)x + ba_0$, which is also in $\mathcal{P}_n$. Hence, the closure property of scalar multiplication holds. Then, if the remaining eight vector space properties hold, $\mathcal{P}_n$ is a vector space under these operations. We verify properties (1), (3), and (4) of the definition and leave the others for you to check.

(1) Commutative Law of Addition: We must show that the order in which two vectors (polynomials) are added makes no difference. Now, by the commutative law of addition for real numbers,

$$(a_n x^n + \cdots + a_1 x + a_0) + (b_n x^n + \cdots + b_1 x + b_0)$$
$$= (a_n + b_n)x^n + \cdots + (a_1 + b_1)x + (a_0 + b_0)$$
$$= (b_n + a_n)x^n + \cdots + (b_1 + a_1)x + (b_0 + a_0)$$
$$= (b_n x^n + \cdots + b_1 x + b_0) + (a_n x^n + \cdots + a_1 x + a_0).$$

(3) Existence of Identity Element for Addition: The zero-degree polynomial $\mathbf{z} = 0x^n + \cdots + 0x + 0$ acts as the additive identity element $\mathbf{0}$. That is, adding $\mathbf{z}$ to any vector $\mathbf{p} = a_n x^n + \cdots + a_1 x + a_0$ does not change the vector:

$$\mathbf{z} + \mathbf{p} = (0 + a_n)x^n + \cdots + (0 + a_1)x + (0 + a_0) = \mathbf{p}.$$

(4) Existence of Additive Inverse: We must show that each vector $\mathbf{p} = a_n x^n + \cdots + a_1 x + a_0$ in $\mathcal{P}_n$ has an additive inverse in $\mathcal{P}_n$. But, the vector $-\mathbf{p} = -(a_n x^n + \cdots + a_1 x + a_0) = (-a_n)x^n + \cdots + (-a_1)x + (-a_0)$ has the property that $\mathbf{p} + [-\mathbf{p}] = \mathbf{z}$, the zero vector, and so $-\mathbf{p}$ acts as the additive inverse of $\mathbf{p}$. Because $-\mathbf{p}$ is also in $\mathcal{P}_n$, we are done. ∎

The vector space in Example 4 is similar to our prototype $\mathbb{R}^n$. For any polynomial in $\mathcal{P}_n$, consider the sequence of its $n+1$ coefficients. This sequence completely describes that polynomial and can be thought of as an $(n+1)$-vector. For example, a polynomial $a_2 x^2 + a_1 x + a_0$ in $\mathcal{P}_2$ can be described by the 3-vector $[a_2, a_1, a_0]$. In this way, the vector space $\mathcal{P}_2$ "resembles" the vector space $\mathbb{R}^3$, and in general, $\mathcal{P}_n$ "resembles" $\mathbb{R}^{n+1}$.

EXAMPLE 5

The set $\mathcal{P}$ of all polynomials (of all degrees) is a vector space under the usual (term-by-term) operations of addition and scalar multiplication (see Exercise 15). ∎

EXAMPLE 6

Let $\mathcal{V}$ be the set of all real-valued functions defined on $\mathbb{R}$. For example, $\mathbf{f}(x) = \arctan(x)$ is in $\mathcal{V}$. We define addition of functions as usual: $\mathbf{h} = \mathbf{f} + \mathbf{g}$ is the function such that $\mathbf{h}(x) = \mathbf{f}(x) + \mathbf{g}(x)$, for every $x \in \mathbb{R}$. Similarly, if $a \in \mathbb{R}$ and $\mathbf{f}$ is in $\mathcal{V}$, we define the scalar multiple $\mathbf{h} = a\mathbf{f}$ to be the function such that $\mathbf{h}(x) = a\mathbf{f}(x)$, for every $x \in \mathbb{R}$. Now, the closure properties hold for $\mathcal{V}$ because sums and scalar multiples of real-valued functions produce real-valued functions. To finish verifying that $\mathcal{V}$ is a vector space, we must check that the remaining eight vector space properties hold.

Suppose that $\mathbf{f}$, $\mathbf{g}$, and $\mathbf{h}$ are in $\mathcal{V}$, and a and b are real numbers.

Property (1): For every x in $\mathbb{R}$, $\mathbf{f}(x)$ and $\mathbf{g}(x)$ are both real numbers. Hence, $\mathbf{f}(x) + \mathbf{g}(x) = \mathbf{g}(x) + \mathbf{f}(x)$ for all $x \in \mathbb{R}$, by the commutative law of addition for real numbers, so each represents the same function of x. Hence, $\mathbf{f} + \mathbf{g} = \mathbf{g} + \mathbf{f}$.

Property (2): For every $x \in \mathbb{R}$, $\mathbf{f}(x) + (\mathbf{g}(x) + \mathbf{h}(x)) = (\mathbf{f}(x) + \mathbf{g}(x)) + \mathbf{h}(x)$, by the associative law of addition for real numbers. Thus, $\mathbf{f} + (\mathbf{g} + \mathbf{h}) = (\mathbf{f} + \mathbf{g}) + \mathbf{h}$.

Property (3): Let $\mathbf{z}$ be the function given by $\mathbf{z}(x) = 0$ for every $x \in \mathbb{R}$. Then, for each x, $\mathbf{z}(x) + \mathbf{f}(x) = 0 + \mathbf{f}(x) = \mathbf{f}(x)$. Hence, $\mathbf{z} + \mathbf{f} = \mathbf{f}$.

Property (4): Given $\mathbf{f}$ in $\mathcal{V}$, define $-\mathbf{f}$ by $[-\mathbf{f}](x) = -(\mathbf{f}(x))$ for every $x \in \mathbb{R}$. Then, for all x, $[-\mathbf{f}](x) + \mathbf{f}(x) = -(\mathbf{f}(x)) + \mathbf{f}(x) = 0$. Therefore, $[-\mathbf{f}] + \mathbf{f} = \mathbf{z}$, the zero vector, and so the additive inverse of $\mathbf{f}$ is also in $\mathcal{V}$.

Properties (5) and (6): For every $x \in \mathbb{R}$, $a(\mathbf{f}(x) + \mathbf{g}(x)) = a\mathbf{f}(x) + a\mathbf{g}(x)$ and $(a + b)\mathbf{f}(x) = a\mathbf{f}(x) + b\mathbf{f}(x)$ by the distributive laws for real numbers of multiplication over addition. Hence, $a(\mathbf{f} + \mathbf{g}) = a\mathbf{f} + a\mathbf{g}$, and $(a + b)\mathbf{f} = a\mathbf{f} + b\mathbf{f}$.

Property (7): For every $x \in \mathbb{R}$, $(ab)\mathbf{f}(x) = a(b\mathbf{f}(x))$ follows from the associative law of multiplication for real numbers. Hence, $(ab)\mathbf{f} = a(b\mathbf{f})$.

Property (8): Since $1 \cdot \mathbf{f}(x) = \mathbf{f}(x)$ for every real number x, we have $1 \cdot \mathbf{f} = \mathbf{f}$ in $\mathcal{V}$. ∎

Two Unusual Vector Spaces

The next two examples place unusual operations on familiar sets to create new vector spaces. In such cases, regardless of how the operations are defined, we sometimes use the symbols $\oplus$ and $\odot$ to denote addition and scalar multiplication, respectively, in order to remind ourselves that these operations are not the "regular" ones. Note that $\oplus$ is defined differently in Examples 7 and 8 (and similarly for $\odot$).

EXAMPLE 7

Let $\mathcal{V}$ be the set $\mathbb{R}^+$ of positive real numbers. This set is not a vector space under the usual operations of addition and scalar multiplication (why?). However, we can define new rules for these operations to make $\mathcal{V}$ a vector space. In what follows, we sometimes think of elements of $\mathbb{R}^+$ as abstract vectors (in which case we use boldface type, such as **v**) or as the values on the positive real number line they represent (in which case we use italics, such as v).

To define "addition" on $\mathcal{V}$, we use *multiplication* of real numbers. That is,

$$\mathbf{v}_1 \oplus \mathbf{v}_2 = v_1 \cdot v_2$$

for every $\mathbf{v}_1$ and $\mathbf{v}_2$ in $\mathcal{V}$, where we use the symbol $\oplus$ for the "addition" operation on $\mathcal{V}$ to emphasize that this is not addition of real numbers. The definition of a vector space states only that vector addition must be a rule for combining two vectors to yield a third vector so that properties (1) through (8) hold. There is no stipulation that vector addition must be at all similar to ordinary addition of real numbers.[2]

We next define "scalar multiplication," $\odot$, on $\mathcal{V}$ by

$$a \odot \mathbf{v} = v^a$$

for every $a \in \mathbb{R}$ and $\mathbf{v} \in \mathcal{V}$.

From the given definitions, we see that if $\mathbf{v}_1$ and $\mathbf{v}_2$ are in $\mathcal{V}$ and a is in $\mathbb{R}$, then both $\mathbf{v}_1 \oplus \mathbf{v}_2$ and $a \odot \mathbf{v}_1$ are in $\mathcal{V}$, thus verifying the two closure properties. To prove the other eight properties, we assume that $\mathbf{v}_1, \mathbf{v}_2, \mathbf{v}_3 \in \mathcal{V}$ and that $a, b \in \mathbb{R}$. We then have the following:

Property (1): $\mathbf{v}_1 \oplus \mathbf{v}_2 = v_1 \cdot v_2 = v_2 \cdot v_1$ (by the commutative law of multiplication for real numbers) $= \mathbf{v}_2 \oplus \mathbf{v}_1$.

Property (2): $\mathbf{v}_1 \oplus (\mathbf{v}_2 \oplus \mathbf{v}_3) = \mathbf{v}_1 \oplus (v_2 \cdot v_3) = v_1 \cdot (v_2 \cdot v_3) = (v_1 \cdot v_2) \cdot v_3$ (by the associative law of multiplication for real numbers) $= (\mathbf{v}_1 \oplus \mathbf{v}_2) \cdot v_3 = (\mathbf{v}_1 \oplus \mathbf{v}_2) \oplus \mathbf{v}_3$.

Property (3): The number 1 in $\mathbb{R}^+$ acts as the zero vector $\mathbf{0}$ in $\mathcal{V}$ (why?).

Property (4): The additive inverse of $\mathbf{v}$ in $\mathcal{V}$ is the positive real number $(1/v)$, because $\mathbf{v} \oplus (1/v) = v \cdot (1/v) = 1$, the zero vector in $\mathcal{V}$.

Property (5): $a \odot (\mathbf{v}_1 \oplus \mathbf{v}_2) = a \odot (v_1 \cdot v_2) = (v_1 \cdot v_2)^a = v_1^a \cdot v_2^a = (a \odot \mathbf{v}_1) \cdot (a \odot \mathbf{v}_2) = (a \odot \mathbf{v}_1) \oplus (a \odot \mathbf{v}_2)$.

Property (6): $(a + b) \odot \mathbf{v} = v^{a+b} = v^a \cdot v^b = (a \odot \mathbf{v}) \cdot (b \odot \mathbf{v}) = (a \odot \mathbf{v}) \oplus (b \odot \mathbf{v})$.

Property (7): $(ab) \odot \mathbf{v} = v^{ab} = \left(v^b\right)^a = (b \odot \mathbf{v})^a = a \odot (b \odot \mathbf{v})$.

Property (8): $1 \odot \mathbf{v} = v^1 = \mathbf{v}$. ∎

EXAMPLE 8 Let $\mathcal{V} = \mathbb{R}^2$, with addition defined by

$$[x, y] \oplus [w, z] = [x + w + 1, \ y + z - 2]$$

and scalar multiplication defined by

$$a \odot [x, y] = [ax + a - 1, \ ay - 2a + 2].$$

---

[2] You might expect the operation $\oplus$ to be called something other than "addition." However, most of our vector space terminology comes from the motivating example of $\mathbb{R}^n$, so the word *addition* is a natural choice for the name of the operation.

The closure properties hold for these operations (why?). In fact, $\mathcal{V}$ forms a vector space because the eight vector properties also hold. We verify properties (2), (3), (4), and (6) and leave the others for you to check.

Property (2):

$$
\begin{aligned}
[x, y] \oplus ([u, v] \oplus [w, z]) &= [x, y] \oplus [u + w + 1, \ v + z - 2] \\
&= [x + u + w + 2, \ y + v + z - 4] \\
&= [x + u + 1, \ y + v - 2] \oplus [w, z] \\
&= ([x, y] \oplus [u, v]) \oplus [w, z].
\end{aligned}
$$

Property (3): The vector $[-1, 2]$ acts as the zero vector, since

$$[x, y] \oplus [-1, 2] = [x + (-1) + 1, \ y + 2 - 2] = [x, y].$$

Property (4): The additive inverse of $[x, y]$ is $[-x - 2, -y + 4]$, because

$$[x, y] \oplus [-x - 2, -y + 4] = [x - x - 2 + 1, \ y - y + 4 - 2] = [-1, 2],$$

the zero vector in $\mathcal{V}$.

Property (6):

$$
\begin{aligned}
(a + b) \odot [x, y] &= [(a + b)x + (a + b) - 1, \ (a + b)y - 2(a + b) + 2] \\
&= [(ax + a - 1) + (bx + b - 1) + 1, \ (ay - 2a + 2) \\
&\quad + (by - 2b + 2) - 2] \\
&= [ax + a - 1, \ ay - 2a + 2] \oplus [bx + b - 1, \ by - 2b + 2] \\
&= (a \odot [x, y]) \oplus (b \odot [x, y]). \qquad\blacksquare
\end{aligned}
$$

Some Elementary Properties of Vector Spaces

The next theorem contains several simple results regarding vector spaces. Although these are obviously true in the most familiar examples, we must prove them in general before we know they hold in every possible vector space.

THEOREM 4.1

Let $\mathcal{V}$ be a vector space. Then, for every vector $\mathbf{v}$ in $\mathcal{V}$ and every real number a, we have

(1) $a\mathbf{0} = \mathbf{0}$ — Any scalar multiple of the zero vector yields the zero vector.

(2) $0\mathbf{v} = \mathbf{0}$ — The scalar zero multiplied by any vector yields the zero vector.

(3) $(-1)\mathbf{v} = -\mathbf{v}$ — The scalar -1 multiplied by any vector yields the additive inverse of that vector.

(4) If $a\mathbf{v} = \mathbf{0}$, then $a = 0$ or $\mathbf{v} = \mathbf{0}$ — If a scalar multiplication yields the zero vector, then either the scalar is zero, or the vector is the zero vector, or both.

Part (3) justifies the notation for the additive inverse in property (4) of the definition of a vector space and shows we do not need to distinguish between $-\mathbf{v}$ and $(-1)\mathbf{v}$.

This theorem must be proved directly from the properties in the definition of a vector space because at this point we have no other known facts about general vector spaces. We prove parts (1), (3), and (4). The proof of part (2) is similar to the proof of part (1) and is left as Exercise 18.

PROOF OF THEOREM 4.1 (abridged):

$$
\begin{aligned}
\textbf{Part (1)}: a\mathbf{0} &= a\mathbf{0} + \mathbf{0} && \text{by property (3)} \\
&= a\mathbf{0} + (a\mathbf{0} + (-[a\mathbf{0}])) && \text{by property (4)} \\
&= (a\mathbf{0} + a\mathbf{0}) + (-[a\mathbf{0}]) && \text{by property (2)} \\
&= a(\mathbf{0} + \mathbf{0}) + (-[a\mathbf{0}]) && \text{by property (5)} \\
&= a\mathbf{0} + (-[a\mathbf{0}]) && \text{by property (3)} \\
&= \mathbf{0}. && \text{by property (4)}
\end{aligned}
$$

$$
\begin{aligned}
\textbf{Part (3)}: -\mathbf{v} &= (-\mathbf{v}) + \mathbf{0} && \text{by property (3)} \\
&= (-\mathbf{v}) + 0\mathbf{v} && \text{by part (2) of Theorem 4.1} \\
&= (-\mathbf{v}) + (1 + (-1))\mathbf{v} && \\
&= (-\mathbf{v}) + (1\mathbf{v} + (-1)\mathbf{v}) && \text{by property (6)} \\
&= ((-\mathbf{v}) + \mathbf{v}) + (-1)\mathbf{v} && \text{by properties (2) and (8)} \\
&= \mathbf{0} + (-1)\mathbf{v} && \text{by property (4)} \\
&= (-1)\mathbf{v}. && \text{by property (3)}
\end{aligned}
$$

Part (4): Assume that $a\mathbf{v} = \mathbf{0}$ and $a \neq 0$. We must show that $\mathbf{v} = \mathbf{0}$. Now,

$$
\begin{aligned}
\mathbf{v} &= 1\mathbf{v} && \text{by property (8)} \\
&= \left(\tfrac{1}{a} \cdot a\right)\mathbf{v} && \text{because } a \neq 0 \\
&= \left(\tfrac{1}{a}\right)(a\mathbf{v}) && \text{by property (7)} \\
&= \left(\tfrac{1}{a}\right)\mathbf{0} && \text{because } a\mathbf{v} = \mathbf{0} \\
&= \mathbf{0}. && \text{by part (1) of Theorem 4.1} \qquad \blacksquare
\end{aligned}
$$

Theorem 4.1 is valid even for unusual vector spaces, such as those in Examples 7 and 8. For instance, part (4) of the theorem claims that, in general, $a\mathbf{v} = \mathbf{0}$ implies $a = 0$ or $\mathbf{v} = \mathbf{0}$. This statement can quickly be verified for the vector space $\mathcal{V} = \mathbb{R}^+$ with operations $\oplus$ and $\odot$ from Example 7. In this case, $a \odot \mathbf{v} = v^a$, and the zero vector $\mathbf{0}$ is the real number 1. Then, part (4) is equivalent here to the true statement that $v^a = 1$ implies $a = 0$ or $v = 1$.

Applying parts (2) and (3) of Theorem 4.1 to an unusual vector space $\mathcal{V}$ gives a quick way of finding the zero vector $\mathbf{0}$ of $\mathcal{V}$ and the additive inverse $-\mathbf{v}$ for any vector $\mathbf{v}$ in $\mathcal{V}$. For instance, in Example 8, we have $\mathcal{V} = \mathbb{R}^2$ with scalar multiplication $a \odot [x, y] = [ax + a - 1, \, ay - 2a + 2]$. To find the zero vector $\mathbf{0}$ in $\mathcal{V}$, we simply multiply the scalar 0 by any general vector $[x, y]$ in $\mathcal{V}$.

$$
\mathbf{0} = 0 \odot [x, y] = [0x + 0 - 1, \, 0y - 2(0) + 2] = [-1, 2].
$$

Similarly, if $[x, y] \in \mathcal{V}$, then $-1 \odot [x, y]$ gives the additive inverse of $[x, y]$.

$$-[x, y] = -1 \odot [x, y] = [-1x + (-1) - 1, -1y - 2(-1) + 2]$$
$$= [-x - 2, -y + 4].$$

Failure of the Vector Space Conditions

We conclude this section by considering some sets that are not vector spaces to see what can go wrong.

EXAMPLE 9   The set Φ of real-valued functions, f, defined on the interval $[0, 1]$ such that $f\left(\frac{1}{2}\right) = 1$, is not a vector space under the usual operations of function addition and scalar multiplication because the closure properties do not hold. If f and g are in Φ, then

$$(f + g)\left(\frac{1}{2}\right) = f\left(\frac{1}{2}\right) + g\left(\frac{1}{2}\right) = 1 + 1 = 2 \neq 1,$$

so $f + g$ is not in Φ. Therefore, Φ is not closed under addition and cannot be a vector space. (Is Φ closed under scalar multiplication?)   ∎

EXAMPLE 10   Let Υ be the set $\mathbb{R}^2$ with operations

$$\mathbf{v}_1 \oplus \mathbf{v}_2 = \mathbf{v}_1 + \mathbf{v}_2 \quad \text{and} \quad c \odot \mathbf{v} = c(\mathbf{Av}), \quad \text{where } \mathbf{A} = \begin{bmatrix} -3 & 1 \\ 5 & -2 \end{bmatrix}.$$

With these operations, Υ is not a vector space. You can verify that Υ is closed under $\oplus$ and $\odot$, but properties (7) and (8) of the definition are not satisfied. For example, property (8) fails since

$$1 \odot \begin{bmatrix} 2 \\ 7 \end{bmatrix} = 1\left(\begin{bmatrix} -3 & 1 \\ 5 & -2 \end{bmatrix} \begin{bmatrix} 2 \\ 7 \end{bmatrix}\right) = 1 \begin{bmatrix} 1 \\ -4 \end{bmatrix} = \begin{bmatrix} 1 \\ -4 \end{bmatrix} \neq \begin{bmatrix} 2 \\ 7 \end{bmatrix}.$$ ∎

Exercises for Section 4.1

Remember: To verify that a given set with its operations is a vector space, you must prove the two closure properties as well as the remaining eight properties in the definition. To show that a set with operations is *not* a vector space, you need only find an example showing that one of the closure properties or one of the remaining eight properties is not satisfied.

1. Rewrite properties (2), (5), (6), and (7) in the definition of a vector space using the symbols $\oplus$ for vector addition and $\odot$ for scalar multiplication. (The notations for real number addition and multiplication should not be changed.)

2. Prove that the set of all scalar multiples of the vector $[1, 3, 2]$ in $\mathbb{R}^3$ forms a vector space with the usual operations on 3-vectors.

3. Verify that the set of polynomials $\mathbf{f}$ in $\mathcal{P}_3$ such that $\mathbf{f}(2) = 0$ forms a vector space with the standard operations.

4. Prove that $\mathbb{R}$ is a vector space using the operations $\oplus$ and $\odot$ given by $\mathbf{x} \oplus \mathbf{y} = (x^3 + y^3)^{1/3}$ and $a \odot \mathbf{x} = (\sqrt[3]{a})x$.

★**5.** Show that the set of singular 2×2 matrices under the usual operations is *not* a vector space.

6. Prove that the set of nonsingular $n \times n$ matrices under the usual operations is *not* a vector space.

7. Show that $\mathbb{R}$, with ordinary addition but with scalar multiplication replaced by $a \odot \mathbf{x} = \mathbf{0}$ for every real number a, is *not* a vector space.

★**8.** Show that the set $\mathbb{R}$, with the usual scalar multiplication but with addition given by $x \oplus y = 2(x + y)$, is *not* a vector space.

9. Show that the set $\mathbb{R}^2$, with the usual scalar multiplication but with vector addition replaced by $[x, y] \oplus [w, z] = [x + w, 0]$, does *not* form a vector space.

10. Let $\mathcal{A} = \mathbb{R}$, with the operations $\oplus$ and $\odot$ given by $\mathbf{x} \oplus \mathbf{y} = (x^5 + y^5)^{1/5}$ and $a \odot \mathbf{x} = a\mathbf{x}$. Determine whether $\mathcal{A}$ is a vector space. Prove your answer.

11. Let $\mathbf{A}$ be a fixed $m \times n$ matrix, and let $\mathbf{B}$ be a fixed m-vector (in $\mathbb{R}^m$). Let $\mathcal{V}$ be the set of solutions $\mathbf{X}$ (in $\mathbb{R}^n$) to the matrix equation $\mathbf{AX} = \mathbf{B}$. Endow $\mathcal{V}$ with the usual n-vector operations.

    (a) Assume $\mathcal{V}$ is nonempty. Show that the closure properties are satisfied in $\mathcal{V}$ if and only if $\mathbf{B} = \mathbf{0}$.
    (b) Explain why properties (1), (2), (5), (6), (7), and (8) in the definition of a vector space have already been proved for $\mathcal{V}$ in Theorem 1.3.
    (c) Prove that property (3) in the definition of a vector space is satisfied if and only if $\mathbf{B} = \mathbf{0}$.
    (d) Explain why property (4) in the definition makes no sense unless property (3) is satisfied. Prove property (4) when $\mathbf{B} = \mathbf{0}$.
    (e) Use parts (a) through (d) of this exercise to determine necessary and sufficient conditions for $\mathcal{V}$ to be a vector space.

12. Let $\mathcal{V}$ be a vector space.

    (a) Prove that the identity element for vector addition in $\mathcal{V}$ is unique. (Hint: Use a proof by contradiction.)
    (b) Let $\mathbf{v} \in \mathcal{V}$. Prove that the additive inverse for $\mathbf{v}$ in $\mathcal{V}$ is unique.

13. The set $\mathbb{R}^2$ with operations $[x, y] \oplus [w, z] = [x + w - 2, y + z + 3]$ and $a \odot [x, y] = [ax - 2a + 2, ay + 3a - 3]$ is a vector space. Use parts (2) and (3) of Theorem 4.1 to find the zero vector $\mathbf{0}$ and the additive inverse of each vector $\mathbf{v} = [x, y]$ for this vector space. Then check your answers.

14. Let $\mathcal{V}$ be a vector space. Prove the following **cancellation laws**:

    (a) If $\mathbf{u}, \mathbf{v}$, and $\mathbf{w}$ are vectors in $\mathcal{V}$ for which $\mathbf{u} + \mathbf{v} = \mathbf{w} + \mathbf{v}$, then $\mathbf{u} = \mathbf{w}$.
    (b) If a and b are scalars and $\mathbf{v} \neq \mathbf{0}$ is a vector in $\mathcal{V}$ with $a\mathbf{v} = b\mathbf{v}$, then $a = b$.
    (c) If $a \neq 0$ is a scalar and $\mathbf{v}, \mathbf{w} \in \mathcal{V}$ with $a\mathbf{v} = a\mathbf{w}$, then $\mathbf{v} = \mathbf{w}$.

15. Prove that the set $\mathcal{P}$ of all polynomials with real coefficients forms a vector space under the usual operations of polynomial (term-by-term) addition and scalar multiplication.

16. Let X be any set, and let $\mathcal{V} = \{$all real-valued functions with domain $X\}$. Prove that $\mathcal{V}$ is a vector space using ordinary addition and scalar multiplication of real-valued functions.

17. Let $\mathbf{v}_1, \ldots, \mathbf{v}_n$ be vectors in a vector space $\mathcal{V}$, and let $a_1, \ldots, a_n$ be any real numbers. Use induction to prove that $\sum_{i=1}^{n} a_i \mathbf{v}_i$ is in $\mathcal{V}$.

18. Prove part (2) of Theorem 4.1.

19. Prove that every nontrivial vector space has an infinite number of elements.

★20. True or False:

  (a) The set $\mathbb{R}^n$ under any operations of "addition" and "scalar multiplication" is a vector space.

  (b) The set of all polynomials of degree 7 is a vector space under the usual operations of addition and scalar multiplication.

  (c) The set of all polynomials of degree ≤ 7 is a vector space under the usual operations of addition and scalar multiplication.

  (d) If $\mathbf{x}$ is a vector in a vector space $\mathcal{V}$, and c is a nonzero scalar, then $c\mathbf{x} = \mathbf{0}$ implies $\mathbf{x} = \mathbf{0}$.

  (e) In a vector space, scalar multiplication by the zero scalar always results in the zero scalar.

  (f) In a vector space, scalar multiplication of a vector $\mathbf{x}$ by -1 always results in the additive inverse of $\mathbf{x}$.

  (g) The set of all real valued functions f on $\mathbb{R}$ such that $f(1) = 0$ is a vector space under the usual operations of addition and scalar multiplication.

4.2 Subspaces

Section 4.1 presented several examples in which two vector spaces share the same addition and scalar multiplication operations, with one as a subset of the other. In fact, most of these examples involve subsets of either $\mathbb{R}^n$, $\mathcal{M}_{mn}$, or the vector space of real-valued functions defined on some set (see Exercise 16 in Section 4.1). As we will see, when a vector space is a subset of a known vector space and has the same operations, it becomes easier to handle. These subsets, called **subspaces**, also provide additional information about the larger vector space.

Definition of a Subspace and Examples

DEFINITION

Let $\mathcal{V}$ be a vector space. Then $\mathcal{W}$ is a **subspace** of $\mathcal{V}$ if and only if $\mathcal{W}$ is a subset of $\mathcal{V}$, and $\mathcal{W}$ is itself a vector space with the same operations as $\mathcal{V}$.

That is, $\mathcal{W}$ is a subspace of $\mathcal{V}$ if and only if $\mathcal{W}$ is a vector space inside $\mathcal{V}$ such that for every a in $\mathbb{R}$ and every $\mathbf{v}$ and $\mathbf{w}$ in $\mathcal{W}$, $\mathbf{v} + \mathbf{w}$ and $a\mathbf{v}$ yield the same vectors when the operations are performed in $\mathcal{W}$ as when they are performed in $\mathcal{V}$.

EXAMPLE 1

Example 3 of Section 4.1 showed that the set of points lying on a plane $\mathcal{W}$ through the origin in $\mathbb{R}^3$ forms a vector space under the usual addition and scalar multiplication in $\mathbb{R}^3$. $\mathcal{W}$ is certainly a subset of $\mathbb{R}^3$. Hence, the vector space $\mathcal{W}$ is a subspace of $\mathbb{R}^3$. ∎

EXAMPLE 2

The set $\mathcal{S}$ of scalar multiples of the vector $[1, 3, 2]$ in $\mathbb{R}^3$ forms a vector space under the usual addition and scalar multiplication in $\mathbb{R}^3$ (see Exercise 2 in Section 4.1). $\mathcal{S}$ is certainly a subset of $\mathbb{R}^3$. Hence, $\mathcal{S}$ is a subspace of $\mathbb{R}^3$. Notice that $\mathcal{S}$ corresponds geometrically to the set of points lying on the line through the origin in $\mathbb{R}^3$ in the direction of the vector $[1, 3, 2]$ (see Figure 4.2). In the same manner, every line through the origin determines a subspace of $\mathbb{R}^3$, namely, the set of scalar multiples of a nonzero vector in the direction of that line. ∎

Figure 4.2

Line containing all scalar multiples of $[1, 3, 2]$

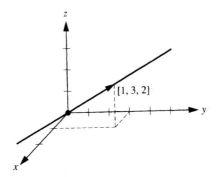

EXAMPLE 3

Let $\mathcal{V}$ be any vector space. Then $\mathcal{V}$ is a subspace of itself (why?). Also, if $\mathcal{W}$ is the subset $\{\mathbf{0}\}$ of $\mathcal{V}$, then $\mathcal{W}$ is a vector space under the same operations as $\mathcal{V}$ (see Example 2 of Section 4.1). Therefore, $\mathcal{W} = \{\mathbf{0}\}$ is a subspace of $\mathcal{V}$. ∎

Although the subspaces $\mathcal{V}$ and $\{\mathbf{0}\}$ of a vector space $\mathcal{V}$ are important, they occasionally complicate matters because they must be considered as special cases in proofs. The subspace $\mathcal{W} = \{\mathbf{0}\}$ is called the **trivial subspace** of $\mathcal{V}$. A vector space containing at least one nonzero vector has at least two distinct subspaces. These are the trivial subspace and the vector space itself. In fact, under the usual operations, $\mathbb{R}$ has only these two subspaces (see Exercise 16).

All subspaces of $\mathcal{V}$ other than $\mathcal{V}$ itself are called **proper subspaces** of $\mathcal{V}$. If we consider Examples 1 to 3 in the context of $\mathbb{R}^3$, we find at least four different types of subspaces of $\mathbb{R}^3$. These are the trivial subspace $\{[0, 0, 0]\} = \{\mathbf{0}\}$, subspaces like Example 2 that can be geometrically represented as a line (thus "resembling" $\mathbb{R}$), subspaces like Example 1 that can be represented as a plane (thus "resembling" $\mathbb{R}^2$), and the subspace $\mathbb{R}^3$ itself.[3] All but the last are proper subspaces. Later we will show that each subspace of $\mathbb{R}^3$ is, in fact, one of these four types. Similarly, we will show later that all subspaces of $\mathbb{R}^n$ "resemble" $\{\mathbf{0}\}$, $\mathbb{R}$, $\mathbb{R}^2$, $\mathbb{R}^3$, ..., $\mathbb{R}^{n-1}$, or $\mathbb{R}^n$.

EXAMPLE 4 Consider the vector spaces (using ordinary function addition and scalar multiplication) in the following chain:

$$\mathcal{P}_0 \subset \mathcal{P}_1 \subset \mathcal{P}_2 \subset \cdots \subset \mathcal{P}$$
$$\subset \{\text{differentiable real-valued functions on } \mathbb{R}\}$$
$$\subset \{\text{continuous real-valued functions on } \mathbb{R}\}$$
$$\subset \{\text{all real-valued functions on } \mathbb{R}\}.$$

Some of these we encountered in Section 4.1, and the rest are discussed in Exercise 7 of this section. Each of these vector spaces is a proper subspace of every vector space after it in the chain (why?). ∎

When Is a Subset a Subspace?

It is important to note that not every subset of a vector space is a subspace. A subset $\mathcal{S}$ of a vector space $\mathcal{V}$ fails to be a subspace of $\mathcal{V}$ if $\mathcal{S}$ does not satisfy the properties of a vector space in its own right or if $\mathcal{S}$ does not use the same operations as $\mathcal{V}$.

EXAMPLE 5 Consider the first quadrant in $\mathbb{R}^2$ — that is, the set Ω of all 2-vectors of the form $[x, y]$ where $x \geq 0$ and $y \geq 0$. This subset Ω of $\mathbb{R}^2$ is not a vector space under the normal operations of $\mathbb{R}^2$ because it is not closed under scalar multiplication. (For example, $[3, 4]$ is in Ω, but $-2 \cdot [3, 4] = [-6, -8]$ is not in Ω.) Therefore, Ω cannot be a subspace of $\mathbb{R}^2$. ∎

EXAMPLE 6 Consider the vector space $\mathbb{R}$ under the usual operations. Let $\mathcal{W}$ be the subset $\mathbb{R}^+$. By Example 7 of Section 4.1, we know that $\mathcal{W}$ is a vector space under the unusual operations $\oplus$ and $\odot$, where $\oplus$ represents multiplication and $\odot$ represents exponentiation. Although $\mathcal{W}$ is a nonempty subset of $\mathbb{R}$ and is itself a vector space, $\mathcal{W}$ is not a subspace of $\mathbb{R}$ because $\mathcal{W}$ and $\mathbb{R}$ do not share the same operations. ∎

The following theorem provides a shortcut for verifying that a (nonempty) subset $\mathcal{W}$ of a vector space is a subspace; if the closure properties hold for $\mathcal{W}$, then the remaining eight vector space properties automatically follow as well.

---

[3] Although some subspaces of $\mathbb{R}^3$ "resemble" $\mathbb{R}$ and $\mathbb{R}^2$ geometrically, note that $\mathbb{R}$ and $\mathbb{R}^2$ are not actually subspaces of $\mathbb{R}^3$ because they are not subsets of $\mathbb{R}^3$.

THEOREM 4.2

Let $\mathcal{V}$ be a vector space, and let $\mathcal{W}$ be a nonempty subset of $\mathcal{V}$ using the same operations. Then $\mathcal{W}$ is a subspace of $\mathcal{V}$ if and only if $\mathcal{W}$ is closed under vector addition and scalar multiplication in $\mathcal{V}$.

Notice that this theorem applies only to *nonempty subsets* of a vector space. Even though the empty set is a subset of every vector space, it is not a subspace of any vector space because it does not contain an additive identity.

PROOF OF THEROEM 4.2

Since this is an "if and only if" statement, the proof has two parts. First we must show that if $\mathcal{W}$ is a subspace of $\mathcal{V}$, then it is closed under the two operations. Now, as a subspace, $\mathcal{W}$ is itself a vector space. Hence, the closure properties hold for $\mathcal{W}$ as they do for any vector space.

For the other part of the proof, we must show that if the closure properties hold for a nonempty subset $\mathcal{W}$ of $\mathcal{V}$, then $\mathcal{W}$ is itself a vector space under the operations in $\mathcal{V}$. That is, we must prove the remaining eight vector space properties hold for $\mathcal{W}$.

Properties (1), (2), (5), (6), (7), and (8) are all true in $\mathcal{W}$ because they are true in $\mathcal{V}$, a known vector space. That is, since these properties hold for all vectors in $\mathcal{V}$, they must be true for all vectors in its subset, $\mathcal{W}$. For example, to prove property (1) for $\mathcal{W}$, let $\mathbf{u}, \mathbf{v} \in \mathcal{W}$. Then,

$$\underbrace{\mathbf{u} + \mathbf{v}}_{\text{addition in } \mathcal{W}} = \underbrace{\mathbf{u} + \mathbf{v}}_{\text{addition in } \mathcal{V}} \quad \text{because } \mathcal{W} \text{ and } \mathcal{V} \text{ share the same operations}$$

$$= \underbrace{\mathbf{v} + \mathbf{u}}_{\text{addition in } \mathcal{V}} \quad \text{because } \mathcal{V} \text{ is a vector space and property (1) holds}$$

$$= \underbrace{\mathbf{v} + \mathbf{u}}_{\text{addition in } \mathcal{W}} \quad \text{because } \mathcal{W} \text{ and } \mathcal{V} \text{ share the same operations.}$$

Next we prove property (3), the existence of an additive identity in $\mathcal{W}$. Because $\mathcal{W}$ is nonempty, we can choose an element $\mathbf{w}_1$ from $\mathcal{W}$. Now $\mathcal{W}$ is closed under scalar multiplication, so $0\mathbf{w}_1$ is in $\mathcal{W}$. However, since this is the same operation as in $\mathcal{V}$, a known vector space, part (2) of Theorem 4.1 implies that $0\mathbf{w}_1 = \mathbf{0}$. Hence, $\mathbf{0}$ is in $\mathcal{W}$. Because $\mathbf{0} + \mathbf{v} = \mathbf{v}$ for all $\mathbf{v}$ in $\mathcal{V}$, it follows that $\mathbf{0} + \mathbf{w} = \mathbf{w}$ for all $\mathbf{w}$ in $\mathcal{W}$. Therefore, $\mathcal{W}$ contains the same additive identity that $\mathcal{V}$ has.

Finally we must prove that property (4), the existence of additive inverses, holds for $\mathcal{W}$. Let $\mathbf{w} \in \mathcal{W}$. Then $\mathbf{w} \in \mathcal{V}$. Part (3) of Theorem 4.1 shows $(-1)\mathbf{w}$ is the additive inverse of $\mathbf{w}$ in $\mathcal{V}$. If we can show that this additive inverse is also in $\mathcal{W}$, we will be done. But since $\mathcal{W}$ is closed under scalar multiplication, $(-1)\mathbf{w} \in \mathcal{W}$. ∎

Checking for Subspaces in $\mathcal{M}_{nn}$ and $\mathbb{R}^n$

In the next three examples, we apply Theorem 4.2 to determine whether several subsets of $\mathcal{M}_{nn}$ and $\mathbb{R}^n$ are subspaces. Assume that $\mathcal{M}_{nn}$ and $\mathbb{R}^n$ have the usual operations.

EXAMPLE 7 Consider $\mathcal{U}_n$, the set of upper triangular $n \times n$ matrices. Since $\mathcal{U}_n$ is nonempty, we may apply Theorem 4.2 to see whether $\mathcal{U}_n$ is a subspace of $\mathcal{M}_{nn}$. Closure of $\mathcal{U}_n$ under vector addition holds because the sum of any two $n \times n$ upper triangular matrices is again upper triangular. The closure property in $\mathcal{U}_n$ for scalar multiplication also holds, since any scalar multiple of an upper triangular matrix is again upper triangular. Hence, $\mathcal{U}_n$ is a subspace of $\mathcal{M}_{nn}$.

Similar arguments show that $\mathcal{L}_n$ (lower triangular $n \times n$ matrices) and $\mathcal{D}_n$ (diagonal $n \times n$ matrices) are also subspaces of $\mathcal{M}_{nn}$. ■

The subspace $\mathcal{D}_n$ of $\mathcal{M}_{nn}$ in Example 7 is the intersection of the subspaces $\mathcal{U}_n$ and $\mathcal{L}_n$. In fact, the intersection of subspaces of a vector space always produces a subspace under the same operations (see Exercise 18).

If either closure property fails to hold for a subset, it cannot be a subspace. For this reason, none of the following subsets of $\mathcal{M}_{nn}$, $n \geq 2$, is a subspace:

(1) the set of nonsingular $n \times n$ matrices
(2) the set of singular $n \times n$ matrices
(3) the set of $n \times n$ matrices in reduced row echelon form.

You should check that the closure property for vector addition fails in each case and that the closure property for scalar multiplication fails in (1) and (3).

EXAMPLE 8 Let $\mathcal{Y}$ be the set of vectors in $\mathbb{R}^4$ of the form $[a, 0, b, 0]$, that is, 4-vectors whose second and fourth coordinates are zero. We prove that $\mathcal{Y}$ is a subspace of $\mathbb{R}^4$ by checking the closure properties.

To prove closure under vector addition, we must add two arbitrary elements of $\mathcal{Y}$ and check that the result has the correct form for a vector in $\mathcal{Y}$. Now, $[a, 0, b, 0] + [c, 0, d, 0] = [(a + c), 0, (b + d), 0]$. The second and fourth coordinates of the sum are zero, so $\mathcal{Y}$ is closed under addition. Similarly, we must prove closure under scalar multiplication. Now, $k[a, 0, b, 0] = [ka, 0, kb, 0]$. Since the second and fourth coordinates of the product are zero, $\mathcal{Y}$ is closed under scalar multiplication. Hence, by Theorem 4.2, $\mathcal{Y}$ is a subspace of $\mathbb{R}^4$. ■

EXAMPLE 9 Let $\mathcal{W}$ be the set of vectors in $\mathbb{R}^3$ of the form $\left[a, \; b, \; \frac{1}{2}a - 2b\right]$, that is, 3-vectors whose third coordinate is half the first coordinate minus twice the second coordinate. We show that $\mathcal{W}$ is a subspace of $\mathbb{R}^3$ by checking the closure properties.

Checking closure under vector addition, we have

$$\left[a, \; b, \; \frac{1}{2}a - 2b\right] + \left[c, \; d, \; \frac{1}{2}c - 2d\right] = \left[a + c, \; b + d, \; \frac{1}{2}a - 2b + \frac{1}{2}c - 2d\right]$$

$$= \left[a + c, \; b + d, \; \frac{1}{2}(a + c) - 2(b + d)\right],$$

which has the required form, since it equals $\left[A, \; B, \; \frac{1}{2}A - 2B\right]$, where $A = a + c$ and $B = b + d$.

Checking closure under scalar multiplication, we get

$$k\left[a,\ b,\ \frac{1}{2}a - 2b\right] = \left[ka,\ kb,\ k\left(\frac{1}{2}a - 2b\right)\right] = \left[ka,\ kb,\ \frac{1}{2}(ka) - 2(kb)\right],$$

which has the required form (why?).

Note that

$$\left[a,\ b,\ \frac{1}{2}a - 2b\right] = a\left[1, 0, \frac{1}{2}\right] + b[0, 1, -2],$$

and so $\mathcal{W}$ consists of the set of all linear combinations of $\left[1, 0, \frac{1}{2}\right]$ and $[0, 1, -2]$. Geometrically, $\mathcal{W}$ is the plane in $\mathbb{R}^3$ through the origin containing the vectors $\left[1, 0, \frac{1}{2}\right]$, and $[0, 1, -2]$, shown in Figure 4.3. This plane is the set of all possible "destinations" using these two directions (starting from the origin). This is the type of subspace of $\mathbb{R}^3$ discussed in Example 1. ∎

Figure 4.3

The plane through the origin containing $\left[1, 0, \frac{1}{2}\right]$, and $[0, 1, -2]$

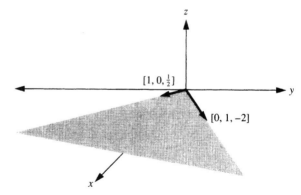

The following subsets of $\mathbb{R}^n$ are not subspaces. In each case, at least one of the two closure properties fails. (Can you determine which ones?)

(1) The set of n-vectors whose first coordinate is nonnegative (in $\mathbb{R}^2$, this set is a half-plane)
(2) The set of unit n-vectors (in $\mathbb{R}^3$, this set is a sphere)
(3) For $n \geq 2$, the set of n-vectors with a zero in at least one coordinate (in $\mathbb{R}^3$, this set is the union of three planes)
(4) The set of n-vectors having all integer coordinates
(5) For $n \geq 2$, the set of all n-vectors whose first two coordinates add up to 3 (in $\mathbb{R}^2$, this is the line $x + y = 3$)

The subsets (2) and (5), which do not contain the additive identity $\mathbf{0}$ of $\mathbb{R}^n$, can quickly be disqualified as subspaces. In general,

If a subset $\mathcal{S}$ of a vector space $\mathcal{V}$ does not contain the zero vector $\mathbf{0}$ of $\mathcal{V}$, then $\mathcal{S}$ is not a subspace of $\mathcal{V}$.

Checking for the presence of the additive identity is usually easy and thus is a fast way to show that certain subsets are not subspaces.

Linear Combinations Remain in a Subspace

As in Chapter 1, we define a **linear combination** of vectors in a general vector space to be a sum of scalar multiples of the vectors. The next theorem asserts that if a finite set of vectors is in a given subspace of a vector space, then so is any linear combination of those vectors.

THEOREM 4.3

Let $\mathcal{W}$ be a subspace of a vector space $\mathcal{V}$, and let $\mathbf{v}_1, \mathbf{v}_2, \ldots, \mathbf{v}_n$ be vectors in $\mathcal{W}$. Then, for any scalars $a_1, a_2, \ldots, a_n$, we have $a_1\mathbf{v}_1 + a_2\mathbf{v}_2 + \cdots + a_n\mathbf{v}_n \in \mathcal{W}$.

Essentially, this theorem points out that a subspace is "closed under linear combinations." That is, when the vectors of a subspace are used to form linear combinations, all possible "destination vectors" remain in the subspace.

PROOF OF THEOREM 4.3

Suppose that $\mathcal{W}$ is a subspace of a vector space $\mathcal{V}$. We give a proof by induction on n.

    Base Step: If $n = 1$, then we must show that if $\mathbf{v}_1 \in \mathcal{W}$ and a_1 is a scalar, then $a_1\mathbf{v}_1 \in \mathcal{W}$. But this is certainly true since the subspace $\mathcal{W}$ is closed under scalar multiplication.

    Inductive Step: Assume that the theorem is true for any linear combination of n vectors in $\mathcal{W}$. We must prove the theorem holds for a linear combination of $n + 1$ vectors. Suppose $\mathbf{v}_1, \mathbf{v}_2, \ldots, \mathbf{v}_n, \mathbf{v}_{n+1}$ are vectors in $\mathcal{W}$, and $a_1, a_2, \ldots, a_n, a_{n+1}$ are scalars. We must show that $a_1\mathbf{v}_1 + a_2\mathbf{v}_2 + \cdots + a_n\mathbf{v}_n + a_{n+1}\mathbf{v}_{n+1} \in \mathcal{W}$. However, by the inductive hypothesis, we know that $a_1\mathbf{v}_1 + a_2\mathbf{v}_2 + \cdots + a_n\mathbf{v}_n \in \mathcal{W}$. Also, $a_{n+1}\mathbf{v}_{n+1} \in \mathcal{W}$, since $\mathcal{W}$ is closed under scalar multiplication. But since $\mathcal{W}$ is also closed under addition, the sum of any two vectors in $\mathcal{W}$ is again in $\mathcal{W}$, so $(a_1\mathbf{v}_1 + a_2\mathbf{v}_2 + \cdots + a_n\mathbf{v}_n) + (a_{n+1}\mathbf{v}_{n+1}) \in \mathcal{W}$.

EXAMPLE 10

In Example 9, we found that the set $\mathcal{W}$ of all vectors of the form $\left[a, \ b, \ \frac{1}{2}a - 2b\right]$ is a subspace of $\mathbb{R}^3$. In particular, $[1, 0, \frac{1}{2}]$ and $[0, 1, -2]$ are in $\mathcal{W}$. By Theorem 4.3, any linear combination of these vectors is also in $\mathcal{W}$. For example, $6[1, 0, \frac{1}{2}] - 5[0, 1, -2] = [6, -5, 13]$ and $-4[1, 0, \frac{1}{2}] + 2[0, 1, -2] = [-4, 2, -6]$ are both in $\mathcal{W}$. Of course, this makes sense geometrically, since $\mathcal{W}$ is a plane through the origin, and any linear combination of vectors in such a plane remains in that plane. ∎

An Eigenspace is a Subspace

We conclude this section by noting that any eigenspace of an $n \times n$ matrix is a subspace of $\mathbb{R}^n$. (In fact, this is why the word "space" appears in the term "eigenspace.")

THEOREM 4.4

Let $\mathbf{A}$ be an $n \times n$ matrix, and let λ be an eigenvalue for $\mathbf{A}$, having eigenspace E_λ. Then E_λ is a subspace of $\mathbb{R}^n$.

PROOF OF THEOREM 4.4

Let λ be an eigenvalue for an $n \times n$ matrix $\mathbf{A}$. By definition, the eigenspace E_λ of λ is the set of all n-vectors $\mathbf{X}$ having the property that $\mathbf{AX} = \lambda\mathbf{X}$, including the zero n-vector. Thus, E_λ is a nonempty subset of $\mathbb{R}^n$. We must show that E_λ is closed under addition and scalar multiplication.

Let $\mathbf{X}_1$, $\mathbf{X}_2$ be any two vectors in E_λ. To show that $\mathbf{X}_1 + \mathbf{X}_2 \in E_\lambda$, we need to verify that $\mathbf{A}(\mathbf{X}_1 + \mathbf{X}_2) = \lambda(\mathbf{X}_1 + \mathbf{X}_2)$. But, $\mathbf{A}(\mathbf{X}_1 + \mathbf{X}_2) = \mathbf{AX}_1 + \mathbf{AX}_2 = \lambda\mathbf{X}_1 + \lambda\mathbf{X}_2 = \lambda(\mathbf{X}_1 + \mathbf{X}_2)$.

Similarly, let $\mathbf{X}$ be a vector in E_λ, and let c be a scalar. We must show that $c\mathbf{X} \in E_\lambda$. But, $\mathbf{A}(c\mathbf{X}) = c(\mathbf{AX}) = c(\lambda\mathbf{X}) = \lambda(c\mathbf{X})$, and so $c\mathbf{X} \in E_\lambda$. Hence, E_λ is a subspace of $\mathbb{R}^n$. ∎

EXAMPLE 11 Consider
$$\mathbf{A} = \begin{bmatrix} 16 & -4 & -2 \\ 3 & 3 & -6 \\ 2 & -8 & 11 \end{bmatrix}.$$

Computing $|x\mathbf{I}_3 - \mathbf{A}|$ produces $p_\mathbf{A}(x) = x^3 - 30x^2 + 225x = x(x-15)^2$. Solving $(0\mathbf{I}_3 - \mathbf{A})\mathbf{X} = \mathbf{0}$ yields $E_0 = \{c[1, 3, 2] \mid c \in \mathbb{R}\}$. Thus, the eigenspace for $\lambda_1 = 0$ is the subspace of $\mathbb{R}^3$ from Example 2. Similarly, solving $(15\mathbf{I}_3 - \mathbf{A}) = \mathbf{0}$ gives $E_{15} = \{a[4, 1, 0] + b[2, 0, 1] \mid a, b \in \mathbb{R}\}$. By Theorem 4.4, E_{15} is also a subspace of $\mathbb{R}^3$. Although it is not obvious, E_{15} is the same subspace of $\mathbb{R}^3$ that we studied in Examples 9 and 10 (see Exercises 14(b) and 14(c)). ∎

Exercises for Section 4.2

Note: From this point onward in the book, use a calculator or available software packages to avoid tedious calculations.

1. Prove or disprove that each given subset of $\mathbb{R}^2$ is a subspace of $\mathbb{R}^2$ under the usual vector operations. (In these problems, a and b represent arbitrary real numbers.)

   ⋆(a) The set of unit 2-vectors
   (b) The set of 2-vectors of the form $[1, a]$
   ⋆(c) The set of 2-vectors of the form $[a, 2a]$
   (d) The set of 2-vectors having a zero in at least one coordinate
   ⋆(e) The set $\{[1, 2]\}$
   (f) The set of 2-vectors whose second coordinate is zero
   ⋆(g) The set of 2-vectors of the form $[a, b]$, where $|a| = |b|$
   (h) The set of points in the plane lying on the line $y = -3x$
   (i) The set of points in the plane lying on the line $y = 7x - 5$
   ⋆(j) The set of points lying on the parabola $y = x^2$

(k) The set of points in the plane lying above the line $y = 2x - 5$

*(l) The set of points in the plane lying inside the circle of radius 1 centered at the origin

2. Prove or disprove that each given subset of $\mathcal{M}_{22}$ is a subspace of $\mathcal{M}_{22}$ under the usual matrix operations. (In these problems, a and b represent arbitrary real numbers.)

*(a) The set of matrices of the form $\begin{bmatrix} a & -a \\ b & 0 \end{bmatrix}$

(b) The set of 2×2 matrices that have at least one row of zeroes

*(c) The set of symmetric 2×2 matrices

(d) The set of nonsingular 2×2 matrices

*(e) The set of 2×2 matrices having the sum of all entries zero

(f) The set of 2×2 matrices having trace zero (Recall that the *trace* of a square matrix is the sum of the main diagonal entries.)

*(g) The set of 2×2 matrices $\mathbf{A}$ such that $\mathbf{A} \begin{bmatrix} 1 & 3 \\ -2 & -6 \end{bmatrix} = \begin{bmatrix} 0 & 0 \\ 0 & 0 \end{bmatrix}$

*(h) The set of 2×2 matrices having the product of all entries zero

3. Prove or disprove that each given subset of $\mathcal{P}_5$ is a subspace of $\mathcal{P}_5$ under the usual operations.

*(a) $\{\mathbf{p} \in \mathcal{P}_5 |$ the coefficient of the first-degree term of $\mathbf{p}$ equals the coefficient of the fifth-degree term of $\mathbf{p}\}$

*(b) $\{\mathbf{p} \in \mathcal{P}_5 | \mathbf{p}(3) = 0\}$

(c) $\{\mathbf{p} \in \mathcal{P}_5 |$ the sum of the coefficients of $\mathbf{p}$ is zero$\}$

(d) $\{\mathbf{p} \in \mathcal{P}_5 | \mathbf{p}(3) = \mathbf{p}(5)\}$

*(e) $\{\mathbf{p} \in \mathcal{P}_5 | \mathbf{p}$ is an odd-degree polynomial (highest-order nonzero term has odd degree)$\}$

(f) $\{\mathbf{p} \in \mathcal{P}_5 | \mathbf{p}$ has a relative maximum at $x = 0\}$

*(g) $\{\mathbf{p} \in \mathcal{P}_5 | \mathbf{p}'(4) = 0$, where $\mathbf{p}'$ is the derivative of $\mathbf{p}\}$

(h) $\{\mathbf{p} \in \mathcal{P}_5 | \mathbf{p}'(4) = 1$, where $\mathbf{p}'$ is the derivative of $\mathbf{p}\}$

4. Show that the set of vectors of the form $[a, b, 0, c, a - 2b + c]$ in $\mathbb{R}^5$ forms a subspace of $\mathbb{R}^5$ under the usual operations.

5. Show that the set of vectors of the form $[2a - 3b, a - 5c, a, 4c - b, c]$ in $\mathbb{R}^5$ forms a subspace of $\mathbb{R}^5$ under the usual operations.

6. (a) Prove that the set of all 3-vectors orthogonal to $[1, -1, 4]$ forms a subspace of $\mathbb{R}^3$.

(b) Is the subspace from part (a) all of $\mathbb{R}^3$, a plane passing through the origin in $\mathbb{R}^3$, or a line passing through the origin in $\mathbb{R}^3$?

7. Show that each of the following sets is a subspace of the vector space of all real-valued functions on the given domain, under the usual operations of function addition and scalar multiplication:

(a) The set of continuous real-valued functions with domain $\mathbb{R}$

(b) The set of differentiable real-valued functions with domain $\mathbb{R}$

(c) The set of all real-valued functions $\mathbf{f}$ defined on the interval $[0, 1]$ such that $\mathbf{f}\left(\frac{1}{2}\right) = 0$ (Compare this vector space with the set in Example 9 of Section 4.1.)

(d) The set of all continuous real-valued functions **f** defined on the interval [0, 1] such that $\int_0^1 \mathbf{f}(x)\, dx = 0$

8. Let $\mathcal{W}$ be the set of differentiable real-valued functions $y = \mathbf{f}(x)$ defined on $\mathbb{R}$ that satisfy the differential equation $3(dy/dx) - 2y = 0$. Show that, under the usual function operations, $\mathcal{W}$ is a subspace of the vector space of all differentiable real-valued functions. (Do not forget to show $\mathcal{W}$ is nonempty.)

9. Show that the set $\mathcal{W}$ of solutions to the differential equation $y'' + 2y' - 9y = 0$ is a subspace of the vector space of all twice-differentiable real-valued functions defined on $\mathbb{R}$. (Do not forget to show that $\mathcal{W}$ is nonempty.)

10. Prove that the set of discontinuous real-valued functions defined on $\mathbb{R}$ (for example, $\mathbf{f}(x) = \begin{cases} 0 \text{ if } x \le 0 \\ 1 \text{ if } x > 0 \end{cases}$) with the usual function operations is not a subspace of the vector space of all real-valued functions with domain $\mathbb{R}$.

11. Let **A** be a fixed $n \times n$ matrix, and let $\mathcal{W}$ be the subset of $\mathcal{M}_{nn}$ of all $n \times n$ matrices that commute with **A** under multiplication (that is, $\mathbf{B} \in \mathcal{W}$ if and only if $\mathbf{AB} = \mathbf{BA}$). Show that $\mathcal{W}$ is a subspace of $\mathcal{M}_{nn}$ under the usual vector space operations. (Do not forget to show that $\mathcal{W}$ is nonempty.)

12. (a) A careful reading of the proof of Theorem 4.2 reveals that only closure under scalar multiplication (not closure under addition) is sufficient to prove the remaining eight vector space properties for $\mathcal{W}$. Explain, nevertheless, why closure under addition is a necessary condition for $\mathcal{W}$ to be a subspace of $\mathcal{V}$.

(b) Show that the set of singular $n \times n$ matrices is closed under scalar multiplication in $\mathcal{M}_{nn}$.

(c) Use parts (a) and (b) to determine which of the eight vector space properties are true for the set of singular $n \times n$ matrices.

(d) Show that the set of singular $n \times n$ matrices is not closed under vector addition and hence is not a subspace of $\mathcal{M}_{nn}$ $(n \ge 2)$.

★(e) Is the set of nonsingular $n \times n$ matrices closed under scalar multiplication? Why or why not?

13. (a) Prove that the set of all points lying on a line passing through the origin in $\mathbb{R}^2$ is a subspace of $\mathbb{R}^2$ (under the usual operations).

(b) Prove that the set of all points in $\mathbb{R}^2$ lying on a line not passing through the origin does not form a subspace of $\mathbb{R}^2$ (under the usual operations).

14. Let $\mathcal{W}$ be the subspace from Examples 9 and 10, and let **A** and E_{15} be as given in Example 11.

(a) Use Theorem 4.2 to prove directly that E_{15} is a subspace of $\mathbb{R}^3$.

(b) Show that $E_{15} \subseteq \mathcal{W}$ by proving that every vector in E_{15} has the form $[a, b, \frac{1}{2}a - 2b]$.

(c) Prove that $\mathcal{W} \subseteq E_{15}$ by showing that every nonzero vector of the form $[a, b, \frac{1}{2}a - 2b]$ is an eigenvector for **A** corresponding to $\lambda_2 = 15$.

★15. Suppose $\mathbf{A}$ is an $n \times n$ matrix and $\lambda \in \mathbb{R}$ is *not* an eigenvalue for $\mathbf{A}$. Determine exactly which vectors are in $S = \{\mathbf{X} \in \mathbb{R}^n \mid \mathbf{AX} = \lambda\mathbf{X}\}$. Is this set a subspace of $\mathbb{R}^n$? Why or why not?

16. Prove that $\mathbb{R}$ (under the usual operations) has no subspaces except $\mathbb{R}$ and $\{\mathbf{0}\}$. (Hint: Let $\mathcal{V}$ be a nontrivial subspace of $\mathbb{R}$, and show that $\mathcal{V} = \mathbb{R}$.)

17. Let $\mathcal{W}$ be a subspace of a vector space $\mathcal{V}$. Show that the set $\mathcal{W}' = \{\mathbf{v} \in \mathcal{V} \mid \mathbf{v} \notin \mathcal{W}\}$ is not a subspace of $\mathcal{V}$.

18. Let $\mathcal{V}$ be a vector space, and let $\mathcal{W}_1$ and $\mathcal{W}_2$ be subspaces of $\mathcal{V}$. Prove that $\mathcal{W}_1 \cap \mathcal{W}_2$ is a subspace of $\mathcal{V}$. (Do not forget to show $\mathcal{W}_1 \cap \mathcal{W}_2$ is nonempty.)

19. Let $\mathcal{V}$ be any vector space, and let $\mathcal{W}$ be a nonempty subset of $\mathcal{V}$.

(a) Prove that $\mathcal{W}$ is a subspace of $\mathcal{V}$ if and only if $a\mathbf{w}_1 + b\mathbf{w}_2$ is an element of $\mathcal{W}$ for every $a, b \in \mathbb{R}$ and every $\mathbf{w}_1, \mathbf{w}_2 \in \mathcal{W}$. (Hint: For one half of the proof, first consider the case where $a = b = 1$ and then the case where $b = 0$ and a is arbitrary.)

(b) Prove that $\mathcal{W}$ is a subspace of $\mathcal{V}$ if and only if $a\mathbf{w}_1 + \mathbf{w}_2$ is an element of $\mathcal{W}$ for every real number a and every $\mathbf{w}_1$ and $\mathbf{w}_2$ in $\mathcal{W}$.

20. Let $\mathcal{W}$ be a nonempty subset of a vector space $\mathcal{V}$, and suppose every linear combination of vectors in $\mathcal{W}$ is also in $\mathcal{W}$. Prove that $\mathcal{W}$ is a subspace of $\mathcal{V}$. (This is the converse of Theorem 4.3.)

21. Let λ be an eigenvalue for an $n \times n$ matrix $\mathbf{A}$. Show that if $\mathbf{X}_1, \ldots, \mathbf{X}_k$ are eigenvectors for $\mathbf{A}$ corresponding to λ, then any linear combination of $\mathbf{X}_1, \ldots, \mathbf{X}_k$ is in E_λ.

★22. True or False:

(a) A nonempty subset $\mathcal{W}$ of a vector space $\mathcal{V}$ is always a subspace of $\mathcal{V}$ under the same operations as those in $\mathcal{V}$.

(b) Every vector space has at least one subspace.

(c) Any plane $\mathcal{W}$ in $\mathbb{R}^3$ is a subspace of $\mathbb{R}^3$ (under the usual operations).

(d) The set of all lower triangular 5×5 matrices is a subspace of $\mathcal{M}_{55}$ (under the usual operations).

(e) The set of all vectors of the form $[0, a, b, 0]$ is a subspace of $\mathbb{R}^4$ (under the usual operations).

(f) If a subset $\mathcal{W}$ of a vector space $\mathcal{V}$ contains the zero vector $\mathbf{0}$ of $\mathcal{V}$, then $\mathcal{W}$ must be a subspace of $\mathcal{V}$ (under the same operations).

(g) Any linear combination of vectors from a subspace $\mathcal{W}$ of a vector space $\mathcal{V}$ must also be in $\mathcal{W}$.

(h) If λ is an eigenvalue for a 4×4 matrix $\mathbf{A}$, then E_λ is a subspace of $\mathbb{R}^4$.

4.3 Span

In this section, we study the concept of linear combinations in more depth. We show that the set of all linear combinations of the vectors in a subset S of $\mathcal{V}$ forms an important subspace of $\mathcal{V}$, called the span of S in $\mathcal{V}$.

Finite Linear Combinations

In Section 4.2, we introduced linear combinations of vectors in a general vector space. We now extend the concept of linear combination to include the possibility of forming sums of scalar multiples from infinite, as well as finite, sets.

DEFINITION

Let S be a nonempty (possibly infinite) subset of a vector space $\mathcal{V}$. Then a vector $\mathbf{v}$ in $\mathcal{V}$ is a **(finite) linear combination of the vectors in** S if and only if there exists a *finite* subset $\{\mathbf{v}_1, \mathbf{v}_2, \ldots, \mathbf{v}_n\}$ of S such that $\mathbf{v} = a_1\mathbf{v}_1 + a_2\mathbf{v}_2 + \cdots + a_n\mathbf{v}_n$ for some real numbers $a_1, \ldots, a_n$.

Examples 1 and 2 below involve a finite set S, while Examples 3 and 4 use an infinite set S. In all these examples, however, only a *finite* number of vectors from S are used at any given time to form linear combinations.

EXAMPLE 1

Consider the subset $S = \{[1, -1, 0], [1, 0, 2], [0, -2, 5]\}$ of $\mathbb{R}^3$. The vector $[1, -2, -2]$ is a linear combination of the vectors in S according to the definition, because $[1, -2, -2] = 2[1, -1, 0] + (-1)[1, 0, 2]$. In this case, the (finite) subset of S used (from the definition) is $\{[1, -1, 0], [1, 0, 2]\}$. However, we could have used all of S to form the linear combination by placing a zero coefficient in front of the remaining vector $[0, -2, 5]$. That is, $[1, -2, -2] = 2[1, -1, 0] + (-1)[1, 0, 2] + 0[0, -2, 5]$. ■

We see from Example 1 that if S is a *finite* subset of a vector space $\mathcal{V}$, any linear combination $\mathbf{v}$ formed using *some* of the vectors in S can always be formed using *all* the vectors in S by placing zero coefficients on the remaining vectors.

A linear combination formed from a set $\{\mathbf{v}\}$ containing a single vector is just a scalar multiple $a\mathbf{v}$ of $\mathbf{v}$, as we see in the next example.

EXAMPLE 2

Let $S = \{[1, -2, 7]\}$, a subset of $\mathbb{R}^3$ containing a single element. Then the only linear combinations that can be formed from S are scalar multiples of $[1, -2, 7]$, such as $[3, -6, 21]$ and $[-4, 8, -28]$. ■

EXAMPLE 3

Consider $\mathcal{P}$, the vector space of polynomials with real coefficients, and let $S = \{1, x^2, x^4, \ldots\}$, the infinite subset of $\mathcal{P}$ consisting of all nonnegative even powers of x (since $x^0 = 1$). We can form linear combinations of vectors in S using any finite subset of S. For example, $\mathbf{p}(x) = 7x^8 - (1/4)x^4 + 10$ is a linear combination formed from S because it is a sum of scalar multiples of elements of a finite subset $\{x^8, x^4, 1\}$ of S. In fact, the possible linear combinations of vectors in S are precisely the polynomials involving only even powers of x. ■

Notice that we cannot use all of the elements in an infinite set S when forming a linear combination because an "infinite" sum would result. This is why a linear combination is frequently called a *finite* linear combination in order to stress that only a finite number of vectors are combined at any time.

EXAMPLE 4

Let $S = \mathcal{U}_2 \cup \mathcal{L}_2$, an infinite subset of $\mathcal{M}_{22}$. (Recall that $\mathcal{U}_2$ and $\mathcal{L}_2$ are, respectively, the sets of upper and lower triangular 2×2 matrices.) The matrix $\mathbf{A} = \begin{bmatrix} 2 & 3 \\ -1 & \frac{1}{2} \end{bmatrix}$ is a linear combination of the elements in S, because

$$\mathbf{A} = \frac{1}{2}\underbrace{\begin{bmatrix} 4 & 6 \\ 0 & 1 \end{bmatrix}}_{\text{in } \mathcal{U}_2} + (-1)\underbrace{\begin{bmatrix} 0 & 0 \\ 1 & 0 \end{bmatrix}}_{\text{in } \mathcal{L}_2}.$$

But there are many other ways to express $\mathbf{A}$ as a finite linear combination of the elements in S. We can add more elements from S with zero coefficients, as in Example 1, but in this case there are further possibilities. For example,

$$\mathbf{A} = 2\underbrace{\begin{bmatrix} 1 & 0 \\ 0 & 0 \end{bmatrix}}_{\text{in } \mathcal{U}_2 \text{ and } \mathcal{L}_2} + 3\underbrace{\begin{bmatrix} 0 & 1 \\ 0 & 0 \end{bmatrix}}_{\text{in } \mathcal{U}_2} + (-1)\underbrace{\begin{bmatrix} 0 & 0 \\ 1 & 0 \end{bmatrix}}_{\text{in } \mathcal{L}_2} + \frac{1}{2}\underbrace{\begin{bmatrix} 0 & 0 \\ 0 & 1 \end{bmatrix}}_{\text{in } \mathcal{U}_2 \text{ and } \mathcal{L}_2}.$$

∎

Definition of the Span of a Set

DEFINITION

Let S be a nonempty subset of a vector space $\mathcal{V}$. Then the **span** of S in $\mathcal{V}$ is the set of all possible (finite) linear combinations of the vectors in S. We use the notation span(S) to denote the span of S in $\mathcal{V}$.

The span of a set S is a generalization of the row space of a matrix; each is just the set of all linear combinations of a set of vectors. In fact, from this definition

The span of the set of rows of a matrix is precisely the row space of the matrix.

We now consider some examples of the span of a subset.

EXAMPLE 5

In Example 3, we found that for $S = \{1, x^2, x^4, \dots\}$ in $\mathcal{P}$, span(S) is the set of all polynomials containing only even-degree terms. This consists of all the "destinations" obtainable by traveling in the "directions" $1, x^2, x^4, \dots$, etc. Thus, we can visualize span(S) as the set of "possible destinations" in the same sense as the row space is the set of "possible destinations" obtainable from the rows of a given matrix. Notice that we may only use a finite number of the possible "directions" to obtain a given "destination." That is, span(S) only contains polynomials, not infinite series. ∎

EXAMPLE 6

Let $S = \mathcal{U}_2 \cup \mathcal{L}_2$ in $\mathcal{M}_{22}$, as in Example 4. Then span(S) $= \mathcal{M}_{22}$ because every 2×2 matrix can be expressed as a finite linear combination of upper and lower triangular matrices, as follows:

$$\begin{bmatrix} a & b \\ c & d \end{bmatrix} = a\begin{bmatrix} 1 & 0 \\ 0 & 0 \end{bmatrix} + b\begin{bmatrix} 0 & 1 \\ 0 & 0 \end{bmatrix} + c\begin{bmatrix} 0 & 0 \\ 1 & 0 \end{bmatrix} + d\begin{bmatrix} 0 & 0 \\ 0 & 1 \end{bmatrix}.$$

Notice that the span of a given set often (but not always) contains many more vectors than the set itself. ∎

Example 6 shows that, when $S = \mathcal{U}_2 \cup \mathcal{L}_2$ and $\mathcal{V} = \mathcal{M}_{22}$, every vector in $\mathcal{V}$ is a linear combination of vectors in S. That is, span(S) $= \mathcal{V}$ itself. When this happens, we say that $\mathcal{V}$ is **spanned by** S or that S **spans** $\mathcal{V}$. Here, we are using span as a *verb* to indicate that the span (noun) of a set S equals $\mathcal{V}$. Thus, $\mathcal{M}_{22}$ is spanned (verb) by $\mathcal{U}_2 \cup \mathcal{L}_2$, since the span (noun) of $\mathcal{U}_2 \cup \mathcal{L}_2$ is $\mathcal{M}_{22}$.

EXAMPLE 7

Note that $\mathbb{R}^3$ is spanned by $S_1 = \{\mathbf{i}, \mathbf{j}, \mathbf{k}\}$, since span($S_1$) $= \mathbb{R}^3$. That is, every 3-vector can be expressed as a linear combination of $\mathbf{i}$, $\mathbf{j}$, and $\mathbf{k}$ (why?). However, $\mathbb{R}^3$ is not spanned by the smaller set $S_2 = \{\mathbf{i}, \mathbf{j}\}$, since span($S_2$) is the xy-plane in $\mathbb{R}^3$ (why?). More generally, $\mathbb{R}^n$ is spanned by the set of standard unit vectors $\{\mathbf{e}_1, \ldots, \mathbf{e}_n\}$. Note that no proper subset of $\{\mathbf{e}_1, \ldots, \mathbf{e}_n\}$ will span $\mathbb{R}^n$. ∎

Span(S) Is the Minimal Subspace Containing S

The next theorem completely characterizes the span.

THEOREM 4.5

Let S be a nonempty subset of a vector space $\mathcal{V}$. Then:
(1) $S \subseteq$ span(S).
(2) Span(S) is a subspace of $\mathcal{V}$ (under the same operations as $\mathcal{V}$).
(3) If $\mathcal{W}$ is a subspace of $\mathcal{V}$ with $S \subseteq \mathcal{W}$, then span(S) $\subseteq \mathcal{W}$.
(4) Span(S) is the smallest subspace of $\mathcal{V}$ containing S.

PROOF OF THEOREM 4.5
Part (1): We must show that each vector $\mathbf{w} \in S$ is also in span(S). But if $\mathbf{w} \in S$, then $\mathbf{w} = 1\mathbf{w}$ is a sum of scalar multiples from the subset $\{\mathbf{w}\}$ of S. Hence, $\mathbf{w} \in$ span(S).

Part (2): Since S is nonempty, part (1) shows that span(S) is nonempty. Therefore, by Theorem 4.2, span(S) is a subspace of $\mathcal{V}$ if we can prove the closure properties hold for span(S).

First, let us verify closure under scalar multiplication. Let $\mathbf{v}$ be in span(S), and let c be a scalar. We must show that $c\mathbf{v} \in$ span(S). Now, since $\mathbf{v} \in$ span(S), a finite subset $\{\mathbf{v}_1, \ldots, \mathbf{v}_n\}$ of S and real numbers $a_1, \ldots, a_n$ exist such that $\mathbf{v} = \sum_{i=1}^{n} a_i \mathbf{v}_i$. Then,

$$c\mathbf{v} = c(a_1\mathbf{v}_1 + \cdots + a_n\mathbf{v}_n) = (ca_1)\mathbf{v}_1 + \cdots + (ca_n)\mathbf{v}_n.$$

Hence, $c\mathbf{v}$ is a linear combination of the finite subset $\{\mathbf{v}_1, \ldots, \mathbf{v}_n\}$ of S, and so $c\mathbf{v} \in$ span(S).

Finally, we show that span(S) is closed under vector addition. Let $\mathbf{u}$ and $\mathbf{v}$ be two vectors in span(S). Then there are finite subsets $\{\mathbf{u}_1, \ldots, \mathbf{u}_k\}$ and $\{\mathbf{v}_1, \ldots, \mathbf{v}_l\}$ of S such that $\mathbf{u} = a_1\mathbf{u}_1 + \cdots + a_k\mathbf{u}_k$ and $\mathbf{v} = b_1\mathbf{v}_1 + \cdots + b_l\mathbf{v}_l$ for some real numbers $a_1, \ldots, a_k, b_1, \ldots, b_l$. The natural thing to do at this point is to combine the expressions for $\mathbf{u}$ and $\mathbf{v}$ by adding corresponding coefficients. However, each of the subsets $\{\mathbf{u}_1, \ldots, \mathbf{u}_k\}$ or $\{\mathbf{v}_1, \ldots, \mathbf{v}_l\}$ may contain elements not found in the other, so it is difficult to match up their coefficients. We need to create a larger set containing all the vectors in both subsets.

Consider the finite subset $X = \{\mathbf{u}_1, \ldots, \mathbf{u}_k\} \cup \{\mathbf{v}_1, \ldots, \mathbf{v}_l\}$ of S. Renaming the elements of X, we can suppose X is the finite subset $\{\mathbf{w}_1, \ldots, \mathbf{w}_m\}$ of S. Hence, there are real numbers $c_1, \ldots, c_m$ and $d_1, \ldots, d_m$ such that $\mathbf{u} = c_1\mathbf{w}_1 + \cdots + c_m\mathbf{w}_m$ and $\mathbf{v} = d_1\mathbf{w}_1 + \cdots + d_m\mathbf{w}_m$. (Note that $c_i = a_j$ if $\mathbf{w}_i = \mathbf{u}_j$ and that $c_i = 0$ if $\mathbf{w}_i \notin \{\mathbf{u}_1, \ldots, \mathbf{u}_k\}$. A similar formula gives the value of each d_i.) Then,

$$\mathbf{u} + \mathbf{v} = \sum_{i=1}^{m} c_i \mathbf{w}_i + \sum_{i=1}^{m} d_i \mathbf{w}_i = \sum_{i=1}^{m} (c_i + d_i)\,\mathbf{w}_i,$$

a linear combination of the elements $\mathbf{w}_i$ in the subset X of S. Thus, $\mathbf{u} + \mathbf{v} \in$ span(S).

Part (3): This part asserts that if S is a subset of a subspace $\mathcal{W}$, then any (finite) linear combination from S is also in $\mathcal{W}$. This is merely a rewording of Theorem 4.3 using the "span" concept. The fact that span(S) cannot contain vectors outside of $\mathcal{W}$ is illustrated in Figure 4.4.

Figure 4.4

(a) Situation that *must* occur if $\mathcal{W}$ is a subspace containing S; (b) situation that *cannot* occur if $\mathcal{W}$ is a subspace containing S

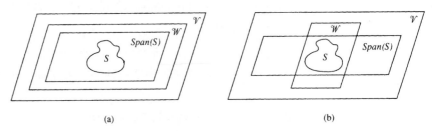

(a)   (b)

Part (4): This is merely a summary of the other three parts. Parts (1) and (2) assert that span(S) is a subspace of $\mathcal{V}$ containing S. But part (3) shows that span(S) is the smallest such subspace because span(S) must be a subset of, and hence smaller than, any other subspace of $\mathcal{V}$ that contains S. ∎

Theorem 4.5 implies that span(S) is created by appending to S precisely those vectors needed to make the closure properties hold. In fact, the whole idea behind span is to "close up" a subset of a vector space to create a subspace.

EXAMPLE 8

Let $\mathbf{v}_1$ and $\mathbf{v}_2$ be any two vectors in $\mathbb{R}^4$. Then, by Theorem 4.5, span($\{\mathbf{v}_1, \mathbf{v}_2\}$) is the smallest subspace of $\mathbb{R}^4$ containing $\mathbf{v}_1$ and $\mathbf{v}_2$. In particular, if $\mathbf{v}_1 = [1, 3, -2, 5]$ and $\mathbf{v}_2 = [0, -4, 3, -1]$, then span($\{\mathbf{v}_1, \mathbf{v}_2\}$) is the subspace of $\mathbb{R}^4$ consisting of all vectors of the form

$$a[1, 3, -2, 5] + b[0, -4, 3, -1] = [a,\ 3a - 4b, -2a + 3b,\ 5a - b].$$

No smaller subspace of $\mathbb{R}^4$ contains $\mathbf{v}_1$ and $\mathbf{v}_2$.   ∎

The following useful result is left for you to prove in Exercise 21.

THEOREM 4.6

Let $\mathcal{V}$ be a vector space, and let S_1 and S_2 be subsets of $\mathcal{V}$ with $S_1 \subseteq S_2$. Then span(S_1) $\subseteq$ span(S_2).

Simplifying Span(S) using Row Reduction

Our next goal is to find a simplified form for the vectors in the span of a given set S. The fact that span is a generalization of the row space concept suggests that we can use results from Chapter 2 involving row spaces to help us compute and simplify span(S). If we form the matrix $\mathbf{A}$ whose rows are the vectors in S, the rows of the reduced row echelon form of $\mathbf{A}$ often give a simpler expression for span(S), since row equivalent matrices have the same row space. Hence, we have the following:

METHOD FOR SIMPLIFYING SPAN(S) USING ROW REDUCTION (SIMPLIFIED SPAN METHOD)

Suppose that S is a finite subset of $\mathbb{R}^n$ containing k vectors, with $k \geq 2$.

Step 1: Form a $k \times n$ matrix $\mathbf{A}$ by using the vectors in S as the rows of $\mathbf{A}$. (Thus, span(S) is the row space of $\mathbf{A}$.)

Step 2: Let $\mathbf{C}$ be the reduced row echelon form matrix for $\mathbf{A}$.

Step 3: Then, a simplified form for span(S) is given by the set of all linear combinations of the *nonzero* rows of $\mathbf{C}$.

EXAMPLE 9

Let S be the subset $\{[1, 4, -1, -5], [2, 8, 5, 4], [-1, -4, 2, 7], [6, 24, -1, -20]\}$ of $\mathbb{R}^4$. By definition, span(S) is the set of all vectors of the form

$$a[1, 4, -1, -5] + b[2, 8, 5, 4] + c[-1, -4, 2, 7] + d[6, 24, -1, -20]$$

for $a, b, c, d \in \mathbb{R}$. We want to use the Simplified Span Method to find a simplified form for the vectors in span(S). We first create

$$\mathbf{A} = \begin{bmatrix} 1 & 4 & -1 & -5 \\ 2 & 8 & 5 & 4 \\ -1 & -4 & 2 & 7 \\ 6 & 24 & -1 & -20 \end{bmatrix},$$

whose rows are the vectors in S. Then, span(S) is the row space of $\mathbf{A}$; that is, the set of all linear combinations of the rows of $\mathbf{A}$.

Next, we simplify the form of the row space of $\mathbf{A}$ by obtaining its reduced row echelon form matrix

$$\mathbf{C} = \begin{bmatrix} 1 & 4 & 0 & -3 \\ 0 & 0 & 1 & 2 \\ 0 & 0 & 0 & 0 \\ 0 & 0 & 0 & 0 \end{bmatrix}.$$

By Theorem 2.8, the row space of $\mathbf{A}$ is the same as the row space of $\mathbf{C}$, which is the set of all 4-vectors of the form

$$a[1, 4, 0, -3] + b[0, 0, 1, 2] = [a, \; 4a, \; b, \; -3a + 2b].$$

Therefore, span(S) = $\{[a, \; 4a, \; b, \; -3a + 2b] \mid a, b \in \mathbb{R}\}$, a subspace of $\mathbb{R}^4$. Note, for example, that the vector $[3, 12, -2, -13]$ is in span(S) ($a = 3, b = -2$). However, the vector $[-2, -8, 4, 6]$ is not in span(S) because the following system has no solutions:

$$\begin{cases} a & = -2 \\ 4a & = -8 \\ b = & 4 \\ -3a + 2b = & 6 \end{cases}.$$
■

EXAMPLE 10

Recall that the eigenspace E_{15} for the matrix $\mathbf{A}$ in Example 11 in Section 4.2 is $E_{15} = \{a[4, 1, 0] + b[2, 0, 1] \mid a, b \in \mathbb{R}\}$. Hence, E_{15} is spanned by $\{[4, 1, 0], [2, 0, 1]\}$. Although the form of E_{15} is already simple, we can obtain an alternative form by using the Simplified Span Method. Row reducing the matrix

$$A = \begin{bmatrix} 4 & 1 & 0 \\ 2 & 0 & 1 \end{bmatrix}, \quad \text{we obtain} \quad \mathbf{C} = \begin{bmatrix} 1 & 0 & \frac{1}{2} \\ 0 & 1 & -2 \end{bmatrix}.$$

Hence, an alternative form for the vectors in E_{15} is $\left\{ a\left[1, 0, \frac{1}{2}\right] + b[0, 1, -2] \; \middle| \; a, b \in \mathbb{R} \right\} = \left\{ [a, b, \frac{1}{2}a - 2b] \; \middle| \; a, b \in \mathbb{R} \right\}$, just as we claimed in Example 11 in Section 4.2. ■

The method used in Examples 9 and 10 works in vector spaces other than $\mathbb{R}^n$, as we see in the next example. This fact will follow from the discussion of isomorphism in Section 5.4. (However, we will not use this fact in proofs of theorems until after Section 5.4.)

EXAMPLE 11

Let S be the subset $\{5x^3 + 2x^2 + 4x - 3, -x^2 + 3x - 7, 2x^3 + 4x^2 - 8x + 5, x^3 + 2x + 5\}$ of $\mathcal{P}_3$. We use the Simplified Span Method to find a simplified form for the vectors in span(S).

Consider the coefficients of each polynomial as the coordinates of a vector in $\mathbb{R}^4$, yielding the corresponding set of vectors $T = \{[5, 2, 4, -3], [0, -1, 3, -7], [2, 4, -8, 5], [1, 0, 2, 5]\}$. Using the method of Example 9, we create the following matrix, whose rows are the vectors in T.

$$\mathbf{A} = \begin{bmatrix} 5 & 2 & 4 & -3 \\ 0 & -1 & 3 & -7 \\ 2 & 4 & -8 & 5 \\ 1 & 0 & 2 & 5 \end{bmatrix}$$

Then span(T) is the row space of the reduced row echelon form of $\mathbf{A}$.

$$\mathbf{C} = \begin{bmatrix} 1 & 0 & 2 & 0 \\ 0 & 1 & -3 & 0 \\ 0 & 0 & 0 & 1 \\ 0 & 0 & 0 & 0 \end{bmatrix}$$

Taking each nonzero row of $\mathbf{C}$ as the coefficients of a polynomial in $\mathcal{P}_3$, we see that

$$\text{span}(S) = \{a(x^3 + 2x) + b(x^2 - 3x) + c(1) \mid a, b, c \in \mathbb{R}\}$$

$$= \{ax^3 + bx^2 + (2a - 3b)x + c \mid a, b, c \in \mathbb{R}\}.$$

∎

A Spanning Set for an Eigenspace

In Section 3.4 we illustrated a method for diagonalizing an $n \times n$ matrix, when possible. In fact, the set S of eigenvectors generated for a given eigenvalue λ using the Diagonalization Method spans the eigenspace E_λ (see Exercise 27). We illustrate this in the following example:

EXAMPLE 12

Let

$$\mathbf{A} = \begin{bmatrix} 0 & -6 & 3 \\ 2 & -13 & 6 \\ 4 & -24 & 11 \end{bmatrix}.$$

A little work yields $p_{\mathbf{A}}(x) = x^3 + 2x^2 + x = x(x+1)^2$. We solve the homogeneous system $(-1\mathbf{I}_3 - \mathbf{A})\mathbf{X} = \mathbf{0}$ to find the eigenspace E_{-1} for $\mathbf{A}$.

Row reducing $[(-\mathbf{I}_3 - \mathbf{A})|\mathbf{0}]$ produces

$$\begin{bmatrix} 1 & -6 & 3 & 0 \\ 0 & 0 & 0 & 0 \\ 0 & 0 & 0 & 0 \end{bmatrix},$$

giving the solution set

$$E_{-1} = \{[6b - 3c, b, c] \mid b, c \in \mathbb{R}\} = \{b[6, 1, 0] + c[-3, 0, 1] \mid b, c \in \mathbb{R}\}.$$

Thus, $E_{-1} = \text{span}(S)$, where $S = \{[6, 1, 0], [-3, 0, 1]\}$. The set S is precisely the set of eigenvectors for $\lambda = -1$ that Step 3 of the Diagonalization Method in Section 3.4 produces (verify!).   ∎

Special Case: The Span of the Empty Set

Until now, our results involving span have specified that the subset S of the vector space $\mathcal{V}$ be nonempty. However, our understanding of span(S) as the smallest subspace of $\mathcal{V}$ containing S allows us to give a meaningful definition for the span of the empty set.

DEFINITION

Span($\{\}$) = $\{\mathbf{0}\}$.

This definition makes sense because the trivial subspace is the smallest subspace of $\mathcal{V}$, hence the smallest one containing the empty set. Thus, Theorem 4.5 is also true when the set S is empty. Similarly, to maintain consistency, we *define* any linear combination of the empty set of vectors to be $\mathbf{0}$. This ensures that the span of the empty set equals the set of all linear combinations of vectors taken from this set.

Exercises for Section 4.3

1. In each of the following cases, use the Simplified Span Method to find a simplified general form for all the vectors in span(S), where S is the given subset of $\mathbb{R}^n$:
   *(a) $S = \{[1, 1, 0], [2, -3, -5]\}$
    (b) $S = \{[3, 1, -2], [-3, -1, 2], [6, 2, -4]\}$
   *(c) $S = \{[1, -1, 1], [2, -3, 3], [0, 1, -1]\}$
    (d) $S = \{[1, 1, 1], [2, 1, 1], [1, 1, 2]\}$
   *(e) $S = \{[1, 3, 0, 1], [0, 0, 1, 1], [0, 1, 0, 1], [1, 5, 1, 4]\}$
    (f) $S = \{[2, -1, 3, 1], [1, -2, 0, -1], [3, -3, 3, 0], [5, -4, 6, 1],$
          $[1, -5, -3, -4]\}$

2. In each case, use the Simplified Span Method to find a simplified general form for all the vectors in span(S), where S is the given subset of $\mathcal{P}_3$:
   *(a) $S = \{x^3 - 1, x^2 - x, x - 1\}$
    (b) $S = \{x^3 + 2x^2, 1 - 4x^2, 12 - 5x^3, x^3 - x^2\}$
   *(c) $S = \{x^3 - x + 5, 3x^3 - 3x + 10, 5x^3 - 5x - 6, 6x - 6x^3 - 13\}$

3. In each case, use the Simplified Span Method to find a simplified general form for all the vectors in span(S), where S is the given subset of $\mathcal{M}_{22}$. (Hint: Rewrite each matrix as a 4-vector.)

$$\star\text{(a)} \quad S = \left\{ \begin{bmatrix} -1 & 1 \\ 0 & 0 \end{bmatrix}, \begin{bmatrix} 0 & 0 \\ 1 & -1 \end{bmatrix}, \begin{bmatrix} -1 & 0 \\ 0 & 1 \end{bmatrix} \right\}$$

$$\text{(b)} \quad S = \left\{ \begin{bmatrix} 4 & -1 \\ 1 & 0 \end{bmatrix}, \begin{bmatrix} 1 & -1 \\ 3 & 0 \end{bmatrix}, \begin{bmatrix} 5 & 1 \\ -7 & 0 \end{bmatrix} \right\}$$

$$\star\text{(c)} \quad S = \left\{ \begin{bmatrix} 1 & -1 \\ 3 & 0 \end{bmatrix}, \begin{bmatrix} 2 & -1 \\ 8 & -1 \end{bmatrix}, \begin{bmatrix} -1 & 4 \\ 4 & -1 \end{bmatrix}, \begin{bmatrix} 3 & -4 \\ 5 & 6 \end{bmatrix} \right\}$$

$\star$**4.** (a) Express the subspace $\mathcal{W}$ of $\mathbb{R}^4$ of all 4-vectors of the form $[a + b, a + c, b + c, c]$ as the row space of a matrix $\mathbf{A}$.

(b) Find the reduced row echelon form matrix $\mathbf{B}$ for $\mathbf{A}$.

(c) Use the matrix $\mathbf{B}$ from part (b) to find a simplified form for the vectors in $\mathcal{W}$.

5. (a) Express the subspace $\mathcal{W}$ of $\mathbb{R}^5$ of all 5-vectors of the form $[2a - 3b, a - 5c, a, 4c - b, c]$ as the row space of a matrix $\mathbf{A}$.

(b) Find the reduced row echelon form matrix $\mathbf{B}$ for $\mathbf{A}$.

(c) Use the matrix $\mathbf{B}$ from part (b) to find a simplified form for the vectors in $\mathcal{W}$.

6. Prove that the set $S = \{[1, 3, -1], [2, 7, -3], [4, 8, -7]\}$ spans $\mathbb{R}^3$.

7. Prove that the set $S = \{[1, -2, 2], [3, -4, -1], [1, -4, 9], [0, 2, -7]\}$ does not span $\mathbb{R}^3$.

8. Show that the set $\{x^2 + x + 1, x + 1, 1\}$ spans $\mathcal{P}_2$.

9. Prove that the set $\{x^2 + 4x - 3, 2x^2 + x + 5, 7x - 11\}$ does not span $\mathcal{P}_2$.

10. (a) Let $S = \{[1, -2, -2], [3, -5, 1], [-1, 1, -5]\}$. Show that $[-4, 5, -13] \in \text{span}(S)$ by expressing it as a linear combination of the vectors in S.

(b) Prove that the set S in part (a) does not span $\mathbb{R}^3$.

$\star$**11.** Consider the subset $S = \{x^3 - 2x^2 + x - 3, 2x^3 - 3x^2 + 2x + 5, 4x^2 + x - 3, 4x^3 - 7x^2 + 4x - 1\}$ of $\mathcal{P}$. Show that $3x^3 - 8x^2 + 2x + 16$ is in span(S) by expressing it as a linear combination of the elements of S.

12. Prove that the set S of all vectors in $\mathbb{R}^4$ that have zeroes in exactly two coordinates spans $\mathbb{R}^4$. (Hint: Find a subset of S that spans $\mathbb{R}^4$.)

13. Let $\mathbf{a}$ be any nonzero element of $\mathbb{R}$. Prove that span($\{\mathbf{a}\}$) $= \mathbb{R}$.

14. $\star$(a) Suppose that S_1 is the set of symmetric 2×2 matrices and that S_2 is the set of skew-symmetric 2×2 matrices. Prove that span($S_1 \cup S_2$) $= \mathcal{M}_{22}$.

(b) State and prove the corresponding statement for $n \times n$ matrices.

15. Consider the subset $S = \{1 + x^2, x + x^3, 3 - 2x + 3x^2 - 12x^3\}$ of $\mathcal{P}$, and let $\mathcal{W} = \left\{ ax^3 + bx^2 + cx + b \mid a, b, c \in \mathbb{R} \right\}$. Show that $\mathcal{W} = \text{span}(S)$.

16. Let $\mathbf{A} = \begin{bmatrix} -9 & -15 & 8 \\ -10 & -14 & 8 \\ -30 & -45 & 25 \end{bmatrix}$.

$\star$(a) Use Step 3 of the Diagonalization Method of Section 3.4 to find a set S of two eigenvectors for $\mathbf{A}$ corresponding to the eigenvalue $\lambda = 1$. Multiply by a scalar to eliminate any fractions in your answers.

(b) Verify that the set S from part (a) spans E_1.

17. Let $S_1 = \{\mathbf{v}_1, \ldots, \mathbf{v}_n\}$ be a nonempty subset of a vector space $\mathcal{V}$. Let $S_2 = \{-\mathbf{v}_1, -\mathbf{v}_2, \ldots, -\mathbf{v}_n\}$. Show that $\text{span}(S_1) = \text{span}(S_2)$.

18. Let $\mathbf{u}$ and $\mathbf{v}$ be two nonzero vectors in $\mathbb{R}^3$, and let $S = \{\mathbf{u}, \mathbf{v}\}$. Show that $\text{span}(S)$ is a line through the origin if $\mathbf{u} = a\mathbf{v}$ for some real number a, but otherwise $\text{span}(S)$ is a plane through the origin.

19. Let $\mathbf{u} = [u_1, u_2, u_3]$, $\mathbf{v} = [v_1, v_2, v_3]$, and $\mathbf{w} = [w_1, w_2, w_3]$ be three vectors in $\mathbb{R}^3$. Show that $S = \{\mathbf{u}, \mathbf{v}, \mathbf{w}\}$ spans $\mathbb{R}^3$ if and only if

$$\begin{vmatrix} u_1 & u_2 & u_3 \\ v_1 & v_2 & v_3 \\ w_1 & w_2 & w_3 \end{vmatrix} \neq 0.$$

(Hint: Consider all of the possible reduced row echelon forms of the corresponding matrix. Using Exercise 18, show that unless the rank is 3, the rows will not span $\mathbb{R}^3$.)

20. Let $S = \{\mathbf{p}_1, \ldots, \mathbf{p}_k\}$ be a finite subset of $\mathcal{P}$. Prove that there is some positive integer n such that $\text{span}(S) \subseteq \mathcal{P}_n$.

▶**21.** Prove Theorem 4.6.

22. (a) Prove that if S is a nonempty subset of a vector space $\mathcal{V}$, then S is a subspace of $\mathcal{V}$ if and only if $\text{span}(S) = S$.

(b) Use part (a) to show that every subspace $\mathcal{W}$ of a vector space $\mathcal{V}$ has a set of vectors that spans $\mathcal{W}$, namely, the set $\mathcal{W}$ itself.

(c) Describe the span of the set of the skew-symmetric matrices in $\mathcal{M}_{33}$.

23. Let S_1 and S_2 be subsets of a vector space $\mathcal{V}$. Prove that $\text{span}(S_1) = \text{span}(S_2)$ if and only if $S_1 \subseteq \text{span}(S_2)$ and $S_2 \subseteq \text{span}(S_1)$.

24. Let S_1 and S_2 be two subsets of a vector space $\mathcal{V}$.

(a) Prove that $\text{span}(S_1 \cap S_2) \subseteq \text{span}(S_1) \cap \text{span}(S_2)$.

⋆(b) Give an example of distinct subsets S_1 and S_2 of $\mathbb{R}^3$ for which the inclusion in part (a) is actually an equality.

⋆(c) Give an example of subsets S_1 and S_2 of $\mathbb{R}^3$ for which the inclusion in part (a) is not an equality.

25. Let S_1 and S_2 be subsets of a vector space $\mathcal{V}$.

(a) Show that $\text{span}(S_1) \cup \text{span}(S_2) \subseteq \text{span}(S_1 \cup S_2)$.

(b) Prove that if $S_1 \subseteq S_2$, then the inclusion in part (a) is an equality.

⋆(c) Give an example of subsets S_1 and S_2 in $\mathcal{P}_5$ for which the inclusion in part (a) is not an equality.

26. Let S be a subset of a vector space $\mathcal{V}$, and let $\mathbf{v} \in \mathcal{V}$. Show that $\text{span}(S) = \text{span}(S \cup \{\mathbf{v}\})$ if and only if $\mathbf{v} \in \text{span}(S)$.

27. Let $\mathbf{A}$ be an $n \times n$ matrix and λ be an eigenvalue for $\mathbf{A}$. Suppose S is the set of eigenvectors for $\mathbf{A}$ corresponding to λ produced by Step 3 of the Diagonalization Method of Section 3.4. Prove that S spans E_λ.

⋆**28.** True or False:

(a) Span(S) is only defined if S is a finite subset of a vector space.

(b) If S is a subset of a vector space $\mathcal{V}$, then $\text{span}(S)$ contains every finite linear combination of vectors in S.

(c) If S is a subset of a vector space $\mathcal{V}$, then span(S) is the smallest set in $\mathcal{V}$ containing S.

(d) If S is a subset of a vector space $\mathcal{V}$, and $\mathcal{W}$ is a subspace of $\mathcal{V}$ containing S, then we must have $\mathcal{W} \subseteq$ span(S).

(e) The row space of a 4×5 matrix $\mathbf{A}$ is a subspace of $\mathbb{R}^4$.

(f) A simplified form for the span of a finite set S of vectors in $\mathbb{R}^n$ can be found by row reducing the matrix whose rows are the vectors of S.

(g) The eigenspace E_λ for an eigenvalue λ of an $n \times n$ matrix $\mathbf{A}$ is the row space of $\lambda \mathbf{I}_n - \mathbf{A}$.

4.4 Linear Independence

In this section, we explore the concept of a linearly independent set of vectors and examine methods for determining whether or not a given set of vectors is linearly independent. We will also see that there are important connections between the concepts of span and linear independence.

Linear Independence and Dependence

DEFINITION

Let S be a subset of a vector space $\mathcal{V}$.

If S has at least two elements, then S is **linearly dependent** if and only if some vector in S can be expressed as a linear combination of the other vectors in S.

If $S = \{\mathbf{v}\}$, a one-element set, then S is **linearly dependent** if and only if $\mathbf{v} = \mathbf{0}$.

The empty set is not linearly dependent.

Finally, the set S is **linearly independent** if and only if S is not linearly dependent.

We can see from this definition that a set S with at least two vectors is linearly dependent whenever at least one vector in S is a "destination" that can be reached using a linear combination of the other vectors in S. Conversely, a set S with at least two vectors is linearly independent if each vector in S represents a "destination" that we cannot reach using a linear combination of the other vectors. Hence, each vector in S represents a new "direction."

EXAMPLE 1 The set of vectors $\{\mathbf{i}, \mathbf{j}, \mathbf{k}\}$ in $\mathbb{R}^3$ is linearly independent. None of the three vectors in this set can be expressed as a linear combination of the other two (why?). More generally, the set $\{\mathbf{e}_1, \ldots, \mathbf{e}_n\}$ in $\mathbb{R}^n$ is linearly independent. ∎

EXAMPLE 2　The set of vectors $S = \{[1, 2, -1], [0, 1, 2], [2, 7, 4]\}$ in $\mathbb{R}^3$ is linearly dependent because it is possible to express some vector in the set S as a linear combination of the others. For example, $[2, 7, 4] = 2[1, 2, -1] + 3[0, 1, 2]$. From a geometric point of view, the fact that $[2, 7, 4]$ can be expressed as a linear combination of the vectors $[1, 2, -1]$ and $[0, 1, 2]$ means that $[2, 7, 4]$ lies in the plane spanned by $[1, 2, -1]$ and $[0, 1, 2]$, assuming that all three vectors have their initial points at the origin (see Figure 4.5). ∎

Figure 4.5

The vector $[2, 7, 4]$ in the plane spanned by $[1, 2, -1]$ and $[0, 1, 2]$

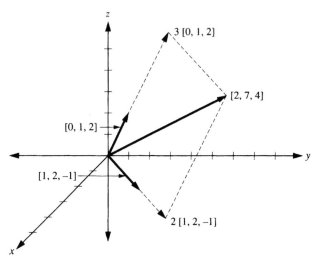

EXAMPLE 3　The set of vectors $S = \{[1, -1, 2], [-3, 3, -6]\}$ in $\mathbb{R}^3$ is linearly dependent since each of the vectors is a linear combination of the other. For example, $[1, -1, 2] = (-\frac{1}{3})[-3, 3, -6]$. ∎

The last example demonstrates that any set of exactly two vectors is linearly dependent precisely when one of the vectors is a scalar multiple of the other, that is, if the vectors are parallel. Similarly, a set of exactly two vectors is linearly independent precisely when neither is a scalar multiple of the other. For example, the subset $\{[3, -8], [2, 5]\}$ is a linearly independent subset of $\mathbb{R}^2$.

EXAMPLE 4　Let $S_1 = \{[3, -1, 4]\}$. Since S_1 contains a single vector and this vector is nonzero, S_1 is a linearly independent subset of $\mathbb{R}^3$. On the other hand, $S_2 = \{[0, 0, 0, 0]\}$ is a linearly dependent subset of $\mathbb{R}^4$. ∎

EXAMPLE 5　Let S be any subset of a vector space $\mathcal{V}$ containing the zero vector $\mathbf{0}$. If S contains no vector other than $\mathbf{0}$, then by definition, S is linearly dependent. If S also contains at least one other nonzero vector $\mathbf{v}$, then note that $\mathbf{0}$ can be expressed as a (finite) linear combination of the remaining vectors in S since $\mathbf{0} = 0 \cdot \mathbf{v}$. Hence, by the definition, S is linearly dependent. Therefore, in all cases, any subset of a vector space that contains the zero vector $\mathbf{0}$ is linearly dependent. ∎

An Alternate Characterization of Linear Independence

If $\mathbf{v}$ is a vector in a set S, we use the notation $S - \{\mathbf{v}\}$ to represent the set of all (other) vectors in S except $\mathbf{v}$. Of course, in the special case where $S = \{\mathbf{v}\}$ itself, the set $S - \{\mathbf{v}\} = \{\}$, the empty set. Notice that a subset S of two or more vectors in a vector space $\mathcal{V}$ is linearly independent precisely when no vector $\mathbf{v}$ in S is in the span of the remaining vectors. That is,

> A set S in a vector space $\mathcal{V}$ is linearly independent if and only if there is no vector $\mathbf{v} \in S$ such that $\mathbf{v} \in \text{span}(S - \{\mathbf{v}\})$.

We can see that this statement holds even in the special cases when $S = \{\mathbf{v}\}$ or $S = \{\}$. In the first case, $\text{span}(S - \{\mathbf{v}\}) = \text{span}(\{\}) = \{\mathbf{0}\}$, as defined in Section 4.3, and so S is linearly independent if and only if $\mathbf{v} \notin \{\mathbf{0}\}$, as expected. On the other hand, if $S = \{\}$, then there certainly is no vector $\mathbf{v}$ in S with the stated property, so S is linearly independent, as before.

Similarly, we have

> A set S in a vector space $\mathcal{V}$ is linearly dependent if and only if there is some vector $\mathbf{v} \in S$ such that $\mathbf{v} \in \text{span}(S - \{\mathbf{v}\})$.

We will find these alternate characterizations of linear independence and dependence useful in future proofs.

An Algebraic Test for Linear Independence

Next, we establish another characterization of linear independence. In addition to its theoretical importance, it also supplies us with a fast method for determining whether a finite subset of a vector space is linearly independent.

THEOREM 4.7

Let $S = \{\mathbf{v}_1, \ldots, \mathbf{v}_n\}$ be a nonempty finite subset of a vector space $\mathcal{V}$. Then S is linearly independent if and only if the equation $a_1\mathbf{v}_1 + \cdots + a_n\mathbf{v}_n = \mathbf{0}$ implies that $a_1 = a_2 = \cdots = a_n = 0$.

That is, a set S is linearly independent if and only if the *only* way the zero vector can be expressed as a (nonempty) linear combination of the vectors in S is by letting all coefficients equal zero.

To prove Theorem 4.7, we prove its contrapositive instead.

> Let $S = \{\mathbf{v}_1, \ldots, \mathbf{v}_n\}$ be a nonempty finite subset of a vector space $\mathcal{V}$. Then S is *linearly dependent* if and only if there are real numbers $a_1, \ldots, a_n$ such that $a_1\mathbf{v}_1 + \cdots + a_n\mathbf{v}_n = \mathbf{0}$, where some $a_i \neq 0$.

PROOF OF THEOREM 4.7

We first complete both halves of the proof assuming S has at least two elements. For one half, assume that S is linearly dependent. Then there is a vector $\mathbf{v}_i$ in S

such that $\mathbf{v}_i \in \mathrm{span}(S - \{\mathbf{v}_i\})$. Without loss of generality, assume $\mathbf{v}_i = \mathbf{v}_1$, that is, $i = 1$. Therefore, there are real numbers $a_2, \ldots, a_n$ such that

$$\mathbf{v}_1 = a_2\mathbf{v}_2 + a_3\mathbf{v}_3 + \cdots + a_n\mathbf{v}_n.$$

Letting $a_1 = -1$, we get $\mathbf{0} = a_1\mathbf{v}_1 + \cdots + a_n\mathbf{v}_n$. Since $a_1 \neq 0$, this result completes the first half of the proof.

To prove the reverse implication, assume that we have coefficients $a_1, \ldots, a_n$ such that $\mathbf{0} = a_1\mathbf{v}_1 + \cdots + a_n\mathbf{v}_n$, with $a_i \neq 0$ for some i. Then,

$$\mathbf{v}_i = \left(-\frac{a_1}{a_i}\right)\mathbf{v}_1 + \cdots + \left(-\frac{a_{i-1}}{a_i}\right)\mathbf{v}_{i-1} + \left(-\frac{a_{i+1}}{a_i}\right)\mathbf{v}_{i+1} + \cdots + \left(-\frac{a_n}{a_i}\right)\mathbf{v}_n,$$

which shows that $\mathbf{v}_i \in \mathrm{span}(S - \{\mathbf{v}_i\})$ and hence that S is linearly dependent.

Finally, in the special case where $S = \{\mathbf{v}_1\}$, we know S is linearly dependent if and only if $\mathbf{v}_1 = \mathbf{0}$, which is certainly true if and only if $a_1\mathbf{v}_1 = \mathbf{0}$ for some $a_1 \neq 0$. ∎

EXAMPLE 6

Consider the subset $S = \{[1, -1, 0, 2], [0, -2, 1, 0], [2, 0, -1, 1]\}$ of $\mathbb{R}^4$. We will investigate whether S is linearly independent using Theorem 4.7.

We proceed by assuming that $a[1, -1, 0, 2] + b[0, -2, 1, 0] + c[2, 0, -1, 1] = [0, 0, 0, 0]$ and solve for a, b, and c to see whether all these coefficients must be zero. That is, we determine whether the following homogeneous system has only the trivial solution:

$$\begin{cases} a & & + 2c = 0 \\ -a & - 2b & = 0 \\ & b & - c = 0 \\ 2a & & + c = 0 \end{cases}.$$

Row reducing

$$\begin{bmatrix} a & b & c \\ 1 & 0 & 2 & 0 \\ -1 & -2 & 0 & 0 \\ 0 & 1 & -1 & 0 \\ 2 & 0 & 1 & 0 \end{bmatrix}, \quad \text{we obtain} \quad \begin{bmatrix} a & b & c \\ 1 & 0 & 0 & 0 \\ 0 & 1 & 0 & 0 \\ 0 & 0 & 1 & 0 \\ 0 & 0 & 0 & 0 \end{bmatrix},$$

which shows that this system has only the trivial solution $a = b = c = 0$. Hence, S is linearly independent. ∎

Using Row Reduction to Test for Linear Independence

Notice that in Example 6 the columns of the matrix to the left of the augmentation bar are just the vectors in S. In general, to test a finite set of vectors in $\mathbb{R}^n$ for linear independence, we row reduce the matrix whose *columns* are the vectors in the set, and then check whether the associated homogeneous system has only the trivial solution. Notice that in practice it is not necessary to include the augmentation bar and the column of zeroes to its right, since this column never changes in the row reduction process. Thus, we have

---

METHOD TO TEST FOR LINEAR INDEPENDENCE USING ROW REDUCTION (INDEPENDENCE TEST METHOD)

Let S be a finite set of vectors in $\mathbb{R}^n$. To determine whether S is linearly independent, perform the following steps:

Step 1: Create the matrix $\mathbf{A}$ whose *columns* are the vectors in S.

Step 2: Find $\mathbf{B}$, the reduced row echelon form of $\mathbf{A}$.

Step 3: If there is a pivot in every column of $\mathbf{B}$, then S is linearly independent. Otherwise, S is linearly dependent.

EXAMPLE 7

Consider the subset $S = \{[3, 1, -1], [-5, -2, 2], [2, 2, -1]\}$ of $\mathbb{R}^3$. Using the Independence Test Method, we row reduce

$$\begin{bmatrix} 3 & -5 & 2 \\ 1 & -2 & 2 \\ -1 & 2 & -1 \end{bmatrix} \text{ to obtain } \begin{bmatrix} 1 & 0 & 0 \\ 0 & 1 & 0 \\ 0 & 0 & 1 \end{bmatrix}.$$

Since we found a pivot in every column, the set S is linearly independent. ∎

EXAMPLE 8

Consider the subset $S = \{[2, 5], [3, 7], [4, -9], [-8, 3]\}$ of $\mathbb{R}^2$. Using the Independence Test Method, we row reduce

$$\begin{bmatrix} 2 & 3 & 4 & -8 \\ 5 & 7 & -9 & 3 \end{bmatrix} \text{ to obtain } \begin{bmatrix} 1 & 0 & -55 & 65 \\ 0 & 1 & 38 & -46 \end{bmatrix}.$$

Since we have no pivots in columns 3 and 4, the set S is linearly dependent. ∎

In the last example, there are more columns than rows in the matrix we row reduced. Hence, there must ultimately be some column without a pivot, since each pivot is in a different row. In such cases, the original set of vectors must be linearly dependent. This motivates the following result, which we ask you to formally prove as Exercise 16:

COROLLARY 4.8

If S is any set in $\mathbb{R}^n$ containing k distinct vectors, where $k > n$, then S is linearly dependent.

The Independence Test Method can be adapted for use on vector spaces other than $\mathbb{R}^n$, as in the next example.

EXAMPLE 9   Consider the following subset of $\mathcal{M}_{22}$:

$$S = \left\{ \begin{bmatrix} 2 & 3 \\ -1 & 4 \end{bmatrix}, \begin{bmatrix} -1 & 0 \\ 1 & 1 \end{bmatrix}, \begin{bmatrix} 6 & -1 \\ 3 & 2 \end{bmatrix}, \begin{bmatrix} -11 & 3 \\ -2 & 2 \end{bmatrix} \right\}.$$

We determine whether S is linearly independent using the Independence Test Method. First, we represent the 2×2 matrices in S as 4-vectors. Placing them in a matrix, using each 4-vector as a column, we get

$$\begin{bmatrix} 2 & -1 & 6 & -11 \\ 3 & 0 & -1 & 3 \\ -1 & 1 & 3 & -2 \\ 4 & 1 & 2 & 2 \end{bmatrix}, \text{ which reduces to } \begin{bmatrix} 1 & 0 & 0 & \frac{1}{2} \\ 0 & 1 & 0 & 3 \\ 0 & 0 & 1 & -\frac{3}{2} \\ 0 & 0 & 0 & 0 \end{bmatrix}.$$

There is no pivot in column 4. Hence, S is linearly dependent. ■

Linear Independence of Infinite Sets

Most cases in which we check for linear independence involve a *finite* set S. However, when S is an infinite subset of a vector space $\mathcal{V}$, linear independence is established by checking that all of its finite subsets are linearly independent. While this is computationally practical only in simple cases, it is an important theoretical extension of Theorem 4.7. We state this generalization in the next theorem, which you are asked to prove as Exercise 25.

> **THEOREM 4.9 (Generalization of Theorem 4.7)**
> Let S be a nonempty subset of a vector space $\mathcal{V}$. Then S is linearly independent if and only if for every *finite* subset $\{\mathbf{v}_1, \ldots, \mathbf{v}_n\}$ of S, the equation $a_1\mathbf{v}_1 + \cdots + a_n\mathbf{v}_n = \mathbf{0}$ implies that $a_i = 0$, for $1 \leq i \leq n$.

EXAMPLE 10   The set $S = \{x, x^3, x^5, \ldots\}$ containing all odd powers of x is a linearly independent subset of $\mathcal{P}$ by Theorem 4.9. This is because the powers of x have distinct exponents. Hence, any finite linear combination of these powers of x can produce the zero polynomial only if all of the coefficients involved are zero. ■

Uniqueness of Expression of a Vector as a Linear Combination

The next theorem serves as the foundation for the rest of this chapter because it gives an even more powerful connection between the concepts of span and linear independence.

THEOREM 4.10

Let S be a nonempty subset of a vector space $\mathcal{V}$. Then S is linearly independent if and only if every vector $\mathbf{v} \in \text{span}(S)$ can be expressed *uniquely* as a finite linear combination of the elements of S, if terms with zero coefficients are ignored.

The phrase "if terms with zero coefficients are ignored" means that two finite linear combinations of a set S are considered the same when their nonzero coefficient terms agree. For example, with $S = \{\mathbf{v}_1, \mathbf{v}_2, \mathbf{v}_3, \mathbf{v}_4\}$, we consider $3\mathbf{v}_1 - 4\mathbf{v}_3$ to be the same linear combination as $3\mathbf{v}_1 + 0\mathbf{v}_2 - 4\mathbf{v}_3 + 0\mathbf{v}_4$.

PROOF OF THEOREM 4.10

Suppose that S is linearly independent, and suppose $\mathbf{v} \in \text{span}(S)$ can be expressed both as $\mathbf{v} = a_1\mathbf{u}_1 + \cdots + a_n\mathbf{u}_k$ and $\mathbf{v} = b_1\mathbf{v}_1 + \cdots + b_k\mathbf{v}_l$, for distinct $\mathbf{u}_1, \ldots, \mathbf{u}_k \in S$ and distinct $\mathbf{v}_1, \ldots, \mathbf{v}_l \in S$, where these expressions differ in at least one nonzero term. Since the $\mathbf{u}_i$'s might not be distinct from the $\mathbf{v}_i$'s, we consider the set $X = \{\mathbf{u}_1, \ldots, \mathbf{u}_k\} \cup \{\mathbf{v}_1, \ldots, \mathbf{v}_l\}$ and label the distinct vectors in X as $\{\mathbf{w}_1, \ldots, \mathbf{w}_m\}$. Then we can express $\mathbf{v} = a_1\mathbf{u}_1 + \cdots + a_n\mathbf{u}_k$ as $\mathbf{v} = c_1\mathbf{w}_1 + \cdots + c_m\mathbf{w}_m$, and also $\mathbf{v} = b_1\mathbf{v}_1 + \cdots + b_k\mathbf{v}_l$ as $\mathbf{v} = d_1\mathbf{w}_1 + \cdots + d_m\mathbf{w}_m$, choosing the scalars c_i, d_i, $1 \le i \le m$, as in the proof of part (2) of Theorem 4.5. Since the original linear combinations for $\mathbf{v}$ are distinct, we know that $c_i \ne d_i$ for some i. Now, $\mathbf{v} - \mathbf{v} = \mathbf{0} = (c_1 - d_1)\mathbf{w}_1 + \cdots + (c_m - d_m)\mathbf{w}_m$. Since $\mathbf{w}_1, \ldots, \mathbf{w}_m \in S$, a linearly independent set, each $c_i - d_i = 0$, by Theorem 4.9. But this is a contradiction since some $c_i \ne d_i$.

Conversely, assume every vector in $\text{span}(S)$ can be uniquely expressed as a linear combination of elements of S. Since $\mathbf{0} \in \text{span}(S)$, there is exactly one linear combination of elements of S that equals $\mathbf{0}$. Thus, if $\{\mathbf{v}_1, \ldots, \mathbf{v}_n\}$ is any finite subset of S, the fact that $\mathbf{0} = 0\mathbf{v}_1 + \cdots + 0\mathbf{v}_n$ means the only possible coefficients of $\mathbf{v}_1, \ldots, \mathbf{v}_n$ here are all zero. Thus, by Theorem 4.9, S is linearly independent. ∎

By Theorem 4.10, S is linearly independent if there is precisely one way of reaching any "destination" in $\text{span}(S)$ using the given "directions" in S!

 Recall the subset $S = \{[1, -1, 0, 2], [0, -2, 1, 0], [2, 0, -1, 1]\}$ of $\mathbb{R}^4$ from Example 6. In that example, we proved that S is linearly independent. Now

$$[11, 1, -6, 10] = 3[1, -1, 0, 2] + (-2)[0, -2, 1, 0] + 4[2, 0, -1, 1]$$

so $[11, 1, -6, 10]$ is in $\text{span}(S)$. Then by Theorem 4.10, this is the *only* possible way to express $[11, 1, -6, 10]$ as a linear combination of the elements in S. ∎

Linear Independence of Eigenvectors

We will prove in Section 5.5 that the set of eigenvectors for an $n \times n$ matrix produced by the Diagonalization Method of Section 3.4 is always linearly independent (also see Exercise 26). Let us assume this for the moment. Now, if the method produces n eigenvectors, then the matrix $\mathbf{P}$ whose columns are these eigenvectors must row reduce to $\mathbf{I}_n$, by the Independence Test Method. This will establish the claim in Section 3.4 that $\mathbf{P}$ is nonsingular.

EXAMPLE 12   Consider the 3×3 matrix

$$\mathbf{A} = \begin{bmatrix} -2 & 12 & -4 \\ -2 & 8 & -2 \\ -3 & 9 & -1 \end{bmatrix}.$$

You are asked to show in Exercise 14 that the Diagonalization Method of Section 3.4 produces the eigenvector $[4, 2, 3]$ for the eigenvalue $\lambda_1 = 1$, and the eigenvectors $[3, 1, 0]$ and $[-1, 0, 1]$ for the eigenvalue $\lambda_2 = 2$. We test their linear independence by row reducing

$$\mathbf{P} = \begin{bmatrix} 4 & 3 & -1 \\ 2 & 1 & 0 \\ 3 & 0 & 1 \end{bmatrix} \text{ to obtain } \begin{bmatrix} 1 & 0 & 0 \\ 0 & 1 & 0 \\ 0 & 0 & 1 \end{bmatrix},$$

thus illustrating that this set of eigenvectors is linearly independent *and* that $\mathbf{P}$ is nonsingular.   ∎

Summary of Results

This section includes several different, but equivalent, descriptions of linearly independent and linearly dependent sets of vectors. Several additional characterizations are described in the exercises. The most important results from both the section and the exercises are summarized in Table 4.1.

Exercises for Section 4.4

*1. In each part, determine by quick inspection whether the given set of vectors is linearly independent. State a reason for your conclusion.
   (a) $\{[0, 1, 1]\}$ (b) $\{[1, 2, -1], [3, 1, -1]\}$
   (c) $\{[1, 2, -5], [-2, -4, 10]\}$ (d) $\{[4, 2, 1], [-1, 3, 7], [0, 0, 0]\}$
   (e) $\{[2, -5, 1], [1, 1, -1], [0, 2, -3], [2, 2, 6]\}$

2. Use the Independence Test Method to determine which of the following sets of vectors are linearly independent:
   *(a) $\{[1, 9, -2], [3, 4, 5], [-2, 5, -7]\}$
   *(b) $\{[2, -1, 3], [4, -1, 6], [-2, 0, 2]\}$
   (c) $\{[12, 4, 1], [-1, 4, -3], [1, -1, 1]\}$
   (d) $\{[5, -2, 3], [-4, 1, -7], [7, -4, -5]\}$
   *(e) $\{[2, 5, -1, 6], [4, 3, 1, 4], [1, -1, 1, -1]\}$
   (f) $\{[1, 3, -2, 4], [3, 11, -2, -2], [2, 8, 3, -9], [3, 11, -8, 5]\}$

3. Use the Independence Test Method to determine which of the following subsets of $\mathcal{P}_2$ are linearly independent:
   *(a) $\{x^2 + x + 1, \ x^2 - 1, \ x^2 + 1\}$

$\rightarrow \forall \vec{v} \in S, \vec{v} \notin span(S - \{\vec{v}\})$

Table 4.1

Equivalent conditions
for a subset S of a
vector space to be
linearly independent or
linearly dependent

Linear Independence of S	Linear Dependence of S	Source
For every $\mathbf{v} \in S$, we have $\mathbf{v} \notin \text{span}(S - \{\mathbf{v}\})$.	There is a $\mathbf{v} \in S$ such that $\mathbf{v} \in \text{span}(S - \{\mathbf{v}\})$.	Alternate Characterization
If $\{\mathbf{v}_1, \ldots, \mathbf{v}_n\} \subseteq S$ and $a_1\mathbf{v}_1 + \cdots + a_n\mathbf{v}_n = \mathbf{0}$, then $a_1 = a_2 = \cdots = a_n = 0$. (The zero vector requires zero coefficients.)	There is a subset $\{\mathbf{v}_1, \ldots, \mathbf{v}_n\}$ of S such that $a_1\mathbf{v}_1 + \cdots + a_n\mathbf{v}_n = \mathbf{0}$, for scalars $a_1, a_2, \ldots, a_n$, with some $a_i \neq 0$. (The zero vector does not require all coefficients to be zero.)	Theorems 4.7 and 4.9
Every vector in span(S) can be uniquely expressed as a linear combination of the vectors in S.	*Some* vector in span(S) can be expressed in more than one way as a linear combination of the vectors in S.	Theorem 4.10
If $S = \{\mathbf{v}_1, \ldots, \mathbf{v}_n\}$, then for each k, $\mathbf{v}_k \notin$ span $(\{\mathbf{v}_1, ..., \mathbf{v}_{k-1}\})$. (Each $\mathbf{v}_k$ is not a linear combination of the previous vectors in S.)	If $S = \{\mathbf{v}_1, \ldots, \mathbf{v}_n\}$, some $\mathbf{v}_k$ can be expressed as $\mathbf{v}_k = a_1\mathbf{v}_1 + \cdots + a_{k-1}\mathbf{v}_{k-1}$. (Some $\mathbf{v}_k$ is a linear combination of the previous vectors in S.)	Exercise 22
For every $\mathbf{v} \in S$, span($S - \{\mathbf{v}\}$) does not contain all the vectors of span(S).	There is some $\mathbf{v} \in S$ such that span($S - \{\mathbf{v}\}$) = span(S).	Exercise 12
Every finite subset of S is linearly independent.	*Some* finite subset of S is linearly dependent.	Exercises 20 and 21

(b) $\{x^2 - x + 3, 2x^2 - 3x - 1, 5x^2 - 9x - 7\}$
⋆(c) $\{2x - 6, 7x + 2, 12x - 7\}$
(d) $\{x^2 + ax + b \mid |a| = |b| = 1\}$

4. Determine which of the following subsets of $\mathcal{P}$ are linearly independent:
⋆(a) $\{x^2 - 1, x^2 + 1, x^2 + x\}$
(b) $\{1 + x^2 - x^3, 2x - 1, x + x^3\}$
⋆(c) $\{4x^2 + 2, x^2 + x - 1, x, x^2 - 5x - 3\}$
(d) $\{3x^3 + 2x + 1, x^3 + x, x - 5, x^3 + x - 10\}$
⋆(e) $\{1, x, x^2, x^3, \ldots\}$
(f) $\{1, 1 + x, 1 + x + x^2, 1 + x + x^2 + x^3, \ldots\}$

5. Show that the following is a linearly dependent subset of $\mathcal{M}_{22}$:

$$\left\{ \begin{bmatrix} 1 & -2 \\ 0 & 1 \end{bmatrix}, \begin{bmatrix} 3 & 2 \\ -6 & 1 \end{bmatrix}, \begin{bmatrix} 4 & -1 \\ -5 & 2 \end{bmatrix}, \begin{bmatrix} 3 & -3 \\ 0 & 0 \end{bmatrix} \right\}.$$

6. Prove that the following is linearly independent in $\mathcal{M}_{32}$:

$$\left\{ \begin{bmatrix} 1 & 2 \\ -1 & 1 \\ 3 & 0 \end{bmatrix}, \begin{bmatrix} 4 & 2 \\ -6 & 1 \\ 0 & 1 \end{bmatrix}, \begin{bmatrix} 0 & 1 \\ 1 & -1 \\ 2 & 2 \end{bmatrix}, \begin{bmatrix} 0 & 7 \\ 5 & 2 \\ -1 & 6 \end{bmatrix} \right\}.$$

7. Let $S = \{[1, 1, 0], [-2, 0, 1]\}$.
   (a) Show that S is a linearly independent subset of $\mathbb{R}^3$.
   ⋆(b) Find a vector $\mathbf{v}$ in $\mathbb{R}^3$ such that $S \cup \{\mathbf{v}\}$ is also linearly independent.
   ⋆(c) Is the vector $\mathbf{v}$ from part (b) unique, or could some other choice for $\mathbf{v}$ have been made? Why or why not?
   ⋆(d) Find a nonzero vector $\mathbf{u}$ in $\mathbb{R}^3$ such that $S \cup \{\mathbf{u}\}$ is linearly dependent.

8. Suppose that S is the subset $\{[2, -1, 0, 5], [1, -1, 2, 0], [-1, 0, 1, 1]\}$ of $\mathbb{R}^4$.
   (a) Show that S is linearly independent.
   (b) Find a linear combination of vectors in S that produces $[-2, 0, 3, -4]$ (an element of span(S)).
   (c) Is there a different linear combination of the elements of S that yields $[-2, 0, 3, -4]$? If so, find one. If not, why not?

9. Consider $S = \{2x^3 - x + 3, 3x^3 + 2x - 2, x^3 - 4x + 8, 4x^3 + 5x - 7\} \subseteq \mathcal{P}_3$.
   (a) Show that S is linearly dependent.
   (b) Show that every three-element subset of S is linearly dependent.
   (c) Explain why every subset of S containing exactly two vectors is linearly independent. (Note: There are six possible two-element subsets.)

10. Let $\mathbf{u} = [u_1, u_2, u_3]$, $\mathbf{v} = [v_1, v_2, v_3]$, $\mathbf{w} = [w_1, w_2, w_3]$ be three vectors in $\mathbb{R}^3$. Show that $S = \{\mathbf{u}, \mathbf{v}, \mathbf{w}\}$ is linearly independent if and only if

$$\begin{vmatrix} u_1 & u_2 & u_3 \\ v_1 & v_2 & v_3 \\ w_1 & w_2 & w_3 \end{vmatrix} \neq 0.$$

(Hint: Consider the transpose and use the Independence Test Method.) (Compare this exercise with Exercise 19 in Section 4.3.)

11. For each of the following vector spaces, find a linearly independent subset S containing exactly four elements:
   ⋆(a) $\mathbb{R}^4$ (b) $\mathbb{R}^5$
   ⋆(c) $\mathcal{P}_3$ (d) $\mathcal{M}_{23}$
   ⋆(e) $\mathcal{V} = $ set of all symmetric matrices in $\mathcal{M}_{33}$.

12. Let S be a subset of a vector space $\mathcal{V}$. Prove that S is linearly dependent if and only if there is a vector $\mathbf{v} \in S$ such that span($S - \{\mathbf{v}\}$) = span(S). (We say that such a vector $\mathbf{v}$ is **redundant** in S because the same set of linear combinations is obtained after $\mathbf{v}$ is removed from S; that is, $\mathbf{v}$ is not needed.)

13. Find a redundant vector in each given linearly dependent set, and show that it satisfies the definition of a redundant vector given in Exercise 12.
   (a) $\{[4, -2, 6, 1], [1, 0, -1, 2], [0, 0, 0, 0], [6, -2, 5, 5]\}$
   ⋆(b) $\{[1, 1, 0, 0], [1, 1, 1, 0], [0, 0, -6, 0]\}$
   (c) $\{[x_1, x_2, x_3, x_4] \in \mathbb{R}^4 \mid x_i = \pm 1, \text{ for each } i\}$

14. Verify that the Diagonalization Method of Section 3.4 produces the eigenvectors given in the text for the matrix $\mathbf{A}$ of Example 12.

15. Let $S_1 = \{\mathbf{v}_1, \ldots, \mathbf{v}_n\}$ be a subset of a vector space $\mathcal{V}$, let c be a nonzero real number, and let $S_2 = \{c\mathbf{v}_1, \ldots, c\mathbf{v}_n\}$. Show that S_1 is linearly independent if and only if S_2 is linearly independent.

▶16. Prove Corollary 4.8. (Hint: Use Theorem 4.7. Construct an appropriate homogeneous system of linear equations, and show that the system has a nontrivial solution.)

17. Let $\mathbf{f}$ be a polynomial with at least two nonzero terms having different degrees. Prove that the set $\{\mathbf{f}(x), x\mathbf{f}'(x)\}$ (where $\mathbf{f}'$ is the derivative of $\mathbf{f}$) is linearly independent in $\mathcal{P}$.

18. Let $\mathcal{V}$ be a vector space, $\mathcal{W}$ a subspace of $\mathcal{V}$, S a linearly independent subset of $\mathcal{W}$, and $\mathbf{v} \in \mathcal{V} - \mathcal{W}$. Prove that $S \cup \{\mathbf{v}\}$ is linearly independent.

19. Let $\mathbf{A}$ be an $n \times m$ matrix, let $S = \{\mathbf{v}_1, \ldots, \mathbf{v}_k\}$ be a finite subset of $\mathbb{R}^m$, and let $T = \{\mathbf{A}\mathbf{v}_1, \ldots, \mathbf{A}\mathbf{v}_k\}$, a subset of $\mathbb{R}^n$.

   (a) Prove that if T is a linearly independent subset of $\mathbb{R}^n$, then S is a linearly independent subset of $\mathbb{R}^m$.
   ⋆(b) Find a matrix $\mathbf{A}$ for which the converse to part (a) is false.
   (c) Show that the converse to part (a) is true if $\mathbf{A}$ is square and nonsingular.

20. Prove that every subset of a linearly independent set is linearly independent.

21. Suppose that S is a subset of a vector space $\mathcal{V}$. Show that if every finite subset of S is linearly independent, then S itself is linearly independent.

22. Suppose $S = \{\mathbf{v}_1, \ldots, \mathbf{v}_n\}$ is a finite subset of a vector space $\mathcal{V}$. Prove that S is linearly independent if and only if $\mathbf{v}_1 \neq \mathbf{0}$ and, for each k with $2 \leq k \leq n$, $\mathbf{v}_k \notin \text{span}(\{\mathbf{v}_1, \ldots, \mathbf{v}_{k-1}\})$. (Hint: Half of the proof is done by contrapositive. For this half, assume that S is linearly dependent, and use an argument similar to the second half of the proof of Theorem 4.7.)

23. Let $\mathbf{f}$ be an nth degree polynomial in $\mathcal{P}$, and let $\mathbf{f}^{(i)}$ be the ith derivative of $\mathbf{f}$. Show that $\{\mathbf{f}, \mathbf{f}^{(1)}, \mathbf{f}^{(2)}, \ldots, \mathbf{f}^{(n)}\}$ is a linearly independent subset of $\mathcal{P}$. (Hint: Reverse the order of the elements, and use Exercise 22.)

24. Let S be a nonempty subset of a vector space $\mathcal{V}$.
   (a) Prove that S is linearly independent if and only if *some* vector $\mathbf{v}$ in span(S) has a unique expression as a linear combination of the vectors in S (ignoring zero coefficients).
   (b) The contrapositive of the statement in part (a) gives a necessary and sufficient condition for S to be linearly dependent. What is this condition?

▶25. Prove Theorem 4.9 by proving its contrapositive, which is the following: Let S be a nonempty subset of a vector space $\mathcal{V}$. Then S is linearly dependent if and only if there is some finite subset $\{\mathbf{v}_1, \mathbf{v}_2, \ldots, \mathbf{v}_n\}$ of S such that $a_1\mathbf{v}_1 + a_2\mathbf{v}_2 + \cdots + a_n\mathbf{v}_n = \mathbf{0}$, for coefficients $a_1, a_2, \ldots, a_n$, with some $a_i \neq 0$. (Hint: Your proof should have the same general outline as the proof of Theorem 4.7. However, it will have several technical differences because S could be infinite.)

26. Suppose $\mathbf{A}$ is an $n \times n$ matrix and that λ is an eigenvalue for $\mathbf{A}$. Let $\{\mathbf{v}_1, \ldots, \mathbf{v}_k\}$ be the set of eigenvectors for $\mathbf{A}$ corresponding to λ generated by Step 3 of the Diagonalization Method in Section 3.4. Prove that S is linearly independent. (Hint: Consider that each $\mathbf{v}_i$ has a 1 in a coordinate in which all the other vectors in S have a 0. Then apply Theorem 4.7.)

27. Suppose T is a linearly independent subset of a vector space $\mathcal{V}$ and that $\mathbf{v} \in \mathcal{V}$ such that $T \cup \{\mathbf{v}\}$ is linearly dependent. Prove that $\mathbf{v} \in \text{span}(T)$.

★28. True or False:

(a) The set $\{[2, -3, 1], [-8, 12, -4]\}$ is a linearly independent subset of $\mathbb{R}^3$.

(b) A set $S = \{\mathbf{v}_1, \mathbf{v}_2, \mathbf{v}_3\}$ in a vector space $\mathcal{V}$ is linearly dependent if $\mathbf{v}_2$ is a linear combination of $\mathbf{v}_1$ and $\mathbf{v}_3$.

(c) A subset $S = \{\mathbf{v}\}$ of a vector space $\mathcal{V}$ is linearly dependent if $\mathbf{v} = \mathbf{0}$.

(d) A subset S of a vector space $\mathcal{V}$ is linearly independent if there is a vector $\mathbf{v} \in S$ such that $\mathbf{v} \in \text{span}(S - \{\mathbf{v}\})$.

(e) If $\{\mathbf{v}_1, \mathbf{v}_2, \ldots, \mathbf{v}_n\}$ is a linearly independent set of vectors in a vector space $\mathcal{V}$, and $a_1\mathbf{v}_1 + a_2\mathbf{v}_2 + \cdots + a_n\mathbf{v}_n = \mathbf{0}$, then $a_1 = a_2 = \cdots = a_n = 0$.

(f) If S is a subset of $\mathbb{R}^4$ containing 6 vectors, then S is linearly dependent.

(g) Let S be a finite nonempty set of vectors in $\mathbb{R}^n$. If the matrix $\mathbf{A}$ whose rows are the vectors in S has n pivots after row reduction, then S is linearly independent.

(h) If $S = \{\mathbf{v}_1, \mathbf{v}_2, \mathbf{v}_3\}$ is a linearly independent subset of a vector space $\mathcal{V}$, then no vector in $\text{span}(S)$ can be expressed as two different linear combinations of $\mathbf{v}_1, \mathbf{v}_2$, and $\mathbf{v}_3$.

(i) If $S = \{\mathbf{v}_1, \mathbf{v}_2\}$ is a subset of a vector space $\mathcal{V}$, and $\mathbf{v}_3 = 3\mathbf{v}_1 - 2\mathbf{v}_2$, then $\{\mathbf{v}_1, \mathbf{v}_2, \mathbf{v}_3\}$ is linearly dependent.

4.5 Basis and Dimension

Suppose that S is a subset of a vector space $\mathcal{V}$ and that $\mathbf{v}$ is some vector in $\mathcal{V}$. We can ask the following two fundamental questions about S and $\mathbf{v}$:

Existence: Is there a linear combination of vectors in S equal to $\mathbf{v}$?

Uniqueness: If so, is this the only such linear combination?

The interplay between existence and uniqueness questions is a pervasive theme throughout mathematics. Answering the existence question is equivalent to determining whether $\mathbf{v} \in \text{span}(S)$. Answering the uniqueness question is equivalent (by Theorem 4.10) to determining whether S is linearly independent.

We are most interested in cases where both existence and uniqueness occur. In this section, we tie together these concepts by examining those subsets of vector spaces that simultaneously span and are linearly independent. Such a subset is called a **basis**.

Definition of Basis

DEFINITION
Let $\mathcal{V}$ be a vector space, and let B be a subset of $\mathcal{V}$. Then B is a **basis** for $\mathcal{V}$ if and only if both of the following are true:
(1) B spans $\mathcal{V}$.
(2) B is linearly independent.

EXAMPLE 1   We show that $B = \{[1, 2, 1], [2, 3, 1], [-1, 2, -3]\}$ is a basis for $\mathbb{R}^3$ by showing that it both spans $\mathbb{R}^3$ and is linearly independent.

First, we use the Simplified Span Method in Section 4.3 to show that B spans $\mathbb{R}^3$. Expressing the vectors in B as rows and row reducing the matrix

$$\begin{bmatrix} 1 & 2 & 1 \\ 2 & 3 & 1 \\ -1 & 2 & -3 \end{bmatrix} \text{ yields } \begin{bmatrix} 1 & 0 & 0 \\ 0 & 1 & 0 \\ 0 & 0 & 1 \end{bmatrix},$$

which proves that $\operatorname{span}(B) = \{a[1, 0, 0] + b[0, 1, 0] + c[0, 0, 1] \mid a, b, c \in \mathbb{R}\} = \mathbb{R}^3$.

Next, we must show that B is linearly independent. Expressing the vectors in B as columns, and using the Independence Test Method in Section 4.4, we row reduce

$$\begin{bmatrix} 1 & 2 & -1 \\ 2 & 3 & 2 \\ 1 & 1 & -3 \end{bmatrix} \text{ to obtain } \begin{bmatrix} 1 & 0 & 0 \\ 0 & 1 & 0 \\ 0 & 0 & 1 \end{bmatrix}.$$

Hence, B is also linearly independent.

Since B spans $\mathbb{R}^3$ and is linearly independent, B is a basis for $\mathbb{R}^3$. (B is not the only basis for $\mathbb{R}^3$, as we show in the next example.) ∎

EXAMPLE 2   The vector space $\mathbb{R}^n$ has $\{\mathbf{e}_1, \ldots, \mathbf{e}_n\}$ as a basis. Although $\mathbb{R}^n$ has other bases as well, the basis $\{\mathbf{e}_1, \ldots, \mathbf{e}_n\}$ is the most useful for general applications and is therefore referred to as the **standard basis** for $\mathbb{R}^n$. Thus, we refer to $\{\mathbf{i}, \mathbf{j}\}$ and $\{\mathbf{i}, \mathbf{j}, \mathbf{k}\}$ as the standard bases for $\mathbb{R}^2$ and $\mathbb{R}^3$, respectively. ∎

Each of our fundamental examples of vector spaces also has a "standard basis."

EXAMPLE 3   The standard basis in $\mathcal{M}_{32}$ is defined as the set

$$\left\{ \begin{bmatrix} 1 & 0 \\ 0 & 0 \\ 0 & 0 \end{bmatrix}, \begin{bmatrix} 0 & 1 \\ 0 & 0 \\ 0 & 0 \end{bmatrix}, \begin{bmatrix} 0 & 0 \\ 1 & 0 \\ 0 & 0 \end{bmatrix}, \begin{bmatrix} 0 & 0 \\ 0 & 1 \\ 0 & 0 \end{bmatrix}, \begin{bmatrix} 0 & 0 \\ 0 & 0 \\ 1 & 0 \end{bmatrix}, \begin{bmatrix} 0 & 0 \\ 0 & 0 \\ 0 & 1 \end{bmatrix} \right\}.$$

More generally, we define the **standard basis** in $\mathcal{M}_{mn}$ to be the set of $m \cdot n$ distinct matrices

$$\left\{ \mathbf{\Psi}_{ij} \mid 1 \le i \le m, 1 \le j \le n \right\},$$

where $\boldsymbol{\Psi}_{ij}$ is the $m \times n$ matrix with 1 in the (i, j) position and zeroes elsewhere. You should check that these $m \cdot n$ matrices are linearly independent and span $\mathcal{M}_{mn}$. In addition to the standard basis, $\mathcal{M}_{mn}$ has many other bases as well. ■

 We define $\{1, x, x^2, x^3\}$ to be the standard basis for $\mathcal{P}_3$. More generally, the **standard basis** for $\mathcal{P}_n$ is defined to be the set $\{1, x, x^2, \ldots, x^n\}$, containing $n + 1$ elements. Similarly, we define the infinite set $\{1, x, x^2, \ldots\}$ to be the **standard basis** for $\mathcal{P}$. Again, note that in each case these sets both span and are linearly independent.

Of course, the polynomial spaces have other bases. For example, the following is also a basis for $\mathcal{P}_4$:

$$\left\{ x^4, \ x^4 - x^3, \ x^4 - x^3 + x^2, \ x^4 - x^3 + x^2 - x, \ x^3 - 1 \right\}.$$

In Exercise 3, you are asked to verify that this is a basis. ■

Two Technical Lemmas

In Examples 1 through 4 we saw that $\mathbb{R}^n$, $\mathcal{P}_n$, and $\mathcal{M}_{mn}$ each have some *finite* set for a basis, while $\mathcal{P}$ has an infinite basis. We will mostly be concerned with those vector spaces that have finite bases. To begin our study of such vector spaces, we first show that if a vector space has *one* basis that is finite, then *all* of its bases are finite, and all have the same size. Proving this requires some effort. We begin with the following lemma:

LEMMA 4.11
Let $\mathcal{V}$ be a vector space spanned by the n-element set $S = \{\mathbf{v}_1, \ldots, \mathbf{v}_n\}$, and let $T = \{\mathbf{w}_1, \ldots, \mathbf{w}_n\}$ be a set of n linearly independent vectors in $\mathcal{V}$. Then $\mathrm{span}(T) = \mathcal{V}$, and hence T is a basis for $\mathcal{V}$.

The proof of this lemma may be difficult on a first reading. To prove the lemma, it is enough to show that T spans $\mathcal{V}$, since we already know that T is linearly independent. The idea behind the proof is to replace each element of S with an element chosen from T, one at a time, and prove that at each step, the new set obtained still spans $\mathcal{V}$. Once all the elements of S have been replaced by elements of T, the final set becomes T itself, and the lemma will be proved. In the actual proof below, we assume that we have already done as many replacements as possible while still maintaining the spanning property. We then show that there is a contradiction if we have not completely replaced S with T.

We use the notation $|S|$ to represent the number of elements in a set S. For example, in the statement of Lemma 4.11, $|S| = |T| = n$.

PROOF OF LEMMA 4.11

Let $\mathcal{V}$, n, S, and T be as in the statement of the lemma. Define

$$A = \{X \mid X \subseteq S \cup T \text{ and } |X| = n \text{ and } \operatorname{span}(X) = \mathcal{V}\}.$$

A is the collection of all subsets of $S \cup T$ of size n that span $\mathcal{V}$. Notice that $S \in A$ because $S \subseteq S \cup T$, $|S| = n$, and S spans $\mathcal{V}$. Our goal is to prove that $T \in A$, for, if so, then T spans $\mathcal{V}$ and we will be done.

Now, because $S \cup T$ is a finite set, it has only a finite number of subsets. Hence, A has only a finite number of elements. A also has at least one element since $S \in A$. Therefore, we can examine each element of A in turn, and choose $Y = \{y_1, y_2, \ldots, y_n\}$ to be an element of A whose intersection with T is the largest possible. (In case of a tie, pick any such element of A.) We will prove that $Y = T$, which will imply that $T \in A$, to complete the proof.

We prove $Y = T$ by contradiction. So, assume $Y \neq T$. Then, since $|Y| = n = |T|$, there must be an element of T that is not in Y. Without loss of generality, assume $\mathbf{w}_1 \notin Y$. Now Y spans $\mathcal{V}$, so $\mathbf{w}_1 \in \operatorname{span}(Y)$. Thus, by the alternate characterization of linear dependence, $Y \cup \{\mathbf{w}_1\}$ is linearly dependent.

Thus, the contrapositive of Theorem 4.7 shows that there is a linear combination

$$c\mathbf{w}_1 + a_1\mathbf{y}_1 + a_2\mathbf{y}_2 + \cdots + a_n\mathbf{y}_n = \mathbf{0},$$

with not all coefficients equal to zero. Now at least one of the nonzero coefficients in this linear combination must belong to some $\mathbf{y}_j \notin T$. (If the nonzero terms only involved vectors in T, then Theorem 4.7 would contradict the linear independence of T.) Without loss of generality, assume $j = 1$; that is, $a_1 \neq 0$ and $\mathbf{y}_1 \notin T$.

Let $Z = Y \cup \{\mathbf{w}_1\} - \{\mathbf{y}_1\}$; that is, $Z = \{\mathbf{w}_1, \mathbf{y}_2, \ldots \mathbf{y}_n\}$, the set obtained by putting $\mathbf{w}_1$ into Y and taking $\mathbf{y}_1$ out. If we can show $Z \in A$, we will have the desired contradiction because Z will be an element of A containing one more element of T than Y does (namely $\mathbf{w}_1$). (This contradicts the assumption that among the sets in A, Y has the most elements of T.)

We now prove $Z \in A$. The definition of Z implies that $|Z| = n$ and $Z \subseteq S \cup T$. We need to show that $\operatorname{span}(Z) = \mathcal{V}$.

Now, rearranging the linear combination involving $\mathbf{w}_1$ and the $\mathbf{y}_i$'s, we get

$$\mathbf{y}_1 = -\frac{c}{a_1}\mathbf{w}_1 - \frac{a_2}{a_1}\mathbf{y}_2 - \cdots - \frac{a_n}{a_1}\mathbf{y}_n.$$

So $\mathbf{y}_1 \in \operatorname{span}(Z)$. Also, $Y - \{\mathbf{y}_1\} \subset Z \subseteq \operatorname{span}(Z)$. Hence, $Y \subseteq \operatorname{span}(Z)$. Theorem 4.5 then shows that $\operatorname{span}(Y) \subseteq \operatorname{span}(Z)$. But Y spans $\mathcal{V}$. Therefore, Z spans $\mathcal{V}$ as well, thus establishing the contradiction. ∎

EXAMPLE 5

Let $S = \{\mathbf{e}_1, \mathbf{e}_2, \mathbf{e}_3, \mathbf{e}_4\} \subseteq \mathbb{R}^4$. Now, S spans $\mathbb{R}^4$. Also, let

$$T = \{[1, -1, 1, 2], [2, 1, 8, 13], [3, -6, -5, -5], [-4, 10, -2, 7]\}.$$

The Independence Test Method proves that T is linearly independent (try it!). Therefore, since $|S| = |T|$, Lemma 4.11 shows that T also spans $\mathbb{R}^4$, and hence is a basis for $\mathbb{R}^4$. ∎

The next lemma, which is very useful, assures us that the size of a linearly independent set is never larger than the size of a spanning set, if the spanning set is finite.

LEMMA 4.12

Let S and T be subsets of a vector space V such that S spans V, S is finite, and T is linearly independent. Then T is finite and $|T| \le |S|$.

PROOF OF LEMMA 4.12

If S is empty, the lemma is certainly true. For $|S| \ge 1$, we give a proof by contradiction. Suppose that T is infinite, or that T is finite, with $|T| > |S| = n$. Then there is some subset $\{\mathbf{w}_1, \dots, \mathbf{w}_{n+1}\}$ of T containing $n + 1$ vectors. Since T is linearly independent and every subset of a linearly independent subset is also linearly independent (see Table 4.1 in Section 4.4), both $\{\mathbf{w}_1, \dots, \mathbf{w}_{n+1}\}$ and $\{\mathbf{w}_1, \dots, \mathbf{w}_n\}$ are linearly independent. But S spans V and $|S| = n$, so Lemma 4.11 implies that $\{\mathbf{w}_1, \dots, \mathbf{w}_n\}$ is a basis for V. Hence, $\mathbf{w}_{n+1} \in V = \text{span}(\{\mathbf{w}_1, \dots, \mathbf{w}_n\})$. This statement contradicts the linear independence of $\{\mathbf{w}_1, \dots, \mathbf{w}_{n+1}\}$, and therefore, $|T| \le |S|$. ∎

Dimension

We can now prove the main result of this section.

THEOREM 4.13

Let V be a vector space, and let B_1 and B_2 be bases for V such that B_1 has finitely many elements. Then B_2 also has finitely many elements, and $|B_1| = |B_2|$.

PROOF OF THEOREM 4.13

Because B_1 and B_2 are bases for V, B_1 spans V and B_2 is linearly independent. Hence, Lemma 4.12 shows that B_2 has finitely many elements and $|B_2| \le |B_1|$. Now, since B_2 is finite, we can reverse the roles of B_1 and B_2 in this argument to show that $|B_1| \le |B_2|$. Therefore, $|B_1| = |B_2|$. ∎

It follows from Theorem 4.13 that if a vector space V has *one* basis containing a finite number of elements, then *every* basis for V is finite, and all bases for V have the same number of elements. This allows us to unambiguously define the **dimension** of such a vector space, as follows:

DEFINITION

Let V be a vector space. If V has a basis B containing a finite number of elements, then V is said to be **finite dimensional**. In this case, the **dimension** of V, $\dim(V)$, is the number of elements in any basis for V. In particular, $\dim(V) = |B|$.

If V has no finite basis, then V is **infinite dimensional**.

| EXAMPLE 6 |

Because $\mathbb{R}^3$ has the (standard) basis $\{\mathbf{i}, \mathbf{j}, \mathbf{k}\}$, the dimension of $\mathbb{R}^3$ is 3. Theorem 4.13 then implies that every other basis for $\mathbb{R}^3$ also has exactly three elements. More generally, $\dim(\mathbb{R}^n) = n$, since $\mathbb{R}^n$ has the basis $\{\mathbf{e}_1, \ldots, \mathbf{e}_n\}$. ∎

| EXAMPLE 7 |

Because the standard basis $\{1, x, x^2, x^3\}$ for $\mathcal{P}_3$ has four elements, $\dim(\mathcal{P}_3) = 4$. Every other basis for $\mathcal{P}_3$, such as $\{x^3 - x, x^2 + x + 1, x^3 + x - 5, 2x^3 + x^2 + x - 3\}$, also has four elements. (Verify that this set is a basis for $\mathcal{P}_3$.)

Also, $\dim(\mathcal{P}_n) = n + 1$, since $\mathcal{P}_n$ has the basis $\{1, x, x^2, \ldots, x^n\}$, containing $n + 1$ elements. Be careful! Many students *erroneously* believe that the dimension of $\mathcal{P}_n$ is n because of the subscript n. ∎

| EXAMPLE 8 |

The standard basis for $\mathcal{M}_{22}$ contains four elements. Hence, $\dim(\mathcal{M}_{22}) = 4$. In general, from the size of the standard basis for $\mathcal{M}_{mn}$, we see that $\dim(\mathcal{M}_{mn}) = m \cdot n$. ∎

| EXAMPLE 9 |

Let $\mathcal{V} = \{\mathbf{0}\}$ be the trivial vector space. Then $\dim(\mathcal{V}) = 0$ because the empty set, which contains no elements, is a basis for $\mathcal{V}$. (Recall that $\text{span}(\{\}) = \{\mathbf{0}\}$ by definition, and that $\{\}$ is defined to be linearly independent.) ∎

| EXAMPLE 10 |

Consider the following subsets of $\mathbb{R}^4$:

$$S_1 = \{[1, 3, 1, 2], [3, 11, 5, 10], [-2, 4, 4, 4]\} \text{ and}$$
$$S_2 = \{[1, 5, -2, 3], [-2, -8, 8, 8], [1, 1, -10, -2], [0, 2, 4, -9], [3, 13, -10, -8]\}.$$

Since $\dim(\mathbb{R}^4) = 4$, $|S_1| = 3$, and $|S_2| = 5$, Theorem 4.13 shows us that neither S_1 nor S_2 is a basis for $\mathbb{R}^4$. In particular, S_1 cannot span $\mathbb{R}^4$ because the standard basis for $\mathbb{R}^4$ would then be a linearly independent set that is larger than S_1, contradicting Lemma 4.12. Similarly, S_2 cannot be linearly independent because the standard basis would be a spanning set that is smaller than S_2, again contradicting Lemma 4.12.

Notice, however, that in this case we can make no conclusions regarding whether S_1 is linearly independent or whether S_2 spans $\mathbb{R}^4$ based solely on the size of these sets. We must check for these properties separately using the Methods of Sections 4.3 and 4.4. ∎

Sizes of Spanning Sets and Linearly Independent Sets

Example 10 illustrates the next result, which summarizes much of what we have learned regarding the sizes of spanning sets and linearly independent sets.

COROLLARY 4.14

Let V be a finite dimensional vector space.

(1) Suppose S is a finite subset of V that spans V. Then $\dim(V) \leq |S|$. Moreover, $|S| = \dim(V)$ if and only if S is a basis for V.

(2) Suppose T is a linearly independent subset of V. Then T is finite and $|T| \leq \dim(V)$. Moreover, $|T| = \dim(V)$ if and only if T is a basis for V.

PROOF OF COROLLARY 4.14

Let B be a basis for V with $|B| = n$. Then $\dim(V) = |B|$, by definition.

Part (1): Since S is a finite spanning set and B is linearly independent, Lemma 4.12 implies that $|B| \leq |S|$, and so $\dim(V) \leq |S|$.

If $|S| = \dim(V)$, we prove that S is a basis for V by contradiction. If S is not a basis, then it is not linearly independent (because it spans). So, by Exercise 12 in Section 4.4 (see Table 4.1), there is a redundant vector in S, that is, a vector $\mathbf{v}$ such that $\text{span}(S - \{\mathbf{v}\}) = \text{span}(S) = V$. But then $S - \{\mathbf{v}\}$ is a spanning set for V having fewer than n elements, contradicting the fact that we just observed that the size of a spanning set is never less than the dimension.

Finally, suppose S is a basis for V. By Theorem 4.13, S is finite, and $|S| = \dim(V)$ by the definition of dimension.

Part (2): Using B as the spanning set S in Lemma 4.12 proves that T is finite and $|T| \leq \dim(V)$.

If $|T| = \dim(V)$, then by using B as the spanning set S in Lemma 4.11, we conclude that T is a basis for V.

Finally, if T is a basis for V, then $|T| = \dim(V)$, by the definition of dimension. ■

EXAMPLE 11   Recall the subset $B = \{[1, 2, 1], [2, 3, 1], [-1, 2, -3]\}$ of $\mathbb{R}^3$ from Example 1. In that example, after showing that B spans $\mathbb{R}^3$, we could have immediately concluded that B is a basis for $\mathbb{R}^3$ without having proved linear independence by using part (1) of Corollary 4.14 because B is a spanning set with $\dim(\mathbb{R}^3) = 3$ elements.

Similarly, consider $T = \{3, x + 5, x^2 - 7x + 12, x^3 + 4\}$, a subset of $\mathcal{P}_3$. T is linearly independent from Exercise 22 in Section 4.4 (see Table 4.1) because each vector in T is not in the span of those before it. Since $|T| = 4 = \dim(\mathcal{P}_3)$, part (2) of Corollary 4.14 shows that T is a basis for $\mathcal{P}_3$. ■

Maximal Linearly Independent Sets and Minimal Spanning Sets

Corollary 4.14 shows that in a finite dimensional vector space, a large enough linearly independent set is a basis, as is a small enough spanning set. The "borderline" size is the dimension of the vector space. No linearly independent sets are larger than this, and no spanning sets are smaller. The next two results illustrate this same principle without explicitly using the dimension. Thus, they are useful in cases in which the dimension is not known or for infinite dimensional vector spaces. Outlines of their proofs are given in Exercises 18 and 19.

THEOREM 4.15
Let V be a vector space with spanning set S (so, span(S) $= V$), and let B be a maximal linearly independent subset of S. Then B is a basis for V.

The phrase, "B is a maximal linearly independent subset of S," means that both of the following are true:

- B is a linearly independent subset of S.

- If $B \subset C \subseteq S$ and $B \neq C$, then C is linearly dependent. *B is as big as it gets

Theorem 4.15 asserts that if there is no way to include another vector from S in B without making B linearly dependent, then B is a basis for span(S) $= V$. The converse to Theorem 4.15 is also true (see Exercise 20).

EXAMPLE 12   Consider the subset $S = \{[1, -2, 1], [3, 1, -2], [5, -3, 0], [5, 4, -5], [0, 0, 0]\}$ of $\mathbb{R}^3$ and the subset $B = \{[1, -2, 1], [5, -3, 0]\}$ of S. We show that B is a maximal linearly independent subset of S and hence, by Theorem 4.15, it is a basis for $V = $ span(S).

Now, B is a linearly independent subset of S. The following equations show that if any of the remaining vectors of S are added to B, the set is no longer linearly independent:

$$[3, 1, -2] = -2[1, -2, 1] + [5, -3, 0]$$
$$[5, 4, -5] = -5[1, -2, 1] + 2[5, -3, 0]$$
$$[0, 0, 0] = 0[1, -2, 1] + 0[5, -3, 0]$$

Thus, B is a maximal linearly independent subset of S and so is a basis for span(S). ∎

Another consequence of Theorem 4.15 is that any vector space V having a finite spanning set S must be finite dimensional. This is because a maximal linearly independent subset of S, which must also be finite, is a basis for V (see Exercise 24).

We also have the following result for spanning sets:

THEOREM 4.16
Let V be a vector space, and let B be a minimal spanning set for V. Then B is a basis for V.

The phrase, "B is a minimal spanning set for V," means that both of the following are true:

- B is a subset of V that spans V.

- If $C \subset B$ and $C \neq B$, then C does not span V.

The converse of Theorem 4.16 is true as well (see Exercise 21).

EXAMPLE 13   Consider the subsets S and B of $\mathbb{R}^3$ given in Example 12. We can use Theorem 4.16 to give another justification that B is a basis for $V = \text{span}(S)$. Recall from Example 12 that every vector in S is a linear combination of vectors in B, so $S \subseteq \text{span}(B)$. This fact along with $B \subseteq S$ and Theorem 4.6 shows that $\text{span}(B) = \text{span}(S) = V$. Also, neither vector in B is a scalar multiple of the other so that neither vector alone can span V (why?). Hence, B is a minimal spanning set for V, and by Theorem 4.16, B is a basis for $\text{span}(S)$.   ∎

Dimension of a Subspace

We conclude this section with the result that every subspace of a finite dimensional vector space is also finite dimensional. This is important because it tells us that the theorems we have developed about finite dimension apply to all *subspaces* of our basic examples $\mathbb{R}^n$, $\mathcal{M}_{mn}$, and $\mathcal{P}_n$.

THEOREM 4.17
Let V be a finite dimensional vector space, and let W be a subspace of V. Then W is also finite dimensional with $\dim(W) \leq \dim(V)$. Moreover, $\dim(W) = \dim(V)$ if and only if $W = V$.

The proof of Theorem 4.17 is left for you to do, with hints, in Exercise 22. The only subtle part of this proof involves showing that W actually has a basis.[4]

EXAMPLE 14   Consider the nested sequence of subspaces of $\mathbb{R}^3$ given by $\{\mathbf{0}\} \subset \{$scalar multiples of $[4, -7, 0]\} \subset xy$-plane $\subset \mathbb{R}^3$. Their respective dimensions are 0, 1, 2, and 3 (why?). Hence, the dimensions of each successive pair of these subspaces satisfy the inequality given in Theorem 4.17.   ∎

EXAMPLE 15   It can be shown that $B = \{x^3+2x^2-4x+18, 3x^2+4x-4, x^3+5x^2-3, 3x+2\}$ is a linearly independent subset of $\mathcal{P}_3$. Therefore, by part (2) of Corollary 4.14, B is a basis for $\mathcal{P}_3$. However, we can also reach the same conclusion from Theorem 4.17. For, $W = \text{span}(B)$ has B as a basis (why?), and hence, $\dim(W) = 4$. But since W is a subspace of $\mathcal{P}_3$ and $\dim(\mathcal{P}_3) = 4$, Theorem 4.17 implies that $W = \mathcal{P}_3$. Hence, B is a basis for $\mathcal{P}_3$.   ∎

---

[4] Although it is true that *every* vector space has a basis, we must be careful here, because we have not proven this. In fact, Theorem 4.17 establishes that every subspace of a finite dimensional vector space *does* have a basis and that this basis is finite. Although every finite dimensional vector space has a finite basis by definition, the proof that every infinite dimensional vector space has a basis requires advanced set theory and is beyond the scope of this text.

Exercises for Section 4.5

1. Prove that each of the following subsets of $\mathbb{R}^4$ is a basis for $\mathbb{R}^4$ by showing both that it spans $\mathbb{R}^4$ and is linearly independent:
    (a) $\{[2, 1, 0, 0], [0, 1, 1, -1], [0, -1, 2, -2], [3, 1, 0, -2]\}$
    (b) $\{[6, 1, 1, -1], [1, 0, 0, 9], [-2, 3, 2, 4], [2, 2, 5, -5]\}$
    (c) $\{[1, 1, 1, 1], [1, 1, 1, -1], [1, 1, -1, -1], [1, -1, -1, -1]\}$
    (d) $\left\{\left[\frac{15}{2}, 5, \frac{12}{5}, 1\right], \left[2, \frac{1}{2}, \frac{3}{4}, 1\right], \left[-\frac{13}{2}, 1, 0, 4\right], \left[\frac{18}{5}, 0, \frac{1}{5}, -\frac{1}{5}\right]\right\}$

2. Prove that the following set is a basis for $\mathcal{M}_{22}$ by showing that it spans $\mathcal{M}_{22}$ and is linearly independent:
$$\left\{\begin{bmatrix} 1 & 4 \\ 2 & 0 \end{bmatrix}, \begin{bmatrix} 0 & 2 \\ 1 & 0 \end{bmatrix}, \begin{bmatrix} -3 & 1 \\ -1 & 0 \end{bmatrix}, \begin{bmatrix} 5 & -2 \\ 0 & -3 \end{bmatrix}\right\}.$$

3. Show that the subset $\{x^4, x^4 - x^3, x^4 - x^3 + x^2, x^4 - x^3 + x^2 - x, x^3 - 1\}$ of $\mathcal{P}_4$ is a basis for $\mathcal{P}_4$.

4. Determine which of the following subsets of $\mathbb{R}^4$ form a basis for $\mathbb{R}^4$:
    ⋆(a) $S = \{[7, 1, 2, 0], [8, 0, 1, -1], [1, 0, 0, -2]\}$
    (b) $S = \{[1, 3, 2, 0], [-2, 0, 6, 7], [0, 6, 10, 7]\}$
    ⋆(c) $S = \{[7, 1, 2, 0], [8, 0, 1, -1], [1, 0, 0, -2], [3, 0, 1, -1]\}$
    (d) $S = \{[1, 3, 2, 0], [-2, 0, 6, 7], [0, 6, 10, 7], [2, 10, -3, 1]\}$
    ⋆(e) $S = \{[1, 2, 3, 2], [1, 4, 9, 3], [6, -2, 1, 4], [3, 1, 2, 1], [10, -9, -15, 6]\}$

5. (a) Show that $B = \{[2, 3, 0, -1], [-1, 1, 1, -1]\}$ is a maximal linearly independent subset of $S = \{[1, 4, 1, -2], [-1, 1, 1, -1], [3, 2, -1, 0], [2, 3, 0, -1]\}$.
    ⋆(b) Calculate dim(span(S)).
    ⋆(c) Does span(S) $= \mathbb{R}^4$? Why or why not?
    (d) Is B a minimal spanning set for span(S)? Why or why not?

6. (a) Show that $B = \{x^3 - x^2 + 2x + 1, 2x^3 + 4x - 7, 3x^3 - x^2 - 6x + 6\}$ is a maximal linearly independent subset of $S = \{x^3 - x^2 + 2x + 1, x - 1, 2x^3 + 4x - 7, x^3 - 3x^2 - 22x + 34, 3x^3 - x^2 - 6x + 6\}$.
    (b) Calculate dim(span(S)).
    (c) Does span(S) $= \mathcal{P}_3$? Why or why not?
    (d) Is B a minimal spanning set for span(S)? Why or why not?

7. Let $\mathcal{W}$ be the solution set to the matrix equation $\mathbf{AX} = \mathbf{O}$, where
$$\mathbf{A} = \begin{bmatrix} 1 & 2 & 1 & 0 & -1 \\ 2 & -1 & 0 & 1 & 3 \\ 1 & -3 & -1 & 1 & 4 \\ 2 & 9 & 4 & -1 & -7 \end{bmatrix}.$$

    (a) Show that $\mathcal{W}$ is a subspace of $\mathbb{R}^5$.
    (b) Find a basis for $\mathcal{W}$.
    (c) Show that dim($\mathcal{W}$)+rank($\mathbf{A}$) $= 5$.

8. Prove that every proper nontrivial subspace of $\mathbb{R}^3$ can be thought of, from a geometric point of view, as either a line through the origin or a plane through the origin.

9. Let $\mathbf{f}$ be a polynomial of degree n. Show that the set $\{\mathbf{f}, \mathbf{f}^{(1)}, \mathbf{f}^{(2)}, \dots, \mathbf{f}^{(n)}\}$ is a basis for $\mathcal{P}_n$ (where $\mathbf{f}^{(i)}$ denotes the ith derivative of $\mathbf{f}$). (Hint: See Exercise 23 in Section 4.4.)

10. (a) Let $\mathbf{A}$ be a 2×2 matrix. Prove that there are real numbers $a_0, a_1, \dots,$ a_4, not all zero, such that $a_4\mathbf{A}^4 + a_3\mathbf{A}^3 + a_2\mathbf{A}^2 + a_1\mathbf{A} + a_0\mathbf{I}_2 = \mathbf{O}_2$. (Hint: You can assume that $\mathbf{A}, \mathbf{A}^2, \mathbf{A}^3, \mathbf{A}^4$, and $\mathbf{I}_2$ are all distinct because if they are not, opposite nonzero coefficients can be chosen for any identical pair to demonstrate that the given statement holds.)

    (b) Suppose $\mathbf{B}$ is an $n \times n$ matrix. Show that there must be a nonzero polynomial $\mathbf{p} \in \mathcal{P}_{n^2}$ such that $\mathbf{p}(\mathbf{B}) = \mathbf{O}_n$.

11. (a) Show that $B = \{(x-2), x(x-2), x^2(x-2), x^3(x-2), x^4(x-2)\}$ is a basis for $V = \{\mathbf{p} \in \mathcal{P}_5 \mid \mathbf{p}(2) = 0\}$.

    $\star$(b) What is $\dim(V)$?

    $\star$(c) Find a basis for $\mathcal{W} = \{\mathbf{p} \in \mathcal{P}_5 \mid \mathbf{p}(2) = \mathbf{p}(3) = 0\}$.

    $\star$(d) Calculate $\dim(\mathcal{W})$.

$\star$**12.** Let V be a finite dimensional vector space.

    (a) Let S be a subset of V with $\dim(V) \le |S|$. Find an example to show that S need not span V.

    (b) Let T be a subset of V with $|T| \le \dim(V)$. Find an example to show that T need not be linearly independent.

13. Let S be a subset of a finite dimensional vector space V such that $|S| = \dim(V)$. If S is not a basis for V, prove that S neither spans V nor is linearly independent.

14. Let V be an n-dimensional vector space, and let S be a subset of V containing exactly n elements. Prove that S spans V if and only if S is linearly independent.

15. Let $\mathbf{A}$ be a nonsingular $n \times n$ matrix, and let B be a basis for $\mathbb{R}^n$.

    (a) Show that $B_1 = \{\mathbf{Av} \mid \mathbf{v} \in B\}$ is also a basis for $\mathbb{R}^n$. (Treat the vectors in B as column vectors.)

    (b) Show that $B_2 = \{\mathbf{vA} \mid \mathbf{v} \in B\}$ is also a basis for $\mathbb{R}^n$. (Treat the vectors in B as row vectors.)

    (c) Letting B be the standard basis for $\mathbb{R}^n$, use the result of part (a) to show that the columns of $\mathbf{A}$ form a basis for $\mathbb{R}^n$.

    (d) Prove that the rows of $\mathbf{A}$ form a basis for $\mathbb{R}^n$.

16. Prove that $\mathcal{P}$ is infinite dimensional by showing that no finite subset S of $\mathcal{P}$ can span $\mathcal{P}$, as follows:

    (a) Let S be a finite subset of $\mathcal{P}$. Show that $S \subseteq \mathcal{P}_n$, for some n.

    (b) Use part (a) to prove that $\text{span}(S) \subseteq \mathcal{P}_n$.

    (c) Conclude that S cannot span $\mathcal{P}$.

17. (a) Prove that if a vector space V has an infinite linearly independent subset, then V is not finite dimensional.

    (b) Use part (a) to prove that any vector space having $\mathcal{P}$ as a subspace is not finite dimensional.

18. The purpose of this exercise is to prove Theorem 4.15. Let $\mathcal{V}$, S, and B be as given in the statement of the theorem. Suppose $B \neq S$, and $\mathbf{w} \in S$ with $\mathbf{w} \notin B$.

    (a) Explain why it is sufficient to prove that B spans $\mathcal{V}$.

    ▶(b) Prove that if $S \subseteq \mathrm{span}(B)$, then B spans $\mathcal{V}$.

    ▶(c) Let $C = B \cup \{\mathbf{w}\}$. Prove that C is linearly dependent.

    (d) Use part (c) to prove that $\mathbf{w} \in \mathrm{span}(B)$. (Also see Exercise 27 in Section 4.4.)

    (e) Tie together all parts to finish the proof.

19. The purpose of this exercise is to prove Theorem 4.16.

    (a) Explain why it is sufficient to prove the following statement: Let S be a spanning set for a vector space $\mathcal{V}$. If S is a minimal spanning set for $\mathcal{V}$, then S is linearly independent.

    ▶(b) State the contrapositive of the statement in part (a).

    ▶(c) Prove the statement from part (b). (Hint: Use Exercise 12 from Section 4.4.)

20. Let B be a basis for a vector space $\mathcal{V}$. Prove that B is a maximal linearly independent subset of $\mathcal{V}$. (Note: You may *not* use $\dim(\mathcal{V})$ in your proof, since $\mathcal{V}$ could be infinite dimensional.)

21. Let B be a basis for a vector space $\mathcal{V}$. Prove that B is a minimal spanning set for $\mathcal{V}$. (Note: You may *not* use $\dim(\mathcal{V})$ in your proof, since $\mathcal{V}$ could be infinite dimensional.)

22. The purpose of this exercise is to prove Theorem 4.17. Let $\mathcal{V}$ and $\mathcal{W}$ be as given in the theorem. Consider the set A of nonnegative integers defined by

$A = \{k \mid \text{a set } T \text{ exists with } T \subseteq \mathcal{W}, |T| = k, \text{ and } T \text{ linearly independent}\}$.

    (a) Prove that $0 \in A$. (Hence, A is nonempty.)

    (b) Prove that $k \in A$ implies $k \leq \dim(\mathcal{V})$. (Hint: Use Corollary 4.14) (Hence, A is finite.)

    ▶(c) Let n be the largest element of A. Let T be a linearly independent subset of $\mathcal{W}$ such that $|T| = n$. Prove T is a maximal linearly independent subset of $\mathcal{W}$.

    ▶(d) Use part (c) and Theorem 4.15 to prove that T is a basis for $\mathcal{W}$.

    (e) Conclude that $\mathcal{W}$ is finite dimensional and use part (b) to show that $\dim(\mathcal{W}) \leq \dim(\mathcal{V})$.

    (f) Prove that if $\dim(\mathcal{W}) = \dim(\mathcal{V})$, then $\mathcal{W} = \mathcal{V}$. (Hint: Let T be a basis for $\mathcal{W}$ and use part (2) of Corollary 4.14 to show that T is also a basis for $\mathcal{V}$.)

    (g) Prove the converse of part (f).

23. Let $\mathcal{V}$ be a subspace of $\mathbb{R}^n$ with $\dim(\mathcal{V}) = n - 1$. (Such a subspace is called a **hyperplane** in $\mathbb{R}^n$.) Prove that there is a nonzero $\mathbf{x} \in \mathbb{R}^n$ such that $\mathcal{V} = \{\mathbf{v} \in \mathbb{R}^n \mid \mathbf{x} \cdot \mathbf{v} = 0\}$. (Hint: Set up a homogeneous system of equations whose coefficient matrix has a basis for $\mathcal{V}$ as its rows. Then notice that this $(n-1) \times n$ system has at least one nontrivial solution, say **x**.)

24. Let $\mathcal{V}$ be a vector space and let S be a finite spanning set for $\mathcal{V}$. Prove that $\mathcal{V}$ is finite dimensional.

$\star$**25.** True or False:

    (a) A set B of vectors in a vector space $\mathcal{V}$ is a basis for $\mathcal{V}$ if B spans $\mathcal{V}$ and B is linearly independent.

    (b) All bases for $\mathcal{P}_4$ have 4 elements.

    (c) $\dim(\mathcal{M}_{43}) = 7$.

    (d) If S is a spanning set for $\mathcal{W}$ and $\dim(\mathcal{W}) = n$, then $|S| \leq n$.

    (e) If T is a linearly independent set in $\mathcal{W}$ and $\dim(\mathcal{W}) = n$, then $|T| = n$.

    (f) If T is a linearly independent set in a finite dimensional vector space $\mathcal{W}$ and S is a finite spanning set for $\mathcal{W}$, then $|T| \leq |S|$.

    (g) If $\mathcal{W}$ is a subspace of a finite dimensional vector space $\mathcal{V}$, then $\dim(\mathcal{W}) < \dim(\mathcal{V})$.

    (h) Every subspace of an infinite dimensional vector space is infinite dimensional.

    (i) If T is a maximal linearly independent set for a vector space $\mathcal{V}$ and S is a minimal spanning set for $\mathcal{V}$, then $S = T$.

    (j) If **A** is a nonsingular 4×4 matrix, then the rows of **A** are a basis for $\mathbb{R}^4$.

4.6 Constructing Special Bases

In this section, we present additional methods for finding a basis for a given finite dimensional vector space, starting with either a spanning set or a linearly independent subset.

Using Row Reduction to Construct a Basis

Recall the Simplified Span Method from Section 4.3. Using that method, we were able to simplify the form of span(S) for a subset S of $\mathbb{R}^n$. This was done by creating a matrix **A** whose rows are the vectors in S, and then row reducing **A** to obtain a reduced row echelon form matrix **C**. We discovered that a simplified form of span(S) is given by the set of all linear combinations of the nonzero rows of **C**. Now, each nonzero row of the matrix **C** has a (pivot) 1 in a column in which all other rows have zeroes, so the nonzero rows of **C** must be linearly independent. Thus, the nonzero rows of **C** not only span S but are linearly independent as well, and so they form a basis for span(S). Therefore, whenever we use the Simplified Span Method on a subset S of $\mathbb{R}^n$, we are actually creating a basis for span(S).

EXAMPLE 1

Let $S = \{[2, -2, 3, 5, 5], [-1, 1, 4, 14, -8], [4, -4, -2, -14, 18], [3, -3, -1, -9, 13]\}$, a subset of $\mathbb{R}^5$. We can use the Simplified Span Method to find a basis B for $V = \text{span}(S)$. We construct the matrix

$$\mathbf{A} = \begin{bmatrix} 2 & -2 & 3 & 5 & 5 \\ -1 & 1 & 4 & 14 & -8 \\ 4 & -4 & -2 & -14 & 18 \\ 3 & -3 & -1 & -9 & 13 \end{bmatrix},$$

whose rows are the vectors in S. The reduced row echelon form matrix for $\mathbf{A}$ is

$$\mathbf{C} = \begin{bmatrix} 1 & -1 & 0 & -2 & 4 \\ 0 & 0 & 1 & 3 & -1 \\ 0 & 0 & 0 & 0 & 0 \\ 0 & 0 & 0 & 0 & 0 \end{bmatrix}.$$

Therefore, the desired basis for V is the set $B = \{[1, -1, 0, -2, 4], [0, 0, 1, 3, -1]\}$ of nonzero rows of $\mathbf{C}$, and $\dim(V) = 2$. ∎

In general, the Simplified Span Method creates a basis of vectors with a simpler form than the original vectors. This is because a reduced row echelon form matrix has the simplest form of all matrices that are row equivalent to it.

This method can also be adapted to vector spaces other than $\mathbb{R}^n$, as in the next example.

EXAMPLE 2 Consider the subset $S = \{x^2 - 3x + 5, \ 3x^3 + 4x - 8, \ 6x^3 - x^2 + 11x - 21, \ 2x^5 - 7x^3 + 5x\}$ of $\mathcal{P}_5$. We use the Simplified Span Method to find a basis for $\mathcal{W} = \text{span}(S)$.

Since S is a subset of $\mathcal{P}_5$ instead of $\mathbb{R}^n$, we must alter our method slightly. We cannot use the polynomials in S themselves as rows of a matrix, so we "peel off" their coefficients to create four 6-vectors, which we use as the rows of the following matrix:

$$\mathbf{A} = \begin{array}{c} \\ \\ \\ \\ \end{array} \begin{matrix} x^5 & x^4 & x^3 & x^2 & x & 1 \\ \begin{bmatrix} 0 & 0 & 0 & 1 & -3 & 5 \\ 0 & 0 & 3 & 0 & 4 & -8 \\ 0 & 0 & 6 & -1 & 11 & -21 \\ 2 & 0 & -7 & 0 & 5 & 0 \end{bmatrix} \end{matrix}.$$

Row reducing this matrix produces

$$\mathbf{C} = \begin{matrix} x^5 & x^4 & x^3 & x^2 & x & 1 \\ \begin{bmatrix} 1 & 0 & 0 & 0 & \frac{43}{6} & -\frac{28}{3} \\ 0 & 0 & 1 & 0 & \frac{4}{3} & -\frac{8}{3} \\ 0 & 0 & 0 & 1 & -3 & 5 \\ 0 & 0 & 0 & 0 & 0 & 0 \end{bmatrix} \end{matrix}.$$

The nonzero rows of $\mathbf{C}$ yield the following three-element basis for $\mathcal{W}$:

$$D = \left\{ x^5 + \tfrac{43}{6}x - \tfrac{28}{3}, \ x^3 + \tfrac{4}{3}x - \tfrac{8}{3}, \ x^2 - 3x + 5 \right\}.$$

Hence, $\dim(\mathcal{W}) = 3$. ∎

Every Spanning Set for a Finite Dimensional Vector Space Contains a Basis

Sometimes, we are interested in reducing a spanning set to a basis by eliminating redundant vectors without changing the form of the original vectors. The next theorem asserts that this is possible; that is, if V is a finite dimensional vector space, then any spanning set of V, finite or infinite, must contain a basis for V.

THEOREM 4.18

If S is a spanning set for a finite dimensional vector space V, then there is a set $B \subseteq S$ that is a basis for V.

The proof of this theorem is very similar to the first part of the proof of Theorem 4.17[5] and is left as Exercise 14.

EXAMPLE 3   Let $S = \{[1, 3, -2], [2, 1, 4], [0, 5, -8], [1, -7, 14]\}$, and let $V = \text{span}(S)$. Theorem 4.18 indicates that some subset of S is a basis for V. Now, the equations

$$[0, 5, -8] = \quad 2[1, 3, -2] - \quad [2, 1, 4] \quad \text{and}$$
$$[1, -7, 14] = -3[1, 3, -2] + 2[2, 1, 4]$$

show that the subset $B = \{[1, 3, -2], [2, 1, 4]\}$ is a maximal linearly independent subset of S (why?). Hence, by Theorem 4.15, B is a basis for V contained in S. ∎

Shrinking a Spanning Set to a Basis Using Row Reduction

As Example 3 illustrates, to find a subset B of a spanning set S that is a basis for span(S), it is necessary to remove enough redundant vectors from S until we are left with a (maximal) linearly independent subset of S. This can be done using the Independence Test Method from Section 4.4. Suppose we row reduce the matrix whose columns are all the vectors in S. Then those vectors of S corresponding to the pivot columns form a linearly independent subset B. This is because if we had row reduced the matrix having just these columns, every column would have had a pivot. Also, no larger subset of S containing B can be linearly independent because reinserting a column corresponding to any of the remaining vectors would result in a nonpivot column after row reduction. Therefore, B is a maximal linearly independent subset of S, and hence is a basis for span(S). This procedure is illustrated in the next two examples.

---

[5] Theorem 4.18 is also true for infinite dimensional vector spaces, but the proof requires advanced topics in set theory beyond the scope of this book.

EXAMPLE 4

Consider the subset $S = \{[1, 2, -1], [3, 6, -3], [4, 1, 2], [0, 0, 0], [-1, 5, -5]\}$ of $\mathbb{R}^3$. We use the Independence Test Method to find a subset B of S that is a basis for $\mathcal{V} = \text{span}(S)$. We form the matrix $\mathbf{A}$ whose columns are the vectors in S, and then row reduce

$$\mathbf{A} = \begin{bmatrix} 1 & 3 & 4 & 0 & -1 \\ 2 & 6 & 1 & 0 & 5 \\ -1 & -3 & 2 & 0 & -5 \end{bmatrix} \quad \text{to obtain} \quad \mathbf{C} = \begin{bmatrix} 1 & 3 & 0 & 0 & 3 \\ 0 & 0 & 1 & 0 & -1 \\ 0 & 0 & 0 & 0 & 0 \end{bmatrix}.$$

Since there are nonzero pivots in the first and third columns of $\mathbf{C}$, we choose $B = \{[1, 2, -1], [4, 1, 2]\}$, the first and third vectors in S. Since $|B| = 2$, $\dim(\mathcal{V}) = 2$. (Hence, S does not span all of $\mathbb{R}^3$.)   ∎

This method can also be adapted to vector spaces other than $\mathbb{R}^n$.

EXAMPLE 5

Let $S = \{x^3 - 3x^2 + 1, 2x^2 + x, 2x^3 + 3x + 2, 4x - 5\} \subseteq \mathcal{P}_3$. We use the Independence Test Method to find a subset B of S that is a basis for $\mathcal{V} = \text{span}(S)$. Let $\mathbf{A}$ be the matrix whose columns are the analogous vectors in $\mathbb{R}^4$ for the given vectors in S. Then

$$\mathbf{A} = \begin{bmatrix} 1 & 0 & 2 & 0 \\ -3 & 2 & 0 & 0 \\ 0 & 1 & 3 & 4 \\ 1 & 0 & 2 & -5 \end{bmatrix}, \quad \text{which reduces to} \quad \mathbf{C} = \begin{bmatrix} 1 & 0 & 2 & 0 \\ 0 & 1 & 3 & 0 \\ 0 & 0 & 0 & 1 \\ 0 & 0 & 0 & 0 \end{bmatrix}.$$

Because we have nonzero pivots in the first, second, and fourth columns of $\mathbf{C}$, we choose $B = \{x^3 - 3x^2 + 1, 2x^2 + x, 4x - 5\}$. These are the first, second, and fourth vectors in S. Then B is the desired basis for $\mathcal{V}$.

Note that $2x^3 + 3x + 2 = 2(x^3 - 3x^2 + 1) + 3(2x^2 + x)$; that is, the third vector in $S = 2$(first vector in S) $+3$(second vector in S). This fact is gleaned from the third column of $\mathbf{C}$.   ∎

The Simplified Span Method and the Independence Test Method for finding a basis are similar enough to cause confusion, so we contrast their various features in the following table:

Table 4.2

Contrasting the Simplified Span Method and Independence Test Method for finding a basis from a given spanning set S

Simplified Span Method	Independence Test Method
The vectors in S become the *rows* of a matrix.	The vectors in S become the *columns* of a matrix.
The basis created is *not* a subset of the spanning set S but contains vectors with a simpler form.	The basis created *is* a subset of the spanning set S.
The nonzero rows of the reduced row echelon form matrix are used as the basis vectors.	The pivot columns of the reduced row echelon form matrix are used to determine which vectors to select from S.

Shrinking an Infinite Spanning Set to a Basis

The Independence Test Method can sometimes be used successfully when the spanning set S is infinite.

EXAMPLE 6

Let $\mathcal{V}$ be the subspace of $\mathcal{M}_{22}$ consisting of all 2×2 symmetric matrices. Let S be the set of nonsingular matrices in $\mathcal{V}$, and let $\mathcal{W} = \text{span}(S) = \text{span}$ ({nonsingular, symmetric 2×2 matrices}). We reduce S to a basis for $\mathcal{W}$ using the Independence Test Method, even though S is infinite. (We prove later that $\mathcal{W} = \mathcal{V}$, and so the basis we construct is actually a basis for $\mathcal{V}$.)

The strategy is to guess a *finite* subset Y of S that spans $\mathcal{W}$. We then use the Independence Test Method on Y to find the desired basis. We try to pick vectors for Y whose forms are as simple as possible to make computation easier. In this case, we choose the set of all nonsingular symmetric 2×2 matrices having only zeroes and ones as entries. That is,

$$Y = \left\{ \begin{bmatrix} 1 & 0 \\ 0 & 1 \end{bmatrix}, \begin{bmatrix} 1 & 1 \\ 1 & 0 \end{bmatrix}, \begin{bmatrix} 0 & 1 \\ 1 & 0 \end{bmatrix}, \begin{bmatrix} 0 & 1 \\ 1 & 1 \end{bmatrix} \right\}.$$

Now, before continuing, we must ensure that $\text{span}(Y) = \mathcal{W}$. That is, we must show every nonsingular symmetric 2×2 matrix is in $\text{span}(Y)$. In fact, we will show every symmetric 2×2 matrix is in $\text{span}(Y)$ by finding real numbers w, x, y, and z so that

$$\begin{bmatrix} a & b \\ b & c \end{bmatrix} = w \begin{bmatrix} 1 & 0 \\ 0 & 1 \end{bmatrix} + x \begin{bmatrix} 1 & 1 \\ 1 & 0 \end{bmatrix} + y \begin{bmatrix} 0 & 1 \\ 1 & 0 \end{bmatrix} + z \begin{bmatrix} 0 & 1 \\ 1 & 1 \end{bmatrix}.$$

Thus, we must prove that the system

$$\begin{cases} w + x & = a \\ x + y + z = b \\ x + y + z = b \\ w \quad\quad + z = c \end{cases}$$

has solutions for w, x, y, and z in terms of a, b, and c. But $w = 0$, $x = a$, $y = b - a - c$, $z = c$ certainly satisfies the system. Hence, $\mathcal{V} \subseteq \text{span}(Y)$. Since $\text{span}(Y) \subseteq \mathcal{V}$, we have $\text{span}(Y) = \mathcal{V} = \mathcal{W}$.

We can now use the Independence Test Method on Y. We express the matrices in Y as corresponding vectors in $\mathbb{R}^4$ and create the matrix with these vectors as columns, as follows:

$$\mathbf{A} = \begin{bmatrix} 1 & 1 & 0 & 0 \\ 0 & 1 & 1 & 1 \\ 0 & 1 & 1 & 1 \\ 1 & 0 & 0 & 1 \end{bmatrix}, \quad \text{which reduces to} \quad \mathbf{C} = \begin{bmatrix} 1 & 0 & 0 & 1 \\ 0 & 1 & 0 & -1 \\ 0 & 0 & 1 & 2 \\ 0 & 0 & 0 & 0 \end{bmatrix}.$$

Then, the desired basis is

$$B = \left\{ \begin{bmatrix} 1 & 0 \\ 0 & 1 \end{bmatrix}, \begin{bmatrix} 1 & 1 \\ 1 & 0 \end{bmatrix}, \begin{bmatrix} 0 & 1 \\ 1 & 0 \end{bmatrix} \right\},$$

the elements of Y corresponding to the pivot columns of C. ∎

The method used in Example 6 is not guaranteed to work when the spanning set S has infinitely many elements because our choice for the finite set Y might not have the same span as S. When this happens, the choice of a larger set Y may lead to success.

Finding a Basis from a Spanning Set by Inspection

Sometimes, a spanning set S for a vector space $\mathcal{V}$ contains vectors whose form is simple enough that it is easier to choose a maximal linearly independent subset of S by inspection than it is to set up a matrix and row reduce. The formal technique presented in the next method resembles a proof by induction in that there is a "base" step followed by an "inductive" step that is repeated until the desired basis is found.[6]

METHOD FOR FINDING A BASIS FROM A SPANNING SET BY INSPECTION (INSPECTION METHOD)

Let S be a finite set of vectors spanning a vector space $\mathcal{V}$.

(1) Base Step: Choose $\mathbf{v}_1 \neq \mathbf{0}$ in S.
   Repeat the following step as many times as possible:

(2) Inductive Step: Assuming $\mathbf{v}_1, \ldots, \mathbf{v}_{k-1}$ have already been chosen from S, choose $\mathbf{v}_k \in S$ such that $\mathbf{v}_k \notin \text{span}(\{\mathbf{v}_1, \ldots, \mathbf{v}_{k-1}\})$.

The final set constructed is a basis for $\mathcal{V}$.

The idea behind this technique is to choose only vectors that are not redundant. The method stops when we run out of vectors to choose in the Inductive Step that are linearly independent of those previously chosen.

The Inspection Method is computationally effective only when you can determine easily which vectors to choose next in the Inductive Step without tedious computations. Otherwise, you should apply the Independence Test Method.

EXAMPLE 7   Let $S = \{[0, 0, 0], [2, -8, 12], [-1, 4, -6], [7, 2, 2]\}$, a subset of $\mathbb{R}^3$. Let $\mathcal{V} = \text{span}(S)$, a subspace of $\mathbb{R}^3$. We use the Inspection Method to find a subset B of S that is a basis for $\mathcal{V}$.

The Base Step is to choose $\mathbf{v}_1$, a nonzero vector in S. So, we skip over the first vector listed in S, $[0, 0, 0]$ and let $\mathbf{v}_1 = [2, -8, 12]$.

Moving on to the Inductive Step, we look for $\mathbf{v}_2$ in S so that $\mathbf{v}_2 \notin \text{span}(\{\mathbf{v}_1\})$. Hence, $\mathbf{v}_2$ may not be a scalar multiple of $\mathbf{v}_1$. Therefore, we may not choose $[-1, 4, -6]$ because $[-1, 4, -6] = -\frac{1}{2}[2, -8, 12]$. Instead, we choose $\mathbf{v}_2 = [7, 2, 2]$.

---

[6] We assume that S has at least one nonzero vector, Otherwise, $\mathcal{V}$ would be the trivial vector space. In this case, the desired basis for $\mathcal{V}$ is the empty set, { }.

At this point, there are no more vectors in S for us to try, so the induction process must terminate here. Therefore, $B = \{\mathbf{v}_1, \mathbf{v}_2\} = \{[2, -8, 12], [7, 2, 2]\}$ is the desired basis for $\mathcal{V}$. Notice that $\mathcal{V} = \text{span}(B)$ is not all of $\mathbb{R}^3$ because $\dim(\mathcal{V}) = 2 \neq \dim\left(\mathbb{R}^3\right)$. (You can verify, for example, that the vector $[1, 0, 0] \in \mathbb{R}^3$ cannot be expressed as a linear combination of the vectors in B and hence is not in $\mathcal{V} = \text{span}(B)$.) ∎

Every Linearly Independent Set in a Finite Dimensional Vector Space Is Contained in Some Basis

Suppose that $T = \{\mathbf{t}_1, \ldots, \mathbf{t}_k\}$ is a linearly independent set of vectors in a finite dimensional vector space $\mathcal{V}$. Because $\mathcal{V}$ is finite dimensional, it has a finite basis, say $A = \{\mathbf{a}_1, \ldots, \mathbf{a}_n\}$. Consider the set $T \cup A$. Now, $T \cup A$ certainly spans $\mathcal{V}$ (since A alone spans $\mathcal{V}$). We can therefore apply the Independence Test Method to $T \cup A$ to produce a basis B for $\mathcal{V}$. If we order the vectors in $T \cup A$ so that all the vectors in T are listed first, then none of these vectors will be eliminated, since no vector in T is a linear combination of vectors listed earlier in T. In this manner we construct a basis B for $\mathcal{V}$ that contains T. We have just proved the following:

THEOREM 4.19

Let T be a linearly independent subset of a finite dimensional vector space $\mathcal{V}$. Then $\mathcal{V}$ has a basis B with $T \subseteq B$.

Compare this result with Theorem 4.18.

We modify slightly the method outlined just before Theorem 4.19 to find a basis for a finite dimensional vector space containing a given linearly independent subset T.

METHOD FOR FINDING A BASIS BY ENLARGING A LINEARLY INDEPENDENT SUBSET (ENLARGING METHOD)

Suppose that $T = \{\mathbf{t}_1, \ldots, \mathbf{t}_k\}$ is a linearly independent subset of a finite dimensional vector space $\mathcal{V}$.

Step 1: Find a finite spanning set $A = \{\mathbf{a}_1, \ldots, \mathbf{a}_n\}$ for $\mathcal{V}$.

Step 2: Form the ordered spanning set $S = \{\mathbf{t}_1, \ldots, \mathbf{t}_k, \mathbf{a}_1, \ldots, \mathbf{a}_n\}$ for $\mathcal{V}$.

Step 3: Use either the Independence Test Method or the Inspection Method on S to produce a subset B of S.

Then B is a basis for $\mathcal{V}$ containing T.

The basis produced by this method is easier to use if the additional vectors in the set A have a simple form. Ideally, we choose A to be the standard basis for $\mathcal{V}$.

EXAMPLE 8   Consider the linearly independent subset $T = \{[2, 0, 4, -12], [0, -1, -3, 9]\}$ of $V = \mathbb{R}^4$. We use the Enlarging Method to find a basis for $\mathbb{R}^4$ that contains T.

Step 1: We choose A to be the standard basis $\{e_1, e_2, e_3, e_4\}$ for $\mathbb{R}^4$.

Step 2: We create

$$S = \{[2, 0, 4, -12], [0, -1, -3, 9], [1, 0, 0, 0], [0, 1, 0, 0], [0, 0, 1, 0], [0, 0, 0, 1]\}.$$

Step 3: The matrix

$$\begin{bmatrix} 2 & 0 & 1 & 0 & 0 & 0 \\ 0 & -1 & 0 & 1 & 0 & 0 \\ 4 & -3 & 0 & 0 & 1 & 0 \\ -12 & 9 & 0 & 0 & 0 & 1 \end{bmatrix} \quad \text{reduces to} \quad \begin{bmatrix} 1 & 0 & 0 & -\frac{3}{4} & 0 & -\frac{1}{12} \\ 0 & 1 & 0 & -1 & 0 & 0 \\ 0 & 0 & 1 & \frac{3}{2} & 0 & \frac{1}{6} \\ 0 & 0 & 0 & 0 & 1 & \frac{1}{3} \end{bmatrix}.$$

Since columns 1, 2, 3, and 5 have nonzero pivots, the Independence Test Method indicates that the set $B = \{[2, 0, 4, -12], [0, -1, -3, 9], [1, 0, 0, 0], [0, 0, 1, 0]\}$ is a basis for $\mathbb{R}^4$ containing T.   ■

In general, we can use the Enlarging Method only when we already know a finite spanning set to use for A. Otherwise, we can make an intelligent guess, just as we did when using the Independence Test Method on an infinite spanning set. However, we must then take care to verify that the resulting set actually spans the vector space.

Exercises for Section 4.6

1. For each of the given subsets S of $\mathbb{R}^5$, find a basis for $V = \text{span}(S)$ using the Simplified Span Method:

   ⋆(a) $S = \{[1, 2, 3, -1, 0], [3, 6, 8, -2, 0], [-1, -1, -3, 1, 1], [-2, -3, -5, 1, 1]\}$

   (b) $S = \{[3, 2, -1, 0, 1], [1, -1, 0, 3, 1], [4, 1, -1, 3, 2], [3, 7, -2, -9, -1], [-1, -4, 1, 6, 1]\}$

   (c) $S = \{[0, 1, 1, 0, 6], [2, -1, 0, -2, 1], [-1, 2, 1, 1, 2], [3, -2, 0, -2, -3], [1, 1, 1, -1, 4], [2, -1, -1, 1, 3]\}$

   ⋆(d) $S = \{[1, 1, 1, 1, 1], [1, 2, 3, 4, 5], [0, 1, 2, 3, 4], [0, 0, 4, 0, -1]\}$

⋆2. Adapt the Simplified Span Method to find a basis for the subspace of $\mathcal{P}_3$ spanned by $S = \{x^3 - 3x^2 + 2, 2x^3 - 7x^2 + x - 3, 4x^3 - 13x^2 + x + 5\}$.

⋆3. Adapt the Simplified Span Method to find a basis for the subspace of $\mathcal{M}_{32}$ spanned by

$$S = \left\{ \begin{bmatrix} 1 & 4 \\ 0 & -1 \\ 2 & 2 \end{bmatrix}, \begin{bmatrix} 2 & 5 \\ 1 & -1 \\ 4 & 9 \end{bmatrix}, \begin{bmatrix} 1 & 7 \\ -1 & -2 \\ 2 & -3 \end{bmatrix}, \begin{bmatrix} 3 & 6 \\ 2 & -1 \\ 6 & 12 \end{bmatrix} \right\}.$$

4. For each given subset S of $\mathbb{R}^3$, find a subset B of S that is a basis for $V = \text{span}(S)$.

$\star$(a) $S = \{[3, 1, -2], [0, 0, 0], [6, 2, -3]\}$

(b) $S = \{[4, 7, 1], [1, 0, 0], [6, 7, 1], [-4, 0, 0]\}$

$\star$(c) $S = \{[1, 3, -2], [2, 1, 4], [3, -6, 18], [0, 1, -1], [-2, 1, -6]\}$

(d) $S = \{[1, 4, -2], [-2, -8, 4], [2, -8, 5], [0, -7, 2]\}$

$\star$(e) $S = \{[3, -2, 2], [1, 2, -1], [3, -2, 7], [-1, -10, 6]\}$

(f) $S = \{[3, 1, 0], [2, -1, 7], [0, 0, 0], [0, 5, -21], [6, 2, 0], [1, 5, 7]\}$

(g) $S = $ the set of all 3-vectors whose second coordinate is zero

$\star$(h) $S = $ the set of all 3-vectors whose second coordinate is -3 times its first coordinate plus its third coordinate

5. For each given subset S of $\mathcal{P}_3$, find a subset B of S that is a basis for $V = \text{span}(S)$.

$\star$(a) $S = \{x^3 - 8x^2 + 1, 3x^3 - 2x^2 + x, 4x^3 + 2x - 10, x^3 - 20x^2 - x + 12, x^3 + 24x^2 + 2x - 13, x^3 + 14x^2 - 7x + 18\}$

(b) $S = \{-2x^3 + x + 2, 3x^3 - x^2 + 4x + 6, 8x^3 + x^2 + 6x + 10, -4x^3 - 3x^2 + 3x + 4, -3x^3 - 4x^2 + 8x + 12\}$

$\star$(c) $S = $ the set of all polynomials in $\mathcal{P}_3$ with a zero constant term

(d) $S = \mathcal{P}_2$

$\star$(e) $S = $ the set of all polynomials in $\mathcal{P}_3$ with the coefficient of the x^2 term equal to the coefficient of the x^3 term

(f) $S = $ the set of all polynomials in $\mathcal{P}_3$ with the coefficient of the x^3 term equal to 8

6. For each given subset S of $\mathcal{M}_{33}$, find a subset B of S that is a basis for $V = \text{span}(S)$.

$\star$(a) $S = \{\mathbf{A} \in \mathcal{M}_{33} | \text{ each } a_{ij} \text{ is either 0 or 1}\}$

(b) $S = \{\mathbf{A} \in \mathcal{M}_{33} | \text{ each } a_{ij} \text{ is either 1 or } -1\}$

$\star$(c) $S = $ the set of all symmetric 3×3 matrices

(d) $S = $ the set of all nonsingular 3×3 matrices

7. Enlarge each of the following linearly independent subsets T of $\mathbb{R}^5$ to a basis B for $\mathbb{R}^5$ containing T:

$\star$(a) $T = \{[1, -3, 0, 1, 4], [2, 2, 1, -3, 1]\}$

(b) $T = \{[1, 1, 1, 1, 1], [0, 1, 1, 1, 1], [0, 0, 1, 1, 1]\}$

$\star$(c) $T = \{[1, 0, -1, 0, 0], [0, 1, -1, 1, 0], [2, 3, -8, -1, 0]\}$

8. Enlarge each of the following linearly independent subsets T of $\mathcal{P}_4$ to a basis B for $\mathcal{P}_4$ that contains T:

$\star$(a) $T = \{x^3 - x^2, x^4 - 3x^3 + 5x^2 - x\}$

(b) $T = \{3x - 2, x^3 - 6x + 4\}$

$\star$(c) $T = \{x^4 - x^3 + x^2 - x + 1, x^3 - x^2 + x - 1, x^2 - x + 1\}$

9. Enlarge each of the following linearly independent subsets T of $\mathcal{M}_{32}$ to a basis B for $\mathcal{M}_{32}$ that contains T:

$\star$(a) $T = \left\{ \begin{bmatrix} 1 & -1 \\ -1 & 1 \\ 0 & 0 \end{bmatrix}, \begin{bmatrix} 0 & 0 \\ 1 & -1 \\ -1 & 1 \end{bmatrix} \right\}$

$$\text{(b) } T = \left\{ \begin{bmatrix} 0 & -2 \\ 1 & 0 \\ -1 & 2 \end{bmatrix}, \begin{bmatrix} 0 & -3 \\ 0 & 1 \\ 3 & -6 \end{bmatrix}, \begin{bmatrix} 0 & 1 \\ 1 & 1 \\ -4 & 8 \end{bmatrix} \right\}$$

$$\star\text{(c) } T = \left\{ \begin{bmatrix} 3 & 0 \\ -1 & 7 \\ 0 & 1 \end{bmatrix}, \begin{bmatrix} -1 & 0 \\ 1 & 3 \\ 0 & -2 \end{bmatrix}, \begin{bmatrix} 2 & 0 \\ 3 & 1 \\ 0 & -1 \end{bmatrix}, \begin{bmatrix} 6 & 0 \\ 0 & 1 \\ 0 & -1 \end{bmatrix} \right\}$$

$\star$**10.** Find a basis for the vector space $\mathcal{U}_4$ consisting of all 4×4 upper triangular matrices.

11. In each case, find the dimension of $\mathcal{V}$ by using an appropriate method to create a basis.

(a) $\mathcal{V} = \text{span}(\{[5, 2, 1, 0, -1], [3, 0, 1, 1, 0], [0, 0, 0, 0, 0],$ $[-2, 4, -2, -4, -2], [0, 12, -4, -10, -6], [-6, 0, -2, -2, 0]\})$, a subspace of $\mathbb{R}^5$

$\star$(b) $\mathcal{V} = \{\mathbf{A} \in \mathcal{M}_{33} | \text{ trace}(\mathbf{A}) = 0\}$, a subspace of $\mathcal{M}_{33}$ (Recall that the trace of a matrix is the sum of the terms on the main diagonal.)

(c) $\mathcal{V} = \text{span}(\{x^4 - x^3 + 2x^2, 2x^4 + x - 5, 2x^3 - 4x^2 + x - 4, 6, x^2 - 1\})$

$\star$(d) $\mathcal{V} = \{\mathbf{p} \in \mathcal{P}_6 | \mathbf{p} = ax^6 - bx^5 + ax^4 - cx^3 + (a + b + c)x^2 - (a - c)x + (3a - 2b + 16c)$, for real numbers a, b, and $c\}$

12. (a) Show that each of these subspaces of $\mathcal{M}_{nn}$ has dimension $(n^2 + n)/2$.

(i) The set of upper triangular $n \times n$ matrices

(ii) The set of lower triangular $n \times n$ matrices

(iii) The set of symmetric $n \times n$ matrices

$\star$(b) What is the dimension of the set of skew-symmetric $n \times n$ matrices?

13. Let $\mathbf{A}$ be an $m \times n$ matrix.

(a) Prove that $S_{\mathbf{A}} = \{\mathbf{X} \in \mathbb{R}^n | \mathbf{A}\mathbf{X} = \mathbf{0}\}$, the solution set of the homogeneous system $\mathbf{A}\mathbf{X} = \mathbf{0}$, is a subspace of $\mathbb{R}^n$.

(b) Prove that $\dim(S_{\mathbf{A}}) + \text{rank}(\mathbf{A}) = n$. (Hint: First consider the case where $\mathbf{A}$ is in reduced row echelon form.)

$\blacktriangleright$**14.** Prove Theorem 4.18. This proof should be similar to the part of the proof for Theorem 4.17 outlined in parts (a), (b), and (c) of Exercise 22 in Section 4.5. However, change the definition of the set A in that exercise so that each set T is a subset of S rather than of $\mathcal{W}$.

15. Let $\mathcal{W}$ be a subspace of a finite dimensional vector space $\mathcal{V}$.

(a) Show that $\mathcal{V}$ has some basis B with a subset B' that is a basis for $\mathcal{W}$.

$\star$(b) If B is any given basis for $\mathcal{V}$, must some subset B' of B be a basis for $\mathcal{W}$? Prove that your answer is correct.

$\star$(c) If B is any given basis for $\mathcal{V}$ and $B' \subseteq B$, is there necessarily a subspace $\mathcal{Y}$ of $\mathcal{V}$ such that B' is a basis for $\mathcal{Y}$? Why or why not?

16. Let $\mathcal{V}$ be a finite dimensional vector space, and let $\mathcal{W}$ be a subspace of $\mathcal{V}$.

(a) Prove that $\mathcal{V}$ has a subspace $\mathcal{W}'$ such that every vector in $\mathcal{V}$ can be uniquely expressed as a sum of a vector in $\mathcal{W}$ and a vector in $\mathcal{W}'$. (In other words, show that there is a subspace $\mathcal{W}'$ so that, for every $\mathbf{v}$ in $\mathcal{V}$, there are unique vectors $\mathbf{w} \in \mathcal{W}$ and $\mathbf{w}' \in \mathcal{W}'$ such that $\mathbf{v} = \mathbf{w} + \mathbf{w}'$.)

★(b) Give an example of a subspace $\mathcal{W}$ of some finite dimensional vector space $\mathcal{V}$ for which the subspace $\mathcal{W}'$ from part (a) is not unique.

17. (a) Let S be a finite subset of $\mathbb{R}^n$. Prove that the Simplified Span Method applied to S produces the standard basis for $\mathbb{R}^n$ if and only if span$(S) = \mathbb{R}^n$.

(b) Let $B \subseteq \mathbb{R}^n$ with $|B| = n$, and let $\mathbf{A}$ be the $n \times n$ matrix whose rows are the vectors in B. Prove that B is a basis for $\mathbb{R}^n$ if and only if $|\mathbf{A}| \neq 0$.

18. Let $\mathbf{A}$ be an $m \times n$ matrix and let S be the set of vectors consisting of the rows of $\mathbf{A}$.

(a) Use the Simplified Span Method to show that dim(span(S)) = rank$(\mathbf{A})$.

(b) Use the Independence Test Method to prove that dim(span(S)) = rank$(\mathbf{A}^T)$.

(c) Use parts (a) and (b) to prove that rank$(\mathbf{A})$ = rank$(\mathbf{A}^T)$. (We will state this formally as Corollary 5.13 in Section 5.3.)

19. Let $\alpha_1, \ldots, \alpha_n$ and $\beta_1, \ldots, \beta_n$ be any real numbers, with $n > 2$. Consider the $n \times n$ matrix $\mathbf{A}$ whose (i, j) term is $a_{ij} = \sin(\alpha_i + \beta_j)$. Prove that $|\mathbf{A}| = 0$. (Hint: Consider $\mathbf{x}_1 = [\sin \beta_1, \sin \beta_2, \ldots, \sin \beta_n]$, $\mathbf{x}_2 = [\cos \beta_1, \cos \beta_2, \ldots, \cos \beta_n]$. Show that the row space of $\mathbf{A} \subseteq$ span$(\{\mathbf{x}_1, \mathbf{x}_2\})$, and hence, dim(row space of $\mathbf{A}$) $< n$.)

★20. True or False:

(a) Given any spanning set S for a finite dimensional vector space $\mathcal{V}$, there is some $B \subseteq S$ that is a basis for $\mathcal{V}$.

(b) Given any linearly independent set T in a finite dimensional vector space $\mathcal{V}$, there is a basis B for $\mathcal{V}$ containing T.

(c) If S is a finite spanning set for $\mathbb{R}^n$, then the Simplified Span Method must produce a subset of S that is a basis for $\mathbb{R}^n$.

(d) If S is a finite spanning set for $\mathbb{R}^n$, then the Independence Test Method produces a subset of S that is a basis for $\mathbb{R}^n$.

(e) If S is a finite spanning set for $\mathbb{R}^n$, then the Inspection Method produces a subset of S that is a basis for $\mathbb{R}^n$.

(f) If T is a linearly independent set in $\mathbb{R}^n$, then the Enlarging Method must produce a subset of T that is a basis for $\mathbb{R}^n$.

(g) Before row reduction, the Simplified Span Method places the vectors of a given spanning set S as columns in a matrix, while the Independence Test Method places the vectors of S as rows.

4.7 Coordinatization

If B is a basis for a vector space $\mathcal{V}$, then we know every vector in $\mathcal{V}$ has a unique expression as a linear combination of the vectors in B. For example, $\{\mathbf{e}_1, \ldots, \mathbf{e}_n\}$ is a basis (the standard one) for $\mathbb{R}^n$, and we are accustomed to writing the vector $[a_1, \ldots, a_n]$ in $\mathbb{R}^n$ as $a_1\mathbf{e}_1 + \cdots + a_n\mathbf{e}_n$. Dealing with the standard basis in $\mathbb{R}^n$ is easy because the coefficients in the linear combination are the same as the coordinates of the vector. However, this is not necessarily true for other bases.

In this section, we develop a process, called coordinatization, for representing any vector in a finite dimensional vector space in terms of its coefficients with respect to a given basis. We also determine how the coordinatization changes whenever we switch bases.

Coordinates with Respect to an Ordered Basis

DEFINITION

An **ordered basis** for a vector space $\mathcal{V}$ is an ordered n-tuple of vectors $(\mathbf{v}_1, \ldots, \mathbf{v}_n)$ such that the set $\{\mathbf{v}_1, \ldots, \mathbf{v}_n\}$ is a basis for $\mathcal{V}$.

In an ordered basis, the elements are written in a specific order. Thus, $(\mathbf{i}, \mathbf{j}, \mathbf{k})$ and $(\mathbf{j}, \mathbf{i}, \mathbf{k})$ are different ordered bases for $\mathbb{R}^3$.

By Theorem 4.10, if $B = (\mathbf{v}_1, \mathbf{v}_2, \ldots, \mathbf{v}_n)$ is an ordered basis for $\mathcal{V}$, then for every vector $\mathbf{w} \in \mathcal{V}$, there are unique scalars $a_1, a_2, \ldots, a_n$ such that $\mathbf{w} = a_1\mathbf{v}_1 + a_2\mathbf{v}_2 + \cdots + a_n\mathbf{v}_n$. We use these scalars $a_1, a_2, \ldots, a_n$ to **coordinatize** the vector $\mathbf{w}$ as follows:

DEFINITION

Let $B = (\mathbf{v}_1, \mathbf{v}_2, \ldots, \mathbf{v}_n)$ be an ordered basis for a vector space $\mathcal{V}$. Suppose that $\mathbf{w} = a_1\mathbf{v}_1 + a_2\mathbf{v}_2 + \cdots + a_n\mathbf{v}_n \in \mathcal{V}$. Then $[\mathbf{w}]_B$, the **coordinatization of w with respect to** B, is the n-vector $[a_1, a_2, \ldots, a_n]$.

The vector $[\mathbf{w}]_B = [a_1, a_2, \ldots, a_n]$ is frequently referred to as "**w expressed in B-coordinates.**" When useful, we will express $[\mathbf{w}]_B$ as a column vector.

EXAMPLE 1 Let $B = (x^3, x^2, x, 1)$, an ordered basis for $\mathcal{P}_3$. Then $[6x^3 - 2x + 18]_B = [6, 0, -2, 18]$, and $[4 - 3x + 9x^2 - 7x^3]_B = [-7, 9, -3, 4]$. ∎

EXAMPLE 2 A general 3×3 skew-symmetric matrix can be expressed as

$$\begin{bmatrix} 0 & a & b \\ -a & 0 & c \\ -b & -c & 0 \end{bmatrix} = a\begin{bmatrix} 0 & 1 & 0 \\ -1 & 0 & 0 \\ 0 & 0 & 0 \end{bmatrix} + b\begin{bmatrix} 0 & 0 & 1 \\ 0 & 0 & 0 \\ -1 & 0 & 0 \end{bmatrix} + c\begin{bmatrix} 0 & 0 & 0 \\ 0 & 0 & 1 \\ 0 & -1 & 0 \end{bmatrix}.$$

Since the last three matrices are linearly independent,

$$B = \left(\begin{bmatrix} 0 & 1 & 0 \\ -1 & 0 & 0 \\ 0 & 0 & 0 \end{bmatrix}, \begin{bmatrix} 0 & 0 & 1 \\ 0 & 0 & 0 \\ -1 & 0 & 0 \end{bmatrix}, \begin{bmatrix} 0 & 0 & 0 \\ 0 & 0 & 1 \\ 0 & -1 & 0 \end{bmatrix} \right)$$

is an ordered basis for the vector space of skew-symmetric 3×3 matrices. Then, for example,

$$\begin{bmatrix} 0 & 2 & -3 \\ -2 & 0 & 1 \\ 3 & -1 & 0 \end{bmatrix}_B = [2, -3, 1], \quad \text{and in general,} \quad \begin{bmatrix} 0 & a & b \\ -a & 0 & c \\ -b & -c & 0 \end{bmatrix}_B = [a, b, c].$$

∎

Notice in Example 2 that the coordinatized vector $[a, b, c]$ is more "compact" than the full matrix of nine entries but still contains the same essential information.

EXAMPLE 3 Let $B = ([4, 2], [1, 3])$ be an ordered basis for $\mathbb{R}^2$. Notice that $[4, 2] = 1[4, 2] + 0[1, 3]$, so $[4, 2]_B = [1, 0]$. Similarly, $[1, 3]_B = [0, 1]$. From a geometric viewpoint, converting to B-coordinates in $\mathbb{R}^2$ results in a new coordinate system in $\mathbb{R}^2$ with $[4, 2]$ and $[1, 3]$ as its "unit" vectors. The new coordinate grid consists of parallelograms whose sides are the vectors in B, as shown in Figure 4.6. For example, $[11, 13]$ equals $[2, 3]$ when expressed in B-coordinates because $[11, 13] = 2[4, 2] + 3[1, 3]$. ∎

Figure 4.6

A B-coordinate grid in $\mathbb{R}^2$: picturing $[11, 3]$ in B-coordinates

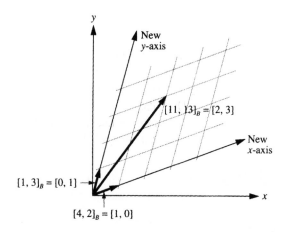

EXAMPLE 4 Consider the subspace $\mathcal{V}$ of $\mathbb{R}^5$ spanned by the ordered basis

$$C = ([-4, 5, -1, 0, -1], [1, -3, 2, 2, 5], [1, -2, 1, 1, 3]).$$

Notice that the vectors in $\mathcal{V}$ can be put into C-coordinates by solving an appropriate system. For example, to find $[-23, 30, -7, -1, -7]_C$, we solve the equation

$$[-23, 30, -7, -1, -7] = a[-4, 5, -1, 0, -1] + b[1, -3, 2, 2, 5] + c[1, -2, 1, 1, 3].$$

The equivalent system is

$$\begin{cases} -4a + b + c = -23 \\ 5a - 3b - 2c = 30 \\ -a + 2b + c = -7 \\ 2b + c = -1 \\ -a + 5b + 3c = -7 \end{cases}.$$

Since the (unique) solution for this system is $a = 6$, $b = -2$, $c = 3$, we see that $[-23, 30, -7, -1, -7]_C = [6, -2, 3]$. Similarly, you can see that

$$\begin{aligned} [-4, 5, -1, 0, -1]_C &= [1, 0, 0], \\ [1, -3, 2, 2, 5]_C &= [0, 1, 0], \quad \text{and} \\ [1, -2, 1, 1, 3]_C &= [0, 0, 1]. \end{aligned}$$

Now, vectors in $\mathbb{R}^5$ that are not in $\text{span}(C)$ cannot be expressed in C-coordinates. For example, the vector $[1, 2, 3, 4, 5]$ is not in $\mathcal{V} = \text{span}(C)$ because the system

$$\begin{cases} -4a + b + c = 1 \\ 5a - 3b - 2c = 2 \\ -a + 2b + c = 3 \\ 2b + c = 4 \\ -a + 5b + 3c = 5 \end{cases}$$

has no solutions. ∎

As part of Example 4, we saw an illustration of the general principle that if $B = (\mathbf{v}_1, \ldots, \mathbf{v}_n)$, then every vector in B has a simple coordinatization. In particular, $[\mathbf{v}_i]_B = \mathbf{e}_i$. You are asked to prove this in Exercise 6.

The following theorem shows that the coordinatization of a vector behaves in a manner similar to the original vector with respect to addition and scalar multiplication:

THEOREM 4.20

Let $B = (\mathbf{v}_1, \ldots, \mathbf{v}_n)$ be an ordered basis for a vector space $\mathcal{V}$. Suppose $\mathbf{w}_1, \ldots, \mathbf{w}_k \in \mathcal{V}$ and $a_1, \ldots, a_k$ are scalars. Then
(1) $[\mathbf{w}_1 + \mathbf{w}_2]_B = [\mathbf{w}_1]_B + [\mathbf{w}_2]_B$
(2) $[a_1\mathbf{w}_1]_B = a_1[\mathbf{w}_1]_B$
(3) $[a_1\mathbf{w}_1 + a_2\mathbf{w}_2 + \cdots + a_k\mathbf{w}_k]_B = a_1[\mathbf{w}_1]_B + a_2[\mathbf{w}_2]_B + \cdots + a_k[\mathbf{w}_k]_B$

Figure 4.7 illustrates part (1) of this theorem. Moving along either path from the upper left to the lower right in the diagram produces the same answer. (Such a picture is called a **commutative diagram**.)

Part (3) asserts that to put a linear combination of vectors in $\mathcal{V}$ into B-coordinates, we can first find the B-coordinates of each vector individually and then calculate the analogous linear combination in $\mathbb{R}^n$. The proof of Theorem 4.20 is left for you to do in Exercise 12.

Figure 4.7

Commutative diagram
involving addition and
coordinatization of
vectors

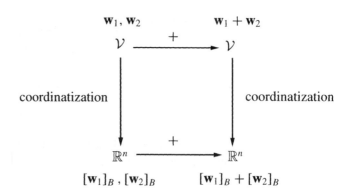

EXAMPLE 5   Recall the subspace $\mathcal{V}$ of $\mathbb{R}^5$ from Example 4 spanned by the ordered basis

$$C = ([-4, 5, -1, 0, -1], [1, -3, 2, 2, 5], [1, -2, 1, 1, 3]).$$

Notice that if we are given the C-coordinates of a 5-vector in $\mathcal{V}$, we can find that vector merely by calculating the appropriate linear combination. For example,

$$
\begin{aligned}
[3, 7, -1] &= 3[1, 0, 0] + 7[0, 1, 0] - 1[0, 0, 1] \\
&= 3[-4, 5, -1, 0, -1]_C + 7[1, -3, 2, 2, 5]_C - [1, -2, 1, 1, 3]_C \\
&= [-12, 15, -3, 0, -3]_C + [7, -21, 14, 14, 35]_C + [-1, 2, -1, -1, -3]_C \\
&\qquad\qquad\qquad\text{by part (2) of Theorem 4.20} \\
&= [-6, -4, 10, 13, 29]_C. \qquad\quad \text{by part (1) of Theorem 4.20}
\end{aligned}
$$

Thus, $[-6, -4, 10, 13, 29]$ is the vector in $\mathcal{V}$ having coordinatization $[3, 7, -1]$ with respect to C. ∎

Using Row Reduction to Coordinatize a Vector

As we saw in Example 4, finding the coordinates of a vector with respect to an ordered basis typically amounts to solving a system of linear equations. Since this is usually done using row reduction, we have the following general method. Although it applies to subspaces of $\mathbb{R}^n$, we can adapt it to other finite dimensional vector spaces, such as $\mathcal{P}_n$ and $\mathcal{M}_{mn}$, as with other techniques we have examined.

METHOD FOR COORDINATIZING A VECTOR WITH RESPECT TO A FINITE ORDERED BASIS (COORDINATIZATION METHOD)

Let V be a nontrivial subspace of $\mathbb{R}^n$, let $B = (\mathbf{v}_1, \ldots, \mathbf{v}_k)$ be an ordered basis for V, and let $\mathbf{v} \in \mathbb{R}^n$. To calculate $[\mathbf{v}]_B$, if possible, perform the following steps:

Step 1: Form an augmented matrix $[\mathbf{A} \mid \mathbf{v}]$ by using the vectors in B as the *columns* of $\mathbf{A}$, in *order*, and using $\mathbf{v}$ as a column on the right.

Step 2: Row reduce $[\mathbf{A} \mid \mathbf{v}]$ to obtain the reduced row echelon form $[\mathbf{C} \mid \mathbf{w}]$.

Step 3: If there is a row of $[\mathbf{C} \mid \mathbf{w}]$ that contains all zeroes on the left and has a nonzero entry on the right, then $\mathbf{v} \notin \text{span}(B) = V$, and coordinatization is not possible. Stop.

Step 4: Otherwise, $\mathbf{v} \in \text{span}(B) = V$. Eliminate all rows consisting entirely of zeroes in $[\mathbf{C} \mid \mathbf{w}]$ to obtain $[\mathbf{D} \mid \mathbf{y}]$. Then, $[\mathbf{v}]_B = \mathbf{y}$, the last column of $[\mathbf{D} \mid \mathbf{y}]$.

EXAMPLE 6

Consider the subspace V of $\mathbb{R}^3$ with ordered basis $B = ([1, 3, 3], [3, 10, 7])$, and let $\mathbf{v}_1 = [3, 11, 5]$. Now

$$\begin{bmatrix} 1 & 3 & 3 \\ 3 & 10 & 11 \\ 3 & 7 & 5 \end{bmatrix} \quad \text{row reduces to} \quad \begin{bmatrix} 1 & 0 & -3 \\ 0 & 1 & 2 \\ 0 & 0 & 0 \end{bmatrix}.$$

Ignoring the final row of zeroes, we obtain $[\mathbf{v}_1]_B = [-3, 2]$.

Similarly, with $\mathbf{v}_2 = [1, 1, -2]$, we find

$$\begin{bmatrix} 1 & 3 & 1 \\ 3 & 10 & 1 \\ 3 & 7 & -2 \end{bmatrix} \quad \text{row reduces to} \quad \begin{bmatrix} 1 & 0 & 7 \\ 0 & 1 & -2 \\ 0 & 0 & -9 \end{bmatrix}.$$

Step 3 applied to the final row shows that $\mathbf{v}_2 \notin V = \text{span}(B)$, and so $[\mathbf{v}_2]_B$ is not defined. ∎

Transition Matrix for Change of Coordinates

Our next goal is to determine how the coordinates of a vector change when we convert from one ordered basis to another.

DEFINITION

Suppose that V is a nontrivial n-dimensional vector space with ordered bases B and C. Let $\mathbf{P}$ be the $n \times n$ matrix whose ith column, for $1 \leq i \leq n$, equals $[\mathbf{b}_i]_C$, where $\mathbf{b}_i$ is the ith basis vector in B. Then $\mathbf{P}$ is called the **transition matrix from B-coordinates to C-coordinates**.

We often refer to the matrix $\mathbf{P}$ in this definition as the "**transition matrix from B to C.**"

EXAMPLE 7   Consider the following ordered bases for $\mathcal{U}_2$:

$$B = \left(\begin{bmatrix} 7 & 3 \\ 0 & 0 \end{bmatrix}, \begin{bmatrix} 1 & 2 \\ 0 & -1 \end{bmatrix}, \begin{bmatrix} 1 & -1 \\ 0 & 1 \end{bmatrix} \right) \quad \text{and} \quad C = \left(\begin{bmatrix} 1 & 0 \\ 0 & 0 \end{bmatrix}, \begin{bmatrix} 0 & 1 \\ 0 & 0 \end{bmatrix}, \begin{bmatrix} 0 & 0 \\ 0 & 1 \end{bmatrix} \right).$$

Calculating the C-coordinates for each vector in B, we obtain

$$\begin{bmatrix} 7 & 3 \\ 0 & 0 \end{bmatrix}_C = \begin{bmatrix} 7 \\ 3 \\ 0 \end{bmatrix}, \quad \begin{bmatrix} 1 & 2 \\ 0 & -1 \end{bmatrix}_C = \begin{bmatrix} 1 \\ 2 \\ -1 \end{bmatrix}, \quad \text{and} \quad \begin{bmatrix} 1 & -1 \\ 0 & 1 \end{bmatrix}_C = \begin{bmatrix} 1 \\ -1 \\ 1 \end{bmatrix}.$$

Thus, the transition matrix from B to C is given by

$$\mathbf{P} = \begin{bmatrix} 7 & 1 & 1 \\ 3 & 2 & -1 \\ 0 & -1 & 1 \end{bmatrix}. \qquad \blacksquare$$

The next theorem shows that if $[\mathbf{v}]_B$ is known, then $[\mathbf{v}]_C$ can be found by using the transition matrix from B to C.

THEOREM 4.21

Suppose that B and C are ordered bases for a nontrivial n-dimensional vector space $\mathcal{V}$, and let $\mathbf{P}$ be an $n \times n$ matrix. Then $\mathbf{P}$ is the transition matrix from B to C if and only if for every $\mathbf{v} \in \mathcal{V}$, $\mathbf{P}[\mathbf{v}]_B = [\mathbf{v}]_C$.

PROOF OF THEOREM 4.21

Let B and C be ordered bases for a vector space $\mathcal{V}$, with $B = (\mathbf{b}_1, \ldots, \mathbf{b}_n)$.

First, suppose $\mathbf{P}$ is the transition matrix from B to C. Let $\mathbf{v} \in \mathcal{V}$. We want to show $\mathbf{P}[\mathbf{v}]_B = [\mathbf{v}]_C$. Suppose $[\mathbf{v}]_B = [a_1, \ldots, a_n]$. Then $\mathbf{v} = a_1 \mathbf{b}_1 + \cdots + a_n \mathbf{b}_n$. Hence,

$$\mathbf{P}[\mathbf{v}]_B = \begin{bmatrix} p_{11} & \cdots & p_{1n} \\ \vdots & \ddots & \vdots \\ p_{n1} & \cdots & p_{nn} \end{bmatrix} \begin{bmatrix} a_1 \\ a_2 \\ \vdots \\ a_n \end{bmatrix}$$

$$= a_1 \begin{bmatrix} p_{11} \\ p_{21} \\ \vdots \\ p_{n1} \end{bmatrix} + a_2 \begin{bmatrix} p_{12} \\ p_{22} \\ \vdots \\ p_{n2} \end{bmatrix} + \cdots + a_n \begin{bmatrix} p_{1n} \\ p_{2n} \\ \vdots \\ p_{nn} \end{bmatrix}.$$

However, **P** is the transition matrix from B to C, so the ith column of **P** equals $[\mathbf{b}_i]_C$. Therefore,

$$
\begin{aligned}
\mathbf{P}[\mathbf{v}]_B &= a_1[\mathbf{b}_1]_C + a_2[\mathbf{b}_2]_C + \cdots + a_n[\mathbf{b}_n]_C \\
&= [a_1\mathbf{b}_1 + a_2\mathbf{b}_2 + \cdots + a_n\mathbf{b}_n]_C \qquad \text{by Theorem 4.20} \\
&= [\mathbf{v}]_C.
\end{aligned}
$$

Conversely, suppose that **P** is an $n \times n$ matrix and that $\mathbf{P}[\mathbf{v}]_B = [\mathbf{v}]_C$ for every $\mathbf{v} \in \mathcal{V}$. We show that **P** is the transition matrix from B to C. By definition, it is enough to show that the ith column of **P** is equal to $[\mathbf{b}_i]_C$. Since $\mathbf{P}[\mathbf{v}]_B = [\mathbf{v}]_C$, for all $\mathbf{v} \in \mathcal{V}$, let $\mathbf{v} = \mathbf{b}_i$. Then since $[\mathbf{v}]_B = \mathbf{e}_i$, we have $\mathbf{P}[\mathbf{v}]_B = \mathbf{P}\mathbf{e}_i = [\mathbf{b}_i]_C$. But $\mathbf{P}\mathbf{e}_i = i$th column of **P**, and we are done. ∎

EXAMPLE 8

Consider the ordered bases $B = ([12, 9], [15, -1])$ and $C = ([3, 4], [6, 1])$ for $\mathbb{R}^2$. Row reducing

$$
\begin{bmatrix} 3 & 6 & | & 12 \\ 4 & 1 & | & 9 \end{bmatrix} \quad \text{to obtain} \quad \begin{bmatrix} 1 & 0 & | & 2 \\ 0 & 1 & | & 1 \end{bmatrix}
$$

gives $[12, 9]_C = [2, 1]$. Similarly, row reducing

$$
\begin{bmatrix} 3 & 6 & | & 15 \\ 4 & 1 & | & -1 \end{bmatrix} \quad \text{to obtain} \quad \begin{bmatrix} 1 & 0 & | & -1 \\ 0 & 1 & | & 3 \end{bmatrix}
$$

yields $[15, -1]_C = [-1, 3]$. Hence, the transition matrix from B to C is

$$
\mathbf{P} = \begin{bmatrix} 2 & -1 \\ 1 & 3 \end{bmatrix}.
$$

Now, consider $\mathbf{v} = [3, 39]$. A quick computation verifies that $[\mathbf{v}]_B = [4, -3]$. Then, by Theorem 4.21,

$$
[\mathbf{v}]_C = \mathbf{P}[\mathbf{v}]_B = \begin{bmatrix} 2 & -1 \\ 1 & 3 \end{bmatrix} \begin{bmatrix} 4 \\ -3 \end{bmatrix} = \begin{bmatrix} 11 \\ -5 \end{bmatrix},
$$

which gives the correct answer for $[\mathbf{v}]_C$. ∎

Calculating the Transition Matrix by Row Reduction

Example 8 illustrates that solving for the columns of the transition matrix involves row reducing a collection of augmented matrices, each of which has the same matrix on the left side of the augmentation bar. Thus, we can accomplish this task more efficiently by performing a single row reduction using a matrix with several columns to the right of the augmentation bar. (We used this technique in Section 2.2 when we solved several similar systems of linear equations simultaneously.) Hence, we have the following:

> **METHOD FOR CALCULATING A TRANSITION MATRIX (TRANSITION MATRIX METHOD)**
>
> To find the transition matrix $\mathbf{P}$ from B to C where B and C are ordered bases for a nontrivial k-dimensional subspace of $\mathbb{R}^n$, use row reduction on
>
> $$\left[\begin{array}{cccc|cccc} \text{1st} & \text{2nd} & & k\text{th} & \text{1st} & \text{2nd} & & k\text{th} \\ \text{vector} & \text{vector} & \cdots & \text{vector} & \text{vector} & \text{vector} & \cdots & \text{vector} \\ \text{in} & \text{in} & & \text{in} & \text{in} & \text{in} & & \text{in} \\ C & C & & C & B & B & & B \end{array}\right]$$
>
> to produce $\left[\begin{array}{c|c} \mathbf{I}_k & \mathbf{P} \\ \hline \text{rows of} & \text{zeroes} \end{array}\right].$

In Exercise 8 you are asked to show that, in the special cases where either B or C is the standard basis in $\mathbb{R}^n$, there are simple expressions for the transition matrix from B to C.

EXAMPLE 9

Recall from Example 4 the subspace $\mathcal{V}$ of $\mathbb{R}^5$ that is spanned by the ordered basis $C = ([-4, 5, -1, 0, -1], [1, -3, 2, 2, 5], [1, -2, 1, 1, 3])$. Using the Simplified Span Method, we find that $B = ([1, 0, -1, 0, 4], [0, 1, -1, 0, 3], [0, 0, 0, 1, 5])$ is also an ordered basis for $\mathcal{V}$. To find the transition matrix from B to C we must solve for the C-coordinates of each vector in B. Using the Transition Matrix Method, we row reduce the augmented matrix

$$\left[\begin{array}{ccc|ccc} -4 & 1 & 1 & 1 & 0 & 0 \\ 5 & -3 & -2 & 0 & 1 & 0 \\ -1 & 2 & 1 & -1 & -1 & 0 \\ 0 & 2 & 1 & 0 & 0 & 1 \\ -1 & 5 & 3 & 4 & 3 & 5 \end{array}\right] \quad \text{to produce} \quad \left[\begin{array}{ccc|ccc} 1 & 0 & 0 & 1 & 1 & 1 \\ 0 & 1 & 0 & -5 & -4 & -3 \\ 0 & 0 & 1 & 10 & 8 & 7 \\ \cancel{0\ 0\ 0} & & & \cancel{0} & \cancel{0} & \cancel{0} \\ \cancel{0\ 0\ 0} & & & \cancel{0} & \cancel{0} & \cancel{0} \end{array}\right].$$

We ignore the last two rows of zeroes. The entries to the right of the bar give us the transition matrix from B to C, namely

$$\mathbf{P} = \left[\begin{array}{ccc} 1 & 1 & 1 \\ -5 & -4 & -3 \\ 10 & 8 & 7 \end{array}\right].$$

Notice that the columns of $\mathbf{P}$ show how to express each vector in B as a linear combination of the vectors in C. For example,

$$[1, 0, -1, 0, 4] = 1[-4, 5, -1, 0, -1] - 5[1, -3, 2, 2, 5] + 10[1, -2, 1, 1, 3].$$

That is, $[1, 0, -1, 0, 4]_C = [1, -5, 10]$. ∎

Composition of Transitions

The next theorem shows that the cumulative effect of two transitions between bases is represented by the product of the transition matrices in *reverse* order.

> **THEOREM 4.22**
> Suppose that B, C, and D are ordered bases for a nontrivial finite dimensional vector space $\mathcal{V}$. Let **P** be the transition matrix from B to C, and let **Q** be the transition matrix from C to D. Then **QP** is the transition matrix from B to D.

The proof of this theorem is left as Exercise 13.

EXAMPLE 10   Consider the ordered bases B and C for $\mathcal{P}_2$ given by

$$B = (-x^2 + 4x + 2,\ 2x^2 - x - 1,\ -x^2 + 2x + 1) \quad \text{and}$$
$$C = (x^2 - 2x - 3,\ 2x^2 - 1,\ x^2 + x + 1).$$

Also consider the standard basis $S = (x^2, x, 1)$ for $\mathcal{P}_2$.

Now, row reducing

$$\left[\begin{array}{rrr|rrr} 1 & 2 & 1 & -1 & 2 & -1 \\ -2 & 0 & 1 & 4 & -1 & 2 \\ -3 & -1 & 1 & 2 & -1 & 1 \end{array}\right] \quad \text{to obtain} \quad \left[\begin{array}{rrr|rrr} 1 & 0 & 0 & -9 & 3 & -5 \\ 0 & 1 & 0 & 11 & -3 & 6 \\ 0 & 0 & 1 & -14 & 5 & -8 \end{array}\right],$$

we see that the transition matrix from B to C is

$$\mathbf{P} = \left[\begin{array}{rrr} -9 & 3 & -5 \\ 11 & -3 & 6 \\ -14 & 5 & -8 \end{array}\right].$$

Because it is simple to express each vector in C in S-coordinates, we can quickly calculate that the transition matrix from C to S is

$$\mathbf{Q} = \left[\begin{array}{rrr} 1 & 2 & 1 \\ -2 & 0 & 1 \\ -3 & -1 & 1 \end{array}\right].$$

Then, by Theorem 4.22, the product

$$\mathbf{QP} = \left[\begin{array}{rrr} 1 & 2 & 1 \\ -2 & 0 & 1 \\ -3 & -1 & 1 \end{array}\right]\left[\begin{array}{rrr} -9 & 3 & -5 \\ 11 & -3 & 6 \\ -14 & 5 & -8 \end{array}\right] = \left[\begin{array}{rrr} -1 & 2 & -1 \\ 4 & -1 & 2 \\ 2 & -1 & 1 \end{array}\right]$$

is the transition matrix from B to S. This matrix is correct, since the columns of **QP** are, in fact, the vectors of B expressed in S-coordinates. ∎

Reversing the Order of Transition

The next theorem shows how to reverse a transition from one basis to another. The proof of this theorem is left as Exercise 14.

THEOREM 4.23

Let B and C be ordered bases for a nontrivial finite dimensional vector space $\mathcal{V}$, and let $\mathbf{P}$ be the transition matrix from B to C. Then $\mathbf{P}$ is nonsingular, and $\mathbf{P}^{-1}$ is the transition matrix from C to B.

EXAMPLE 11

Let us consider again the vector space $\mathcal{V}$ of Examples 4 and 9, with ordered bases

$$B = ([1, 0, -1, 0, 4], [0, 1, -1, 0, 3], [0, 0, 0, 1, 5]) \quad \text{and}$$
$$C = ([-4, 5, -1, 0, -1], [1, -3, 2, 2, 5], [1, -2, 1, 1, 3]).$$

In Example 9, we calculated the transition matrix $\mathbf{P}$ from B to C. Now, by Theorem 4.23, $\mathbf{P}^{-1}$ is the transition matrix from C to B. Using

$$\mathbf{P} = \begin{bmatrix} 1 & 1 & 1 \\ -5 & -4 & -3 \\ 10 & 8 & 7 \end{bmatrix}, \quad \text{we calculate} \quad \mathbf{P}^{-1} = \begin{bmatrix} -4 & 1 & 1 \\ 5 & -3 & -2 \\ 0 & 2 & 1 \end{bmatrix}.$$

A direct computation of the transition matrix from C to B using row reduction will produce the same result. (Try it! The simple form of the vectors in B make this a quick calculation.) ∎

Let us return to the situation in Example 10 and use the inverses of the transition matrices to find the B-coordinates of a polynomial in $\mathcal{P}_2$.

EXAMPLE 12

Consider again the bases B, C, and S in Example 10 and the transition matrices $\mathbf{P}$ from B to C and $\mathbf{Q}$ from C to S. The transition matrices from C to B and from S to C, respectively, are

$$\mathbf{P}^{-1} = \begin{bmatrix} -6 & -1 & 3 \\ 4 & 2 & -1 \\ 13 & 3 & -6 \end{bmatrix} \quad \text{and} \quad \mathbf{Q}^{-1} = \begin{bmatrix} 1 & -3 & 2 \\ -1 & 4 & -3 \\ 2 & -5 & 4 \end{bmatrix}.$$

Now,

$$[\mathbf{v}]_B = \mathbf{P}^{-1}[\mathbf{v}]_C = \mathbf{P}^{-1}\left(\mathbf{Q}^{-1}[\mathbf{v}]_S\right) = \left(\mathbf{P}^{-1}\mathbf{Q}^{-1}\right)[\mathbf{v}]_S,$$

and so $\mathbf{P}^{-1}\mathbf{Q}^{-1}$ acts as the transition matrix from S to B (see Figure 4.8). For example, if $\mathbf{v} = x^2 + 7x + 3$, then

$$[\mathbf{v}]_B = \left(\mathbf{P}^{-1}\mathbf{Q}^{-1}\right)[\mathbf{v}]_S$$

$$= \begin{bmatrix} -6 & -1 & 3 \\ 4 & 2 & -1 \\ 13 & 3 & -6 \end{bmatrix} \begin{bmatrix} 1 & -3 & 2 \\ -1 & 4 & -3 \\ 2 & -5 & 4 \end{bmatrix} \begin{bmatrix} 1 \\ 7 \\ 3 \end{bmatrix} = \begin{bmatrix} 3 \\ 1 \\ -2 \end{bmatrix}.$$

∎

Figure 4.8

Transition matrices
used to convert
between B-, C-, and
S-coordinates in $\mathcal{P}_2$

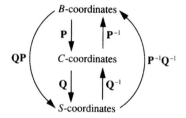

Diagonalization and the Transition Matrix

The matrix $\mathbf{P}$ obtained in the process of diagonalizing an $n \times n$ matrix turns out to be a transition matrix between two different bases for $\mathbb{R}^n$, as we see in our final example.

EXAMPLE 13 Consider

$$\mathbf{A} = \begin{bmatrix} 14 & -15 & -30 \\ 6 & -7 & -12 \\ 3 & -3 & -7 \end{bmatrix}.$$

A quick calculation produces $p_{\mathbf{A}}(x) = x^3 - 3x - 2 = (x - 2)(x + 1)^2$. Row reducing $(2\mathbf{I}_3 - \mathbf{A})$ yields the eigenvector $\mathbf{v}_1 = [5, 2, 1]$. The set $\{\mathbf{v}_1\}$ is a basis for the eigenspace E_2. Similarly, we row reduce $(-1\mathbf{I}_3 - \mathbf{A})$ to obtain the eigenvectors $\mathbf{v}_2 = [1, 1, 0]$ and $\mathbf{v}_3 = [2, 0, 1]$. The set $\{\mathbf{v}_2, \mathbf{v}_3\}$ forms a basis for the eigenspace E_{-1}.

Let $B = (\mathbf{v}_1, \mathbf{v}_2, \mathbf{v}_3)$. These vectors are linearly independent (see the remarks before Example 12 in Section 4.4), and thus B is a basis for $\mathbb{R}^3$ by Corollary 4.14. Let S be the standard basis. Then, the transition matrix $\mathbf{P}$ from B to S is given by the matrix whose columns are the vectors in B.

$$\mathbf{P} = \begin{bmatrix} 5 & 1 & 2 \\ 2 & 1 & 0 \\ 1 & 0 & 1 \end{bmatrix}; \quad \text{and therefore,} \quad \mathbf{P}^{-1} = \begin{bmatrix} 1 & -1 & -2 \\ -2 & 3 & 4 \\ -1 & 1 & 3 \end{bmatrix}$$

is the transition matrix from S to B.

Notice that the transition matrix $\mathbf{P}$ is precisely the matrix $\mathbf{P}$ created by the Diagonalization Method of Section 3.4! Recall from Section 3.4 that $\mathbf{P}^{-1}\mathbf{A}\mathbf{P}$ is a diagonal matrix $\mathbf{D}$ with the eigenvalues of $\mathbf{A}$ on the main diagonal: namely,

$$\mathbf{D} = \begin{bmatrix} 2 & 0 & 0 \\ 0 & -1 & 0 \\ 0 & 0 & -1 \end{bmatrix}.$$

We can understand this relationship between $\mathbf{A}$ and $\mathbf{D}$ more fully from a "change of coordinates" perspective. In fact, if $\mathbf{v}$ is any vector in $\mathbb{R}^3$ expressed in standard coordinates, we claim that $\mathbf{D}[\mathbf{v}]_B = [\mathbf{A}\mathbf{v}]_B$. That is, multiplication by $\mathbf{D}$ working in B-coordinates corresponds to first multiplying by $\mathbf{A}$ in standard coordinates, and then converting to B-coordinates (see Figure 4.9).

Why does this relationship hold? Well,

$$\mathbf{D}[\mathbf{v}]_B = (\mathbf{P}^{-1}\mathbf{A}\mathbf{P})[\mathbf{v}]_B = (\mathbf{P}^{-1}\mathbf{A})\mathbf{P}[\mathbf{v}]_B = \mathbf{P}^{-1}\mathbf{A}[\mathbf{v}]_S = \mathbf{P}^{-1}(\mathbf{A}\mathbf{v}) = [\mathbf{A}\mathbf{v}]_B$$

because multiplication by $\mathbf{P}$ and $\mathbf{P}^{-1}$ perform the appropriate transitions between B- and S-coordinates. Thus, we can think of $\mathbf{D}$ as being the "B-coordinates version" of $\mathbf{A}$. By using a basis of eigenvectors we have converted to a new coordinate system in which multiplication by $\mathbf{A}$ has been replaced with multiplication by a diagonal matrix, which is much easier to work with because of its simpler form. ∎

Figure 4.9

Multiplication by $\mathbf{A}$ in standard coordinates corresponds to multiplication by $\mathbf{D}$ in B-coordinates

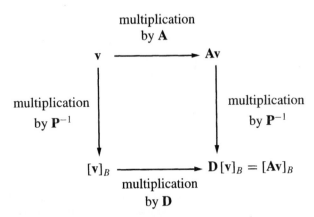

Example 13 illustrates the following general principle:

When the Diagonalization Method of Section 3.4 is successfully performed on a matrix $\mathbf{A}$, the matrix $\mathbf{P}$ obtained is the transition matrix from B-coordinates to standard coordinates, where B is an ordered basis for $\mathbb{R}^n$ consisting of eigenvectors for $\mathbf{A}$.

◆ **Application**: You have now covered the prerequisites for Section 8.7, "Rotation of Axes."
◆ **Supplemental Material:** You have now covered the prerequisites for Section 10.2, "Function Spaces."

Exercises for Section 4.7

1. In each part, let B represent an ordered basis for a subspace $\mathcal{V}$ of $\mathbb{R}^n$, $\mathcal{P}_n$, or $\mathcal{M}_{mn}$. Find $[\mathbf{v}]_B$, for the given $\mathbf{v} \in \mathcal{V}$.

   ⋆(a) $B = ([1, -4, 1], [5, -7, 2], [0, -4, 1])$; $\mathbf{v} = [2, -1, 0]$

   (b) $B = ([4, 6, 0, 1], [5, 1, -1, 0], [0, 15, 1, 3], [1, 5, 0, 1])$; $\mathbf{v} = [0, -9, 1, -2]$

   ⋆(c) $B = ([2, 3, 1, -2, 2], [4, 3, 3, 1, -1], [1, 2, 1, -1, 1])$; $\mathbf{v} = [7, -4, 5, 13, -13]$

   (d) $B = ([-3, 1, -2, 5, -1], [6, 1, 2, -1, 0], [9, 2, 1, -4, 2], [3, 1, 0, -2, 1])$; $\mathbf{v} = [3, 16, -12, 41, -7]$

   ⋆(e) $B = (3x^2 - x + 2, x^2 + 2x - 3, 2x^2 + 3x - 1)$; $\mathbf{v} = 13x^2 - 5x + 20$

   (f) $B = (4x^2 + 3x - 1, 2x^2 - x + 4, x^2 - 2x + 3)$; $\mathbf{v} = -5x^2 - 17x + 20$

$\star$(g) $B = (2x^3 - x^2 + 3x - 1, \ x^3 + 2x^2 - x + 3, \ -3x^3 - x^2 + x + 1)$;
$\mathbf{v} = 8x^3 + 11x^2 - 9x + 11$

$\star$(h) $B = \left(\begin{bmatrix} 1 & -2 \\ 0 & 1 \end{bmatrix}, \begin{bmatrix} 2 & -1 \\ 1 & 0 \end{bmatrix}, \begin{bmatrix} 1 & -1 \\ 3 & 1 \end{bmatrix} \right)$; $\mathbf{v} = \begin{bmatrix} -3 & -2 \\ 0 & 3 \end{bmatrix}$

(i) $B = \left(\begin{bmatrix} -2 & 3 \\ 0 & 2 \end{bmatrix}, \begin{bmatrix} 1 & 1 \\ -1 & 2 \end{bmatrix}, \begin{bmatrix} 0 & -3 \\ 2 & 1 \end{bmatrix} \right)$; $\mathbf{v} = \begin{bmatrix} -8 & 35 \\ -14 & 8 \end{bmatrix}$

$\star$(j) $B = \left(\begin{bmatrix} 1 & 3 & -1 \\ 2 & 1 & 4 \end{bmatrix}, \begin{bmatrix} -3 & 1 & 7 \\ 1 & 2 & 5 \end{bmatrix} \right)$; $\mathbf{v} = \begin{bmatrix} 11 & 13 & -19 \\ 8 & 1 & 10 \end{bmatrix}$

2. In each part, ordered bases B and C are given for a subspace of $\mathbb{R}^n$, $\mathcal{P}_n$, or $\mathcal{M}_{mn}$. Find the transition matrix from B to C.

$\star$(a) $B = ([1, 0, 0], [0, 1, 0], [0, 0, 1])$; $C = ([1, 5, 1], [1, 6, -6], [1, 3, 14])$

(b) $B = ([1, 0, -1], [10, 5, 4], [2, 1, 1])$; $C = ([1, 0, 2], [5, 2, 5], [2, 1, 2])$

$\star$(c) $B = (2x^2 + 3x - 1, 8x^2 + x + 1, x^2 + 6)$; $C = (x^2 + 3x + 1, 3x^2 + 4x + 1, 10x^2 + 17x + 5)$

$\star$(d) $B = \left(\begin{bmatrix} 1 & 3 \\ 5 & 1 \end{bmatrix}, \begin{bmatrix} 2 & 1 \\ 0 & 4 \end{bmatrix}, \begin{bmatrix} 3 & 1 \\ 1 & 0 \end{bmatrix}, \begin{bmatrix} 0 & 2 \\ -4 & 1 \end{bmatrix} \right)$;

$C = \left(\begin{bmatrix} -1 & 1 \\ 3 & -1 \end{bmatrix}, \begin{bmatrix} 1 & 0 \\ 0 & 1 \end{bmatrix}, \begin{bmatrix} 3 & -4 \\ -7 & 4 \end{bmatrix}, \begin{bmatrix} 1 & -1 \\ -2 & 1 \end{bmatrix} \right)$

(e) $B = ([1, 3, -2, 0, 1, 4], [-6, 2, 7, -5, -11, -14])$;
$C = ([3, 1, -4, 2, 5, 8], [4, 0, -5, 3, 7, 10])$

$\star$(f) $B = (6x^4 + 20x^3 + 7x^2 + 19x - 4, \ x^4 + 5x^3 + 7x^2 - x + 6, \ 5x^3 + 17x^2 - 10x + 19)$; $C = (x^4 + 3x^3 + 4x - 2, \ 2x^4 + 7x^3 + 4x^2 + 3x + 1, \ 2x^4 + 5x^3 - 3x^2 + 8x - 7)$

(g) $B = \left(\begin{bmatrix} -1 & 4 \\ 8 & 4 \\ -9 & 0 \end{bmatrix}, \begin{bmatrix} 3 & 7 \\ 2 & 10 \\ -7 & 3 \end{bmatrix}, \begin{bmatrix} -9 & -1 \\ 20 & 0 \\ 3 & 1 \end{bmatrix} \right)$;

$C = \left(\begin{bmatrix} 1 & 2 \\ 0 & 4 \\ 1 & 2 \end{bmatrix}, \begin{bmatrix} 2 & 3 \\ -1 & 5 \\ -1 & 2 \end{bmatrix}, \begin{bmatrix} -1 & 2 \\ 5 & 3 \\ -2 & 1 \end{bmatrix} \right)$

3. Draw the B-coordinate grid in $\mathbb{R}^2$ as in Example 3, where $B = ([3, 2], [-2, 1])$. Plot the point $(2, 6)$. Convert this point to B-coordinates, and show that it is at the proper place on the B-coordinate grid.

4. In each part of this exercise, ordered bases B, C, and D are given for $\mathbb{R}^n$ or $\mathcal{P}_n$. Calculate the following independently:

   (i) The transition matrix $\mathbf{P}$ from B to C

   (ii) The transition matrix $\mathbf{Q}$ from C to D

   (iii) The transition matrix $\mathbf{T}$ from B to D
   Then verify Theorem 4.22 by showing that $\mathbf{T} = \mathbf{QP}$.

$\star$(a) $B = ([3, 1], [7, 2])$; $C = ([3, 7], [2, 5])$; $D = ([5, 2], [2, 1])$

(b) $B = ([8, 1, 0], [2, 11, 5], [-1, 2, 1])$; $C = ([2, 11, 5], [-1, 2, 1], [8, 1, 0])$; $D = ([-1, 2, 1], [2, 11, 5], [8, 1, 0])$

$\star$(c) $B = (x^2 + 2x + 2, 3x^2 + 7x + 8, 3x^2 + 9x + 13)$; $C = (x^2 + 4x + 1, 2x^2 + x, x^2)$; $D = (7x^2 - 3x + 2, x^2 + 7x - 3, x^2 - 2x + 1)$

(d) $B = (4x^3 + x^2 + 5x + 2, 2x^3 - 2x^2 + x + 1, 3x^3 - x^2 + 7x + 3,$
$x^3 - x^2 + 2x + 1)$; $C = (x^3 + x + 3, x^2 + 2x - 1, x^3 + 2x^2 + 6x + 6,$
$3x^3 - x^2 + 6x + 36)$; $D = (x^3, x^2, x, 1)$

5. In each part of this exercise, an ordered basis B is given for a subspace $\mathcal{V}$ of $\mathbb{R}^n$. Perform the following steps:

   (i) Use the Simplified Span Method to find a second ordered basis C.
   (ii) Find the transition matrix $\mathbf{P}$ from B to C.
   (iii) Use Theorem 4.23 to find the transition matrix $\mathbf{Q}$ from C to B.
   (iv) For the given vector $\mathbf{v} \in \mathcal{V}$, independently calculate $[\mathbf{v}]_B$ and $[\mathbf{v}]_C$.
   (v) Check your answer to step (iv) by using $\mathbf{Q}$ and $[\mathbf{v}]_C$ to calculate $[\mathbf{v}]_B$.

   ★(a) $B = ([1, -4, 1, 2, 1], [6, -24, 5, 8, 3], [3, -12, 3, 6, 2])$; $\mathbf{v} = [2, -8, -2, -12, 3]$

   (b) $B = ([1, -5, 2, 0, -4], [3, -14, 9, 2, -3], [1, -4, 5, 3, 7])$; $\mathbf{v} = [2, -9, 7, 5, 7]$

   ★(c) $B = ([3, -1, 4, 6], [6, 7, -3, -2], [-4, -3, 3, 4], [-2, 0, 1, 2])$; $\mathbf{v} = [10, 14, 3, 12]$

6. Let $B = (\mathbf{v}_1, \ldots, \mathbf{v}_n)$ be an ordered basis for a vector space $\mathcal{V}$. Prove that for each i, $[\mathbf{v}_i]_B = \mathbf{e}_i$.

7. ★(a) Let $\mathbf{u} = [-5, 9, -1]$, $\mathbf{v} = [3, -9, 2]$, and $\mathbf{w} = [2, -5, 1]$. Find the transition matrix from the ordered basis $B = (\mathbf{u}, \mathbf{v}, \mathbf{w})$ to each of the following ordered bases: $C_1 = (\mathbf{v}, \mathbf{w}, \mathbf{u})$, $C_2 = (\mathbf{w}, \mathbf{u}, \mathbf{v})$, $C_3 = (\mathbf{u}, \mathbf{w}, \mathbf{v})$, $C_4 = (\mathbf{v}, \mathbf{u}, \mathbf{w})$, $C_5 = (\mathbf{w}, \mathbf{v}, \mathbf{u})$.

   (b) Let B be an ordered basis for an n-dimensional vector space $\mathcal{V}$. Let C be another ordered basis for $\mathcal{V}$ with the same vectors as B but rearranged in a different order. Prove that the transition matrix from B to C is obtained by rearranging rows of $\mathbf{I}_n$ in exactly the same fashion.

8. Let B and C be ordered bases for $\mathbb{R}^n$.

   (a) Show that if B is the standard basis in $\mathbb{R}^n$, then the transition matrix from B to C is given by

   $$\begin{bmatrix} \text{1st} & \text{2nd} & & n\text{th} \\ \text{vector} & \text{vector} & \cdots & \text{vector} \\ \text{in} & \text{in} & & \text{in} \\ C & C & & C \end{bmatrix}^{-1}.$$

   (b) Show that if C is the standard basis in $\mathbb{R}^n$, then the transition matrix from B to C is given by

   $$\begin{bmatrix} \text{1st} & \text{2nd} & & n\text{th} \\ \text{vector} & \text{vector} & \cdots & \text{vector} \\ \text{in} & \text{in} & & \text{in} \\ B & B & & B \end{bmatrix}.$$

9. Let B and C be ordered bases for $\mathbb{R}^n$. Let $\mathbf{P}$ be the matrix whose columns are the vectors in B and let $\mathbf{Q}$ be the matrix whose columns are the

vectors in C. Prove that the transition matrix from B to C equals $\mathbf{Q}^{-1}\mathbf{P}$. (Hint: Use Exercise 8.)

★**10.** Consider the ordered basis $B = ([-2, 1, 3], [1, 0, 2], [-13, 5, 10])$ for $\mathbb{R}^3$. Suppose that C is another ordered basis for $\mathbb{R}^3$ and that the transition matrix from B to C is given by

$$\begin{bmatrix} 1 & 9 & -1 \\ 2 & 13 & -11 \\ -1 & -8 & 3 \end{bmatrix}.$$

Find C. (Hint: Use Exercise 9.)

11. (a) Verify all of the computations in Example 13, including the computation of $p_{\mathbf{A}}(x)$, the eigenvectors $\mathbf{v}_1$, $\mathbf{v}_2$, and $\mathbf{v}_3$, the transition matrix $\mathbf{P}$, and its inverse $\mathbf{P}^{-1}$. Check that $\mathbf{D} = \mathbf{P}^{-1}\mathbf{A}\mathbf{P}$.

★(b) Let $\mathbf{v} = [1, 4, -2]$. With B, $\mathbf{A}$, and $\mathbf{D}$ as in Example 13, compute $\mathbf{D}[\mathbf{v}]_B$ and $[\mathbf{A}\mathbf{v}]_B$ independently, without using multiplication by the matrices $\mathbf{P}$ or $\mathbf{P}^{-1}$ in that example. Compare your results.

▶**12.** Prove Theorem 4.20. (Hint: Use a proof by induction for part (3).)

▶**13.** Prove Theorem 4.22. (Hint: Use Theorem 4.21.)

14. Prove Theorem 4.23. (Hint: Let $\mathbf{Q}$ be the transition matrix from C to B. Prove that $\mathbf{QP} = \mathbf{I}$ by using Theorems 4.21 and 4.22.)

★**15.** True or False:

(a) For the ordered bases $B = (\mathbf{i}, \mathbf{j}, \mathbf{k})$ and $C = (\mathbf{j}, \mathbf{k}, \mathbf{i})$ for $\mathbb{R}^3$, we have $[\mathbf{v}]_B = [\mathbf{v}]_C$ for each $\mathbf{v} \in \mathbb{R}^3$.

(b) If B is a finite ordered basis for $\mathcal{V}$ and $\mathbf{b}_i$ is the ith vector in B, then $[\mathbf{b}_i]_B = \mathbf{e}_i$.

(c) If $B = (\mathbf{b}_1, ..., \mathbf{b}_n)$ and $C = (\mathbf{c}_1, ..., \mathbf{c}_n)$ are ordered bases for a vector space $\mathcal{V}$, then the ith column of the transition matrix $\mathbf{Q}$ from C to B is $[\mathbf{c}_i]_B$.

(d) If B and C are ordered bases for a finite dimensional vector space $\mathcal{V}$ and $\mathbf{P}$ is the transition matrix from B to C, then $\mathbf{P}[\mathbf{v}]_C = [\mathbf{v}]_B$ for every vector $\mathbf{v} \in \mathcal{V}$.

(e) If B, C, and D are finite ordered bases for a vector space $\mathcal{V}$, $\mathbf{P}$ is the transition matrix from B to C, and $\mathbf{Q}$ is the transition matrix from C to D, then $\mathbf{PQ}$ is the transition matrix from B to D.

(f) If B and C are ordered bases for a finite dimensional vector space $\mathcal{V}$ and if $\mathbf{P}$ is the transition matrix from B to C, then $\mathbf{P}$ is nonsingular.

(g) If the Diagonalization Method is used to find a diagonal form $\mathbf{D}$ for an $n \times n$ matrix $\mathbf{A}$, then $\mathbf{D}$ is the transition matrix from standard coordinates to an ordered basis of eigenvectors for $\mathbf{A}$.

Chapter 5

Linear Transformations

Transforming Space

Although a vector can be used to indicate a particular type of movement, actual vectors themselves are essentially static, unchanging objects. For example, if we represent the edges of a particular image on a computer screen by vectors, then these vectors are fixed in place. However, when we want to move or alter the image in some way, such as rotating it about a point on the screen, we need a function to calculate the new position for each of the original vectors.

This suggests that we need another "tool" in our arsenal — functions that move a given set of vectors in a prescribed "linear" manner. Such functions are called linear transformations. Just as we saw in Chapter 4 that general vector spaces are abstract generalizations of $\mathbb{R}^n$, we will find in this chapter that linear transformations are the corresponding abstract generalization of matrix multiplication.

In this chapter, we study functions that map the vectors in one vector space to those in another. We concentrate on a special class of these functions, known as linear transformations. The formal definition of a linear transformation is introduced in Section 5.1 along with several of its fundamental properties. In Section 5.2, we show that the effect of any linear transformation is equivalent to multiplication by a corresponding matrix. In Section 5.3, we examine an important relationship between the dimensions of the domain and the range of a linear transformation, known as the Dimension Theorem. In Section 5.4, we use linear transformations to establish that all n-dimensional vector spaces are in some sense equivalent. Finally, in Section 5.5, we return to the topic of eigenvalues and eigenvectors to study them in the context of linear transformations.

5.1 Introduction to Linear Transformations

In this section, we introduce linear transformations and examine their elementary properties.

Functions

If you are not familiar with the terms *domain, codomain, range, image,* and *pre-image* in the context of functions, read Appendix B before proceeding. The following example illustrates some of these terms:

EXAMPLE 1   Let $f \colon \mathcal{M}_{23} \to \mathcal{M}_{22}$ be given by

$$f\left(\begin{bmatrix} a & b & c \\ d & e & f \end{bmatrix}\right) = \begin{bmatrix} a & b \\ 0 & 0 \end{bmatrix}.$$

Then f is a function that maps one vector space to another. The domain of f is $\mathcal{M}_{23}$, the codomain of f is $\mathcal{M}_{22}$, and the range of f is the set of all 2×2 matrices with second row entries equal to zero. The image of $\begin{bmatrix} 1 & 2 & 3 \\ 4 & 5 & 6 \end{bmatrix}$ under f is $\begin{bmatrix} 1 & 2 \\ 0 & 0 \end{bmatrix}$. The matrix $\begin{bmatrix} 1 & 2 & 10 \\ 11 & 12 & 13 \end{bmatrix}$ is one of the pre-images of $\begin{bmatrix} 1 & 2 \\ 0 & 0 \end{bmatrix}$ under f. Also, the image under f of the set S of all matrices of the form $\begin{bmatrix} 7 & * & * \\ * & * & * \end{bmatrix}$ (where "$*$" represents any real number) is the set $f(S)$ containing all matrices of the form $\begin{bmatrix} 7 & * \\ 0 & 0 \end{bmatrix}$. Finally, the pre-image under f of the set T of all matrices of the form $\begin{bmatrix} a & a+2 \\ 0 & 0 \end{bmatrix}$ is the set $f^{-1}(T)$ consisting of all matrices of the form $\begin{bmatrix} a & a+2 & * \\ * & * & * \end{bmatrix}$.   ■

Linear Transformations

DEFINITION

Let $\mathcal{V}$ and $\mathcal{W}$ be vector spaces, and let $f: \mathcal{V} \to \mathcal{W}$ be a function from $\mathcal{V}$ to $\mathcal{W}$. (That is, for each vector $\mathbf{v} \in \mathcal{V}$, $f(\mathbf{v})$ denotes exactly one vector of $\mathcal{W}$.) Then f is a **linear transformation** if and only if both of the following are true:

(1) $f(\mathbf{v}_1 + \mathbf{v}_2) = f(\mathbf{v}_1) + f(\mathbf{v}_2)$, for all $\mathbf{v}_1, \mathbf{v}_2 \in \mathcal{V}$
(2) $f(c\mathbf{v}) = cf(\mathbf{v})$, for all $c \in \mathbb{R}$ and all $\mathbf{v} \in \mathcal{V}$.

Properties (1) and (2) insist that the operations of addition and scalar multiplication give the same result on vectors whether the operations are performed before f is applied (in $\mathcal{V}$) or after f is applied (in $\mathcal{W}$). Thus, a linear transformation is a function between vector spaces that "preserves" the operations that give structure to the spaces.

To determine whether a given function f from a vector space $\mathcal{V}$ to a vector space $\mathcal{W}$ is a linear transformation, we need only verify properties (1) and (2) in the definition, as in the next three examples.

EXAMPLE 2

Consider the mapping $f: \mathcal{M}_{mn} \to \mathcal{M}_{nm}$, given by $f(\mathbf{A}) = \mathbf{A}^T$ for any $m \times n$ matrix $\mathbf{A}$. We will show that f is a linear transformation.

(1) We must show that $f(\mathbf{A}_1 + \mathbf{A}_2) = f(\mathbf{A}_1) + f(\mathbf{A}_2)$, for matrices $\mathbf{A}_1$, $\mathbf{A}_2 \in \mathcal{M}_{mn}$. However, $f(\mathbf{A}_1 + \mathbf{A}_2) = (\mathbf{A}_1 + \mathbf{A}_2)^T = \mathbf{A}_1^T + \mathbf{A}_2^T$ (by part (2) of Theorem 1.12) $= f(\mathbf{A}_1) + f(\mathbf{A}_2)$.

(2) We must show that $f(c\mathbf{A}) = cf(\mathbf{A})$, for all $c \in \mathbb{R}$ and for all $\mathbf{A} \in \mathcal{M}_{mn}$. However, $f(c\mathbf{A}) = (c\mathbf{A})^T = c\left(\mathbf{A}^T\right)$ (by part (3) of Theorem 1.12) $= cf(\mathbf{A})$.

Hence, f is a linear transformation. ∎

EXAMPLE 3

Consider the function $g: \mathcal{P}_n \to \mathcal{P}_{n-1}$ given by $g(\mathbf{p}) = \mathbf{p}'$, the derivative of $\mathbf{p}$. We will show that g is a linear transformation.

(1) We must show that $g(\mathbf{p}_1 + \mathbf{p}_2) = g(\mathbf{p}_1) + g(\mathbf{p}_2)$, for all $\mathbf{p}_1, \mathbf{p}_2 \in \mathcal{P}_n$. Now, $g(\mathbf{p}_1 + \mathbf{p}_2) = (\mathbf{p}_1 + \mathbf{p}_2)'$. From calculus we know that the derivative of a sum is the sum of the derivatives, so $(\mathbf{p}_1 + \mathbf{p}_2)' = \mathbf{p}_1' + \mathbf{p}_2' = g(\mathbf{p}_1) + g(\mathbf{p}_2)$.

(2) We must show that $g(c\mathbf{p}) = cg(\mathbf{p})$, for all $c \in \mathbb{R}$ and $\mathbf{p} \in \mathcal{P}_n$. Now, $g(c\mathbf{p}) = (c\mathbf{p})'$. Again, from calculus we know that the derivative of a constant times a function is equal to the constant times the derivative of the function, so $(c\mathbf{p})' = c(\mathbf{p}') = cg(\mathbf{p})$.

Hence, g is a linear transformation. ∎

EXAMPLE 4

Let $\mathcal{V}$ be a finite dimensional vector space, and let B be an ordered basis for $\mathcal{V}$. Then every element $\mathbf{v} \in \mathcal{V}$ has its coordinatization $[\mathbf{v}]_B$ with respect to B. Consider the mapping $f: \mathcal{V} \to \mathbb{R}^n$ given by $f(\mathbf{v}) = [\mathbf{v}]_B$. We will show that f is a linear transformation.

Let $\mathbf{v}_1, \mathbf{v}_2 \in \mathcal{V}$. By Theorem 4.20, $[\mathbf{v}_1 + \mathbf{v}_2]_B = [\mathbf{v}_1]_B + [\mathbf{v}_2]_B$. Hence,

$$f(\mathbf{v}_1 + \mathbf{v}_2) = [\mathbf{v}_1 + \mathbf{v}_2]_B = [\mathbf{v}_1]_B + [\mathbf{v}_2]_B = f(\mathbf{v}_1) + f(\mathbf{v}_2).$$

Next, let $c \in \mathbb{R}$ and $\mathbf{v} \in \mathcal{V}$. Again by Theorem 4.20, $[c\mathbf{v}]_B = c\,[\mathbf{v}]_B$. Hence,

$$f(c\mathbf{v}) = [c\mathbf{v}]_B = c\,[\mathbf{v}]_B = cf(\mathbf{v}).$$

Thus, f is a linear transformation from $\mathcal{V}$ to $\mathbb{R}^n$.   ∎

Not every function between vector spaces is a linear transformation. For example, consider the function $h: \mathbb{R}^2 \to \mathbb{R}^2$ given by $h([x, y]) = [x + 1, y - 2] = [x, y] + [1, -2]$. In this case, h merely adds $[1, -2]$ to each vector $[x, y]$ (see Figure 5.1). This type of mapping is called a **translation**. However, h is not a linear transformation. To show that it is not, we have to produce a counterexample to verify that either property (1) or property (2) of the definition fails. Property (1) fails since $h([1, 2] + [3, 4]) = h([4, 6]) = [5, 4]$, while $h([1, 2]) + h([3, 4]) = [2, 0] + [4, 2] = [6, 2]$.

Figure 5.1

A translation in $\mathbb{R}^2$

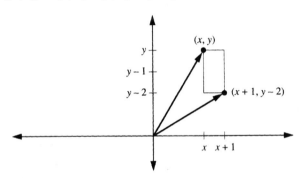

In general, when given a function f between vector spaces, we do not always know right away whether f is a linear transformation. If we suspect that either property (1) or (2) does not hold for f, then we look for a counterexample.

Linear Operators and Some Geometric Examples

An important type of linear transformation is one that maps a vector space to itself.

DEFINITION

Let $\mathcal{V}$ be a vector space. A **linear operator** on $\mathcal{V}$ is a linear transformation whose domain and codomain are both $\mathcal{V}$.

EXAMPLE 5   If $\mathcal{V}$ is any vector space, then the mapping $i: \mathcal{V} \to \mathcal{V}$ given by $i(\mathbf{v}) = \mathbf{v}$ for all $\mathbf{v} \in \mathcal{V}$ is a linear operator, known as the **identity linear operator**. Also, the constant mapping $z: \mathcal{V} \to \mathcal{V}$ given by $z(\mathbf{v}) = \mathbf{0}_\mathcal{V}$, is a linear operator known as the **zero linear operator** (see Exercise 2).   ∎

The next few examples exhibit important geometric operators. In these examples, assume that all vectors begin at the origin.

EXAMPLE 6

Reflections: Consider the mapping $f \colon \mathbb{R}^3 \to \mathbb{R}^3$ given by $f([a_1, a_2, a_3]) = [a_1, a_2, -a_3]$. This mapping "reflects" the vector $[a_1, a_2, a_3]$ through the xy-plane, which acts like a "mirror" (see Figure 5.2). Now, since

$$
\begin{aligned}
f([a_1, a_2, a_3] + [b_1, b_2, b_3]) &= f([a_1 + b_1, a_2 + b_2, a_3 + b_3]) \\
&= [a_1 + b_1, a_2 + b_2, -(a_3 + b_3)] \\
&= [a_1, a_2, -a_3] + [b_1, b_2, -b_3] \\
&= f([a_1, a_2, a_3]) + f([b_1, b_2, b_3]), \quad \text{and}
\end{aligned}
$$

$$
\begin{aligned}
f(c[a_1, a_2, a_3]) &= f([ca_1, ca_2, ca_3]) = [ca_1, ca_2, -ca_3] \\
&= c[a_1, a_2, -a_3] = cf([a_1, a_2, a_3]),
\end{aligned}
$$

we see that f is a linear operator. Similarly, reflection through the xz-plane or the yz-plane is also a linear operator on $\mathbb{R}^3$ (see Exercise 4). ∎

Figure 5.2

Reflection in $\mathbb{R}^3$
through the xy-plane

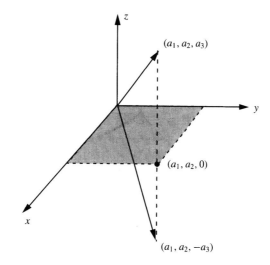

EXAMPLE 7

Contractions and Dilations: Consider the mapping $g \colon \mathbb{R}^n \to \mathbb{R}^n$ given by scalar multiplication by k, where $k \in \mathbb{R}$; that is, $g(\mathbf{v}) = k\mathbf{v}$, for $\mathbf{v} \in \mathbb{R}^n$. The function g is a linear operator (see Exercise 3). If $|k| > 1$, g represents a **dilation** (lengthening) of the vectors in $\mathbb{R}^n$; if $|k| < 1$, g represents a **contraction** (shrinking). ∎

EXAMPLE 8

Projections: Consider the mapping $h \colon \mathbb{R}^3 \to \mathbb{R}^3$ given by $h([a_1, a_2, a_3]) = [a_1, a_2, 0]$. This mapping takes each vector in $\mathbb{R}^3$ to a corresponding vector in the xy-plane (see Figure 5.3). Similarly, consider the mapping $j \colon \mathbb{R}^4 \to \mathbb{R}^4$ given by $j([a_1, a_2, a_3, a_4]) = [0, a_2, 0, a_4]$. This mapping takes each vector in $\mathbb{R}^4$ to a corresponding vector whose first and third coordinates are zero. The functions h and j are both linear operators (see Exercise 5). Such mappings, where at least one of the coordinates is "zeroed out," are examples of **projection mappings**. You can verify that all such mappings are linear operators. (Other types of projection mappings are illustrated in Exercises 6 and 7.) ∎

Figure 5.3

Projection of $[a_1, a_2, a_3]$
to the xy-plane

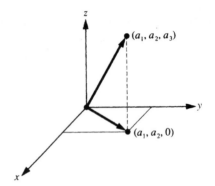

| EXAMPLE 9 |

Rotations: Let θ be a fixed angle in $\mathbb{R}^2$, and let $l: \mathbb{R}^2 \to \mathbb{R}^2$ be given by

$$l\left(\begin{bmatrix} x \\ y \end{bmatrix}\right) = \begin{bmatrix} \cos\theta & -\sin\theta \\ \sin\theta & \cos\theta \end{bmatrix}\begin{bmatrix} x \\ y \end{bmatrix} = \begin{bmatrix} x\cos\theta - y\sin\theta \\ x\sin\theta + y\cos\theta \end{bmatrix}.$$

In Exercise 9 you are asked to show that l rotates $[x, y]$ counterclockwise through the angle θ (see Figure 5.4).

Now, let $\mathbf{v}_1 = [x_1, y_1]$ and $\mathbf{v}_2 = [x_2, y_2]$ be two vectors in $\mathbb{R}^2$. Then,

$$\begin{aligned} l(\mathbf{v}_1 + \mathbf{v}_2) &= \begin{bmatrix} \cos\theta & -\sin\theta \\ \sin\theta & \cos\theta \end{bmatrix}(\mathbf{v}_1 + \mathbf{v}_2) \\ &= \begin{bmatrix} \cos\theta & -\sin\theta \\ \sin\theta & \cos\theta \end{bmatrix}\mathbf{v}_1 + \begin{bmatrix} \cos\theta & -\sin\theta \\ \sin\theta & \cos\theta \end{bmatrix}\mathbf{v}_2 \\ &= l(\mathbf{v}_1) + l(\mathbf{v}_2). \end{aligned}$$

Similarly, $l(c\mathbf{v}) = cl(\mathbf{v})$, for any $c \in \mathbb{R}$ and $\mathbf{v} \in \mathbb{R}^2$. Hence, l is a linear operator. ∎

Beware! Not all geometric operations are linear operators. Recall that the translation function is not a linear operator!

Figure 5.4

Counterclockwise
rotation of $[x, y]$
through an angle θ in $\mathbb{R}^2$

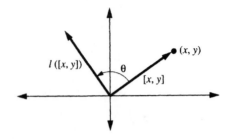

Multiplication Transformation

The linear operator in Example 9 is actually a special case of the next example, which shows that multiplication by an $m \times n$ matrix is always a linear transformation from $\mathbb{R}^n$ to $\mathbb{R}^m$.

EXAMPLE 10 Let $\mathbf{A}$ be a given $m \times n$ matrix. We show that the function $f: \mathbb{R}^n \to \mathbb{R}^m$ defined by $f(\mathbf{x}) = \mathbf{A}\mathbf{x}$, for all $\mathbf{x} \in \mathbb{R}^n$, is a linear transformation. Let $\mathbf{x}_1, \mathbf{x}_2 \in \mathbb{R}^n$. Then $f(\mathbf{x}_1 + \mathbf{x}_2) = \mathbf{A}(\mathbf{x}_1 + \mathbf{x}_2) = \mathbf{A}\mathbf{x}_1 + \mathbf{A}\mathbf{x}_2 = f(\mathbf{x}_1) + f(\mathbf{x}_2)$. Also, let $\mathbf{x} \in \mathbb{R}^n$ and $c \in \mathbb{R}$. Then, $f(c\mathbf{x}) = \mathbf{A}(c\mathbf{x}) = c(\mathbf{A}\mathbf{x}) = cf(\mathbf{x})$. ∎

For a specific example of the multiplication transformation, consider the matrix $\mathbf{A} = \begin{bmatrix} -1 & 4 & 2 \\ 5 & 6 & -3 \end{bmatrix}$. The mapping given by

$$f\left(\begin{bmatrix} x_1 \\ x_2 \\ x_3 \end{bmatrix}\right) = \begin{bmatrix} -1 & 4 & 2 \\ 5 & 6 & -3 \end{bmatrix}\begin{bmatrix} x_1 \\ x_2 \\ x_3 \end{bmatrix} = \begin{bmatrix} -x_1 + 4x_2 + 2x_3 \\ 5x_1 + 6x_2 - 3x_3 \end{bmatrix}$$

is a linear transformation from $\mathbb{R}^3$ to $\mathbb{R}^2$. In the next section, we will show that the converse of the result in Example 10 also holds; every linear transformation from $\mathbb{R}^n$ to $\mathbb{R}^m$ is equivalent to multiplication by an appropriate $m \times n$ matrix.

Elementary Properties of Linear Transformations

We now prove some basic properties of linear transformations. From here on, we usually use italicized capital letters, such as "L," to represent linear transformations.

THEOREM 5.1

Let $\mathcal{V}$ and $\mathcal{W}$ be vector spaces, and let $L: \mathcal{V} \to \mathcal{W}$ be a linear transformation. Let $\mathbf{0}_{\mathcal{V}}$ be the zero vector in $\mathcal{V}$ and $\mathbf{0}_{\mathcal{W}}$ be the zero vector in $\mathcal{W}$. Then
(1) $L(\mathbf{0}_{\mathcal{V}}) = \mathbf{0}_{\mathcal{W}}$
(2) $L(-\mathbf{v}) = -L(\mathbf{v})$, for all $\mathbf{v} \in \mathcal{V}$
(3) $L(a_1\mathbf{v}_1 + a_2\mathbf{v}_2 + \cdots + a_n\mathbf{v}_n) = a_1 L(\mathbf{v}_1) + a_2 L(\mathbf{v}_2) + \cdots + a_n L(\mathbf{v}_n)$,
for all $a_1, \ldots, a_n \in \mathbb{R}$, and $\mathbf{v}_1, \ldots, \mathbf{v}_n \in \mathcal{V}$, for $n \geq 2$.

PROOF OF THEOREM 5.1

Part (1):

$$\begin{aligned} L(\mathbf{0}_{\mathcal{V}}) &= L(0\mathbf{0}_{\mathcal{V}}) && \text{part (2) of Theorem 4.1, in } \mathcal{V} \\ &= 0L(\mathbf{0}_{\mathcal{V}}) && \text{property (2) of linear transformation} \\ &= \mathbf{0}_{\mathcal{W}} && \text{part (2) of Theorem 4.1, in } \mathcal{W} \end{aligned}$$

Part (2):

$$\begin{aligned} L(-\mathbf{v}) &= L(-1\mathbf{v}) && \text{part (3) of Theorem 4.1, in } \mathcal{V} \\ &= -1(L(\mathbf{v})) && \text{property (2) of linear transformation} \\ &= -L(\mathbf{v}) && \text{part (3) of Theorem 4.1, in } \mathcal{W} \end{aligned}$$

Part (3): (abridged) This part is proved by induction. We prove the Base Step ($n = 2$) here and leave the Inductive Step as Exercise 29. For the Base

Step, we must show that $L(a_1\mathbf{v}_1 + a_2\mathbf{v}_2) = a_1 L(\mathbf{v}_1) + a_2 L(\mathbf{v}_2)$. But,

$$
\begin{aligned}
L(a_1\mathbf{v}_1 + a_2\mathbf{v}_2) &= L(a_1\mathbf{v}_1) + L(a_2\mathbf{v}_2) \quad \text{property (1) of linear transformation}\\
&= a_1 L(\mathbf{v}_1) + a_2 L(\mathbf{v}_2) \quad \text{property (2) of linear transformation.}
\end{aligned}
$$

■

The next theorem asserts that the composition $L_2 \circ L_1$ of linear transformations L_1 and L_2 is again a linear transformation (see Appendix B for a review of composition of functions).

THEOREM 5.2

Let $\mathcal{V}_1, \mathcal{V}_2$, and $\mathcal{V}_3$ be vector spaces. Let $L_1: \mathcal{V}_1 \to \mathcal{V}_2$ and $L_2: \mathcal{V}_2 \to \mathcal{V}_3$ be linear transformations. Then $L_2 \circ L_1: \mathcal{V}_1 \to \mathcal{V}_3$ given by $(L_2 \circ L_1)(\mathbf{v}) = L_2(L_1(\mathbf{v}))$, for all $\mathbf{v} \in \mathcal{V}_1$, is a linear transformation.

PROOF OF THEOREM 5.2

(abridged) To show that $L_2 \circ L_1$ is a linear transformation, we must show that for all $c \in \mathbb{R}$ and $\mathbf{v}, \mathbf{v}_1, \mathbf{v}_2 \in \mathcal{V}$

$$
\begin{aligned}
(L_2 \circ L_1)(\mathbf{v}_1 + \mathbf{v}_2) &= (L_2 \circ L_1)(\mathbf{v}_1) + (L_2 \circ L_1)(\mathbf{v}_2)\\
\text{and} \qquad (L_2 \circ L_1)(c\mathbf{v}) &= c(L_2 \circ L_1)(\mathbf{v}).
\end{aligned}
$$

The first property holds since

$$
\begin{aligned}
(L_2 \circ L_1)(\mathbf{v}_1 + \mathbf{v}_2) &= L_2(L_1(\mathbf{v}_1 + \mathbf{v}_2))\\
&= L_2(L_1(\mathbf{v}_1) + L_1(\mathbf{v}_2)) \qquad \text{because } L_1 \text{ is a linear}\\
&\qquad\qquad\qquad\qquad\qquad\qquad \text{transformation}\\
&= L_2(L_1(\mathbf{v}_1)) + L_2(L_1(\mathbf{v}_2)) \qquad \text{because } L_2 \text{ is a linear}\\
&\qquad\qquad\qquad\qquad\qquad\qquad \text{transformation}\\
&= (L_2 \circ L_1)(\mathbf{v}_1) + (L_2 \circ L_1)(\mathbf{v}_2).
\end{aligned}
$$

We leave the proof of the second property as Exercise 33.   ■

EXAMPLE 11

Let L_1 represent the rotation of vectors in $\mathbb{R}^2$ through a fixed angle θ (as in Example 9), and let L_2 represent the reflection of vectors in $\mathbb{R}^2$ through the x-axis. That is, if $\mathbf{v} = [v_1, v_2]$, then

$$
L_1(\mathbf{v}) = \begin{bmatrix} \cos\theta & -\sin\theta \\ \sin\theta & \cos\theta \end{bmatrix} \begin{bmatrix} v_1 \\ v_2 \end{bmatrix} \quad \text{and} \quad L_2(\mathbf{v}) = \begin{bmatrix} v_1 \\ -v_2 \end{bmatrix}.
$$

Because L_1 and L_2 are both linear transformations, Theorem 5.2 asserts that

$$
L_2(L_1(\mathbf{v})) = L_2\left(\begin{bmatrix} v_1\cos\theta - v_2\sin\theta \\ v_1\sin\theta + v_2\cos\theta \end{bmatrix}\right) = \begin{bmatrix} v_1\cos\theta - v_2\sin\theta \\ -v_1\sin\theta - v_2\cos\theta \end{bmatrix}
$$

is also a linear transformation. $L_2 \circ L_1$ represents a rotation of $\mathbf{v}$ through θ followed by a reflection through the x-axis.   ■

Theorem 5.2 generalizes naturally to more than two linear transformations. That is, if $L_1, L_2, \ldots, L_k$ are linear transformations and the composition $L_k \circ \cdots \circ L_2 \circ L_1$ makes sense, then $L_k \circ \cdots \circ L_2 \circ L_1$ is also a linear transformation.

Linear Transformations and Subspaces

The final theorem of this section assures us that, under a linear transformation $L: \mathcal{V} \to \mathcal{W}$, subspaces of $\mathcal{V}$ "correspond" to subspaces of $\mathcal{W}$ and vice versa.

THEOREM 5.3

Let $L: \mathcal{V} \to \mathcal{W}$ be a linear transformation.

(1) If $\mathcal{V}'$ is a subspace of $\mathcal{V}$, then $L(\mathcal{V}') = \{L(\mathbf{v}) \mid \mathbf{v} \in \mathcal{V}'\}$, the image of $\mathcal{V}'$ in $\mathcal{W}$, is a subspace of $\mathcal{W}$. In particular, the range of L is a subspace of $\mathcal{W}$.

(2) If $\mathcal{W}'$ is a subspace of $\mathcal{W}$, then $L^{-1}(\mathcal{W}') = \{\mathbf{v} \mid L(\mathbf{v}) \in \mathcal{W}'\}$, the pre-image of $\mathcal{W}'$ in $\mathcal{V}$, is a subspace of $\mathcal{V}$.

We prove part (1) and leave part (2) as Exercise 31.

PROOF OF THEOREM 5.3

Part (1): Suppose that $L: \mathcal{V} \to \mathcal{W}$ is a linear transformation and that $\mathcal{V}'$ is a subspace of $\mathcal{V}$. Now, $L\left(\mathcal{V}'\right)$, the image of $\mathcal{V}'$ in $\mathcal{V}$ (see Figure 5.5), is certainly nonempty (why?). Hence, to show that $L\left(\mathcal{V}'\right)$ is a subspace of $\mathcal{W}$, we must prove that $L\left(\mathcal{V}'\right)$ is closed under addition and scalar multiplication.

First, suppose that $\mathbf{w}_1, \mathbf{w}_2 \in L\left(\mathcal{V}'\right)$. Then, by definition of $L\left(\mathcal{V}'\right)$, we have $\mathbf{w}_1 = L(\mathbf{v}_1)$ and $\mathbf{w}_2 = L(\mathbf{v}_2)$, for some $\mathbf{v}_1, \mathbf{v}_2 \in \mathcal{V}'$. Then, $\mathbf{w}_1 + \mathbf{w}_2 = L(\mathbf{v}_1) + L(\mathbf{v}_2) = L(\mathbf{v}_1 + \mathbf{v}_2)$ because L is a linear transformation. However, since $\mathcal{V}'$ is a subspace of $\mathcal{V}$, $(\mathbf{v}_1 + \mathbf{v}_2) \in \mathcal{V}'$. Thus, $(\mathbf{w}_1 + \mathbf{w}_2)$ is the image of $(\mathbf{v}_1 + \mathbf{v}_2) \in \mathcal{V}'$, and so $(\mathbf{w}_1 + \mathbf{w}_2) \in L\left(\mathcal{V}'\right)$. Hence, $L\left(\mathcal{V}'\right)$ is closed under addition.

Next, suppose that $c \in \mathbb{R}$ and $\mathbf{w} \in L\left(\mathcal{V}'\right)$. By definition of $L\left(\mathcal{V}'\right)$, $\mathbf{w} = L(\mathbf{v})$, for some $\mathbf{v} \in \mathcal{V}'$. Then, $c\mathbf{w} = cL(\mathbf{v}) = L(c\mathbf{v})$ since L is a linear transformation. Now, $c\mathbf{v} \in \mathcal{V}'$, because $\mathcal{V}'$ is a subspace of $\mathcal{V}$. Thus, $c\mathbf{w}$ is the image of $c\mathbf{v} \in \mathcal{V}'$, and so $c\mathbf{w} \in L\left(\mathcal{V}'\right)$. Hence, $L\left(\mathcal{V}'\right)$ is closed under scalar multiplication. ∎

Figure 5.5

Subspaces of $\mathcal{V}$ correspond to subspaces of $\mathcal{W}$ under a linear transformation $L: \mathcal{V} \to \mathcal{W}$.

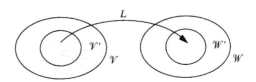

| EXAMPLE 12 | Let $L: \mathcal{M}_{22} \to \mathbb{R}^3$ where $L\left(\begin{bmatrix} a & b \\ c & d \end{bmatrix}\right) = [b, 0, c]$. L is a linear transformation |

(verify!). By Theorem 5.3, the range of any linear transformation is a subspace of the codomain. Hence, the range of $L = \{[b, 0, c] \mid b, c \in \mathbb{R}\}$ is a subspace of $\mathbb{R}^3$.

Also, consider the subspace $\mathcal{U}_2 = \left\{ \begin{bmatrix} a & b \\ 0 & d \end{bmatrix} \mid a, b, d \in \mathbb{R} \right\}$ of $\mathcal{M}_{22}$. Then the image of $\mathcal{U}_2$ under L is $\{[b, 0, 0] \mid b \in \mathbb{R}\}$. This image is a subspace of $\mathbb{R}^3$, as Theorem 5.3 asserts. Finally, consider the subspace $W = \{[b, e, 2b] \mid b, e \in \mathbb{R}\}$ of $\mathbb{R}^3$. The pre-image of W consists of all matrices in $\mathcal{M}_{22}$ of the form $\begin{bmatrix} a & b \\ 2b & d \end{bmatrix}$. Notice that this pre-image is a subspace of $\mathcal{M}_{22}$, as claimed by Theorem 5.3. ∎

Exercises for Section 5.1

1. Determine which of the following functions are linear transformations. Prove that your answers are correct. Which are linear operators?

   ⋆(a) $f: \mathbb{R}^2 \to \mathbb{R}^2$ given by $f([x, y]) = [3x - 4y, -x + 2y]$

   ⋆(b) $h: \mathbb{R}^4 \to \mathbb{R}^4$ given by $h([x_1, x_2, x_3, x_4]) = [x_1 + 2, x_2 - 1, x_3, -3]$

   (c) $k: \mathbb{R}^3 \to \mathbb{R}^3$ given by $k([x_1, x_2, x_3]) = [x_2, x_3, x_1]$

   ⋆(d) $l: \mathcal{M}_{22} \to \mathcal{M}_{22}$ given by $l\left(\begin{bmatrix} a & b \\ c & d \end{bmatrix}\right) = \begin{bmatrix} a - 2c + d & 3b - c \\ -4a & b + c - 3d \end{bmatrix}$

   (e) $n: \mathcal{M}_{22} \to \mathbb{R}$ given by $n\left(\begin{bmatrix} a & b \\ c & d \end{bmatrix}\right) = ad - bc$

   ⋆(f) $r: \mathcal{P}_3 \to \mathcal{P}_2$ given by $r(ax^3 + bx^2 + cx + d) = (\sqrt[3]{a})x^2 - b^2x + c$

   (g) $s: \mathbb{R}^3 \to \mathbb{R}^3$ given by $s([x_1, x_2, x_3]) = [\cos x_1, \sin x_2, e^{x_3}]$

   ⋆(h) $t: \mathcal{P}_3 \to \mathbb{R}$ given by $t(a_3x^3 + a_2x^2 + a_1x + a_0) = a_3 + a_2 + a_1 + a_0$

   (i) $u: \mathbb{R}^4 \to \mathbb{R}$ given by $u([x_1, x_2, x_3, x_4]) = |x_2|$

   ⋆(j) $v: \mathcal{P}_2 \to \mathbb{R}$ given by $v(ax^2 + bx + c) = abc$

   (k) $g: \mathcal{M}_{32} \to \mathcal{P}_4$ given by $g\left(\begin{bmatrix} a_{11} & a_{12} \\ a_{21} & a_{22} \\ a_{31} & a_{32} \end{bmatrix}\right) = a_{11}x^4 - a_{21}x^2 + a_{31}$

   ⋆(l) $e: \mathbb{R}^2 \to \mathbb{R}$ given by $e([x, y]) = \sqrt{x^2 + y^2}$

2. Let $\mathcal{V}$ and $\mathcal{W}$ be vector spaces.

   (a) Show that the identity mapping $i: \mathcal{V} \to \mathcal{V}$ given by $i(\mathbf{v}) = \mathbf{v}$, for all $\mathbf{v} \in \mathcal{V}$, is a linear operator.

   (b) Show that the zero mapping $z: \mathcal{V} \to \mathcal{W}$ given by $z(\mathbf{v}) = \mathbf{0}_{\mathcal{W}}$, for all $\mathbf{v} \in \mathcal{V}$, is a linear transformation.

3. Let k be a fixed scalar in $\mathbb{R}$. Show that the mapping $f: \mathbb{R}^n \to \mathbb{R}^n$ given by $f([x_1, x_2, \ldots, x_n]) = k[x_1, x_2, \ldots, x_n]$ is a linear operator.

4. (a) Show that $f: \mathbb{R}^3 \to \mathbb{R}^3$ given by $f([x, y, z]) = [-x, y, z]$ (reflection of a vector through the yz-plane) is a linear operator.

   (b) What mapping from $\mathbb{R}^3$ to $\mathbb{R}^3$ would reflect a vector through the xz-plane? Is it a linear operator? Why or why not?

(c) What mapping from $\mathbb{R}^2$ to $\mathbb{R}^2$ would reflect a vector through the y-axis? through the x-axis? Are these linear operators? Why or why not?

5. Show that the projection mappings $h: \mathbb{R}^3 \to \mathbb{R}^3$ given by $h([a_1, a_2, a_3]) = [a_1, a_2, 0]$ and $j: \mathbb{R}^4 \to \mathbb{R}^4$ given by $j([a_1, a_2, a_3, a_4]) = [0, a_2, 0, a_4]$ are linear operators.

6. The mapping $f: \mathbb{R}^n \to \mathbb{R}$ given by $f([x_1, x_2, \ldots, x_i, \ldots, x_n]) = x_i$ is another type of projection mapping. Show that f is a linear transformation.

7. Let $\mathbf{x}$ be a fixed nonzero vector in $\mathbb{R}^3$. Show that the mapping $g: \mathbb{R}^3 \to \mathbb{R}^3$ given by $g(\mathbf{y}) = \mathbf{proj_x y}$ is a linear operator.

8. Let $\mathbf{x}$ be a fixed vector in $\mathbb{R}^n$. Prove that $L: \mathbb{R}^n \to \mathbb{R}$ given by $L(\mathbf{y}) = \mathbf{x} \cdot \mathbf{y}$ is a linear transformation.

9. Let θ be a fixed angle in the xy-plane. Show that the linear operator $L: \mathbb{R}^2 \to \mathbb{R}^2$ given by $L\left(\begin{bmatrix} x \\ y \end{bmatrix}\right) = \begin{bmatrix} \cos\theta & -\sin\theta \\ \sin\theta & \cos\theta \end{bmatrix} \begin{bmatrix} x \\ y \end{bmatrix}$ rotates the vector $[x, y]$ counterclockwise through the angle θ in the plane. (Hint: Consider the vector $[x', y']$, obtained by rotating $[x, y]$ counterclockwise through the angle θ. Let $r = \sqrt{x^2 + y^2}$. Then $x = r\cos\alpha$ and $y = r\sin\alpha$, where α is the angle shown in Figure 5.6. Notice that $x' = r(\cos(\theta + \alpha))$ and $y' = r(\sin(\theta + \alpha))$. Then show that $L([x, y]) = [x', y']$.)

Figure 5.6

The vectors $[x, y]$ and $[x', y']$

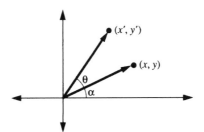

10. (a) Explain why the mapping $L: \mathbb{R}^3 \to \mathbb{R}^3$ given by

$$L\left(\begin{bmatrix} x \\ y \\ z \end{bmatrix}\right) = \begin{bmatrix} \cos\theta & -\sin\theta & 0 \\ \sin\theta & \cos\theta & 0 \\ 0 & 0 & 1 \end{bmatrix} \begin{bmatrix} x \\ y \\ z \end{bmatrix}$$

is a linear operator.

(b) Show that the mapping L in part (a) rotates every vector in $\mathbb{R}^3$ about the z-axis through an angle of θ (as measured relative to the xy-plane).

$\star$(c) What matrix should be multiplied times $[x, y, z]$ to create the linear operator that rotates $\mathbb{R}^3$ about the y-axis through an angle θ (relative to the xz-plane)?

11. Shears: Let $f_1, f_2: \mathbb{R}^2 \to \mathbb{R}^2$ be given by

$$f_1\left(\begin{bmatrix} x \\ y \end{bmatrix}\right) = \begin{bmatrix} 1 & k \\ 0 & 1 \end{bmatrix} \begin{bmatrix} x \\ y \end{bmatrix} = \begin{bmatrix} x + ky \\ y \end{bmatrix}$$

and

$$f_2\left(\begin{bmatrix} x \\ y \end{bmatrix}\right) = \begin{bmatrix} 1 & 0 \\ k & 1 \end{bmatrix}\begin{bmatrix} x \\ y \end{bmatrix} = \begin{bmatrix} x \\ kx + y \end{bmatrix}.$$

The mapping f_1 is called a **shear in the x-direction with factor** k; f_2 is called a **shear in the y-direction with factor** k. The effect of these functions (for $k > 1$) on the vector $[2, 3]$ is shown in Figure 5.7. Show that f_1 and f_2 are linear operators directly, without using Example 10.

Figure 5.7

(a) Shear in the x-direction; (b) Shear in the y-direction (both for $k > 0$)

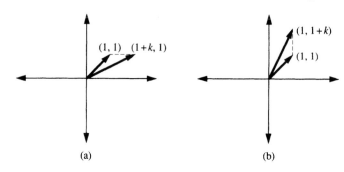

(a)   (b)

12. Let $f\colon \mathcal{M}_{nn} \to \mathbb{R}$ be given by $f(\mathbf{A}) = \text{trace}(\mathbf{A})$. (The trace is defined in Exercise 14 of Section 1.4.) Prove that f is a linear transformation.

13. Show that the mappings $g, h\colon \mathcal{M}_{nn} \to \mathcal{M}_{nn}$ given by $g(\mathbf{A}) = \mathbf{A} + \mathbf{A}^T$ and $h(\mathbf{A}) = \mathbf{A} - \mathbf{A}^T$ are linear operators on $\mathcal{M}_{nn}$.

14. (a) Show that if $\mathbf{p} \in \mathcal{P}_n$, then the (indefinite integral) function $f\colon \mathcal{P}_n \to \mathcal{P}_{n+1}$, where $f(\mathbf{p})$ is the vector $\int \mathbf{p}(x)\, dx$ with zero constant term, is a linear transformation.

   (b) Show that if $\mathbf{p} \in \mathcal{P}_n$, then the (definite integral) function $g\colon \mathcal{P}_n \to \mathbb{R}$ given by $g(\mathbf{p}) = \int_a^b \mathbf{p}\, dx$ is a linear transformation, for any fixed $a, b \in \mathbb{R}$.

15. Let $\mathcal{V}$ be the vector space of all functions f from $\mathbb{R}$ to $\mathbb{R}$ that are infinitely differentiable (that is, for which $f^{(n)}$, the nth derivative of f, exists for every integer $n \geq 1$). Use induction and Theorem 5.2 to show that for any given integer $k \geq 1$, $L\colon \mathcal{V} \to \mathcal{V}$ given by $L(f) = f^{(k)}$ is a linear operator.

16. Consider the function $f\colon \mathcal{M}_{nn} \to \mathcal{M}_{nn}$ given by $f(\mathbf{A}) = \mathbf{BA}$, where $\mathbf{B}$ is some fixed $n \times n$ matrix. Show that f is a linear operator.

17. Let $\mathbf{B}$ be a fixed nonsingular matrix in $\mathcal{M}_{nn}$. Show that the mapping $f\colon \mathcal{M}_{nn} \to \mathcal{M}_{nn}$ given by $f(\mathbf{A}) = \mathbf{B}^{-1}\mathbf{AB}$ is a linear operator.

18. Let a be a fixed real number.
   (a) Let $L\colon \mathcal{P}_n \to \mathbb{R}$ be given by $L(\mathbf{p}(x)) = \mathbf{p}(a)$. (That is, L evaluates polynomials in $\mathcal{P}_n$ at $x = a$.) Show that L is a linear transformation.
   (b) Let $L\colon \mathcal{P}_n \to \mathcal{P}_n$ be given by $L(\mathbf{p}(x)) = \mathbf{p}(x + a)$. (For example, when a is positive, L shifts the graph of $\mathbf{p}(x)$ to the *left* by a units.) Prove that L is a linear operator.

19. Let $\mathbf{A}$ be a fixed matrix in $\mathcal{M}_{nn}$. Define $f\colon \mathcal{P}_n \to \mathcal{M}_{nn}$ by

$$f(a_n x^n + a_{n-1} x^{n-1} + \cdots + a_1 x + a_0) = a_n \mathbf{A}^n + a_{n-1} \mathbf{A}^{n-1} + \cdots + a_1 \mathbf{A} + a_0 \mathbf{I}_n.$$

Show that f is a linear transformation.

20. Let $\mathcal{V}$ be the unusual vector space from Example 7 in Section 4.1. Show that $L: \mathcal{V} \to \mathbb{R}$ given by $L(x) = \ln(x)$ is a linear transformation.

21. Let $\mathcal{V}$ be a vector space, and let $\mathbf{x} \neq \mathbf{0}$ be a fixed vector in $\mathcal{V}$. Prove that the translation function $f: \mathcal{V} \to \mathcal{V}$ given by $f(\mathbf{v}) = \mathbf{v} + \mathbf{x}$ is not a linear transformation.

22. Show that if $\mathbf{A}$ is a fixed matrix in $\mathcal{M}_{mn}$ and $\mathbf{y} \neq \mathbf{0}$ is a fixed vector in $\mathbb{R}^m$, then the mapping $f: \mathbb{R}^n \to \mathbb{R}^m$ given by $f(\mathbf{x}) = \mathbf{Ax} + \mathbf{y}$ is not a linear transformation by showing that part (1) of Theorem 5.1 fails for f.

23. Prove that $f: \mathcal{M}_{33} \to \mathbb{R}$ given by $f(\mathbf{A}) = |\mathbf{A}|$ is not a linear transformation. (A similar result is true for $\mathcal{M}_{nn}$, for $n > 1$.)

24. Suppose $L_1: \mathcal{V} \to \mathcal{W}$ is a linear transformation and $L_2: \mathcal{V} \to \mathcal{W}$ is defined by $L_2(\mathbf{v}) = L_1(2\mathbf{v})$. Show that L_2 is a linear transformation.

25. Suppose $L: \mathbb{R}^3 \to \mathbb{R}^3$ is a linear operator and $L([1, 0, 0]) = [-2, 1, 0]$, $L([0, 1, 0]) = [3, -2, 1]$, and $L([0, 0, 1]) = [0, -1, 3]$. Find $L([-3, 2, 4])$. Give a formula for $L([x, y, z])$, for any $[x, y, z] \in \mathbb{R}^3$.

⋆26. Suppose $L: \mathbb{R}^2 \to \mathbb{R}^2$ is a linear operator and $L(\mathbf{i} + \mathbf{j}) = \mathbf{i} - 3\mathbf{j}$ and $L(-2\mathbf{i} + 3\mathbf{j}) = -4\mathbf{i} + 2\mathbf{j}$. Express $L(\mathbf{i})$ and $L(\mathbf{j})$ as linear combinations of $\mathbf{i}$ and $\mathbf{j}$.

27. Let $L: \mathcal{V} \to \mathcal{W}$ be a linear transformation. Show that $L(\mathbf{x} - \mathbf{y}) = L(\mathbf{x}) - L(\mathbf{y})$, for all vectors $\mathbf{x}, \mathbf{y} \in \mathcal{V}$.

28. Part (3) of Theorem 5.1 assures us that if $L: \mathcal{V} \to \mathcal{W}$ is a linear transformation, then $L(a\mathbf{v}_1 + b\mathbf{v}_2) = aL(\mathbf{v}_1) + bL(\mathbf{v}_2)$, for all $\mathbf{v}_1, \mathbf{v}_2 \in \mathcal{V}$ and all $a, b \in \mathbb{R}$. Prove that the converse of this statement is true. (Hint: Consider two cases: first $a = b = 1$ and then $b = 0$.)

29. Finish the proof of part (3) of Theorem 5.1 by doing the Inductive Step.

30. (a) Suppose that $L: \mathcal{V} \to \mathcal{W}$ is a linear transformation. Show that if $\{L(\mathbf{v}_1), L(\mathbf{v}_2), \ldots, L(\mathbf{v}_n)\}$ is a linearly independent set of n distinct vectors in $\mathcal{W}$, for some vectors $\mathbf{v}_1, \ldots, \mathbf{v}_n \in \mathcal{V}$, then $\{\mathbf{v}_1, \mathbf{v}_2, \ldots, \mathbf{v}_n\}$ is a linearly independent set in $\mathcal{V}$.

⋆(b) Find a counterexample to the converse of part (a).

▶31. Finish the proof of Theorem 5.3 by showing that if $L: \mathcal{V} \to \mathcal{W}$ is a linear transformation and $\mathcal{W}'$ is a subspace of $\mathcal{W}$ with pre-image $L^{-1}(\mathcal{W}')$, then $L^{-1}(\mathcal{W}')$ is a subspace of $\mathcal{V}$.

32. Show that every linear operator $L: \mathbb{R} \to \mathbb{R}$ has the form $L(\mathbf{x}) = c\mathbf{x}$, for some $c \in \mathbb{R}$.

33. Finish the proof of Theorem 5.2 by proving property (2) of a linear transformation for $L_2 \circ L_1$.

34. Let $L_1, L_2: \mathcal{V} \to \mathcal{W}$ be linear transformations. Define $(L_1 \oplus L_2): \mathcal{V} \to \mathcal{W}$ by $(L_1 \oplus L_2)(\mathbf{v}) = L_1(\mathbf{v}) + L_2(\mathbf{v})$ (where the latter addition takes place in $\mathcal{W}$). Also define $(c \odot L_1): \mathcal{V} \to \mathcal{W}$ by $(c \odot L_1)(\mathbf{v}) = c(L_1(\mathbf{v}))$ (where the latter scalar multiplication takes place in $\mathcal{W}$).

(a) Show that $(L_1 \oplus L_2)$ and $(c \odot L_1)$ are linear transformations.

(b) Use the results in part (a) above and part (b) of Exercise 2 to show that the set of all linear transformations from $\mathcal{V}$ to $\mathcal{W}$ is a vector space under the operations $\oplus$ and $\odot$.

35. Let $L: \mathbb{R}^2 \to \mathbb{R}^2$ be a nonzero linear operator. Show that L maps a line to either a line or a point.

★36. True or False:

(a) If $L: \mathcal{V} \to \mathcal{W}$ is a function between vector spaces for which $L(c\mathbf{v}) = cL(\mathbf{v})$, then L is a linear transformation.

(b) If $\mathcal{V}$ is an n-dimensional vector space with ordered basis B, then $L: \mathcal{V} \to \mathbb{R}^n$ given by $L(\mathbf{v}) = [\mathbf{v}]_B$ is a linear transformation.

(c) The function $L: \mathbb{R}^3 \to \mathbb{R}^3$ given by $L([x, y, z]) = [x+1, \ y-2, \ z+3]$ is a linear operator.

(d) If $\mathbf{A}$ is a 4×3 matrix, then $L(\mathbf{v}) = \mathbf{A}\mathbf{v}$ is a linear transformation from $\mathbb{R}^4$ to $\mathbb{R}^3$.

(e) A linear transformation from $\mathcal{V}$ to $\mathcal{W}$ always maps $\mathbf{0}_{\mathcal{V}}$ to $\mathbf{0}_{\mathcal{W}}$.

(f) If $M_1: \mathcal{V} \to \mathcal{W}$ and $M_2: \mathcal{W} \to \mathcal{X}$ are linear transformations, then $M_1 \circ M_2$ is a well-defined linear transformation.

(g) If $L: \mathcal{V} \to \mathcal{W}$ is a linear transformation, then the image of any subspace of $\mathcal{V}$ is a subspace of $\mathcal{W}$.

(h) If $L: \mathcal{V} \to \mathcal{W}$ is a linear transformation, then the pre-image of $\{\mathbf{0}_{\mathcal{W}}\}$ is a subspace of $\mathcal{V}$.

5.2 The Matrix of a Linear Transformation

In this section we show that the behavior of any linear transformation $L: \mathcal{V} \to \mathcal{W}$ is determined by its effect on a basis for $\mathcal{V}$. In particular, when $\mathcal{V}$ and $\mathcal{W}$ are finite dimensional and ordered bases for $\mathcal{V}$ and $\mathcal{W}$ are chosen, we can obtain a matrix corresponding to L that is useful in computing images under L. Finally, we investigate how the matrix for L changes as the bases for $\mathcal{V}$ and $\mathcal{W}$ change.

A Linear Transformation Is Determined by Its Action on a Basis

If the action of a linear transformation $L: \mathcal{V} \to \mathcal{W}$ on a basis for $\mathcal{V}$ is known, then the action of L can be computed for all elements of $\mathcal{V}$, as we see in the next example.

EXAMPLE 1 You can quickly verify that

$$B = ([0, 4, 0, 1], \ [-2, 5, 0, 2], \ [-3, 5, 1, 1], \ [-1, 2, 0, 1])$$

is an ordered basis for $\mathbb{R}^4$. Now suppose that $L: \mathbb{R}^4 \to \mathbb{R}^3$ is a linear transformation for which

$$L([0, 4, 0, 1]) \ = \ [3, 1, 2], \qquad L([-2, 5, 0, 2]) = [2, -1, 1],$$
$$L([-3, 5, 1, 1]) \ = \ [-4, 3, 0], \quad \text{and} \quad L([-1, 2, 0, 1]) = [6, 1, -1].$$

We can use the values of L on B to compute L for other vectors in $\mathbb{R}^4$. For example, let $\mathbf{v} = [-4, 14, 1, 4]$. By using row reduction, we see that $[\mathbf{v}]_B =$

$[2, -1, 1, 3]$ (verify!). So,

$$
\begin{aligned}
L(\mathbf{v}) &= L\left(2\,[0,4,0,1] - 1\,[-2,5,0,2] + 1\,[-3,5,1,1] + 3\,[-1,2,0,1]\right) \\
&= 2L\left([0,4,0,1]\right) - 1L\left([-2,5,0,2]\right) + 1L\left([-3,5,1,1]\right) \\
&\quad + 3L\left([-1,2,0,1]\right) \\
&= 2[3,1,2] - [2,-1,1] + [-4,3,0] + 3[6,1,-1] \\
&= [18,9,0].
\end{aligned}
$$

In general, if $\mathbf{v} \in \mathbb{R}^4$ and $[\mathbf{v}]_B = [k_1, k_2, k_3, k_4]$, then

$$
\begin{aligned}
L(\mathbf{v}) &= k_1[3,1,2] + k_2[2,-1,1] + k_3[-4,3,0] + k_4[6,1,-1] \\
&= [3k_1 + 2k_2 - 4k_3 + 6k_4,\ k_1 - k_2 + 3k_3 + k_4,\ 2k_1 + k_2 - k_4].
\end{aligned}
$$

Thus, we have derived a general formula for L from its effect on the basis B. ∎

Example 1 illustrates the next theorem.

THEOREM 5.4

Let $B = (\mathbf{v}_1, \mathbf{v}_2, \ldots, \mathbf{v}_n)$ be an ordered basis for a vector space $\mathcal{V}$. Let $\mathcal{W}$ be a vector space, and let $\mathbf{w}_1, \mathbf{w}_2, \ldots, \mathbf{w}_n$ be any n vectors in $\mathcal{W}$. Then there is a unique linear transformation $L\colon \mathcal{V} \to \mathcal{W}$ such that $L(\mathbf{v}_1) = \mathbf{w}_1$, $L(\mathbf{v}_2) = \mathbf{w}_2, \ldots, L(\mathbf{v}_n) = \mathbf{w}_n$.

PROOF OF THEOREM 5.4

(abridged) Let $B = (\mathbf{v}_1, \mathbf{v}_2, \ldots, \mathbf{v}_n)$ be an ordered basis for $\mathcal{V}$, and let $\mathbf{v} \in \mathcal{V}$. Then $\mathbf{v} = a_1\mathbf{v}_1 + \cdots + a_n\mathbf{v}_n$, for some unique a_i's in $\mathbb{R}$. Let $\mathbf{w}_1, \ldots, \mathbf{w}_n$ be any vectors in $\mathcal{W}$. Define $L\colon \mathcal{V} \to \mathcal{W}$ by $L(\mathbf{v}) = a_1\mathbf{w}_1 + a_2\mathbf{w}_2 + \cdots + a_n\mathbf{w}_n$. Notice that $L(\mathbf{v})$ is well defined since the a_i's are unique.

To show that L is a linear transformation, we must prove that $L(\mathbf{x}_1 + \mathbf{x}_2) = L(\mathbf{x}_1) + L(\mathbf{x}_2)$ and $L(c\mathbf{x}_1) = cL(\mathbf{x}_1)$, for all $\mathbf{x}_1, \mathbf{x}_2 \in \mathcal{V}$ and all $c \in \mathbb{R}$. Suppose that $\mathbf{x}_1 = d_1\mathbf{v}_1 + \cdots + d_n\mathbf{v}_n$ and $\mathbf{x}_2 = e_1\mathbf{v}_1 + \cdots + e_n\mathbf{v}_n$. Then, by definition of L, $L(\mathbf{x}_1) = d_1\mathbf{w}_1 + \cdots + d_n\mathbf{w}_n$ and $L(\mathbf{x}_2) = e_1\mathbf{w}_n + \cdots + e_n\mathbf{w}_n$. However,

$$
\begin{aligned}
\mathbf{x}_1 + \mathbf{x}_2 &= (d_1 + e_1)\mathbf{v}_1 + \cdots + (d_n + e_n)\mathbf{v}_n, \\
\text{so,} \quad L(\mathbf{x}_1 + \mathbf{x}_2) &= (d_1 + e_1)\mathbf{w}_1 + \cdots + (d_n + e_n)\mathbf{w}_n,
\end{aligned}
$$

again by definition of L. Hence, $L(\mathbf{x}_1) + L(\mathbf{x}_2) = L(\mathbf{x}_1 + \mathbf{x}_2)$.

Similarly, suppose $\mathbf{x} \in \mathcal{V}$, and $\mathbf{x} = t_1\mathbf{v}_1 + \cdots + t_n\mathbf{v}_n$. Then, $c\mathbf{x} = ct_1\mathbf{v}_1 + \cdots + ct_n\mathbf{v}_n$, and so $L(c\mathbf{x}) = ct_1\mathbf{w}_1 + \cdots + ct_n\mathbf{w}_n = cL(\mathbf{x})$. Hence, L is a linear transformation.

Finally, the proof of the uniqueness assertion is straightforward and is left as Exercise 30. ∎

The Matrix of a Linear Transformation

Our next goal is to show that every linear transformation on a finite dimensional vector space can be expressed as a matrix multiplication. This will allow us to solve problems involving linear transformations by performing matrix multiplications, which can easily be done by computer. As we will see, the matrix for a linear transformation is determined by the ordered bases B and C chosen for the domain and codomain, respectively. Our goal is to find a matrix that takes the B-coordinates of a vector in the domain to the C-coordinates of its image vector in the codomain.

Recall the linear transformation $L: \mathbb{R}^4 \to \mathbb{R}^3$ with the ordered basis B for $\mathbb{R}^4$ from Example 1. For $\mathbf{v} \in \mathbb{R}^4$, we let $[\mathbf{v}]_B = [k_1, k_2, k_3, k_4]$, and obtained the following formula for L:

$$L(\mathbf{v}) = [3k_1 + 2k_2 - 4k_3 + 6k_4, \ k_1 - k_2 + 3k_3 + k_4, \ 2k_1 + k_2 - k_4].$$

Now, to keep matters simple, we select the standard basis $C = (\mathbf{e}_1, \mathbf{e}_2, \mathbf{e}_3)$ for the codomain $\mathbb{R}^3$, so that the C-coordinates of vectors in the codomain are the same as the vectors themselves. (That is, $L(\mathbf{v}) = [L(\mathbf{v})]_C$, since C is the standard basis.) Then this formula for L takes the B-coordinates of each vector in the domain to the C-coordinates of its image vector in the codomain. Now, notice that if

$$\mathbf{A}_{BC} = \begin{bmatrix} 3 & 2 & -4 & 6 \\ 1 & -1 & 3 & 1 \\ 2 & 1 & 0 & -1 \end{bmatrix}, \quad \text{then } \mathbf{A}_{BC} \begin{bmatrix} k_1 \\ k_2 \\ k_3 \\ k_4 \end{bmatrix} = \begin{bmatrix} 3k_1 + 2k_2 - 4k_3 + 6k_4 \\ k_1 - k_2 + 3k_3 + k_4 \\ 2k_1 + k_2 - k_4 \end{bmatrix}.$$

Hence, the matrix $\mathbf{A}$ contains all of the information needed for carrying out the linear transformation L with respect to the chosen bases B and C.

A similar process can be used for any linear transformation between finite dimensional vector spaces.

THEOREM 5.5

Let $\mathcal{V}$ and $\mathcal{W}$ be nontrivial vector spaces, with $\dim(\mathcal{V}) = n$ and $\dim(\mathcal{W}) = m$. Let $B = (\mathbf{v}_1, \mathbf{v}_2, \ldots, \mathbf{v}_n)$ and $C = (\mathbf{w}_1, \mathbf{w}_2, \ldots, \mathbf{w}_m)$ be ordered bases for $\mathcal{V}$ and $\mathcal{W}$, respectively. Let $L: \mathcal{V} \to \mathcal{W}$ be a linear transformation. Then there is a unique $m \times n$ matrix $\mathbf{A}_{BC}$ such that $\mathbf{A}_{BC}[\mathbf{v}]_B = [L(\mathbf{v})]_C$, for all $\mathbf{v} \in \mathcal{V}$. (That is, $\mathbf{A}_{BC}$ times the coordinatization of $\mathbf{v}$ with respect to B gives the coordinatization of $L(\mathbf{v})$ with respect to C.)

Furthermore, for $1 \le i \le n$, the ith column of $\mathbf{A}_{BC} = [L(\mathbf{v}_i)]_C$.

Theorem 5.5 asserts that once ordered bases for $\mathcal{V}$ and $\mathcal{W}$ have been selected, *each linear transformation $L: \mathcal{V} \to \mathcal{W}$ is equivalent to multiplication by a unique corresponding matrix*. The matrix $\mathbf{A}_{BC}$ in this theorem is known as the **matrix of the linear transformation L with respect to the ordered bases B (for $\mathcal{V}$) and C (for $\mathcal{W}$)**. Theorem 5.5 also says that the matrix $\mathbf{A}_{BC}$ is computed as follows: find the image of each domain basis element $\mathbf{v}_i$ in turn, and then express these images in C-coordinates to get the respective columns of $\mathbf{A}_{BC}$.

The subscripts B and C on $\mathbf{A}$ are sometimes omitted when the bases being used are clear from context. Beware! If different ordered bases are chosen for $\mathcal{V}$ or $\mathcal{W}$, the matrix for the linear transformation will probably change.

PROOF OF THEOREM 5.5

Consider the $m \times n$ matrix $\mathbf{A}_{BC}$ whose ith column equals $[L(\mathbf{v}_i)]_C$, for $1 \leq i \leq n$. Let $\mathbf{v} \in \mathcal{V}$. We first prove that $\mathbf{A}_{BC}[\mathbf{v}]_B = [L(\mathbf{v})]_C$.

Suppose that $[\mathbf{v}]_B = [k_1, k_2, \ldots, k_n]$. Then $\mathbf{v} = k_1\mathbf{v}_1 + k_2\mathbf{v}_2 + \cdots + k_n\mathbf{v}_n$, and $L(\mathbf{v}) = k_1 L(\mathbf{v}_1) + k_2 L(\mathbf{v}_2) + \cdots + k_n L(\mathbf{v}_n)$, by Theorem 5.1. Hence,

$$
\begin{aligned}
[L(\mathbf{v})]_C &= [k_1 L(\mathbf{v}_1) + k_2 L(\mathbf{v}_2) + \cdots + k_n L(\mathbf{v}_n)]_C \\
&= k_1 [L(\mathbf{v}_1)]_C + k_2 [L(\mathbf{v}_2)]_C + \cdots + k_n [L(\mathbf{v}_n)]_C \quad \text{by Theorem 4.20} \\
&= k_1(\text{1st column of } \mathbf{A}_{BC}) + k_2(\text{2nd column of } \mathbf{A}_{BC}) \\
&\quad + \cdots + k_n(\text{nth column of } \mathbf{A}_{BC}) \\
&= \mathbf{A}_{BC} \begin{bmatrix} k_1 \\ k_2 \\ \vdots \\ k_n \end{bmatrix} = \mathbf{A}_{BC}[\mathbf{v}]_B.
\end{aligned}
$$

To complete the proof, we need to establish the uniqueness of $\mathbf{A}_{BC}$. Suppose that $\mathbf{H}$ is an $m \times n$ matrix such that $\mathbf{H}[\mathbf{v}]_B = [L(\mathbf{v})]_C$ for all $\mathbf{v} \in \mathcal{V}$. We will show that $\mathbf{H} = \mathbf{A}_{BC}$. It is enough to show that the ith column of $\mathbf{H}$ equals the ith column of $\mathbf{A}_{BC}$, for $1 \leq i \leq n$. Consider the ith vector, $\mathbf{v}_i$, of the ordered basis B for $\mathcal{V}$. Since $[\mathbf{v}_i]_B = \mathbf{e}_i$, we have: ith column of $\mathbf{H} = \mathbf{H}\mathbf{e}_i = \mathbf{H}[\mathbf{v}_i]_B = [L(\mathbf{v}_i)]_C$, and this is the ith column of $\mathbf{A}_{BC}$. ∎

Notice that in the special case where the codomain $\mathcal{W}$ is $\mathbb{R}^m$, and the basis C for $\mathcal{W}$ is the standard basis, Theorem 5.5 asserts that the ith column of $\mathbf{A}_{BC}$ is simply $L(\mathbf{v}_i)$ itself (why?).

EXAMPLE 2

Table 5.1 lists the matrices corresponding to some geometric linear operators on $\mathbb{R}^3$, with respect to the standard basis. The columns of each matrix are quickly calculated using Theorem 5.5, since we simply find the images $L(\mathbf{e}_1)$, $L(\mathbf{e}_2)$, and $L(\mathbf{e}_3)$ of the domain basis elements $\mathbf{e}_1$, $\mathbf{e}_2$, and $\mathbf{e}_3$. (Each image is equal to its coordinatization in the codomain since we are using the standard basis for the codomain as well.)

Once the matrix for each transformation is calculated, we can easily find the image of any vector using matrix multiplication. For example, to find the effect of the reflection L_1 in Table 5.1 on the vector $[3, -4, 2]$, we simply multiply by the matrix for L_1 to get

$$
\begin{bmatrix} 1 & 0 & 0 \\ 0 & 1 & 0 \\ 0 & 0 & -1 \end{bmatrix} \begin{bmatrix} 3 \\ -4 \\ 2 \end{bmatrix} = \begin{bmatrix} 3 \\ -4 \\ -2 \end{bmatrix}.
$$
■

EXAMPLE 3

We will find the matrix for the linear transformation $L: \mathcal{P}_3 \to \mathbb{R}^3$ given by

$$L(a_3 x^3 + a_2 x^2 + a_1 x + a_0) = [a_0 + a_1, 2a_2, a_3 - a_0]$$

with respect to the standard ordered bases $B = (x^3, x^2, x, 1)$ for $\mathcal{P}_3$ and $C = (\mathbf{e}_1, \mathbf{e}_2, \mathbf{e}_3)$ for $\mathbb{R}^3$. We first need to find $L(\mathbf{v})$, for each $\mathbf{v} \in B$. By definition of

Table 5.1

Matrices for several geometric linear operators on $\mathbb{R}^3$

Transformation	Formula	Matrix
Reflection (through xy-plane)	$L_1\left(\begin{bmatrix} a_1 \\ a_2 \\ a_3 \end{bmatrix}\right) = \begin{bmatrix} a_1 \\ a_2 \\ -a_3 \end{bmatrix}$	$\begin{array}{ccc} L_1(\mathbf{e}_1) & L_1(\mathbf{e}_2) & L_1(\mathbf{e}_3) \end{array}$ $\begin{bmatrix} 1 & 0 & 0 \\ 0 & 1 & 0 \\ 0 & 0 & -1 \end{bmatrix}$
Contraction or dilation	$L_2\left(\begin{bmatrix} a_1 \\ a_2 \\ a_3 \end{bmatrix}\right) = \begin{bmatrix} ca_1 \\ ca_2 \\ ca_3 \end{bmatrix}$, for $c \in \mathbb{R}$	$\begin{array}{ccc} L_2(\mathbf{e}_1) & L_2(\mathbf{e}_2) & L_2(\mathbf{e}_3) \end{array}$ $\begin{bmatrix} c & 0 & 0 \\ 0 & c & 0 \\ 0 & 0 & c \end{bmatrix}$
Projection (onto xy-plane)	$L_3\left(\begin{bmatrix} a_1 \\ a_2 \\ a_3 \end{bmatrix}\right) = \begin{bmatrix} a_1 \\ a_2 \\ 0 \end{bmatrix}$	$\begin{array}{ccc} L_3(\mathbf{e}_1) & L_3(\mathbf{e}_2) & L_3(\mathbf{e}_3) \end{array}$ $\begin{bmatrix} 1 & 0 & 0 \\ 0 & 1 & 0 \\ 0 & 0 & 0 \end{bmatrix}$
Rotation (about z-axis through angle θ relative to the xy-plane)	$L_4\left(\begin{bmatrix} a_1 \\ a_2 \\ a_3 \end{bmatrix}\right) = \begin{bmatrix} a_1\cos\theta - a_2\sin\theta \\ a_1\sin\theta + a_2\cos\theta \\ a_3 \end{bmatrix}$	$\begin{array}{ccc} L_4(\mathbf{e}_1) & L_4(\mathbf{e}_2) & L_4(\mathbf{e}_3) \end{array}$ $\begin{bmatrix} \cos\theta & -\sin\theta & 0 \\ \sin\theta & \cos\theta & 0 \\ 0 & 0 & 1 \end{bmatrix}$
Shear (in the z-direction with factor k) (analog of Exercise 11 in Section 5.1)	$L_5\left(\begin{bmatrix} a_1 \\ a_2 \\ a_3 \end{bmatrix}\right) = \begin{bmatrix} a_1 + ka_3 \\ a_2 + ka_3 \\ a_3 \end{bmatrix}$	$\begin{array}{ccc} L_5(\mathbf{e}_1) & L_5(\mathbf{e}_2) & L_5(\mathbf{e}_3) \end{array}$ $\begin{bmatrix} 1 & 0 & k \\ 0 & 1 & k \\ 0 & 0 & 1 \end{bmatrix}$

L, we have

$$L(x^3) = [0, 0, 1], \quad L(x^2) = [0, 2, 0], \quad L(x) = [1, 0, 0], \quad \text{and} \quad L(1) = [1, 0, -1].$$

Since we are using the standard basis C for $\mathbb{R}^3$, each of these images in $\mathbb{R}^3$ is its own C-coordinatization. Then by Theorem 5.5, the matrix $\mathbf{A}_{BC}$ for L is the matrix whose columns are these images; that is,

$$\mathbf{A}_{BC} = \begin{array}{cccc} L(x^3) & L(x^2) & L(x) & L(1) \end{array} \\ \begin{bmatrix} 0 & 0 & 1 & 1 \\ 0 & 2 & 0 & 0 \\ 1 & 0 & 0 & -1 \end{bmatrix}.$$

We will compute $L(5x^3 - x^2 + 3x + 2)$ using this matrix. Now, $[5x^3 - x^2 + 3x + 2]_B = [5, -1, 3, 2]$. Hence, multiplication by $\mathbf{A}_{BC}$ gives

$$\left[L(5x^3 - x^2 + 3x + 2)\right]_C = \begin{bmatrix} 0 & 0 & 1 & 1 \\ 0 & 2 & 0 & 0 \\ 1 & 0 & 0 & -1 \end{bmatrix} \begin{bmatrix} 5 \\ -1 \\ 3 \\ 2 \end{bmatrix} = \begin{bmatrix} 5 \\ -2 \\ 3 \end{bmatrix}.$$

Since C is the standard basis for $\mathbb{R}^3$, we have $L(5x^3 - x^2 + 3x + 2) = [5, -2, 3]$, which can be quickly verified to be the correct answer. ∎

EXAMPLE 4 We will find the matrix for the same linear transformation $L: \mathcal{P}_3 \to \mathbb{R}^3$ of Example 3 with respect to the different ordered bases

$$D = (x^3 + x^2,\ x^2 + x,\ x + 1,\ 1)$$
$$\text{and} \quad E = ([-2, 1, -3], [1, -3, 0], [3, -6, 2]).$$

You should verify that D and E are bases for $\mathcal{P}_3$ and $\mathbb{R}^3$, respectively.

We first need to find $L(\mathbf{v})$, for each $\mathbf{v} \in D$. By definition of L, we have $L(x^3 + x^2) = [0, 2, 1]$, $L(x^2 + x) = [1, 2, 0]$, $L(x + 1) = [2, 0, -1]$, and $L(1) = [1, 0, -1]$. Now we must find the coordinatization of each of these images in terms of the basis E for $\mathbb{R}^3$. Since we must solve for the coordinates of many vectors, it is quicker to use the transition matrix $\mathbf{Q}$ from the standard basis C for $\mathbb{R}^3$ to the basis E. From Theorem 4.23, $\mathbf{Q}$ is the inverse of the matrix whose columns are the vectors in E; that is,

$$\mathbf{Q} = \begin{bmatrix} -2 & 1 & 3 \\ 1 & -3 & -6 \\ -3 & 0 & 2 \end{bmatrix}^{-1} = \begin{bmatrix} -6 & -2 & 3 \\ 16 & 5 & -9 \\ -9 & -3 & 5 \end{bmatrix}.$$

Now, multiplying $\mathbf{Q}$ by each of the images, we get

$$\left[L(x^3 + x^2) \right]_E = \mathbf{Q} \begin{bmatrix} 0 \\ 2 \\ 1 \end{bmatrix} = \begin{bmatrix} -1 \\ 1 \\ -1 \end{bmatrix}, \quad \left[L(x^2 + x) \right]_E = \mathbf{Q} \begin{bmatrix} 1 \\ 2 \\ 0 \end{bmatrix} = \begin{bmatrix} -10 \\ 26 \\ -15 \end{bmatrix},$$

$$[L(x + 1)]_E = \mathbf{Q} \begin{bmatrix} 2 \\ 0 \\ -1 \end{bmatrix} = \begin{bmatrix} -15 \\ 41 \\ -23 \end{bmatrix}, \quad \text{and} \quad [L(1)]_E = \mathbf{Q} \begin{bmatrix} 1 \\ 0 \\ -1 \end{bmatrix} = \begin{bmatrix} -9 \\ 25 \\ -14 \end{bmatrix}.$$

By Theorem 5.5, the matrix $\mathbf{A}_{DE}$ for L is the matrix whose columns are these products.

$$\mathbf{A}_{DE} = \begin{bmatrix} -1 & -10 & -15 & -9 \\ 1 & 26 & 41 & 25 \\ -1 & -15 & -23 & -14 \end{bmatrix}$$

We will compute $L(5x^3 - x^2 + 3x + 2)$ using this matrix. We must first find the representation for $5x^3 - x^2 + 3x + 2$ in terms of the basis D. Solving $5x^3 - x^2 + 3x + 2 = a(x^3 + x^2) + b(x^2 + x) + c(x + 1) + d(1)$ for a, b, c, and d, we get the unique solution $a = 5$, $b = -6$, $c = 9$, and $d = -7$ (verify!). Hence, $\left[5x^3 - x^2 + 3x + 2\right]_D = [5, -6, 9, -7]$. Then

$$\left[L(5x^3 - x^2 + 3x + 2) \right]_E = \begin{bmatrix} -1 & -10 & -15 & -9 \\ 1 & 26 & 41 & 25 \\ -1 & -15 & -23 & -14 \end{bmatrix} \begin{bmatrix} 5 \\ -6 \\ 9 \\ -7 \end{bmatrix} = \begin{bmatrix} -17 \\ 43 \\ -24 \end{bmatrix}.$$

This answer represents a coordinate vector in terms of the basis E, and so

$$L(5x^3 - x^2 + 3x + 2) = -17 \begin{bmatrix} -2 \\ 1 \\ -3 \end{bmatrix} + 43 \begin{bmatrix} 1 \\ -3 \\ 0 \end{bmatrix} - 24 \begin{bmatrix} 3 \\ -6 \\ 2 \end{bmatrix} = \begin{bmatrix} 5 \\ -2 \\ 3 \end{bmatrix},$$

which agrees with the answer in Example 3. ∎

Finding the New Matrix for a Linear Transformation after a Change of Basis

The next theorem indicates precisely how the matrix for a linear transformation changes when we alter the bases for the domain and codomain.

THEOREM 5.6

Let V and W be two nontrivial finite dimensional vector spaces with ordered bases B and C, respectively. Let $L: V \to W$ be a linear transformation with matrix $\mathbf{A}_{BC}$ in terms of bases B and C. Suppose that D and E are other ordered bases for V and W, respectively. Let $\mathbf{P}$ be the transition matrix from B to D, and let $\mathbf{Q}$ be the transition matrix from C to E. Then the matrix $\mathbf{A}_{DE}$ for L in terms of bases D and E is given by $\mathbf{A}_{DE} = \mathbf{Q}\mathbf{A}_{BC}\mathbf{P}^{-1}$.

The situation in Theorem 5.6 is summarized in Figure 5.8.

Figure 5.8

Relationship between matrices $\mathbf{A}_{BC}$ and $\mathbf{A}_{DE}$ for a linear transformation under a change of basis

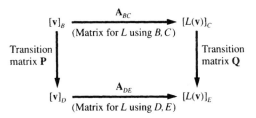

PROOF OF THEOREM 5.6

For all $\mathbf{v} \in V$,

$$\mathbf{A}_{BC}[\mathbf{v}]_B = [L(\mathbf{v})]_C \qquad \text{by Theorem 5.5}$$
$$\Rightarrow \quad \mathbf{Q}\mathbf{A}_{BC}[\mathbf{v}]_B = \mathbf{Q}[L(\mathbf{v})]_C$$
$$\Rightarrow \quad \mathbf{Q}\mathbf{A}_{BC}[\mathbf{v}]_B = [L(\mathbf{v})]_E \qquad \begin{array}{l}\text{because } \mathbf{Q} \text{ is the transition} \\ \text{matrix from } C \text{ to } E\end{array}$$
$$\Rightarrow \quad \mathbf{Q}\mathbf{A}_{BC}\mathbf{P}^{-1}[\mathbf{v}]_D = [L(\mathbf{v})]_E \qquad \begin{array}{l}\text{because } \mathbf{P}^{-1} \text{ is the transition} \\ \text{matrix from } D \text{ to } B\end{array}$$

However, $\mathbf{A}_{DE}$ is the *unique* matrix such that $\mathbf{A}_{DE}[\mathbf{v}]_D = [L(\mathbf{v})]_E$, for all $\mathbf{v} \in V$. Hence, $\mathbf{A}_{DE} = \mathbf{Q}\mathbf{A}_{BC}\mathbf{P}^{-1}$. ∎

Theorem 5.6 gives us an alternate method for finding the matrix of a linear transformation with respect to one pair of bases when the matrix for another pair of bases is known.

EXAMPLE 5   Recall the linear transformation $L: \mathcal{P}_3 \to \mathbb{R}^3$ from Examples 3 and 4, given by

$$L(a_3x^3 + a_2x^2 + a_1x + a_0) = [a_0 + a_1, 2a_2, a_3 - a_0].$$

Example 3 shows that the matrix for L using the standard bases B (for $\mathcal{P}_3$) and C (for $\mathbb{R}^3$) is

$$\mathbf{A}_{BC} = \begin{bmatrix} 0 & 0 & 1 & 1 \\ 0 & 2 & 0 & 0 \\ 1 & 0 & 0 & -1 \end{bmatrix}.$$

Also, in Example 4, we computed directly to find the matrix $\mathbf{A}_{DE}$ for the ordered bases $D = (x^3 + x^2, x^2 + x, x + 1, 1)$ for $\mathcal{P}_3$ and $E = ([-2, 1, -3], [1, -3, 0], [3, -6, 2])$ for $\mathbb{R}^3$. Instead, we now use Theorem 5.6 to calculate $\mathbf{A}_{DE}$. Recall from Example 4 that the transition matrix $\mathbf{Q}$ from bases C to E is

$$\mathbf{Q} = \begin{bmatrix} -6 & -2 & 3 \\ 16 & 5 & -9 \\ -9 & -3 & 5 \end{bmatrix}.$$

Also, the transition matrix $\mathbf{P}^{-1}$ from bases D to B is

$$\mathbf{P}^{-1} = \begin{bmatrix} 1 & 0 & 0 & 0 \\ 1 & 1 & 0 & 0 \\ 0 & 1 & 1 & 0 \\ 0 & 0 & 1 & 1 \end{bmatrix}. \quad \text{(Verify!)}$$

Hence,

$$\mathbf{A}_{DE} = \mathbf{Q}\mathbf{A}_{BC}\mathbf{P}^{-1} = \begin{bmatrix} -6 & -2 & 3 \\ 16 & 5 & -9 \\ -9 & -3 & 5 \end{bmatrix} \begin{bmatrix} 0 & 0 & 1 & 1 \\ 0 & 2 & 0 & 0 \\ 1 & 0 & 0 & -1 \end{bmatrix} \begin{bmatrix} 1 & 0 & 0 & 0 \\ 1 & 1 & 0 & 0 \\ 0 & 1 & 1 & 0 \\ 0 & 0 & 1 & 1 \end{bmatrix}$$

$$= \begin{bmatrix} -1 & -10 & -15 & -9 \\ 1 & 26 & 41 & 25 \\ -1 & -15 & -23 & -14 \end{bmatrix},$$

which agrees with the result obtained for $\mathbf{A}_{DE}$ in Example 4. ∎

Similar Matrices

Suppose L is a linear operator on a finite dimensional vector space $\mathcal{V}$. If B is a basis for $\mathcal{V}$, then there is some matrix $\mathbf{A}_{BB}$ for L with respect to B. Also, if C is another basis for $\mathcal{V}$, then there is some matrix $\mathbf{A}_{CC}$ for L with respect to C. Let $\mathbf{P}$ be the transition matrix from B to C (see Figure 5.9). Notice that by Theorem 5.6 we have $\mathbf{A}_{CC} = \mathbf{P}\mathbf{A}_{BB}\mathbf{P}^{-1}$, which leads us to the following definition:

Figure 5.9

Relationship between matrices $\mathbf{A}_{BB}$ and $\mathbf{A}_{CC}$ for a linear operator under a change of basis

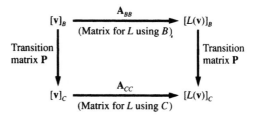

DEFINITION

Let $\mathbf{X}$ and $\mathbf{Y}$ be $n \times n$ matrices. Then $\mathbf{X}$ is **similar** to $\mathbf{Y}$ if and only if $\mathbf{Y} = \mathbf{P}\mathbf{X}\mathbf{P}^{-1}$, for some nonsingular $n \times n$ matrix $\mathbf{P}$.

Notice that if $\mathbf{X}$ is similar to $\mathbf{Y}$, then $\mathbf{Y} = \mathbf{PXP}^{-1}$, and so $\mathbf{X} = \mathbf{P}^{-1}\mathbf{YP} = \mathbf{P}^{-1}\mathbf{Y}\left(\mathbf{P}^{-1}\right)^{-1}$, which means that $\mathbf{Y}$ is also similar to $\mathbf{X}$. Thus, we usually speak of such matrices $\mathbf{X}$ and $\mathbf{Y}$ as being similar to each other.

The discussion before the definition shows that if two matrices $\mathbf{A}_{BB}$ and $\mathbf{A}_{CC}$ represent the same linear operator, but with respect to two different bases, then $\mathbf{A}_{BB}$ is similar to $\mathbf{A}_{CC}$. In fact, the converse is also true (see Exercise 25).

EXAMPLE 6

Consider the linear operator $L: \mathbb{R}^3 \to \mathbb{R}^3$ whose matrix with respect to the standard basis B for $\mathbb{R}^3$ is

$$\mathbf{A}_{BB} = \frac{1}{7}\begin{bmatrix} 6 & 3 & -2 \\ 3 & -2 & 6 \\ -2 & 6 & 3 \end{bmatrix}.$$

We will use eigenvectors to find another basis D for $\mathbb{R}^3$ so that with respect to D, L has a much simpler matrix representation. Now, $p_{\mathbf{A}_{BB}}(x) = |x\mathbf{I}_3 - \mathbf{A}_{BB}| = x^3 - x^2 - x + 1 = (x-1)^2(x+1)$ (verify!).

By row reducing $(1\mathbf{I}_3 - \mathbf{A}_{BB})$ and $(-1\mathbf{I}_3 - \mathbf{A}_{BB})$ we find the basis $\{[3, 1, 0], [-2, 0, 1]\}$ for the eigenspace E_1 for $\mathbf{A}_{BB}$ and the basis $\{[1, -3, 2]\}$ for the eigenspace E_{-1} for $\mathbf{A}_{BB}$. (Again, verify!) A quick check verifies that $D = \{[3, 1, 0], [-2, 0, 1], [1, -3, 2]\}$ is a basis for $\mathbb{R}^3$ consisting of eigenvectors for $\mathbf{A}_{BB}$.

Next, recall that $\mathbf{A}_{DD}$ is similar to $\mathbf{A}_{BB}$. In particular, $\mathbf{A}_{DD} = \mathbf{PA}_{BB}\mathbf{P}^{-1}$, where $\mathbf{P}$ is the transition matrix from B to D. Now, the matrix whose columns are the vectors in D is the transition matrix from D to the standard basis B. This matrix plays the role of $\mathbf{P}^{-1}$, and so we let

$$\mathbf{P}^{-1} = \begin{bmatrix} 3 & -2 & 1 \\ 1 & 0 & -3 \\ 0 & 1 & 2 \end{bmatrix}, \quad \text{and obtain} \quad \mathbf{P} = \frac{1}{14}\begin{bmatrix} 3 & 5 & 6 \\ -2 & 6 & 10 \\ 1 & -3 & 2 \end{bmatrix}$$

for the transition matrix from B to D. Then,

$$\mathbf{A}_{DD} = \mathbf{PA}_{BB}\mathbf{P}^{-1} = \begin{bmatrix} 1 & 0 & 0 \\ 0 & 1 & 0 \\ 0 & 0 & -1 \end{bmatrix},$$

a diagonal matrix with the eigenvalues 1 and -1 on the main diagonal. (Compare this with Theorem 3.14, where the roles of $\mathbf{P}$ and $\mathbf{P}^{-1}$ are reversed.)

Written in this form, the operator L is more comprehensible. Compare $\mathbf{A}_{DD}$ to the matrix for a reflection through the xy-plane given in Table 5.1. Now, because D is not the standard basis for $\mathbb{R}^3$, L is *not* a reflection through the xy-plane. But we can show that L is a reflection of all vectors in $\mathbb{R}^3$ through the plane formed by the two basis vectors for E_1 (that is, the plane is the eigenspace E_1 itself). By the uniqueness assertion in Theorem 5.4, it is enough to show that L acts as a reflection through the plane E_1 for each of the three basis vectors of D.

Since $[3, 1, 0]$ and $[-2, 0, 1]$ are in the plane E_1, we need to show that L "reflects" these vectors to themselves. But this is true since $L([3, 1, 0]) = 1[3, 1, 0] = [3, 1, 0]$, and similarly for $[-2, 0, 1]$. Finally, notice that $[1, -3, 2]$ is orthogonal to the plane E_1 (since it is orthogonal to both $[3, 1, 0]$ and $[-2, 0, 1]$). Therefore, we need to show that L "reflects" this vector to its opposite. But, $L([1, -3, 2]) = -1[1, -3, 2] = -[1, -3, 2]$, and we are done. Hence, L is a reflection through the plane E_1. ∎

Because the matrix $\mathbf{A}_{DD}$ in Example 6 is diagonal, it follows that $p_{\mathbf{A}_{DD}}(x) = (x - 1)^2(x + 1)$, which equals $p_{\mathbf{A}_{BB}}(x)$. This illustrates the following theorem, which is left as Exercise 21:

THEOREM 5.7

If $\mathbf{A}$ and $\mathbf{B}$ are similar $n \times n$ matrices, then $p_{\mathbf{A}}(x) = p_{\mathbf{B}}(x)$.

Matrix for the Composition of Linear Transformations

Our final theorem for this section shows how to find the corresponding matrix for the composition of linear transformations. The proof is left as Exercise 15.

THEOREM 5.8

Let $\mathcal{V}_1, \mathcal{V}_2,$ and $\mathcal{V}_3$ be nontrivial finite dimensional vector spaces with ordered bases $B, C,$ and D, respectively. Let $L_1: \mathcal{V}_1 \to \mathcal{V}_2$ be a linear transformation with matrix $\mathbf{A}_{BC}$ with respect to bases B and C, and let $L_2: \mathcal{V}_2 \to \mathcal{V}_3$ be a linear transformation with matrix $\mathbf{A}_{CD}$ with respect to bases C and D. Then the matrix for the composite linear transformation $L_2 \circ L_1: \mathcal{V}_1 \to \mathcal{V}_3$ with respect to bases B and D is the product $\mathbf{A}_{CD}\mathbf{A}_{BC}$.

Theorem 5.8 can be generalized to compositions of several linear transformations, as in the next example.

EXAMPLE 7

Let $L_1, L_2, \ldots, L_5$ be the geometric linear operators on $\mathbb{R}^3$ given in Table 5.1. Let $\mathbf{A}_1, \ldots, \mathbf{A}_5$ be the matrices for these operators using the standard basis for $\mathbb{R}^3$. Then, the matrix for the composition $L_4 \circ L_5$ is

$$\mathbf{A}_4\mathbf{A}_5 = \begin{bmatrix} \cos\theta & -\sin\theta & 0 \\ \sin\theta & \cos\theta & 0 \\ 0 & 0 & 1 \end{bmatrix} \begin{bmatrix} 1 & 0 & k \\ 0 & 1 & k \\ 0 & 0 & 1 \end{bmatrix} = \begin{bmatrix} \cos\theta & -\sin\theta & k\cos\theta - k\sin\theta \\ \sin\theta & \cos\theta & k\sin\theta + k\cos\theta \\ 0 & 0 & 1 \end{bmatrix}.$$

Similarly, the matrix for the composition $L_2 \circ L_3 \circ L_1 \circ L_5$ is

$$\mathbf{A}_2\mathbf{A}_3\mathbf{A}_1\mathbf{A}_5 = \begin{bmatrix} c & 0 & 0 \\ 0 & c & 0 \\ 0 & 0 & c \end{bmatrix} \begin{bmatrix} 1 & 0 & 0 \\ 0 & 1 & 0 \\ 0 & 0 & 0 \end{bmatrix} \begin{bmatrix} 1 & 0 & 0 \\ 0 & 1 & 0 \\ 0 & 0 & -1 \end{bmatrix} \begin{bmatrix} 1 & 0 & k \\ 0 & 1 & k \\ 0 & 0 & 1 \end{bmatrix} = \begin{bmatrix} c & 0 & kc \\ 0 & c & kc \\ 0 & 0 & 0 \end{bmatrix}.$$

∎

♦ **Supplemental Material:** You have now covered the prerequisites for Section 7.3, "Complex Vector Spaces."

♦ **Application**: You have now covered the prerequisites for Section 8.8, "Computer Graphics."

Exercises for Section 5.2

1. Verify that the correct matrix is given for each of the geometric linear operators in Table 5.1.

2. For each of the following linear transformations $L: V \rightarrow W$, find the matrix for L with respect to the standard bases for V and W.

   ⋆(a) $L: \mathbb{R}^3 \rightarrow \mathbb{R}^3$ given by $L([x, y, z]) = [-6x + 4y - z, -2x + 3y - 5z, 3x - y + 7z]$

   (b) $L: \mathbb{R}^4 \rightarrow \mathbb{R}^2$ given by $L([x, y, z, w]) = [3x - 5y + z - 2w, 5x + y - 2z + 8w]$

   ⋆(c) $L: \mathcal{P}_3 \rightarrow \mathbb{R}^3$ given by $L(ax^3 + bx^2 + cx + d) = [4a - b + 3c + 3d, a + 3b - c + 5d, -2a - 7b + 5c - d]$

   (d) $L: \mathcal{P}_3 \rightarrow \mathcal{M}_{22}$ given by

$$L(ax^3 + bx^2 + cx + d) = \begin{bmatrix} -3a - 2c & -b + 4d \\ 4b - c + 3d & -6a - b + 2d \end{bmatrix}$$

3. For each of the following linear transformations $L: V \rightarrow W$, find the matrix $\mathbf{A}_{BC}$ for L with respect to the given bases B for V and C for W using the method of Theorem 5.5:

   ⋆(a) $L: \mathbb{R}^3 \rightarrow \mathbb{R}^2$ given by $L([x, y, z]) = [-2x + 3z, x + 2y - z]$ with $B = ([1, -3, 2], [-4, 13, -3], [2, -3, 20])$ and $C = ([-2, -1], [5, 3])$

   (b) $L: \mathbb{R}^2 \rightarrow \mathbb{R}^3$ given by $L([x, y]) = [13x - 9y, -x - 2y, -11x + 6y]$ with $B = ([2, 3], [-3, -4])$ and $C = ([-1, 2, 2], [-4, 1, 3], [1, -1, -1])$

   ⋆(c) $L: \mathbb{R}^2 \rightarrow \mathcal{P}_2$ given by $L([a, b]) = (-a + 5b)x^2 + (3a - b)x + 2b$ with $B = ([5, 3], [3, 2])$ and $C = (3x^2 - 2x, -2x^2 + 2x - 1, x^2 - x + 1)$

   (d) $L: \mathcal{M}_{22} \rightarrow \mathbb{R}^3$ given by $L\left(\begin{bmatrix} a & b \\ c & d \end{bmatrix} \right) = [a - c + 2d, 2a + b - d, -2c + d]$ with $B = \left(\begin{bmatrix} 2 & 5 \\ 2 & -1 \end{bmatrix}, \begin{bmatrix} -2 & -2 \\ 0 & 1 \end{bmatrix}, \begin{bmatrix} -3 & -4 \\ 1 & 2 \end{bmatrix}, \begin{bmatrix} -1 & -3 \\ 0 & 1 \end{bmatrix} \right)$ and $C = ([7, 0, -3], [2, -1, -2], [-2, 0, 1])$

   ⋆(e) $L: \mathcal{P}_2 \rightarrow \mathcal{M}_{23}$ given by

$$L(ax^2 + bx + c) = \begin{bmatrix} -a & 2b + c & 3a - c \\ a + b & c & -2a + b - c \end{bmatrix}$$

with $B = (-5x^2 - x - 1, -6x^2 + 3x + 1, 2x + 1)$ and $C = \left(\begin{bmatrix} 1 & 0 & 0 \\ 0 & 0 & 0 \end{bmatrix}, \begin{bmatrix} 0 & -1 & 0 \\ 0 & 0 & 0 \end{bmatrix}, \begin{bmatrix} 0 & 1 & 1 \\ 0 & 0 & 0 \end{bmatrix}, \begin{bmatrix} 0 & 0 & 0 \\ -1 & 0 & 0 \end{bmatrix}, \begin{bmatrix} 0 & 0 & 0 \\ 0 & 1 & 1 \end{bmatrix}, \begin{bmatrix} 0 & 0 & 0 \\ 0 & 0 & 1 \end{bmatrix} \right)$

4. In each case, find the matrix $\mathbf{A}_{DE}$ for the given linear transformation $L: \mathcal{V} \rightarrow \mathcal{W}$ with respect to the given bases D and E by first finding the matrix for L with respect to the standard bases B and C for $\mathcal{V}$ and $\mathcal{W}$, respectively, and then using the method of Theorem 5.6.

⋆(a) $L: \mathbb{R}^3 \rightarrow \mathbb{R}^3$ given by $L([a, b, c]) = [-2a + b, -b - c, a + 3c]$ with $D = ([15, -6, 4], [2, 0, 1], [3, -1, 1])$ and $E = ([1, -3, 1], [0, 3, -1], [2, -2, 1])$

⋆(b) $L: \mathcal{M}_{22} \rightarrow \mathbb{R}^2$ given by

$$
L\left(\begin{bmatrix} a & b \\ c & d \end{bmatrix}\right) = [6a - b + 3c - 2d, -2a + 3b - c + 4d]
$$

with

$$
D = \left(\begin{bmatrix} 2 & 1 \\ 0 & 1 \end{bmatrix}, \begin{bmatrix} 0 & 2 \\ 1 & 1 \end{bmatrix}, \begin{bmatrix} 1 & 1 \\ 2 & 1 \end{bmatrix}, \begin{bmatrix} 1 & 1 \\ 1 & 1 \end{bmatrix}\right)
$$

and $E = ([-2, 5], [-1, 2])$

(c) $L: \mathcal{M}_{22} \rightarrow \mathcal{P}_2$ given by

$$
L\left(\begin{bmatrix} a & b \\ c & d \end{bmatrix}\right) = (b - c)x^2 + (3a - d)x + (4a - 2c + d)
$$

with

$$
D = \left(\begin{bmatrix} 3 & -4 \\ 1 & -1 \end{bmatrix}, \begin{bmatrix} -2 & 1 \\ 1 & 1 \end{bmatrix}, \begin{bmatrix} 2 & -2 \\ 1 & -1 \end{bmatrix}, \begin{bmatrix} -2 & 1 \\ 0 & 1 \end{bmatrix}\right)
$$

and $E = (2x - 1, -5x^2 + 3x - 1, x^2 - 2x + 1)$

5. Verify that the same matrix is obtained for L in Exercise 3(d) by first finding the matrix for L with respect to the standard bases and then using the method of Theorem 5.6.

6. In each case, find the matrix $\mathbf{A}_{BB}$ for each of the given linear operators $L: \mathcal{V} \rightarrow \mathcal{V}$ with respect to the given basis B by using the method of Theorem 5.5. Then, check your answer by calculating the matrix for L using the standard basis and applying the method of Theorem 5.6.

⋆(a) $L: \mathbb{R}^2 \rightarrow \mathbb{R}^2$ given by $L([x, y]) = [2x - y, x - 3y]$ with $B = ([4, -1], [-7, 2])$

⋆(b) $L: \mathcal{P}_2 \rightarrow \mathcal{P}_2$ given by $L(ax^2 + bx + c) = (b - 2c)x^2 + (2a + c)x + (a - b - c)$ with $B = (2x^2 + 2x - 1, x, -3x^2 - 2x + 1)$

(c) $L: \mathcal{M}_{22} \rightarrow \mathcal{M}_{22}$ given by

$$
L\left(\begin{bmatrix} a & b \\ c & d \end{bmatrix}\right) = \begin{bmatrix} 2a - c + d & a - b \\ -3b - 2d & -a - 2c + 3d \end{bmatrix}
$$

with

$$
B = \left(\begin{bmatrix} -2 & -1 \\ 0 & 1 \end{bmatrix}, \begin{bmatrix} 3 & 1 \\ 0 & -1 \end{bmatrix}, \begin{bmatrix} -2 & 0 \\ 0 & 1 \end{bmatrix}, \begin{bmatrix} 1 & -1 \\ 1 & -1 \end{bmatrix}\right)
$$

7. ★(a) Let $L: \mathcal{P}_3 \rightarrow \mathcal{P}_2$ be given by $L(\mathbf{p}) = \mathbf{p}'$, for $\mathbf{p} \in \mathcal{P}_3$. Find the matrix for L with respect to the standard bases for $\mathcal{P}_3$ and $\mathcal{P}_2$. Use this matrix to calculate $L(4x^3 - 5x^2 + 6x - 7)$ by matrix multiplication.

(b) Let $L: \mathcal{P}_2 \rightarrow \mathcal{P}_3$ be the indefinite integral linear transformation; that is, $L(\mathbf{p})$ is the vector $\int \mathbf{p}(x)\, dx$ with zero constant term. Find the matrix for L with respect to the standard bases for $\mathcal{P}_2$ and $\mathcal{P}_3$. Use this matrix to calculate $L(2x^2 - x + 5)$ by matrix multiplication.

8. Let $L: \mathbb{R}^2 \rightarrow \mathbb{R}^2$ be the linear operator that performs a counterclockwise rotation through an angle of $\frac{\pi}{6}$ radians ($30°$).

★(a) Find the matrix for L with respect to the standard basis for $\mathbb{R}^2$.

(b) Find the matrix for L with respect to the basis $B = ([4, -3], [3, -2])$.

9. Let $L: \mathcal{M}_{23} \rightarrow \mathcal{M}_{32}$ be given by $L(\mathbf{A}) = \mathbf{A}^T$.

(a) Find the matrix for L with respect to the standard bases.

★(b) Find the matrix for L with respect to the bases

$$B = \left(\begin{bmatrix} 1 & 0 & 0 \\ 0 & 0 & 0 \end{bmatrix}, \begin{bmatrix} 0 & 1 & -1 \\ 0 & 0 & 0 \end{bmatrix}, \begin{bmatrix} 0 & 1 & 0 \\ 0 & 0 & 0 \end{bmatrix}, \begin{bmatrix} 0 & 0 & 0 \\ -1 & 0 & 0 \end{bmatrix}, \right.$$

$$\left. \begin{bmatrix} 0 & 0 & 0 \\ 0 & -1 & -1 \end{bmatrix}, \begin{bmatrix} 0 & 0 & 0 \\ 0 & 0 & 1 \end{bmatrix} \right) \text{ for } \mathcal{M}_{23}, \text{ and}$$

$$C = \left(\begin{bmatrix} 1 & 1 \\ 0 & 0 \\ 0 & 0 \end{bmatrix}, \begin{bmatrix} 1 & -1 \\ 0 & 0 \\ 0 & 0 \end{bmatrix}, \begin{bmatrix} 0 & 0 \\ 1 & 1 \\ 0 & 0 \end{bmatrix}, \begin{bmatrix} 0 & 0 \\ 1 & -1 \\ 0 & 0 \end{bmatrix}, \begin{bmatrix} 0 & 0 \\ 0 & 0 \\ 1 & 1 \end{bmatrix}, \right.$$

$$\left. \begin{bmatrix} 0 & 0 \\ 0 & 0 \\ 1 & -1 \end{bmatrix} \right) \text{ for } \mathcal{M}_{32}.$$

★10. Let B be a basis for $\mathcal{V}_1$, C be a basis for $\mathcal{V}_2$, and D be a basis for $\mathcal{V}_3$. Suppose $L_1: \mathcal{V}_1 \rightarrow \mathcal{V}_2$ and $L_2: \mathcal{V}_2 \rightarrow \mathcal{V}_3$ are represented, respectively, by the matrices

$$\mathbf{A}_{BC} = \begin{bmatrix} -2 & 3 & -1 \\ 4 & 0 & -2 \end{bmatrix} \quad \text{and} \quad \mathbf{A}_{CD} = \begin{bmatrix} 4 & -1 \\ 2 & 0 \\ -1 & -3 \end{bmatrix}.$$

Find the matrix $\mathbf{A}_{BD}$ representing the composition $L_2 \circ L_1: \mathcal{V}_1 \rightarrow \mathcal{V}_3$.

11. Let $L_1: \mathbb{R}^3 \rightarrow \mathbb{R}^4$ be given by $L_1([x, y, z]) = [x - y - z, 2y + 3z, x + 3y, -2x + z]$, and let $L_2: \mathbb{R}^4 \rightarrow \mathbb{R}^2$ be given by $L_2([x, y, z, w]) = [2y - 2z + 3w, x - z + w]$.

(a) Find the matrices for L_1 and L_2 with respect to the standard bases in each case.

(b) Find the matrix for $L_2 \circ L_1$ with respect to the standard bases for $\mathbb{R}^3$ and $\mathbb{R}^2$ using Theorem 5.8.

(c) Check your answer to part (b) by computing $(L_2 \circ L_1)([x, y, z])$ and finding the matrix for $L_2 \circ L_1$ directly from this result.

12. Let $\mathbf{A} = \begin{bmatrix} \cos\theta & -\sin\theta \\ \sin\theta & \cos\theta \end{bmatrix}$, the matrix representing the counterclockwise rotation of $\mathbb{R}^2$ about the origin through an angle θ.

(a) Use Theorem 5.8 to show that

$$\mathbf{A}^2 = \begin{bmatrix} \cos 2\theta & -\sin 2\theta \\ \sin 2\theta & \cos 2\theta \end{bmatrix}.$$

(b) Generalize the result of part (a) to show that for any integer $n \geq 1$

$$\mathbf{A}^n = \begin{bmatrix} \cos n\theta & -\sin n\theta \\ \sin n\theta & \cos n\theta \end{bmatrix}.$$

13. Let $B = (\mathbf{v}_1, \mathbf{v}_2, \ldots, \mathbf{v}_n)$ be an ordered basis for a vector space $\mathcal{V}$. Find the matrix with respect to B for each of the following linear operators $L: \mathcal{V} \to \mathcal{V}$:

★(a) $L(\mathbf{v}) = \mathbf{v}$, for all $\mathbf{v} \in \mathcal{V}$ (identity linear operator)
   (b) $L(\mathbf{v}) = \mathbf{0}$, for all $\mathbf{v} \in \mathcal{V}$ (zero linear operator)
★(c) $L(\mathbf{v}) = c\mathbf{v}$, for all $\mathbf{v} \in \mathcal{V}$, and for some fixed $c \in \mathbb{R}$ (scalar linear operator)
   (d) $L: \mathcal{V} \to \mathcal{V}$ given by $L(\mathbf{v}_1) = \mathbf{v}_2$, $L(\mathbf{v}_2) = \mathbf{v}_3, \ldots, L(\mathbf{v}_{n-1}) = \mathbf{v}_n$, $L(\mathbf{v}_n) = \mathbf{v}_1$ (forward replacement of basis vectors)
★(e) $L: \mathcal{V} \to \mathcal{V}$ given by $L(\mathbf{v}_1) = \mathbf{v}_n$, $L(\mathbf{v}_2) = \mathbf{v}_1, \ldots, L(\mathbf{v}_{n-1}) = \mathbf{v}_{n-2}$, $L(\mathbf{v}_n) = \mathbf{v}_{n-1}$ (reverse replacement of basis vectors)

14. Let $L: \mathbb{R}^n \to \mathbb{R}$ be a linear transformation. Prove that there is a vector $\mathbf{x}$ in $\mathbb{R}^n$ such that $L(\mathbf{y}) = \mathbf{x} \cdot \mathbf{y}$ for all $\mathbf{y} \in \mathbb{R}^n$.

▶**15.** Prove Theorem 5.8.

16. Let $L: \mathbb{R}^3 \to \mathbb{R}^3$ be given by $L([x, y, z]) = [-4y - 13z, -6x + 5y + 6z, 2x - 2y - 3z]$.

(a) What is the matrix for L with respect to the standard basis for $\mathbb{R}^3$?
(b) What is the matrix for L with respect to the basis

$$B = ([-1, -6, 2], [3, 4, -1], [-1, -3, 1])?$$

(c) What does your answer to part (b) tell you about the vectors in B? Explain.

17. Let $\mathbf{A}$, $\mathbf{B}$, and $\mathbf{C}$ be $n \times n$ matrices. Show that if $\mathbf{A}$ is similar to $\mathbf{B}$ and $\mathbf{B}$ is similar to $\mathbf{C}$, then $\mathbf{A}$ is similar to $\mathbf{C}$.

18. Let $\mathbf{A}$ and $\mathbf{B}$ be similar $n \times n$ matrices. Prove each of the following:

(a) $\mathbf{A}^k$ is similar to $\mathbf{B}^k$, for each positive integer k.
(b) $\mathbf{A}^T$ is similar to $\mathbf{B}^T$.
(c) $|\mathbf{A}| = |\mathbf{B}|$.
(d) $\mathbf{A}$ is nonsingular if and only if $\mathbf{B}$ is nonsingular.
(e) If $\mathbf{A}$ and $\mathbf{B}$ are both nonsingular, then $\mathbf{A}^{-1}$ is similar to $\mathbf{B}^{-1}$.
(f) $\mathbf{A} + \mathbf{I}_n$ is similar to $\mathbf{B} + \mathbf{I}_n$.
(g) Trace$(\mathbf{A}) = $ trace$(\mathbf{B})$. (Hint: Use Exercise 22(c) in Section 1.5.)

19. Show that if $\mathbf{A}$ is a nonsingular $n \times n$ matrix and $\mathbf{B}$ is any $n \times n$ matrix, then $\mathbf{AB}$ is similar to $\mathbf{BA}$.

20. Show that a 2×2 matrix similar only to itself must be a diagonal matrix with both main diagonal entries equal.

▶**21.** Prove Theorem 5.7.

22. In Example 6, verify that $p_{\mathbf{A}_{BB}}(x) = (x-1)^2(x+1)$, $\{[3, 1, 0], [-2, 0, 1]\}$ is a basis for the eigenspace E_1, $\{[1, -3, 2]\}$ is a basis for the eigenspace E_{-1}, the transition matrices $\mathbf{P}^{-1}$ and $\mathbf{P}$ are as indicated, and, finally, $\mathbf{A}_{DD} = \mathbf{PA}_{BB}\mathbf{P}^{-1}$ is a diagonal matrix with entries 1, 1, and -1, respectively, on the main diagonal.

23. Let $L: \mathbb{R}^3 \to \mathbb{R}^3$ be the linear operator whose matrix with respect to the standard basis B for $\mathbb{R}^3$ is

$$\mathbf{A}_{BB} = \frac{1}{9} \begin{bmatrix} 8 & 2 & 2 \\ 2 & 5 & -4 \\ 2 & -4 & 5 \end{bmatrix}.$$

⋆(a) Calculate and factor $p_{\mathbf{A}_{BB}}(x)$. (Be sure to incorporate $\frac{1}{9}$ correctly into your calculations.)

⋆(b) Solve for a basis for each eigenspace for L. Combine these to form a basis C for $\mathbb{R}^3$.

⋆(c) Find the transition matrix $\mathbf{P}$ from B to C.

(d) Calculate $\mathbf{A}_{CC}$ using $\mathbf{A}_{BB}$, $\mathbf{P}$ and $\mathbf{P}^{-1}$.

(e) Use $\mathbf{A}_{CC}$ to give a geometric description of the operator L, as was done in Example 6.

24. Let L be a linear operator on a vector space $\mathcal{V}$ with ordered basis $B = (\mathbf{v}_1, \ldots, \mathbf{v}_n)$. Suppose that k is a nonzero real number, and let C be the ordered basis $(k\mathbf{v}_1, \ldots, k\mathbf{v}_n)$ for $\mathcal{V}$. Show that $\mathbf{A}_{BB} = \mathbf{A}_{CC}$.

25. Let $\mathcal{V}$ be an n-dimensional vector space, and let $\mathbf{X}$ and $\mathbf{Y}$ be similar $n \times n$ matrices. Prove that there is a linear operator $L: \mathcal{V} \to \mathcal{V}$ and bases B and C such that $\mathbf{X}$ is the matrix for L with respect to B and $\mathbf{Y}$ is the matrix for L with respect to C. (Hint: Suppose that $\mathbf{Y} = \mathbf{PXP}^{-1}$. Choose any basis B for $\mathcal{V}$. Then create the linear operator $L: \mathcal{V} \to \mathcal{V}$ whose matrix with respect to B is $\mathbf{X}$. Let $\mathbf{v}_i$ be the vector so that $[\mathbf{v}_i]_B = i$th column of $\mathbf{P}^{-1}$. Define C to be $(\mathbf{v}_1, \ldots, \mathbf{v}_n)$. Prove that C is a basis for $\mathcal{V}$. Then show that $\mathbf{P}$ is the transition matrix from B to C and that $\mathbf{Y}$ is the matrix for L with respect to C.)

26. Let $B = ([a, b], [c, d])$ be a basis for $\mathbb{R}^2$. Then $ad - bc \neq 0$ (why?). Let $L: \mathbb{R}^2 \to \mathbb{R}^2$ be a linear operator such that $L([a, b]) = [c, d]$ and $L([c, d]) = [a, b]$. Show that the matrix for L with respect to the standard basis for $\mathbb{R}^2$ is

$$\frac{1}{ad - bc} \begin{bmatrix} cd - ab & a^2 - c^2 \\ d^2 - b^2 & ab - cd \end{bmatrix}.$$

27. Let $L: \mathbb{R}^2 \to \mathbb{R}^2$ be the linear transformation where $L(\mathbf{v})$ is the reflection of $\mathbf{v}$ through the line $y = mx$. (Assume that the initial point of $\mathbf{v}$ is the origin.) Show that the matrix for L with respect to the standard basis for $\mathbb{R}^2$ is

$$\frac{1}{1 + m^2} \begin{bmatrix} 1 - m^2 & 2m \\ 2m & m^2 - 1 \end{bmatrix}.$$

(Hint: Use Exercise 19 in Section 1.2.)

28. Find the set of all matrices with respect to the standard basis for $\mathbb{R}^2$ for all linear operators that

(a) Take all vectors of the form $[0, y]$ to vectors of the form $[0, y']$
(b) Take all vectors of the form $[x, 0]$ to vectors of the form $[x', 0]$
(c) Satisfy both parts (a) and (b) simultaneously

29. Let V and W be finite dimensional vector spaces, and let $\mathcal{Y}$ be a subspace of V. Suppose that $L: \mathcal{Y} \rightarrow W$ is a linear transformation. Prove that there is a linear transformation $L': V \rightarrow W$ such that $L'(\mathbf{y}) = L(\mathbf{y})$ for every $\mathbf{y} \in \mathcal{Y}$. ($L'$ is called an **extension** of L to V.)

▶**30.** Prove the uniqueness assertion in Theorem 5.4. (Hint: Let $\mathbf{v}$ be any vector in V. Show that there is only one possible answer for $L(\mathbf{v})$ by expressing $L(\mathbf{v})$ as a linear combination of the $\mathbf{w}_i$'s.)

⋆**31.** True or False:

(a) If $L: V \rightarrow W$ is a linear transformation, and $B = (\mathbf{v}_1, \mathbf{v}_2, \ldots, \mathbf{v}_n)$ is an ordered basis for V, then for any $\mathbf{v} \in V$, $L(\mathbf{v})$ can be computed if $L(\mathbf{v}_1), L(\mathbf{v}_2), \ldots, L(\mathbf{v}_n)$ are known.
(b) There is a unique linear transformation $L: \mathbb{R}^3 \rightarrow \mathcal{P}_3$ such that $L([1, 0, 0]) = x^3 - x^2$, $L([0, 1, 0]) = x^3 - x^2$, and $L([0, 0, 1]) = x^3 - x^2$.
(c) If V is a finite dimensional vector space and $L: V \rightarrow W$ is a linear transformation, then there is a unique matrix $\mathbf{A}$ corresponding to L.
(d) If $L: V \rightarrow W$ is a linear transformation and B is a (finite) ordered basis for V, and C is a (finite) ordered basis for W, then $[\mathbf{v}]_B = \mathbf{A}_{BC}[L(\mathbf{v})]_C$.
(e) If $L: V \rightarrow W$ is a linear transformation and $B = (\mathbf{v}_1, \mathbf{v}_2, \ldots, \mathbf{v}_n)$ is an ordered basis for V, and C is a finite ordered basis for W, then the ith column of $\mathbf{A}_{BC}$ is $[L(\mathbf{v}_i)]_C$.
(f) The matrix for the projection of $\mathbb{R}^3$ onto the xz-plane (with respect to the standard basis) is $\begin{bmatrix} 1 & 0 & 0 \\ 0 & 1 & 0 \\ 0 & 0 & 0 \end{bmatrix}$.
(g) If $L: V \rightarrow W$ is a linear transformation, and B and D are (finite) ordered bases for V, and C and E are (finite) ordered bases for W, then $\mathbf{A}_{DE}\mathbf{P} = \mathbf{Q}\mathbf{A}_{BC}$, where $\mathbf{P}$ is the transition matrix from B to D, and $\mathbf{Q}$ is the transition matrix from C to E.
(h) If $L: V \rightarrow V$ is a linear operator on a finite dimensional vector space, and B and D are ordered bases for V, then $\mathbf{A}_{BB}$ is similar to $\mathbf{A}_{DD}$.
(i) Similar square matrices have identical characteristic polynomials.
(j) If $L_1, L_2: \mathbb{R}^2 \rightarrow \mathbb{R}^2$ are linear transformations with matrices $\begin{bmatrix} 1 & 2 \\ 3 & 4 \end{bmatrix}$ and $\begin{bmatrix} 0 & 1 \\ 1 & 0 \end{bmatrix}$, respectively, with respect to the standard basis, then the matrix for $L_2 \circ L_1$ with respect to the standard basis equals $\begin{bmatrix} 1 & 2 \\ 3 & 4 \end{bmatrix}\begin{bmatrix} 0 & 1 \\ 1 & 0 \end{bmatrix}$.

5.3 The Dimension Theorem

In this section, we introduce two special subspaces associated with a linear transformation $L: \mathcal{V} \to \mathcal{W}$: the kernel of L (a subspace of $\mathcal{V}$) and the range of L (a subspace of $\mathcal{W}$). We illustrate techniques for calculating bases for both the kernel and range and show their dimensions are related to the rank of any matrix for the linear transformation. Finally, we show that any matrix and its transpose have the same rank.

Kernel and Range

DEFINITION

Let $L: \mathcal{V} \to \mathcal{W}$ be a linear transformation. The **kernel** of L, denoted by $\ker(L)$, is the subset of all vectors in $\mathcal{V}$ that map to $\mathbf{0}_{\mathcal{W}}$. That is, $\ker(L) = \{\mathbf{v} \in \mathcal{V} | L(\mathbf{v}) = \mathbf{0}_{\mathcal{W}}\}$. The **range** of L, or, $\text{range}(L)$, is the subset of all vectors in $\mathcal{W}$ that are the image of some vector in $\mathcal{V}$. That is, $\text{range}(L) = \{L(\mathbf{v}) | \mathbf{v} \in \mathcal{V}\}$.

Remember that the kernel[1] is a subset of the *domain* and that the range is a subset of the *codomain*. Since the kernel of $L: \mathcal{V} \to \mathcal{W}$ is the pre-image of the subspace $\{\mathbf{0}_{\mathcal{W}}\}$ of $\mathcal{W}$, it must be a subspace of $\mathcal{V}$ by Theorem 5.3. That theorem also assures us that the range of L is a subspace of $\mathcal{W}$. Hence, we have

---

THEOREM 5.9

If $L: \mathcal{V} \to \mathcal{W}$ is a linear transformation, then the kernel of L is a subspace of $\mathcal{V}$ and the range of L is a subspace of $\mathcal{W}$.

EXAMPLE 1

Projection: For $n \geq 3$, consider the linear operator $L: \mathbb{R}^n \to \mathbb{R}^n$ given by $L([a_1, a_2, \ldots, a_n]) = [a_1, a_2, 0, \ldots, 0]$. Now, $\ker(L)$ consists of those elements of the domain that map to $[0, 0, \ldots, 0]$, the zero vector of the codomain. Hence, for vectors in the kernel, $a_1 = a_2 = 0$, but $a_3, \ldots, a_n$ can have any values. Thus,

$$\ker(L) = \{[0, 0, a_3, \ldots, a_n] | a_3, \ldots, a_n \in \mathbb{R}\}.$$

Notice that $\ker(L)$ is a subspace of the domain and that $\dim(\ker(L)) = n - 2$, because the standard basis vectors $\mathbf{e}_3, \ldots, \mathbf{e}_n$ of $\mathbb{R}^n$ span $\ker(L)$.

Also, $\text{range}(L)$ consists of those elements of the codomain that are images of domain elements. Hence, $\text{range}(L) = \{[a_1, a_2, 0, \cdots, 0] | a_1, a_2 \in \mathbb{R}\}$. Notice that $\text{range}(L)$ is a subspace of the codomain and that $\dim(\text{range}(L)) = 2$, since the standard basis vectors $\mathbf{e}_1$ and $\mathbf{e}_2$ span $\text{range}(L)$. ∎

---

[1] Some textbooks refer to the kernel of L as the **nullspace** of L.

EXAMPLE 2

Differentiation: Consider the linear transformation $L: \mathcal{P}_3 \to \mathcal{P}_2$ given by $L(ax^3+bx^2+cx+d) = 3ax^2+2bx+c$. Now, $\ker(L)$ consists of the polynomials in $\mathcal{P}_3$ that map to the zero polynomial in $\mathcal{P}_2$. However, if $3ax^2 + 2bx + c = 0$, we must have $a = b = c = 0$. Hence, $\ker(L) = \left\{ 0x^3 + 0x^2 + 0x + d \,\middle|\, d \in \mathbb{R} \right\}$; that is, $\ker(L)$ is just the subset of $\mathcal{P}_3$ of all constant polynomials. Notice that $\ker(L)$ is a subspace of $\mathcal{P}_3$ and that $\dim(\ker(L)) = 1$ because the single polynomial "1" spans $\ker(L)$.

Also, $\text{range}(L)$ consists of all polynomials in the codomain $\mathbb{R}^2$ of the form $3ax^2 + 2bx + c$. Since every polynomial $Ax^2 + Bx + C$ of degree 2 or less can be expressed in this form (take $a = A/3,\ b = B/2,\ c = C$), $\text{range}(L)$ is all of $\mathcal{P}_2$. Therefore, $\text{range}(L)$ is a subspace of $\mathcal{P}_2$, and $\dim(\text{range}(L)) = 3$. ∎

EXAMPLE 3

Rotation: Recall that the linear transformation $L: \mathbb{R}^2 \to \mathbb{R}^2$ given by

$$L\left(\begin{bmatrix} x \\ y \end{bmatrix} \right) = \begin{bmatrix} \cos\theta & -\sin\theta \\ \sin\theta & \cos\theta \end{bmatrix} \begin{bmatrix} x \\ y \end{bmatrix},$$

for some (fixed) angle θ, represents the counterclockwise rotation of any vector $[x, y]$ with initial point at the origin through the angle θ.

Now, $\ker(L)$ consists of all vectors in the domain $\mathbb{R}^2$ that map to $[0,0]$ in the codomain $\mathbb{R}^2$. However, only $[0, 0]$ itself is rotated by L to the zero vector. Hence, $\ker(L) = \{[0, 0]\}$. Notice that $\ker(L)$ is a subspace of $\mathbb{R}^2$, and $\dim(\ker(L)) = 0$.

Also, $\text{range}(L)$ is all of the codomain $\mathbb{R}^2$ because every nonzero vector $\mathbf{v}$ in $\mathbb{R}^2$ is the image of the vector of the same length at the angle θ *clockwise* from $\mathbf{v}$. Thus, $\text{range}(L) = \mathbb{R}^2$ and so, $\text{range}(L)$ is a subspace of $\mathbb{R}^2$ with $\dim(\text{range}(L)) = 2$. ∎

The Dimension Theorem

Notice that in Examples 1, 2, and 3, the dimensions of the kernel and the range add up to the dimension of the domain. This is no coincidence.

THEOREM 5.10 (Dimension Theorem)

If $L: \mathcal{V} \to \mathcal{W}$ is a linear transformation and $\mathcal{V}$ is finite dimensional, then $\text{range}(L)$ is finite dimensional, and
$$\dim(\ker(L)) + \dim(\text{range}(L)) = \dim(\mathcal{V}).$$

PROOF OF THEOREM 5.10

Let $L: \mathcal{V} \to \mathcal{W}$ be a linear transformation with $\mathcal{V}$ finite dimensional. Now, $\ker(L)$ is a subspace of $\mathcal{V}$, so by Theorem 4.17, $\ker(L)$ is finite dimensional and $\dim(\ker(L)) \leq \dim(\mathcal{V})$. Suppose that $\dim(\ker(L)) = s$. Let $\{\mathbf{k}_1, \ldots, \mathbf{k}_s\}$ be a basis for $\ker(L)$, and using Theorem 4.19, let $\mathbf{q}_1, \ldots, \mathbf{q}_t$ be vectors in $\mathcal{V}$ such that $B = \{\mathbf{k}_1, \ldots, \mathbf{k}_s, \mathbf{q}_1, \ldots, \mathbf{q}_t\}$ is a basis for $\mathcal{V}$. Hence, $\dim(\mathcal{V}) = s + t$.

We claim that $C = \{L(\mathbf{q}_1), \ldots, L(\mathbf{q}_t)\}$ consists of t distinct vectors and that C is a basis for range(L). If we can show this, we will be done, since we will have dim(range(L)) $= t$, which is finite, and that dim($\mathcal{V}$) $= s + t =$ dim(ker(L)) + dim(range(L)).

Now, range(L) is a subspace of $\mathcal{W}$. If $\mathbf{v} \in \mathcal{V}$, then $\mathbf{v} = a_1\mathbf{k}_1 + \cdots + a_s\mathbf{k}_s + b_1\mathbf{q}_1 + \cdots + b_t\mathbf{q}_t$ for some $a_1, \ldots, a_s, b_1, \ldots, b_t \in \mathbb{R}$. Hence

$$
\begin{aligned}
L(\mathbf{v}) &= L(a_1\mathbf{k}_1 + \cdots + a_s\mathbf{k}_s + b_1\mathbf{q}_1 + \cdots + b_t\mathbf{q}_t) \\
&= a_1 L(\mathbf{k}_1) + \cdots + a_s L(\mathbf{k}_s) + b_1 L(\mathbf{q}_1) + \cdots + b_t L(\mathbf{q}_t) \\
&= a_1\mathbf{0} + \cdots + a_s\mathbf{0} + b_1 L(\mathbf{q}_1) + \cdots + b_t L(\mathbf{q}_t) \\
&= b_1 L(\mathbf{q}_1) + \cdots + b_t L(\mathbf{q}_t).
\end{aligned}
$$

Thus, range(L) is spanned by $C = \{L(\mathbf{q}_1), \ldots, L(\mathbf{q}_t)\}$.

Next, we show that C is linearly independent. Suppose $c_1 L(\mathbf{q}_1) + \cdots + c_t L(\mathbf{q}_t) = 0$. Then $L(c_1\mathbf{q}_1 + \cdots + c_t\mathbf{q}_t) = 0$, and so $c_1\mathbf{q}_1 + \cdots + c_t\mathbf{q}_t \in$ ker(L). Therefore, there exist $d_1, \ldots, d_s \in \mathbb{R}$ such that

$$
c_1\mathbf{q}_1 + \cdots + c_t\mathbf{q}_t = d_1\mathbf{k}_1 + \cdots + d_s\mathbf{k}_s,
$$

implying $d_1\mathbf{k}_1 + \cdots + d_s\mathbf{k}_s - c_1\mathbf{q}_1 - \cdots - c_t\mathbf{q}_t = 0$. But the fact that $B = \{\mathbf{k}_1, \ldots, \mathbf{k}_s, \mathbf{q}_1, \ldots, \mathbf{q}_t\}$ is linearly independent shows that $d_1 = \cdots = d_s = c_1 = \cdots = c_t = 0$. Since all of the c_i's are zero, C is linearly independent.

Finally, we need to show the vectors in C are distinct. But, if C did not consist of t *distinct* vectors, we could have had $c_1 L(\mathbf{q}_1) + \cdots + c_t L(\mathbf{q}_t) = 0$ with not every c_i equal to zero (why?). Hence, we are done.   ∎

Figure 5.10

Images of basis elements in the Dimension Theorem

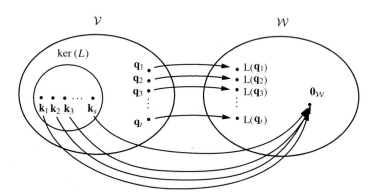

An immediate corollary of the Dimension Theorem is the following:

COROLLARY 5.11

If $L\colon \mathcal{V} \to \mathcal{W}$ is a linear transformation and $\mathcal{V}$ is finite dimensional, then dim(ker(L)) $\leq$ dim($\mathcal{V}$) and dim(range(L)) $\leq$ dim($\mathcal{V}$).

Beware! If $L\colon \mathcal{V} \to \mathcal{W}$ is a linear transformation, it is not necessarily true that dim($\mathcal{W}$) $\leq$ dim($\mathcal{V}$). In fact, let $\mathcal{V} = \mathbb{R}^3$ and $\mathcal{W} = \mathbb{R}^4$ and consider the linear transformation $L: \mathbb{R}^3 \to \mathbb{R}^4$ given by $L([x, y, z]) = [x, y, z, 0]$. Then $\dim(\mathcal{W}) = \dim\left(\mathbb{R}^4\right) = 4 > 3 = \dim\left(\mathbb{R}^3\right) = \dim(\mathcal{V})$.

EXAMPLE 4 Consider $L: \mathcal{M}_{nn} \to \mathcal{M}_{nn}$ given by $L(\mathbf{A}) = \mathbf{A} + \mathbf{A}^T$. Now, $\ker(L) = \{\mathbf{A} \in \mathcal{M}_{nn} \mid \mathbf{A} + \mathbf{A}^T = \mathbf{O}_n\}$. However, $\mathbf{A} + \mathbf{A}^T = \mathbf{O}_n$ implies that $\mathbf{A} = -\mathbf{A}^T$. Hence, $\ker(L)$ is precisely the set of all skew-symmetric $n \times n$ matrices.

The range of L is the set of all matrices $\mathbf{B}$ of the form $\mathbf{A} + \mathbf{A}^T$ for some $n \times n$ matrix $\mathbf{A}$. However, if $\mathbf{B} = \mathbf{A} + \mathbf{A}^T$, then $\mathbf{B}^T = \left(\mathbf{A} + \mathbf{A}^T\right)^T = \mathbf{A}^T + \mathbf{A} = \mathbf{B}$, so $\mathbf{B}$ is symmetric. Thus, $\text{range}(L) \subseteq \{\text{symmetric } n \times n \text{ matrices}\}$.

Next, if $\mathbf{B}$ is a symmetric $n \times n$ matrix, then $L(\frac{1}{2}\mathbf{B}) = \frac{1}{2}L(\mathbf{B}) = \frac{1}{2}(\mathbf{B} + \mathbf{B}^T) = \frac{1}{2}(\mathbf{B} + \mathbf{B}) = \mathbf{B}$, and so $\mathbf{B} \in \text{range}(L)$, thus proving $\{\text{symmetric } n \times n \text{ matrices}\} \subseteq \text{range}(L)$. Hence, $\text{range}(L)$ is the set of all symmetric $n \times n$ matrices.

In Exercise 12 of Section 4.6, we found that $\dim(\{\text{skew-symmetric } n \times n \text{ matrices}\}) = (n^2-n)/2$ and that $\dim(\{\text{symmetric } n \times n \text{ matrices}\}) = \left(n^2 + n\right)/2$. Notice that the Dimension Theorem holds here, since $\dim(\ker(L)) + \dim(\text{range}(L)) = (n^2 - n)/2 + \left(n^2 + n\right)/2 = (n^2 - n + n^2 + n)/2 = n^2 = \dim\left(\mathcal{M}_{nn}\right)$. ∎

Finding the Kernel from the Matrix of a Linear Transformation

Consider the linear transformation $L: \mathbb{R}^n \to \mathbb{R}^m$ given by $L(\mathbf{X}) = \mathbf{AX}$, where $\mathbf{A}$ is a (fixed) $m \times n$ matrix and $\mathbf{X} \in \mathbb{R}^n$. Now, $\ker(L)$ is the subspace of all vectors $\mathbf{X}$ in the domain $\mathbb{R}^n$ that are solutions of the homogeneous system $\mathbf{AX} = \mathbf{O}$. If $\mathbf{B}$ is the reduced row echelon form matrix for $\mathbf{A}$, we find a basis for $\ker(L)$ by solving for particular solutions to the system $\mathbf{BX} = \mathbf{O}$ by systematically setting each independent variable equal to 1 in turn, while setting the others equal to 0. (You should be familiar with this process from the Diagonalization Method for finding eigenvectors in Section 3.4.) Thus, $\dim(\ker(L))$ equals the number of independent variables in the system $\mathbf{BX} = \mathbf{O}$.

We present an example of this technique.

EXAMPLE 5 Let $L : \mathbb{R}^5 \to \mathbb{R}^4$ be given by $L(\mathbf{X}) = \mathbf{AX}$, where

$$\mathbf{A} = \begin{bmatrix} 8 & 4 & 16 & 32 & 0 \\ 4 & 2 & 10 & 22 & -4 \\ -2 & -1 & -5 & -11 & 7 \\ 6 & 3 & 15 & 33 & -7 \end{bmatrix}.$$

To solve for $\ker(L)$, we first row reduce $\mathbf{A}$ to

$$\mathbf{B} = \begin{bmatrix} 1 & \frac{1}{2} & 0 & -2 & 0 \\ 0 & 0 & 1 & 3 & 0 \\ 0 & 0 & 0 & 0 & 1 \\ 0 & 0 & 0 & 0 & 0 \end{bmatrix}.$$

The homogeneous system $\mathbf{BX} = \mathbf{O}$ has independent variables x_2 and x_4, and

$$\begin{cases} x_1 &=& -\frac{1}{2}x_2 &+& 2x_4 \\ x_3 &=& & -& 3x_4 \\ x_5 &=& & & 0 \end{cases}.$$

We construct two particular solutions, first by setting $x_2 = 1$ and $x_4 = 0$ to obtain $\mathbf{v}_1 = [-\frac{1}{2}, 1, 0, 0, 0]$, and then setting $x_2 = 0$ and $x_4 = 1$, yielding $\mathbf{v}_2 = [2, 0, -3, 1, 0]$. The set $\{\mathbf{v}_1, \mathbf{v}_2\}$ forms a basis for $\ker(L)$, and thus, $\dim(\ker(L)) = 2$, the number of independent variables. The entire subspace $\ker(L)$ consists of all linear combinations of the basis vectors; that is,

$$\ker(L) = \{a\mathbf{v}_1 + b\mathbf{v}_2 \mid a, b \in \mathbb{R}\} = \left\{ \left[-\frac{1}{2}a + 2b, a, -3b, b, 0 \right] \Big| \, a, b \in \mathbb{R} \right\}.$$

Finally, note that we could have eliminated fractions in this basis, just as we did with eigenvectors in Section 3.4 by replacing $\mathbf{v}_1$ with $2\mathbf{v}_1 = [-1, 2, 0, 0, 0]$. ∎

Example 5 illustrates the following general technique:

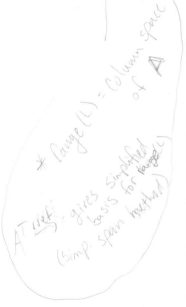

> **METHOD FOR FINDING A BASIS FOR THE KERNEL OF A LINEAR TRANSFORMATION (KERNEL METHOD)**
>
> Let $L: \mathbb{R}^n \to \mathbb{R}^m$ be a linear transformation given by $\mathbf{L}(\mathbf{X}) = \mathbf{AX}$ for some $m \times n$ matrix $\mathbf{A}$. To find a basis for $\ker(L)$, perform the following steps:
>
> **Step 1:** Find $\mathbf{B}$, the reduced row echelon form of $\mathbf{A}$.
>
> **Step 2:** Solve for one particular solution for each independent variable in the homogeneous system $\mathbf{BX} = \mathbf{O}$. The ith such solution, $\mathbf{v}_i$, is found by setting the ith independent variable equal to 1 and setting all other independent variables equal to 0.
>
> **Step 3:** The set $\{\mathbf{v}_1, \ldots, \mathbf{v}_k\}$ is a basis for $\ker(L)$. (We can replace any $\mathbf{v}_i$ with $c\mathbf{v}_i$, where $c \neq 0$, to eliminate fractions.)

The method for finding a basis for $\ker(L)$ is practically identical to Step 3 of the Diagonalization Method of Section 3.4, in which we create a basis of eigenvectors for the eigenspace E_λ for a matrix $\mathbf{A}$. This is to be expected, since E_λ is really the kernel of the linear transformation L whose matrix is $(\lambda \mathbf{I}_n - \mathbf{A})$.

Finding the Range from the Matrix of a Linear Transformation

Next, we determine a method for finding a basis for the range of $L: \mathbb{R}^n \to \mathbb{R}^m$ given by $L(\mathbf{X}) = \mathbf{AX}$. Now by Theorem 5.4, a linear transformation is completely determined by the images of a basis for the domain. In fact, from the proof of Theorem 5.10, $\mathrm{range}(L)$ is spanned by these images. Thus, $\{L(\mathbf{e}_1), \ldots, L(\mathbf{e}_n)\}$ spans $\mathrm{range}(L)$. But $L(\mathbf{e}_i)$ is the ith column of $\mathbf{A}$. Hence, the columns of $\mathbf{A}$ span range(L). That is, range(L) = span({columns of $\mathbf{A}$}). Applying the Independence Test Method to this set of columns gives us the following general technique:

METHOD FOR FINDING A BASIS FOR THE RANGE OF A LINEAR TRANSFORMATION (RANGE METHOD)

Let $L: \mathbb{R}^n \rightarrow \mathbb{R}^m$ be a linear transformation given by $\mathbf{L}(\mathbf{X}) = \mathbf{AX}$, for some $m \times n$ matrix $\mathbf{A}$. To find a basis for range(L), perform the following steps:

Step 1: Find $\mathbf{B}$, the reduced row echelon form of $\mathbf{A}$.

Step 2: Form the set of those columns of $\mathbf{A}$ whose corresponding columns in $\mathbf{B}$ have nonzero pivots. This set is a basis for range(L).

EXAMPLE 6

Consider the linear transformation $L: \mathbb{R}^5 \longrightarrow \mathbb{R}^4$ given in Example 5. After row reducing the matrix $\mathbf{A}$ for L we obtained a matrix $\mathbf{B}$ in reduced row echelon form having nonzero pivots in columns 1, 3, and 5. Hence, columns 1, 3, and 5 of $\mathbf{A}$ form a basis for range(L). In particular, we get the basis {[8, 4, −2, 6], [16, 10, −5, 15], [0, −4, 7, −7]}, and so dim(range(L)) = 3. ∎

From Examples 5 and 6, we see that dim(ker(L))+dim(range(L)) = 2+3 = 5 = dim($\mathbb{R}^5$) = dim(domain(L)), for the given linear transformation L. We can understand why this works by examining our methods for calculating bases for the kernel and range. For ker(L), we get one basis vector for each independent variable, which corresponds to a nonpivot column of $\mathbf{A}$ after row reducing. For range(L), we get one basis vector for each pivot column of $\mathbf{A}$. Together, these account for the total number of columns of $\mathbf{A}$, which is the dimension of the domain.

The fact that the number of nonzero pivots of $\mathbf{A}$ equals the number of nonzero rows in the reduced row echelon form matrix for $\mathbf{A}$ shows that dim(range(L)) = rank($\mathbf{A}$). This result is stated in the following theorem, which also holds when bases other than the standard bases are used. The proof of this, as well as a generalization of this result to other vector spaces, is left as Exercises 17 and 18.

THEOREM 5.12

If $L: \mathbb{R}^n \rightarrow \mathbb{R}^m$ is a linear transformation with matrix $\mathbf{A}$ with respect to any bases for $\mathbb{R}^n$ and $\mathbb{R}^m$, then

(1) dim(range(L)) = rank($\mathbf{A}$)

(2) dim(ker(L)) = n−rank($\mathbf{A}$).

Rank of the Transpose

We can use the Range Method to prove the following result[2]:

COROLLARY 5.13
If $\mathbf{A}$ is any matrix, then rank$(\mathbf{A}) =$ rank$(\mathbf{A}^T)$.

PROOF OF COROLLARY 5.13
Let $\mathbf{A}$ be an $m \times n$ matrix. Consider the linear transformation $L\colon \mathbb{R}^n \to \mathbb{R}^m$ with associated matrix $\mathbf{A}$ (using the standard bases). By the Range Method, range(L) is the span of the column vectors of $\mathbf{A}$. Hence, range(L) is the span of the row vectors of $\mathbf{A}^T$; that is, range(L) is the row space of $\mathbf{A}^T$. Thus, dim(range(L)) = rank$(\mathbf{A}^T)$, by the Simplified Span Method. But by Theorem 5.12, dim(range(L)) = rank$(\mathbf{A})$. Hence, rank$(\mathbf{A}) =$ rank$(\mathbf{A}^T)$. ∎

EXAMPLE 7
Let $\mathbf{A}$ be the matrix from Examples 5 and 6. We calculated its reduced row echelon form $\mathbf{B}$ in Example 5 and found it has 3 nonzero rows. Hence, rank$(\mathbf{A}) = 3$. Now,

$$\mathbf{A}^T = \begin{bmatrix} 8 & 4 & -2 & 6 \\ 4 & 2 & -1 & 3 \\ 16 & 10 & -5 & 15 \\ 32 & 22 & -11 & 33 \\ 0 & -4 & 7 & -7 \end{bmatrix} \text{ row reduces to } \begin{bmatrix} 1 & 0 & 0 & 0 \\ 0 & 1 & 0 & \frac{7}{5} \\ 0 & 0 & 1 & -\frac{1}{5} \\ 0 & 0 & 0 & 0 \\ 0 & 0 & 0 & 0 \end{bmatrix},$$

showing that rank$(\mathbf{A}^T) = 3$ as well. ∎

In some textbooks, rank$(\mathbf{A})$ is called the **row rank** of $\mathbf{A}$ and rank$(\mathbf{A}^T)$ is called the **column rank** of $\mathbf{A}$. Thus, Corollary 5.13 asserts that the row rank of $\mathbf{A}$ equals the column rank of $\mathbf{A}$.

Recall that rank$(\mathbf{A}) =$ dim(row space of $\mathbf{A}$). Analogous to the concept of row space, we define the **column space** of a matrix $\mathbf{A}$ as the span of the columns of $\mathbf{A}$. In Corollary 5.13, we observed that if $L\colon \mathbb{R}^n \to \mathbb{R}^m$ with $L(\mathbf{X}) = \mathbf{AX}$ (using the standard bases), then range(L) = span({columns of $\mathbf{A}$}) = column space of $\mathbf{A}$, and so dim(range(L)) = dim(column space of $\mathbf{A}$) = rank$(\mathbf{A}^T)$. With this new terminology, Corollary 5.13 asserts that dim(row space of $\mathbf{A}$) = dim(column space of $\mathbf{A}$). Be careful! This statement does not imply that these *spaces* are equal, only that their *dimensions* are equal. In fact, unless $\mathbf{A}$ is square, they contain vectors of different sizes. Notice that in Examples 5 through 7, the row space of $\mathbf{A}$ is a subspace of $\mathbb{R}^5$, but the column space of $\mathbf{A}$ is a subspace of $\mathbb{R}^4$.

---

[2] In Exercise 18 of Section 4.6 you were asked to prove Corollary 5.13 by essentially the same method given here, only using different notation.

Exercises for Section 5.3

1. Let $L\colon \mathbb{R}^3 \to \mathbb{R}^3$ be given by

$$L\left(\begin{bmatrix} x_1 \\ x_2 \\ x_3 \end{bmatrix}\right) = \begin{bmatrix} 5 & 1 & -1 \\ -3 & 0 & 1 \\ 1 & -1 & -1 \end{bmatrix} \begin{bmatrix} x_1 \\ x_2 \\ x_3 \end{bmatrix}.$$

$\star$(a) Is $[1, -2, 3]$ in $\ker(L)$? Why or why not?
   (b) Is $[2, -1, 4]$ in $\ker(L)$? Why or why not?
$\star$(c) Is $[2, -1, 4]$ in $\operatorname{range}(L)$? Why or why not?
   (d) Is $[-16, 12, -8]$ in $\operatorname{range}(L)$? Why or why not?

2. Let $L\colon \mathcal{P}_3 \to \mathcal{P}_3$ be given by $L(ax^3+bx^2+cx+d) = 2cx^3-(a+b)x+(d-c)$.

$\star$(a) Is $x^3 - 5x^2 + 3x - 6$ in $\ker(L)$? Why or why not?
   (b) Is $4x^3 - 4x^2$ in $\ker(L)$? Why or why not?
$\star$(c) Is $8x^3 - x - 1$ in $\operatorname{range}(L)$? Why or why not?
   (d) Is $4x^3 - 3x^2 + 7$ in $\operatorname{range}(L)$? Why or why not?

3. For each of the following linear transformations $L\colon \mathcal{V} \to \mathcal{W}$, find a basis for $\ker(L)$ and a basis for $\operatorname{range}(L)$. Verify that $\dim(\ker(L)) + \dim(\operatorname{range}(L)) = \dim(\mathcal{V})$.

$\star$(a) $L\colon \mathbb{R}^3 \to \mathbb{R}^3$ given by

$$L\left(\begin{bmatrix} x_1 \\ x_2 \\ x_3 \end{bmatrix}\right) = \begin{bmatrix} 1 & -1 & 5 \\ -2 & 3 & -13 \\ 3 & -3 & 15 \end{bmatrix} \begin{bmatrix} x_1 \\ x_2 \\ x_3 \end{bmatrix}$$

(b) $L\colon \mathbb{R}^3 \to \mathbb{R}^4$ given by

$$L\left(\begin{bmatrix} x_1 \\ x_2 \\ x_3 \end{bmatrix}\right) = \begin{bmatrix} 4 & -2 & 8 \\ 7 & 1 & 5 \\ -2 & -1 & 0 \\ 3 & -2 & 7 \end{bmatrix} \begin{bmatrix} x_1 \\ x_2 \\ x_3 \end{bmatrix}$$

(c) $L\colon \mathbb{R}^3 \to \mathbb{R}^2$ given by

$$L\left(\begin{bmatrix} x_1 \\ x_2 \\ x_3 \end{bmatrix}\right) = \begin{bmatrix} 3 & 2 & 11 \\ 2 & 1 & 8 \end{bmatrix} \begin{bmatrix} x_1 \\ x_2 \\ x_3 \end{bmatrix}$$

$\star$(d) $L\colon \mathbb{R}^4 \to \mathbb{R}^5$ given by

$$L\left(\begin{bmatrix} x_1 \\ x_2 \\ x_3 \\ x_4 \end{bmatrix}\right) = \begin{bmatrix} -14 & -8 & -10 & 2 \\ -4 & -1 & 1 & -2 \\ -6 & 2 & 12 & -10 \\ 3 & -7 & -24 & 17 \\ 4 & 2 & 2 & 0 \end{bmatrix} \begin{bmatrix} x_1 \\ x_2 \\ x_3 \\ x_4 \end{bmatrix}$$

4. For each of the following linear transformations $L: \mathcal{V} \to \mathcal{W}$, find a basis for ker($L$) and a basis for range($L$), and verify that dim(ker(L)) + dim(range(L)) = dim($\mathcal{V}$):

★(a) $L: \mathbb{R}^3 \to \mathbb{R}^2$ given by $L([x_1, x_2, x_3]) = [0, x_2]$

(b) $L: \mathbb{R}^2 \to \mathbb{R}^3$ given by $L([x_1, x_2]) = [x_1, x_1 + x_2, x_2]$

(c) $L: \mathcal{M}_{22} \to \mathcal{M}_{32}$ given by $L\left(\begin{bmatrix} a_{11} & a_{12} \\ a_{21} & a_{22} \end{bmatrix}\right) = \begin{bmatrix} 0 & -a_{12} \\ -a_{21} & 0 \\ 0 & 0 \end{bmatrix}$

★(d) $L: \mathcal{P}_4 \to \mathcal{P}_2$ given by $L(ax^4 + bx^3 + cx^2 + dx + e) = cx^2 + dx + e$

(e) $L: \mathcal{P}_2 \to \mathcal{P}_3$ given by $L(ax^2 + bx + c) = cx^3 + bx^2 + ax$

★(f) $L: \mathbb{R}^3 \to \mathbb{R}^3$ given by $L([x_1, x_2, x_3]) = [x_1, 0, x_1 - x_2 + x_3]$

★(g) $L: \mathcal{M}_{22} \to \mathcal{M}_{22}$ given by $L(\mathbf{A}) = \mathbf{A}^T$

(h) $L: \mathcal{M}_{33} \to \mathcal{M}_{33}$ given by $L(\mathbf{A}) = \mathbf{A} - \mathbf{A}^T$

★(i) $L: \mathcal{P}_2 \to \mathbb{R}^2$ given by $L(\mathbf{p}) = \left[\mathbf{p}(1), \mathbf{p}'(1)\right]$

(j) $L: \mathcal{P}_4 \to \mathbb{R}^3$ given by $L(\mathbf{p}) = [\mathbf{p}(-1), \mathbf{p}(0), \mathbf{p}(1)]$

5. (a) Suppose that $L: \mathcal{V} \to \mathcal{W}$ is the linear transformation given by $L(\mathbf{v}) = \mathbf{0}_{\mathcal{W}}$, for all $\mathbf{v} \in \mathcal{V}$. What is ker($L$)? What is range($L$)?

(b) Suppose that $L: \mathcal{V} \to \mathcal{V}$ is the linear transformation given by $L(\mathbf{v}) = \mathbf{v}$, for all $\mathbf{v} \in \mathcal{V}$. What is ker($L$)? What is range($L$)?

★6. Consider the mapping $L: \mathcal{M}_{33} \to \mathbb{R}$ given by $L(\mathbf{A}) = \text{trace}(\mathbf{A})$ (see Exercise 14 in Section 1.4). Show that L is a linear transformation. What is ker(L)? What is range(L)? Calculate dim(ker(L)) and dim(range(L)).

7. Let $\mathcal{V}$ be a vector space with fixed basis $B = \{\mathbf{v}_1, \ldots, \mathbf{v}_n\}$. Define $L: \mathcal{V} \to \mathcal{V}$ by $L(\mathbf{v}_1) = \mathbf{v}_2$, $L(\mathbf{v}_2) = \mathbf{v}_3, \ldots$, $L(\mathbf{v}_{n-1}) = \mathbf{v}_n$, $L(\mathbf{v}_n) = \mathbf{v}_1$. Find range($L$). What is ker($L$)?

★8. Consider $L: \mathcal{P}_2 \to \mathcal{P}_4$ given by $L(\mathbf{p}) = x^2\mathbf{p}$. What is ker($L$)? What is range($L$)? Verify that dim(ker($L$)) + dim(range($L$)) = dim($\mathcal{P}_2$).

9. Consider $L: \mathcal{P}_4 \to \mathcal{P}_2$ given by $L(\mathbf{p}) = \mathbf{p}''$. What is ker($L$)? What is range($L$)? Verify that dim(ker($L$)) + dim(range($L$)) = dim($\mathcal{P}_4$).

★10. Consider $L: \mathcal{P}_n \to \mathcal{P}_n$ given by $L(\mathbf{p}) = \mathbf{p}^{(k)}$ (the kth derivative of $\mathbf{p}$), where $k \le n$. What is dim(ker(L))? What is dim(range(L))? What happens when $k > n$?

11. Let a be a fixed real number. Consider $L: \mathcal{P}_n \to \mathbb{R}$ given by $L(\mathbf{p}(x)) = \mathbf{p}(a)$ (that is, the evaluation of $\mathbf{p}$ at $x = a$). (Recall from Exercise 18 in Section 5.1 that L is a linear transformation.) Show that $\{x - a, x^2 - a^2, \ldots, x^n - a^n\}$ is a basis for ker(L). (Hint: What is range(L)?)

★12. Suppose that $L: \mathbb{R}^n \to \mathbb{R}^n$ is a linear operator given by $L(\mathbf{X}) = \mathbf{AX}$, where $|\mathbf{A}| \ne 0$. What is ker(L)? What is range(L)?

13. Let $\mathcal{V}$ be a finite dimensional vector space, and let $L: \mathcal{V} \to \mathcal{V}$ be a linear operator. Show that ker(L) = $\{\mathbf{0}_{\mathcal{V}}\}$ if and only if range(L) = $\mathcal{V}$.

14. Let $L: \mathcal{V} \to \mathcal{W}$ be a linear transformation. Prove directly that ker(L) is a subspace of $\mathcal{V}$ and that range(L) is a subspace of $\mathcal{W}$ using Theorem 4.2; that is, without invoking Theorem 5.9.

15. Let $L_1: \mathcal{V} \to \mathcal{W}$ and $L_2: \mathcal{W} \to \mathcal{X}$ be linear transformations.

   (a) Show that $\ker(L_1) \subseteq \ker(L_2 \circ L_1)$.

   (b) Show that $\mathrm{range}(L_2 \circ L_1) \subseteq \mathrm{range}(L_2)$.

   (c) If $\mathcal{V}$ is finite dimensional, prove that $\dim(\mathrm{range}(L_2 \circ L_1)) \leq \dim(\mathrm{range}(L_1))$.

★16. Give an example of a linear operator $L: \mathbb{R}^2 \to \mathbb{R}^2$ such that $\ker(L) = \mathrm{range}(L)$.

17. Let $L : \mathbb{R}^n \to \mathbb{R}^m$ be given by $L(\mathbf{X}) = \mathbf{AX}$ for a fixed $m \times n$ matrix $\mathbf{A}$.

   (a) Prove that the Range Method correctly produces a basis for $\mathrm{range}(L)$.

   (b) Use part (a) to prove that $\dim(\mathrm{range}(L)) = \mathrm{rank}(\mathbf{A})$.

   (c) Use part (b) and the Dimension Theorem to prove $\dim(\ker(L)) = n - \mathrm{rank}(\mathbf{A})$.

   (d) Prove that the Kernel Method correctly produces a basis for $\ker(L)$. (Hint: First show the vectors created by the method are linearly independent. Then use part (c) and apply part (2) of Corollary 4.14.)

18. Let $L : \mathcal{V} \to \mathcal{W}$ be a linear transformation between finite dimensional vector spaces. Suppose that the $m \times n$ matrix $\mathbf{A}$ is the matrix for L with respect to bases B and C. Let $T : \mathbb{R}^n \to \mathbb{R}^m$ be the linear transformation given by $T(\mathbf{X}) = \mathbf{AX}$.

   (a) Prove that $\dim(\ker(L)) = \dim(\ker(T))$.

   (b) Prove that $\dim(\mathrm{range}(L)) = \dim(\mathrm{range}(T))$. (Hint: Use part (a) and the Dimension Theorem.)

   (c) With $\mathcal{V} = \mathbb{R}^n$ and $\mathcal{W} = \mathbb{R}^m$, use parts (a) and (b) and Exercise 17 to prove Theorem 5.12. (Notice that Theorem 5.12 allows *any* bases to be used for $\mathbb{R}^n$ and $\mathbb{R}^m$, while Exercise 17 assumes that the standard bases are used.)

   ★(d) State a theorem that generalizes Theorem 5.12, and prove it using Theorem 5.12 and parts (a) and (b) of this exercise.

19. Let $\mathbf{A}$ be an $m \times n$ matrix and let $\mathbf{P}$ and $\mathbf{Q}$ be nonsingular matrices of size $m \times m$ and $n \times n$, respectively.

   (a) Use Theorem 5.12 to prove that $\mathrm{rank}(\mathbf{A}) = \mathrm{rank}(\mathbf{PA})$.

   (b) Use Theorem 5.12 to prove that $\mathrm{rank}(\mathbf{A}) = \mathrm{rank}(\mathbf{AQ})$.

   (c) Give an alternate proof for part (b) using part (a) and Corollary 5.13.

★20. True or False:

   (a) If $L: \mathcal{V} \to \mathcal{W}$ is a linear transformation, then $\ker(L) = \{L(\mathbf{v}) \mid \mathbf{v} \in \mathcal{V}\}$.

   (b) If $L: \mathcal{V} \to \mathcal{W}$ is a linear transformation, then $\mathrm{range}(L)$ is a subspace of $\mathcal{V}$.

   (c) If $L: \mathcal{V} \to \mathcal{W}$ is a linear transformation and $\dim(\mathcal{V}) = n$, then $\dim(\ker(L)) = n - \dim(\mathrm{range}(L))$.

   (d) If $L: \mathcal{V} \to \mathcal{W}$ is a linear transformation and $\dim(\mathcal{V}) = 5$ and $\dim(\mathcal{W}) = 3$, then the Dimension Theorem implies that $\dim(\ker(L)) = 2$.

   (e) If $L: \mathbb{R}^n \to \mathbb{R}^m$ is a linear transformation and $L(\mathbf{X}) = \mathbf{AX}$, then $\dim(\ker(L))$ equals the number of nonpivot columns in the reduced row echelon form matrix for $\mathbf{A}$.

(f) If $L: \mathbb{R}^n \to \mathbb{R}^m$ is a linear transformation and $L(\mathbf{X}) = \mathbf{AX}$, then $\dim(\text{range}(L)) = n - \text{rank}(\mathbf{A})$.

(g) If $\mathbf{A}$ is a 5×5 matrix, and $\text{rank}(\mathbf{A}) = 2$, then $\text{rank}(\mathbf{A}^T) = 3$.

(h) If $\mathbf{A}$ is any matrix, then the row space of $\mathbf{A}$ equals the column space of $\mathbf{A}$.

5.4 Isomorphism

In this section, we examine methods for determining whether two vector spaces are equivalent, or *isomorphic*. Isomorphism is important because if certain algebraic results are true in one of two isomorphic vector spaces, corresponding results hold true in the other as well.

We first study the following two special types of linear transformations: *one-to-one* and *onto*. We will see that if there is a linear transformation from a vector space $\mathcal{V}$ to a vector space $\mathcal{W}$ that is both one-to-one and onto, then $\mathcal{V}$ and $\mathcal{W}$ are isomorphic vector spaces.

One-to-One and Onto Linear Transformations

One-to-one functions and onto functions are defined and discussed in Appendix B. In particular, Appendix B contains the usual methods for proving that a given function is, or is not, one-to-one or onto. Now, we are interested primarily in linear transformations, so we restate the definitions of *one-to-one* and *onto* specifically as they apply to this type of function.

DEFINITION

Let $L: \mathcal{V} \to \mathcal{W}$ be a linear transformation.

(1) L is **one-to-one** if and only if distinct vectors in $\mathcal{V}$ have different images in $\mathcal{W}$. That is, L is **one-to-one** if and only if, for all $\mathbf{v}_1, \mathbf{v}_2 \in \mathcal{V}$, $L(\mathbf{v}_1) = L(\mathbf{v}_2)$ implies $\mathbf{v}_1 = \mathbf{v}_2$.

(2) L is **onto** if and only if every vector in the codomain $\mathcal{W}$ is the image of some vector in the domain $\mathcal{V}$. That is, L is **onto** if and only if, for every $\mathbf{w} \in \mathcal{W}$, there is some $\mathbf{v} \in \mathcal{V}$ such that $L(\mathbf{v}) = \mathbf{w}$.

Notice that the two descriptions of a one-to-one linear transformation given in this definition are really contrapositives of each other.

EXAMPLE 1   **Rotation:** Recall the rotation linear operator $L: \mathbb{R}^2 \to \mathbb{R}^2$ from Example 9 in Section 5.1 given by $L(\mathbf{v}) = \mathbf{Av}$, where $\mathbf{A} = \begin{bmatrix} \cos\theta & -\sin\theta \\ \sin\theta & \cos\theta \end{bmatrix}$. We will show that L is *both one-to-one and onto*.

To show that L is one-to-one, we take any two arbitrary vectors $\mathbf{v}_1$ and $\mathbf{v}_2$ in the domain $\mathbb{R}^2$, assume that $L(\mathbf{v}_1) = L(\mathbf{v}_2)$, and prove that $\mathbf{v}_1 = \mathbf{v}_2$. Now, if $L(\mathbf{v}_1) = L(\mathbf{v}_2)$, then $\mathbf{Av}_1 = \mathbf{Av}_2$. Because $\mathbf{A}$ is nonsingular, we can multiply both sides on the left by $\mathbf{A}^{-1}$ to obtain $\mathbf{v}_1 = \mathbf{v}_2$. Hence, L is one-to-one.

To show that L is onto, we must take any arbitrary vector $\mathbf{w}$ in the codomain $\mathbb{R}^2$ and show that there is some vector $\mathbf{v}$ in the domain $\mathbb{R}^2$ that maps to $\mathbf{w}$. Recall that multiplication by $\mathbf{A}^{-1}$ undoes the action of multiplication by $\mathbf{A}$, and so it must represent a *clockwise* rotation through the angle θ. Hence, we can find a pre-image for $\mathbf{w}$ by rotating it *clockwise* through the angle θ; that is, consider $\mathbf{v} = \mathbf{A}^{-1}\mathbf{w} \in \mathbb{R}^2$. When we apply L to $\mathbf{v}$, we rotate it *counterclockwise* through the same angle θ: $L(\mathbf{v}) = \mathbf{A}(\mathbf{A}^{-1}\mathbf{w}) = \mathbf{w}$, thus obtaining the original vector $\mathbf{w}$. Since $\mathbf{v}$ is in the domain and $\mathbf{v}$ maps to $\mathbf{w}$ under L, L is onto. ∎

EXAMPLE 2

Differentiation: Consider the linear transformation $L: \mathcal{P}_3 \to \mathcal{P}_2$ given by $L(\mathbf{p}) = \mathbf{p}'$. We will show that L is *onto but not one-to-one*.

To show that L is not one-to-one, we must find two different vectors $\mathbf{p}_1$ and $\mathbf{p}_2$ in the domain $\mathcal{P}_3$ that have the same image. Consider $\mathbf{p}_1 = x + 1$ and $\mathbf{p}_2 = x + 2$. Since $L(\mathbf{p}_1) = L(\mathbf{p}_2) = 1$, L is not one-to-one.

To show that L is onto, we must take an arbitrary vector $\mathbf{q}$ in $\mathcal{P}_2$ and find some vector $\mathbf{p}$ in $\mathcal{P}_3$ such that $L(\mathbf{p}) = \mathbf{q}$. Consider the vector $\mathbf{p} = \int \mathbf{q}(x)\,dx$ with zero constant term. Because $L(\mathbf{p}) = \mathbf{q}$, we see that L is onto. ∎

If in Example 2 we had used $\mathcal{P}_3$ for the codomain instead of $\mathcal{P}_2$, the linear transformation would not have been onto because x^3 would have no pre-image (why?). This provides an example of a linear transformation that is neither one-to-one nor onto. Also, Exercise 7 illustrates a linear transformation that is one-to-one but not onto. These examples, together with Examples 1 and 2, show that the concepts of one-to-one and onto are independent of each other; that is, there are linear transformations that have either property with or without the other.

Theorem B.1 in Appendix B shows that the composition of one-to-one linear transformations is one-to-one, and similarly, the composition of onto linear transformations is onto.

The Kernel of a One-to-One Linear Transformation

The next theorem gives an alternate way of proving that a linear transformation is one-to-one.

THEOREM 5.14

Let $\mathcal{V}$ and $\mathcal{W}$ be vector spaces, and let $L: \mathcal{V} \to \mathcal{W}$ be a linear transformation. Then L is one-to-one if and only if $\ker(L) = \{\mathbf{0}_\mathcal{V}\}$ (or, equivalently, if and only if $\dim(\ker(L)) = 0$).

Thus, a linear transformation whose kernel contains a nonzero vector cannot be one-to-one.

PROOF OF THEOREM 5.14

First suppose that L is one-to-one, and let $\mathbf{v} \in \ker(L)$. We must show that $\mathbf{v} = \mathbf{0}_\mathcal{V}$. Now, $L(\mathbf{v}) = \mathbf{0}_\mathcal{W}$. However, by Theorem 5.1, $L(\mathbf{0}_\mathcal{V}) = \mathbf{0}_\mathcal{W}$. Because $L(\mathbf{v}) = L(\mathbf{0}_\mathcal{V})$ and L is one-to-one, we must have $\mathbf{v} = \mathbf{0}_\mathcal{V}$.

Conversely, suppose that $\ker(L) = \{\mathbf{0}_\mathcal{V}\}$. We must show that L is one-to-one. Let $\mathbf{v}_1, \mathbf{v}_2 \in \mathcal{V}$, with $L(\mathbf{v}_1) = L(\mathbf{v}_2)$. We must show that $\mathbf{v}_1 = \mathbf{v}_2$. Now, $L(\mathbf{v}_1) - L(\mathbf{v}_2) = \mathbf{0}_\mathcal{W}$, implying that $L(\mathbf{v}_1 - \mathbf{v}_2) = \mathbf{0}_\mathcal{W}$. Hence, $\mathbf{v}_1 - \mathbf{v}_2 \in \ker(L)$, by definition of the kernel. Since $\ker(L) = \{\mathbf{0}_\mathcal{V}\}$, $\mathbf{v}_1 - \mathbf{v}_2 = \mathbf{0}_\mathcal{V}$ and so $\mathbf{v}_1 = \mathbf{v}_2$. ∎

EXAMPLE 3   Consider the linear transformation $L: \mathcal{M}_{22} \to \mathcal{M}_{23}$ given by $L\left(\begin{bmatrix} a & b \\ c & d \end{bmatrix}\right) = \begin{bmatrix} a-b & 0 & c-d \\ c+d & a+b & 0 \end{bmatrix}$. If $\begin{bmatrix} a & b \\ c & d \end{bmatrix} \in \ker(L)$, then $a-b = c-d = c+d = a+b = 0$. Solving these equations yields $a = b = c = d = 0$, and so $\ker(L)$ contains only the zero matrix $\begin{bmatrix} 0 & 0 \\ 0 & 0 \end{bmatrix}$. By Theorem 5.14, L is one-to-one.

On the other hand, consider $M: \mathcal{M}_{23} \to \mathcal{M}_{22}$ given by $M\left(\begin{bmatrix} a & b & c \\ d & e & f \end{bmatrix}\right) = \begin{bmatrix} a+b & a+c \\ d+e & d+f \end{bmatrix}$. Notice that $M\left(\begin{bmatrix} 1 & -1 & -1 \\ 1 & -1 & -1 \end{bmatrix}\right) = \begin{bmatrix} 0 & 0 \\ 0 & 0 \end{bmatrix}$, and so $\ker(M)$ contains a nonzero vector. Therefore, M is not one-to-one. ∎

Suppose that $\mathcal{V}$ and $\mathcal{W}$ are finite dimensional vector spaces and $L: \mathcal{V} \to \mathcal{W}$ is a linear transformation. If $\dim(\mathcal{V}) = \dim(\mathcal{W})$, the next result asserts that we need only check that L is *either* one-to-one or onto to know that L has the other property as well.

COROLLARY 5.15

Let $\mathcal{V}$ and $\mathcal{W}$ be finite dimensional vector spaces with $\dim(\mathcal{V}) = \dim(\mathcal{W})$. Let $L: \mathcal{V} \to \mathcal{W}$ be a linear transformation. Then L is one-to-one if and only if L is onto.

PROOF OF COROLLARY 5.15

Let $\mathcal{V}$ and $\mathcal{W}$ be finite dimensional vector spaces with $\dim(\mathcal{V}) = \dim(\mathcal{W})$, and let $L: \mathcal{V} \to \mathcal{W}$ be a linear transformation. Then

L is one-to-one	$\Leftrightarrow \dim(\ker(L)) = 0$	by Theorem 5.14
	$\Leftrightarrow \dim(\mathcal{V}) = \dim(\mathrm{range}(L))$	by the Dimension Theorem
	$\Leftrightarrow \dim(\mathcal{W}) = \dim(\mathrm{range}(L))$	because $\dim(\mathcal{V}) = \dim(\mathcal{W})$
	$\Leftrightarrow L$ is onto.	by Theorem 4.17

∎

EXAMPLE 4   Consider $L: \mathcal{P}_2 \to \mathbb{R}^3$ given by $L(\mathbf{p}) = [\mathbf{p}(0), \mathbf{p}(1), \mathbf{p}(2)]$. Now, $\dim(\mathcal{P}_2) = \dim(\mathbb{R}^3) = 3$. Hence, by Corollary 5.15, if L is either one-to-one or onto, it has the other property as well.

We will show that L is one-to-one using Theorem 5.14. If $\mathbf{p} \in \ker(L)$, then $L(\mathbf{p}) = \mathbf{0}$, and so $\mathbf{p}(0) = \mathbf{p}(1) = \mathbf{p}(2) = 0$. Hence $\mathbf{p}$ is a polynomial of degree

≤ 2 touching the x-axis at $x = 0$, $x = 1$, and $x = 2$. Since the graph of $\mathbf{p}$ must be either a parabola or a line, it cannot touch the x-axis at three distinct points unless its graph is the line $y = 0$. That is, $\mathbf{p} = \mathbf{0}$ in $\mathcal{P}_2$. Therefore, $\ker(L) = \{\mathbf{0}\}$, and L is one-to-one.

Now, by Corollary 5.15, L is onto. Thus, given any 3-vector $[a, b, c]$, there is some $\mathbf{p} \in \mathcal{P}_2$ such that $\mathbf{p}(0) = a$, $\mathbf{p}(1) = b$, and $\mathbf{p}(2) = c$. (This example is generalized further in Exercise 29.) ■

Isomorphisms: Invertible Linear Transformations

We restate here the definition from Appendix B for the inverse of a function as it applies to linear transformations.

DEFINITION

Let $L: \mathcal{V} \to \mathcal{W}$ be a linear transformation. Then L is an **invertible linear transformation** if and only if there is a function $M: \mathcal{W} \to \mathcal{V}$ such that $(M \circ L)(\mathbf{v}) = \mathbf{v}$, for all $\mathbf{v} \in \mathcal{V}$, and $(L \circ M)(\mathbf{w}) = \mathbf{w}$, for all $\mathbf{w} \in \mathcal{W}$. Such a function M is called an **inverse** of L.

If the inverse M of $L: \mathcal{V} \to \mathcal{W}$ exists, then it is unique by Theorem B.3 and is usually denoted by $L^{-1}: \mathcal{W} \to \mathcal{V}$.

DEFINITION

A linear transformation $L: \mathcal{V} \to \mathcal{W}$ that is both one-to-one and onto is called an **isomorphism** from $\mathcal{V}$ to $\mathcal{W}$.

The next result shows that the previous two definitions actually refer to the same class of linear transformations.

THEOREM 5.16

Let $L: \mathcal{V} \to \mathcal{W}$ be a linear transformation. Then L is an isomorphism if and only if L is an invertible linear transformation. Moreover, if L is invertible, then L^{-1} is also a linear transformation.

Notice that Theorem 5.16 also asserts that whenever L is an isomorphism, L^{-1} is an isomorphism as well because L^{-1} is an invertible linear transformation (with L as its inverse).

PROOF OF THEOREM 5.16

The "if and only if" part of Theorem 5.16 follows directly from Theorem B.2. Thus, we only need to prove the last assertion in Theorem 5.16. That is, suppose $L: \mathcal{V} \to \mathcal{W}$ is invertible (and thus, an isomorphism) with inverse L^{-1}. We need to prove L^{-1} is a linear transformation. To do this, we must show both of the following properties hold:

(1) $L^{-1}(\mathbf{w}_1 + \mathbf{w}_2) = L^{-1}(\mathbf{w}_1) + L^{-1}(\mathbf{w}_2)$, for all $\mathbf{w}_1, \mathbf{w}_2 \in \mathcal{W}$

(2) $L^{-1}(c\mathbf{w}) = cL^{-1}(\mathbf{w})$, for all $c \in \mathbb{R}$, and for all $\mathbf{w} \in \mathcal{W}$.

Property (1): Because L is an isomorphism, L is one-to-one. Hence, if we can show that $L(L^{-1}(\mathbf{w}_1 + \mathbf{w}_2)) = L(L^{-1}(\mathbf{w}_1) + L^{-1}(\mathbf{w}_2))$, we will be done. But,

$$\begin{aligned} L(L^{-1}(\mathbf{w}_1) + L^{-1}(\mathbf{w}_2)) &= L(L^{-1}(\mathbf{w}_1)) + L(L^{-1}(\mathbf{w}_2)) \\ &= \mathbf{w}_1 + \mathbf{w}_2 \\ &= L(L^{-1}(\mathbf{w}_1 + \mathbf{w}_2)). \end{aligned}$$

Property (2): Again, because L is an isomorphism, L is one-to-one. Hence, if we can show that $L(L^{-1}(c\mathbf{w})) = L(cL^{-1}(\mathbf{w}))$, we will be done. But,

$$\begin{aligned} L(cL^{-1}(\mathbf{w})) &= cL(L^{-1}(\mathbf{w})) \\ &= c\mathbf{w} \\ &= L(L^{-1}(c\mathbf{w})). \end{aligned}$$

Because both properties (1) and (2) hold, L^{-1} is a linear transformation. ∎

EXAMPLE 5   Recall the rotation linear operator $L: \mathbb{R}^2 \to \mathbb{R}^2$ with

$$L\left(\begin{bmatrix} x \\ y \end{bmatrix}\right) = \begin{bmatrix} \cos\theta & -\sin\theta \\ \sin\theta & \cos\theta \end{bmatrix} \begin{bmatrix} x \\ y \end{bmatrix}$$

given in Example 1. We have shown that L is both one-to-one and onto. Hence, L is an isomorphism and has an inverse, L^{-1}. Because L represents a *counterclockwise* rotation of vectors through the angle θ, then L^{-1} must represent a *clockwise* rotation through the angle θ, as we saw in Example 1. Equivalently, L^{-1} can be thought of as a *counterclockwise* rotation through the angle $-\theta$. Thus,

$$L^{-1}\left(\begin{bmatrix} x \\ y \end{bmatrix}\right) = \begin{bmatrix} \cos(-\theta) & -\sin(-\theta) \\ \sin(-\theta) & \cos(-\theta) \end{bmatrix} \begin{bmatrix} x \\ y \end{bmatrix} = \begin{bmatrix} \cos\theta & \sin\theta \\ -\sin\theta & \cos\theta \end{bmatrix} \begin{bmatrix} x \\ y \end{bmatrix}.$$

Of course, L^{-1} is also an isomorphism. ∎

The next theorem gives a simple method for determining whether a linear transformation between finite dimensional vector spaces is an isomorphism.

THEOREM 5.17

Let $\mathcal{V}$ and $\mathcal{W}$ both be nontrivial n-dimensional vector spaces with ordered bases B and C, respectively, and let $L: \mathcal{V} \to \mathcal{W}$ be a linear transformation. Then L is an isomorphism if and only if the matrix representation $\mathbf{A}_{BC}$ for L with respect to B and C is nonsingular.

To prove one half of Theorem 5.17, let $\mathbf{A}_{BC}$ be the matrix for L with respect to B and C, and let $\mathbf{D}_{CB}$ be the matrix for L^{-1} with respect to C and B. Theorem 5.8 then shows that $\mathbf{D}_{CB}\mathbf{A}_{BC} = \mathbf{I}_n$, and so $\mathbf{A}_{BC}$ is nonsingular. An analogous observation proves the converse. We leave the details as Exercise 15.

EXAMPLE 6 Consider $L: \mathbb{R}^3 \to \mathbb{R}^3$ given by $L(\mathbf{v}) = \mathbf{A}\mathbf{v}$, where

$$\mathbf{A} = \begin{bmatrix} 1 & 0 & 3 \\ 0 & 1 & 3 \\ 0 & 0 & 1 \end{bmatrix}.$$

Now, $\mathbf{A}$ is nonsingular ($|\mathbf{A}| = 1 \neq 0$). Hence, by Theorem 5.17, L is an isomorphism. Geometrically, L represents a shear in the z-direction (see Table 5.1). ∎

Theorem B.4 in Appendix B shows that the composition of isomorphisms results in an isomorphism. In particular, the inverse of the composition $L_2 \circ L_1$ is $L_1^{-1} \circ L_2^{-1}$. That is, the transformations must be undone in *reverse* order to arrive at the correct inverse. (Compare this with part (3) of Theorem 2.11 for matrix multiplication.)

Isomorphic Vector Spaces

DEFINITION

Let $\mathcal{V}$ and $\mathcal{W}$ be vector spaces. Then $\mathcal{V}$ is **isomorphic** to $\mathcal{W}$, denoted $\mathcal{V} \cong \mathcal{W}$, if and only if there exists an isomorphism $L: \mathcal{V} \to \mathcal{W}$.

If $\mathcal{V} \cong \mathcal{W}$, there is some isomorphism $L: \mathcal{V} \to \mathcal{W}$. Then by Theorem 5.16, $L^{-1}: \mathcal{W} \to \mathcal{V}$ is also an isomorphism, so $\mathcal{W} \cong \mathcal{V}$. Hence, we usually speak of such $\mathcal{V}$ and $\mathcal{W}$ as being *isomorphic to each other*.

Also notice that if $\mathcal{V} \cong \mathcal{W}$ and $\mathcal{W} \cong \mathcal{X}$, then there are isomorphisms $L_1: \mathcal{V} \to \mathcal{W}$ and $L_2: \mathcal{W} \to \mathcal{X}$. But then $L_2 \circ L_1: \mathcal{V} \to \mathcal{X}$ is an isomorphism, and so $\mathcal{V} \cong \mathcal{X}$. In other words, two vector spaces such as $\mathcal{V}$ and $\mathcal{X}$ that are both isomorphic to the same vector space $\mathcal{W}$ are isomorphic to each other.

EXAMPLE 7 Consider $L_1: \mathbb{R}^4 \to \mathcal{P}_3$ given by $L_1([a, b, c, d]) = ax^3 + bx^2 + cx + d$ and $L_2: \mathcal{M}_{22} \to \mathcal{P}_3$ given by $L_2\left(\begin{bmatrix} a & b \\ c & d \end{bmatrix}\right) = ax^3 + bx^2 + cx + d$. L_1 and L_2 are certainly both isomorphisms. Hence, $\mathbb{R}^4 \cong \mathcal{P}_3$ and $\mathcal{M}_{22} \cong \mathcal{P}_3$. Thus, the composition $L_2^{-1} \circ L_1: \mathbb{R}^4 \to \mathcal{M}_{22}$ is also an isomorphism, and so $\mathbb{R}^4 \cong \mathcal{M}_{22}$. Notice that all of these vector spaces have dimension 4. ∎

Next, we show that finite dimensional vector spaces $\mathcal{V}$ and $\mathcal{W}$ must have the same dimension for an isomorphism to exist between them.

THEOREM 5.18

Suppose $\mathcal{V} \cong \mathcal{W}$ and $\mathcal{V}$ is finite dimensional. Then $\mathcal{W}$ is finite dimensional and $\dim(\mathcal{V}) = \dim(\mathcal{W})$.

PROOF OF THEOREM 5.18

Since $\mathcal{V} \cong \mathcal{W}$, there is an isomorphism $L: \mathcal{V} \rightarrow \mathcal{W}$. Since L is onto, $\mathcal{W} = $ range(L). Hence, $\mathcal{W}$ is finite dimensional by the Dimension Theorem.

Because L is one-to-one, $\dim(\ker(L)) = 0$ by Theorem 5.14. Then, by the Dimension Theorem, $\dim(\mathcal{V}) = \dim(\text{range}(L)) = \dim(\mathcal{W})$, since L is onto. ∎

Theorem 5.18 implies that there is no possible isomorphism from, say, $\mathbb{R}^3$ to $\mathcal{P}_4$ or from $\mathcal{M}_{22}$ to $\mathbb{R}^3$, because the dimensions of the spaces do not agree. In addition, the corresponding matrix for any isomorphism $L: \mathcal{V} \rightarrow \mathcal{W}$ (with respect to any bases for $\mathcal{V}$ and $\mathcal{W}$) must be a square matrix, since $\dim(\mathcal{V}) = \dim(\mathcal{W})$.

All n-Dimensional Vector Spaces Are Isomorphic

Example 7 hints that any two finite dimensional vector spaces of the same dimension are isomorphic. This result, which is one of the most important in all linear algebra, is a corollary of the next theorem.

THEOREM 5.19

If $\mathcal{V}$ is any n-dimensional vector space, then $\mathcal{V} \cong \mathbb{R}^n$.

PROOF OF THEOREM 5.19

Suppose that $\mathcal{V}$ is a vector space with $\dim(\mathcal{V}) = n$. If we can find an isomorphism $L: \mathcal{V} \rightarrow \mathbb{R}^n$, then $\mathcal{V} \cong \mathbb{R}^n$, and we will be done. Let $B = (\mathbf{v}_1, \ldots, \mathbf{v}_n)$ be an ordered basis for $\mathcal{V}$. Consider the mapping $L(\mathbf{v}) = [\mathbf{v}]_B$, for all $\mathbf{v} \in \mathcal{V}$. Now, L is a linear transformation by Example 4 in Section 5.1. Also,

$$\mathbf{v} \in \ker(L) \iff [\mathbf{v}]_B = [0, \ldots, 0] \iff \mathbf{v} = 0\mathbf{v}_1 + \cdots + 0\mathbf{v}_n \iff \mathbf{v} = \mathbf{0}.$$

Hence, $\ker(L) = \{\mathbf{0}_\mathcal{V}\}$, and L is one-to-one. Since $\dim(\mathcal{V}) = \dim(\mathbb{R}^n)$, L is onto by Corollary 5.15. Hence, L is an isomorphism. ∎

In particular, Theorem 5.19 tells us that $\mathcal{P}_n \cong \mathbb{R}^{n+1}$ and that $\mathcal{M}_{mn} \cong \mathbb{R}^{mn}$. Also, the proof of Theorem 5.19 illustrates that coordinatization of vectors in an n-dimensional vector space $\mathcal{V}$ automatically gives an isomorphism of $\mathcal{V}$ with $\mathbb{R}^n$.

By the remarks before Example 7, Theorem 5.19 implies the following converse of Theorem 5.18:

COROLLARY 5.20

Any two n-dimensional vector spaces $\mathcal{V}$ and $\mathcal{W}$ are isomorphic. That is, if $\dim(\mathcal{V}) = \dim(\mathcal{W})$, then $\mathcal{V} \cong \mathcal{W}$.

For example, suppose that V and W are both vector spaces with $\dim(V) = \dim(W) = 47$. Then by Corollary 5.20, $V \cong W$ and by Theorem 5.19, $V \cong W \cong \mathbb{R}^{47}$.

Vector Space Properties Inherited through Isomorphism

Whenever two vector spaces V and W are isomorphic, the isomorphism $L: V \rightarrow W$ carries certain properties of the vectors in V over to their corresponding vectors in W. (Similarly, L^{-1} carries properties of vectors in W back to V.) Essentially, the elements of W behave "identically" to those of V, except that their "names" are different. For example, linearly independent subsets are carried over to linearly independent subsets, spanning sets to spanning sets, and bases to bases. These and other properties shared by isomorphic vector spaces are presented in Exercises 20 through 23.

We have been using the coordinatization isomorphism between $\mathcal{P}_n$ and $\mathbb{R}^{n+1}$ as well as between $\mathcal{M}_{mn}$ and $\mathbb{R}^{mn}$ in an intuitive way throughout the book to apply computational techniques that we developed in $\mathbb{R}^k$ to these other vector spaces. In particular, whenever we used the Simplified Span Method or the Independence Test Method on a vector space other than $\mathbb{R}^n$, we were essentially relying on an appropriate isomorphism to validate the process.

The next example shows in detail how the result in Theorem 5.12 can be extended to other vector spaces by using appropriate isomorphisms from the domain and codomain to $\mathbb{R}^n$ and $\mathbb{R}^m$, respectively[3]. (Recall from Theorem 5.12 that if $L: \mathbb{R}^n \rightarrow \mathbb{R}^m$ is a linear transformation with matrix $\mathbf{A}$, we have $\dim(\text{range}(L)) = \text{rank}(\mathbf{A})$ and $\dim(\ker(L)) = n - \text{rank}(\mathbf{A})$.)

EXAMPLE 8 Suppose that $L: \mathcal{P}_4 \rightarrow \mathcal{M}_{22}$ is a linear transformation with matrix

$$\mathbf{A} = \begin{bmatrix} 3 & -2 & 1 & 13 & -9 \\ -3 & 3 & -1 & -16 & 11 \\ 2 & -2 & 1 & 11 & -8 \\ -5 & 3 & -3 & -22 & 17 \end{bmatrix},$$

using the standard bases for $\mathcal{P}_4$ and $\mathcal{M}_{22}$. We will find a basis for $\ker(L)$ and a basis for $\text{range}(L)$.

Now, $\mathcal{P}_4 \cong \mathbb{R}^5$ and $\mathcal{M}_{22} \cong \mathbb{R}^4$, and it is simpler to "replace" vectors in $\mathcal{P}_4$ and $\mathcal{M}_{22}$ with their coordinatizations with respect to $\mathbb{R}^5$ and $\mathbb{R}^4$, respectively. That is, we replace every polynomial $a_4 x^4 + a_3 x^3 + a_2 x^2 + a_1 x + a_0$ in $\mathcal{P}_4$ with its coordinatization $[a_4, a_3, a_2, a_1, a_0]$ in $\mathbb{R}^5$ and every matrix $\begin{bmatrix} a & b \\ c & d \end{bmatrix}$ in $\mathcal{M}_{22}$ with its coordinatization $[a, b, c, d]$ in $\mathbb{R}^4$. Hence, for all practical purposes, we have replaced the given mapping $L: \mathcal{P}_4 \rightarrow \mathcal{M}_{22}$ with an equivalent transformation $T: \mathbb{R}^5 \rightarrow \mathbb{R}^4$. Then $\mathbf{A}$ is also the matrix for this equivalent mapping T (why?).

---

[3] Also see part (d) of Exercise 18 in Section 5.3.

Now the reduced row echelon form for $\mathbf{A}$ is

$$\mathbf{B} = \begin{bmatrix} 1 & 0 & 0 & 2 & -1 \\ 0 & 1 & 0 & -3 & 2 \\ 0 & 0 & 1 & 1 & -2 \\ 0 & 0 & 0 & 0 & 0 \end{bmatrix}.$$

Table 5.2

Conditions on linear transformations that are one-to-one, onto, or isomorphisms

Let $L: \mathcal{V} \to \mathcal{W}$ be a linear transformation, and let B be a basis for $\mathcal{V}$.

L is *one-to-one*

$\Leftrightarrow$	$\ker(L) = \{\mathbf{0}_{\mathcal{V}}\}$	Theorem 5.14
$\Leftrightarrow$	$\dim(\ker(L)) = 0$	Theorem 5.14
$\Leftrightarrow$	the image of every linearly independent set in $\mathcal{V}$ is linearly independent in $\mathcal{W}$	Exercise 20

L is *onto*

$\Leftrightarrow$	$\text{range}(L) = \mathcal{W}$	Definition
$\Leftrightarrow$	$\dim(\text{range}(L)) = \dim(\mathcal{W})$	Theorem 4.17*
$\Leftrightarrow$	the image of *every* spanning set for $\mathcal{V}$ is a spanning set for $\mathcal{W}$	Exercise 21
$\Leftrightarrow$	the image of *some* spanning set for $\mathcal{V}$ is a spanning set for $\mathcal{W}$	Exercise 21

L is an *isomorphism*

$\Leftrightarrow$	L is both one-to-one and onto	Definition
$\Leftrightarrow$	L is invertible (that is, $L^{-1}: \mathcal{W} \to \mathcal{V}$ exists)	Theorem 5.16
$\Leftrightarrow$	the matrix for L (with respect to *every* pair of ordered bases for $\mathcal{V}$ and $\mathcal{W}$) is nonsingular	Theorem 5.17*
$\Leftrightarrow$	the matrix for L (with respect to *some* pair of ordered bases for $\mathcal{V}$ and $\mathcal{W}$) is nonsingular	Theorem 5.17*
$\Leftrightarrow$	the images of vectors in B are distinct and $L(B)$ is a basis for $\mathcal{W}$	Exercise 22
$\Leftrightarrow$	L is one-to-one and $\dim(\mathcal{V}) = \dim(\mathcal{W})$	Corollary 5.15*
$\Leftrightarrow$	L is onto and $\dim(\mathcal{V}) = \dim(\mathcal{W})$	Corollary 5.15*

Furthermore, if $L: \mathcal{V} \to \mathcal{W}$ is an isomorphism, then

(1)	$\dim(\mathcal{V}) = \dim(\mathcal{W})$	Theorem 5.18*
(2)	L^{-1} is an isomorphism from $\mathcal{W}$ to $\mathcal{V}$	Theorem 5.16

*True only in the finite dimensional case

By Theorem 5.12, $\dim(\text{range}(T)) = \text{rank}(\mathbf{A}) = \text{rank}(\mathbf{B}) = 3$, and $\dim(\ker(T)) = \dim(\mathbb{R}^5) - \text{rank}(\mathbf{A}) = 5 - 3 = 2$. Hence, $\dim(\text{range}(L)) = 3$ and $\dim(\ker(L)) = 2$ as well.

We can find bases for range(L) and ker(L) by first finding bases for range(T) and ker(T). By taking the columns of $\mathbf{A}$ that correspond to pivot columns of $\mathbf{B}$, we have the following:

$$\text{Basis for range}(T) = \{[3, -3, 2, -5], [-2, 3, -2, 3], [1, -1, 1, -3]\}.$$

By replacing these coordinatization vectors in $\mathbb{R}^4$ with their corresponding 2×2 matrices in $\mathcal{M}_{22}$, we have the following:

$$\text{Basis for range}(L) = \left\{ \begin{bmatrix} 3 & -3 \\ 2 & -5 \end{bmatrix}, \begin{bmatrix} -2 & 3 \\ -2 & 3 \end{bmatrix}, \begin{bmatrix} 1 & -1 \\ 1 & -3 \end{bmatrix} \right\}.$$

Also, ker(T) is the solution set of $\mathbf{BX} = \mathbf{O}$, which is

$$\{a[-2, 3, -1, 1, 0] + b[1, -2, 2, 0, 1] \mid a, b \in \mathbb{R}\}.$$

Hence, a basis for ker(T) is $\{[-2, 3, -1, 1, 0], [1, -2, 2, 0, 1]\}$. By replacing these coordinatization vectors in $\mathbb{R}^5$ with their corresponding polynomials in $\mathcal{P}_4$, we find that a basis for ker(L) is $\{-2x^4 + 3x^3 - x^2 + x, \ x^4 - 2x^3 + 2x^2 + 1\}$. ∎

So far, we have proved many important results concerning the concepts of one-to-one, onto, and isomorphism. For convenience, these and other useful properties from the exercises are summarized in Table 5.2.

Exercises for Section 5.4

1. Which of the following linear transformations are one-to-one? Which are onto? Which are isomorphisms? Justify your answers without using row reduction.

   ★(a) $L: \mathbb{R}^3 \to \mathbb{R}^4$ given by $L([x, y, z]) = [y, z, -y, 0]$

   (b) $L: \mathbb{R}^3 \to \mathbb{R}^2$ given by $L([x, y, z]) = [x + y, y + z]$

   ★(c) $L: \mathbb{R}^3 \to \mathbb{R}^3$ given by $L([x, y, z]) = [2x, x + y + z, -y]$

   (d) $L: \mathcal{P}_3 \to \mathcal{P}_2$ given by $L(ax^3 + bx^2 + cx + d) = ax^2 + bx + c$

   ★(e) $L: \mathcal{P}_2 \to \mathcal{P}_2$ given by $L(ax^2 + bx + c) = (a+b)x^2 + (b+c)x + (a+c)$

   (f) $L: \mathcal{M}_{22} \to \mathcal{M}_{22}$ given by $L\left(\begin{bmatrix} a & b \\ c & d \end{bmatrix} \right) = \begin{bmatrix} d & b+c \\ b-c & a \end{bmatrix}$

   ★(g) $L: \mathcal{M}_{23} \to \mathcal{M}_{22}$ given by $L\left(\begin{bmatrix} a & b & c \\ d & e & f \end{bmatrix} \right) = \begin{bmatrix} a & -c \\ 2e & d+f \end{bmatrix}$

   ★(h) $L: \mathcal{P}_2 \to \mathcal{M}_{22}$ given by $L(ax^2 + bx + c) = \begin{bmatrix} a+c & 0 \\ b-c & -3a \end{bmatrix}$

2. Which of the following linear transformations are one-to-one? Which are onto? Which are isomorphisms? Justify your answers by using row reduction to determine the dimensions of the kernel and range.

   ★(a) $L: \mathbb{R}^2 \to \mathbb{R}^2$ given by $L\left(\begin{bmatrix} x_1 \\ x_2 \end{bmatrix} \right) = \begin{bmatrix} -4 & -3 \\ 2 & 2 \end{bmatrix} \begin{bmatrix} x_1 \\ x_2 \end{bmatrix}$

$\star$(b) $L: \mathbb{R}^2 \to \mathbb{R}^3$ given by $L\left(\begin{bmatrix} x_1 \\ x_2 \end{bmatrix}\right) = \begin{bmatrix} -3 & 4 \\ -6 & 9 \\ 7 & -8 \end{bmatrix} \begin{bmatrix} x_1 \\ x_2 \end{bmatrix}$

$\star$(c) $L: \mathbb{R}^3 \to \mathbb{R}^3$ given by $L\left(\begin{bmatrix} x_1 \\ x_2 \\ x_3 \end{bmatrix}\right) = \begin{bmatrix} -7 & 4 & -2 \\ 16 & -7 & 2 \\ 4 & -3 & 2 \end{bmatrix} \begin{bmatrix} x_1 \\ x_2 \\ x_3 \end{bmatrix}$

(d) $L: \mathbb{R}^4 \to \mathbb{R}^3$ given by $L\left(\begin{bmatrix} x_1 \\ x_2 \\ x_3 \\ x_4 \end{bmatrix}\right) = \begin{bmatrix} -5 & 3 & 1 & 18 \\ -2 & 1 & 1 & 6 \\ -7 & 3 & 4 & 19 \end{bmatrix} \begin{bmatrix} x_1 \\ x_2 \\ x_3 \\ x_4 \end{bmatrix}$

3. In each of the following cases, the matrix for a linear transformation with respect to some ordered bases for the domain and codomain is given. Which of these linear transformations are one-to-one? Which are onto? Which are isomorphisms? Justify your answers by using row reduction to determine the dimensions of the kernel and range.

$\star$(a) $L: \mathcal{P}_2 \to \mathcal{P}_2$ having matrix $\begin{bmatrix} 1 & -3 & 0 \\ -4 & 13 & -1 \\ 8 & -25 & 2 \end{bmatrix}$

(b) $L: \mathcal{M}_{22} \to \mathcal{M}_{22}$ having matrix $\begin{bmatrix} 6 & -9 & 2 & 8 \\ 10 & -6 & 12 & 4 \\ -3 & 3 & -4 & -4 \\ 8 & -9 & 9 & 11 \end{bmatrix}$

$\star$(c) $L: \mathcal{M}_{22} \to \mathcal{P}_3$ having matrix $\begin{bmatrix} 2 & 3 & -1 & 1 \\ 5 & 2 & -4 & 7 \\ 1 & 7 & 1 & -4 \\ -2 & 19 & 7 & -19 \end{bmatrix}$

$\star$**4.** (a) Suppose that $L: \mathbb{R}^6 \to \mathcal{P}_5$ is a linear transformation and that L is not onto. Is L one-to-one? Why or why not?

(b) Suppose that $L: \mathcal{M}_{22} \to \mathcal{P}_3$ is a linear transformation and that L is not one-to-one. Is L onto? Why or why not?

5. Suppose that $m > n$.

(a) Show there is no onto linear transformation from $\mathbb{R}^n$ to $\mathbb{R}^m$.

(b) Show there is no one-to-one linear transformation from $\mathbb{R}^m$ to $\mathbb{R}^n$.

6. Let $\mathbf{A}$ be a fixed $n \times n$ matrix, and consider $L: \mathcal{M}_{nn} \to \mathcal{M}_{nn}$ given by $L(\mathbf{B}) = \mathbf{AB} - \mathbf{BA}$.

(a) Show that L is not one-to-one. (Hint: Consider $L(\mathbf{I}_n)$.)

(b) Use part (a) to show that L is not onto.

7. Define $L: \mathcal{U}_3 \to \mathcal{M}_{33}$ by $L(\mathbf{A}) = \frac{1}{2}(\mathbf{A} + \mathbf{A}^T)$. Prove that L is one-to-one but is *not* onto.

8. Each part of this exercise gives matrices for linear operators L_1 and L_2 on $\mathbb{R}^3$ with respect to the standard basis. For each part, do the following:

(i) Show that L_1 and L_2 are isomorphisms.

(ii) Find L_1^{-1} and L_2^{-1}.

(iii) Calculate $L_2 \circ L_1$ directly.

(iv) Calculate $(L_2 \circ L_1)^{-1}$ by inverting the appropriate matrix.

(v) Calculate $L_1^{-1} \circ L_2^{-1}$ directly from your answer to (ii) and verify that the answer agrees with the result you obtained in (iv).

★(a) L_1:
$$\begin{bmatrix} 0 & -2 & 1 \\ 0 & -1 & 0 \\ 1 & 0 & 0 \end{bmatrix}, \quad L_2: \begin{bmatrix} 1 & 0 & 0 \\ -2 & 0 & 1 \\ 0 & -3 & 0 \end{bmatrix}$$

(b) L_1:
$$\begin{bmatrix} -4 & 0 & 1 \\ 0 & 1 & 0 \\ 1 & 2 & 0 \end{bmatrix}, \quad L_2: \begin{bmatrix} 0 & 3 & -1 \\ 1 & 0 & 0 \\ 0 & -2 & 1 \end{bmatrix}$$

★(c) L_1:
$$\begin{bmatrix} -9 & 2 & 1 \\ -6 & 1 & 1 \\ 5 & 0 & -2 \end{bmatrix}, \quad L_2: \begin{bmatrix} -4 & 2 & 1 \\ -3 & 1 & 0 \\ -5 & 2 & 1 \end{bmatrix}$$

9. Show that $L: \mathcal{M}_{mn} \to \mathcal{M}_{nm}$ given by $L(\mathbf{A}) = \mathbf{A}^T$ is an isomorphism.

10. Let $\mathbf{A}$ be a fixed nonsingular $n \times n$ matrix.

(a) Show that $L_1: \mathcal{M}_{nn} \to \mathcal{M}_{nn}$ given by $L_1(\mathbf{B}) = \mathbf{AB}$ is an isomorphism. (Hint: Be sure to show first that L_1 is a linear operator.)

(b) Show that $L_2: \mathcal{M}_{nn} \to \mathcal{M}_{nn}$ given by $L_2(\mathbf{B}) = \mathbf{ABA}^{-1}$ is an isomorphism.

11. Show that $L: \mathcal{P}_n \to \mathcal{P}_n$ given by $L(\mathbf{p}) = \mathbf{p} + \mathbf{p}'$ is an isomorphism. (Hint: First show that L is a linear operator.)

12. Let $R: \mathbb{R}^2 \to \mathbb{R}^2$ be the operator that reflects a vector through the line $y = x$; that is, $R([a, b]) = [b, a]$.

★(a) Find the matrix for R with respect to the standard basis for $\mathbb{R}^2$.

(b) Show that R is an isomorphism.

(c) Prove that $R^{-1} = R$ using the matrix from part (a).

(d) Give a geometric explanation for the result in part (c).

13. Prove that the change of basis process is essentially an isomorphism; that is, if B and C are two different finite bases for a vector space $\mathcal{V}$, with $\dim(\mathcal{V}) = n$, then the mapping $L: \mathbb{R}^n \to \mathbb{R}^n$ given by $L([\mathbf{v}]_B) = [\mathbf{v}]_C$ is an isomorphism. (Hint: First show that L is a linear operator.)

14. Let $\mathcal{V}$, $\mathcal{W}$, and $\mathcal{X}$ be vector spaces. Let $L_1: \mathcal{V} \to \mathcal{W}$ and $L_2: \mathcal{V} \to \mathcal{W}$ be linear transformations. Let $M: \mathcal{W} \to \mathcal{X}$ be an isomorphism. If $M \circ L_1 = M \circ L_2$, show that $L_1 = L_2$.

▶**15.** Prove Theorem 5.17.

16. (a) Explain why $\mathcal{M}_{mn} \cong \mathcal{M}_{nm}$.

(b) Explain why $\mathcal{P}_{4n+3} \cong \mathcal{M}_{4,n+1}$.

(c) Explain why the subspace of upper triangular matrices in $\mathcal{M}_{nn}$ is isomorphic to $\mathbb{R}^{n(n+1)/2}$. Is the subspace still isomorphic to $\mathbb{R}^{n(n+1)/2}$ if *upper* is replaced by *lower*?

17. Let $\mathcal{V}$ be a vector space. Show that a linear operator $L: \mathcal{V} \to \mathcal{V}$ is an isomorphism if and only if $L \circ L$ is an isomorphism.

18. Let $\mathcal{V}$ be a nontrivial vector space. Suppose that $L: \mathcal{V} \to \mathcal{V}$ is a linear operator.

(a) If $L \circ L$ is the zero transformation, show that L is not an isomorphism.

(b) If $L \circ L = L$ and L is not the identity transformation, show that L is not an isomorphism.

19. Let $L: \mathbb{R}^n \to \mathbb{R}^n$ be a linear operator with matrix $\mathbf{A}$ (using the standard basis for $\mathbb{R}^n$). Prove that L is an isomorphism if and only if the columns of $\mathbf{A}$ are linearly independent.

20. Let $L: V \longrightarrow W$ be a linear transformation between vector spaces.

(a) Suppose T is a linearly independent subset of V. Prove that if L is one-to-one, then $L(T)$ is linearly independent in W.

(b) Suppose that for every linearly independent set T in V, $L(T)$ is linearly independent in W. Prove that L is one-to-one. (Hint: Prove $\ker(L) = \{\mathbf{0}_V\}$ using a proof by contradiction.)

21. Let $L: V \longrightarrow W$ be a linear transformation between vector spaces, and let S be a spanning set for V.

(a) Prove that if L is onto, then $L(S)$ spans W.

(b) Show that if $L(S)$ spans W, then L is onto.

22. Let $L: V \longrightarrow W$ be a linear transformation between vector spaces, and let B be a basis for V.

(a) Show that if L is an isomorphism, then $L(B)$ is a basis for W. (Hint: Use Exercises 20(a) and 21(a).)

(b) Prove that if $L(B)$ is a basis for W, and the images of vectors in B are distinct, then L is an isomorphism. (Hint: Use Exercise 21(b) to show L is onto. Then show $\ker(L) = \{\mathbf{0}_V\}$ using a proof by contradiction.)

(c) Define $T : \mathbb{R}^3 \longrightarrow \mathbb{R}^2$ by $T(\mathbf{X}) = \begin{bmatrix} 3 & 5 & 3 \\ 1 & 2 & 1 \end{bmatrix} \mathbf{X}$, and let B be the standard basis in $\mathbb{R}^3$. Show that $T(B)$ is a basis for $\mathbb{R}^2$, but T is not an isomorphism.

(d) Explain why part (c) does not provide a counterexample to part (b).

23. Let $L: V \longrightarrow W$ be an isomorphism between finite dimensional vector spaces, and let B be a basis for V. Show that for all $\mathbf{v} \in V$, $[\mathbf{v}]_B = [L(\mathbf{v})]_{L(B)}$. (Hint: Use the fact from Exercise 22(a) that $L(B)$ is a basis for W.)

24. Let $L: V \longrightarrow W$ be an isomorphism between finite dimensional vector spaces, and let $T: W \longrightarrow X$ be a linear transformation between vector spaces.

(a) Prove that $\text{range}(T) = \text{range}(T \circ L)$.

(b) Show that $\dim(\ker(T)) = \dim(\ker(T \circ L))$. (Hint: Use part (a).)

25. Let $T: V \longrightarrow W$ be a linear transformation between finite dimensional vector spaces, and let $L: W \longrightarrow X$ be an isomorphism between vector spaces.

(a) Prove that $\ker(T) = \ker(L \circ T)$.

(b) Show that $\dim(\text{range}(T)) = \dim(\text{range}(L \circ T))$. (Hint: Use part (a).)

26. Recall the linear transformation $L: \mathcal{P}_4 \longrightarrow \mathcal{M}_{22}$ and associated matrix **A** from Example 8. Define $F_1: \mathcal{P}_4 \longrightarrow \mathbb{R}^5$ and $F_2: \mathcal{M}_{22} \longrightarrow \mathbb{R}^4$ to be the coordinatization isomorphisms alluded to in the example.

★(a) Express the mapping $T: \mathbb{R}^5 \longrightarrow \mathbb{R}^4$ from Example 8 in terms of L, F_1, and F_2.

(b) Use part (a) and the definitions of L, F_1, and F_2 to show that the matrix for T with respect to the standard bases for $\mathbb{R}^5$ and $\mathbb{R}^4$ is the matrix **A** given in the example.

27. We show in this exercise that any isomorphism from $\mathbb{R}^2$ to $\mathbb{R}^2$ is the composition of certain types of reflections, contractions/dilations, and shears. (See Exercise 11 in Section 5.1 for the definition of a shear.) Note that if $a \neq 0$,

$$\begin{bmatrix} a & b \\ c & d \end{bmatrix} = \begin{bmatrix} a & 0 \\ 0 & 1 \end{bmatrix} \begin{bmatrix} 1 & 0 \\ c & 1 \end{bmatrix} \begin{bmatrix} 1 & 0 \\ 0 & \frac{ad-bc}{a} \end{bmatrix} \begin{bmatrix} 1 & \frac{b}{a} \\ 0 & 1 \end{bmatrix},$$

and if $c \neq 0$,

$$\begin{bmatrix} a & b \\ c & d \end{bmatrix} = \begin{bmatrix} 0 & 1 \\ 1 & 0 \end{bmatrix} \begin{bmatrix} c & 0 \\ 0 & 1 \end{bmatrix} \begin{bmatrix} 1 & 0 \\ a & 1 \end{bmatrix} \begin{bmatrix} 1 & 0 \\ 0 & \frac{bc-ad}{c} \end{bmatrix} \begin{bmatrix} 1 & \frac{d}{c} \\ 0 & 1 \end{bmatrix}.$$

(a) Use the given equations to show that every nonsingular 2×2 matrix can be expressed as a product of matrices, each of which is in one of the following forms:

$$\begin{bmatrix} k & 0 \\ 0 & 1 \end{bmatrix}, \begin{bmatrix} 1 & 0 \\ 0 & k \end{bmatrix}, \begin{bmatrix} 1 & 0 \\ k & 1 \end{bmatrix}, \begin{bmatrix} 1 & k \\ 0 & 1 \end{bmatrix}, \text{ or } \begin{bmatrix} 0 & 1 \\ 1 & 0 \end{bmatrix}.$$

(b) Show that when $k \geq 0$, multiplying either of the first two matrices in part (a) times the vector $[x, y]$ represents a contraction/dilation along the x-coordinate or the y-coordinate.

(c) Show that when $k < 0$, multiplying either of the first two matrices in part (a) times the vector $[x, y]$ represents a contraction/dilation along the x-coordinate or the y-coordinate, followed by a reflection through one of the axes. $\left(\text{Hint: } \begin{bmatrix} k & 0 \\ 0 & 1 \end{bmatrix} = \begin{bmatrix} -1 & 0 \\ 0 & 1 \end{bmatrix} \begin{bmatrix} -k & 0 \\ 0 & 1 \end{bmatrix}. \right)$

(d) Explain why multiplying either of the third or fourth matrices in part (a) times $[x, y]$ represents a shear.

(e) Explain why multiplying the last matrix in part (a) times $[x, y]$ represents a reflection through the line $y = x$.

(f) Using parts (a) through (e), show that any isomorphism from $\mathbb{R}^2$ to $\mathbb{R}^2$ is the composition of a finite number of the following linear operators: reflection through an axis, reflection through $y = x$, contraction/dilation of the x- or y-coordinate, shear in the x- or y-direction.

28. Express the linear transformation $L: \mathbb{R}^2 \to \mathbb{R}^2$ that rotates the plane $45°$ in a counterclockwise direction as a composition of the transformations described in part (f) of Exercise 27.

29. (a) Let x_1, x_2, x_3 be distinct real numbers. Use an argument similar to that in Example 4 to show that for any given $a, b, c \in \mathbb{R}$, there is a polynomial $\mathbf{p} \in \mathcal{P}_2$ such that $\mathbf{p}(x_1) = a$, $\mathbf{p}(x_2) = b$, and $\mathbf{p}(x_3) = c$.

(b) For each choice of $x_1, x_2, x_3, a, b, c \in \mathbb{R}$, show that the polynomial $\mathbf{p}$ from part (a) is unique.

(c) Recall from algebra that a nonzero polynomial of degree n can have at most n roots. Use this fact to prove that if $x_1, \ldots, x_{n+1} \in \mathbb{R}$, with $x_1, \ldots, x_{n+1}$ distinct, then for any given $a_1, \ldots, a_{n+1} \in \mathbb{R}$, there is a unique polynomial $\mathbf{p} \in \mathcal{P}_n$ such that $\mathbf{p}(x_1) = a_1$, $\mathbf{p}(x_2) = a_2, \ldots, \mathbf{p}(x_n) = a_n$, and $\mathbf{p}(x_{n+1}) = a_{n+1}$.

30. Define $L: \mathcal{P} \longrightarrow \mathcal{P}$ by $L(\mathbf{p}(x)) = x\mathbf{p}(x)$.

(a) Show that L is one-to-one but not onto.

(b) Explain why L does not contradict Corollary 5.15.

★31. True or False:

(a) A linear transformation $L: \mathcal{V} \longrightarrow \mathcal{W}$ is one-to-one if for all $\mathbf{v}_1, \mathbf{v}_2 \in \mathcal{V}$, $\mathbf{v}_1 = \mathbf{v}_2$ implies $L(\mathbf{v}_1) = L(\mathbf{v}_2)$.

(b) A linear transformation $L: \mathcal{V} \longrightarrow \mathcal{W}$ is onto if for all $\mathbf{v} \in \mathcal{V}$, there is some $\mathbf{w} \in \mathcal{W}$ such that $L(\mathbf{v}) = \mathbf{w}$.

(c) A linear transformation $L: \mathcal{V} \longrightarrow \mathcal{W}$ is one-to-one if $\ker(L)$ contains no vectors other than $\mathbf{0}_\mathcal{V}$.

(d) If $L: \mathcal{V} \longrightarrow \mathcal{W}$ is a linear transformation, then L is one-to-one if and only if L is onto.

(e) If the inverse L^{-1} of a linear transformation L exists, then L^{-1} is also a linear transformation.

(f) A linear transformation is an isomorphism if and only if it is invertible.

(g) If $L: \mathcal{V} \longrightarrow \mathcal{V}$ is a linear operator, and the matrix for L with respect to the finite basis B for $\mathcal{V}$ is $\mathbf{A}_{BB}$, then L is an isomorphism if and only if $|\mathbf{A}_{BB}| = 0$.

(h) $\mathbb{R}^{28} \cong \mathcal{P}_{27} \cong \mathcal{M}_{74}$.

5.5 Diagonalization of Linear Operators

In Section 3.4, we examined a method for diagonalizing certain square matrices. Now, we generalize this process to diagonalize linear operators.

Eigenvalues, Eigenvectors, and Eigenspaces for Linear Operators

We define eigenvalues and eigenvectors for linear operators in a manner analogous to their definitions for matrices.

DEFINITION

Let $L: \mathcal{V} \to \mathcal{V}$ be a linear operator. A real number λ is said to be an **eigenvalue** of L if and only if there is a nonzero vector $\mathbf{v} \in \mathcal{V}$ such that $L(\mathbf{v}) = \lambda \mathbf{v}$. Also, any nonzero vector $\mathbf{v}$ such that $L(\mathbf{v}) = \lambda \mathbf{v}$ is said to be an **eigenvector** for L corresponding to the eigenvalue λ.

If L is a linear operator on $\mathbb{R}^n$ given by multiplication by a square matrix $\mathbf{A}$ (that is, $L(\mathbf{v}) = \mathbf{Av}$), then the eigenvalues and eigenvectors for L are merely the eigenvalues and eigenvectors of the matrix $\mathbf{A}$, since $L(\mathbf{v}) = \lambda\mathbf{v}$ if and only if $\mathbf{Av} = \lambda\mathbf{v}$. Hence, all of the results regarding eigenvalues and eigenvectors for matrices in Section 3.4 apply to this type of operator. Let us now consider an example involving a different type of linear operator.

EXAMPLE 1

Consider $L\colon \mathcal{M}_{nn} \longrightarrow \mathcal{M}_{nn}$ given by $L(\mathbf{A}) = \mathbf{A} + \mathbf{A}^T$. Then every nonzero $n \times n$ symmetric matrix $\mathbf{S}$ is an eigenvector for L corresponding to the eigenvalue $\lambda_1 = 2$ because $L(\mathbf{S}) = \mathbf{S} + \mathbf{S}^T = \mathbf{S} + \mathbf{S}$ (since $\mathbf{S}$ is symmetric) $= 2\mathbf{S}$. Similarly, every nonzero skew-symmetric $n \times n$ matrix $\mathbf{V}$ is an eigenvector for L corresponding to the eigenvalue $\lambda_2 = 0$ because $L(\mathbf{V}) = \mathbf{V} + \mathbf{V}^T = \mathbf{V} + (-\mathbf{V}) = \mathbf{O}_{nn} = 0\mathbf{V}$. ∎

We now define an eigenspace for a linear operator.

DEFINITION

Let $L\colon \mathcal{V} \to \mathcal{V}$ be a linear operator on $\mathcal{V}$. Let λ be an eigenvalue for L. Then E_λ, the **eigenspace of** λ, is defined to be the set of all eigenvectors for L corresponding to λ, together with the zero vector $\mathbf{0}_{\mathcal{V}}$ of $\mathcal{V}$. That is, $E_\lambda = \{\mathbf{v} \in \mathcal{V} \mid L(\mathbf{v}) = \lambda\mathbf{v}\}$.

Just as the eigenspace of an $n \times n$ matrix is a subspace of $\mathbb{R}^n$ (see Theorem 4.4), the eigenspace of a linear operator $L\colon \mathcal{V} \to \mathcal{V}$ is a subspace of the vector space $\mathcal{V}$. This can be proved directly by showing that the eigenspace is nonempty and closed under vector addition and scalar multiplication, and then applying Theorem 4.2.

EXAMPLE 2

Recall the operator $L : \mathcal{M}_{nn} \longrightarrow \mathcal{M}_{nn}$ from Example 1 given by $L(\mathbf{A}) = \mathbf{A} + \mathbf{A}^T$. We have already seen that the eigenspace E_2 for L contains all symmetric $n \times n$ matrices. In fact, these are the only elements of E_2 because

$$L(\mathbf{A}) = 2\mathbf{A} \implies \mathbf{A} + \mathbf{A}^T = 2\mathbf{A} \implies \mathbf{A} + \mathbf{A}^T = \mathbf{A} + \mathbf{A} \implies \mathbf{A}^T = \mathbf{A}.$$

Hence $E_2 = \{\text{symmetric } n \times n \text{ matrices}\}$, which we know to be a subspace of $\mathcal{M}_{nn}$ having dimension $n(n+1)/2$.

Similarly, the eigenspace $E_0 = \{\text{skew-symmetric } n \times n \text{ matrices}\}$. ∎

The Characteristic Polynomial of a Linear Operator

Frequently, we analyze a linear operator L on a finite dimensional vector space $\mathcal{V}$ by looking at its matrix with respect to some basis for $\mathcal{V}$. In particular, to solve for the eigenvalues of L, we first find an ordered basis B for $\mathcal{V}$, and then solve for the matrix representation $\mathbf{A}$ of L with respect to B. For this matrix $\mathbf{A}$, we have $[L(\mathbf{v})]_B = \mathbf{A}[\mathbf{v}]_B$. Thus, finding the eigenvalues of $\mathbf{A}$ gives the eigenvalues of L.

EXAMPLE 3

Let $L\colon \mathbb{R}^2 \to \mathbb{R}^2$ be the linear operator given by $L([a, b]) = [b, a]$, that is, a reflection about the line $y = x$. We will calculate the eigenvalues for L two ways — first, using the standard basis for $\mathbb{R}^2$, and then, using a nonstandard basis.

Since $L(\mathbf{i}) = \mathbf{j}$ and $L(\mathbf{j}) = \mathbf{i}$, the matrix for L with respect to the standard basis is

$$\mathbf{A} = \begin{bmatrix} 0 & 1 \\ 1 & 0 \end{bmatrix}. \text{ Then } p_{\mathbf{A}}(x) = \begin{vmatrix} x & -1 \\ -1 & x \end{vmatrix} = x^2 - 1 = (x-1)(x+1).$$

Hence, the eigenvalues for $\mathbf{A}$ (and L) are $\lambda_1 = 1$ and $\lambda_2 = -1$. Solving the homogeneous system $(1\mathbf{I}_2 - \mathbf{A})\mathbf{v} = \mathbf{0}$ yields $\mathbf{v}_1 = [1, 1]$ as an eigenvector corresponding to $\lambda_1 = 1$. Similarly, we obtain $\mathbf{v}_2 = [1, -1]$, for $\lambda_2 = -1$.

Notice that this result makes sense geometrically. The vector $\mathbf{v}_1$ runs parallel to the line of reflection and thus L leaves $\mathbf{v}_1$ unchanged; $L(\mathbf{v}_1) = \lambda_1 \mathbf{v}_1 = \mathbf{v}_1$. On the other hand, $\mathbf{v}_2$ is perpendicular to the axis of reflection, and so L reverses its direction; $L(\mathbf{v}_2) = \lambda_2 \mathbf{v}_2 = -\mathbf{v}_2$.

Now, instead of using the standard basis in $\mathbb{R}^2$, let us find the matrix representation of L with respect to $B = (\mathbf{v}_1, \mathbf{v}_2)$. Since $[L(\mathbf{v}_1)]_B = [1, 0]$ and $[L(\mathbf{v}_2)]_B = [0, -1]$ (why?), the matrix for L with respect to B is

$$\overset{\displaystyle [L(\mathbf{v}_1)]_B \ \ [L(\mathbf{v}_2)]_B}{\mathbf{D} = \begin{bmatrix} 1 & 0 \\ 0 & -1 \end{bmatrix},}$$

a diagonal matrix with the eigenvalues for L on the main diagonal. Notice that

$$p_{\mathbf{D}}(x) = \begin{vmatrix} x - 1 & 0 \\ 0 & x + 1 \end{vmatrix} = (x-1)(x+1) = p_{\mathbf{A}}(x),$$

giving us (of course) the same eigenvalues $\lambda_1 = 1$ and $\lambda_2 = -1$ for L.   ■

Example 3 illustrates how two different matrix representations for the same linear operator (using different ordered bases) produce the same characteristic polynomial. Theorems 5.6 and 5.7 together show that this is true in general. Therefore, we can define the characteristic polynomial of a linear operator as follows, without concern about which particular ordered basis is used:

DEFINITION

Let L be a linear operator on a finite dimensional vector space $\mathcal{V}$. Suppose $\mathbf{A}$ is the matrix representation of L with respect to some ordered basis for $\mathcal{V}$. Then the **characteristic polynomial of** L, $p_L(x)$, is defined to be $p_{\mathbf{A}}(x)$.

EXAMPLE 4	Consider $L\colon \mathcal{P}_2 \to \mathcal{P}_2$ determined by $L(\mathbf{p}(x)) = x^2\mathbf{p}''(x) + (3x-2)\mathbf{p}'(x) + 5\mathbf{p}(x)$. You can check that $L(x^2) = 13x^2 - 4x$, $L(x) = 8x - 2$, and $L(1) = 5$. Thus, the matrix representation of L with respect to the standard basis $S = (x^2, x, 1)$ is

$$\mathbf{A} = \begin{bmatrix} 13 & 0 & 0 \\ -4 & 8 & 0 \\ 0 & -2 & 5 \end{bmatrix}.$$

Hence,

$$p_L(x) = p_{\mathbf{A}}(x) = \begin{vmatrix} x-13 & 0 & 0 \\ 4 & x-8 & 0 \\ 0 & 2 & x-5 \end{vmatrix} = (x-13)(x-8)(x-5),$$

since this is the determinant of a lower triangular matrix. The eigenvalues of L are the roots of $p_L(x)$, namely $\lambda_1 = 13$, $\lambda_2 = 8$, and $\lambda_3 = 5$. ∎

Criterion for Diagonalization

Given a linear operator L on a finite dimensional vector space $\mathcal{V}$, our goal is to find a basis B for $\mathcal{V}$ such that the matrix for L with respect to B is diagonal, as in Example 3. But, just as every square matrix cannot be diagonalized, neither can every linear operator.

DEFINITION

A linear operator L on a finite dimensional vector space $\mathcal{V}$ is **diagonalizable** if and only if the matrix representation of L with respect to some ordered basis for $\mathcal{V}$ is a diagonal matrix.

The next result indicates precisely which linear operators are diagonalizable.

THEOREM 5.21

Let L be a linear operator on an n-dimensional vector space $\mathcal{V}$. Then L is diagonalizable if and only if there is a set of n linearly independent eigenvectors for L.

PROOF OF THEOREM 5.21

Suppose that L is diagonalizable. Then there is an ordered basis $B = (\mathbf{v}_1, \ldots, \mathbf{v}_n)$ for $\mathcal{V}$ such that the matrix representation for L with respect to B is a diagonal matrix $\mathbf{D}$. Now, B is a linearly independent set. If we can show that each vector $\mathbf{v}_i$ in B, for $1 \le i \le n$, is an eigenvector corresponding to some eigenvalue for L, then B will be a set of n linearly independent eigenvectors for L. Now, for each $\mathbf{v}_i$, we have $[L(\mathbf{v}_i)]_B = \mathbf{D}[\mathbf{v}_i]_B = \mathbf{D}\mathbf{e}_i = d_{ii}\mathbf{e}_i = d_{ii}[\mathbf{v}_i]_B = [d_{ii}\mathbf{v}_i]_B$, where d_{ii} is the (i, i) entry of $\mathbf{D}$. Since coordinatization of vectors with respect to B is an isomorphism, we have $L(\mathbf{v}_i) = d_{ii}\mathbf{v}_i$, and so each $\mathbf{v}_i$ is an eigenvector for L corresponding to the eigenvalue d_{ii}.

Conversely, suppose that $B = \{\mathbf{w}_1, \ldots, \mathbf{w}_n\}$ is a set of n linearly independent eigenvectors for L, corresponding to the (not necessarily distinct) eigenvalues $\lambda_1, \ldots, \lambda_n$, respectively. Since B contains $n = \dim(\mathcal{V})$ linearly independent vectors, B is a basis for $\mathcal{V}$, by part (2) of Corollary 4.14. We show that the matrix $\mathbf{A}$ for L with respect to B is, in fact, diagonal. Now, for $1 \leq i \leq n$,

$$i\text{th column of } \mathbf{A} = [L(\mathbf{w}_i)]_B = [\lambda_i \mathbf{w}_i]_B = \lambda_i [\mathbf{w}_i]_B = \lambda_i \mathbf{e}_i.$$

Thus, $\mathbf{A}$ is a diagonal matrix, and so L is diagonalizable.   ■

EXAMPLE 5

In Example 3, $L: \mathbb{R}^2 \to \mathbb{R}^2$ was defined by $L([a, b]) = [b, a]$. In that example, we found a set of two linearly independent eigenvectors for L, namely $\mathbf{v}_1 = [1, 1]$ and $\mathbf{v}_2 = [1, -1]$. Since $\dim(\mathbb{R}^2) = 2$, Theorem 5.21 indicates that L is diagonalizable. In fact, in Example 3, we computed the matrix for L with respect to the ordered basis $(\mathbf{v}_1, \mathbf{v}_2)$ for $\mathbb{R}^2$ to be the diagonal matrix $\begin{bmatrix} 1 & 0 \\ 0 & -1 \end{bmatrix}$.   ■

EXAMPLE 6

Consider the linear operator $L: \mathbb{R}^2 \to \mathbb{R}^2$ that rotates the plane counterclockwise through an angle of $\frac{\pi}{4}$. Now, every nonzero vector $\mathbf{v}$ is moved to $L(\mathbf{v})$, which is not parallel to $\mathbf{v}$, since $L(\mathbf{v})$ forms a $45°$ angle with $\mathbf{v}$. Hence, L has *no* eigenvectors, and so a set of two linearly independent eigenvectors cannot be found for L. Therefore, by Theorem 5.21, L is not diagonalizable.   ■

Linear Independence of Eigenvectors

Theorem 5.21 asserts that finding enough *linearly independent* eigenvectors is crucial to the diagonalization process. The next theorem gives a condition under which a set of eigenvectors is guaranteed to be linearly independent.

THEOREM 5.22

Let L be a linear operator on a vector space $\mathcal{V}$, and let $\lambda_1, \ldots, \lambda_t$ be distinct eigenvalues for L. If $\mathbf{v}_1, \ldots, \mathbf{v}_t$ are eigenvectors for L corresponding to $\lambda_1, \ldots, \lambda_t$, respectively, then the set $\{\mathbf{v}_1, \ldots, \mathbf{v}_t\}$ is linearly independent. That is, eigenvectors corresponding to distinct eigenvalues are linearly independent.

PROOF OF THEOREM 5.22

We proceed by induction on t.

Base Step: Suppose that $t = 1$. Any eigenvector $\mathbf{v}_1$ for λ_1 is nonzero, so $\{\mathbf{v}_1\}$ is linearly independent.

Inductive Step: Let $\lambda_1, \ldots, \lambda_{k+1}$ be distinct eigenvalues for L, and let $\mathbf{v}_1, \ldots, \mathbf{v}_{k+1}$ be corresponding eigenvectors. Our inductive hypothesis is that the set $\{\mathbf{v}_1, \ldots, \mathbf{v}_k\}$ is linearly independent. We must prove that $\{\mathbf{v}_1, \ldots, \mathbf{v}_k, \mathbf{v}_{k+1}\}$ is linearly independent. Suppose that $a_1\mathbf{v}_1 + \cdots + a_k\mathbf{v}_k + a_{k+1}\mathbf{v}_{k+1} = \mathbf{0}_{\mathcal{V}}$. Showing that $a_1 = a_2 = \cdots = a_k = a_{k+1} = 0$ will finish the proof by Theorem 4.7. Now,

$$
\begin{aligned}
L(a_1\mathbf{v}_1 + \cdots + a_k\mathbf{v}_k + a_{k+1}\mathbf{v}_{k+1}) &= L(\mathbf{0}_{\mathcal{V}}) \\
\implies \quad a_1 L(\mathbf{v}_1) + \cdots + a_k L(\mathbf{v}_k) + a_{k+1}L(\mathbf{v}_{k+1}) &= L(\mathbf{0}_{\mathcal{V}}) \\
\implies \quad a_1\lambda_1\mathbf{v}_1 + \cdots + a_k\lambda_k\mathbf{v}_k + a_{k+1}\lambda_{k+1}\mathbf{v}_{k+1} &= \mathbf{0}_{\mathcal{V}}.
\end{aligned}
$$

Multiplying both sides of the original equation $a_1\mathbf{v}_1 + \cdots + a_k\mathbf{v}_k + a_{k+1}\mathbf{v}_{k+1} = \mathbf{0}_{\mathcal{V}}$ by λ_{k+1} yields

$$
a_1\lambda_{k+1}\mathbf{v}_1 + \cdots + a_k\lambda_{k+1}\mathbf{v}_k + a_{k+1}\lambda_{k+1}\mathbf{v}_{k+1} = \mathbf{0}_{\mathcal{V}}.
$$

Subtracting the last two equations containing λ_{k+1} gives

$$
a_1(\lambda_1 - \lambda_{k+1})\mathbf{v}_1 + \cdots + a_k(\lambda_k - \lambda_{k+1})\mathbf{v}_k = \mathbf{0}_{\mathcal{V}}.
$$

Hence, our inductive hypothesis, together with Theorem 4.7, implies that

$$
a_1(\lambda_1 - \lambda_{k+1}) = \cdots = a_k(\lambda_k - \lambda_{k+1}) = 0.
$$

Since the eigenvalues $\lambda_1, \ldots, \lambda_{k+1}$ are distinct, none of the factors $\lambda_i - \lambda_{k+1}$ in these equations can equal zero, for $1 \le i \le k$. Thus, $a_1 = a_2 = \cdots = a_k = 0$. Finally, plugging these values into the earlier equation $a_1\mathbf{v}_1 + \cdots + a_k\mathbf{v}_k + a_{k+1}\mathbf{v}_{k+1} = \mathbf{0}_{\mathcal{V}}$ gives $a_{k+1}\mathbf{v}_{k+1} = \mathbf{0}_{\mathcal{V}}$. Since $\mathbf{v}_{k+1} \ne \mathbf{0}_{\mathcal{V}}$, we must have $a_{k+1} = 0$ as well. ∎

EXAMPLE 7

Consider the linear operator $L: \mathbb{R}^3 \to \mathbb{R}^3$ given by $L(\mathbf{x}) = \mathbf{Ax}$, where

$$
\mathbf{A} = \begin{bmatrix} 31 & -14 & -92 \\ -50 & 28 & 158 \\ 18 & -9 & -55 \end{bmatrix}.
$$

It can be shown that the characteristic polynomial for $\mathbf{A}$ is $p_{\mathbf{A}}(x) = x^3 - 4x^2 + x + 6 = (x+1)(x-2)(x-3)$. Hence, the eigenvalues for $\mathbf{A}$ are $\lambda_1 = -1$, $\lambda_2 = 2$, and $\lambda_3 = 3$. A quick check verifies that $[2, -2, 1]$, $[10, 1, 3]$, and $[1, 2, 0]$ are eigenvectors, respectively, for the distinct eigenvalues λ_1, λ_2, and λ_3. Therefore, by Theorem 5.22, the set $B = \{[2, -2, 1], [10, 1, 3], [1, 2, 0]\}$ is linearly independent (verify!). In fact, since $\dim(\mathbb{R}^3) = 3$, this set B is a basis for $\mathbb{R}^3$.

Also note that L is diagonalizable by Theorem 5.21, since there are 3 linearly independent eigenvectors for L and $\dim(\mathbb{R}^3) = 3$. In fact, the matrix for L with respect to B is

$$
\mathbf{D} = \begin{bmatrix} -1 & 0 & 0 \\ 0 & 2 & 0 \\ 0 & 0 & 3 \end{bmatrix}.
$$

This can be verified by computing $\mathbf{D} = \mathbf{P}^{-1}\mathbf{A}\mathbf{P}$, where

$$\mathbf{P} = \begin{bmatrix} 2 & 10 & 1 \\ -2 & 1 & 2 \\ 1 & 3 & 0 \end{bmatrix}$$

is the transition matrix from B-coordinates to standard coordinates, that is, the matrix whose columns are the vectors in B (see Exercise 8(b) in Section 4.7). ∎

As illustrated in Example 7, Theorems 5.21 and 5.22 combine to prove the following:

COROLLARY 5.23

If L is a linear operator on an n-dimensional vector space and L has n distinct eigenvalues, then L is diagonalizable.

The converse to this corollary is false, since it is possible to get n linearly independent eigenvectors from fewer than n eigenvalues (see Exercise 6).

The proof of the following generalization of Theorem 5.22 is left as Exercises 15 and 16.

THEOREM 5.24

Let $L : \mathcal{V} \longrightarrow \mathcal{V}$ be a linear operator on a finite dimensional vector space $\mathcal{V}$, and let $B_1, B_2, \ldots, B_k$ be bases for eigenspaces $E_{\lambda_1}, \ldots, E_{\lambda_k}$ for L, where $\lambda_1, \ldots, \lambda_k$ are distinct eigenvalues for L. Then $B_i \cap B_j = \emptyset$ for $1 \leq i < j \leq k$, and $B_1 \cup B_2 \cup \cdots \cup B_k$ is a linearly independent subset of $\mathcal{V}$.

This theorem asserts that for a given operator on a finite dimensional vector space, the bases for distinct eigenspaces are disjoint, and the union of two or more bases from distinct eigenspaces always constitutes a linearly independent set.

 Consider the linear operator $L: \mathbb{R}^4 \to \mathbb{R}^4$ given by $L(\mathbf{x}) = \mathbf{A}\mathbf{x}$, for the matrix $\mathbf{A}$ in Example 5 of Section 3.4; namely,

$$\mathbf{A} = \begin{bmatrix} -4 & 7 & 1 & 4 \\ 6 & -16 & -3 & -9 \\ 12 & -27 & -4 & -15 \\ -18 & 43 & 7 & 24 \end{bmatrix}.$$

In that example, we showed there were precisely three eigenvalues for $\mathbf{A}$ (and hence, for L): $\lambda_1 = -1$, $\lambda_2 = 2$, and $\lambda_3 = 0$. In the row reduction of $[(-1)\mathbf{I}_4 - \mathbf{A} \mid \mathbf{0}]$ in that example, we found two independent variables, and so $\dim(\mathbf{E}_{\lambda_1}) = 2$. We also discovered eigenvectors $\mathbf{X}_1 = [-2, -1, 1, 0]$ and $\mathbf{X}_2 = [-1, -1, 0, 1]$ for λ_1. Therefore, $\{\mathbf{X}_1, \mathbf{X}_2\}$ is a basis for $\mathbf{E}_{\lambda_1}$. Similarly, we can verify that $\dim(\mathbf{E}_{\lambda_2}) = \dim(\mathbf{E}_{\lambda_3}) = 1$. We found eigenvector $\mathbf{X}_3 = [1, -2, -4, 6]$ for λ_2, and eigenvector $\mathbf{X}_4 = [1, -3, -3, 7]$ for λ_3. Thus $\{\mathbf{X}_3\}$ is a basis for $\mathbf{E}_{\lambda_2}$, and $\{\mathbf{X}_4\}$ is a basis for $\mathbf{E}_{\lambda_3}$. Now, by Theorem 5.24, the union $\{\mathbf{X}_1, \mathbf{X}_2, \mathbf{X}_3, \mathbf{X}_4\}$ of these bases is a linearly independent subset of $\mathbb{R}^4$. Of course, since $\dim(\mathbb{R}^4) = 4$, $\{\mathbf{X}_1, \mathbf{X}_2, \mathbf{X}_3, \mathbf{X}_4\}$ is also a basis for $\mathbb{R}^4$. Hence, by Theorem 5.21, L is diagonalizable. ∎

Method for Diagonalizing a Linear Operator

Theorem 5.24 suggests a method for diagonalizing a given linear operator $L: \mathcal{V} \to \mathcal{V}$, when possible. This method, outlined below, illustrates how to find a basis B so that the matrix for L with respect to B is diagonal. In the case where $\mathcal{V} = \mathbb{R}^n$ and the standard basis is used, we simply apply the Diagonalization Method of Section 3.4 to the matrix for L to find a basis for $\mathcal{V}$. In other cases, we first need to choose a basis C for $\mathcal{V}$. Next we find the matrix for L with respect to C, and then use the Diagonalization Method on this matrix to obtain a basis Z of eigenvectors in $\mathbb{R}^n$. Finally, the desired basis B for $\mathcal{V}$ consists of the vectors in $\mathcal{V}$ whose coordinatization with respect to C are the vectors in Z.

METHOD FOR DIAGONALIZING A LINEAR OPERATOR (IF POSSIBLE) (GENERALIZED DIAGONALIZATION METHOD)

Let $L: \mathcal{V} \to \mathcal{V}$ be a linear operator on an n-dimensional vector space $\mathcal{V}$.

Step 1: Find a basis C for $\mathcal{V}$ (if $\mathcal{V} = \mathbb{R}^n$, we can use the standard basis), and calculate the matrix representation $\mathbf{A}$ of L with respect to C.

Step 2: Apply the Diagonalization Method of Section 3.4 to $\mathbf{A}$ in order to obtain all of the eigenvalues $\lambda_1, \ldots, \lambda_k$ of $\mathbf{A}$ and a basis in $\mathbb{R}^n$ for each eigenspace E_{λ_i} of $\mathbf{A}$ (by solving an appropriate homogeneous system if necessary). If the union of the bases of the E_{λ_i} contains fewer than n elements, then L is not diagonalizable, and we stop. Otherwise, let $Z = (\mathbf{w}_1 \ldots, \mathbf{w}_n)$ be an ordered basis for $\mathbb{R}^n$ consisting of the union of the bases for the E_{λ_i}.

Step 3: Reverse the C-coordinatization isomorphism on the vectors in Z to obtain an ordered basis $B = (\mathbf{v}_1 \ldots, \mathbf{v}_n)$ for $\mathcal{V}$; that is, $[\mathbf{v}_i]_C = \mathbf{w}_i$.

The matrix representation for L with respect to B is the diagonal matrix $\mathbf{D}$ whose (i, i) entry d_{ii} is the eigenvalue for L corresponding to $\mathbf{v}_i$. In most practical situations, the transition matrix $\mathbf{P}$ from B- to C-coordinates is useful; $\mathbf{P}$ is the $n \times n$ matrix whose columns are $[\mathbf{v}_1]_C, \ldots, [\mathbf{v}_n]_C$; that is, $\mathbf{w}_1, \mathbf{w}_2, \ldots, \mathbf{w}_n$. Note that $\mathbf{D} = \mathbf{P}^{-1}\mathbf{AP}$.

Step 2 of this method uses the fact that if there are n vectors in the union of the bases for the distinct eigenspaces, then this union is a linearly independent set of vectors (and hence, a basis for $\mathbb{R}^n$). This follows from Theorem 5.24.

If we have a linear operator on $\mathbb{R}^n$ and use the standard basis for C, then the C-coordinatization isomorphism in this method is merely the identity mapping. In this case, Steps 1 and 3 are a lot easier to perform, as we see in the next example.

EXAMPLE 9

We use the method outlined above to diagonalize the operator $L: \mathbb{R}^4 \to \mathbb{R}^4$ given by $L(\mathbf{v}) = \mathbf{Av}$, where

$$\mathbf{A} = \begin{bmatrix} 5 & 0 & -8 & 8 \\ 8 & 1 & -16 & 16 \\ -4 & 0 & 9 & -8 \\ -8 & 0 & 16 & -15 \end{bmatrix}.$$

Step 1: Since $\mathcal{V} = \mathbb{R}^4$, we let C be the standard basis for $\mathbb{R}^4$. Then no additional work needs to be done here, since the matrix representation for L with respect to C is simply $\mathbf{A}$ itself.

Step 2: We apply the Diagonalization Method of Section 3.4 to $\mathbf{A}$. A lengthy computation produces the characteristic polynomial

$$p_{\mathbf{A}}(x) = x^4 - 6x^2 + 8x - 3 = (x-1)^3(x+3).$$

Thus, the eigenvalues for $\mathbf{A}$ are $\lambda_1 = 1$ and $\lambda_2 = -3$.

To obtain a basis for the eigenspace E_{λ_1}, we row reduce

$$[1\mathbf{I}_4 - \mathbf{A} \mid \mathbf{0}] = \begin{bmatrix} -4 & 0 & 8 & -8 & \mid & 0 \\ -8 & 0 & 16 & -16 & \mid & 0 \\ 4 & 0 & -8 & 8 & \mid & 0 \\ 8 & 0 & -16 & 16 & \mid & 0 \end{bmatrix} \quad \text{to obtain} \quad \begin{bmatrix} 1 & 0 & -2 & 2 & \mid & 0 \\ 0 & 0 & 0 & 0 & \mid & 0 \\ 0 & 0 & 0 & 0 & \mid & 0 \\ 0 & 0 & 0 & 0 & \mid & 0 \end{bmatrix}.$$

There are 3 independent variables, so $\dim(E_{\lambda_1}) = 3$. As in Section 3.4, we set each independent variable in turn to 1, while setting the others equal to 0. This yields the linearly independent eigenvectors $\mathbf{w}_1 = [0, 1, 0, 0]$, $\mathbf{w}_2 = [2, 0, 1, 0]$, and $\mathbf{w}_3 = [-2, 0, 0, 1]$. Thus, $\{\mathbf{w}_1, \mathbf{w}_2, \mathbf{w}_3\}$ is a basis for E_{λ_1}. A similar procedure yields $\dim(E_{\lambda_2}) = 1$, and the eigenvector $\mathbf{w}_4 = [1, 2, -1, -2]$ for E_{λ_2}. Also, $\{\mathbf{w}_4\}$ is a basis for E_{λ_2}. Since $\dim(\mathcal{V}) = 4$ and since we obtained 4 eigenvectors overall from the Diagonalization Method, L is diagonalizable. We form the union $Z = \{\mathbf{w}_1, \mathbf{w}_2, \mathbf{w}_3, \mathbf{w}_4\}$ of the bases for E_{λ_1} and E_{λ_2}.

Step 3: Since C is the standard basis for $\mathbb{R}^4$ and the C-coordinatization isomorphism is the identity mapping, no additional work needs to be done here. We simply let $B = (\mathbf{v}_1, \mathbf{v}_2, \mathbf{v}_3, \mathbf{v}_4)$ where $\mathbf{v}_1 = \mathbf{w}_1$, $\mathbf{v}_2 = \mathbf{w}_2$, $\mathbf{v}_3 = \mathbf{w}_3$, and $\mathbf{v}_4 = \mathbf{w}_4$. That is, $B = ([0, 1, 0, 0], [2, 0, 1, 0], [-2, 0, 0, 1], [1, 2, -1, -2])$. B is an ordered basis for $\mathcal{V} = \mathbb{R}^4$.

Notice that the matrix representation of L with respect to B is the 4×4 diagonal matrix $\mathbf{D}$ with each d_{ii} equal to the eigenvalue for $\mathbf{v}_i$, for $1 \le i \le 4$. In particular,

$$\mathbf{D} = \begin{bmatrix} 1 & 0 & 0 & 0 \\ 0 & 1 & 0 & 0 \\ 0 & 0 & 1 & 0 \\ 0 & 0 & 0 & -3 \end{bmatrix}.$$

Also, the transition matrix **P** from B-coordinates to standard coordinates is formed by using $\mathbf{v}_1$, $\mathbf{v}_2$, $\mathbf{v}_3$, and $\mathbf{v}_4$ as columns. Hence,

$$\mathbf{P} = \begin{bmatrix} 0 & 2 & -2 & 1 \\ 1 & 0 & 0 & 2 \\ 0 & 1 & 0 & -1 \\ 0 & 0 & 1 & -2 \end{bmatrix}, \quad \text{and its inverse is} \quad \mathbf{P}^{-1} = \begin{bmatrix} 2 & 1 & -4 & 4 \\ -1 & 0 & 3 & -2 \\ -2 & 0 & 4 & -3 \\ -1 & 0 & 2 & -2 \end{bmatrix}.$$

You should verify that $\mathbf{P}^{-1}\mathbf{A}\mathbf{P} = \mathbf{D}$. ∎

In the next example, the linear operator is not originally defined as a matrix multiplication, and so Steps 1 and 3 of the process require additional work.

EXAMPLE 10

Let $L: \mathcal{P}_3 \to \mathcal{P}_3$ be given by $L(\mathbf{p}(x)) = x\mathbf{p}'(x) + \mathbf{p}(x+1)$. We want to find an ordered basis B for $\mathcal{P}_3$ such that the matrix representation of L with respect to B is diagonal.

Step 1: Let $C = (x^3, x^2, x, 1)$, the standard basis for $\mathcal{P}_3$. We need the matrix for L with respect to C. Calculating directly, we get

$$\begin{array}{rcll} L(x^3) & = & x(3x^2) + (x+1)^3 & = & 4x^3 + 3x^2 + 3x + 1, \\ L(x^2) & = & x(2x) + (x+1)^2 & = & 3x^2 + 2x + 1, \\ L(x) & = & x(1) + (x+1) & = & 2x + 1, \\ \text{and} \quad L(1) & = & x(0) + 1 & = & 1. \end{array}$$

Thus, the matrix for L with respect to C is

$$\mathbf{A} = \begin{bmatrix} 4 & 0 & 0 & 0 \\ 3 & 3 & 0 & 0 \\ 3 & 2 & 2 & 0 \\ 1 & 1 & 1 & 1 \end{bmatrix}.$$

Step 2: We now apply the Diagonalization Method of Section 3.4 to **A**. The characteristic polynomial of **A** is $p_{\mathbf{A}}(x) = (x-4)(x-3)(x-2)(x-1)$ since **A** is lower triangular. Thus, the eigenvalues for **A** are $\lambda_1 = 4$, $\lambda_2 = 3$, $\lambda_3 = 2$, and $\lambda_4 = 1$. Solving for a basis for each eigenspace of **A** gives: basis for $E_{\lambda_1} = \{[6, 18, 27, 17]\}$, basis for $E_{\lambda_2} = \{[0, 2, 4, 3]\}$, basis for $E_{\lambda_3} = \{[0, 0, 1, 1]\}$, and basis for $E_{\lambda_4} = \{[0, 0, 0, 1]\}$. Since $\dim(\mathcal{P}_3) = 4$ and since we obtained 4 distinct eigenvectors, L is diagonalizable. The union

$$Z = \{[6, 18, 27, 17], [0, 2, 4, 3], [0, 0, 1, 1], [0, 0, 0, 1]\}$$

of these eigenspaces is a linearly independent set by Theorem 5.24, and hence, Z is a basis for $\mathbb{R}^4$.

Step 3: Reversing the C-coordinatization isomorphism on the vectors in Z yields the ordered basis $B = (\mathbf{v}_1, \mathbf{v}_2, \mathbf{v}_3, \mathbf{v}_4)$ for $\mathcal{P}_3$, where $\mathbf{v}_1 = 6x^3 + 18x^2 + 27x + 17$, $\mathbf{v}_2 = 2x^2 + 4x + 3$, $\mathbf{v}_3 = x + 1$, and $\mathbf{v}_4 = 1$. The diagonal matrix

$$\mathbf{D} = \begin{bmatrix} 4 & 0 & 0 & 0 \\ 0 & 3 & 0 & 0 \\ 0 & 0 & 2 & 0 \\ 0 & 0 & 0 & 1 \end{bmatrix}$$

is the matrix representation of L in B-coordinates and has the eigenvalues of L appearing on the main diagonal. Finally, the transition matrix $\mathbf{P}$ from B-coordinates to C-coordinates is

$$\mathbf{P} = \begin{bmatrix} 6 & 0 & 0 & 0 \\ 18 & 2 & 0 & 0 \\ 27 & 4 & 1 & 0 \\ 17 & 3 & 1 & 1 \end{bmatrix}.$$

It can quickly be verified that $\mathbf{D} = \mathbf{P}^{-1}\mathbf{AP}$. ■

Geometric and Algebraic Multiplicity

As we have seen, the number of eigenvectors in a basis for each eigenspace is crucial in determining whether a given linear operator is diagonalizable, and so we often need to consider the dimension of each eigenspace.

DEFINITION

Let L be a linear operator on a finite dimensional vector space, and let λ be an eigenvalue for L. Then the dimension of the eigenspace E_λ is called the **geometric multiplicity of** λ.

EXAMPLE 11

In Example 9 we studied a linear operator on $\mathbb{R}^4$ having eigenvalues $\lambda_1 = 1$ and $\lambda_2 = -3$. In that example, we found $\dim(E_{\lambda_1}) = 3$ and $\dim(E_{\lambda_2}) = 1$. Hence, the geometric multiplicity of λ_1 is 3 and the geometric multiplicity of λ_2 is 1. ■

We define the algebraic multiplicity of a linear operator in a manner analogous to the matrix-related definition in Section 3.4.

DEFINITION

Let L be a linear operator on a finite dimensional vector space, and let λ be an eigenvalue for L. Suppose that $(x - \lambda)^k$ is the highest power of $(x - \lambda)$ that divides $p_L(x)$. Then k is called the **algebraic multiplicity of** λ.

In Section 3.4 we suggested, but did not prove, the following relationship between the algebraic and geometric multiplicities of an eigenvalue.

THEOREM 5.25

Let L be a linear operator on a finite dimensional vector space $\mathcal{V}$, and let λ be an eigenvalue for L. Then

$$1 \le \text{(geometric multiplicity of } \lambda) \le \text{(algebraic multiplicity of } \lambda).$$

The proof of Theorem 5.25 uses the following lemma:

LEMMA 5.26

Let $\mathbf{A}$ be an $n \times n$ matrix symbolically represented by $\mathbf{A} = \begin{bmatrix} \mathbf{B} & \mathbf{C} \\ \mathbf{O} & \mathbf{D} \end{bmatrix}$, where $\mathbf{B}$ is an $m \times m$ submatrix, $\mathbf{C}$ is an $m \times (n-m)$ submatrix, $\mathbf{O}$ is an $(n-m) \times m$ zero submatrix, and $\mathbf{D}$ is an $(n-m) \times (n-m)$ submatrix. Then, $|\mathbf{A}| = |\mathbf{B}| \cdot |\mathbf{D}|$.

Lemma 5.26 follows from Exercise 14 in Section 3.2. (We suggest you complete that exercise if you have not already done so.)

PROOF OF THEOREM 5.25

Let $\mathcal{V}$, L, and λ be as given in the statement of the theorem, and let k represent the geometric multiplicity of λ. By definition, the eigenspace E_λ must contain at least one nonzero vector, and thus $k = \dim(E_\lambda) \geq 1$. Thus, the first inequality in the theorem is proved.

Next, choose a basis $\{\mathbf{v}_1, \ldots, \mathbf{v}_k\}$ for E_λ and expand it to an ordered basis $B = (\mathbf{v}_1, \ldots, \mathbf{v}_k, \mathbf{v}_{k+1}, \ldots, \mathbf{v}_n)$ for $\mathcal{V}$. Let $\mathbf{A}$ be the matrix representation for L with respect to B. Notice that for $1 \leq i \leq k$, the ith column of $\mathbf{A} = [L(\mathbf{v}_i)]_B = [\lambda \mathbf{v}_i]_B = \lambda [\mathbf{v}_i]_B = \lambda \mathbf{e}_i$. Thus, $\mathbf{A}$ has the form

$$\mathbf{A} = \begin{bmatrix} \lambda \mathbf{I}_k & \mathbf{C} \\ \mathbf{O} & \mathbf{D} \end{bmatrix},$$

where $\mathbf{C}$ is a $k \times (n-k)$ submatrix, $\mathbf{O}$ is an $(n-k) \times k$ zero submatrix, and $\mathbf{D}$ is an $(n-k) \times (n-k)$ submatrix.

The form of $\mathbf{A}$ makes it straightforward to calculate the characteristic polynomial of L:

$$
\begin{aligned}
p_L(x) = p_{\mathbf{A}}(x) = |x\mathbf{I}_n - \mathbf{A}| &= \left| x\mathbf{I}_n - \begin{bmatrix} \lambda \mathbf{I}_k & \mathbf{C} \\ \mathbf{O} & \mathbf{D} \end{bmatrix} \right| \\
&= \begin{vmatrix} (x - \lambda)\mathbf{I}_k & -\mathbf{C} \\ \mathbf{O} & x\mathbf{I}_{n-k} - \mathbf{D} \end{vmatrix} \\
&= |(x - \lambda)\mathbf{I}_k| \cdot |x\mathbf{I}_{n-k} - \mathbf{D}| \qquad \text{by Lemma 5.26} \\
&= (x - \lambda)^k \cdot p_{\mathbf{D}}(x).
\end{aligned}
$$

Let l be the number of factors of $x - \lambda$ in $p_{\mathbf{D}}(x)$. (Note that $l \geq 0$, with $l = 0$ if $p_{\mathbf{D}}(\lambda) \neq 0$.) Then, altogether, $(x - \lambda)^{k+l}$ is the largest power of $x - \lambda$ that divides $p_L(x)$. Hence,

geometric multiplicity of $\lambda = k \leq k + l =$ algebraic multiplicity of λ.

■

EXAMPLE 12

Consider the linear operator $L: \mathbb{R}^4 \to \mathbb{R}^4$ given by

$$L\left(\begin{bmatrix} x_1 \\ x_2 \\ x_3 \\ x_4 \end{bmatrix}\right) = \begin{bmatrix} 5 & 2 & 0 & 1 \\ -2 & 1 & 0 & -1 \\ 4 & 4 & 3 & 2 \\ 16 & 0 & -8 & -5 \end{bmatrix} \begin{bmatrix} x_1 \\ x_2 \\ x_3 \\ x_4 \end{bmatrix}.$$

In Exercise 3(a), you are asked to verify that $p_L(x) = (x-3)^3(x+5)$. Thus, the eigenvalues for L are $\lambda_1 = 3$ and $\lambda_2 = -5$. Notice that the algebraic multiplicity of λ_1 is 3 and the algebraic multiplicity of λ_2 is 1.

Next we find the eigenspaces of λ_1 and λ_2 by solving appropriate homogeneous systems. Let $\mathbf{A}$ be the matrix for L. For $\lambda_1 = 3$, we solve $(3\mathbf{I}_4 - \mathbf{A})\mathbf{v} = \mathbf{0}$ by row reducing

$$\begin{bmatrix} -2 & -2 & 0 & -1 & | & 0 \\ 2 & 2 & 0 & 1 & | & 0 \\ -4 & -4 & 0 & -2 & | & 0 \\ -16 & 0 & 8 & 8 & | & 0 \end{bmatrix} \quad \text{to obtain} \quad \begin{bmatrix} 1 & 0 & -\frac{1}{2} & -\frac{1}{2} & | & 0 \\ 0 & 1 & \frac{1}{2} & 1 & | & 0 \\ 0 & 0 & 0 & 0 & | & 0 \\ 0 & 0 & 0 & 0 & | & 0 \end{bmatrix}.$$

Thus, a basis for E_3 is $\{[1, -1, 2, 0], [1, -2, 0, 2]\}$, and so the geometric multiplicity of λ_1 is 2, which is less than its algebraic multiplicity.

In Exercise 3(b), you are asked to solve an appropriate system to show that the eigenspace for $\lambda_2 = -5$ has dimension 1, with $\{[1, -1, 2, -8]\}$ being a basis for E_{-5}. Thus, the geometric multiplicity of λ_2 is 1. Hence, the geometric and algebraic multiplicities of λ_2 are actually equal. ∎

The eigenvalue λ_2 in Example 12 also illustrates the principle that if the algebraic multiplicity of an eigenvalue is 1, then its geometric multiplicity must also be 1. This follows immediately from Theorem 5.25.

Multiplicities and Diagonalization

Theorem 5.25 gives us a way to use algebraic and geometric multiplicities to determine whether a linear operator is diagonalizable. Let $L: V \longrightarrow V$ be a linear operator, with $\dim(V) = n$. Then $p_L(x)$ has degree n. Therefore, the sum of the algebraic multiplicities for all eigenvalues can be at most n. Now, for L to be diagonalizable, L must have n linearly independent eigenvectors by Theorem 5.21. This can only happen if the sum of the geometric multiplicities of all eigenvalues for L equals n. Theorem 5.25 then forces the geometric multiplicity of every eigenvalue to equal its algebraic multiplicity (why?). We have therefore proven the following alternative characterization of diagonalizability:

THEOREM 5.27

Let $L: V \to V$ be a linear operator with $\dim(V) = n$. Then L is diagonal-izable if and only if the sum of the algebraic multiplicities over all eigenvalues of L equals n and the geometric multiplicity of each eigenvalue equals its algebraic multiplicity.

Theorem 5.27 gives another justification that the operator L on $\mathbb{R}^4$ in Example 9 is diagonalizable. The eigenvalues $\lambda_1 = 1$ and $\lambda_2 = -3$ have algebraic multiplicities 3 and 1, respectively, and $3 + 1 = 4 = \dim(\mathbb{R}^4)$. Also, the eigenvalues respectively have geometric multiplicities 3 and 1, which equal their algebraic multiplicities. These conditions ensure L is diagonalizable, as we demonstrated in that example.

EXAMPLE 13

Theorem 5.27 shows the operator on $\mathbb{R}^4$ in Example 12 is not diagonalizable because the geometric multiplicity of $\lambda_1 = 3$ is 2, while its algebraic multiplicity is 3.   ∎

EXAMPLE 14

Let $L : \mathbb{R}^3 \to \mathbb{R}^3$ be a rotation about the z-axis through an angle of $\frac{\pi}{3}$. Then the matrix for L with respect to the standard basis is

$$\mathbf{A} = \begin{bmatrix} \frac{1}{2} & -\frac{\sqrt{3}}{2} & 0 \\ \frac{\sqrt{3}}{2} & \frac{1}{2} & 0 \\ 0 & 0 & 1 \end{bmatrix},$$

as described in Table 5.1. Using $\mathbf{A}$, we calculate $p_L(x) = x^3 - 2x^2 + 2x - 1 = (x - 1)(x^2 - x + 1)$, where the quadratic factor has no real roots. Therefore, $\lambda = 1$ is the only eigenvalue, and its algebraic multiplicity is 1. Hence, by Theorem 5.27, L is not diagonalizable because the sum of the algebraic multiplicities of its eigenvalues equals 1, which is less than $\dim(\mathbb{R}^3) = 3$.   ∎

The Cayley-Hamilton Theorem

We conclude this section with an interesting relationship between a matrix and its characteristic polynomial. If $p(x) = a_n x^n + a_{n-1} x^{n-1} \cdots + a_1 x + a_0$ is any polynomial and $\mathbf{A}$ is an $n \times n$ matrix, we define $p(\mathbf{A})$ to be the $n \times n$ matrix given by $p(\mathbf{A}) = a_n \mathbf{A}^n + a_{n-1} \mathbf{A}^{n-1} + \cdots + a_1 \mathbf{A} + a_0 \mathbf{I}_n$.

THEOREM 5.28 (Cayley-Hamilton Theorem)
Let $\mathbf{A}$ be an $n \times n$ matrix, and let $p_{\mathbf{A}}(x)$ be its characteristic polynomial. Then $p_{\mathbf{A}}(\mathbf{A}) = \mathbf{O}_n$.

The Cayley-Hamilton Theorem is an important result in advanced linear algebra. We have placed its proof in Appendix A for the interested reader.

EXAMPLE 15

Let $\mathbf{A} = \begin{bmatrix} 3 & 2 \\ 4 & -1 \end{bmatrix}$. Then $p_{\mathbf{A}}(x) = x^2 - 2x - 11$ (verify!). The Cayley-Hamilton Theorem states that $p_{\mathbf{A}}(\mathbf{A}) = \mathbf{O}_2$. To check this, note that

$$p_{\mathbf{A}}(\mathbf{A}) = \mathbf{A}^2 - 2\mathbf{A} - 11\mathbf{I}_2 = \begin{bmatrix} 17 & 4 \\ 8 & 9 \end{bmatrix} - \begin{bmatrix} 6 & 4 \\ 8 & -2 \end{bmatrix} - \begin{bmatrix} 11 & 0 \\ 0 & 11 \end{bmatrix} = \begin{bmatrix} 0 & 0 \\ 0 & 0 \end{bmatrix}.$$

∎

♦ **Application:** You have now covered the prerequisites for Section 8.9, "Differential Equations."

♦ **Numerical Method:** You have now covered the prerequisites for Section 9.3 "The Power Method for Finding Eigenvalues."

Exercises for Section 5.5

1. For each of the following, let L be a linear operator on $\mathbb{R}^n$ represented by the given matrix with respect to the standard basis. Find all eigenvalues for L, and find a basis for the eigenspace corresponding to each eigenvalue. Compare the geometric and algebraic multiplicities of each eigenvalue.

   ⋆(a) $\begin{bmatrix} 2 & 1 \\ 0 & 2 \end{bmatrix}$ (b) $\begin{bmatrix} 3 & 0 \\ 4 & 2 \end{bmatrix}$

   ⋆(c) $\begin{bmatrix} 0 & 1 & 1 \\ -1 & 4 & -1 \\ -1 & 5 & -2 \end{bmatrix}$ ⋆(d) $\begin{bmatrix} 2 & 0 & 0 \\ 4 & -3 & -6 \\ -4 & 5 & 8 \end{bmatrix}$

   (e) $\begin{bmatrix} 7 & 1 & 2 \\ -11 & -2 & -3 \\ -24 & -3 & -7 \end{bmatrix}$ (f)] $\begin{bmatrix} -13 & 10 & 12 & 19 \\ 1 & 5 & 7 & -2 \\ -2 & -1 & -1 & 3 \\ -9 & 8 & 10 & 13 \end{bmatrix}$

2. Each of the following represents a linear operator L on a vector space $\mathcal{V}$. Let C be the standard basis in each case, and let $\mathbf{A}$ be the matrix representation of L with respect to C. Follow Steps 1 and 2 of the Generalized Diagonalization Method to determine whether L is diagonalizable. If L is diagonalizable, finish the method by performing Step 3. In particular, find the following:

   (i) An ordered basis B for $\mathcal{V}$ consisting of eigenvectors for L

   (ii) The diagonal matrix $\mathbf{D}$ that is the matrix representation of L with respect to B

   (iii) The transition matrix $\mathbf{P}$ from B to C

   Finally, check your work by verifying that $\mathbf{D} = \mathbf{P}^{-1}\mathbf{AP}$.

   (a) $L: \mathbb{R}^4 \to \mathbb{R}^4$ given by $L([x_1, x_2, x_3, x_4]) = [x_2, x_1, x_4, x_3]$

   ⋆(b) $L: \mathcal{P}_2 \to \mathcal{P}_2$ given by $L(\mathbf{p}(x)) = (x-1)\mathbf{p}'(x)$

   (c) $L: \mathcal{P}_2 \to \mathcal{P}_2$ given by $L(\mathbf{p}(x)) = x^2\mathbf{p}''(x) + \mathbf{p}'(x) - 3\mathbf{p}(x)$

   ⋆(d) $L: \mathcal{P}_2 \to \mathcal{P}_2$ given by $L(\mathbf{p}(x)) = (x-3)^2\mathbf{p}''(x) + x\mathbf{p}'(x) - 5\mathbf{p}(x)$

   ⋆(e) $L: \mathbb{R}^2 \to \mathbb{R}^2$ such that L is the counterclockwise rotation about the origin through an angle of $\frac{\pi}{3}$ radians

   (f) $L: \mathcal{M}_{22} \to \mathcal{M}_{22}$ given by $L(\mathbf{K}) = \mathbf{K}^T$

   (g) $L: \mathcal{M}_{22} \to \mathcal{M}_{22}$ given by $L(\mathbf{K}) = \mathbf{K} - \mathbf{K}^T$

   ⋆(h) $L: \mathcal{M}_{22} \to \mathcal{M}_{22}$ given by $L(\mathbf{K}) = \begin{bmatrix} -4 & 3 \\ -10 & 7 \end{bmatrix}\mathbf{K}$

3. Consider the linear operator $L: \mathbb{R}^4 \to \mathbb{R}^4$ from Example 12.

   (a) Verify that $p_L(x) = (x-3)^3(x+5) = x^4 - 4x^3 - 18x^2 + 108x - 135$. (Hint: Use a cofactor expansion along the third column.)

   (b) Show that $\{[1, -1, 2, -8]\}$ is a basis for the eigenspace E_{-5} for L by solving an appropriate homogeneous system.

4. Let $L: \mathcal{P}_2 \rightarrow \mathcal{P}_2$ be the translation operator given by $L(\mathbf{p}(x)) = \mathbf{p}(x + a)$, for some (fixed) real number a.

⋆(a) Find all eigenvalues for L when $a = 1$, and find a basis for each eigenspace.

   (b) Find all eigenvalues for L when a is an arbitrary nonzero number, and find a basis for each eigenspace.

5. Let $\mathbf{A}$ be an $n \times n$ upper triangular matrix with all main diagonal entries equal. Show that $\mathbf{A}$ is diagonalizable if and only if $\mathbf{A}$ is a diagonal matrix.

6. Explain why Examples 8 and 9 provide counterexamples to the converse of Corollary 5.23.

⋆**7.** (a) Give an example of a 3×3 upper triangular matrix having an eigenvalue λ with algebraic multiplicity 3 and geometric multiplicity 2.

   (b) Give an example of a 3×3 upper triangular matrix, one of whose eigenvalues has algebraic multiplicity 2 and geometric multiplicity 2.

8. (a) Suppose that L is a linear operator on a finite dimensional vector space. Prove L is an isomorphism if and only if 0 is not an eigenvalue for L.

   (b) Let L be an isomorphism from a vector space to itself. Suppose that λ is an eigenvalue for L having eigenvector $\mathbf{v}$. Prove that $\mathbf{v}$ is an eigenvector for L^{-1} corresponding to the eigenvalue $1/\lambda$.

9. Let L be a linear operator on a finite dimensional vector space $\mathcal{V}$, and let B be an ordered basis for $\mathcal{V}$. Also, let $\mathbf{A}$ be the matrix for L with respect to B. Assume that $\mathbf{A}$ is a diagonalizable matrix. Prove that there is an ordered basis C for $\mathcal{V}$ such that the matrix representation of L with respect to C is diagonal and hence that L is a diagonalizable operator.

10. Let $\mathbf{A}$ be an $n \times n$ matrix. Suppose that $(\mathbf{v}_1, \ldots, \mathbf{v}_n)$ is a basis for $\mathbb{R}^n$ of eigenvectors for $\mathbf{A}$ with corresponding eigenvalues $\lambda_1, \lambda_2, \ldots, \lambda_n$. Show that $|\mathbf{A}| = \lambda_1 \lambda_2 \cdots \lambda_n$.

11. Let L be a linear operator on an n-dimensional vector space, with $\{\lambda_1, \ldots, \lambda_k\}$ equal to the set of all distinct eigenvalues for L. Show that $\Sigma_{i=1}^{k}$(geometric multiplicity of λ_i) $\leq n$.

12. Let L be a linear operator on a finite dimensional vector space $\mathcal{V}$. Show that if L is diagonalizable, then every root of $p_L(x)$ is real.

13. Let $\mathbf{A}$ and $\mathbf{B}$ be commuting $n \times n$ matrices.

   (a) Show that if λ is an eigenvalue for $\mathbf{A}$ and $\mathbf{v} \in E_\lambda$ (the eigenspace for $\mathbf{A}$ associated with λ), then $\mathbf{Bv} \in E_\lambda$.

   (b) Prove that if $\mathbf{A}$ has n distinct eigenvalues, then $\mathbf{B}$ is diagonalizable.

14. (a) Let $\mathbf{A}$ be a fixed 2×2 matrix with distinct eigenvalues λ_1 and λ_2. Show that the linear operator $L: \mathcal{M}_{22} \rightarrow \mathcal{M}_{22}$ given by $L(\mathbf{K}) = \mathbf{AK}$ is diagonalizable with eigenvalues λ_1 and λ_2, each having multiplicity 2. (Hint: Use eigenvectors for $\mathbf{A}$ to help create eigenvectors for L.)

   (b) Generalize part (a) as follows: Let $\mathbf{A}$ be a fixed diagonalizable $n \times n$ matrix with distinct eigenvalues $\lambda_1, \ldots, \lambda_k$. Show that the linear operator $L: \mathcal{M}_{nn} \rightarrow \mathcal{M}_{nn}$ given by $L(\mathbf{K}) = \mathbf{AK}$ is diagonalizable with eigenvalues $\lambda_1, \ldots, \lambda_k$. In addition, show that, for each i, the geometric multiplicity of λ_i for L is n times the geometric multiplicity of λ_i for $\mathbf{A}$.

▶**15.** Let $L: \mathcal{V} \to \mathcal{V}$ be a linear operator on a finite dimensional vector space $\mathcal{V}$. Suppose that λ_1 and λ_2 are distinct eigenvalues for L and that B_1 and B_2 are bases for the eigenspaces E_{λ_1} and E_{λ_2} for L. Prove that $B_1 \cap B_2$ is empty.

▶**16.** Let $L: \mathcal{V} \to \mathcal{V}$ be a linear operator on a finite dimensional vector space $\mathcal{V}$. Suppose that $\lambda_1, \ldots, \lambda_n$ are distinct eigenvalues for L and that $B_i = \{\mathbf{v}_{i1}, \ldots, \mathbf{v}_{ik_i}\}$ is a basis for the eigenspace E_{λ_i}, for $1 \leq i \leq n$. The goal of this exercise is to show that $B = \cup_{i=1}^{n} B_i$ is linearly independent. Suppose that $\Sigma_{i=1}^{n} \Sigma_{j=1}^{k_i} a_{ij} \mathbf{v}_{ij} = \mathbf{0}$.

(a) Let $\mathbf{u}_i = \Sigma_{j=1}^{k_i} a_{ij} \mathbf{v}_{ij}$. Show that $\mathbf{u}_i \in E_{\lambda_i}$.

(b) Note that $\Sigma_{i=1}^{n} \mathbf{u}_i = \mathbf{0}$. Use Theorem 5.22 and Theorem 4.7 to show that $\mathbf{u}_i = \mathbf{0}$, for $1 \leq i \leq n$.

(c) Conclude that $a_{ij} = 0$, for $1 \leq i \leq n$ and $1 \leq j \leq k_i$.

(d) Explain why parts (a) through (c) prove that B is linearly independent.

★**17.** True or False:

(a) If $L: \mathcal{V} \to \mathcal{V}$ is a linear operator and λ is an eigenvalue for L, then $E_\lambda = \{\lambda L(\mathbf{v}) \mid \mathbf{v} \in \mathcal{V}\}$.

(b) If L is a linear operator on a finite dimensional vector space $\mathcal{V}$ and $\mathbf{A}$ is a matrix for L with respect to some ordered basis for $\mathcal{V}$, then $p_L(x) = p_{\mathbf{A}}(x)$.

(c) If $\dim(\mathcal{V}) = 5$, a linear operator L on $\mathcal{V}$ is diagonalizable when L has 5 linearly independent eigenvectors.

(d) Eigenvectors for a given linear operator L are linearly independent if and only if they correspond to distinct eigenvalues of L.

(e) If L is a linear operator on a finite dimensional vector space, then the union of bases for distinct eigenspaces for L is a linearly independent set.

(f) If $L: \mathbb{R}^6 \to \mathbb{R}^6$ is a diagonalizable linear operator, then the union of bases for all the distinct eigenspaces of L is actually a basis for $\mathbb{R}^6$.

(g) If L is a diagonalizable linear operator on a finite dimensional vector space $\mathcal{V}$, the Generalized Diagonalization Method produces a basis B for $\mathcal{V}$ so that the matrix for L with respect to B is diagonal.

(h) If L is a linear operator on a finite dimensional vector space $\mathcal{V}$ and λ is an eigenvalue for L, then the algebraic multiplicity of λ is never greater than the geometric multiplicity of λ.

(i) If $\dim(\mathcal{V}) = 7$ and $L: \mathcal{V} \to \mathcal{V}$ is a linear operator, then L is diagonalizable whenever the sum of the algebraic multiplicities of all the eigenvalues equals 7.

(j) If $\mathbf{A} = \begin{bmatrix} 1 & 2 \\ 0 & 4 \end{bmatrix}$, then $(1\mathbf{I}_2 - \mathbf{A})(4\mathbf{I}_2 - \mathbf{A}) = \mathbf{O}_2$.

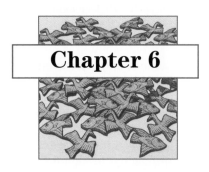

Chapter 6

Orthogonality

Geometry Is Never Pointless

Linear algebra exists at the crossroads between algebra and geometry. Yet, in our study of abstract vector spaces in Chapters 4 and 5, we usually concentrated on the algebra involved at the expense of the geometry. But the underlying geometry is important also. For example, a linear transformation on $\mathbb{R}^n$ can cause the distance between the images of two points to be different than the original distance between the points. Angles between a pair of vectors and their images can differ as well.

In Chapter 1, we noted that the geometric properties of $\mathbb{R}^n$, such as orthogonality, are derived from the length and dot product of vectors. In this chapter, we enhance our understanding of these properties and operations by reexamining them in the light of the more general vector space properties of Chapters 4 and 5. This new level of understanding will put additional applications within our reach.

In our study of general vector spaces and linear transformations in Chapters 4 and 5, we avoided the dot product because it is not defined in every vector space. Therefore, we could not discuss lengths of vectors or angles in general vector spaces as we can in $\mathbb{R}^n$. In this chapter, we restrict our attention to $\mathbb{R}^n$ and present some additional structures and properties related to the dot product.

In Section 6.1, we examine special bases for $\mathbb{R}^n$ whose vectors are mutually orthogonal. In Section 6.2, we introduce orthogonal complements of subspaces of $\mathbb{R}^n$. Finally, in Section 6.3, we use orthogonality to diagonalize any symmetric matrix.

6.1 Orthogonal Bases and the Gram-Schmidt Process

In this section, we investigate orthogonality of vectors in more detail. Our main goal is the Gram-Schmidt Process, a method for constructing a basis of mutually orthogonal vectors for any nontrivial subspace of $\mathbb{R}^n$.

Orthogonal and Orthonormal Vectors

DEFINITION

Let $\{\mathbf{v}_1, \mathbf{v}_2, \ldots, \mathbf{v}_k\}$ be a subset of k distinct vectors of $\mathbb{R}^n$. Then $\{\mathbf{v}_1, \mathbf{v}_2, \ldots, \mathbf{v}_k\}$ is an **orthogonal set of vectors** if and only if the dot product of any two distinct vectors in this set is zero, that is, if and only if $\mathbf{v}_i \cdot \mathbf{v}_j = 0$, for $1 \leq i, j \leq k, i \neq j$. Also, $\{\mathbf{v}_1, \mathbf{v}_2, \ldots, \mathbf{v}_k\}$ is an **orthonormal set of vectors** if and only if it is an orthogonal set and all its vectors are unit vectors (that is, $\|\mathbf{v}_i\| = 1$, for $1 \leq i \leq k$).

In particular, any set containing a single vector is orthogonal, and any set containing a single unit vector is orthonormal.

EXAMPLE 1

In $\mathbb{R}^3$, $\{\mathbf{i}, \mathbf{j}, \mathbf{k}\}$ is an orthogonal set because $\mathbf{i} \cdot \mathbf{j} = \mathbf{j} \cdot \mathbf{k} = \mathbf{k} \cdot \mathbf{i} = 0$. In fact, this is an orthonormal set since we also have $\|\mathbf{i}\| = \|\mathbf{j}\| = \|\mathbf{k}\| = 1$.

In $\mathbb{R}^4$, $\{[1, 0, -1, 0], [3, 0, 3, 0]\}$ is an orthogonal set because $[1, 0, -1, 0] \cdot [3, 0, 3, 0] = 0$. If we normalize each vector (that is, divide each of these vectors by its length), we create the orthonormal set of vectors

$$\left\{ \left[\frac{1}{\sqrt{2}}, 0, -\frac{1}{\sqrt{2}}, 0 \right], \left[\frac{1}{\sqrt{2}}, 0, \frac{1}{\sqrt{2}}, 0 \right] \right\}. \qquad \blacksquare$$

The next theorem is proved in the same manner as Result 7 in Section 1.3.

THEOREM 6.1

Let $T = \{\mathbf{v}_1, \ldots, \mathbf{v}_k\}$ be an orthogonal set of nonzero vectors in $\mathbb{R}^n$. Then T is a linearly independent set.

Notice that the orthogonal sets in Example 1 are indeed linearly independent.

Orthogonal and Orthonormal Bases

Theorem 6.1 assures us that any orthogonal set of nonzero vectors in $\mathbb{R}^n$ is linearly independent, so any such set forms a basis for some subspace of $\mathbb{R}^n$.

DEFINITION

A basis B for a subspace $\mathcal{W}$ of $\mathbb{R}^n$ is an **orthogonal basis** for $\mathcal{W}$ if and only if B is an orthogonal set. Similarly, a basis B for $\mathcal{W}$ is an **orthonormal basis** for $\mathcal{W}$ if and only if B is an orthonormal set.

The following corollary follows immediately from Theorem 6.1:

COROLLARY 6.2

If B is an orthogonal set of n nonzero vectors in $\mathbb{R}^n$, then B is an orthogonal basis for $\mathbb{R}^n$. Similarly, if B is an orthonormal set of n vectors in $\mathbb{R}^n$, then B is an orthonormal basis for $\mathbb{R}^n$.

EXAMPLE 2

Consider the following subset of $\mathbb{R}^3$: $\{[1, 0, -1], [-1, 4, -1], [2, 1, 2]\}$. Because every pair of distinct vectors in this set is orthogonal (verify!), this is an orthogonal set. By Corollary 6.2, this is also an orthogonal basis for $\mathbb{R}^3$. Normalizing each vector, we obtain the following orthonormal basis for $\mathbb{R}^3$:

$$\left\{ \left[\frac{1}{\sqrt{2}}, 0, -\frac{1}{\sqrt{2}} \right], \left[-\frac{1}{3\sqrt{2}}, \frac{4}{3\sqrt{2}}, -\frac{1}{3\sqrt{2}} \right], \left[\frac{2}{3}, \frac{1}{3}, \frac{2}{3} \right] \right\}.$$

■

One of the advantages of using an orthogonal or orthonormal basis is that it is easy to coordinatize vectors with respect to that basis.

THEOREM 6.3

If $B = (\mathbf{v}_1, \mathbf{v}_2, \ldots, \mathbf{v}_k)$ is a nonempty orthogonal ordered basis for a subspace $\mathcal{W}$ of $\mathbb{R}^n$, and if $\mathbf{v}$ is any vector in $\mathcal{W}$, then

$$[\mathbf{v}]_B = \left[\frac{(\mathbf{v} \cdot \mathbf{v}_1)}{(\mathbf{v}_1 \cdot \mathbf{v}_1)}, \frac{(\mathbf{v} \cdot \mathbf{v}_2)}{(\mathbf{v}_2 \cdot \mathbf{v}_2)}, \ldots, \frac{(\mathbf{v} \cdot \mathbf{v}_k)}{(\mathbf{v}_k \cdot \mathbf{v}_k)} \right] = \left[\frac{(\mathbf{v} \cdot \mathbf{v}_1)}{||\mathbf{v}_1||^2}, \frac{(\mathbf{v} \cdot \mathbf{v}_2)}{||\mathbf{v}_2||^2}, \ldots, \frac{(\mathbf{v} \cdot \mathbf{v}_k)}{||\mathbf{v}_k||^2} \right].$$

In particular, if B is an orthonormal ordered basis for $\mathcal{W}$, then
$[\mathbf{v}]_B = [\mathbf{v} \cdot \mathbf{v}_1, \mathbf{v} \cdot \mathbf{v}_2, \ldots, \mathbf{v} \cdot \mathbf{v}_k]$.

PROOF OF THEOREM 6.3

Suppose that $[\mathbf{v}]_B = [a_1, a_2, \ldots, a_k]$, where $a_1, a_2, \ldots, a_k \in \mathbb{R}$. We must show that $a_i = (\mathbf{v} \cdot \mathbf{v}_i)/(\mathbf{v}_i \cdot \mathbf{v}_i)$, for $1 \le i \le k$. Now, $\mathbf{v} = a_1\mathbf{v}_1 + a_2\mathbf{v}_2 + \cdots + a_k\mathbf{v}_k$. Hence,

$$
\begin{aligned}
\mathbf{v} \cdot \mathbf{v}_i &= (a_1\mathbf{v}_1 + a_2\mathbf{v}_2 + \cdots + a_i\mathbf{v}_i + \cdots + a_k\mathbf{v}_k) \cdot \mathbf{v}_i \\
&= a_1(\mathbf{v}_1 \cdot \mathbf{v}_i) + a_2(\mathbf{v}_2 \cdot \mathbf{v}_i) + \cdots + a_i(\mathbf{v}_i \cdot \mathbf{v}_i) + \cdots + a_k(\mathbf{v}_k \cdot \mathbf{v}_i) \\
&= a_1(0) + a_2(0) + \cdots + a_i(\mathbf{v}_i \cdot \mathbf{v}_i) + \cdots + a_k(0) \quad \text{because } B \text{ is orthogonal} \\
&= a_i(\mathbf{v}_i \cdot \mathbf{v}_i).
\end{aligned}
$$

Thus, $a_i = (\mathbf{v} \cdot \mathbf{v}_i)/(\mathbf{v}_i \cdot \mathbf{v}_i) = (\mathbf{v} \cdot \mathbf{v}_i)/\|\mathbf{v}_i\|^2$. In the special case when B is orthonormal, $\|\mathbf{v}_i\| = 1$, and so $a_i = \mathbf{v} \cdot \mathbf{v}_i$. ∎

EXAMPLE 3

Consider the ordered orthogonal basis $B = (\mathbf{v}_1, \mathbf{v}_2, \mathbf{v}_3)$ for $\mathbb{R}^3$ from Example 2, where $\mathbf{v}_1 = [1, 0, -1]$, $\mathbf{v}_2 = [-1, 4, -1]$, and $\mathbf{v}_3 = [2, 1, 2]$. Let $\mathbf{v} = [-1, 5, 3]$. We will use Theorem 6.3 to find $[\mathbf{v}]_B$.

Now, $\mathbf{v} \cdot \mathbf{v}_1 = -4$, $\mathbf{v} \cdot \mathbf{v}_2 = 18$, and $\mathbf{v} \cdot \mathbf{v}_3 = 9$. Also, $\mathbf{v}_1 \cdot \mathbf{v}_1 = 2$, $\mathbf{v}_2 \cdot \mathbf{v}_2 = 18$, and $\mathbf{v}_3 \cdot \mathbf{v}_3 = 9$. Hence,

$$
[\mathbf{v}]_B = \left[\frac{(\mathbf{v} \cdot \mathbf{v}_1)}{(\mathbf{v}_1 \cdot \mathbf{v}_1)}, \frac{(\mathbf{v} \cdot \mathbf{v}_2)}{(\mathbf{v}_2 \cdot \mathbf{v}_2)}, \ldots, \frac{(\mathbf{v} \cdot \mathbf{v}_k)}{(\mathbf{v}_k \cdot \mathbf{v}_k)} \right] = \left[\frac{-4}{2}, \frac{18}{18}, \frac{9}{9} \right] = [-2, 1, 1].
$$

Similarly, suppose $C = (\mathbf{w}_1, \mathbf{w}_2, \mathbf{w}_3)$ is the ordered orthonormal basis for $\mathbb{R}^3$ from Example 2; that is, $\mathbf{w}_1 = \left[\frac{1}{\sqrt{2}}, 0, -\frac{1}{\sqrt{2}} \right]$, $\mathbf{w}_2 = \left[-\frac{1}{3\sqrt{2}}, \frac{4}{3\sqrt{2}}, -\frac{1}{3\sqrt{2}} \right]$, and $\mathbf{w}_3 = \left[\frac{2}{3}, \frac{1}{3}, \frac{2}{3} \right]$. Again, let $\mathbf{v} = [-1, 5, 3]$. Then $\mathbf{v} \cdot \mathbf{v}_1 = -2\sqrt{2}$, $\mathbf{v} \cdot \mathbf{v}_2 = 3\sqrt{2}$, and $\mathbf{v} \cdot \mathbf{v}_3 = 3$. By Theorem 6.3, $[\mathbf{v}]_C = \left[-2\sqrt{2}, 3\sqrt{2}, 3 \right]$. These coordinates can be verified by checking that

$$
[-1, 5, 3] = -2\sqrt{2} \left[\frac{1}{\sqrt{2}}, 0, -\frac{1}{\sqrt{2}} \right] + 3\sqrt{2} \left[-\frac{1}{3\sqrt{2}}, \frac{4}{3\sqrt{2}}, -\frac{1}{3\sqrt{2}} \right] + 3 \left[\frac{2}{3}, \frac{1}{3}, \frac{2}{3} \right].
$$

■

The Gram-Schmidt Process: Finding an Orthogonal Basis for a Subspace of $\mathbb{R}^n$

We have just seen that it is convenient to work with an orthogonal basis whenever possible. Now, suppose $\mathcal{W}$ is a subspace of $\mathbb{R}^n$ with basis $B = \{\mathbf{w}_1, \ldots, \mathbf{w}_k\}$. There is a straightforward way to replace B with an orthogonal basis for $\mathcal{W}$. This is known as the Gram-Schmidt Process.

GRAM-SCHMIDT PROCESS

Let $\{\mathbf{w}_1, \ldots, \mathbf{w}_k\}$ be a linearly independent subset of $\mathbb{R}^n$. We create a new set $\{\mathbf{v}_1, \ldots, \mathbf{v}_k\}$ of vectors as follows:

Let $\mathbf{v}_1 = \mathbf{w}_1.$

Let $\mathbf{v}_2 = \mathbf{w}_2 - \left(\dfrac{\mathbf{w}_2 \cdot \mathbf{v}_1}{\mathbf{v}_1 \cdot \mathbf{v}_1}\right) \mathbf{v}_1.$

Let $\mathbf{v}_3 = \mathbf{w}_3 - \left(\dfrac{\mathbf{w}_3 \cdot \mathbf{v}_1}{\mathbf{v}_1 \cdot \mathbf{v}_1}\right) \mathbf{v}_1 - \left(\dfrac{\mathbf{w}_3 \cdot \mathbf{v}_2}{\mathbf{v}_2 \cdot \mathbf{v}_2}\right) \mathbf{v}_2.$

$$\vdots$$

Let $\mathbf{v}_k = \mathbf{w}_k - \left(\dfrac{\mathbf{w}_k \cdot \mathbf{v}_1}{\mathbf{v}_1 \cdot \mathbf{v}_1}\right) \mathbf{v}_1 - \left(\dfrac{\mathbf{w}_k \cdot \mathbf{v}_2}{\mathbf{v}_2 \cdot \mathbf{v}_2}\right) \mathbf{v}_2 - \cdots - \left(\dfrac{\mathbf{w}_k \cdot \mathbf{v}_{k-1}}{\mathbf{v}_{k-1} \cdot \mathbf{v}_{k-1}}\right) \mathbf{v}_{k-1}.$

Then $\{\mathbf{v}_1, \ldots, \mathbf{v}_k\}$ is an orthogonal basis for $\text{span}(\{\mathbf{w}_1, \ldots, \mathbf{w}_k\}).$

The justification that the Gram-Schmidt Process is valid is given in the following theorem:

THEOREM 6.4

Let $B = \{\mathbf{w}_1, \ldots, \mathbf{w}_k\}$ be a basis for a subspace $\mathcal{W}$ of $\mathbb{R}^n$. Then the set $T = \{\mathbf{v}_1, \ldots, \mathbf{v}_K\}$ obtained by applying the Gram-Schmidt Process to B is an orthogonal basis for $\mathcal{W}$.

Hence, any nontrivial subspace $\mathcal{W}$ of $\mathbb{R}^n$ has an orthogonal basis.

PROOF OF THEOREM 6.4

Let $\mathcal{W}$, B, and T be as given in the statement of the theorem. To prove that T is an orthogonal basis for $\mathcal{W}$, we must prove three statements about T.

(1) $T \subseteq \mathcal{W}$.

(2) Every vector in T is nonzero.

(3) T is an orthogonal set.

Theorem 6.1 will then show that T is linearly independent, and since $|T| = k = \dim(\mathcal{W})$, T is an orthogonal basis for $\mathcal{W}$.

We proceed by induction, proving for each i, $1 \leq i \leq k$, that

(1′) $\{\mathbf{v}_1, \ldots, \mathbf{v}_i\} \subseteq \text{span}(\{\mathbf{w}_1, \ldots, \mathbf{w}_i\})$,

(2′) $\mathbf{v}_i \neq \mathbf{0}$,

(3′) $\{\mathbf{v}_1, \ldots, \mathbf{v}_i\}$ is an orthogonal set.

Obviously, once the induction is complete, properties (1), (2), and (3) will be established for T, and the theorem will be proved.

Base Step: Since $\mathbf{v}_1 = \mathbf{w}_1 \in B$, it is clear that $\{\mathbf{v}_1\} \subseteq \text{span}(\{\mathbf{w}_1\})$, $\mathbf{v}_1 \neq \mathbf{0}$, and $\{\mathbf{v}_1\}$ is an orthogonal set.

Inductive Step: The inductive hypothesis asserts that $\{\mathbf{v}_1, \ldots, \mathbf{v}_i\}$ is an orthogonal subset of $\text{span}(\{\mathbf{w}_1, \ldots, \mathbf{w}_i\})$ consisting of nonzero vectors. We need to prove $(1')$, $(2')$, and $(3')$ for $\{\mathbf{v}_1, \ldots, \mathbf{v}_{i+1}\}$.

To establish $(1')$, we only need to prove that $\mathbf{v}_{i+1} \in \text{span}(\{\mathbf{w}_1, \ldots, \mathbf{w}_{i+1}\})$, since we already know from the inductive hypothesis that $\{\mathbf{v}_1, \ldots, \mathbf{v}_i\}$ is a subset of $\text{span}(\{\mathbf{w}_1, \ldots, \mathbf{w}_i\})$, and hence of $\text{span}(\{\mathbf{w}_1, \ldots, \mathbf{w}_{i+1}\})$. But by definition, $\mathbf{v}_{i+1}$ is a linear combination of $\mathbf{w}_{i+1}$ and $\mathbf{v}_1, \ldots, \mathbf{v}_i$, all of which are in $\text{span}(\{\mathbf{w}_1, \ldots, \mathbf{w}_{i+1}\})$. Hence, $\mathbf{v}_{i+1} \in \text{span}(\{\mathbf{w}_1, \ldots, \mathbf{w}_{i+1}\})$.

To prove $(2')$, we assume that $\mathbf{v}_{i+1} = \mathbf{0}$ and produce a contradiction. Now, from the definition of $\mathbf{v}_{i+1}$, if $\mathbf{v}_{i+1} = \mathbf{0}$ we have

$$\mathbf{w}_{i+1} = \left(\frac{\mathbf{w}_{i+1} \cdot \mathbf{v}_1}{\mathbf{v}_1 \cdot \mathbf{v}_1} \right) \mathbf{v}_1 + \left(\frac{\mathbf{w}_{i+1} \cdot \mathbf{v}_2}{\mathbf{v}_2 \cdot \mathbf{v}_2} \right) \mathbf{v}_2 + \cdots + \left(\frac{\mathbf{w}_{i+1} \cdot \mathbf{v}_i}{\mathbf{v}_i \cdot \mathbf{v}_i} \right) \mathbf{v}_i.$$

But then $\mathbf{w}_{i+1} \in \text{span}(\{\mathbf{v}_1, \ldots, \mathbf{v}_i\}) \subseteq \text{span}(\{\mathbf{w}_1, \ldots, \mathbf{w}_i\})$, from the inductive hypothesis. This result contradicts the fact that B is a linearly independent set. Therefore, $\mathbf{v}_{i+1} \neq \mathbf{0}$.

Finally, we need to prove $(3')$. By the inductive hypothesis, $\{\mathbf{v}_1, \ldots, \mathbf{v}_i\}$ is an orthogonal set. Hence, we only need to show that $\mathbf{v}_{i+1}$ is orthogonal to each of $\mathbf{v}_1, \ldots, \mathbf{v}_i$. Now,

$$\mathbf{v}_{i+1} = \mathbf{w}_{i+1} - \left(\frac{\mathbf{w}_{i+1} \cdot \mathbf{v}_1}{\mathbf{v}_1 \cdot \mathbf{v}_1} \right) \mathbf{v}_1 - \left(\frac{\mathbf{w}_{i+1} \cdot \mathbf{v}_2}{\mathbf{v}_2 \cdot \mathbf{v}_2} \right) \mathbf{v}_2 - \cdots - \left(\frac{\mathbf{w}_{i+1} \cdot \mathbf{v}_i}{\mathbf{v}_i \cdot \mathbf{v}_i} \right) \mathbf{v}_i.$$

Notice that

$$
\begin{aligned}
\mathbf{v}_{i+1} \cdot \mathbf{v}_1 &= \mathbf{w}_{i+1} \cdot \mathbf{v}_1 - \left(\frac{\mathbf{w}_{i+1} \cdot \mathbf{v}_1}{\mathbf{v}_1 \cdot \mathbf{v}_1} \right) (\mathbf{v}_1 \cdot \mathbf{v}_1) - \left(\frac{\mathbf{w}_{i+1} \cdot \mathbf{v}_2}{\mathbf{v}_2 \cdot \mathbf{v}_2} \right) (\mathbf{v}_2 \cdot \mathbf{v}_1) \\
&\quad - \cdots - \left(\frac{\mathbf{w}_{i+1} \cdot \mathbf{v}_i}{\mathbf{v}_i \cdot \mathbf{v}_i} \right) (\mathbf{v}_i \cdot \mathbf{v}_1) \\
&= \mathbf{w}_{i+1} \cdot \mathbf{v}_1 - \left(\frac{\mathbf{w}_{i+1} \cdot \mathbf{v}_1}{\mathbf{v}_1 \cdot \mathbf{v}_1} \right) (\mathbf{v}_1 \cdot \mathbf{v}_1) - \left(\frac{\mathbf{w}_{i+1} \cdot \mathbf{v}_2}{\mathbf{v}_2 \cdot \mathbf{v}_2} \right) (0) \\
&\quad - \cdots - \left(\frac{\mathbf{w}_{i+1} \cdot \mathbf{v}_i}{\mathbf{v}_i \cdot \mathbf{v}_i} \right) (0) \qquad \text{inductive hypothesis} \\
&= \mathbf{w}_{i+1} \cdot \mathbf{v}_1 - \mathbf{w}_{i+1} \cdot \mathbf{v}_1 = 0.
\end{aligned}
$$

Similar arguments show that $\mathbf{v}_{i+1} \cdot \mathbf{v}_2 = \mathbf{v}_{i+1} \cdot \mathbf{v}_3 = \cdots = \mathbf{v}_{i+1} \cdot \mathbf{v}_i = 0$. Hence, $\{\mathbf{v}_1, \ldots, \mathbf{v}_{i+1}\}$ is an orthogonal set. This finishes the Inductive Step, completing the proof of the theorem. ∎

Once we have an orthogonal basis for a subspace $\mathcal{W}$ of $\mathbb{R}^n$, we can easily convert it to an orthonormal basis for $\mathcal{W}$ by normalizing each vector. Also, a little thought will convince you that if any of the newly created vectors $\mathbf{v}_i$ in the Gram-Schmidt Process is replaced with a nonzero scalar multiple of itself, the proof of Theorem 6.4 still holds. Hence, in applying the Gram-Schmidt

Process, we can often replace the $\mathbf{v}_i$'s we create with appropriate multiples to avoid fractions. The next example illustrates these techniques.

EXAMPLE 4

You can verify that $B = \{[2, 1, 0, -1], [1, 0, 2, -1], [0, -2, 1, 0]\}$ is a linearly independent set in $\mathbb{R}^4$. Let $\mathcal{W} = \text{span}(B)$. Now, B is not an orthogonal basis for $\mathcal{W}$, but we will apply the Gram-Schmidt Process to replace B with an orthogonal basis. Let $\mathbf{w}_1 = [2, 1, 0, -1]$, $\mathbf{w}_2 = [1, 0, 2, -1]$, and $\mathbf{w}_3 = [0, -2, 1, 0]$. Beginning the Gram-Schmidt Process, we obtain $\mathbf{v}_1 = \mathbf{w}_1 = [2, 1, 0, -1]$ and

$$\mathbf{v}_2 = \mathbf{w}_2 - \left(\frac{\mathbf{w}_2 \cdot \mathbf{v}_1}{\mathbf{v}_1 \cdot \mathbf{v}_1}\right) \mathbf{v}_1$$

$$= [1, 0, 2, -1] - \left(\frac{[1, 0, 2, -1] \cdot [2, 1, 0, -1]}{[2, 1, 0, -1] \cdot [2, 1, 0, -1]}\right) [2, 1, 0, -1]$$

$$= [1, 0, 2, -1] - \left(\frac{3}{6}\right) [2, 1, 0, -1] = \left[0, -\frac{1}{2}, 2, -\frac{1}{2}\right].$$

To avoid fractions, we replace this vector with an appropriate scalar multiple. Multiplying by 2, we get $\mathbf{v}_2 = [0, -1, 4, -1]$. Notice that $\mathbf{v}_2$ is orthogonal to $\mathbf{v}_1$. Finally,

$$\mathbf{v}_3 = \mathbf{w}_3 - \left(\frac{\mathbf{w}_3 \cdot \mathbf{v}_1}{\mathbf{v}_1 \cdot \mathbf{v}_1}\right) \mathbf{v}_1 - \left(\frac{\mathbf{w}_3 \cdot \mathbf{v}_2}{\mathbf{v}_2 \cdot \mathbf{v}_2}\right) \mathbf{v}_2$$

$$= [0, -2, 1, 0] - \left(\frac{[0, -2, 1, 0] \cdot [2, 1, 0, -1]}{[2, 1, 0, -1] \cdot [2, 1, 0, -1]}\right) [2, 1, 0, -1]$$

$$\qquad - \left(\frac{[0, -2, 1, 0] \cdot [0, -1, 4, -1]}{[0, -1, 4, -1] \cdot [0, -1, 4, -1]}\right) [0, -1, 4, -1]$$

$$= [0, -2, 1, 0] - \left(\frac{-2}{6}\right) [2, 1, 0, -1] - \left(\frac{6}{18}\right) [0, -1, 4, -1]$$

$$= \left[\frac{2}{3}, -\frac{4}{3}, -\frac{1}{3}, 0\right].$$

To avoid fractions, we multiply this vector by 3, yielding $\mathbf{v}_3 = [2, -4, -1, 0]$. Notice that $\mathbf{v}_3$ is orthogonal to both $\mathbf{v}_1$ and $\mathbf{v}_2$. Hence,

$$\{\mathbf{v}_1, \mathbf{v}_2, \mathbf{v}_3\} = \{[2, 1, 0, -1], [0, -1, 4, -1], [2, -4, -1, 0]\}$$

is an orthogonal basis for $\mathcal{W}$. To find an orthonormal basis for $\mathcal{W}$, we normalize $\mathbf{v}_1$, $\mathbf{v}_2$, and $\mathbf{v}_3$ to obtain

$$\left\{\left[\frac{2}{\sqrt{6}}, \frac{1}{\sqrt{6}}, 0, -\frac{1}{\sqrt{6}}\right], \left[0, -\frac{1}{3\sqrt{2}}, \frac{4}{3\sqrt{2}}, -\frac{1}{3\sqrt{2}}\right], \left[\frac{2}{\sqrt{21}}, -\frac{4}{\sqrt{21}}, -\frac{1}{\sqrt{21}}, 0\right]\right\}.$$

∎

Suppose $T = \{\mathbf{w}_1, \ldots, \mathbf{w}_k\}$ is an orthogonal set of nonzero vectors in a subspace $\mathcal{W}$ of $\mathbb{R}^n$. By Theorem 6.1, T is linearly independent. Hence, by

Theorem 4.19, we can enlarge T to an ordered basis $(\mathbf{w}_1, \ldots, \mathbf{w}_k, \mathbf{w}_{k+1}, \ldots, \mathbf{w}_l)$ for $\mathcal{W}$. Applying the Gram-Schmidt Process to this enlarged basis gives an ordered orthogonal basis $B = (\mathbf{v}_1, \ldots, \mathbf{v}_k, \mathbf{v}_{k+1}, \ldots, \mathbf{v}_l)$ for $\mathcal{W}$. However, because $(\mathbf{w}_1, \ldots, \mathbf{w}_k)$ is already orthogonal, the first k vectors, $\mathbf{v}_1, \ldots, \mathbf{v}_k$, created by the Gram-Schmidt Process will be equal to $\mathbf{w}_1, \ldots, \mathbf{w}_k$, respectively (why?). Hence, B is an ordered orthogonal basis for $\mathcal{W}$ that contains T. Similarly, if the original set $T = \{\mathbf{w}_1, \ldots, \mathbf{w}_k\}$ is *orthonormal*, T can be enlarged to an *orthonormal* basis for $\mathcal{W}$ (why?). These remarks prove the following:

THEOREM 6.5

Let $\mathcal{W}$ be a subspace of $\mathbb{R}^n$. Then any orthogonal set of nonzero vectors in $\mathcal{W}$ is contained in (can be enlarged to) an orthogonal basis for $\mathcal{W}$. Similarly, any orthonormal set of vectors in $\mathcal{W}$ is contained in an orthonormal basis for $\mathcal{W}$.

EXAMPLE 5

We will find an orthogonal basis B for $\mathbb{R}^4$ that contains the orthogonal set $T = \{[2, 1, 0, -1], [0, -1, 4, -1], [2, -4, -1, 0]\}$ from Example 4. To enlarge T to a basis for $\mathbb{R}^4$, we row reduce

$$\begin{bmatrix} 2 & 0 & 2 & 1 & 0 & 0 & 0 \\ 1 & -1 & -4 & 0 & 1 & 0 & 0 \\ 0 & 4 & -1 & 0 & 0 & 1 & 0 \\ -1 & -1 & 0 & 0 & 0 & 0 & 1 \end{bmatrix} \text{ to obtain } \begin{bmatrix} 1 & 0 & 0 & 0 & \frac{1}{18} & -\frac{2}{9} & -\frac{17}{18} \\ 0 & 1 & 0 & 0 & -\frac{1}{18} & \frac{2}{9} & -\frac{1}{18} \\ 0 & 0 & 1 & 0 & -\frac{2}{9} & -\frac{1}{9} & -\frac{2}{9} \\ 0 & 0 & 0 & 1 & \frac{1}{3} & \frac{2}{3} & \frac{7}{3} \end{bmatrix}.$$

Hence, the Enlarging Method from Section 4.6 shows that $\{[2, 1, 0, -1], [0, -1, 4, -1], [2, -4, -1, 0], [1, 0, 0, 0]\}$ is a basis for $\mathbb{R}^4$. Now, we use the Gram-Schmidt Process to convert this basis to an orthogonal basis for $\mathbb{R}^4$.

Let $\mathbf{w}_1 = [2, 1, 0, -1]$, $\mathbf{w}_2 = [0, -1, 4, -1]$, $\mathbf{w}_3 = [2, -4, -1, 0]$, and $\mathbf{w}_4 = [1, 0, 0, 0]$. The first few steps of the Gram-Schmidt Process give $\mathbf{v}_1 = \mathbf{w}_1$, $\mathbf{v}_2 = \mathbf{w}_2$, and $\mathbf{v}_3 = \mathbf{w}_3$ (why?). Finally,

$$\mathbf{v}_4 = \mathbf{w}_4 - \left(\frac{\mathbf{w}_4 \cdot \mathbf{v}_1}{\mathbf{v}_1 \cdot \mathbf{v}_1}\right)\mathbf{v}_1 - \left(\frac{\mathbf{w}_4 \cdot \mathbf{v}_2}{\mathbf{v}_2 \cdot \mathbf{v}_2}\right)\mathbf{v}_2 - \left(\frac{\mathbf{w}_4 \cdot \mathbf{v}_3}{\mathbf{v}_3 \cdot \mathbf{v}_3}\right)\mathbf{v}_3$$

$$= [1, 0, 0, 0] - \left(\frac{[1, 0, 0, 0] \cdot [2, 1, 0, -1]}{[2, 1, 0, -1] \cdot [2, 1, 0, -1]}\right)[2, 1, 0, -1]$$

$$- \left(\frac{[1, 0, 0, 0] \cdot [0, -1, 4, -1]}{[0, -1, 4, -1] \cdot [0, -1, 4, -1]}\right)[0, -1, 4, -1]$$

$$- \left(\frac{[1, 0, 0, 0] \cdot [2, -4, -1, 0]}{[2, -4, -1, 0] \cdot [2, -4, -1, 0]}\right)[2, -4, -1, 0]$$

$$= [1, 0, 0, 0] - \frac{1}{3}[2, 1, 0, -1] - \frac{2}{21}[2, -4, -1, 0] = \left[\frac{1}{7}, \frac{1}{21}, \frac{2}{21}, \frac{1}{3}\right].$$

To avoid fractions, we multiply this vector by 21 to obtain $\mathbf{v}_4 = [3, 1, 2, 7]$. Notice that $\mathbf{v}_4$ is orthogonal to $\mathbf{v}_1$, $\mathbf{v}_2$, and $\mathbf{v}_3$. Hence, $\{\mathbf{v}_1, \mathbf{v}_2, \mathbf{v}_3, \mathbf{v}_4\}$ is an orthogonal basis for $\mathbb{R}^4$ containing T. ∎

Orthogonal Matrices

DEFINITION

A nonsingular (square) matrix $\mathbf{A}$ is **orthogonal** if and only if $\mathbf{A}^T = \mathbf{A}^{-1}$.

Notice that if $\mathbf{A}$ is orthogonal, then $|\mathbf{A}| = \pm 1$ because $|\mathbf{A}^T| = |\mathbf{A}^{-1}| \Rightarrow |\mathbf{A}| = 1/|\mathbf{A}| \Rightarrow |\mathbf{A}|^2 = 1 \Rightarrow |\mathbf{A}| = \pm 1$. (Beware! The converse is not true — if $|\mathbf{A}| = \pm 1$, then $\mathbf{A}$ is not necessarily orthogonal.)

It is straightforward to prove that if $\mathbf{A}$ is orthogonal, then $\mathbf{A}^{-1} = \mathbf{A}^T$ is also orthogonal (see Exercise 11(a)). It also follows that if $\mathbf{A}$ and $\mathbf{B}$ are orthogonal matrices of the same size, then $\mathbf{AB}$ is orthogonal (see Exercise 11(b)).

The next theorem characterizes all orthogonal matrices.

THEOREM 6.6

Let $\mathbf{A}$ be an $n \times n$ matrix. Then $\mathbf{A}$ is orthogonal
    (1) if and only if the rows of $\mathbf{A}$ form an orthonormal basis for $\mathbb{R}^n$
    (2) if and only if the columns of $\mathbf{A}$ form an orthonormal basis for $\mathbb{R}^n$.

Theorem 6.6 suggests that it is probably more appropriate to refer to orthogonal matrices as "orthonormal matrices." Unfortunately, the term *orthogonal matrix* has become traditional usage in linear algebra.

PROOF OF THEOREM 6.6

(abridged) We prove half of part (1) and leave the rest as Exercise 17.

Suppose that $\mathbf{A}$ is an orthogonal $n \times n$ matrix. Then we have $\mathbf{AA}^T = \mathbf{I}_n$ (why?). Hence, for $1 \le i, j \le n$ with $i \ne j$, we have [ith row of $\mathbf{A}$] $\cdot$ [jth column of $\mathbf{A}^T$] $= 0$. Therefore, [ith row of $\mathbf{A}$] $\cdot$ [jth row of $\mathbf{A}$] $= 0$, which shows that distinct rows of $\mathbf{A}$ are orthogonal. Again, because $\mathbf{AA}^T = \mathbf{I}_n$, for each i, $1 \le i \le n$, we have [ith row of $\mathbf{A}$] $\cdot$ [ith column of $\mathbf{A}^T$] $= 1$. But then [ith row of $\mathbf{A}$] $\cdot$ [ith row of $\mathbf{A}$] $= 1$, which shows that each row of $\mathbf{A}$ is a unit vector. Thus, the n rows of $\mathbf{A}$ form an orthonormal set, and hence, an orthonormal basis for $\mathbb{R}^n$. ∎

$\mathbf{I}_n$ is obviously an orthogonal matrix, for any $n \ge 1$. In the next example, we show how Theorem 6.6 can be used to find other orthogonal matrices.

EXAMPLE 6 Consider the orthonormal basis $\{\mathbf{v}_1, \mathbf{v}_2, \mathbf{v}_3\}$ for $\mathbb{R}^3$ from Example 2, where

$$\mathbf{v}_1 = \left[\frac{1}{\sqrt{2}}, 0, -\frac{1}{\sqrt{2}}\right], \mathbf{v}_2 = \left[-\frac{1}{3\sqrt{2}}, \frac{4}{3\sqrt{2}}, -\frac{1}{3\sqrt{2}}\right], \text{ and } \mathbf{v}_3 = \left[\frac{2}{3}, \frac{1}{3}, \frac{2}{3}\right].$$

By parts (1) and (2) of Theorem 6.6, respectively,

$$\mathbf{A} = \begin{bmatrix} \frac{1}{\sqrt{2}} & 0 & -\frac{1}{\sqrt{2}} \\ -\frac{1}{3\sqrt{2}} & \frac{4}{3\sqrt{2}} & -\frac{1}{3\sqrt{2}} \\ \frac{2}{3} & \frac{1}{3} & \frac{2}{3} \end{bmatrix} \quad \text{and} \quad \mathbf{A}^T = \begin{bmatrix} \frac{1}{\sqrt{2}} & -\frac{1}{3\sqrt{2}} & \frac{2}{3} \\ 0 & \frac{4}{3\sqrt{2}} & \frac{1}{3} \\ -\frac{1}{\sqrt{2}} & -\frac{1}{3\sqrt{2}} & \frac{2}{3} \end{bmatrix}$$

are both orthogonal matrices. You can verify that both $\mathbf{A}$ and $\mathbf{A}^T$ are orthogonal by checking that $\mathbf{A}\mathbf{A}^T = \mathbf{I}_3$. ∎

One important example of orthogonal matrices is given in the next theorem.

THEOREM 6.7
Let B and C be ordered orthonormal bases for $\mathbb{R}^n$. Then the transition matrix from B to C is an orthogonal matrix.

In Exercise 20 you are asked to prove a partial converse as well as a generalization of Theorem 6.7.

PROOF OF THEOREM 6.7
Let S be the standard basis for $\mathbb{R}^n$. The matrix $\mathbf{P}$ whose columns are the vectors in B is the transition matrix from B to S. Similarly, the matrix $\mathbf{Q}$, whose columns are the vectors in C, is the transition matrix from C to S. Both $\mathbf{P}$ and $\mathbf{Q}$ are orthogonal matrices by part (2) of Theorem 6.6. But then $\mathbf{Q}^{-1}$ is also orthogonal. Now, by Theorems 4.22 and 4.23, $\mathbf{Q}^{-1}\mathbf{P}$ is the transition matrix from B to C (see Figure 6.1), and $\mathbf{Q}^{-1}\mathbf{P}$ is orthogonal because the product of orthogonal matrices is orthogonal (see Exercise 11(b)). ∎

Figure 6.1

Visualizing $\mathbf{Q}^{-1}\mathbf{P}$ as the transition matrix from B to C

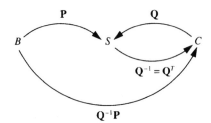

EXAMPLE 7   Consider the following ordered orthonormal bases for $\mathbb{R}^2$:

$$B = \left(\left[\frac{\sqrt{2}}{2}, \frac{\sqrt{2}}{2} \right], \left[\frac{\sqrt{2}}{2}, -\frac{\sqrt{2}}{2} \right] \right) \quad \text{and} \quad C = \left(\left[\frac{\sqrt{3}}{2}, \frac{1}{2} \right], \left[-\frac{1}{2}, \frac{\sqrt{3}}{2} \right] \right).$$

By Theorem 6.7, the transition matrix from B to C is orthogonal. To verify this, we can use Theorem 6.3 to obtain

$$\left[\frac{\sqrt{2}}{2}, \frac{\sqrt{2}}{2}\right]_C = \left[\frac{\sqrt{6}+\sqrt{2}}{4}, \frac{\sqrt{6}-\sqrt{2}}{4}\right] \quad \text{and}$$

$$\left[\frac{\sqrt{2}}{2}, -\frac{\sqrt{2}}{2}\right]_C = \left[\frac{\sqrt{6}-\sqrt{2}}{4}, \frac{-\sqrt{6}-\sqrt{2}}{4}\right].$$

Hence, the transition matrix from B to C is

$$\mathbf{A} = \frac{1}{4}\left[\begin{array}{cc} \sqrt{6}+\sqrt{2} & \sqrt{6}-\sqrt{2} \\ \sqrt{6}-\sqrt{2} & -\sqrt{6}-\sqrt{2} \end{array}\right].$$

Because $\mathbf{AA}^T = \mathbf{I}_2$ (verify!), $\mathbf{A}$ is an orthogonal matrix. ■

The final theorem of this section can be used to prove that multiplying two n-vectors by an orthogonal matrix does not change the angle between them (see Exercise 18).

THEOREM 6.8

Let $\mathbf{A}$ be an $n \times n$ orthogonal matrix, and let $\mathbf{v}$ and $\mathbf{w}$ be vectors in $\mathbb{R}^n$. Then $\mathbf{v} \cdot \mathbf{w} = \mathbf{Av} \cdot \mathbf{Aw}$.

PROOF OF THEOREM 6.8

Notice that the dot product $\mathbf{x} \cdot \mathbf{y}$ of two column vectors $\mathbf{x}$ and $\mathbf{y}$ can be written in matrix multiplication form as $\mathbf{x}^T\mathbf{y}$. Let $\mathbf{v}, \mathbf{w} \in \mathbb{R}^n$, and let $\mathbf{A}$ be an $n \times n$ orthogonal matrix. Then

$$\mathbf{v} \cdot \mathbf{w} = \mathbf{v}^T\mathbf{w} = \mathbf{v}^T\mathbf{I}_n\mathbf{w} = \mathbf{v}^T\mathbf{A}^T\mathbf{Aw} = (\mathbf{Av})^T\mathbf{Aw} = \mathbf{Av} \cdot \mathbf{Aw}. \quad ■$$

Exercises for Section 6.1

1. Which of the following sets of vectors are orthogonal? Which are orthonormal?

   ⋆(a) $\{[3, -2], [4, 6]\}$

   (b) $\left\{\left[-\frac{1}{\sqrt{5}}, \frac{2}{\sqrt{5}}\right], \left[\frac{2}{\sqrt{5}}, \frac{1}{\sqrt{5}}\right]\right\}$

   ⋆(c) $\left\{\left[\frac{3}{\sqrt{13}}, -\frac{2}{\sqrt{13}}\right], \left[\frac{1}{\sqrt{10}}, -\frac{3}{\sqrt{10}}\right]\right\}$

   (d) $\left\{\left[\frac{1}{3}, \frac{2}{3}, \frac{2}{3}\right], \left[\frac{2}{3}, \frac{1}{3}, -\frac{2}{3}\right], \left[\frac{2}{3}, -\frac{2}{3}, \frac{1}{3}\right]\right\}$

   (e) $\left\{\left[\frac{3}{5}, 0, -\frac{4}{5}\right]\right\}$

   ⋆(f) $[2, -3, 1, 2], [-1, 2, 8, 0], [6, -1, 1, -8]\}$

   (g) $\left\{\left[\frac{1}{4}, \frac{1}{4}, \frac{1}{4}, -\frac{1}{2}, \frac{3}{4}\right], \left[\frac{1}{6}, \frac{1}{6}, -\frac{1}{2}, \frac{2}{3}, \frac{1}{2}\right]\right\}$

2. Which of the following matrices are orthogonal?

★(a) $\begin{bmatrix} \frac{\sqrt{3}}{2} & \frac{1}{2} \\ -\frac{1}{2} & \frac{\sqrt{3}}{2} \end{bmatrix}$

(b) $\begin{bmatrix} 3 & -2 \\ 2 & 3 \end{bmatrix}$

★(c) $\begin{bmatrix} 3 & 0 & 10 \\ -1 & 3 & 3 \\ 3 & 1 & -9 \end{bmatrix}$

(d) $\begin{bmatrix} \frac{2}{15} & \frac{5}{15} & \frac{14}{15} \\ \frac{10}{15} & \frac{10}{15} & -\frac{5}{15} \\ \frac{11}{15} & -\frac{10}{15} & \frac{2}{15} \end{bmatrix}$

★(e) $\begin{bmatrix} \frac{2}{3} & \frac{2}{3} & 0 & \frac{1}{3} \\ \frac{2}{3} & -\frac{2}{3} & -\frac{1}{3} & 0 \\ \frac{1}{3} & 0 & \frac{2}{3} & -\frac{2}{3} \\ 0 & \frac{1}{3} & -\frac{2}{3} & -\frac{2}{3} \end{bmatrix}$

3. In each case, verify that the given ordered basis B is orthonormal. Then, for the given $\mathbf{v}$, find $[\mathbf{v}]_B$, using the method of Theorem 6.3.

★(a) $\mathbf{v} = [-2, 3]$, $B = \left(\left[-\frac{\sqrt{3}}{2}, \frac{1}{2} \right], \left[\frac{1}{2}, \frac{\sqrt{3}}{2} \right] \right)$

(b) $\mathbf{v} = [4, -1, 2]$, $B = \left(\left[\frac{3}{7}, -\frac{6}{7}, -\frac{2}{7} \right], \left[\frac{2}{7}, \frac{3}{7}, -\frac{6}{7} \right], \left[\frac{6}{7}, \frac{2}{7}, \frac{3}{7} \right] \right)$

★(c) $\mathbf{v} = [8, 4, -3, 5]$, $B = \left(\left[\frac{1}{2}, -\frac{1}{2}, \frac{1}{2}, \frac{1}{2} \right], \left[\frac{3}{2\sqrt{3}}, \frac{1}{2\sqrt{3}}, -\frac{1}{2\sqrt{3}}, -\frac{1}{2\sqrt{3}} \right], \right.$
$\left. \left[0, \frac{2}{\sqrt{6}}, \frac{1}{\sqrt{6}}, \frac{1}{\sqrt{6}} \right], \left[0, 0, -\frac{1}{\sqrt{2}}, \frac{1}{\sqrt{2}} \right] \right)$

4. Each of the following represents a basis for a subspace of $\mathbb{R}^n$, for some n. Use the Gram-Schmidt Process to find an orthogonal basis for the subspace.

★(a) $\{[5, -1, 2], [2, -1, -4]\}$ in $\mathbb{R}^3$

(b) $\{[2, -1, 3, 1], [-3, 0, -1, 4]\}$ in $\mathbb{R}^4$

★(c) $\{[2, 1, 0, -1], [1, 1, 1, -1], [1, -2, 1, 1]\}$ in $\mathbb{R}^4$

5. Enlarge each of the following orthogonal sets to an orthogonal basis for $\mathbb{R}^n$. (Avoid fractions by using appropriate scalar multiples.)

★(a) $\{[2, 2, -3]\}$

(b) $\{[1, -4, 3]\}$

★(c) $\{[1, -3, 1], [2, 5, 13]\}$

(d) $\{[3, 1, -2], [5, -3, 6]\}$

★(e) $\{[2, 1, -2, 1]\}$

(f) $\{[2, 1, 0, -3], [0, 3, 2, 1]\}$

6. Let $\mathcal{W} = \{[a, b, c, d, e] \mid a + b + c + d + e = 0\}$, a subspace of $\mathbb{R}^5$. Let $T = \{[-2, -1, 4, -2, 1], [4, -3, 0, -2, 1]\}$, an orthogonal subset of $\mathcal{W}$. Enlarge T to an orthogonal basis for $\mathcal{W}$. (Hint: Use the fact that $B = \{[1, -1, 0, 0, 0], [0, 1, -1, 0, 0], [0, 0, 1, -1, 0], [0, 0, 0, 1, -1]\}$ is a basis for $\mathcal{W}$.)

7. It can be shown (although we do not prove it here) that the linear operator represented by a 3×3 orthogonal matrix with determinant 1 (with respect to the standard basis) always represents a rotation about some axis in $\mathbb{R}^3$ and that the axis of rotation is parallel to an eigenvector corresponding to the eigenvalue $\lambda = 1$. Verify that each of the following matrices is orthogonal with determinant 1, and thereby represents a rotation about an axis in $\mathbb{R}^3$. Solve in each case for a vector in the direction of the axis of rotation.

$\star$(a) $\frac{1}{11} \begin{bmatrix} 2 & 6 & -9 \\ -9 & 6 & 2 \\ 6 & 7 & 6 \end{bmatrix}$

(b) $\frac{1}{17} \begin{bmatrix} 12 & 1 & 12 \\ 8 & 12 & -9 \\ -9 & 12 & 8 \end{bmatrix}$

$\star$(c) $\frac{1}{7} \begin{bmatrix} 6 & 2 & 3 \\ 3 & -6 & -2 \\ 2 & 3 & -6 \end{bmatrix}$

(d) $\frac{1}{15} \begin{bmatrix} 2 & 14 & 5 \\ 10 & -5 & 10 \\ 11 & 2 & -10 \end{bmatrix}$

8. (a) Show that if $\{\mathbf{v}_1, \ldots, \mathbf{v}_k\}$ is an orthogonal set in $\mathbb{R}^n$ and $c_1, \ldots, c_k$ are nonzero scalars, then $\{c_1\mathbf{v}_1, \ldots, c_k\mathbf{v}_k\}$ is also an orthogonal set.

   $\star$(b) Is part (a) still true if *orthogonal* is replaced by *orthonormal* everywhere?

9. Suppose that $\{\mathbf{u}_1, \ldots, \mathbf{u}_n\}$ is an orthonormal basis for $\mathbb{R}^n$.

   (a) If $\mathbf{v}, \mathbf{w} \in \mathbb{R}^n$, show that

   $$\mathbf{v} \cdot \mathbf{w} = (\mathbf{v} \cdot \mathbf{u}_1)(\mathbf{w} \cdot \mathbf{u}_1) + (\mathbf{v} \cdot \mathbf{u}_2)(\mathbf{w} \cdot \mathbf{u}_2) + \cdots + (\mathbf{v} \cdot \mathbf{u}_n)(\mathbf{w} \cdot \mathbf{u}_n).$$

   (b) If $\mathbf{v} \in \mathbb{R}^n$, use part (a) to prove **Parseval's Equality**,

   $$\|\mathbf{v}\|^2 = (\mathbf{v} \cdot \mathbf{u}_1)^2 + (\mathbf{v} \cdot \mathbf{u}_2)^2 + \cdots + (\mathbf{v} \cdot \mathbf{u}_n)^2.$$

10. Let $\{\mathbf{u}_1, \ldots, \mathbf{u}_k\}$ be an orthonormal set of vectors in $\mathbb{R}^n$. For any vector $\mathbf{v} \in \mathbb{R}^n$, prove **Bessel's Inequality**,

   $$(\mathbf{v} \cdot \mathbf{u}_1)^2 + \cdots + (\mathbf{v} \cdot \mathbf{u}_k)^2 \leq \|\mathbf{v}\|^2.$$

   (Hint: Let $\mathcal{W}$ be the subspace spanned by $\{\mathbf{u}_1, \ldots, \mathbf{u}_k\}$. Enlarge $\{\mathbf{u}_1, \ldots, \mathbf{u}_k\}$ to an orthonormal basis for $\mathbb{R}^n$. Then use Theorem 6.3.) (Bessel's Inequality is a generalization of Parseval's Equality, which appears in Exercise 9.)

11. Suppose that $\mathbf{A}$ and $\mathbf{B}$ are $n \times n$ orthogonal matrices.

   (a) Show that $\mathbf{A}^{-1} (= \mathbf{A}^T)$ is an orthogonal matrix.

   (b) Prove that $\mathbf{AB}$ is an orthogonal matrix.

12. Let $\mathbf{A}$ be an $n \times n$ matrix with $\mathbf{A}^2 = \mathbf{I}_n$. Prove that $\mathbf{A}$ is symmetric if and only if $\mathbf{A}$ is orthogonal.

13. Show that if n is odd and $\mathbf{A}$ is an orthogonal $n \times n$ matrix, then $\mathbf{A}$ is not skew-symmetric. (Hint: Suppose $\mathbf{A}$ is both orthogonal and skew-symmetric. Show that $\mathbf{A}^2 = -\mathbf{I}_n$, and then use determinants.)

14. If $\mathbf{A}$ is an $n \times n$ orthogonal matrix with $|\mathbf{A}| = -1$, show that $\mathbf{A} + \mathbf{I}_n$ has no inverse. (Hint: Show that $\mathbf{A} + \mathbf{I}_n = \mathbf{A}(\mathbf{A} + \mathbf{I}_n)^T$, and then use determinants.)

15. Suppose that $\mathbf{A}$ is a 3×3 upper triangular orthogonal matrix. Show that $\mathbf{A}$ is diagonal and that all main diagonal entries of $\mathbf{A}$ equal ± 1. (Note: This result is true for any $n \times n$ upper triangular orthogonal matrix.)

16. (a) If $\mathbf{u}$ is any unit vector in $\mathbb{R}^n$, explain why there exists an $n \times n$ orthogonal matrix with $\mathbf{u}$ as its first row. (Hint: Consider Theorem 6.5.)

⋆(b) Find an orthogonal matrix whose first row is $\frac{1}{\sqrt{6}}[1, 2, 1]$.

17. ▶Finish the proof of Theorem 6.6.

18. Suppose that $\mathbf{A}$ is an $n \times n$ orthogonal matrix.

(a) Prove that for every $\mathbf{v} \in \mathbb{R}^n$, $\|\mathbf{v}\| = \|\mathbf{Av}\|$.

(b) Prove that for all $\mathbf{v}, \mathbf{w} \in \mathbb{R}^n$, the angle between $\mathbf{v}$ and $\mathbf{w}$ equals the angle between $\mathbf{Av}$ and $\mathbf{Aw}$.

19. Let B be an ordered orthonormal basis for a k-dimensional subspace $\mathcal{V}$ of $\mathbb{R}^n$. Prove that for all $\mathbf{v}_1, \mathbf{v}_2 \in \mathcal{V}$, $\mathbf{v}_1 \cdot \mathbf{v}_2 = [\mathbf{v}_1]_B \cdot [\mathbf{v}_2]_B$, where the first dot product takes place in $\mathbb{R}^n$ and the second takes place in $\mathbb{R}^k$. (Hint: Let $B = (\mathbf{b}_1, \ldots, \mathbf{b}_k)$, and express $\mathbf{v}_1$ and $\mathbf{v}_2$ as linear combinations of the vectors in B. Substitute these linear combinations in the left side of $\mathbf{v}_1 \cdot \mathbf{v}_2 = [\mathbf{v}_1]_B \cdot [\mathbf{v}_2]_B$ and simplify. Then use the same linear combinations to express $\mathbf{v}_1$ and $\mathbf{v}_2$ in B-coordinates to calculate the right side.)

20. Prove each of the following statements related to Theorem 6.7. (Hint: Use the result of Exercise 19 in proving parts (b) and (c).)

(a) Let B be an orthonormal basis for $\mathbb{R}^n$, C be a basis for $\mathbb{R}^n$, and $\mathbf{P}$ be the transition matrix from B to C. If $\mathbf{P}$ is an orthogonal matrix, then C is an orthonormal basis for $\mathbb{R}^n$.

(b) Let $\mathcal{V}$ be a subspace of $\mathbb{R}^n$, and let B and C be orthonormal bases for $\mathcal{V}$. Then the transition matrix from B to C is an orthogonal matrix.

(c) Let $\mathcal{V}$ be a subspace of $\mathbb{R}^n$, B be an orthonormal basis for $\mathcal{V}$, C be a basis for $\mathcal{V}$, and $\mathbf{P}$ be the transition matrix from B to C. If $\mathbf{P}$ is an orthogonal matrix, then C is an orthonormal basis for $\mathcal{V}$.

⋆**21.** True or False:

(a) Any subset of $\mathbb{R}^n$ containing $\mathbf{0}$ is automatically an orthogonal set of vectors.

(b) The standard basis in $\mathbb{R}^n$ is an orthonormal set of vectors.

(c) If $B = (\mathbf{u}_1, \mathbf{u}_2, \ldots, \mathbf{u}_n)$ is an ordered orthonormal basis for $\mathbb{R}^n$, and $\mathbf{v} \in \mathbb{R}^n$, then $[\mathbf{v}]_B = [\mathbf{v} \cdot \mathbf{u}_1, \mathbf{v} \cdot \mathbf{u}_2, \ldots, \mathbf{v} \cdot \mathbf{u}_n]$.

(d) The Gram-Schmidt Process can be used to enlarge any linearly independent set $\{\mathbf{w}_1, \mathbf{w}_2, \ldots, \mathbf{w}_k\}$ in $\mathbb{R}^n$ to an orthogonal basis $\{\mathbf{w}_1, \mathbf{w}_2, \ldots, \mathbf{w}_k, \mathbf{w}_{k+1}, \ldots, \mathbf{w}_n\}$ for $\mathbb{R}^n$.

(e) If $\mathcal{W}$ is a nontrivial subspace of $\mathbb{R}^n$, then an orthogonal basis for $\mathcal{W}$ exists.

(f) If $\mathbf{A}$ is a square matrix, and $\mathbf{A}^T\mathbf{A} = \mathbf{I}_n$, then $\mathbf{A}$ is orthogonal.

(g) If $\mathbf{A}$ and $\mathbf{B}$ are orthogonal $n \times n$ matrices, then $\mathbf{BA}$ is orthogonal and $|\mathbf{BA}| = \pm 1$.

(h) If either the rows or columns of $\mathbf{A}$ form an orthogonal basis for $\mathbb{R}^n$, then $\mathbf{A}$ is orthogonal.

(i) If $\mathbf{A}$ is an orthogonal matrix and R is a type (III) row operation, then $R(\mathbf{A})$ is an orthogonal matrix.

(j) If $\mathbf{P}$ is the transition matrix from B to C, where B and C are ordered orthonormal bases for $\mathbb{R}^n$, then $\mathbf{P}$ is orthogonal.

6.2 Orthogonal Complements

For each subspace $\mathcal{W}$ of $\mathbb{R}^n$, there is a corresponding subspace of $\mathbb{R}^n$ consisting of the vectors orthogonal to all vectors in $\mathcal{W}$, called the orthogonal complement of $\mathcal{W}$. In this section, we study many elementary properties of orthogonal complements and investigate the orthogonal projection of a vector onto a subspace of $\mathbb{R}^n$.

Orthogonal Complements

DEFINITION

Let $\mathcal{W}$ be a subspace of $\mathbb{R}^n$. The **orthogonal complement**, $\mathcal{W}^\perp$, of $\mathcal{W}$ in $\mathbb{R}^n$ is the set of all vectors $\mathbf{x} \in \mathbb{R}^n$ with the property that $\mathbf{x} \cdot \mathbf{w} = 0$, for all $\mathbf{w} \in \mathcal{W}$. That is, $\mathcal{W}^\perp$ contains those vectors of $\mathbb{R}^n$ orthogonal to every vector in $\mathcal{W}$.

The proof of the next theorem is left as Exercise 18.

THEOREM 6.9

If $\mathcal{W}$ is a subspace of $\mathbb{R}^n$, then $\mathbf{v} \in \mathcal{W}^\perp$ if and only if $\mathbf{v}$ is orthogonal to every vector in a spanning set for $\mathcal{W}$.

EXAMPLE 1

Consider the subspace $\mathcal{W} = \{\,[a, b, 0] \mid a, b \in \mathbb{R}\}$ of $\mathbb{R}^3$. Now, $\mathcal{W}$ is spanned by $\{[1, 0, 0], [0, 1, 0]\}$. By Theorem 6.9, a vector $[x, y, z]$ is in $\mathcal{W}^\perp$, the orthogonal complement of $\mathcal{W}$, if and only if it is orthogonal to both $[1, 0, 0]$ and $[0, 1, 0]$ (why?) — that is, if and only if $x = y = 0$. Hence, $\mathcal{W}^\perp = \{\,[0, 0, z] \mid z \in \mathbb{R}\}$. Notice that $\mathcal{W}^\perp$ is a subspace of $\mathbb{R}^3$ of dimension 1 and that $\dim(\mathcal{W}) + \dim\left(\mathcal{W}^\perp\right) = \dim\left(\mathbb{R}^3\right)$. ∎

EXAMPLE 2

Consider the subspace $\mathcal{W} = \{a[-3, 2, 4] \mid a \in \mathbb{R}\}$ of $\mathbb{R}^3$. Since $\{[-3, 2, 4]\}$ spans $\mathcal{W}$, Theorem 6.9 tells us that the orthogonal complement $\mathcal{W}^\perp$ of $\mathcal{W}$ is the set of all vectors $[x, y, z]$ in $\mathbb{R}^3$ such that $[x, y, z] \cdot [-3, 2, 4] = 0$. That is, $\mathcal{W}^\perp$ is precisely the set of all vectors $[x, y, z]$ lying in the plane $-3x + 2y + 4z = 0$. Notice that $\mathcal{W}^\perp$ is a subspace of $\mathbb{R}^3$ of dimension 2 and that $\dim(\mathcal{W}) + \dim\left(\mathcal{W}^\perp\right) = \dim(\mathbb{R}^3)$. ∎

EXAMPLE 3

The orthogonal complement of $\mathbb{R}^n$ itself is just the trivial subspace $\{\mathbf{0}\}$, since $\mathbf{0}$ is the only vector orthogonal to all of $\mathbf{e}_1, \mathbf{e}_2, \ldots, \mathbf{e}_n \in \mathbb{R}^n$ (why?).

Conversely, the orthogonal complement of the trivial subspace in $\mathbb{R}^n$ is all of $\mathbb{R}^n$ because every vector in $\mathbb{R}^n$ is orthogonal to the zero vector.

Hence, $\{\mathbf{0}\}$ and $\mathbb{R}^n$ itself are orthogonal complements of each other in $\mathbb{R}^n$. Notice that the dimensions of these two subspaces add up to $\dim(\mathbb{R}^n)$. ∎

Properties of Orthogonal Complements

Examples 1, 2, and 3 suggest that the orthogonal complement $\mathcal{W}^\perp$ of a subspace $\mathcal{W}$ is a subspace of $\mathbb{R}^n$. This result is part of the next theorem.

THEOREM 6.10

Let $\mathcal{W}$ be a subspace of $\mathbb{R}^n$. Then $\mathcal{W}^\perp$ is a subspace of $\mathbb{R}^n$, and $\mathcal{W} \cap \mathcal{W}^\perp = \{\mathbf{0}\}$.

PROOF OF THEOREM 6.10

$\mathcal{W}^\perp$ is nonempty because $\mathbf{0} \in \mathcal{W}^\perp$ (why?). Thus, to show that $\mathcal{W}^\perp$ is a subspace, we need only verify the closure properties for $\mathcal{W}^\perp$.

Suppose $\mathbf{x}_1, \mathbf{x}_2 \in \mathcal{W}^\perp$. We want to show $\mathbf{x}_1 + \mathbf{x}_2 \in \mathcal{W}^\perp$. However, for all $\mathbf{w} \in \mathcal{W}$, $(\mathbf{x}_1 + \mathbf{x}_2) \cdot \mathbf{w} = (\mathbf{x}_1 \cdot \mathbf{w}) + (\mathbf{x}_2 \cdot \mathbf{w}) = 0 + 0 = 0$, since $\mathbf{x}_1, \mathbf{x}_2 \in \mathcal{W}^\perp$. Hence, $\mathbf{x}_1 + \mathbf{x}_2 \in \mathcal{W}^\perp$. Next, suppose that $\mathbf{x} \in \mathcal{W}^\perp$ and $c \in \mathbb{R}$. We want to show that $c\mathbf{x} \in \mathcal{W}^\perp$. However, for all $\mathbf{w} \in \mathcal{W}$, $(c\mathbf{x}) \cdot \mathbf{w} = c(\mathbf{x} \cdot \mathbf{w}) = c(0) = 0$, since $\mathbf{x} \in \mathcal{W}^\perp$. Hence, $c\mathbf{x} \in \mathcal{W}^\perp$. Thus, $\mathcal{W}^\perp$ is a subspace of $\mathbb{R}^n$.

Finally, suppose $\mathbf{w} \in \mathcal{W} \cap \mathcal{W}^\perp$. Then $\mathbf{w} \in \mathcal{W}$ and $\mathbf{w} \in \mathcal{W}^\perp$, so $\mathbf{w}$ is orthogonal to itself. Hence, $\mathbf{w} \cdot \mathbf{w} = 0$, and so $\mathbf{w} = \mathbf{0}$. ∎

The next theorem shows how we can obtain an orthogonal basis for $\mathcal{W}^\perp$.

THEOREM 6.11

Let $\mathcal{W}$ be a subspace of $\mathbb{R}^n$. Let $\{\mathbf{v}_1, \ldots, \mathbf{v}_k\}$ be an orthogonal basis for $\mathcal{W}$ contained in an orthogonal basis $\{\mathbf{v}_1, \ldots, \mathbf{v}_k, \mathbf{v}_{k+1}, \ldots, \mathbf{v}_n\}$ for $\mathbb{R}^n$. Then $\{\mathbf{v}_{k+1}, \ldots, \mathbf{v}_n\}$ is an orthogonal basis for $\mathcal{W}^\perp$.

PROOF OF THEOREM 6.11

Let $\{\mathbf{v}_1, \ldots, \mathbf{v}_n\}$ be an orthogonal basis for $\mathbb{R}^n$, with $\mathcal{W} = \text{span}(\{\mathbf{v}_1, \ldots, \mathbf{v}_k\})$. Let $\mathcal{X} = \text{span}(\{\mathbf{v}_{k+1}, \ldots, \mathbf{v}_n\})$. Since $\{\mathbf{v}_{k+1}, \ldots, \mathbf{v}_n\}$ is linearly independent (why?), it is a basis for $\mathcal{W}^\perp$ if $\mathcal{X} = \mathcal{W}^\perp$. We will show that $\mathcal{X} \subseteq \mathcal{W}^\perp$ and $\mathcal{W}^\perp \subseteq \mathcal{X}$.

To show $\mathcal{X} \subseteq \mathcal{W}^\perp$, we must prove that any vector $\mathbf{x}$ of the form $d_{k+1}\mathbf{v}_{k+1} + \cdots + d_n\mathbf{v}_n$ (for some scalars $d_{k+1}, \ldots, d_n$) is orthogonal to every vector $\mathbf{w} \in \mathcal{W}$. Now, if $\mathbf{w} \in \mathcal{W}$, then $\mathbf{w} = c_1\mathbf{v}_1 + \cdots + c_k\mathbf{v}_k$, for some scalars $c_1, \ldots, c_k$. Hence,

$$\mathbf{x} \cdot \mathbf{w} = (d_{k+1}\mathbf{v}_{k+1} + \cdots + d_n\mathbf{v}_n) \cdot (c_1\mathbf{v}_1 + \cdots + c_k\mathbf{v}_k),$$

which equals zero when expanded because each vector in $\{\mathbf{v}_{k+1}, \ldots, \mathbf{v}_n\}$ is orthogonal to every vector in $\{\mathbf{v}_1, \ldots, \mathbf{v}_k\}$. Hence, $\mathbf{x} \in \mathcal{W}^\perp$, and so $\mathcal{X} \subseteq \mathcal{W}^\perp$.

To show $\mathcal{W}^\perp \subseteq \mathcal{X}$, we must show that any vector $\mathbf{x}$ in $\mathcal{W}^\perp$ is also in span $(\{\mathbf{v}_{k+1}, \ldots, \mathbf{v}_n\})$. Let $\mathbf{x} \in \mathcal{W}^\perp$. Since $\{\mathbf{v}_1, \ldots, \mathbf{v}_n\}$ is an orthogonal basis for $\mathbb{R}^n$, Theorem 6.3 tells us that

$$\mathbf{x} = \frac{(\mathbf{x} \cdot \mathbf{v}_1)}{(\mathbf{v}_1 \cdot \mathbf{v}_1)}\mathbf{v}_1 + \cdots + \frac{(\mathbf{x} \cdot \mathbf{v}_k)}{(\mathbf{v}_k \cdot \mathbf{v}_k)}\mathbf{v}_k + \frac{(\mathbf{x} \cdot \mathbf{v}_{k+1})}{(\mathbf{v}_{k+1} \cdot \mathbf{v}_{k+1})}\mathbf{v}_{k+1} + \cdots + \frac{(\mathbf{x} \cdot \mathbf{v}_n)}{(\mathbf{v}_n \cdot \mathbf{v}_n)}\mathbf{v}_n.$$

However, since each of $\mathbf{v}_1, \ldots, \mathbf{v}_k$ is in $\mathcal{W}$, we know that $\mathbf{x} \cdot \mathbf{v}_1 = \cdots = \mathbf{x} \cdot \mathbf{v}_k = 0$. Hence,

$$\mathbf{x} = \frac{(\mathbf{x} \cdot \mathbf{v}_{k+1})}{(\mathbf{v}_{k+1} \cdot \mathbf{v}_{k+1})} \mathbf{v}_{k+1} + \cdots + \frac{(\mathbf{x} \cdot \mathbf{v}_n)}{(\mathbf{v}_n \cdot \mathbf{v}_n)} \mathbf{v}_n,$$

and so $\mathbf{x} \in \text{span}(\{\mathbf{v}_{k+1}, \ldots, \mathbf{v}_n\})$. Thus, $\mathcal{W}^\perp \subseteq \mathcal{X}$. ∎

EXAMPLE 4

Consider the subspace $\mathcal{W} = \text{span}(\{[2, -1, 0, 1], [-1, 3, 1, -1]\})$ of $\mathbb{R}^4$. We want to find an orthogonal basis for $\mathcal{W}^\perp$. We start by finding an orthogonal basis for $\mathcal{W}$.

Let $\mathbf{w}_1 = [2, -1, 0, 1]$ and $\mathbf{w}_2 = [-1, 3, 1, -1]$. Performing the Gram-Schmidt Process yields $\mathbf{v}_1 = \mathbf{w}_1 = [2, -1, 0, 1]$ and $\mathbf{v}_2 = \mathbf{w}_2 - ((\mathbf{w}_2 \cdot \mathbf{v}_1)/(\mathbf{v}_1 \cdot \mathbf{v}_1))\mathbf{v}_1 = [1, 2, 1, 0]$. Hence, $\{\mathbf{v}_1, \mathbf{v}_2\} = \{[2, -1, 0, 1], [1, 2, 1, 0]\}$ is an orthogonal basis for $\mathcal{W}$.

We now expand this basis for $\mathcal{W}$ to a basis for all of $\mathbb{R}^4$ using the Enlarging Method of Section 4.6. Row reducing

$$\begin{bmatrix} 2 & 1 & 1 & 0 & 0 & 0 \\ -1 & 2 & 0 & 1 & 0 & 0 \\ 0 & 1 & 0 & 0 & 1 & 0 \\ 1 & 0 & 0 & 0 & 0 & 1 \end{bmatrix} \quad \text{yields} \quad \begin{bmatrix} 1 & 0 & 0 & 0 & 0 & 1 \\ 0 & 1 & 0 & 0 & 1 & 0 \\ 0 & 0 & 1 & 0 & -1 & -2 \\ 0 & 0 & 0 & 1 & -2 & 1 \end{bmatrix}.$$

Thus, $\{\mathbf{v}_1, \mathbf{v}_2, \mathbf{w}_3, \mathbf{w}_4\}$ is a basis for $\mathbb{R}^4$, where $\mathbf{w}_3 = [1, 0, 0, 0]$ and $\mathbf{w}_4 = [0, 1, 0, 0]$. Applying the Gram-Schmidt Process to $\{\mathbf{v}_1, \mathbf{v}_2, \mathbf{w}_3, \mathbf{w}_4\}$, we replace $\mathbf{w}_3$ and $\mathbf{w}_4$, respectively, with $\mathbf{v}_3 = [1, 0, -1, -2]$ and $\mathbf{v}_4 = [0, 1, -2, 1]$ (verify!). Then $\{\mathbf{v}_1, \mathbf{v}_2, \mathbf{v}_3, \mathbf{v}_4\}$ is an orthogonal basis for $\mathbb{R}^4$. Since $\{\mathbf{v}_1, \mathbf{v}_2\}$ is an orthogonal basis for $\mathcal{W}$, Theorem 6.11 tells us that $\{\mathbf{v}_3, \mathbf{v}_4\} = \{[1, 0, -1, -2], [0, 1, -2, 1]\}$ is an orthogonal basis for $\mathcal{W}^\perp$. ∎

The following is an important corollary of Theorem 6.11, which was illustrated in Examples 1, 2, and 3:

COROLLARY 6.12

Let $\mathcal{W}$ be a subspace of $\mathbb{R}^n$. Then $\dim(\mathcal{W}) + \dim\left(\mathcal{W}^\perp\right) = n = \dim(\mathbb{R}^n)$.

PROOF OF COROLLARY 6.12

Let $\mathcal{W}$ be a subspace of $\mathbb{R}^n$ of dimension k. By Theorem 6.4, $\mathcal{W}$ has an orthogonal basis $\{\mathbf{v}_1, \ldots, \mathbf{v}_k\}$. By Theorem 6.5, we can expand this basis for $\mathcal{W}$ to an orthogonal basis $\{\mathbf{v}_1, \ldots, \mathbf{v}_k, \mathbf{v}_{k+1}, \ldots, \mathbf{v}_n\}$ for all of $\mathbb{R}^n$. Then, by Theorem 6.11, $\{\mathbf{v}_{k+1}, \ldots, \mathbf{v}_n\}$ is a basis for $\mathcal{W}^\perp$, and so $\dim\left(\mathcal{W}^\perp\right) = n - k$. Hence, $\dim(\mathcal{W}) + \dim(\mathcal{W}^\perp) = n$. ∎

EXAMPLE 5

If $\mathcal{W}$ is a one-dimensional subspace of $\mathbb{R}^n$, then Corollary 6.12 asserts that $\dim\left(\mathcal{W}^\perp\right) = n - 1$. For example, in $\mathbb{R}^2$, the one-dimensional subspace

$\mathcal{W} = \text{span}(\{[a, b]\})$, where $[a, b] \neq [0, 0]$, has a one-dimensional orthogonal complement. In fact, $\mathcal{W}^\perp = \text{span}(\{[b, -a]\})$ (see Figure 6.2(a)). That is, $\mathcal{W}^\perp$ is the set of all vectors on the line through the origin perpendicular to $[a, b]$.

In $\mathbb{R}^3$, the one-dimensional subspace $\mathcal{W} = \text{span}(\{[a, b, c]\})$, where $[a, b, c] \neq [0, 0, 0]$, has a two-dimensional orthogonal complement. A little thought will convince you that $\mathcal{W}^\perp$ is the plane through the origin perpendicular to $[a, b, c]$; that is, the plane $ax + by + cz = 0$ (see Figure 6.2(b)). ∎

Figure 6.2

(a) The orthogonal complement of $\mathcal{W} = \text{span}(\{[a, b]\})$ in $\mathbb{R}^2$, a line through the origin perpendicular to $[a, b]$, when $[a, b] \neq [0, 0]$; (b) the orthogonal complement of $\mathcal{W} = \text{span}(\{[a, b, c]\})$ in $\mathbb{R}^3$, a plane through the origin perpendicular to $[a, b, c]$, when $[a, b, c] \neq [0, 0, 0]$

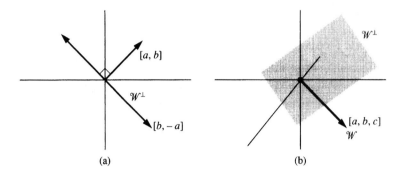

(a)   (b)

If $\mathcal{W}$ is a subspace of $\mathbb{R}^n$, Corollary 6.12 indicates that the dimensions of $\mathcal{W}$ and $\mathcal{W}^\perp$ add up to n. For this reason, many students get the mistaken impression that every vector in $\mathbb{R}^n$ lies either in $\mathcal{W}$ or in $\mathcal{W}^\perp$. But $\mathcal{W}$ and $\mathcal{W}^\perp$ are not "setwise" complements of each other; a more accurate depiction is given in Figure 6.3. For example, recall the subspace $\mathcal{W} = \{[a, b, 0] \mid a, b \in \mathbb{R}\}$ of Example 1. We showed that $\mathcal{W}^\perp = \{[0, 0, z] \mid z \in \mathbb{R}\}$. Yet $[1, 1, 1]$ is in neither $\mathcal{W}$ nor $\mathcal{W}^\perp$, even though $\dim(\mathcal{W}) + \dim(\mathcal{W}^\perp) = \dim(\mathbb{R}^3)$. In this case, $\mathcal{W}$ is the xy-plane and $\mathcal{W}^\perp$ is the z-axis.

Figure 6.3

Symbolic depiction of $\mathcal{W}$ and $\mathcal{W}^\perp$

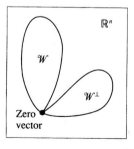

The next corollary asserts that each subspace $\mathcal{W}$ of $\mathbb{R}^n$ is, in fact, the orthogonal complement of $\mathcal{W}^\perp$. Hence, $\mathcal{W}$ and $\mathcal{W}^\perp$ are orthogonal complements of each other. The proof is left as Exercise 19.

> **COROLLARY 6.13**
>
> Let $\mathcal{W}$ be a subspace of $\mathbb{R}^n$. Then $\left(\mathcal{W}^\perp\right)^\perp = \mathcal{W}$.

Orthogonal Projection onto a Subspace

Next, we present the Projection Theorem, a generalization of Theorem 1.10. Recall from Theorem 1.10 that every nonzero vector in $\mathbb{R}^n$ can be decomposed into the sum of two vectors, one parallel to a given vector $\mathbf{a}$ and another orthogonal to $\mathbf{a}$.

> **THEOREM 6.14 (Projection Theorem)**
>
> Let $\mathcal{W}$ be a subspace of $\mathbb{R}^n$. Then every vector $\mathbf{v} \in \mathbb{R}^n$ can be expressed in a unique way as $\mathbf{w}_1 + \mathbf{w}_2$, where $\mathbf{w}_1 \in \mathcal{W}$ and $\mathbf{w}_2 \in \mathcal{W}^\perp$.

PROOF OF THEOREM 6.14

Let $\mathcal{W}$ be a subspace of $\mathbb{R}^n$, and let $\mathbf{v} \in \mathbb{R}^n$. We first show that $\mathbf{v}$ can be expressed as $\mathbf{w}_1 + \mathbf{w}_2$, where $\mathbf{w}_1 \in \mathcal{W}$, $\mathbf{w}_2 \in \mathcal{W}^\perp$. Then we will show that there is a unique pair $\mathbf{w}_1, \mathbf{w}_2$ for each $\mathbf{v}$.

Let $\{\mathbf{u}_1, \ldots, \mathbf{u}_k\}$ be an orthonormal basis for $\mathcal{W}$. Expand $\{\mathbf{u}_1, \ldots, \mathbf{u}_k\}$ to an orthonormal basis $\{\mathbf{u}_1, \ldots, \mathbf{u}_k, \mathbf{u}_{k+1}, \ldots, \mathbf{u}_n\}$ for $\mathbb{R}^n$. Then by Theorem 6.3, $\mathbf{v} = (\mathbf{v} \cdot \mathbf{u}_1)\mathbf{u}_1 + \cdots + (\mathbf{v} \cdot \mathbf{u}_n)\mathbf{u}_n$. Let $\mathbf{w}_1 = (\mathbf{v} \cdot \mathbf{u}_1)\mathbf{u}_1 + \cdots + (\mathbf{v} \cdot \mathbf{u}_k)\mathbf{u}_k$ and $\mathbf{w}_2 = (\mathbf{v} \cdot \mathbf{u}_{k+1})\mathbf{u}_{k+1} + \cdots + (\mathbf{v} \cdot \mathbf{u}_n)\mathbf{u}_n$. Clearly, $\mathbf{v} = \mathbf{w}_1 + \mathbf{w}_2$. Also, Theorem 6.11 implies that $\mathbf{w}_1 \in \mathcal{W}$ and $\mathbf{w}_2$ is in $\mathcal{W}^\perp$.

Finally, we want to show uniqueness of decomposition. Suppose that $\mathbf{v} = \mathbf{w}_1 + \mathbf{w}_2$ and $\mathbf{v} = \mathbf{w}_1' + \mathbf{w}_2'$, where $\mathbf{w}_1, \mathbf{w}_1' \in \mathcal{W}$ and $\mathbf{w}_2, \mathbf{w}_2' \in \mathcal{W}^\perp$. We want to show that $\mathbf{w}_1 = \mathbf{w}_1'$ and $\mathbf{w}_2 = \mathbf{w}_2'$. Now, $\mathbf{w}_1 - \mathbf{w}_1' = \mathbf{w}_2' - \mathbf{w}_2$ (why?). Also, $\mathbf{w}_1 - \mathbf{w}_1' \in \mathcal{W}$, but $\mathbf{w}_2' - \mathbf{w}_2 \in \mathcal{W}^\perp$. Thus, $\mathbf{w}_1 - \mathbf{w}_1' = \mathbf{w}_2' - \mathbf{w}_2 \in \mathcal{W} \cap \mathcal{W}^\perp$. By Theorem 6.10, $\mathbf{w}_1 - \mathbf{w}_1' = \mathbf{w}_2' - \mathbf{w}_2 = \mathbf{0}$. Hence, $\mathbf{w}_1 = \mathbf{w}_1'$ and $\mathbf{w}_2 = \mathbf{w}_2'$. ∎

We give a special name to the vector $\mathbf{w}_1$ in the proof of Theorem 6.14.

DEFINITION

Let $\mathcal{W}$ be a subspace of $\mathbb{R}^n$ with orthonormal basis $\{\mathbf{u}_1, \ldots, \mathbf{u}_k\}$, and let $\mathbf{v} \in \mathbb{R}^n$. Then the **orthogonal projection of $\mathbf{v}$ onto** $\mathcal{W}$ is the vector

$$\mathbf{proj}_{\mathcal{W}}\mathbf{v} = (\mathbf{v} \cdot \mathbf{u}_1)\mathbf{u}_1 + \cdots + (\mathbf{v} \cdot \mathbf{u}_k)\mathbf{u}_k.$$

Notice that the choice of orthonormal basis for $\mathcal{W}$ in this definition does not matter. This is because if $\mathbf{v}$ is any vector in $\mathbb{R}^n$, Theorem 6.14 asserts there is a unique expression $\mathbf{w}_1 + \mathbf{w}_2$ for $\mathbf{v}$ with $\mathbf{w}_1 \in \mathcal{W}$, $\mathbf{w}_2 \in \mathcal{W}^\perp$, and we see from the proof of the theorem that $\mathbf{w}_1 = \mathbf{proj}_{\mathcal{W}}\mathbf{v}$. Hence, if $\{\mathbf{z}_1, \ldots, \mathbf{z}_k\}$ is any other orthonormal basis for $\mathcal{W}$, then $\mathbf{proj}_{\mathcal{W}}\mathbf{v}$ is equal to $(\mathbf{v} \cdot \mathbf{z}_1)\mathbf{z}_1 + \cdots + (\mathbf{v} \cdot \mathbf{z}_k)\mathbf{z}_k$ as well. This fact is illustrated in the next example.

Consider the orthonormal subset

$$B = \{\mathbf{u}_1, \mathbf{u}_2\} = \left\{ \left[\frac{8}{9}, -\frac{1}{9}, -\frac{4}{9} \right], \left[\frac{4}{9}, \frac{4}{9}, \frac{7}{9} \right] \right\}$$

of $\mathbb{R}^3$, and let $\mathcal{W} = \text{span}(B)$. Notice that B is an orthonormal basis for $\mathcal{W}$. Also consider the orthogonal set $S = \{[4, 1, 1], [4, -5, -11]\}$. Now since

$$[4, 1, 1] = 3 \left[\frac{8}{9}, -\frac{1}{9}, -\frac{4}{9} \right] + 3 \left[\frac{4}{9}, \frac{4}{9}, \frac{7}{9} \right]$$

$$\text{and } [4, -5, -11] = 9 \left[\frac{8}{9}, -\frac{1}{9}, -\frac{4}{9} \right] - 9 \left[\frac{4}{9}, \frac{4}{9}, \frac{7}{9} \right],$$

S is an orthogonal subset of $\mathcal{W}$. Since $|S| = \dim(\mathcal{W})$, S is also an orthogonal basis for $\mathcal{W}$. Hence, after normalizing the vectors in S, we obtain the following second orthonormal basis for $\mathcal{W}$:

$$C = \{\mathbf{z}_1, \mathbf{z}_2\} = \left\{ \left[\frac{4}{3\sqrt{2}}, \frac{1}{3\sqrt{2}}, \frac{1}{3\sqrt{2}} \right], \left[\frac{4}{9\sqrt{2}}, -\frac{5}{9\sqrt{2}}, -\frac{11}{9\sqrt{2}} \right] \right\}.$$

Let $\mathbf{v} = [1, 2, 3]$. We will verify that the same vector for $\text{proj}_{\mathcal{W}}\mathbf{v}$ is obtained whether $B = \{\mathbf{u}_1, \mathbf{u}_2\}$ or $C = \{\mathbf{z}_1, \mathbf{z}_2\}$ is used as the orthonormal basis for $\mathcal{W}$. Now, using B yields

$$(\mathbf{v} \cdot \mathbf{u}_1)\,\mathbf{u}_1 + (\mathbf{v} \cdot \mathbf{u}_2)\,\mathbf{u}_2 = -\frac{2}{3} \left[\frac{8}{9}, -\frac{1}{9}, -\frac{4}{9} \right] + \frac{11}{3} \left[\frac{4}{9}, \frac{4}{9}, \frac{7}{9} \right] = \left[\frac{28}{27}, \frac{46}{27}, \frac{85}{27} \right].$$

Similarly, using C gives

$$(\mathbf{v} \cdot \mathbf{z}_1)\,\mathbf{z}_1 + (\mathbf{v} \cdot \mathbf{z}_2)\,\mathbf{z}_2 = \frac{3}{\sqrt{2}} \left[\frac{4}{3\sqrt{2}}, \frac{1}{3\sqrt{2}}, \frac{1}{3\sqrt{2}} \right]$$

$$+ \left(-\frac{13}{3\sqrt{2}} \right) \left[\frac{4}{9\sqrt{2}}, -\frac{5}{9\sqrt{2}}, -\frac{11}{9\sqrt{2}} \right]$$

$$= \left[\frac{28}{27}, \frac{46}{27}, \frac{85}{27} \right].$$

Hence, with either orthonormal basis we obtain $\text{proj}_{\mathcal{W}}\mathbf{v} = \left[\frac{28}{27}, \frac{46}{27}, \frac{85}{27} \right]$. ∎

The proof of Theorem 6.14 illustrates the following:

> If $\mathcal{W}$ is a subspace of $\mathbb{R}^n$ and $\mathbf{v} \in \mathbb{R}^n$, then $\mathbf{v}$ can be expressed as $\mathbf{w}_1 + \mathbf{w}_2$, where $\mathbf{w}_1 = \text{proj}_{\mathcal{W}}\mathbf{v} \in \mathcal{W}$ and $\mathbf{w}_2 = \mathbf{v} - \text{proj}_{\mathcal{W}}\mathbf{v} \in \mathcal{W}^\perp$. Moreover, $\mathbf{w}_2$ can also be expressed as $\text{proj}_{\mathcal{W}^\perp}\mathbf{v}$.

The vector $\mathbf{w}_1$ is the generalization of the projection vector $\text{proj}_{\mathbf{a}}\mathbf{b}$ from Section 1.2 (see Exercise 17).

EXAMPLE 7

Let $\mathcal{W}$ be the subspace of $\mathbb{R}^3$ whose vectors (beginning at the origin) lie in the plane $\mathcal{L}$ with equation $2x + y + z = 0$. Let $\mathbf{v} = [-6, 10, 5]$. (Notice that $\mathbf{v} \notin \mathcal{W}$.) We will find $\mathbf{proj}_{\mathcal{W}}\mathbf{v}$.

First, notice that $[1, 0, -2]$ and $[0, 1, -1]$ are two linearly independent vectors in $\mathcal{W}$. (To find the first vector, choose $x = 1$, $y = 0$, and for the other, let $x = 0$ and $y = 1$.) Using the Gram-Schmidt Process on these vectors, we obtain the orthogonal basis $\{[1, 0, -2], [-2, 5, -1]\}$ for $\mathcal{W}$ (verify!). After normalizing, we have the orthonormal basis $\{\mathbf{u}_1, \mathbf{u}_2\}$ for $\mathcal{W}$, where

$$\mathbf{u}_1 = \left[\frac{1}{\sqrt{5}}, 0, -\frac{2}{\sqrt{5}}\right] \quad \text{and} \quad \mathbf{u}_2 = \left[-\frac{2}{\sqrt{30}}, \frac{5}{\sqrt{30}}, -\frac{1}{\sqrt{30}}\right].$$

Now,

$$\mathbf{proj}_{\mathcal{W}}\mathbf{v} = (\mathbf{v} \cdot \mathbf{u}_1)\,\mathbf{u}_1 + (\mathbf{v} \cdot \mathbf{u}_2)\,\mathbf{u}_2$$

$$= -\frac{16}{\sqrt{5}}\left[\frac{1}{\sqrt{5}}, 0, -\frac{2}{\sqrt{5}}\right] + \frac{57}{\sqrt{30}}\left[-\frac{2}{\sqrt{30}}, \frac{5}{\sqrt{30}}, -\frac{1}{\sqrt{30}}\right]$$

$$= \left[-\frac{16}{5}, 0, \frac{32}{5}\right] + \left[-\frac{114}{30}, \frac{285}{30}, -\frac{57}{30}\right]$$

$$= \left[-7, \frac{19}{2}, \frac{9}{2}\right].$$

Notice that this vector is in $\mathcal{W}$. Finally, $\mathbf{v} - \mathbf{proj}_{\mathcal{W}}\mathbf{v} = \left[1, \frac{1}{2}, \frac{1}{2}\right]$, which is indeed in $\mathcal{W}^{\perp}$ because it is orthogonal to both $\mathbf{u}_1$ and $\mathbf{u}_2$ (verify!). Hence, we have decomposed $\mathbf{v} = [-6, 10, 5]$ as the sum of two vectors $\left[-7, \frac{19}{2}, \frac{9}{2}\right]$ and $\left[1, \frac{1}{2}, \frac{1}{2}\right]$, where the first is in $\mathcal{W}$ and the second is in $\mathcal{W}^{\perp}$. ∎

We can think of the orthogonal projection vector $\mathbf{proj}_{\mathcal{W}}\mathbf{v}$ in Example 7 as the "shadow" that $\mathbf{v}$ casts on the plane $\mathcal{L}$ as light falls directly onto $\mathcal{L}$ from a light source above and parallel to $\mathcal{L}$. This concept is illustrated in Figure 6.4.

Figure 6.4

The orthogonal projection vector $\left[-7, \frac{19}{2}, \frac{9}{2}\right]$ of $\mathbf{v} = [-6, 10, 5]$ onto the plane $2x + y + z = 0$, pictured as a shadow cast by $\mathbf{v}$ from a light source above and parallel to the plane

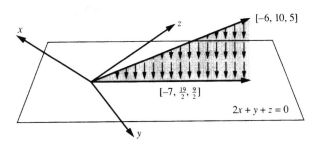

There are two special cases of the Projection Theorem. First, if $\mathbf{v} \in \mathcal{W}$, then $\mathbf{proj}_{\mathcal{W}}\mathbf{v}$ simply equals $\mathbf{v}$ itself. Also, if $\mathbf{v} \in \mathcal{W}^{\perp}$, then $\mathbf{proj}_{\mathcal{W}}\mathbf{v}$ equals $\mathbf{0}$. These results are left as Exercise 13.

The next theorem assures us that orthogonal projection onto a subspace of $\mathbb{R}^n$ is a linear operator on $\mathbb{R}^n$. The proof is left as Exercise 20.

THEOREM 6.15

Let $\mathcal{W}$ be a subspace of $\mathbb{R}^n$. Then the mapping $L: \mathbb{R}^n \to \mathbb{R}^n$ given by $L(\mathbf{v}) = \mathbf{proj}_{\mathcal{W}}\mathbf{v}$ is a linear operator with $\ker(L) = \mathcal{W}^\perp$.

Application: Orthogonal Projections and Reflections in $\mathbb{R}^3$

From Theorem 6.15, an orthogonal projection onto a plane through the origin in $\mathbb{R}^3$ is a linear operator on $\mathbb{R}^3$. We can use eigenvectors and the Generalized Diagonalization Method to find the matrix for such an operator with respect to the standard basis.

EXAMPLE 8   Let $L : \mathbb{R}^3 \longrightarrow \mathbb{R}^3$ be the orthogonal projection onto the plane $\mathcal{W} = \{[x, y, z] \mid 4x - 7y + 4z = 0\}$. To find the matrix for L with respect to the standard basis, we first find bases for $\mathcal{W}$ and $\mathcal{W}^\perp$, which as we will see, are actually bases for the eigenspaces of L.

Since $[4, -7, 4] \cdot [x, y, z] = 0$ for every vector in $\mathcal{W}$, $\mathbf{v}_1 = [4, -7, 4] \in \mathcal{W}^\perp$. Since $\dim(\mathcal{W}) = 2$, we have $\dim(\mathcal{W}^\perp) = 1$ by Corollary 6.12 and so $\{\mathbf{v}_1\}$ is a basis for $\mathcal{W}^\perp$. Notice that $\mathcal{W}^\perp = \ker(L)$ (by Theorem 6.15), and so $\mathcal{W}^\perp =$ the eigenspace E_0 for L. Hence, $\{\mathbf{v}_1\}$ is actually a basis for E_0.

Next, let $\mathbf{v}_2 = [1, 0, -1]$ and $\mathbf{v}_3 = [0, 1, \frac{7}{4}]$. Then $\{\mathbf{v}_2, \mathbf{v}_3\}$ is a linearly independent subset of $\mathcal{W}$. (Do you see how we chose $\mathbf{v}_2$ and $\mathbf{v}_3$?) Hence, $\{\mathbf{v}_2, \mathbf{v}_3\}$ is a basis for $\mathcal{W}$, since $\dim(\mathcal{W}) = 2$. But since every vector in the plane $\mathcal{W}$ is mapped to itself by L, $\mathcal{W} =$ the eigenspace E_1 for L. Thus, $\{\mathbf{v}_2, \mathbf{v}_3\}$ is a basis for E_1. The union $\{\mathbf{v}_1, \mathbf{v}_2, \mathbf{v}_3\}$ of the bases for E_0 and E_1 is a linearly independent set of three vectors for $\mathbb{R}^3$ by Theorem 5.24, and so L is diagonalizable.

Now, by the Generalized Diagonalization Method of Section 5.5, if $\mathbf{A}$ is the matrix for L with respect to the standard basis, then $\mathbf{P}^{-1}\mathbf{AP} = \mathbf{D}$, where $\mathbf{P}$ is the matrix whose columns are the eigenvectors $\mathbf{v}_1$, $\mathbf{v}_2$, and $\mathbf{v}_3$, and $\mathbf{D}$ is the diagonal matrix with the eigenvalues 0, 1, and 1 on the main diagonal. Hence, we compute $\mathbf{P}^{-1}$, and use $\mathbf{A} = \mathbf{PDP}^{-1}$ to obtain

$$\mathbf{A} = \begin{bmatrix} 4 & 1 & 0 \\ -7 & 0 & 1 \\ 4 & -1 & \frac{7}{4} \end{bmatrix} \begin{bmatrix} 0 & 0 & 0 \\ 0 & 1 & 0 \\ 0 & 0 & 1 \end{bmatrix} \begin{bmatrix} \frac{4}{81} & -\frac{7}{81} & \frac{4}{81} \\ \frac{65}{81} & \frac{28}{81} & -\frac{16}{81} \\ \frac{28}{81} & \frac{32}{81} & \frac{28}{81} \end{bmatrix} = \frac{1}{81} \begin{bmatrix} 65 & 28 & -16 \\ 28 & 32 & 28 \\ -16 & 28 & 65 \end{bmatrix},$$

which is the matrix for L with respect to the standard basis.   ∎

The technique used in Example 8 can be used to find the matrix with respect to the standard basis for the orthogonal projection onto any plane through the origin in $\mathbb{R}^3$. In particular, for the plane $ax + by + cz = 0$, let $\mathbf{v}_1 = [a, b, c]$, a vector orthogonal to the plane. Next, choose $\mathbf{v}_2$ and $\mathbf{v}_3$ to be any linearly independent pair of vectors in the plane. Then the matrix $\mathbf{A}$ for the projection with respect to the standard basis is $\mathbf{A} = \mathbf{PDP}^{-1}$, where $\mathbf{P}$ is the matrix whose columns are $\mathbf{v}_1$, $\mathbf{v}_2$, and $\mathbf{v}_3$, in any order, and $\mathbf{D}$ is the diagonal matrix with the eigenvalues 0, 1, and 1 in a corresponding order on the main diagonal. That is, the column containing eigenvalue 0 in $\mathbf{D}$ must correspond to the column in $\mathbf{P}$ containing $\mathbf{v}_1$.

Similarly, we can reverse the process to determine whether a given 3×3 matrix $\mathbf{A}$ represents an orthogonal projection onto a plane through the origin. Such a matrix must diagonalize to the diagonal matrix $\mathbf{D}$ having eigenvalues 0, 1, and 1, respectively, on the main diagonal, and the matrix $\mathbf{P}$ such that $\mathbf{A} = \mathbf{PDP}^{-1}$ must have the property that the column of $\mathbf{P}$ corresponding to the eigenvalue 0 be orthogonal to the other two columns of $\mathbf{P}$.

EXAMPLE 9

The matrix

$$\mathbf{A} = \begin{bmatrix} 18 & -6 & -30 \\ -25 & 10 & 45 \\ 17 & -6 & -29 \end{bmatrix}$$

has eigenvalues 0, 1, and -2 (verify!). Since there is an eigenvalue other than 0 or 1, $\mathbf{A}$ can not represent an orthogonal projection onto a plane through the origin.

Similarly, you can verify that

$$\mathbf{A}_1 = \begin{bmatrix} -3 & 1 & -1 \\ 16 & -3 & 4 \\ 28 & -7 & 8 \end{bmatrix} \quad \text{diagonalizes to} \quad \mathbf{D}_1 = \begin{bmatrix} 0 & 0 & 0 \\ 0 & 1 & 0 \\ 0 & 0 & 1 \end{bmatrix}.$$

Now, $\mathbf{D}_1$ clearly has the proper form. However, the transition matrix $\mathbf{P}_1$ used in the diagonalization is found to be

$$\mathbf{P}_1 = \begin{bmatrix} -1 & 0 & 1 \\ 4 & -1 & -6 \\ 7 & -1 & -10 \end{bmatrix}.$$

Since the first column of $\mathbf{P}_1$ (corresponding to eigenvalue 0) is not orthogonal to the other two columns of $\mathbf{P}_1$, $\mathbf{A}_1$ does not represent an orthogonal projection onto a plane through the origin.

In contrast, the matrix

$$\mathbf{A}_2 = \frac{1}{14} \begin{bmatrix} 5 & -3 & -6 \\ -3 & 13 & -2 \\ -6 & -2 & 10 \end{bmatrix} \quad \text{diagonalizes to} \quad \mathbf{D}_2 = \begin{bmatrix} 0 & 0 & 0 \\ 0 & 1 & 0 \\ 0 & 0 & 1 \end{bmatrix}$$

with transition matrix

$$\mathbf{P}_2 = \begin{bmatrix} 3 & -4 & 1 \\ 1 & 2 & -1 \\ 2 & 5 & -1 \end{bmatrix}.$$

Now, $\mathbf{D}_2$ has the correct form, as does $\mathbf{P}_2$, since the first column of $\mathbf{P}_2$ is orthogonal to both other columns. Hence, $\mathbf{A}_2$ represents an orthogonal projection onto a plane through the origin in $\mathbb{R}^3$. In fact, it is the orthogonal projection onto the plane $3x + y + 2z = 0$, that is, all $[x, y, z]$ orthogonal to the first column of $\mathbf{P}_2$. ∎

We can analyze linear operators that are **orthogonal reflections** through a plane through the origin in $\mathbb{R}^3$ in a manner similar to the techniques we used for orthogonal projections[1]. However, the vector $\mathbf{v}_1$ orthogonal to the plane now corresponds to the eigenvalue $\lambda_1 = -1$ (instead of $\lambda_1 = 0$), since $\mathbf{v}_1$ reflects through the plane into $-\mathbf{v}_1$.

| EXAMPLE 10 |

Consider the orthogonal reflection R through the plane $5x - y + 3z = 0$. The matrix for R with respect to the standard basis for $\mathbb{R}^3$ is $\mathbf{A} = \mathbf{PDP}^{-1}$, where $\mathbf{D}$ has the eigenvalues -1, 1, and 1 on the main diagonal, and where the first column of $\mathbf{P}$ is orthogonal to the plane, and the other two columns of $\mathbf{P}$ are linearly independent vectors in the plane. Hence,

$$\mathbf{A} = \mathbf{PDP}^{-1} = \begin{bmatrix} 5 & 1 & 0 \\ -1 & 0 & 1 \\ 3 & -\frac{5}{3} & \frac{1}{3} \end{bmatrix} \begin{bmatrix} -1 & 0 & 0 \\ 0 & 1 & 0 \\ 0 & 0 & 1 \end{bmatrix} \begin{bmatrix} \frac{1}{7} & -\frac{1}{35} & \frac{3}{35} \\ \frac{2}{7} & \frac{1}{7} & -\frac{3}{7} \\ \frac{1}{7} & \frac{34}{35} & \frac{3}{35} \end{bmatrix} = \begin{bmatrix} -\frac{3}{7} & \frac{2}{7} & -\frac{6}{7} \\ \frac{2}{7} & \frac{33}{35} & \frac{6}{35} \\ -\frac{6}{7} & \frac{6}{35} & \frac{17}{35} \end{bmatrix}.$$

∎

Application: Distance from a Point to a Subspace

DEFINITION

Let $\mathcal{W}$ be a subspace of $\mathbb{R}^n$, and assume all vectors in $\mathcal{W}$ have initial point at the origin. Let P be any point in n-dimensional space. Then the **minimum distance** from P to $\mathcal{W}$ is the shortest distance between P and the terminal point of any vector in $\mathcal{W}$.

The next theorem gives a formula for the minimum distance, and its proof is left as Exercise 23.

THEOREM 6.16

Let $\mathcal{W}$ be a subspace of $\mathbb{R}^n$, and let P be a point in n-dimensional space. If $\mathbf{v}$ is the vector from the origin to P, then the minimum distance from P to $\mathcal{W}$ is $\|\mathbf{v} - \mathbf{proj}_{\mathcal{W}}\mathbf{v}\|$.

Notice that if S is the terminal point of $\mathbf{proj}_{\mathcal{W}}\mathbf{v}$, then $\|\mathbf{v} - \mathbf{proj}_{\mathcal{W}}\mathbf{v}\|$ represents the distance from P to S. Therefore, Theorem 6.16 can be interpreted as saying that no other vector in $\mathcal{W}$ is closer to $\mathbf{v}$ than $\mathbf{proj}_{\mathcal{W}}\mathbf{v}$; that

---

[1] All of the reflection operators we have studied earlier in this text are, in fact, *orthogonal* reflections.

is, the norm of the difference between $\mathbf{v}$ and $\mathbf{proj}_{\mathcal{W}}\mathbf{v}$ is less than or equal to the norm of the difference between $\mathbf{v}$ and any other vector in $\mathcal{W}$. In fact, it can be shown that if $\mathbf{w}$ is a vector in $\mathcal{W}$ equally close to $\mathbf{v}$, then $\mathbf{w}$ must equal $\mathbf{proj}_{\mathcal{W}}\mathbf{v}$.[2]

Figure 6.5

The minimum distance from P to $\mathcal{W}$, $\|\mathbf{v} - \mathbf{proj}_{\mathcal{W}}\mathbf{v}\|$

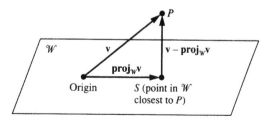

EXAMPLE 11

Consider the subspace $\mathcal{W}$ of $\mathbb{R}^3$ from Example 7, whose vectors lie in the plane $2x + y + z = 0$. In that example, for $\mathbf{v} = [-6, 10, 5]$, we calculated that $\mathbf{v} - \mathbf{proj}_{\mathcal{W}}\mathbf{v} = \left[1, \frac{1}{2}, \frac{1}{2}\right]$. Hence, the minimum distance from $P = (-6, 10, 5)$ to $\mathcal{W}$ is $\|\mathbf{v} - \mathbf{proj}_{\mathcal{W}}\mathbf{v}\| = \sqrt{1^2 + \left(\frac{1}{2}\right)^2 + \left(\frac{1}{2}\right)^2} = \sqrt{\frac{3}{2}} \approx 1.2247$. ∎

♦ **Application:** You have now covered the prerequisites for Section 8.10, "Least-Squares Solutions for Inconsistent Systems."

Exercises for Section 6.2

1. For each of the following subspaces $\mathcal{W}$ of $\mathbb{R}^n$, find a basis for $\mathcal{W}^{\perp}$, and verify Corollary 6.12:
   ⋆(a) In $\mathbb{R}^2$, $\mathcal{W} = \text{span}(\{[3, -2]\})$
   (b) In $\mathbb{R}^3$, $\mathcal{W} = \text{span}(\{[1, -2, 1]\})$
   ⋆(c) In $\mathbb{R}^3$, $\mathcal{W} = \text{span}(\{[1, 4, -2], [2, 1, -1]\})$
   (d) In $\mathbb{R}^3$, $\mathcal{W} = $ the plane $3x - y + 4z = 0$
   ⋆(e) In $\mathbb{R}^3$, $\mathcal{W} = $ the plane $-2x + 5y - z = 0$
   ⋆(f) In $\mathbb{R}^4$, $\mathcal{W} = \text{span}(\{[1, -1, 0, 2], [0, 1, 2, -1]\})$
   (g) In $\mathbb{R}^4$, $\mathcal{W} = \{[x, y, z, w] \mid 3x - 2y + 4z + w = 0\}$
2. For each of the following subspaces $\mathcal{W}$ of $\mathbb{R}^n$ and for the given $\mathbf{v} \in \mathbb{R}^n$, find $\mathbf{proj}_{\mathcal{W}}\mathbf{v}$, and decompose $\mathbf{v}$ into $\mathbf{w}_1 + \mathbf{w}_2$, where $\mathbf{w}_1 \in \mathcal{W}$ and $\mathbf{w}_2 \in \mathcal{W}^{\perp}$. (Hint: You may need to find an orthonormal basis for $\mathcal{W}$ first.)
   ⋆(a) In $\mathbb{R}^3$, $\mathcal{W} = \text{span}(\{[1, -2, -1], [3, -1, 0]\})$, $\mathbf{v} = [-1, 3, 2]$
   ⋆(b) In $\mathbb{R}^3$, $\mathcal{W} = $ the plane $2x - 2y + z = 0$, $\mathbf{v} = [1, -4, 3]$
   (c) In $\mathbb{R}^3$, $\mathcal{W} = \text{span}(\{[-1, 3, 2]\})$, $\mathbf{v} = [2, 2, -3]$
   (d) In $\mathbb{R}^4$, $\mathcal{W} = \text{span}(\{[2, -1, 1, 0], [1, -1, 2, 2]\})$, $\mathbf{v} = [-1, 3, 3, 2]$
3. Let $\mathbf{v} = [a, b, c]$. If $\mathcal{W}$ is the xy-plane, verify that $\mathbf{proj}_{\mathcal{W}}\mathbf{v} = [a, b, 0]$.

---

[2]This statement, in a slightly different form, is proved as part of Theorem 8.8 in Section 8.10.

4. In each of the following, find the minimum distance between the given point P and the given subspace W of $\mathbb{R}^n$:

   ⋆(a) $P = (-2, 3, 1)$, $W = \text{span}(\{[-1, 4, 4], [2, -1, 0]\})$ in $\mathbb{R}^3$

   (b) $P = (4, -1, 2)$, $W = \text{span}(\{[-2, 3, -3]\})$ in $\mathbb{R}^3$

   (c) $P = (2, 3, -3, 1)$, $W = \text{span}(\{[-1, 2, -1, 1], [2, -1, 1, -1]\})$ in $\mathbb{R}^4$

   ⋆(d) $P = (-1, 4, -2, 2)$, $W = \{[x, y, z, w] \mid 2x - 3z + 2w = 0\}$ in $\mathbb{R}^4$

5. In each part, let L be the linear operator on $\mathbb{R}^3$ with the given matrix representation with respect to the standard basis. Determine whether L is

   (i) An orthogonal projection onto a plane through the origin

   (ii) An orthogonal reflection through a plane through the origin

   (iii) Neither

   Also, if L is of type (i) or (ii), state the equation of the plane.

   ⋆(a) $\dfrac{1}{11} \begin{bmatrix} 2 & -3 & -3 \\ -3 & 10 & -1 \\ -3 & -1 & 10 \end{bmatrix}$

   (b) $\dfrac{1}{9} \begin{bmatrix} 7 & 4 & 4 \\ 4 & 1 & -8 \\ 4 & -8 & 1 \end{bmatrix}$

   (c) $\dfrac{1}{3} \begin{bmatrix} 11 & 49 & -77 \\ -18 & -66 & 99 \\ -10 & -35 & 52 \end{bmatrix}$

   ⋆(d) $\dfrac{1}{15} \begin{bmatrix} 7 & -2 & -14 \\ -4 & 14 & -7 \\ -12 & -3 & -6 \end{bmatrix}$

⋆6. Let $L: \mathbb{R}^3 \to \mathbb{R}^3$ be the orthogonal projection onto the plane $2x - y + 2z = 0$. Use eigenvalues and eigenvectors to find the matrix representation of L with respect to the standard basis.

7. Let $L: \mathbb{R}^3 \to \mathbb{R}^3$ be the orthogonal reflection through the plane $3x - y + 2z = 0$. Use eigenvalues and eigenvectors to find the matrix representation of L with respect to the standard basis.

8. Let $L: \mathbb{R}^3 \to \mathbb{R}^3$ be the orthogonal projection onto the plane $2x + y + z = 0$ from Example 7.

   (a) Use eigenvalues and eigenvectors to find the matrix representation of L with respect to the standard basis.

   (b) Use the matrix in part (a) to confirm the computation in Example 7 that $L([-6, 10, 5]) = [-7, \frac{19}{2}, \frac{9}{2}]$.

9. Find the characteristic polynomial for each of the given linear operators. (Hint: This requires almost no computation.)

   ⋆(a) $L: \mathbb{R}^3 \to \mathbb{R}^3$, where L is the orthogonal projection onto the plane $4x - 3y + 2z = 0$

   (b) $L: \mathbb{R}^3 \to \mathbb{R}^3$, where L is the orthogonal projection onto the line through the origin spanned by $[4, -1, 3]$

⋆(c) $L: \mathbb{R}^3 \to \mathbb{R}^3$, where L is the orthogonal reflection through the plane $3x + 5y - z = 0$

10. In each of the following, find the matrix representation of the operator $L: \mathbb{R}^n \to \mathbb{R}^n$ given by $L(\mathbf{v}) = \mathbf{proj}_{\mathcal{W}}\mathbf{v}$, with respect to the standard basis for $\mathbb{R}^n$:

⋆(a) In $\mathbb{R}^3$, $\mathcal{W} = \operatorname{span}(\{[2, -1, 1], [1, 0, -3]\})$
(b) In $\mathbb{R}^3$, $\mathcal{W}$ = the plane $3x - 2y + 2z = 0$
⋆(c) In $\mathbb{R}^4$, $\mathcal{W} = \operatorname{span}(\{[1, 2, 1, 0], [-1, 0, -2, 1]\})$
(d) In $\mathbb{R}^4$, $\mathcal{W} = \operatorname{span}(\{[-3, -1, 1, 2]\})$

11. Prove that if $\mathcal{W}_1$ and $\mathcal{W}_2$ are subspaces of $\mathbb{R}^n$ with $\mathcal{W}_1^\perp = \mathcal{W}_2^\perp$, then $\mathcal{W}_1 = \mathcal{W}_2$.

12. Prove that if $\mathcal{W}_1$ and $\mathcal{W}_2$ are subspaces of $\mathbb{R}^n$ with $\mathcal{W}_1 \subseteq \mathcal{W}_2$, then $\mathcal{W}_2^\perp \subseteq \mathcal{W}_1^\perp$.

13. Let $\mathcal{W}$ be a subspace of $\mathbb{R}^n$.

(a) Show that if $\mathbf{v} \in \mathcal{W}$, then $\mathbf{proj}_{\mathcal{W}}\mathbf{v} = \mathbf{v}$.
(b) Show that if $\mathbf{v} \in \mathcal{W}^\perp$, then $\mathbf{proj}_{\mathcal{W}}\mathbf{v} = \mathbf{0}$.

14. Let $\mathcal{W}$ be a subspace of $\mathbb{R}^n$. Suppose that $\mathbf{v}$ is a nonzero vector with initial point at the origin and terminal point P. Prove that $\mathbf{v} \in \mathcal{W}^\perp$ if and only if the minimum distance between P and $\mathcal{W}$ is $\|\mathbf{v}\|$.

15. Let $\mathcal{W}$ be a subspace of $\mathbb{R}^n$, and let $\mathbf{v}_1$ and $\mathbf{v}_2$ be vectors in $\mathbb{R}^n$. Suppose that $\mathbf{p}_1 = \mathbf{proj}_{\mathcal{W}}\mathbf{v}_1$ and $\mathbf{p}_2 = \mathbf{proj}_{\mathcal{W}}\mathbf{v}_2$.

(a) What is $\mathbf{proj}_{\mathcal{W}}(\mathbf{v}_1 + \mathbf{v}_2)$? Prove your answer.
(b) If $c \in \mathbb{R}$, what is $\mathbf{proj}_{\mathcal{W}}(c\mathbf{v}_1)$? Prove your answer.

16. We can represent matrices in $\mathcal{M}_{nn}$ as n^2-vectors by using their coordinatization with respect to the standard basis in $\mathcal{M}_{nn}$. Use this technique to prove that the orthogonal complement of the subspace $\mathcal{V}$ of symmetric matrices in $\mathcal{M}_{nn}$ is the subspace $\mathcal{W}$ of $n \times n$ skew-symmetric matrices. (Hint: First show that $\mathcal{W} \subseteq \mathcal{V}^\perp$. Then prove equality by showing that $\dim(\mathcal{W}) = n^2 - \dim(\mathcal{V})$.)

17. Show that if $\mathcal{W}$ is a one-dimensional subspace of $\mathbb{R}^n$ spanned by $\mathbf{a}$ and if $\mathbf{b} \in \mathbb{R}^n$, then the value of $\mathbf{proj}_{\mathcal{W}}\mathbf{b}$ agrees with the definition for $\mathbf{proj}_{\mathbf{a}}\mathbf{b}$ in Section 1.2.

▶18. Prove Theorem 6.9.

19. Prove Corollary 6.13. (Hint: First show that $\mathcal{W} \subseteq (\mathcal{W}^\perp)^\perp$. Then use Corollary 6.12 to show that $\dim(\mathcal{W}) = \dim\left((\mathcal{W}^\perp)^\perp\right)$, and apply Theorem 4.17.)

▶20. Prove Theorem 6.15. (Hint: To prove $\ker(L) = \mathcal{W}^\perp$, first show that $\operatorname{range}(L) = \mathcal{W}$. Hence, $\dim(\ker(L)) = n - \dim(\mathcal{W}) = \dim\left(\mathcal{W}^\perp\right)$ (why?). Finally, show $\mathcal{W}^\perp \subseteq \ker(L)$, and apply Theorem 4.17.)

21. Let $L: \mathbb{R}^n \to \mathbb{R}^m$ be a linear transformation with matrix $\mathbf{A}$ (with respect to the standard basis). Show that $\ker(L)$ is the orthogonal complement of the row space of $\mathbf{A}$.

22. Let $L: \mathbb{R}^n \to \mathbb{R}^m$ be a linear transformation. Consider the mapping $T :$ $(\ker(L))^{\perp} \to \mathbb{R}^m$ given by $T(\mathbf{v}) = L(\mathbf{v})$, for all $\mathbf{v} \in (\ker(L))^{\perp}$. ($T$ is the **restriction** of L to $(\ker(L))^{\perp}$.) Prove that T is one-to-one.

▶**23.** Prove Theorem 6.16. (Hint: Suppose that T is any point in $\mathcal{W}$ and $\mathbf{w}$ is the vector from the origin to T. We need to show that $\|\mathbf{v} - \mathbf{w}\| \geq \|\mathbf{v} - \mathbf{proj}_{\mathcal{W}}\mathbf{v}\|$; that is, the distance from P to T is at least as large as the distance from P to the terminal point of $\mathbf{proj}_{\mathcal{W}}\mathbf{v}$. Let $\mathbf{a} = \mathbf{v} - \mathbf{proj}_{\mathcal{W}}\mathbf{v}$ and $\mathbf{b} = (\mathbf{proj}_{\mathcal{W}}\mathbf{v}) - \mathbf{w}$. Show that $\mathbf{a} \in \mathcal{W}^{\perp}$, $\mathbf{b} \in \mathcal{W}$, and $\|\mathbf{v} - \mathbf{w}\|^2 = \|\mathbf{a}\|^2 + \|\mathbf{b}\|^2$.)

24. Let $\mathcal{L}$ be a subspace of $\mathbb{R}^n$, and let $\mathcal{W}$ be a subspace of $\mathcal{L}$. We define the **orthogonal complement of** $\mathcal{W}$ **in** $\mathcal{L}$ to be the set of all vectors in $\mathcal{L}$ that are orthogonal to every vector in $\mathcal{W}$.

   (a) Prove that the orthogonal complement of $\mathcal{W}$ in $\mathcal{L}$ is a subspace of $\mathcal{L}$.

   (b) Prove that the dimensions of $\mathcal{W}$ and its orthogonal complement in $\mathcal{L}$ add up to the dimension of $\mathcal{L}$. (Hint: Let B be an orthonormal basis for $\mathcal{W}$. First enlarge B to an orthonormal basis for $\mathcal{L}$, and then enlarge this basis to an orthonormal basis for $\mathbb{R}^n$.)

★**25.** True or False:

   (a) If $\mathcal{W}$ is a subspace of $\mathbb{R}^n$, then $\mathcal{W}^{\perp} = \{\mathbf{x} \in \mathbb{R}^n \mid \mathbf{x} \cdot \mathbf{w} = 0 \text{ for all } \mathbf{w} \in \mathcal{W}\}$.

   (b) If $\mathcal{W}$ is a subspace of $\mathbb{R}^n$ and every vector in a basis for $\mathcal{W}$ is orthogonal to $\mathbf{v}$, then $\mathbf{v} \in \mathcal{W}^{\perp}$.

   (c) If $\mathcal{W}$ is a subspace of $\mathbb{R}^n$, then $\mathcal{W} \cap \mathcal{W}^{\perp} = \{\,\}$.

   (d) If $\mathcal{W}$ is a subspace of $\mathbb{R}^7$, and $\{\mathbf{b}_1, \mathbf{b}_2, \ldots, \mathbf{b}_7\}$ is a basis for $\mathbb{R}^7$ and $\{\mathbf{b}_1, \mathbf{b}_2, \mathbf{b}_3, \mathbf{b}_4\}$ is a basis for $\mathcal{W}$, then $\{\mathbf{b}_5, \mathbf{b}_6, \mathbf{b}_7\}$ is a basis for $\mathcal{W}^{\perp}$.

   (e) If $\mathcal{W}$ is a subspace of $\mathbb{R}^5$, then $\dim(\mathcal{W}^{\perp}) = 5 - \dim(\mathcal{W})$.

   (f) If $\mathcal{W}$ is a subspace of $\mathbb{R}^n$, then every vector $\mathbf{v} \in \mathbb{R}^n$ lies in $\mathcal{W}$ or $\mathcal{W}^{\perp}$.

   (g) The orthogonal complement of the orthogonal complement of a subspace $\mathcal{W}$ of $\mathbb{R}^n$ is $\mathcal{W}$ itself.

   (h) The orthogonal complement of a plane through the origin in $\mathbb{R}^3$ is a line through the origin perpendicular to the plane.

   (i) The mapping $L: \mathbb{R}^n \to \mathbb{R}^n$ given by $L(\mathbf{v}) = \mathbf{proj}_{\mathcal{W}}\mathbf{v}$, where $\mathcal{W}$ is a given subspace of $\mathbb{R}^n$, has $\mathcal{W}^{\perp}$ as its kernel.

   (j) The matrix for an orthogonal projection onto a plane through the origin in $\mathbb{R}^3$ diagonalizes to a matrix with $-1, 1, 1$ on the main diagonal.

   (k) If $\mathcal{W}$ is a subspace of $\mathbb{R}^n$, and $\mathbf{v} \in \mathbb{R}^n$, then the minimum distance from $\mathcal{V}$ to $\mathcal{W}$ is $\|\mathbf{proj}_{\mathcal{W}^{\perp}}\mathbf{v}\|$.

   (l) If $\mathbf{v} \in \mathbb{R}^n$, and $\mathcal{W}$ is a subspace of $\mathbb{R}^n$, then $\mathbf{v} = \mathbf{proj}_{\mathcal{W}}\mathbf{v} + \mathbf{proj}_{\mathcal{W}^{\perp}}\mathbf{v}$.

6.3 Orthogonal Diagonalization

In this section, we determine which linear operators on $\mathbb{R}^n$ have an orthonormal basis B of eigenvectors. Such operators are said to be orthogonally diagonalizable. For this type of operator, the transition matrix **P** from B-coordinates to standard coordinates is an orthogonal matrix. Such a change of basis preserves much of the geometric structure of $\mathbb{R}^n$, including lengths of vectors and the angles between them. Essentially, then, an orthogonally diagonalizable operator is one for which we can find a diagonal form while keeping certain important geometric properties of the operator.

We begin by defining symmetric operators and studying their properties. Then we show that these operators are precisely the ones that are orthogonally diagonalizable. Also, we present a method for orthogonally diagonalizing an operator analogous to the Generalized Diagonalization Method in Section 5.5.

Symmetric Operators

DEFINITION

Let $\mathcal{V}$ be a subspace of $\mathbb{R}^n$. A linear operator $L: \mathcal{V} \to \mathcal{V}$ is a **symmetric operator** on $\mathcal{V}$ if and only if $L(\mathbf{v}_1) \cdot \mathbf{v}_2 = \mathbf{v}_1 \cdot L(\mathbf{v}_2)$, for every $\mathbf{v}_1, \mathbf{v}_2 \in \mathcal{V}$.

EXAMPLE 1   The operator L on $\mathbb{R}^3$ given by $L([a, b, c]) = [b, a, -c]$ is symmetric since

$$L([a, b, c]) \cdot [d, e, f] = [b, a, -c] \cdot [d, e, f] = bd + ae - cf$$
$$\text{and} \quad [a, b, c] \cdot L([d, e, f]) = [a, b, c] \cdot [e, d, -f] = ae + bd - cf.$$

∎

You can verify that the matrix representation for the operator L in Example 1 with respect to the standard basis is

$$\begin{bmatrix} 0 & 1 & 0 \\ 1 & 0 & 0 \\ 0 & 0 & -1 \end{bmatrix},$$

a symmetric matrix. The next theorem asserts that an operator on a subspace $\mathcal{V}$ of $\mathbb{R}^n$ is symmetric if and only if its matrix representation with respect to any orthonormal basis for $\mathcal{V}$ is symmetric.

THEOREM 6.17

Let $\mathcal{V}$ be a nontrivial subspace of $\mathbb{R}^n$, L be a linear operator on $\mathcal{V}$, B be an ordered orthonormal basis for $\mathcal{V}$, and **A** be the matrix for L with respect to B. Then L is a symmetric operator if and only if **A** is a symmetric matrix.

Theorem 6.17 gives a quick way of recognizing symmetric operators just by looking at their matrix representations. Such operators occur frequently in applications. (For example, see Section 10.3, "Quadratic Forms.") The proof of Theorem 6.17 is long, and so we have placed it in Appendix A for the interested reader.

A Symmetric Operator Always Has an Eigenvalue

The following lemma is needed for the proof of Theorem 6.19, the main theorem of this section:

LEMMA 6.18

Let L be a symmetric operator on a nontrivial subspace $\mathcal{V}$ of $\mathbb{R}^n$. Then L has at least one eigenvalue.

Simpler proofs of Lemma 6.18 exist than the one given below, but they involve complex vector spaces, which will not be discussed until Section 7.3. Nevertheless, the following proof is interesting, since it brings together a variety of topics already developed as well as some familiar theorems from algebra.

PROOF OF LEMMA 6.18

Suppose that L is a symmetric operator on a nontrivial subspace $\mathcal{V}$ of $\mathbb{R}^n$ with $\dim(\mathcal{V}) = k$. Let B be an orthonormal basis for $\mathcal{V}$. By Theorem 6.17, the matrix representation $\mathbf{A}$ for L with respect to B is a symmetric matrix.

Let $p_{\mathbf{A}}(x) = x^k + \alpha_{k-1}x^{k-1} + \cdots + \alpha_1 x + \alpha_0$ be the characteristic polynomial for $\mathbf{A}$. From algebra, we know that $p_{\mathbf{A}}(x)$ can be factored into a product of linear terms and irreducible (nonfactorable) quadratic terms. Since $k \times k$ matrices follow the same laws of algebra as real numbers, with the exception of the commutative law for multiplication, and since $\mathbf{A}$ commutes with itself and $\mathbf{I}_k$, it follows that the polynomial $p_{\mathbf{A}}(\mathbf{A}) = \mathbf{A}^k + \alpha_{k-1}\mathbf{A}^{k-1} + \cdots + \alpha_1\mathbf{A} + \alpha_0\mathbf{I}_k$ can also be factored into linear and irreducible quadratic factors. Hence, $p_{\mathbf{A}}(\mathbf{A}) = \mathbf{F}_1\mathbf{F}_2 \cdots \mathbf{F}_j$, where each factor $\mathbf{F}_i$ is either of the form $a_i\mathbf{A} + b_i\mathbf{I}_k$, with $a_i \neq 0$, or of the form $a_i\mathbf{A}^2 + b_i\mathbf{A} + c_i\mathbf{I}_k$, with $a_i \neq 0$ and $b_i^2 - 4a_ic_i < 0$ (since the latter condition makes this quadratic irreducible).

Now, by the Cayley-Hamilton Theorem, $p_{\mathbf{A}}(\mathbf{A}) = \mathbf{F}_1\mathbf{F}_2 \cdots \mathbf{F}_j = \mathbf{O}_k$. Hence, the determinant $\left|\mathbf{F}_1\mathbf{F}_2 \cdots \mathbf{F}_j\right| = 0$. Since $\left|\mathbf{F}_1\mathbf{F}_2 \cdots \mathbf{F}_j\right| = |\mathbf{F}_1||\mathbf{F}_2| \cdots |\mathbf{F}_j|$, some $\mathbf{F}_i$ must have a zero determinant. There are two possible cases.

Case 1: Suppose that $\mathbf{F}_i = a_i\mathbf{A} + b_i\mathbf{I}_k$. Since $|a_i\mathbf{A} + b_i\mathbf{I}_k| = 0$, there is a nonzero vector $\mathbf{u}$ with $(a_i\mathbf{A} + b_i\mathbf{I}_k)\mathbf{u} = \mathbf{0}$. Thus, $\mathbf{A}\mathbf{u} = -(b_i/a_i)\mathbf{u}$, and $\mathbf{u}$ is an eigenvector for $\mathbf{A}$ with eigenvalue $-b_i/a_i$. Hence, $-b_i/a_i$ is an eigenvalue for L.

Case 2: Suppose that $\mathbf{F}_i = a_i\mathbf{A}^2 + b_i\mathbf{A} + c_i\mathbf{I}_k$. We show that this case cannot occur by exhibiting a contradiction. As in Case 1, there is a nonzero vector $\mathbf{u}$ with $(a_i\mathbf{A}^2 + b_i\mathbf{A} + c_i\mathbf{I}_k)\mathbf{u} = \mathbf{0}$. Completing the square yields

$$a_i\left(\left(\mathbf{A} + \frac{b_i}{2a_i}\mathbf{I}_k\right)^2 - \left(\frac{b_i^2 - 4a_ic_i}{4a_i^2}\right)\mathbf{I}_k\right)\mathbf{u} = \mathbf{0}.$$

Let $\mathbf{C} = \mathbf{A} + (b_i/2a_i)\mathbf{I}_k$ and $d = -(b_i^2 - 4a_i c_i)/4a_i^2$. Then $\mathbf{C}$ is a symmetric matrix since it is the sum of symmetric matrices, and $d > 0$ since $b_i^2 - 4a_i c_i < 0$. These substitutions simplify the preceding equation to $a_i(\mathbf{C}^2 + d\mathbf{I}_k)\mathbf{u} = \mathbf{0}$, or $\mathbf{C}^2\mathbf{u} = -d\mathbf{u}$. Thus,

$$
\begin{aligned}
0 &\leq (\mathbf{Cu}) \cdot (\mathbf{Cu}) \\
&= \mathbf{u} \cdot (\mathbf{C}^2\mathbf{u}), && \text{since } \mathbf{C} \text{ is symmetric} \\
&= \mathbf{u} \cdot (-d\mathbf{u}) \\
&= -d(\mathbf{u} \cdot \mathbf{u}) < 0, && \text{since } d > 0 \text{ and } \mathbf{u} \neq \mathbf{0}.
\end{aligned}
$$

However, $0 < 0$ is a contradiction. Hence, Case 2 cannot occur. ∎

EXAMPLE 2 The operator $L([a, b, c]) = [b, a, -c]$ on $\mathbb{R}^3$ is symmetric, as shown in Example 1. Lemma 6.18 then states that L has at least one eigenvalue. In fact, L has two eigenvalues, which are $\lambda_1 = 1$ and $\lambda_2 = -1$. The eigenspaces E_{λ_1} and E_{λ_2} have bases $\{[1, 1, 0]\}$ and $\{[1, -1, 0], [0, 0, 1]\}$, respectively. ∎

Orthogonally Diagonalizable Operators

We know that a linear operator L on a finite dimensional vector space $\mathcal{V}$ can be diagonalized if we can find a basis for $\mathcal{V}$ consisting of eigenvectors for L. We now examine the special case where the basis of eigenvectors is *orthonormal*.

DEFINITION

Let $\mathcal{V}$ be a nontrivial subspace of $\mathbb{R}^n$, and let $L: \mathcal{V} \to \mathcal{V}$ be a linear operator. Then L is an **orthogonally diagonalizable operator** if and only if there is an ordered orthonormal basis B for $\mathcal{V}$ such that the matrix for L with respect to B is a diagonal matrix.

A $k \times k$ matrix $\mathbf{A}$ is an **orthogonally diagonalizable matrix** if and only if there is an orthogonal matrix $\mathbf{P}$ such that $\mathbf{D} = \mathbf{P}^{-1}\mathbf{AP}$ is a diagonal matrix.

These two definitions are related. In fact, L is an orthogonally diagonalizable operator if and only if the matrix for L with respect to any orthonormal basis is orthogonally diagonalizable. To see this, suppose L is an orthogonally diagonalizable operator on a nontrivial subspace $\mathcal{V}$ of $\mathbb{R}^n$, and B is an orthonormal basis such that the matrix for L with respect to B is $\mathbf{D}$, a diagonal matrix. By a generalization of Theorem 6.7 (see Exercise 20 in Section 6.1), the transition matrix $\mathbf{P}$ between B and any other orthonormal basis C for $\mathcal{V}$ is orthogonal. Then if $\mathbf{A}$ is the matrix for L with respect to C, we have $\mathbf{D} = \mathbf{P}^{-1}\mathbf{AP}$, and thus $\mathbf{A}$ is orthogonally diagonalizable. By reversing this reasoning, we see that the converse is also true.

Equivalence of Symmetric and Orthogonally Diagonalizable Operators

We are now ready to show that symmetric operators and orthogonally diagonalizable operators are really the same.

THEOREM 6.19

Let V be a nontrivial subspace of $\mathbb{R}^n$, and let L be a linear operator on V. Then L is orthogonally diagonalizable if and only if L is symmetric.

PROOF OF THEOREM 6.19

Suppose that L is a linear operator on a nontrivial subspace V of $\mathbb{R}^n$.

First, we show that if L is orthogonally diagonalizable, then L is symmetric. Suppose L is orthogonally diagonalizable. Then, by definition, there is an ordered orthonormal basis B for V such that the matrix representation $\mathbf{A}$ for L with respect to B is diagonal. Since every diagonal matrix is also symmetric, L is a symmetric operator by Theorem 6.17.

To finish the proof, we must show that if L is a symmetric operator, then L is orthogonally diagonalizable. Suppose L is symmetric. If L has an ordered orthonormal basis B consisting entirely of eigenvectors of L, then, clearly, the matrix for L with respect to B is a diagonal matrix (having the eigenvalues corresponding to the eigenvectors in B along its main diagonal), and then L is orthogonally diagonalizable. Therefore, our goal is to find an orthonormal basis of eigenvectors for L. We give a proof by induction on $\dim(V)$.

Base Step: Assume that $\dim(V) = 1$. Normalize any nonzero vector in V to obtain a unit vector $\mathbf{u} \in V$. Then, $\{\mathbf{u}\}$ is an orthonormal basis for V. Since $L(\mathbf{u}) \in V$ and $\{\mathbf{u}\}$ is a basis for V, we must have $L(\mathbf{u}) = \lambda\mathbf{u}$, for some real number λ, and so λ is an eigenvalue for L. Hence, $\{\mathbf{u}\}$ is an orthonormal basis of eigenvectors for V, thus completing the Base Step.

Inductive Step: The inductive hypothesis is as follows:

If W is a subspace of $\mathbb{R}^n$ with dimension k, and T is any symmetric operator on W, then W has an orthonormal basis of eigenvectors for T.

We must prove the following:

If V is a subspace of $\mathbb{R}^n$ with dimension $k + 1$, and L is a symmetric operator on V, then V has an orthonormal basis of eigenvectors for L.

Now, L has at least one eigenvalue λ, by Lemma 6.18. Take any eigenvector for L corresponding to λ and normalize it to create a unit eigenvector $\mathbf{v}$. Let $\mathcal{Y} = \text{span}(\{\mathbf{v}\})$. Now, we want to enlarge $\{\mathbf{v}\}$ to an orthonormal basis of eigenvectors for L in V.

Our goal is to find a subspace $\mathcal{W}$ of $\mathcal{V}$ of dimension k that is orthogonal to $\mathcal{Y}$, together with a symmetric operator on $\mathcal{W}$. We can then invoke the inductive hypothesis to find the remaining orthonormal basis vectors for L.

Since $\dim(\mathcal{V}) = k+1$, we can use the Gram-Schmidt Process to find vectors $\mathbf{v}_1, \ldots, \mathbf{v}_k$ such that $\{\mathbf{v}, \mathbf{v}_1, \ldots, \mathbf{v}_k\}$ is an orthonormal basis for $\mathcal{V}$ containing $\mathbf{v}$. Since $\mathbf{v}_1, \ldots, \mathbf{v}_k$ are orthogonal to $\mathbf{v}$, we have $\{\mathbf{v}_1, \ldots, \mathbf{v}_k\} \subseteq \mathcal{Y}^\perp \cap \mathcal{V}$, the orthogonal complement of $\mathcal{Y}$ in $\mathcal{V}$ (see Exercise 24 in Section 6.2). Let $\mathcal{W} = \mathcal{Y}^\perp \cap \mathcal{V}$. Since $\{\mathbf{v}_1, \ldots, \mathbf{v}_k\}$ is a linearly independent subset of $\mathcal{W}$, $\dim(\mathcal{W}) \geq k$. But $\mathbf{v} \notin \mathcal{W}$ implies $\dim(\mathcal{W}) < \dim(\mathcal{V}) = k+1$, and so $\dim(\mathcal{W}) = k$.

Next, we claim that for every $\mathbf{w} \in \mathcal{W}$, we have $L(\mathbf{w}) \in \mathcal{W}$. For,

$$
\begin{aligned}
\mathbf{v} \cdot L(\mathbf{w}) &= L(\mathbf{v}) \cdot \mathbf{w} && \text{since } L \text{ is symmetric} \\
&= (\lambda \mathbf{v}) \cdot \mathbf{w} && \text{since } \lambda \text{ is an eigenvalue for } L \\
&= \lambda(\mathbf{v} \cdot \mathbf{w}) = \lambda(0) = 0,
\end{aligned}
$$

which shows that $L(\mathbf{w})$ is orthogonal to $\mathbf{v}$ and hence is in $\mathcal{W}$. Therefore, we can define a linear operator $T \colon \mathcal{W} \to \mathcal{W}$ by $T(\mathbf{w}) = L(\mathbf{w})$. ($T$ is the **restriction** of L to $\mathcal{W}$.) Now, T is a symmetric operator on $\mathcal{W}$ since, for every $\mathbf{w}_1, \mathbf{w}_2 \in \mathcal{W}$,

$$
\begin{aligned}
T(\mathbf{w}_1) \cdot \mathbf{w}_2 &= L(\mathbf{w}_1) \cdot \mathbf{w}_2 && \text{definition of } T \\
&= \mathbf{w}_1 \cdot L(\mathbf{w}_2) && \text{since } L \text{ is symmetric} \\
&= \mathbf{w}_1 \cdot T(\mathbf{w}_2). && \text{definition of } T.
\end{aligned}
$$

Since $\dim(\mathcal{W}) = k$, the inductive hypothesis implies that $\mathcal{W}$ has an orthonormal basis $\{\mathbf{u}_1, \ldots, \mathbf{u}_k\}$ of eigenvectors for T. Then, by definition of T, $\{\mathbf{u}_1, \ldots, \mathbf{u}_k\}$ is also a set of eigenvectors for L, all of which are orthogonal to $\mathbf{v}$ (since they are in $\mathcal{W}$). Hence, $B = \{\mathbf{v}, \mathbf{u}_1, \ldots, \mathbf{u}_k\}$ is an orthonormal basis for $\mathcal{V}$ of eigenvectors for L, and we have finished the proof of the Inductive Step. ∎

Method for Orthogonally Diagonalizing a Linear Operator

We now present a method for orthogonally diagonalizing a symmetric operator, based on Theorem 6.19. You should compare this to the method for diagonalizing a linear operator given in Section 5.5. Notice that the following method assumes that eigenvectors of a symmetric operator corresponding to distinct eigenvalues are orthogonal. The proof of this is left as Exercise 11.

METHOD FOR ORTHOGONALLY DIAGONALIZING A SYMMETRIC OPERATOR (ORTHOGONAL DIAGONALIZATION METHOD)

Let $L: \mathcal{V} \to \mathcal{V}$ be a symmetric operator on a subspace $\mathcal{V}$ of $\mathbb{R}^n$, with $\dim(\mathcal{V}) = k$.

Step 1: Find an ordered orthonormal basis C for $\mathcal{V}$ (if $\mathcal{V} = \mathbb{R}^n$, we can use the standard basis), and calculate the matrix representation $\mathbf{A}$ for L with respect to C (which should be a $k \times k$ symmetric matrix).

Step 2: (a) Apply the Diagonalization Method of Section 3.4 to $\mathbf{A}$ in order to obtain all of the eigenvalues $\lambda_1, \ldots, \lambda_k$ of $\mathbf{A}$, and a basis in $\mathbb{R}^k$ for each eigenspace E_{λ_i} of $\mathbf{A}$ (by solving an appropriate homogeneous system if necessary).

(b) Perform the Gram-Schmidt Process on the basis for each E_{λ_i} from Step 2(a), and then normalize to get an orthonormal basis for each E_{λ_i}.

(c) Let $Z = (\mathbf{z}_1 \ldots, \mathbf{z}_k)$ be an ordered basis for $\mathbb{R}^k$ consisting of the union of the orthonormal bases for the E_{λ_i}.

Step 3: Reverse the C-coordinatization isomorphism on the vectors in Z to obtain an ordered orthonormal basis $B = (\mathbf{v}_1 \ldots, \mathbf{v}_k)$ for $\mathcal{V}$; that is, $[\mathbf{v}_i]_C = \mathbf{z}_i$.

The matrix representation for L with respect to B is the diagonal matrix $\mathbf{D}$, where d_{ii} is the eigenvalue for L corresponding to $\mathbf{v}_i$. In most practical situations, the transition matrix $\mathbf{P}$ from B- to C-coordinates is useful. $\mathbf{P}$ is the $k \times k$ matrix whose columns are $[\mathbf{v}_1]_C, \ldots, [\mathbf{v}_k]_C$, that is, the vectors $\mathbf{z}_1, \ldots, \mathbf{z}_k$ in Z. Note that $\mathbf{P}$ is an orthogonal matrix, and $\mathbf{D} = \mathbf{P}^{-1}\mathbf{A}\mathbf{P} = \mathbf{P}^T\mathbf{A}\mathbf{P}$.

The following example illustrates this method.

EXAMPLE 3   Consider the operator $L: \mathbb{R}^4 \to \mathbb{R}^4$ given by $L(\mathbf{v}) = \mathbf{A}\mathbf{v}$, where

$$\mathbf{A} = \frac{1}{7}\begin{bmatrix} 15 & -21 & -3 & -5 \\ -21 & 35 & -7 & 0 \\ -3 & -7 & 23 & 15 \\ -5 & 0 & 15 & 39 \end{bmatrix}.$$

L is clearly symmetric, since its matrix $\mathbf{A}$ with respect to the standard basis C for $\mathbb{R}^4$ is symmetric. We find an orthonormal basis B such that the matrix for L with respect to B is diagonal.

Step 1: We have already seen that $\mathbf{A}$ is the matrix for L with respect to the standard basis C for $\mathbb{R}^4$.

Step 2: (a) A lengthy calculation yields

$$p_{\mathbf{A}}(x) = x^4 - 16x^3 + 77x^2 - 98x = x(x-2)(x-7)^2,$$

giving eigenvalues $\lambda_1 = 0$, $\lambda_2 = 2$, and $\lambda_3 = 7$. Solving the appropriate homogeneous systems to find bases for the eigenspaces produces

$$\text{Basis for } E_{\lambda_1} = \{[3, 2, 1, 0]\}$$
$$\text{Basis for } E_{\lambda_2} = \{[1, 0, -3, 2]\}$$
$$\text{Basis for } E_{\lambda_3} = \{[-2, 3, 0, 1], [3, -5, 1, 0]\}.$$

(b) There is no need to perform the Gram-Schmidt Process on the bases for E_{λ_1} and E_{λ_2}, since each of these eigenspaces is one-dimensional. Normalizing the basis vectors yields

$$\text{Orthonormal basis for } E_{\lambda_1} = \left\{ \frac{1}{\sqrt{14}}[3, 2, 1, 0] \right\}$$

$$\text{Orthonormal basis for } E_{\lambda_2} = \left\{ \frac{1}{\sqrt{14}}[1, 0, -3, 2] \right\}.$$

Let us label the vectors in these bases as $\mathbf{z}_1$, $\mathbf{z}_2$, respectively. However, we must perform the Gram-Schmidt Process on the basis for E_{λ_3}. Let $\mathbf{w}_1 = \mathbf{v}_1 = [-2, 3, 0, 1]$ and $\mathbf{w}_2 = [3, -5, 1, 0]$. Then

$$\mathbf{v}_2 = [3, -5, 1, 0] - \left(\frac{[3, -5, 1, 0] \cdot [-2, 3, 0, 1]}{[-2, 3, 0, 1] \cdot [-2, 3, 0, 1]} \right) [-2, 3, 0, 1] = \left[0, -\frac{1}{2}, 1, \frac{3}{2} \right].$$

Finally, normalizing $\mathbf{v}_1$ and $\mathbf{v}_2$, we obtain

$$\text{Orthonormal basis for } E_{\lambda_3} = \left\{ \frac{1}{\sqrt{14}}[-2, 3, 0, 1], \frac{1}{\sqrt{14}}[0, -1, 2, 3] \right\}.$$

Let us label the vectors in this basis as $\mathbf{z}_3$, $\mathbf{z}_4$, respectively.

(c) We let $Z = (\mathbf{z}_1, \mathbf{z}_2, \mathbf{z}_3, \mathbf{z}_4) =$

$$\left(\frac{1}{\sqrt{14}}[3, 2, 1, 0], \frac{1}{\sqrt{14}}[1, 0, -3, 2], \frac{1}{\sqrt{14}}[-2, 3, 0, 1], \frac{1}{\sqrt{14}}[0, -1, 2, 3] \right)$$

be the union of the orthonormal bases for E_{λ_1}, E_{λ_2}, and E_{λ_3}.

Step 3: Since C is the standard basis for $\mathbb{R}^4$, the C-coordinatization isomorphism is the identity mapping, so $\mathbf{v}_1 = \mathbf{z}_1$, $\mathbf{v}_2 = \mathbf{z}_2$, $\mathbf{v}_3 = \mathbf{z}_3$, and $\mathbf{v}_4 = \mathbf{z}_4$ here, and $B = (\mathbf{v}_1, \mathbf{v}_2, \mathbf{v}_3, \mathbf{v}_4)$ is an ordered orthonormal basis for $\mathbb{R}^4$. The matrix representation $\mathbf{D}$ of L with respect to B is

$$\mathbf{D} = \begin{bmatrix} \lambda_1 & 0 & 0 & 0 \\ 0 & \lambda_2 & 0 & 0 \\ 0 & 0 & \lambda_3 & 0 \\ 0 & 0 & 0 & \lambda_3 \end{bmatrix} = \begin{bmatrix} 0 & 0 & 0 & 0 \\ 0 & 2 & 0 & 0 \\ 0 & 0 & 7 & 0 \\ 0 & 0 & 0 & 7 \end{bmatrix}.$$

The transition matrix $\mathbf{P}$ from B to C is the *orthogonal* matrix

$$\mathbf{P} = \frac{1}{\sqrt{14}} \begin{bmatrix} 3 & 1 & -2 & 0 \\ 2 & 0 & 3 & -1 \\ 1 & -3 & 0 & 2 \\ 0 & 2 & 1 & 3 \end{bmatrix}.$$

You can verify that $\mathbf{P}^{-1}\mathbf{A}\mathbf{P} = \mathbf{P}^T\mathbf{A}\mathbf{P} = \mathbf{D}$. ∎

We conclude by examining a symmetric operator whose domain is a proper subspace of $\mathbb{R}^n$.

EXAMPLE 4 Consider the operators L_1, L_2, and L_3 on $\mathbb{R}^3$ given by

L_1: orthogonal projection onto the plane $x + y + z = 0$
L_2: orthogonal projection onto the plane $x + y - z = 0$
L_3: orthogonal projection onto the xy-plane (that is, $z = 0$).

Let $L: \mathbb{R}^3 \to \mathbb{R}^3$ be given by $L = L_3 \circ L_2 \circ L_1$, and let $\mathcal{V}$ be the xy-plane in $\mathbb{R}^3$. Then, since range$(L_3) = \mathcal{V}$, we see that range$(L) \subseteq \mathcal{V}$. Thus, restricting the domain of L to $\mathcal{V}$, we can think of L as a linear operator on $\mathcal{V}$. We will show that L is a symmetric operator on $\mathcal{V}$ and orthogonally diagonalize L.

Step 1: Choose $C = ([1, 0, 0], [0, 1, 0])$ as an ordered orthonormal basis for $\mathcal{V}$. We need to calculate the matrix representation $\mathbf{A}$ of L with respect to C. Using the orthonormal basis $\left\{ \frac{1}{\sqrt{2}}[1, -1, 0], \frac{1}{\sqrt{6}}[1, 1, -2] \right\}$ for the plane $x + y + z = 0$, the orthonormal basis $\left\{ \frac{1}{\sqrt{2}}[1, -1, 0], \frac{1}{\sqrt{6}}[1, 1, 2] \right\}$ for the plane $x + y - z = 0$, and the orthonormal basis C for the xy-plane, we can use the method of Example 7 in Section 6.2 to compute the required orthogonal projections.

$$L([1, 0, 0]) = L_3(L_2(L_1([1, 0, 0]))) = L_3\left(L_2\left(\left[\frac{2}{3}, -\frac{1}{3}, -\frac{1}{3}\right]\right)\right)$$

$$= L_3\left(\left[\frac{4}{9}, -\frac{5}{9}, -\frac{1}{9}\right]\right) = \frac{1}{9}[4, -5, 0]$$

$$\text{and} \quad L([0, 1, 0]) = L_3(L_2(L_1([0, 1, 0]))) = L_3\left(L_2\left(\left[-\frac{1}{3}, \frac{2}{3}, -\frac{1}{3}\right]\right)\right)$$

$$= L_3\left(\left[-\frac{5}{9}, \frac{4}{9}, -\frac{1}{9}\right]\right) = \frac{1}{9}[-5, 4, 0].$$

Expressing these vectors in C-coordinates, we see that the matrix representation of L with respect to C is $\mathbf{A} = \frac{1}{9}\begin{bmatrix} 4 & -5 \\ -5 & 4 \end{bmatrix}$, a symmetric matrix. Thus, by Theorem 6.17, L is a symmetric operator on $\mathcal{V}$.[3] Hence, L is, indeed, orthogonally diagonalizable.

Step 2: (a) The characteristic polynomial for $\mathbf{A}$ is $p_{\mathbf{A}}(x) = x^2 - \frac{8}{9}x - \frac{1}{9} = (x - 1)\left(x + \frac{1}{9}\right)$, giving eigenvalues $\lambda_1 = 1$ and $\lambda_2 = -\frac{1}{9}$. Solving the appropriate homogeneous systems to find bases for these eigenspaces yields

Basis for $E_{\lambda_1} = \{[1, -1]\}$, Basis for $E_{\lambda_2} = \{[1, 1]\}$.

Notice that we expressed the bases in C-coordinates.

---

[3] You can easily verify that L is *not* a symmetric operator on all of $\mathbb{R}^3$, even though it is symmetric on the subspace $\mathcal{V}$.

(b) Since the eigenspaces are one-dimensional, there is no need to perform the Gram-Schmidt Process on the bases for E_{λ_1} and E_{λ_2}. Normalizing the basis vectors produces

$$\text{Orthonormal basis for } E_{\lambda_1} = \left\{ \frac{1}{\sqrt{2}}[1, -1] \right\}$$

$$\text{Orthonormal basis for } E_{\lambda_2} = \left\{ \frac{1}{\sqrt{2}}[1, 1] \right\}.$$

Let us denote these vectors as $\mathbf{z}_1$, $\mathbf{z}_2$, respectively.

(c) Let $Z = (\mathbf{z}_1, \mathbf{z}_2)$ be the union of the (ordered) orthonormal bases for E_{λ_1} and E_{λ_2}.

Step 3: Reversing the C-coordinatization isomorphism on Z, we obtain $\mathbf{v}_1 = \frac{1}{\sqrt{2}}[1, -1, 0]$, and $\mathbf{v}_2 = \frac{1}{\sqrt{2}}[1, 1, 0]$, respectively. Thus, an ordered orthonormal basis in $\mathbb{R}^3$ for $\mathcal{V}$ is $B = (\mathbf{v}_1, \mathbf{v}_2)$. The matrix $\mathbf{D} = \begin{bmatrix} 1 & 0 \\ 0 & -\frac{1}{9} \end{bmatrix}$ is the matrix representation for L with respect to B. The transition matrix $\mathbf{P} = \left(1/\sqrt{2}\right) \begin{bmatrix} 1 & 1 \\ -1 & 1 \end{bmatrix}$ from B to C is the orthogonal matrix whose columns are the vectors in B expressed in C-coordinates. You can verify that $\mathbf{P}^{-1}\mathbf{AP} = \mathbf{P}^T\mathbf{AP} = \mathbf{D}$. ∎

♦ **Supplemental Material**: You have now covered the prerequisites for Section 7.4, "Orthogonality in $\mathbb{C}^n$;" Section 7.5, "Inner Product Spaces;" and Section 10.3, "Quadratic Forms."

♦ **Application**: You have now covered the prerequisites for Section 8.11, "Max-Min Problems in $\mathbb{R}^3$ and the Hessian Matrix."

Exercises for Section 6.3

Note: Use a calculator or computer (when needed) in solving for eigenvalues and eigenvectors and performing the Gram-Schmidt Process.

1. Determine which of the following linear operators are symmetric. Explain why each is, or is not, symmetric.

★(a) $L: \mathbb{R}^2 \to \mathbb{R}^2$ given by $L\left(\begin{bmatrix} x \\ y \end{bmatrix}\right) = \begin{bmatrix} 3x + 2y \\ 2x + 5y \end{bmatrix}$

(b) $L: \mathbb{R}^2 \to \mathbb{R}^2$ given by $L\left(\begin{bmatrix} x \\ y \end{bmatrix}\right) = \begin{bmatrix} 5x - 7y \\ 7x + 6y \end{bmatrix}$

(c) $L: \mathbb{R}^3 \to \mathbb{R}^3$ given by the orthogonal projection onto the plane $x + y + z = 0$

★(d) $L: \mathbb{R}^3 \to \mathbb{R}^3$ given by the orthogonal projection onto the plane $ax + by + cz = 0$

★(e) $L: \mathbb{R}^3 \to \mathbb{R}^3$ given by a counterclockwise rotation through an angle of $\frac{\pi}{3}$ radians about the line through the origin in the direction $[1, 1, -1]$

(f) $L: \mathbb{R}^3 \to \mathbb{R}^3$ given by the orthogonal reflection through the plane $4x - 3y + 5z = 0$

★(g) $L: \mathbb{R}^4 \to \mathbb{R}^4$ given by $L = L_1^{-1} \circ L_2 \circ L_1$ where $L_1: \mathbb{R}^4 \to \mathcal{M}_{22}$ is given by $L_1([a, b, c, d]) = \begin{bmatrix} a & b \\ c & d \end{bmatrix}$, and $L_2: \mathcal{M}_{22} \to \mathcal{M}_{22}$ is given by

$$L_2(\mathbf{K}) = \begin{bmatrix} 4 & 3 \\ 3 & 9 \end{bmatrix} \mathbf{K}$$

2. In each part, find a symmetric matrix having the given eigenvalues and the given bases for their associated eigenspaces.

   ★(a) $\lambda_1 = 1, \lambda_2 = -1, E_{\lambda_1} = \text{span}\left(\left\{\frac{1}{5}[3, 4]\right\}\right), E_{\lambda_2} = \text{span}\left(\left\{\frac{1}{5}[4, -3]\right\}\right)$

   (b) $\lambda_1 = 0, \lambda_2 = 1, \lambda_3 = 2, E_{\lambda_1} = \text{span}\left(\left\{\frac{1}{11}[6, 2, -9]\right\}\right), E_{\lambda_2} = \text{span}$
   $\left(\left\{\frac{1}{11}[7, 6, 6]\right\}\right), E_{\lambda_3} = \text{span}\left(\left\{\frac{1}{11}[6, -9, 2]\right\}\right)$

   (c) $\lambda_1 = 1, \lambda_2 = 2, E_{\lambda_1} = \text{span}(\{[6, 3, 2], [8, -3, 5]\}), E_{\lambda_2} = \text{span}$
   $(\{[3, -2, -6]\})$

   ★(d) $\lambda_1 = -1, \lambda_2 = 1, E_{\lambda_1} = \text{span}(\{[12, 3, 4, 0], [12, -1, 7, 12]\}), E_{\lambda_2} = \text{span}(\{[-3, 12, 0, 4], [-2, 24, -12, 11]\})$

3. In each part of this exercise, the matrix $\mathbf{A}$ with respect to the standard basis for a symmetric linear operator on $\mathbb{R}^n$ is given. Orthogonally diagonalize each operator by following Steps 2 and 3 of the method given in the text. Your answers should include the ordered orthonormal basis B, the orthogonal matrix $\mathbf{P}$, and the diagonal matrix $\mathbf{D}$. Check your work by verifying that $\mathbf{D} = \mathbf{P}^{-1}\mathbf{AP}$. (Hint: In (e), $p_\mathbf{A}(x) = (x - 2)^2(x + 3)(x - 5)$.)

   ★(a) $\mathbf{A} = \begin{bmatrix} 144 & -60 \\ -60 & 25 \end{bmatrix}$

   (b) $\mathbf{A} = \frac{1}{25}\begin{bmatrix} 39 & 48 \\ 48 & 11 \end{bmatrix}$

   ★(c) $\mathbf{A} = \frac{1}{9}\begin{bmatrix} 17 & 8 & -4 \\ 8 & 17 & -4 \\ -4 & -4 & 11 \end{bmatrix}$

   (d) $\mathbf{A} = \frac{1}{27}\begin{bmatrix} -13 & -40 & -16 \\ -40 & 176 & -124 \\ -16 & -124 & -1 \end{bmatrix}$

   ★(e) $\mathbf{A} = \frac{1}{14}\begin{bmatrix} 23 & 0 & 15 & -10 \\ 0 & 31 & -6 & -9 \\ 15 & -6 & -5 & 48 \\ -10 & -9 & 48 & 35 \end{bmatrix}$

   (f) $\mathbf{A} = \begin{bmatrix} 3 & 4 & 12 \\ 4 & -12 & 3 \\ 12 & 3 & -4 \end{bmatrix}$

   ★(g) $\mathbf{A} = \begin{bmatrix} 11 & 2 & -10 \\ 2 & 14 & 5 \\ -10 & 5 & -10 \end{bmatrix}$

4. In each part of this exercise, use the Orthogonal Diagonalization Method on the given symmetric linear operator L, defined on a subspace $\mathcal{V}$ of $\mathbb{R}^n$. Your answers should include the ordered orthonormal basis C for $\mathcal{V}$, the matrix $\mathbf{A}$ for L with respect to C, the ordered orthonormal basis B for $\mathcal{V}$, the orthogonal matrix $\mathbf{P}$, and the diagonal matrix $\mathbf{D}$. Check your work by verifying that $\mathbf{D} = \mathbf{P}^{-1}\mathbf{AP}$.

   ★(a) $L: \mathcal{V} \to \mathcal{V}$, where $\mathcal{V}$ is the plane $6x + 10y - 15z = 0$ in $\mathbb{R}^3$, $L([-10, 15, 6]) = [50, -18, 8]$, and $L([15, 6, 10]) = [-5, 36, 22]$

(b) $L: \mathcal{V} \to \mathcal{V}$, where $\mathcal{V}$ is the subspace of $\mathbb{R}^4$ spanned by $\{[1, -1, 1, 1], [-1, 1, 1, 1], [1, 1, 1, -1]\}$ and L is given by

$$L\left(\begin{bmatrix} w \\ x \\ y \\ z \end{bmatrix}\right) = \begin{bmatrix} 1 & -2 & 1 & 1 \\ -1 & 2 & 0 & 2 \\ 2 & 2 & 1 & 2 \\ 1 & 1 & 1 & -2 \end{bmatrix} \begin{bmatrix} w \\ x \\ y \\ z \end{bmatrix}$$

5. In each case, use orthogonal diagonalization to find a symmetric matrix **A** such that

   ⋆(a) $\mathbf{A}^3 = \dfrac{1}{25} \begin{bmatrix} 119 & -108 \\ -108 & 56 \end{bmatrix}$ (b) $\mathbf{A}^2 = \begin{bmatrix} 481 & -360 \\ -360 & 964 \end{bmatrix}$

   (c) $\mathbf{A}^2 = \begin{bmatrix} 17 & 16 & -16 \\ 16 & 41 & -32 \\ -16 & -32 & 41 \end{bmatrix}$

⋆**6.** Give an example of a 3×3 matrix that is diagonalizable but not orthogonally diagonalizable.

⋆**7.** Find the diagonal matrix **D** to which $\begin{bmatrix} a & b \\ b & c \end{bmatrix}$ is similar by an orthogonal change of coordinates. (Hint: Think! The full method for orthogonal diagonalization is not needed.)

8. Let L be a symmetric linear operator on a subspace $\mathcal{V}$ of $\mathbb{R}^n$.

   (a) If 1 is the only eigenvalue for L, prove that L is the identity operator.

   ⋆(b) What must be true about L if zero is its only eigenvalue? Prove it.

9. Let L_1 and L_2 be symmetric operators on $\mathbb{R}^n$. Prove that $L_2 \circ L_1$ is symmetric if and only if $L_2 \circ L_1 = L_1 \circ L_2$.

10. Two $n \times n$ matrices **A** and **B** are said to be **orthogonally similar** if and only if there is an orthogonal matrix **P** such that $\mathbf{A} = \mathbf{PBP}^{-1}$. Prove the following statements are equivalent for $n \times n$ symmetric matrices **A** and **B**:

    (i) **A** and **B** are similar.

    (ii) **A** and **B** have the same characteristic polynomial.

    (iii) **A** and **B** are orthogonally similar.

    (Hint: Show that (i) $\Rightarrow$ (ii) $\Rightarrow$ (iii) $\Rightarrow$ (i).)

11. Let L be a symmetric operator on a subspace $\mathcal{V}$ of $\mathbb{R}^n$. Suppose that λ_1 and λ_2 are distinct eigenvalues for L with corresponding eigenvectors $\mathbf{v}_1$ and $\mathbf{v}_2$. Prove that $\mathbf{v}_1 \perp \mathbf{v}_2$. (Hint: Use the definition of a symmetric operator to show that $(\lambda_2 - \lambda_1)(\mathbf{v}_1 \cdot \mathbf{v}_2) = 0$.)

12. Let **A** be an $n \times n$ symmetric matrix. Prove that **A** is orthogonal if and only if all eigenvalues for **A** are either 1 or -1. (Hint: For one half of the proof, use Theorem 6.8. For the other half, orthogonally diagonalize to help calculate $\mathbf{A}^2 = \mathbf{AA}^T$.)

⋆**13.** True or False:

   (a) If $\mathcal{V}$ is a nontrivial subspace of $\mathbb{R}^n$, a linear operator L on $\mathcal{V}$ with the property $\mathbf{v}_1 \cdot L(\mathbf{v}_2) = L(\mathbf{v}_1) \cdot \mathbf{v}_2$ for every $\mathbf{v}_1, \mathbf{v}_2 \in \mathcal{V}$ has at least one eigenvalue.

(b) A symmetric operator on a nontrivial subspace $\mathcal{V}$ of $\mathbb{R}^n$ has a symmetric matrix with respect to any ordered basis for $\mathcal{V}$.

(c) If a linear operator L on a nontrivial subspace $\mathcal{V}$ of $\mathbb{R}^n$ is symmetric, then the matrix for L with respect to any ordered orthonormal basis for $\mathcal{V}$ is symmetric.

(d) A linear operator L on a nontrivial subspace $\mathcal{V}$ of $\mathbb{R}^n$ is symmetric if and only if the matrix for L with respect to some ordered orthonormal basis for $\mathcal{V}$ is diagonal.

(e) Let L be a symmetric linear operator on a nontrivial subspace of $\mathbb{R}^n$ having matrix $\mathbf{A}$ with respect to an ordered orthonormal basis. In using the Orthogonal Diagonalization Method on L, the transition matrix $\mathbf{P}$ and the diagonal matrix $\mathbf{D}$ obtained from this process have the property that $\mathbf{A} = \mathbf{PDP}^T$.

(f) The orthogonal matrix $\mathbf{P}$ in the equation $\mathbf{D} = \mathbf{P}^{-1}\mathbf{AP}$ for a symmetric matrix $\mathbf{A}$ and diagonal matrix $\mathbf{D}$ is the transition matrix from some ordered orthonormal basis to standard coordinates.

Chapter 7

Complex Vector Spaces and General Inner Products

A Complex Situation

Until now, we have kept our theory of linear algebra within the real number system. But many practical mathematical problems, especially in physics and electronics, involve square roots of negative numbers (that is, complex numbers). For example, modern theories of heat transfer, fluid flow, damped harmonic oscillation, alternating current circuit theory, quantum mechanics, and relativity — all beyond the scope of this text — depend on the use of complex quantities. Therefore, our next goal is to extend many of our previous results to the realm of complex numbers.

One excellent reason for generalizing to the complex number system is that we can take advantage of the Fundamental Theorem of Algebra, which states that every nth degree polynomial can be factored completely when complex roots are permitted. In particular, we will see how this permits us to find additional (non-real) solutions to eigenvalue problems.

In this chapter, we extend many previous results to more complicated algebraic structures. In Section 7.1, we study $\mathbb{C}^n$, the set of complex n-vectors, and consider its similarities to and differences from $\mathbb{R}^n$. In Section 7.2, we examine properties of the eigenspaces of matrices with complex entries. Section 7.3 compares the properties of general complex vector spaces and linear transformations with their real counterparts. In Section 7.4, we study the complex analogs of the Gram-Schmidt Process and orthogonal matrices. Finally, in Section 7.5, we discuss inner product spaces, which possess an additional operation analogous to the dot product on $\mathbb{R}^n$.

Section 7.1 can be covered any time after finishing Section 1.5. Each remaining section of this chapter depends on those before it. In addition, Section 7.2 assumes Section 3.4 as a prerequisite, while Section 7.3 assumes Section 5.2 as a prerequisite. Finally, Sections 7.4 and 7.5 have Section 6.3 as a prerequisite. Section 7.5 could be covered without going through Sections 7.1 through 7.4 if attention is only paid to real inner products.

We use the complex number system throughout this chapter, and we assume you are familiar with its basic operations. For quick reference, Appendix C lists the definition of a complex number and the rules for complex addition, multiplication, conjugation, magnitude, and reciprocal.

7.1 Complex n-Vectors and Matrices

Prerequisite: Section 1.5, Matrix Multiplication

Until now, our scalars and entries in vectors and matrices have always been real numbers. In this section, however, we use the complex numbers to define and study complex n-vectors and matrices, emphasizing their differences with real vectors and matrices from Chapter 1.

Complex n-Vectors

DEFINITION

A **complex n-vector** is an ordered sequence (or ordered n-tuple) of n complex numbers. The set of all complex n-vectors is denoted by $\mathbb{C}^n$.

For example, $[3 - 2i, 4 + 3i, -i]$ is a vector in $\mathbb{C}^3$. We often write $\mathbf{z} = [z_1, z_2, \ldots, z_n]$ (where $z_1, z_2, \ldots, z_n \in \mathbb{C}$) to represent an arbitrary vector in $\mathbb{C}^n$.

For complex vectors, we usually need to extend our definition of **scalar** to include complex numbers instead of only real numbers. In what follows, it will always be clear from the context whether we are using complex scalars or real scalars.

Scalar multiplication and addition of complex vectors are defined coordinate-wise, just as for real vectors. For example, $(-2 + i)[4 + i, -1 - 2i] + [-3 - 2i, -2 + i] = [-9 + 2i, 4 + 3i] + [-3 - 2i, -2 + i] = [-12, 2 + 4i]$. You can verify that all the properties in Theorem 1.3 carry over to complex vectors (with real or complex scalars).

The **complex conjugate** of a vector $\mathbf{z} = [z_1, z_2, \ldots, z_n] \in \mathbb{C}^n$ is defined, using the complex conjugate operation, to be $\overline{\mathbf{z}} = [\overline{z_1}, \overline{z_2}, \ldots, \overline{z_n}]$. For example, if $\mathbf{z} = [3 - 2i, -5 - 4i, -2i]$, then $\overline{\mathbf{z}} = [3 + 2i, -5 + 4i, 2i]$.

We define the complex dot product of two vectors as follows:

DEFINITION

Let $\mathbf{z} = [z_1, z_2, \ldots, z_n]$ and $\mathbf{w} = [w_1, w_2, \ldots, w_n]$ be vectors in $\mathbb{C}^n$. The **complex dot (inner) product** of $\mathbf{z}$ and $\mathbf{w}$ is given by

$\mathbf{z} \cdot \mathbf{w} = z_1 \overline{w_1} + z_2 \overline{w_2} + \cdots + z_n \overline{w_n}.$

Notice that if $\mathbf{z}$ and $\mathbf{w}$ are both real vectors, then $\mathbf{z} \cdot \mathbf{w}$ is the familiar dot product in $\mathbb{R}^n$. The next example illustrates the complex dot product.

EXAMPLE 1

Let $\mathbf{z} = [3 - 2i, -2 + i, -4 - 3i]$ and $\mathbf{w} = [-2 + 4i, 5 - i, -2i]$. Then

$$
\begin{aligned}
\mathbf{z} \cdot \mathbf{w} &= (3 - 2i)\overline{(-2 + 4i)} + (-2 + i)\overline{(5 - i)} + (-4 - 3i)\overline{(-2i)} \\
&= (3 - 2i)(-2 - 4i) + (-2 + i)(5 + i) + (-4 - 3i)(+2i) \\
&= -19 - 13i.
\end{aligned}
$$

However,

$$
\begin{aligned}
\mathbf{w} \cdot \mathbf{z} &= (-2 + 4i)\overline{(3 - 2i)} + (5 - i)\overline{(-2 + i)} + (-2i)\overline{(-4 - 3i)} \\
&= -19 + 13i.
\end{aligned}
$$

Notice that $\mathbf{z} \cdot \mathbf{w} = \overline{\mathbf{w} \cdot \mathbf{z}}$. (This is true in general, as we will see shortly.) ∎

Now, if $\mathbf{z} = [z_1, \ldots, z_n]$, then $\mathbf{z} \cdot \mathbf{z} = z_1 \overline{z_1} + \cdots + z_n \overline{z_n} = |z_1|^2 + \cdots + |z_n|^2$, a nonnegative real number. We define the **length** of a complex vector $\mathbf{z} = [z_1, \ldots, z_n]$ as $\|\mathbf{z}\| = \sqrt{\mathbf{z} \cdot \mathbf{z}}$. For example, if $\mathbf{z} = [3 - i, -2i, 4 + 3i]$, then

$$
\|\mathbf{z}\| = \sqrt{(3 - i)(3 + i) + (-2i)(2i) + (4 + 3i)(4 - 3i)} = \sqrt{10 + 4 + 25} = \sqrt{39}.
$$

As with real *n*-vectors, a complex vector having length 1 is called a **unit vector**.

The following theorem lists the most important properties of the complex dot product. You are asked to prove parts of this theorem in Exercise 2. Notice the use of the complex conjugate in parts (1) and (5).

THEOREM 7.1

Let $\mathbf{z}_1$, $\mathbf{z}_2$, and $\mathbf{z}_3$ be vectors in $\mathbb{C}^n$, and let $k \in \mathbb{C}$ be any scalar. Then

(1) $\mathbf{z}_1 \cdot \mathbf{z}_2 = \overline{\mathbf{z}_2 \cdot \mathbf{z}_1}$ Conjugate-Commutativity of
                                        Complex Dot Product

(2) $\mathbf{z}_1 \cdot \mathbf{z}_1 = \|\mathbf{z}_1\|^2 \geq 0$ Relationships between Complex
(3) $\mathbf{z}_1 \cdot \mathbf{z}_1 = 0$ if and only if $\mathbf{z}_1 = \mathbf{0}$ Dot Product and Length

(4) $k(\mathbf{z}_1 \cdot \mathbf{z}_2) = (k\mathbf{z}_1) \cdot \mathbf{z}_2$ Relationships between Scalar
(5) $\overline{k}(\mathbf{z}_1 \cdot \mathbf{z}_2) = \mathbf{z}_1 \cdot (k\mathbf{z}_2)$ Multiplication and Complex
                                        Dot Product

(6) $\mathbf{z}_1 \cdot (\mathbf{z}_2 + \mathbf{z}_3) = (\mathbf{z}_1 \cdot \mathbf{z}_2) + (\mathbf{z}_1 \cdot \mathbf{z}_3)$ Distributive Laws of Complex
(7) $(\mathbf{z}_1 + \mathbf{z}_2) \cdot \mathbf{z}_3 = (\mathbf{z}_1 \cdot \mathbf{z}_3) + (\mathbf{z}_2 \cdot \mathbf{z}_3)$ Dot Product over Addition

---

Unfortunately, we cannot define the angle between two complex n-vectors as we did in Section 1.2 for real vectors, since the complex dot product is not necessarily a real number and hence $\frac{\mathbf{z} \cdot \mathbf{w}}{\|\mathbf{z}\| \, \|\mathbf{w}\|}$ does not always represent the cosine of an angle.

Complex Matrices

DEFINITION

An $m \times n$ **complex matrix** is a rectangular array of complex numbers arranged in m rows and n columns. The set of all $m \times n$ complex matrices is denoted as $\mathcal{M}_{mn}^{\mathbb{C}}$, or **complex** $\mathcal{M}_{mn}$.

Addition and scalar multiplication of matrices are defined entrywise in the usual manner, and the properties in Theorem 1.11 also hold for complex matrices.

We next define multiplication of complex matrices. Beware! Complex matrices are multiplied the same way as real matrices. We do not take complex conjugates of entries in the second matrix as we do with entries in the second vector for the complex dot product.

DEFINITION

If $\mathbf{Z}$ is an $m \times n$ matrix and $\mathbf{W}$ is an $n \times r$ matrix, then $\mathbf{ZW}$ is the $m \times r$ matrix whose (i, j) entry equals

$$(\mathbf{ZW})_{ij} = z_{i1}w_{1j} + z_{i2}w_{2j} + \cdots + z_{in}w_{nj}.$$

EXAMPLE 2   Let $\mathbf{Z} = \begin{bmatrix} 1-i & 2i & -2+i \\ -3i & 3-2i & -1-i \end{bmatrix}$ and $\mathbf{W} = \begin{bmatrix} -2i & 1-4i \\ -1+3i & 2-3i \\ -2+i & -4+i \end{bmatrix}$. Then

the $(1, 1)$ entry of $\mathbf{ZW}$ is

$$(1-i)(-2i) + (2i)(-1+3i) + (-2+i)(-2+i)$$
$$= -2i - 2 - 2i - 6 + 3 - 4i = -5 - 8i.$$

You can verify that the entire product is $\mathbf{ZW} = \begin{bmatrix} -5-8i & 10-7i \\ 12i & -7-13i \end{bmatrix}$. ∎

The familiar properties of matrix multiplication carry over to the complex case.

The **complex conjugate** $\overline{\mathbf{Z}}$ of a complex matrix $\mathbf{Z} = [z_{ij}]$ is the matrix whose (i, j) entry is $\overline{z_{ij}}$. The **transpose** $\mathbf{Z}^T$ of an $m \times n$ complex matrix $\mathbf{Z} = [z_{ij}]$ is the $n \times m$ matrix whose (j, i) entry is z_{ij}. You can verify that $(\overline{\mathbf{Z}})^T = \overline{(\mathbf{Z}^T)}$ for any complex matrix $\mathbf{Z}$, and so we can define the **conjugate transpose** $\mathbf{Z}^*$ of a complex matrix to be

$$\mathbf{Z}^* = \left(\overline{\mathbf{Z}} \right)^T = \overline{(\mathbf{Z}^T)}.$$

EXAMPLE 3   If $\mathbf{Z} = \begin{bmatrix} 2-3i & -i & 5 \\ 4i & 1+2i & -2-4i \end{bmatrix}$, then $\overline{\mathbf{Z}} = \begin{bmatrix} 2+3i & i & 5 \\ -4i & 1-2i & -2+4i \end{bmatrix}$, and

$$\mathbf{Z}^* = \left(\overline{\mathbf{Z}} \right)^T = \begin{bmatrix} 2+3i & -4i \\ i & 1-2i \\ 5 & -2+4i \end{bmatrix}.$$ ∎

The following theorem lists the most important properties of the complex conjugate and conjugate transpose operations:

THEOREM 7.2

Let $\mathbf{Z}$ and $\mathbf{Y}$ be $m \times n$ complex matrices, let $\mathbf{W}$ be an $n \times p$ complex matrix, and let $k \in \mathbb{C}$. Then

(1) $\overline{\left(\overline{\mathbf{Z}} \right)} = \mathbf{Z}$, and $(\mathbf{Z}^*)^* = \mathbf{Z}$
(2) $(\mathbf{Z} + \mathbf{Y})^* = \mathbf{Z}^* + \mathbf{Y}^*$
(3) $(k\mathbf{Z})^* = \overline{k}(\mathbf{Z}^*)$
(4) $\overline{\mathbf{ZW}} = \overline{\mathbf{Z}}\,\overline{\mathbf{W}}$
(5) $(\mathbf{ZW})^* = \mathbf{W}^*\mathbf{Z}^*$

Note the use of $\bar{k}$ in part (3). The proof of this theorem is straightforward, and parts of it are left as Exercise 4. We also have the following useful result:

THEOREM 7.3

If $\mathbf{A}$ is any $n \times n$ complex matrix and $\mathbf{z}$ and $\mathbf{w}$ are complex n-vectors, then $(\mathbf{Az}) \cdot \mathbf{w} = \mathbf{z} \cdot (\mathbf{A}^*\mathbf{w})$.

Compare the following proof of Theorem 7.3 with that of Theorem 6.8.

PROOF OF THEOREM 7.3

$$(\mathbf{Az}) \cdot \mathbf{w} = (\mathbf{Az})^T \overline{\mathbf{w}} = \mathbf{z}^T \mathbf{A}^T \overline{\mathbf{w}} = \mathbf{z}^T \overline{(\mathbf{A}^*\mathbf{w})} = \mathbf{z} \cdot (\mathbf{A}^*\mathbf{w}). \qquad \blacksquare$$

Hermitian, Skew-Hermitian, and Normal Matrices

Real symmetric and skew-symmetric matrices have complex analogs.

DEFINITION

Let $\mathbf{Z}$ be a square complex matrix. Then $\mathbf{Z}$ is **Hermitian** if and only if $\mathbf{Z}^* = \mathbf{Z}$, and $\mathbf{Z}$ is **skew-Hermitian** if and only if $\mathbf{Z}^* = -\mathbf{Z}$.

Notice that an $n \times n$ complex matrix $\mathbf{Z}$ is Hermitian if and only if $z_{ij} = \overline{z_{ji}}$, for $1 \le i, j \le n$. When $i = j$, we have $z_{ii} = \overline{z_{ii}}$ for all i, and so *all main diagonal entries of a Hermitian matrix are real*. Similarly, $\mathbf{Z}$ is skew-Hermitian if and only if $z_{ij} = -\overline{z_{ji}}$ for $1 \le i, j \le n$. When $i = j$, we have $z_{ii} = -\overline{z_{ii}}$ for all i, and so *all main diagonal entries of a skew-Hermitian matrix are pure imaginary*.

EXAMPLE 4   Consider the matrix

$$\mathbf{H} = \begin{bmatrix} 3 & 2-i & 1-2i \\ 2+i & -1 & -3i \\ 1+2i & 3i & 4 \end{bmatrix}.$$

Notice that

$$\overline{\mathbf{H}} = \begin{bmatrix} 3 & 2+i & 1+2i \\ 2-i & -1 & 3i \\ 1-2i & -3i & 4 \end{bmatrix}, \text{ and so } \mathbf{H}^* = \left(\overline{\mathbf{H}}\right)^T = \begin{bmatrix} 3 & 2-i & 1-2i \\ 2+i & -1 & -3i \\ 1+2i & 3i & 4 \end{bmatrix}.$$

Since $\mathbf{H}^* = \mathbf{H}$, $\mathbf{H}$ is Hermitian. Similarly, you can verify that the matrix

$$\mathbf{K} = \begin{bmatrix} -2i & 5+i & -1-3i \\ -5+i & i & 6 \\ 1-3i & -6 & 3i \end{bmatrix}$$

is skew-Hermitian. $\qquad \blacksquare$

Some other useful results concerning Hermitian and skew-Hermitian matrices are left for you to prove in Exercises 6, 7, and 8.

Another very important type of complex matrix is the following:

DEFINITION

A square complex matrix $\mathbf{Z}$ is **normal** if and only if $\mathbf{ZZ}^* = \mathbf{Z}^*\mathbf{Z}$.

The next theorem gives two important classes of normal matrices.

THEOREM 7.4

If $\mathbf{Z}$ is a Hermitian or skew-Hermitian matrix, then $\mathbf{Z}$ is normal.

The proof is left as Exercise 9. The next example gives a normal matrix that is neither Hermitian nor skew-Hermitian, thus illustrating that the converse to Theorem 7.4 is false.

EXAMPLE 5

Consider $\mathbf{Z} = \begin{bmatrix} 1 - 2i & -i \\ 1 & 2 - 3i \end{bmatrix}$. Now, $\mathbf{Z}^* = \begin{bmatrix} 1 + 2i & 1 \\ i & 2 + 3i \end{bmatrix}$, and so

$$\mathbf{ZZ}^* = \begin{bmatrix} 1 - 2i & -i \\ 1 & 2 - 3i \end{bmatrix} \begin{bmatrix} 1 + 2i & 1 \\ i & 2 + 3i \end{bmatrix} = \begin{bmatrix} 6 & 4 - 4i \\ 4 + 4i & 14 \end{bmatrix}.$$

Also, $\mathbf{Z}^*\mathbf{Z} = \begin{bmatrix} 1 + 2i & 1 \\ i & 2 + 3i \end{bmatrix} \begin{bmatrix} 1 - 2i & -i \\ 1 & 2 - 3i \end{bmatrix} = \begin{bmatrix} 6 & 4 - 4i \\ 4 + 4i & 14 \end{bmatrix}.$

Since $\mathbf{ZZ}^* = \mathbf{Z}^*\mathbf{Z}$, $\mathbf{Z}$ is normal. ∎

In Exercise 10 you are asked to prove that a matrix $\mathbf{Z}$ is normal if and only if $\mathbf{Z} = \mathbf{H}_1 + i\mathbf{H}_2$, where $\mathbf{H}_1$ and $\mathbf{H}_2$ are Hermitian matrices such that $\mathbf{H}_1\mathbf{H}_2 = \mathbf{H}_2\mathbf{H}_1$. For example, the normal matrix $\mathbf{Z}$ from Example 5 equals

$$\begin{bmatrix} 1 & \frac{1}{2} - \frac{1}{2}i \\ \frac{1}{2} + \frac{1}{2}i & 2 \end{bmatrix} + i \begin{bmatrix} -2 & -\frac{1}{2} + \frac{1}{2}i \\ -\frac{1}{2} - \frac{1}{2}i & -3 \end{bmatrix}.$$

Exercises for Section 7.1

1. Perform the following computations involving complex vectors.

   ⋆(a) $[2 + i, 3, -i] + [-1 + 3i, -2 + i, 6]$

   ⋆(b) $(-8 + 3i)[4i, 2 - 3i, -7 + i]$

   (c) $\overline{[5 - i, 2 + i, -3i]}$

   ⋆(d) $\overline{(-4)[6 - 3i, 7 - 2i, -8i]}$

   ⋆(e) $[-2 + i, 5 - 2i, 3 + 4i] \cdot [1 + i, 4 - 3i, -6i]$

   (f) $[5 + 2i, 6i, -2 + i] \cdot [3 - 6i, 8 + i, 1 - 4i]$

2. (a) Prove parts (1) and (2) of Theorem 7.1.

  (b) Prove part (5) of Theorem 7.1.

3. Perform the computations below with the following matrices:

$$\mathbf{A} = \begin{bmatrix} 2+5i & -4+i \\ -3-6i & 8-3i \end{bmatrix} \qquad \mathbf{B} = \begin{bmatrix} 9-i & -3i \\ 5+2i & 4+3i \end{bmatrix}$$

$$\mathbf{C} = \begin{bmatrix} 1+i & -2i & 6+4i \\ 0 & 3+i & 5 \\ -10i & 0 & 7-3i \end{bmatrix} \qquad \mathbf{D} = \begin{bmatrix} 5-i & -i & -3 \\ 2+3i & 0 & -4+i \end{bmatrix}$$

$\star$(a) $\mathbf{A} + \mathbf{B}$ (b) $\overline{\mathbf{C}}$

$\star$(c) $\mathbf{C}^*$ $\star$(d) $(-3i)\mathbf{D}$

  (e) $\mathbf{A} - \mathbf{B}^T$ $\star$(f) $\mathbf{AB}$

  (g) $\mathbf{D}\,(\mathbf{C}^*)$ (h) $\mathbf{B}^2$

$\star$(i) $\mathbf{C}^T\mathbf{D}^*$ (j) $(\mathbf{C}^*)^2$

4. (a) Prove part (3) of Theorem 7.2.

  ▶(b) Prove part (5) of Theorem 7.2.

$\star$**5.** Determine which of the following matrices are Hermitian or skew-Hermitian.

(a) $\begin{bmatrix} -4i & 6-2i & 8 \\ -6-2i & 0 & -2-i \\ -8 & 2-i & 5i \end{bmatrix}$ (b) $\begin{bmatrix} 2+3i & 6i & 1+i \\ -6i & 4 & 8-3i \\ 1-i & -8+3i & 5i \end{bmatrix}$

(c) $\begin{bmatrix} 2 & 0 & 0 \\ 0 & -3 & 0 \\ 0 & 0 & 4 \end{bmatrix}$ (d) $\begin{bmatrix} 5i & 0 & 0 \\ 0 & -2i & 0 \\ 0 & 0 & 6i \end{bmatrix}$

(e) $\begin{bmatrix} 2 & -2i & 2 \\ 2i & -2 & -2i \\ 2 & 2i & -2 \end{bmatrix}$

6. Let $\mathbf{Z}$ be any square complex matrix.

  (a) Prove that $\mathbf{H} = \frac{1}{2}(\mathbf{Z} + \mathbf{Z}^*)$ is a Hermitian matrix and $\mathbf{K} = \frac{1}{2}(\mathbf{Z} - \mathbf{Z}^*)$ is skew-Hermitian.

  (b) Prove that $\mathbf{Z}$ can be expressed uniquely as the sum of a Hermitian matrix $\mathbf{H}$ and a skew-Hermitian matrix $\mathbf{K}$. (Hint: Use part (a).)

7. Let $\mathbf{H}$ be an $n \times n$ Hermitian matrix.

  (a) Suppose $\mathbf{J}$ is an $n \times n$ Hermitian matrix. Prove that $\mathbf{HJ}$ is Hermitian if and only if $\mathbf{HJ} = \mathbf{JH}$.

  (b) Prove that $\mathbf{H}^k$ is Hermitian for all integers $k \geq 1$. (Hint: Use part (a) and a proof by induction.)

  (c) Prove that $\mathbf{P}^*\mathbf{HP}$ is Hermitian for any $n \times n$ complex matrix $\mathbf{P}$.

8. Prove that for any complex matrix $\mathbf{A}$, both $\mathbf{AA}^*$ and $\mathbf{A}^*\mathbf{A}$ are Hermitian.

▶**9.** Prove Theorem 7.4.

10. Let $\mathbf{Z}$ be a square complex matrix. Prove that $\mathbf{Z}$ is normal if and only if there exist two Hermitian matrices $\mathbf{H}_1$ and $\mathbf{H}_2$ such that $\mathbf{Z} = \mathbf{H}_1 + i\mathbf{H}_2$ and $\mathbf{H}_1\mathbf{H}_2 = \mathbf{H}_2\mathbf{H}_1$. (Hint: If $\mathbf{Z}$ is normal, let $\mathbf{H}_1 = (\mathbf{Z} + \mathbf{Z}^*)/2$.)

*11. True or False:
   (a) The dot product of two complex n-vectors is always a real number.
   (b) The (i, j) entry of the product $\mathbf{ZW}$ is the complex dot product of the ith row of $\mathbf{Z}$ with the jth column of $\mathbf{W}$.
   (c) The complex conjugate of the transpose of $\mathbf{Z}$ is equal to the transpose of the complex conjugate of $\mathbf{Z}$.
   (d) If $\mathbf{v}_1, \mathbf{v}_2 \in \mathbb{C}^n$ and $k \in \mathbb{C}$, then $k(\mathbf{v}_1 \cdot \mathbf{v}_2) = (k\mathbf{v}_1) \cdot \mathbf{v}_2 = \mathbf{v}_1 \cdot (k\mathbf{v}_2)$.
   (e) Every Hermitian matrix is symmetric.
   (f) The transpose of a skew-Hermitian matrix is normal.

7.2 Complex Eigenvalues and Complex Eigenvectors

Prerequisite: Section 3.4, Eigenvalues and Diagonalization

In this section, we consider row reduction and determinants using complex numbers and matrices and then extend the concept of eigenvalues and eigenvectors to complex $n \times n$ matrices.

Complex Linear Systems and Determinants

The Gaussian elimination and Gauss-Jordan row reduction methods are both used to solve complex systems of linear equations just as described in Sections 2.1 and 2.2 for real linear systems. However, the arithmetic involved is typically more tedious.

EXAMPLE 1   Let us solve the system

$$\begin{cases} (2 - 3i)w & + & (19 + 4i)z & = & -35 + 59i \\ (2 + i)w & + & (-4 + 13i)z & = & -40 - 30i \\ (1 - i)w & + & (9 + 6i)z & = & -32 + 25i \end{cases}$$

using Gaussian elimination. We begin with the augmented matrix

$$\left[\begin{array}{cc|c} 2 - 3i & 19 + 4i & -35 + 59i \\ 2 + i & -4 + 13i & -40 - 30i \\ 1 - i & 9 + 6i & -32 + 25i \end{array} \right].$$

Performing the row operations

$$\langle 1 \rangle \quad \leftarrow \quad \frac{1}{2 - 3i} \langle 1 \rangle, \quad \text{or,} \quad \langle 1 \rangle \leftarrow \left(\frac{2}{13} + \frac{3}{13}i \right) \langle 1 \rangle,$$

$$\langle 2 \rangle \quad \leftarrow \quad -(2 + i) \langle 1 \rangle + \langle 2 \rangle,$$

$$\text{and} \quad \langle 3 \rangle \quad \leftarrow \quad -(1 - i) \langle 1 \rangle + \langle 3 \rangle \qquad \qquad \text{yields}$$

$$\left[\begin{array}{cc|c} 1 & 2 + 5i & -19 + i \\ 0 & -3 + i & -1 - 13i \\ 0 & 2 + 3i & -14 + 5i \end{array} \right].$$

Continuing with

$$\langle 2 \rangle \quad \leftarrow \quad \frac{1}{-3+i}\,\langle 2 \rangle, \quad \text{or,} \quad \langle 2 \rangle \leftarrow \left(-\frac{3}{10} - \frac{1}{10}i\right)\langle 2 \rangle,$$

and $\langle 3 \rangle \quad \leftarrow \quad -(2+3i)\,\langle 2 \rangle + \langle 3 \rangle \qquad$ produces

$$\begin{bmatrix} 1 & 2+5i & -19+i \\ 0 & 1 & -1+4i \\ 0 & 0 & 0 \end{bmatrix}.$$

Hence,

$$w + (2+5i)z = -19 + i, \quad \text{and}$$
$$z = -1 + 4i.$$

Thus, $w = -19 + i - (2+5i)(-1+4i) = 3 - 2i$. Therefore, the unique solution to the system is $(w, z) = (3 - 2i, -1 + 4i)$.   ∎

All of our results for real matrices involving reduced row echelon form, rank, row spaces, homogeneous systems, and inverse matrices carry over to complex matrices. Similarly, determinants of complex matrices are computed in the same manner as for real matrices, and the following results, which we state without proof, are true:

THEOREM 7.5
Let **W** and **Z** be complex $n \times n$ matrices. Then
  (1) $|\mathbf{WZ}| = |\mathbf{W}||\mathbf{Z}|$
  (2) $|\mathbf{W}| = |\mathbf{W}^T|$
  (3) $|\overline{\mathbf{W}}| = |\mathbf{W}^*| = \overline{|\mathbf{W}|}$
  (4) $|\mathbf{W}| \neq 0$ iff **W** is nonsingular if and only if $\text{rank}(\mathbf{W}) = n$

In addition, all the equivalences in Table 3.1 also hold for complex $n \times n$ matrices.

Complex Eigenvalues and Complex Eigenvectors

If **A** is an $n \times n$ complex matrix, then $\lambda \in \mathbb{C}$ is an **eigenvalue** for **A** if and only if there is a nonzero vector $\mathbf{v} \in \mathbb{C}^n$ such that $\mathbf{Av} = \lambda\mathbf{v}$. As before, the nonzero vector **v** is called an **eigenvector** for **A** associated with λ. The **characteristic polynomial** of **A**, defined as $p_\mathbf{A}(x) = |x\mathbf{I}_n - \mathbf{A}|$, is used to find the eigenvalues of **A**, just as in Section 3.4.

EXAMPLE 2   For the matrix

$$\mathbf{A} = \begin{bmatrix} -4+7i & 2+i & 7+7i \\ 1-3i & 1-i & -3-i \\ 5+4i & 1-2i & 7-5i \end{bmatrix},$$

we have

$$x\mathbf{I}_3 - \mathbf{A} = \begin{bmatrix} x + 4 - 7i & -2 - i & -7 - 7i \\ -1 + 3i & x - 1 + i & 3 + i \\ -5 - 4i & -1 + 2i & x - 7 + 5i \end{bmatrix}.$$

After some calculation, you can verify that $p_\mathbf{A}(x) = |x\mathbf{I}_3 - \mathbf{A}| = x^3 - (4 + i)x^2 + (5 + 5i)x - (6 + 6i)$. You can also check that $p_\mathbf{A}(x)$ factors as $(x - (1 - i))(x - 2i)(x - 3)$. Hence, the eigenvalues of $\mathbf{A}$ are $\lambda_1 = 1 - i$, $\lambda_2 = 2i$, and $\lambda_3 = 3$. To find an eigenvector for λ_1, we look for a nontrivial solution $\mathbf{v}$ of the system $((1 - i)\mathbf{I}_3 - \mathbf{A})\mathbf{v} = \mathbf{0}$. Hence, we row reduce

$$\begin{bmatrix} 5 - 8i & -2 - i & -7 - 7i & | & 0 \\ -1 + 3i & 0 & 3 + i & | & 0 \\ -5 - 4i & -1 + 2i & -6 + 4i & | & 0 \end{bmatrix} \text{ to obtain } \begin{bmatrix} 1 & 0 & -i & | & 0 \\ 0 & 1 & i & | & 0 \\ 0 & 0 & 0 & | & 0 \end{bmatrix}.$$

Thus, we get the eigenvector $[i, -i, 1]$ corresponding to λ_1. A similar analysis shows that $[3i, -i, 2]$ is an eigenvector corresponding to λ_2, and $[i, 0, 1]$ is an eigenvector corresponding to λ_3. ∎

Diagonalizable Complex Matrices

We say a complex matrix $\mathbf{A}$ is **diagonalizable** if and only if there is a non-singular complex matrix $\mathbf{P}$ such that $\mathbf{P}^{-1}\mathbf{A}\mathbf{P} = \mathbf{D}$ is a diagonal matrix. Just as with real matrices, the matrix $\mathbf{P}$ has eigenvectors for $\mathbf{A}$ as its columns, and the diagonal matrix $\mathbf{D}$ has the eigenvalues for $\mathbf{A}$ on its main diagonal, with d_{ii} being an eigenvalue corresponding to the eigenvector that is the ith column of $\mathbf{P}$. The six step Method for diagonalizing a matrix given in Section 3.4 works just as well for complex matrices.

Algebraic Multiplicity of an Eigenvalue

The algebraic multiplicity of an eigenvalue of a complex matrix is defined just as for real matrices – that is, k is the algebraic multiplicity of an eigenvalue λ for a matrix $\mathbf{A}$ if and only if $(x - \lambda)^k$ is the highest power of $(x - \lambda)$ that divides $p_\mathbf{A}(x)$. However, an important property of complex polynomials makes the situation for complex matrices a bit different than for real matrices. In particular, the **Fundamental Theorem of Algebra** states that any complex polynomial of degree n factors into a product of n *linear* factors. Thus, for every $n \times n$ matrix $\mathbf{A}$, $p_\mathbf{A}(x)$ can be expressed as a product of n *linear* factors. Therefore, the algebraic multiplicities of the eigenvalues of $\mathbf{A}$ must add up to n. This eliminates one of the two reasons that some real matrices are not diagonalizable. However, there are still some complex matrices that are not diagonalizable, as we will see later in Example 4.

EXAMPLE 3   Consider the matrix

$$\mathbf{A} = \begin{bmatrix} \cos\theta & -\sin\theta \\ \sin\theta & \cos\theta \end{bmatrix}$$

from Example 7 in Section 3.4 for a fixed value of θ such that $\sin\theta \neq 0$. In that example, we computed $p_\mathbf{A}(x) = x^2 - 2(\cos\theta)x + 1$, which factors into complex linear factors as $p_\mathbf{A}(x) = (x - (\cos\theta + i\sin\theta))(x - (\cos\theta - i\sin\theta))$. Thus, the two complex eigenvalues for $\mathbf{A}$ are $\lambda_1 = \cos\theta + i\sin\theta$ and $\lambda_2 = \cos\theta - i\sin\theta$.[1] Row reducing $\lambda_1 \mathbf{I}_2 - \mathbf{A}$ yields $\begin{bmatrix} 1 & -i \\ 0 & 0 \end{bmatrix}$, thus giving the eigenvector $[i, 1]$. Similarly, row reducing $\lambda_2 \mathbf{I}_2 - \mathbf{A}$ produces the eigenvector $[-i, 1]$. Hence, $\mathbf{P} = \begin{bmatrix} i & -i \\ 1 & 1 \end{bmatrix}$. You can verify that

$$\begin{aligned} \mathbf{P}^{-1}\mathbf{A}\mathbf{P} &= \left(\frac{1}{2}\begin{bmatrix} -i & 1 \\ i & 1 \end{bmatrix}\right)\begin{bmatrix} \cos\theta & -\sin\theta \\ \sin\theta & \cos\theta \end{bmatrix}\begin{bmatrix} i & -i \\ 1 & 1 \end{bmatrix} \\ &= \begin{bmatrix} \cos\theta + i\sin\theta & 0 \\ 0 & \cos\theta - i\sin\theta \end{bmatrix} = \mathbf{D}. \end{aligned}$$

For example, if $\theta = \frac{\pi}{6}$, then $\mathbf{A} = \frac{1}{2}\begin{bmatrix} \sqrt{3} & -1 \\ 1 & \sqrt{3} \end{bmatrix}$ and $\mathbf{D} = \frac{1}{2}\begin{bmatrix} \sqrt{3}+i & 0 \\ 0 & \sqrt{3}-i \end{bmatrix}$. Note that the eigenvectors for $\mathbf{A}$ are independent of θ, and hence so is the matrix $\mathbf{P}$. However, $\mathbf{D}$ and the eigenvalues of $\mathbf{A}$ change as θ changes. ∎

This example illustrates how a real matrix could be diagonalizable when thought of as a complex matrix, even though it is not diagonalizable when considered as a real matrix.

Nondiagonalizable Complex Matrices

It is still possible for a complex matrix to be nondiagonalizable. This occurs whenever the number of eigenvectors for a given eigenvalue produced in Step 3 of the diagonalization process is less than the algebraic multiplicity of that eigenvalue.

EXAMPLE 4   Consider the matrix

$$\mathbf{A} = \begin{bmatrix} -3-15i & -6+25i & 43+18i \\ 2-2i & -4+i & 1+8i \\ 2-5i & -7+6i & 9+14i \end{bmatrix},$$

---

[1] We could have solved for λ_1 and λ_2 by using the quadratic formula instead of factoring.

whose characteristic polynomial is $p_{\mathbf{A}}(x) = x^3 - 2x^2 + x = x(x-1)^2$. The eigenvalue $\lambda_1 = 1$ has algebraic multiplicity 2. However, $\lambda_1 \mathbf{I}_2 - \mathbf{A} = \mathbf{I}_2 - \mathbf{A}$ row reduces to

$$
\begin{bmatrix}
1 & 0 & -\frac{3}{2} - \frac{7}{2}i \\
0 & 1 & \frac{1}{2} + \frac{1}{2}i \\
0 & 0 & 0
\end{bmatrix}.
$$

Hence, Step 3 produces only one eigenvector, namely $\left[\frac{3}{2} - \frac{7}{2}i, -\frac{1}{2} - \frac{1}{2}i, 1\right]$. Since the number of eigenvectors produced for λ_1 is less than the algebraic multiplicity of λ_1, $\mathbf{A}$ cannotbe diagonalized.   ∎

Exercises for Section 7.2

1. Give the complete solution set for each of the following complex linear systems:

   ⋆(a) $\begin{cases} (3+i)w & + & (5+5i)z & = & 29+33i \\ (1+i)w & + & (6-2i)z & = & 30-12i \end{cases}$

   (b) $\begin{cases} (1+2i)x & + & (-1+3i)y & + & (9+3i)z & = & 18+46i \\ (2+3i)x & + & (-1+5i)y & + & (15+5i)z & = & 30+76i \\ (5-2i)x & + & (7+3i)y & + & (11-20i)z & = & 120-25i \end{cases}$

   ⋆(c) $\begin{cases} 3ix & + & (-6+3i)y & + & (12+18i)z & = & -51+9i \\ (3+2i)x & + & (1+7i)y & + & (25-2i)z & = & -13+56i \\ (1+i)x & + & 2iy & + & (9+i)z & = & -7+17i \end{cases}$

   (d) $\begin{cases} (1+3i)w & + & 10iz & = & -46-38i \\ (4+2i)w & + & (12+13i)z & = & -111 \end{cases}$

   ⋆(e) $\begin{cases} (3-2i)w & + & (12+5i)z & = & 3+11i \\ (5+4i)w & + & (-2+23i)z & = & -14+15i \end{cases}$

   (f) $\begin{cases} (2-i)x & + & (1-3i)y & + & (21+2i)z & = & -14-13i \\ (1+2i)x & + & (6+2i)y & + & (3+46i)z & = & 24-27i \end{cases}$

2. In each part, compute the determinant of the given matrix $\mathbf{A}$, determine whether $\mathbf{A}$ is nonsingular, and then calculate $|\mathbf{A}^*|$ to verify that $|\mathbf{A}^*| = \overline{|\mathbf{A}|}$.

   (a) $\mathbf{A} = \begin{bmatrix} 2+i & -3+2i \\ 4-3i & 1+8i \end{bmatrix}$ ⋆(b) $\mathbf{A} = \begin{bmatrix} i & 2 & 5i \\ 1+i & 1-i & i \\ 4 & -2 & 2-i \end{bmatrix}$

   (c) $\mathbf{A} = \begin{bmatrix} 0 & i & 0 & 1 \\ -i & 0 & 0 & 0 \\ 0 & -1 & 2 & 1 \\ 1 & 0 & 3i & 4i \end{bmatrix}$

3. For each of the following matrices, find all eigenvalues and express each eigenspace as a set of linear combinations of particular eigenvectors:

★(a) $\begin{bmatrix} 4+3i & -1-3i \\ 8-2i & -5-2i \end{bmatrix}$
     (b) $\begin{bmatrix} 11 & 2 & -7 \\ 0 & 6 & -5 \\ 10 & 2 & -6 \end{bmatrix}$

★(c) $\begin{bmatrix} 4+3i & -4-2i & 4+7i \\ 2-4i & -2+5i & 7-4i \\ -4-2i & 4+2i & -4-6i \end{bmatrix}$
     (d) $\begin{bmatrix} -i & 2i & -1+2i \\ 1 & -1+i & -i \\ -2+i & 2-i & 3+2i \end{bmatrix}$

4. ★(a) Explain why the matrix $\mathbf{A}$ in part (a) of Exercise 3 is diagonalizable. Find a nonsingular $\mathbf{P}$ and diagonal $\mathbf{D}$ such that $\mathbf{P}^{-1}\mathbf{AP} = \mathbf{D}$.
   (b) Show that the matrix in part (d) of Exercise 3 is not diagonalizable.
   (c) Show that the matrix from part (b) of Exercise 3 is diagonalizable as a complex matrix, but not as a real matrix.

5. Give a convincing argument that if the algebraic multiplicity of every eigenvalue of a complex $n \times n$ matrix is 1, then the matrix is diagonalizable.

★6. True or False:
   (a) If $\mathbf{A}$ is a 4×4 complex matrix whose second row is i times its first row, then $|\mathbf{A}| = 0$.
   (b) The algebraic multiplicity of any eigenvalue of an $n \times n$ complex matrix must equal n.
   (c) Every real $n \times n$ matrix is diagonalizable when thought of as a complex matrix.
   (d) The Fundamental Theorem of Algebra guarantees that every nth degree complex polynomial has n distinct roots.

7.3 Complex Vector Spaces

Prerequisite: Section 5.2, The Matrix of a Linear Transformation

In this section, we examine complex vector spaces and their similarities and differences with real vector spaces. We also discuss linear transformations from one complex vector space to another.

Complex Vector Spaces

We define **complex vector spaces** exactly the same way that we defined real vector spaces in Section 4.1, except that the set of scalars is enlarged to allow the use of complex numbers rather than just real numbers. Naturally, $\mathbb{C}^n$ is an example (in fact, the most important one) of a complex vector space. Also, under regular addition and complex scalar multiplication, both $\mathcal{M}_{mn}^{\mathbb{C}}$ and $\mathcal{P}_n^{\mathbb{C}}$ (polynomials of degree $\leq n$ with complex coefficients) are complex vector spaces (see Exercise 1).

The concepts of subspace, span, linear independence, basis, and dimension for real vector spaces carry over to complex vector spaces in an analogous way.

All of the results in Chapter 4 have complex counterparts. In particular, if $\mathcal{W}$ is any subspace of a finite n-dimensional complex vector space (for example, $\mathbb{C}^n$), then $\mathcal{W}$ has a finite basis, and $\dim(\mathcal{W}) \leq n$.

Because every real scalar is also a complex number, every complex vector space is also a real vector space. Therefore, we must be careful about whether a vector space is being considered as a *real* or a *complex* vector space, that is, whether complex scalars are to be used or just real scalars. For example, $\mathbb{C}^3$ is both a real vector space and a complex vector space. As a *real* vector space, $\mathbb{C}^3$ has $\{[1, 0, 0], [i, 0, 0], [0, 1, 0], [0, i, 0], [0, 0, 1], [0, 0, i]\}$ as a basis and $\dim(\mathbb{C}^3) = 6$. But as a *complex* vector space, $\mathbb{C}^3$ has $\{[1, 0, 0], [0, 1, 0], [0, 0, 1]\}$ as a basis (since i can now be used as a scalar) and so $\dim(\mathbb{C}^3) = 3$. In general, $\dim(\mathbb{C}^n) = 2n$ as a *real* vector space, but $\dim(\mathbb{C}^n) = n$ as a *complex* vector space. In Exercise 6, you are asked to prove that if $\mathcal{V}$ is an n-dimensional complex vector space, then $\mathcal{V}$ is a $2n$-dimensional real vector space.[2]

As usual, we let $\mathbf{e}_i = [1, 0, 0, \ldots, 0], \mathbf{e}_2 = [0, 1, 0, \ldots, 0], \ldots, \mathbf{e}_n = [0, 0, 0, \ldots, 1]$ represent the **standard basis vectors** for the complex vector space $\mathbb{C}^n$.

Coordinatization in a complex vector space is done in the usual manner, as the following example indicates:

EXAMPLE 1

Consider the subspace $\mathcal{W}$ of the complex vector space $\mathbb{C}^4$ spanned by the vectors $\mathbf{x}_1 = [1 + i, 3, 0, -2i]$ and $\mathbf{x}_2 = [-i, 1 - i, 3i, 1 + 2i]$. Since these vectors are linearly independent (why?), the set $B = (\mathbf{x}_1, \mathbf{x}_2)$ is an ordered basis for $\mathcal{W}$ and $\dim(\mathcal{W}) = 2$. The linear combination $\mathbf{z} = (1 - i)\mathbf{x}_1 + 3\mathbf{x}_2$ of these basis vectors is equal to

$$\begin{aligned} \mathbf{z} &= (1 - i)\mathbf{x}_1 + 3\mathbf{x}_2 = [2, 3 - 3i, 0, -2 - 2i] + [-3i, 3 - 3i, 9i, 3 + 6i] \\ &= [2 - 3i, 6 - 6i, 9i, 1 + 4i]. \end{aligned}$$

Of course, the coordinatization of $\mathbf{z}$ with respect to B is $[\mathbf{z}]_B = [1 - i, 3]$. ∎

Linear Transformations

Linear transformations between complex vector spaces are defined just as for real vector spaces, except that complex scalars are used in the rule $L(k\mathbf{v}) = kL(\mathbf{v})$. The properties of complex linear transformations are completely analogous to those for linear transformations between real vector spaces.

Now every complex vector space is also a real vector space. Therefore, if $\mathcal{V}$ and $\mathcal{W}$ are complex vector spaces, and $L: \mathcal{V} \to \mathcal{W}$ is a complex linear transformation, then L is also a real linear transformation when we consider $\mathcal{V}$ and $\mathcal{W}$ to be real vector spaces. Beware! The converse is not true. It is possible to have a real linear transformation $T: \mathcal{V} \to \mathcal{W}$ that is not a complex linear transformation, as in the next example.

---

[2] The two different dimensions are sometimes distinguished by calling them the **real dimension** and the **complex dimension**.

EXAMPLE 2   Let $T: \mathbb{C}^2 \to \mathbb{C}^2$ be given by $T([z_1, z_2]) = [\overline{z_2}, \overline{z_1}]$. Then T is a real linear transformation because it satisfies the two properties, as follows:

(1) If $k \in \mathbb{R}$, then $T(k[z_1, z_2]) = T([kz_1, kz_2]) = [\overline{kz_2}, \overline{kz_1}] = [\overline{k}\overline{z_2}, \overline{k}\overline{z_1}] = [k\overline{z_2}, k\overline{z_1}] = k[\overline{z_2}, \overline{z_1}] = kT([z_1, z_2])$.

(2) $T([z_1, z_2] + [z_3, z_4]) = T([z_1 + z_3, z_2 + z_4]) = [\overline{z_2 + z_4}, \overline{z_1 + z_3}] = [\overline{z_2} + \overline{z_4}, \overline{z_1} + \overline{z_3}] = [\overline{z_2}, \overline{z_1}] + [\overline{z_4}, \overline{z_3}] = T([z_1, z_2]) + T([z_3, z_4])$.

However, T is not a complex linear transformation. Consider $T(i[1, i]) = T([i, -1]) = [-1, -i]$, while $iT([1, i]) = i[-i, 1] = [1, i]$ instead. Hence, T is not a complex linear transformation.   ∎

Exercises for Section 7.3

1. (a) Show that the set $\mathcal{P}_n^{\mathbb{C}}$ of all polynomials of degree $\le n$ under addition and complex scalar multiplication is a complex vector space.
    (b) Show that the set $\mathcal{M}_{mn}^{\mathbb{C}}$ of all $m \times n$ complex matrices under addition and complex scalar multiplication is a complex vector space.

2. Determine which of the following subsets of the complex vector space $\mathbb{C}^3$ are linearly independent. Also, in each case find the dimension of the span of the subset.

    (a) $\{[2 + i, -i, 3], [-i, 3 + i, -1]\}$
    ⋆(b) $\{[2 + i, -i, 3], [-3 + 6i, 3, 9i]\}$
    (c) $\{[3 - i, 1 + 2i, -i], [1 + i, -2, 4 + i], [1 - 3i, 5 + 2i, -8 - 3i]\}$
    ⋆(d) $\{[3 - i, 1 + 2i, -i], [1 + i, -2, 4 + i], [3 + i, -2 + 5i, 3 - 8i]\}$

3. Repeat Exercise 2 considering $\mathbb{C}^3$ as a *real* vector space. (Hint: First coordinatize the given vectors with respect to the basis $\{[1, 0, 0], [i, 0, 0], [0, 1, 0], [0, i, 0], [0, 0, 1], [0, 0, i]\}$ for $\mathbb{C}^3$. This essentially replaces the original vectors with vectors in $\mathbb{R}^6$, a more intuitive setting.)

4. (a) Show that $B = ([2i, -1+3i, 4], [3+i, -2, 1-i], [-3+5i, 2i, -5+3i])$ is an ordered basis for the complex vector space $\mathbb{C}^3$.
    ⋆(b) Let $\mathbf{z} = [3 - i, -5 - 5i, 7 + i]$. For the ordered basis B in part (a), find $[\mathbf{z}]_B$.

⋆5. With $\mathbb{C}^2$ as a real vector space, give an ordered basis for $\mathbb{C}^2$ and a matrix with respect to this basis for the linear transformation $L: \mathbb{C}^2 \to \mathbb{C}^2$ given by $L([z_1, z_2]) = [\overline{z_2}, \overline{z_1}]$. (Hint: What is the dimension of $\mathbb{C}^2$ as a real vector space?)

6. Let $\mathcal{V}$ be an n-dimensional complex vector space with basis $\{\mathbf{v}_1, \mathbf{v}_2, \ldots, \mathbf{v}_n\}$. Prove that $\{\mathbf{v}_1, i\mathbf{v}_1, \mathbf{v}_2, i\mathbf{v}_2, \ldots, \mathbf{v}_n, i\mathbf{v}_n\}$ is a basis for $\mathcal{V}$ when considered as a real vector space.

7. Prove that not every real vector space can be considered to be a complex vector space. (Hint: Consider $\mathbb{R}^3$ and Exercise 6.)

⋆8. Give the matrix with respect to the standard bases for the linear transformation $L: \mathbb{C}^2 \to \mathbb{C}^3$ (considered as complex vector spaces) such that $L([1+i, -1+3i]) = [3-i, 5, -i]$ and $L([1-i, 1+2i]) = [2+i, 1-3i, 3]$.

*9. True or False:
  (a) Every linearly independent subset of a complex vector space $\mathcal{V}$ is contained in a basis for $\mathcal{V}$.
  (b) The function $L: \mathbb{C} \to \mathbb{C}$ given by $L(z) = \bar{z}$ is a complex linear transformation.
  (c) If $\mathcal{V}$ is an n-dimensional complex vector space with ordered basis B, then $L: \mathcal{V} \to \mathbb{C}^n$ given by $L(\mathbf{v}) = [\mathbf{v}]_B$ is a complex linear transformation.
  (d) Every complex subspace of a finite dimensional complex vector space has even (complex) dimension.

7.4 Orthogonality in $\mathbb{C}^n$

Prerequisite: Section 6.3, Orthogonal Diagonalization

In this section, we study orthogonality and the Gram-Schmidt Process in $\mathbb{C}^n$, and the complex analog of orthogonal diagonalization.

Orthogonal Bases and the Gram-Schmidt Process

DEFINITION

A subset $\{\mathbf{v}_1, \mathbf{v}_2, \ldots, \mathbf{v}_n\}$ of vectors of $\mathbb{C}^n$ is **orthogonal** if and only if the *complex* dot product of any two distinct vectors in the set is zero. An orthogonal set of vectors in $\mathbb{C}^n$ is **orthonormal** if and only if each vector in the set is a unit vector.

As with real vector spaces, any set of orthogonal nonzero vectors in a complex vector space is linearly independent. The Gram-Schmidt Process for finding an orthogonal basis extends to the complex case, as in the next example.

EXAMPLE 1

We find an orthogonal basis for the complex vector space $\mathbb{C}^3$ containing $\mathbf{w}_1 = [i, 1+i, 1]$. First, we use the Enlarging Method of Section 4.6 to find a basis for $\mathbb{C}^3$ containing $\mathbf{w}_1$. Row reducing

$$\begin{bmatrix} i & 1 & 0 & 0 \\ 1+i & 0 & 1 & 0 \\ 1 & 0 & 0 & 1 \end{bmatrix} \quad \text{to obtain} \quad \begin{bmatrix} 1 & 0 & 0 & 1 \\ 0 & 1 & 0 & -i \\ 0 & 0 & 1 & -1-i \end{bmatrix}$$

shows that if $\mathbf{w}_2 = [1, 0, 0]$ and $\mathbf{w}_3 = [0, 1, 0]$, then $\{\mathbf{w}_1, \mathbf{w}_2, \mathbf{w}_3\}$ is a basis for $\mathbb{C}^3$.

Let $\mathbf{v}_1 = \mathbf{w}_1$. Following the steps of the Gram-Schmidt Process, we obtain

$$\mathbf{v}_2 = \mathbf{w}_2 - \left(\frac{\mathbf{w}_2 \cdot \mathbf{v}_1}{\mathbf{v}_1 \cdot \mathbf{v}_1} \right) \mathbf{v}_1 = [1, 0, 0] - \left(\frac{-i}{4} \right) [i, 1+i, 1].$$

Multiplying by 4 to avoid fractions, we get

$$\mathbf{v}_2 = [4, 0, 0] + i[i, 1 + i, 1] = [3, -1 + i, i].$$

Continuing, we get

$$
\begin{aligned}
\mathbf{v}_3 &= \mathbf{w}_3 - \left(\frac{\mathbf{w}_3 \cdot \mathbf{v}_1}{\mathbf{v}_1 \cdot \mathbf{v}_1}\right)\mathbf{v}_1 - \left(\frac{\mathbf{w}_3 \cdot \mathbf{v}_2}{\mathbf{v}_2 \cdot \mathbf{v}_2}\right)\mathbf{v}_2 \\
&= [0, 1, 0] - \left(\frac{1 - i}{4}\right)[i, 1 + i, 1] - \left(\frac{-1 - i}{12}\right)[3, -1 + i, 1].
\end{aligned}
$$

Multiplying by 12 to avoid fractions, we get

$$\mathbf{v}_3 = [0, 12, 0] + 3(-1 + i)[i, 1 + i, 1] + (1 + i)[3, -1 + i, i] = [0, 4, -4 + 4i].$$

We can divide by 4 to avoid multiples, and so finally get $\mathbf{v}_3 = [0, 1, -1 + i]$. Hence, $\{\mathbf{v}_1, \mathbf{v}_2, \mathbf{v}_3\} = \{[i, 1 + i, 1], [3, -1 + i, i], [0, 1, -1 + i]\}$ is an orthogonal basis for $\mathbb{C}^3$ containing $\mathbf{w}_1$. (You should verify that $\mathbf{v}_1$, $\mathbf{v}_2$, and $\mathbf{v}_3$ are mutually orthogonal.)

We can normalize $\mathbf{v}_1$, $\mathbf{v}_2$, and $\mathbf{v}_3$ to obtain the following orthonormal basis for $\mathbb{C}^3$:

$$
\left\{\left[\frac{i}{2}, \frac{1 + i}{2}, \frac{1}{2}\right], \left[\frac{3}{2\sqrt{3}}, \frac{-1 + i}{2\sqrt{3}}, \frac{i}{2\sqrt{3}}\right], \left[0, \frac{1}{\sqrt{3}}, \frac{-1 + i}{\sqrt{3}}\right]\right\}.
$$

■

Recall that the complex dot product is not symmetric. Hence, in Example 1 we were careful in the Gram-Schmidt Process to compute the dot products $\mathbf{w}_2 \cdot \mathbf{v}_1$, $\mathbf{w}_3 \cdot \mathbf{v}_1$, and $\mathbf{w}_3 \cdot \mathbf{v}_2$ in the correct order. If we had computed $\mathbf{v}_1 \cdot \mathbf{w}_2$, $\mathbf{v}_1 \cdot \mathbf{w}_3$, and $\mathbf{v}_2 \cdot \mathbf{w}_3$ instead, the vectors obtained would not be orthogonal.

Unitary Matrices

We now examine the complex analog of orthogonal matrices.

DEFINITION

A nonsingular (square) complex matrix $\mathbf{A}$ is **unitary** if and only if $\mathbf{A}^* = \mathbf{A}^{-1}$ (that is, if and only if $(\overline{\mathbf{A}})^T = \mathbf{A}^{-1}$).

It follows immediately that every unitary matrix is a normal matrix (why?).

EXAMPLE 2   For

$$
\mathbf{A} = \begin{bmatrix} \frac{1-i}{\sqrt{3}} & 0 & \frac{i}{\sqrt{3}} \\ \frac{-1+i}{\sqrt{15}} & \frac{3}{\sqrt{15}} & \frac{2i}{\sqrt{15}} \\ \frac{1-i}{\sqrt{10}} & \frac{2}{\sqrt{10}} & \frac{-2i}{\sqrt{10}} \end{bmatrix}, \text{ we have } \mathbf{A}^* = (\overline{\mathbf{A}})^T = \begin{bmatrix} \frac{1+i}{\sqrt{3}} & \frac{-1-i}{\sqrt{15}} & \frac{1+i}{\sqrt{10}} \\ 0 & \frac{3}{\sqrt{15}} & \frac{2}{\sqrt{10}} \\ -\frac{i}{\sqrt{3}} & -\frac{2i}{\sqrt{15}} & \frac{2i}{\sqrt{10}} \end{bmatrix}
$$

A quick calculation shows that $\mathbf{A}\mathbf{A}^* = \mathbf{I}_3$ (verify!), so $\mathbf{A}$ is unitary.   ■

It is straightforward to show that if a matrix $\mathbf{A}$ is unitary, then so is $\mathbf{A}^{-1}$, and also that the absolute value of $|\mathbf{A}|$ equals 1; that is, $\Big| |\mathbf{A}| \Big| = 1$ (see Exercise 4).

Also, the product of two unitary matrices of the same size is unitary (see Exercise 5). The next two theorems are the analogs of Theorems 6.6 and 6.7. They are left for you to prove in Exercises 7 and 8. You should verify that the unitary matrix of Example 2 satisfies Theorem 7.6.

THEOREM 7.6

Let $\mathbf{A}$ be an $n \times n$ complex matrix. Then $\mathbf{A}$ is unitary
(1) if and only if the rows of $\mathbf{A}$ form an orthonormal basis for $\mathbb{C}^n$
(2) if and only if the columns of $\mathbf{A}$ form an orthonormal basis for $\mathbb{C}^n$.

THEOREM 7.7

Let B and C be ordered orthonormal bases for $\mathbb{C}^n$. Then the transition matrix from B to C is a unitary matrix.

Unitarily Diagonalizable Matrices

We now consider the complex analog of orthogonal diagonalization.

DEFINITION

An $n \times n$ complex matrix $\mathbf{A}$ is **unitarily diagonalizable** if and only if there is a unitary matrix $\mathbf{P}$ such that $\mathbf{P}^{-1}\mathbf{AP}$ is diagonal.

 Consider the matrix

$$\mathbf{P} = \frac{1}{3}\begin{bmatrix} -2i & 2 & 1 \\ 2i & 1 & 2 \\ 1 & -2i & 2i \end{bmatrix}.$$

Notice that $\mathbf{P}$ is a unitary matrix, since the columns of $\mathbf{P}$ form an orthonormal basis for $\mathbb{C}^3$.

Next, consider the matrix

$$\mathbf{A} = \frac{1}{3}\begin{bmatrix} -1+3i & 2+2i & -2 \\ 2+2i & 2i & -2i \\ 2 & 2i & 1+4i \end{bmatrix}.$$

Now, $\mathbf{A}$ is unitarily diagonalizable because

$$\mathbf{P}^{-1}\mathbf{AP} = \mathbf{P}^*\mathbf{AP} = \begin{bmatrix} -1 & 0 & 0 \\ 0 & 2i & 0 \\ 0 & 0 & 1+i \end{bmatrix},$$

a diagonal matrix. ∎

We saw in Section 6.3 that a matrix is orthogonally diagonalizable if and only if it is symmetric. The following theorem characterizes unitarily diagonalizable matrices:

THEOREM 7.8

An $n \times n$ matrix $\mathbf{A}$ is unitarily diagonalizable if and only if $\mathbf{A}$ is normal.

A quick calculation shows that the matrix $\mathbf{A}$ in Example 3 is normal (see Exercise 9).

EXAMPLE 4   Let $\mathbf{A} = \begin{bmatrix} -48 + 18i & -24 + 36i \\ 24 - 36i & -27 + 32i \end{bmatrix}$. A direct computation of $\mathbf{AA}^*$ and $\mathbf{A}^*\mathbf{A}$ shows that $\mathbf{A}$ is normal (verify!). Therefore, $\mathbf{A}$ is unitarily diagonalizable by Theorem 7.8. After some calculation, you can verify that the eigenvalues of $\mathbf{A}$ are $\lambda_1 = 50i$ and $\lambda_2 = -75$. Hence, $\mathbf{A}$ is unitarily diagonalizable to $\mathbf{D} = \begin{bmatrix} 50i & 0 \\ 0 & -75 \end{bmatrix}$.

In fact, λ_1 and λ_2 have associated eigenvectors $\mathbf{v}_1 = \left[\frac{3}{5}, -\frac{4}{5}i \right]$ and $\mathbf{v}_2 = \left[-\frac{4}{5}i, \frac{3}{5} \right]$. Since $\{\mathbf{v}_1, \mathbf{v}_2\}$ is an orthonormal set, the matrix $\mathbf{P} = \begin{bmatrix} \frac{3}{5} & -\frac{4}{5}i \\ -\frac{4}{5}i & \frac{3}{5} \end{bmatrix}$, whose columns are $\mathbf{v}_1$ and $\mathbf{v}_2$, is a unitary matrix, and $\mathbf{P}^{-1}\mathbf{AP} = \mathbf{P}^*\mathbf{AP} = \mathbf{D}$. ∎

Self-Adjoint Operators and Hermitian Matrices

An immediate corollary of Theorems 7.4 and 7.8 is

COROLLARY 7.9

If $\mathbf{A}$ is a Hermitian matrix, then $\mathbf{A}$ is unitarily diagonalizable.

We can prove even more about Hermitian matrices. First, we introduce some new terminology. If linear operators L and M on $\mathbb{C}^n$ have the property $L(\mathbf{x}) \cdot \mathbf{y} = \mathbf{x} \cdot M(\mathbf{y})$ for all $\mathbf{x}, \mathbf{y} \in \mathbb{C}^n$, then M is called an **adjoint** of L. Now, suppose that $L: \mathbb{C}^n \to \mathbb{C}^n$ is the linear operator $L(\mathbf{x}) = \mathbf{Ax}$, where $\mathbf{A}$ is an $n \times n$ matrix, and let $L^*: \mathbb{C}^n \to \mathbb{C}^n$ be given by $L^*(\mathbf{x}) = \mathbf{A}^*\mathbf{x}$. By Theorem 7.3, $(L(\mathbf{x})) \cdot \mathbf{y} = \mathbf{x} \cdot (L^*(\mathbf{y}))$ for all $\mathbf{x}, \mathbf{y} \in \mathbb{C}^n$, and so L^* is an adjoint of L.

Now, if $\mathbf{A}$ is a Hermitian matrix, then $\mathbf{A} = \mathbf{A}^*$, and so $L = L^*$. Thus, $(L(\mathbf{x})) \cdot \mathbf{y} = \mathbf{x} \cdot (L(\mathbf{y}))$ for all $\mathbf{x}, \mathbf{y} \in \mathbb{C}^n$. Such an operator is called **self-adjoint**, since it is its own adjoint. It can be shown that every self-adjoint operator on $\mathbb{C}^n$ has a Hermitian matrix representation with respect to any orthonormal basis. Self-adjoint operators are the complex analogs of the symmetric operators in Section 6.3. Corollary 7.9 asserts that all self-adjoint operators are unitarily diagonalizable. The converse to Corollary 7.9 is *not* true because there are unitarily diagonalizable (= normal) matrices that are not Hermitian. This differs from the situation with linear operators on real vector spaces where the analog of the converse of Corollary 7.9 *is* true; that is, every orthogonally diagonalizable linear operator is symmetric.

The final theorem of this section shows that any diagonal matrix representation for a self-adjoint operator has all real entries.

THEOREM 7.10

All eigenvalues of a Hermitian matrix are real.

PROOF OF THEOREM 7.10

Let λ be an eigenvalue for a Hermitian matrix $\mathbf{A}$, and let $\mathbf{u}$ be a unit eigenvector for λ. Then $\lambda = \lambda \|\mathbf{u}\|^2 = \lambda(\mathbf{u} \cdot \mathbf{u}) = (\lambda\mathbf{u}) \cdot \mathbf{u} = (\mathbf{A}\mathbf{u}) \cdot \mathbf{u} = \mathbf{u} \cdot (\mathbf{A}\mathbf{u})$ (by Theorem 7.3) $= \mathbf{u} \cdot \lambda\mathbf{u} = \overline{\lambda}(\mathbf{u} \cdot \mathbf{u})$ (by part (5) of Theorem 7.1) $= \overline{\lambda}$. Hence, λ is real. ∎

EXAMPLE 5   Consider the Hermitian matrix

$$\mathbf{A} = \begin{bmatrix} 17 & -24 + 8i & -24 - 32i \\ -24 - 8i & 53 & 4 + 12i \\ -24 + 32i & 4 - 12i & 11 \end{bmatrix}.$$

By Theorem 7.10, all eigenvalues of $\mathbf{A}$ are real. It can be shown that these eigenvalues are $\lambda_1 = 27$, $\lambda_2 = -27$, and $\lambda_3 = 81$. By Corollary 7.9, $\mathbf{A}$ is unitarily diagonalizable. In fact, the unitary matrix

$$\mathbf{P} = \frac{1}{9} \begin{bmatrix} 4 & 6 - 2i & -3 - 4i \\ 6 + 2i & 1 & 2 + 6i \\ -3 + 4i & 2 - 6i & 4 \end{bmatrix}$$

has the property that $\mathbf{P}^{-1}\mathbf{A}\mathbf{P}$ is the diagonal matrix with eigenvalues λ_1, λ_2, and λ_3 on the main diagonal (verify!). ∎

Every real symmetric matrix $\mathbf{A}$ can be thought of as a complex Hermitian matrix. Now $p_{\mathbf{A}}(x)$ must have at least one complex root. But by Theorem 7.10, this eigenvalue for $\mathbf{A}$ must be real. This gives us a shorter proof of Lemma 6.18 in Section 6.3. (We did not use this method of proof in Section 6.3 since it entails complex numbers.)

Exercises for Section 7.4

1. Determine whether the following sets of vectors are orthogonal.
  ⋆(a) In $\mathbb{C}^2$: $\{[1 + 2i, -3 - i], [4 - 2i, 3 + i]\}$
  (b) In $\mathbb{C}^3$: $\{[1 - i, -1 + i, 1 - i], [i, -2i, 2i]\}$

$\star$(c) In $\mathbb{C}^3$: $\{[2i, -1, i], [1, -i, -1], [0, 1, i]\}$

(d) In $\mathbb{C}^4$: $\{[1, i, -1, 1+i], [4, -i, 1, -1-i], [0, 3, -i, -1+i]\}$

2. Suppose $\{\mathbf{z}_1, \ldots, \mathbf{z}_k\}$ is an orthonormal subset of $\mathbb{C}^n$, and $c_1, \ldots, c_k \in \mathbb{C}$ with $|c_i| = 1$ for $1 \leq i \leq k$. Prove that $\{c_1\mathbf{z}_1, \ldots, c_k\mathbf{z}_k\}$ is an orthonormal subset of $\mathbb{C}^n$.

$\star$**3.** (a) Use the Gram-Schmidt Process to find an orthogonal basis for $\mathbb{C}^3$ containing $[1+i, i, 1]$.

(b) Find a 3×3 unitary matrix having a multiple of $[1+i, i, 1]$ as its first row.

4. Prove that if $\mathbf{A}$ is a unitary matrix, then $\mathbf{A}^{-1}$ is unitary and the absolute value of $|\mathbf{A}|$ equals 1.

5. Prove that the product of two unitary matrices of the same size is a unitary matrix.

6. (a) Prove that a complex matrix $\mathbf{A}$ is unitary if and only if $\overline{\mathbf{A}}$ is unitary.

(b) Let $\mathbf{A}$ be a unitary matrix. Prove that $\mathbf{A}^k$ is unitary for all integers $k \geq 1$.

(c) Let $\mathbf{A}$ be a unitary matrix. Prove that $\mathbf{A}^2 = \mathbf{I}_n$ if and only if $\mathbf{A}$ is Hermitian.

7. ▶(a) Without using Theorem 7.6, prove that $\mathbf{A}$ is a unitary matrix if and only if $\mathbf{A}^T$ is unitary.

(b) Prove Theorem 7.6. (Hint: First prove part (1) of Theorem 7.6, and then use part (a) of this exercise to prove part (2). Modify the proof of Theorem 6.6. For instance, when $i \neq j$, to show that the ith row of $\mathbf{A}$ is orthogonal to the jth column of $\mathbf{A}$, we must show that the *complex* dot product of the ith row of $\mathbf{A}$ with the jth column of $\mathbf{A}$ equals zero.)

8. Prove Theorem 7.7. (Hint: Modify the proof of Theorem 6.7.)

9. Show that the matrix $\mathbf{A}$ in Example 3 is normal.

10. (a) Show that the linear operator $L: \mathbb{C}^2 \to \mathbb{C}^2$ given by
$$L\left(\begin{bmatrix} z_1 \\ z_2 \end{bmatrix}\right) = \begin{bmatrix} 1-6i & -10-2i \\ 2-10i & 5 \end{bmatrix}\begin{bmatrix} z_1 \\ z_2 \end{bmatrix}$$ is unitarily diagonalizable.

$\star$(b) If $\mathbf{A}$ is the matrix for L (with respect to the standard basis for $\mathbb{C}^2$), find a unitary matrix $\mathbf{P}$ such that $\mathbf{P}^{-1}\mathbf{AP}$ is diagonal.

11. (a) Show that the following matrix is unitarily diagonalizable:
$$\mathbf{A} = \begin{bmatrix} -4+5i & 2+2i & 4+4i \\ 2+2i & -1+8i & -2-2i \\ 4+4i & -2-2i & -4+5i \end{bmatrix}.$$

(b) Find a unitary matrix $\mathbf{P}$ such that $\mathbf{P}^{-1}\mathbf{AP}$ is diagonal.

12. (a) Let $\mathbf{A}$ be a unitary matrix. Show that $|\lambda| = 1$ for every eigenvalue λ of $\mathbf{A}$. (Hint: Suppose $\mathbf{Az} = \lambda\mathbf{z}$, for some $\mathbf{z}$. Use Theorem 7.3 to calculate $\mathbf{Az} \cdot \mathbf{Az}$ two different ways to show that $\lambda\overline{\lambda} = 1$.)

(b) Prove that a unitary matrix $\mathbf{A}$ is Hermitian if and only if the eigenvalues of $\mathbf{A}$ are 1 and/or -1.

*13. Verify directly that all of the eigenvalues of the following Hermitian matrix are real:

$$\begin{bmatrix} 1 & 2+i & 1-2i \\ 2-i & -3 & -i \\ 1+2i & i & 2 \end{bmatrix}.$$

14. (a) Prove that if **A** is normal and has real eigenvalues, then **A** is Hermitian. (Hint: Use Theorem 7.8 to express **A** as **PDP*** for some unitary **P** and diagonal **D**. Calculate **A***.)

   (b) Prove that if **A** is normal and all eigenvalues have absolute value equal to 1, then **A** is unitary. (Hint: With **A** = **PDP*** as in part (a), show **DD*** = **I** and use this to calculate **AA***.)

*15. True or False:

   (a) Every Hermitian matrix is unitary.
   (b) Every orthonormal basis for $\mathbb{R}^n$ is also an orthonormal basis for $\mathbb{C}^n$.
   (c) An $n \times n$ complex matrix **A** is unitarily diagonalizable if and only if there is a unitary matrix **P** such that **PAP*** is diagonal.
   (d) If the columns of an $n \times n$ matrix **A** form an orthonormal basis for $\mathbb{C}^n$, then the rows of **A** also form an orthonormal basis for $\mathbb{C}^n$.
   (e) If **A** is the matrix with respect to the standard basis for a linear operator L on $\mathbb{C}^n$, then **A**^{T} is the matrix for the adjoint of L with respect to the standard basis.

7.5 Inner Product Spaces

Prerequisite: Section 6.3, Orthogonal Diagonalization

In $\mathbb{R}^n$ and $\mathbb{C}^n$, we have the dot product along with the operations of vector addition and scalar multiplication. In other vector spaces, we can often create a similar type of product, known as an inner product.

Inner Products

DEFINITION

Let $\mathcal{V}$ be a real [complex] vector space with operations + and $\cdot$, and let $\langle \, , \, \rangle$ be an operation that assigns to each pair of vectors $\mathbf{x}, \mathbf{y} \in \mathcal{V}$ a real [complex] number, denoted $\langle \mathbf{x}, \mathbf{y} \rangle$. Then $\langle \, , \, \rangle$ is a **real [complex] inner product** for $\mathcal{V}$ if and only if the following properties hold for all $\mathbf{x}, \mathbf{y} \in \mathcal{V}$ and all $k \in \mathbb{R}$ [$k \in \mathbb{C}$]:

   (1) $\langle \mathbf{x}, \mathbf{x} \rangle$ is always real, and $\langle \mathbf{x}, \mathbf{x} \rangle \geq 0$
   (2) $\langle \mathbf{x}, \mathbf{x} \rangle = 0$ if and only if $\mathbf{x} = \mathbf{0}$
   (3) $\langle \mathbf{x}, \mathbf{y} \rangle = \langle \mathbf{y}, \mathbf{x} \rangle \quad \left[\langle \mathbf{x}, \mathbf{y} \rangle = \overline{\langle \mathbf{y}, \mathbf{x} \rangle} \right]$
   (4) $\langle \mathbf{x} + \mathbf{y}, \mathbf{z} \rangle = \langle \mathbf{x}, \mathbf{z} \rangle + \langle \mathbf{y}, \mathbf{z} \rangle$
   (5) $\langle k\mathbf{x}, \mathbf{y} \rangle = k \langle \mathbf{x}, \mathbf{y} \rangle$

A vector space together with a real [complex] inner product operation is known as a **real [complex] inner product space**.

EXAMPLE 1

Consider the real vector space $\mathbb{R}^n$. Let $\mathbf{x} = [x_1, \ldots, x_n]$ and $\mathbf{y} = [y_1, \ldots, y_n]$ be vectors in $\mathbb{R}^n$. By Theorem 1.5, the operation $\langle \mathbf{x}, \mathbf{y} \rangle = \mathbf{x} \cdot \mathbf{y} = x_1 y_1 + \cdots + x_n y_n$ (usual real dot product) is a real inner product (verify!). Hence, $\mathbb{R}^n$ together with the dot product is a real inner product space.

Similarly, let $\mathbf{x} = [x_1, \ldots, x_n]$ and $\mathbf{y} = [y_1, \ldots, y_n]$ be vectors in the complex vector space $\mathbb{C}^n$. By Theorem 7.1, the operation $\langle \mathbf{x}, \mathbf{y} \rangle = \mathbf{x} \cdot \mathbf{y} = x_1 \overline{y_1} + \cdots + x_n \overline{y_n}$ (usual complex dot product) is an inner product on $\mathbb{C}^n$. Thus, $\mathbb{C}^n$ together with the complex dot product is a complex inner product space. ∎

EXAMPLE 2

Consider the real vector space $\mathbb{R}^2$. For $\mathbf{x} = [x_1, x_2]$ and $\mathbf{y} = [y_1, y_2]$ in $\mathbb{R}^2$, define $\langle \mathbf{x}, \mathbf{y} \rangle = x_1 y_1 - x_1 y_2 - x_2 y_1 + 2 x_2 y_2$. We verify the five properties in the definition of an inner product space.

Property (1): $\langle \mathbf{x}, \mathbf{x} \rangle = x_1 x_1 - x_1 x_2 - x_2 x_1 + 2 x_2 x_2 = x_1^2 - 2 x_1 x_2 + x_2^2 + x_2^2 = (x_1 - x_2)^2 + x_2^2 \geq 0$.

Property (2): $\langle \mathbf{x}, \mathbf{x} \rangle = 0$ exactly when $x_1 = x_2 = 0$ (that is, when $\mathbf{x} = \mathbf{0}$).

Property (3): $\langle \mathbf{y}, \mathbf{x} \rangle = y_1 x_1 - y_1 x_2 - y_2 x_1 + 2 y_2 x_2 = x_1 y_1 - x_1 y_2 - x_2 y_1 + 2 x_2 y_2 = \langle \mathbf{x}, \mathbf{y} \rangle$.

Property (4): Let $\mathbf{z} = [z_1, z_2]$. Then

$$\langle \mathbf{x} + \mathbf{y}, \mathbf{z} \rangle = (x_1 + y_1) z_1 - (x_1 + y_1) z_2 - (x_2 + y_2) z_1 + 2(x_2 + y_2) z_2$$
$$= x_1 z_1 + y_1 z_1 - x_1 z_2 - y_1 z_2 - x_2 z_1 - y_2 z_1 + 2 x_2 z_2 + 2 y_2 z_2$$
$$= (x_1 z_1 - x_1 z_2 - x_2 z_1 + 2 x_2 z_2) + (y_1 z_1 - y_1 z_2 - y_2 z_1 + 2 y_2 z_2)$$
$$= \langle \mathbf{x}, \mathbf{z} \rangle + \langle \mathbf{y}, \mathbf{z} \rangle.$$

Property (5): $\langle k\mathbf{x}, \mathbf{y} \rangle = (k x_1) y_1 - (k x_1) y_2 - (k x_2) y_1 + 2(k x_2) y_2 = k(x_1 y_1 - x_1 y_2 - x_2 y_1 + 2 x_2 y_2) = k \langle \mathbf{x}, \mathbf{y} \rangle$.

Hence, $\langle \, , \, \rangle$ is a real inner product on $\mathbb{R}^2$, and $\mathbb{R}^2$ together with this operation $\langle \, , \, \rangle$ is a real inner product space. ∎

EXAMPLE 3

Consider the real vector space $\mathbb{R}^n$. Let $\mathbf{A}$ be a nonsingular $n \times n$ real matrix. Let $\mathbf{x}, \mathbf{y} \in \mathbb{R}^n$ and define $\langle \mathbf{x}, \mathbf{y} \rangle = (\mathbf{Ax}) \cdot (\mathbf{Ay})$ (the usual dot product of $\mathbf{Ax}$ and $\mathbf{Ay}$). It can be shown (see Exercise 1) that $\langle \, , \, \rangle$ is a real inner product on $\mathbb{R}^n$, and so $\mathbb{R}^n$ together with this operation $\langle \, , \, \rangle$ is a real inner product space. ∎

EXAMPLE 4

Consider the real vector space $\mathcal{P}_n$. Let $\mathbf{p}_1 = a_n x^n + \cdots + a_1 x + a_0$ and $\mathbf{p}_2 = b_n x^n + \cdots + b_1 x + b_0$ be in $\mathcal{P}_n$. Define $\langle \mathbf{p}_1, \mathbf{p}_2 \rangle = a_n b_n + \cdots + a_1 b_1 + a_0 b_0$. It can be shown (see Exercise 2) that $\langle \, , \, \rangle$ is a real inner product on $\mathcal{P}_n$, and so $\mathcal{P}_n$ together with this operation $\langle \, , \, \rangle$ is a real inner product space. ∎

EXAMPLE 5

Let $a, b \in \mathbb{R}$, with $a < b$, and consider the real vector space $\mathcal{V}$ of all real-valued continuous functions defined on the interval $[a, b]$ (for example, polynomials, $\sin x$, e^x). Let $\mathbf{f}, \mathbf{g} \in \mathcal{V}$. Define $\langle \mathbf{f}, \mathbf{g} \rangle = \int_a^b \mathbf{f}(t) \mathbf{g}(t) \, dt$. It can be shown (see Exercise 3) that $\langle \, , \, \rangle$ is a real inner product on $\mathcal{V}$, and so $\mathcal{V}$ together with this operation $\langle \, , \, \rangle$ is a real inner product space.

Analogously, the operation $\langle \mathbf{f}, \mathbf{g} \rangle = \int_a^b \mathbf{f}(t)\overline{\mathbf{g}(t)}\, dt$ makes the complex vector space of all complex-valued continuous functions on $[a, b]$ into a complex inner product space. ∎

Of course, not every operation is an inner product. For example, for the vectors $\mathbf{x} = [x_1, x_2]$ and $\mathbf{y} = [y_1, y_2]$ in $\mathbb{R}^2$, consider the operation $\langle \mathbf{x}, \mathbf{y} \rangle = x_1^2 + y_1^2$. Now, with $\mathbf{x} = \mathbf{y} = [1, 0]$, we have $\langle 2\mathbf{x}, \mathbf{y} \rangle = 2^2 + 1^2 = 5$, but $2\langle \mathbf{x}, \mathbf{y} \rangle = 2(1^2 + 1^2) = 4$. Thus, property (5) fails to hold.

The next theorem lists some useful results for inner product spaces.

THEOREM 7.11

Let $\mathcal{V}$ be a real [complex] inner product space with inner product $\langle \, , \, \rangle$. Then for all $\mathbf{x}, \mathbf{y} \in \mathcal{V}$ and all $k \in \mathbb{R}$ $[k \in \mathbb{C}]$, we have

    (1) $\langle \mathbf{0}, \mathbf{x} \rangle = \langle \mathbf{x}, \mathbf{0} \rangle = 0$.

    (2) $\langle \mathbf{x}, \mathbf{y} + \mathbf{z} \rangle = \langle \mathbf{x}, \mathbf{y} \rangle + \langle \mathbf{x}, \mathbf{z} \rangle$.

    (3) $\langle \mathbf{x}, k\mathbf{y} \rangle = k \langle \mathbf{x}, \mathbf{y} \rangle$ $[\langle \mathbf{x}, k\mathbf{y} \rangle = \overline{k} \langle \mathbf{x}, \mathbf{y} \rangle]$.

Note the use of $\overline{k}$ in part (3) for complex vector spaces. The proof of this theorem is straightforward, and parts are left for you to do in Exercise 5.

Length, Distance, and Angles in Inner Product Spaces

The next definition extends the concept of the length of a vector to any inner product space.

DEFINITION

If $\mathbf{x}$ is a vector in an inner product space, then the **norm (length)** of $\mathbf{x}$ is $\|\mathbf{x}\| = \sqrt{\langle \mathbf{x}, \mathbf{x} \rangle}$.

This definition yields a nonnegative real number for $\|\mathbf{x}\|$ since by definition, $\langle \mathbf{x}, \mathbf{x} \rangle$ is always real and nonnegative for any vector $\mathbf{x}$. Also note that this definition agrees with the earlier definition of length in $\mathbb{R}^n$ based on the usual dot product in $\mathbb{R}^n$. We also have the following result:

THEOREM 7.12

Let $\mathcal{V}$ be a real [complex] inner product space, with $\mathbf{x} \in \mathcal{V}$. Let $k \in \mathbb{R}$ $[k \in \mathbb{C}]$. Then, $\|k\mathbf{x}\| = |k|\, \|\mathbf{x}\|$.

The proof of this theorem is left for you to do in Exercise 6.

As before, we say that a vector of length 1 in an inner product space is a **unit vector**. For instance, in the inner product space of Example 4, the polynomial $\mathbf{p} = \frac{\sqrt{2}}{2}x + \frac{\sqrt{2}}{2}$ is a unit vector since $\|\mathbf{p}\| = \sqrt{\langle \mathbf{p}, \mathbf{p} \rangle} = \sqrt{\left(\frac{\sqrt{2}}{2}\right)^2 + \left(\frac{\sqrt{2}}{2}\right)^2} = 1$.

We define the distance between two vectors in the general inner product space setting as we did for $\mathbb{R}^n$.

DEFINITION

Let $\mathbf{x}, \mathbf{y} \in \mathcal{V}$, an inner product space. Then the **distance between x and y** is $\|\mathbf{x} - \mathbf{y}\|$.

| EXAMPLE 6 | Consider the real vector space $\mathcal{V}$ of real continuous functions from Example 5, with $a = 0$ and $b = \pi$. That is, $\langle \mathbf{f}, \mathbf{g} \rangle = \int_0^\pi \mathbf{f}(t)\mathbf{g}(t)\,dt$ for all $\mathbf{f}, \mathbf{g} \in \mathcal{V}$. Let $\mathbf{f} = \cos t$ and $\mathbf{g} = \sin t$. Then the distance between $\mathbf{f}$ and $\mathbf{g}$ is

$$\|\mathbf{f} - \mathbf{g}\| = \sqrt{\langle \cos t - \sin t, \ \cos t - \sin t \rangle} = \sqrt{\int_0^\pi (\cos t - \sin t)^2 \, dt}$$

$$= \sqrt{\int_0^\pi \left(\cos^2 t - 2\cos t \sin t + \sin^2 t \right) \, dt}$$

$$= \sqrt{\int_0^\pi (1 - \sin 2t) \, dt} = \sqrt{\left(t + \frac{1}{2} \cos 2t \right)\bigg|_0^\pi} = \sqrt{\pi}.$$

Hence, the distance between $\cos t$ and $\sin t$ is $\sqrt{\pi}$ under this inner product. ∎

The next theorem shows that some other familiar results from the ordinary dot product carry over to the general inner product.

THEOREM 7.13

Let $\mathbf{x}, \mathbf{y} \in \mathcal{V}$, an inner product space, with inner product $\langle \, , \, \rangle$. Then

(1) $|\langle \mathbf{x}, \mathbf{y} \rangle| \leq \|\mathbf{x}\| \, \|\mathbf{y}\|$ Cauchy-Schwarz Inequality
(2) $\|\mathbf{x} + \mathbf{y}\| \leq \|\mathbf{x}\| + \|\mathbf{y}\|$ Triangle Inequality.

The proofs of these statements are analogous to the proofs for the ordinary dot product and are left for you to do in Exercise 11.

From the Cauchy-Schwarz Inequality, we have $-1 \leq \langle \mathbf{x}, \mathbf{y} \rangle / (\|\mathbf{x}\| \, \|\mathbf{y}\|) \leq 1$, for any nonzero vectors $\mathbf{x}$ and $\mathbf{y}$ in a *real* inner product space. Hence, we can make the following definition:

DEFINITION

Let $\mathbf{x}, \mathbf{y} \in \mathcal{V}$, a *real* inner product space. Then the **angle between x and y** is the angle θ from 0 to π such that $\cos \theta = \langle \mathbf{x}, \mathbf{y} \rangle / (\|\mathbf{x}\| \, \|\mathbf{y}\|)$.

EXAMPLE 7 Consider again the inner product space of Example 6, where $\langle \mathbf{f}, \mathbf{g} \rangle = \int_0^\pi \mathbf{f}(t)\mathbf{g}(t)\, dt$. Let $\mathbf{f} = t$ and $\mathbf{g} = \sin t$. Then $\langle \mathbf{f}, \mathbf{g} \rangle = \int_0^\pi t \sin t\, dt$. Using integration by parts, we get $(-t \cos t)|_0^\pi + \int_0^\pi \cos t\, dt = \pi + (\sin t)|_0^\pi = \pi$. Also, $\|\mathbf{f}\|^2 = \langle \mathbf{f}, \mathbf{f} \rangle = \int_0^\pi (\mathbf{f}(t))^2\, dt = \int_0^\pi t^2\, dt = \left(t^3/3 \right)\big|_0^\pi = \pi^3/3$, and so $\|\mathbf{f}\| = \sqrt{\pi^3/3}$. Similarly, $\|\mathbf{g}\|^2 = \langle \mathbf{g}, \mathbf{g} \rangle = \int_0^\pi (\mathbf{g}(t))^2\, dt = \int_0^\pi \sin^2 t\, dt = \int_0^\pi \frac{1}{2}(1 - \cos 2t)\, dt = \left(\frac{1}{2}t - \frac{1}{4}\sin 2t \right)\Big|_0^\pi = \pi/2$, and so $\|\mathbf{g}\| = \sqrt{\pi/2}$. Hence, the cosine of the angle θ between t and $\sin t$ equals $\langle \mathbf{f}, \mathbf{g} \rangle / (\|\mathbf{f}\| \, \|\mathbf{g}\|) = \pi / \left(\sqrt{\pi^3/3}\sqrt{\pi/2} \right) = \sqrt{6}/\pi \approx 0.78$. Hence, $\theta \approx 0.68$ radians ($38.8°$). ∎

Orthogonality in Inner Product Spaces

We next define orthogonal vectors in a general inner product space setting and show that nonzero orthogonal vectors are linearly independent.

DEFINITION

A subset $\{\mathbf{x}_1, \ldots, \mathbf{x}_n\}$ of vectors in an inner product space $\mathcal{V}$ with inner product $\langle \, , \, \rangle$ is **orthogonal** if and only if $\langle \mathbf{x}_i, \mathbf{x}_j \rangle = 0$ for $1 \le i, j \le n$, with $i \ne j$. Also, an orthogonal set of vectors in $\mathcal{V}$ is **orthonormal** if and only if each vector in the set is a unit vector.

The next theorem is the analog of Theorem 6.1, and its proof is left for you to do in Exercise 15.

THEOREM 7.14

If $\mathcal{V}$ is an inner product space and T is an orthogonal set of nonzero vectors in $\mathcal{V}$, then T is a linearly independent set.

EXAMPLE 8 Consider again the inner product space $\mathcal{V}$ of Example 5 of real continuous functions with inner product $\langle \mathbf{f}, \mathbf{g} \rangle = \int_a^b \mathbf{f}(t)\mathbf{g}(t)\, dt$, with $a = -\pi$ and $b = \pi$. The set $\{1, \cos t, \sin t\}$ is an orthogonal set in $\mathcal{V}$ since each of the following definite integrals equals zero (verify!):

$$\int_{-\pi}^\pi (1)\cos t\, dt, \qquad \int_{-\pi}^\pi (1)\sin t\, dt, \qquad \int_{-\pi}^\pi (\cos t)(\sin t)\, dt.$$

Also, note that $\|1\|^2 = \langle 1, 1 \rangle = \int_{-\pi}^\pi (1)(1)dt = 2\pi$, $\|\cos t\|^2 = \langle \cos t, \cos t \rangle = \int_{-\pi}^\pi \cos^2 t\, dt = \pi$ (why?), and $\|\sin t\|^2 = \langle \sin t, \sin t \rangle = \int_{-\pi}^\pi \sin^2 t\, dt = \pi$ (why?). Therefore, the set

$$\left\{ \frac{1}{\sqrt{2\pi}}, \frac{\cos t}{\sqrt{\pi}}, \frac{\sin t}{\sqrt{\pi}} \right\}$$

is an orthonormal set in $\mathcal{V}$. ∎

Example 8 can be generalized. The set $\{1, \cos t, \sin t, \cos 2t, \sin 2t, \cos 3t,$ $\sin 3t, \ldots\}$ is an orthogonal set (see Exercise 16) and therefore linearly independent by Theorem 7.14. The functions in this set are important in the theory of partial differential equations. It can be shown that every continuously differentiable function on the interval $[-\pi, \pi]$ can be represented as the (infinite) sum of constant multiples of these functions. Such a sum is known as the **Fourier series** of the function.

A basis for an inner product space V is an **orthogonal [orthonormal] basis** if the vectors in the basis form an orthogonal [orthonormal] set.

EXAMPLE 9

Consider again the inner product space P_n with the inner product of Example 4; that is, if $\mathbf{p}_1 = a_n x^n + \cdots + a_1 x + a_0$ and $\mathbf{p}_2 = b_n x^n + \cdots + b_1 x + b_0$ are in P_n, then $\langle \mathbf{p}_1, \mathbf{p}_2 \rangle = a_n b_n + \cdots + a_1 b_1 + a_0 b_0$. Now, $\{x^n, x^{n-1}, \ldots, x, 1\}$ is an orthogonal basis for P_n with this inner product, since $\langle x^k, x^l \rangle = 0$, for $0 \leq k, l \leq n$, with $k \neq l$ (why?). Since $\|x^k\| = \sqrt{\langle x^k, x^k \rangle} = 1$, for all k, $0 \leq k \leq n$ (why?), the set $\{x^n, x^{n-1}, \ldots, x, 1\}$ is also an orthonormal basis for this inner product space. ∎

A proof analogous to that of Theorem 6.3 gives us the next theorem (see Exercise 17).

THEOREM 7.15

If $B = (\mathbf{v}_1, \mathbf{v}_2, \ldots, \mathbf{v}_k)$ is an orthogonal ordered basis for a subspace W of an inner product space V, and if $\mathbf{v}$ is any vector in W, then

$$[\mathbf{v}]_B = \left[\frac{\langle \mathbf{v}, \mathbf{v}_1 \rangle}{\langle \mathbf{v}_1, \mathbf{v}_1 \rangle}, \frac{\langle \mathbf{v}, \mathbf{v}_2 \rangle}{\langle \mathbf{v}_2, \mathbf{v}_2 \rangle}, \ldots, \frac{\langle \mathbf{v}, \mathbf{v}_k \rangle}{\langle \mathbf{v}_k, \mathbf{v}_k \rangle} \right].$$

In particular, if B is an orthonormal ordered basis for W, then $[\mathbf{v}]_B = [\langle \mathbf{v}, \mathbf{v}_1 \rangle, \langle \mathbf{v}, \mathbf{v}_2 \rangle, \ldots, \langle \mathbf{v}, \mathbf{v}_k \rangle]$.

EXAMPLE 10

Recall the inner product space $\mathbb{R}^2$ in Example 2, with inner product given as follows: if $\mathbf{x} = [x_1, x_2]$ and $\mathbf{y} = [y_1, y_2]$, then $\langle \mathbf{x}, \mathbf{y} \rangle = x_1 y_1 - x_1 y_2 - x_2 y_1 + 2 x_2 y_2$. An ordered orthogonal basis for this space is $B = (\mathbf{v}_1, \mathbf{v}_2) = ([2, 1], [0, 1])$ (verify!). Recall from Example 2 that $\langle \mathbf{x}, \mathbf{x} \rangle = (x_1 - x_2)^2 + x_2^2$. Thus, $\langle \mathbf{v}_1, \mathbf{v}_1 \rangle = (2 - 1)^2 + 1^2 = 2$, and $\langle \mathbf{v}_2, \mathbf{v}_2 \rangle = (0 - 1)^2 + 1^2 = 2$.

Next, suppose that $\mathbf{v} = [a, b]$ is any vector in $\mathbb{R}^2$. Now, $\langle \mathbf{v}, \mathbf{v}_1 \rangle = \langle [a, b], [2, 1] \rangle = (a)(2) - (a)(1) - (b)(2) + 2(b)(1) = a$. Also, $\langle \mathbf{v}, \mathbf{v}_2 \rangle = \langle [a, b], [0, 1] \rangle = (a)(0) - (a)(1) - (b)(0) + 2(b)(1) = -a + 2b$. Then,

$$[\mathbf{v}]_B = \left[\frac{\langle \mathbf{v}, \mathbf{v}_1 \rangle}{\langle \mathbf{v}_1, \mathbf{v}_1 \rangle}, \frac{\langle \mathbf{v}, \mathbf{v}_2 \rangle}{\langle \mathbf{v}_2, \mathbf{v}_2 \rangle} \right] = \left[\frac{a}{2}, \frac{-a + 2b}{2} \right].$$

Notice that $\frac{a}{2}[2, 1] + \left(\frac{-a + 2b}{2} \right)[0, 1]$ does equal $[a, b] = \mathbf{v}$. ∎

The Generalized Gram-Schmidt Process

We can generalize the Gram-Schmidt Process of Section 6.1 to any inner product space. That is, we can replace any linearly independent set of k vectors with an orthogonal set of k vectors that spans the same subspace.

GENERALIZED GRAM-SCHMIDT PROCESS

Let $\{\mathbf{w}_1, \ldots, \mathbf{w}_k\}$ be a linearly independent subset of an inner product space V, with inner product $\langle\,,\,\rangle$. We create a new set $\{\mathbf{v}_1, \ldots, \mathbf{v}_k\}$ of vectors as follows:

Let $\mathbf{v}_1 = \mathbf{w}_1$.

Let $\mathbf{v}_2 = \mathbf{w}_2 - \left(\dfrac{\langle\mathbf{w}_2, \mathbf{v}_1\rangle}{\langle\mathbf{v}_1, \mathbf{v}_1\rangle}\right)\mathbf{v}_1$.

Let $\mathbf{v}_3 = \mathbf{w}_3 - \left(\dfrac{\langle\mathbf{w}_3, \mathbf{v}_1\rangle}{\langle\mathbf{v}_1, \mathbf{v}_1\rangle}\right)\mathbf{v}_1 - \left(\dfrac{\langle\mathbf{w}_3, \mathbf{v}_2\rangle}{\langle\mathbf{v}_2, \mathbf{v}_2\rangle}\right)\mathbf{v}_2$.

$\vdots$

Let $\mathbf{v}_k = \mathbf{w}_k - \left(\dfrac{\langle\mathbf{w}_k, \mathbf{v}_1\rangle}{\langle\mathbf{v}_1, \mathbf{v}_1\rangle}\right)\mathbf{v}_1 - \left(\dfrac{\langle\mathbf{w}_k, \mathbf{v}_2\rangle}{\langle\mathbf{v}_2, \mathbf{v}_2\rangle}\right)\mathbf{v}_2 - \cdots - \left(\dfrac{\langle\mathbf{w}_k, \mathbf{v}_{k-1}\rangle}{\langle\mathbf{v}_{k-1}, \mathbf{v}_{k-1}\rangle}\right)\mathbf{v}_{k-1}$.

A proof similar to that of Theorem 6.4 (see Exercise 21) gives

---

THEOREM 7.16

Let $B = \{\mathbf{w}_1, \ldots, \mathbf{w}_k\}$ be a basis for a finite dimensional inner product space V. Then the set $\{\mathbf{v}_1, \ldots, \mathbf{v}_k\}$ obtained by applying the Generalized Gram-Schmidt Process to B is an orthogonal basis for V.

Hence, every nontrivial finite dimensional inner product space has an orthogonal basis.

EXAMPLE 11 Recall the inner product space V from Example 5 of real continuous functions using $a = -1$ and $b = 1$; that is, with inner product $\langle\mathbf{f}, \mathbf{g}\rangle = \int_{-1}^{1} \mathbf{f}(t)\mathbf{g}(t)\, dt$. Now, $\{1, t, t^2, t^3\}$ is a linearly independent set in V. We use this set to find four orthogonal vectors in V.

Let $\mathbf{w}_1 = 1$, $\mathbf{w}_2 = t$, $\mathbf{w}_3 = t^2$, and $\mathbf{w}_4 = t^3$. Using the Generalized Gram-Schmidt Process, we start with $\mathbf{v}_1 = \mathbf{w}_1 = 1$ and obtain

$$\mathbf{v}_2 = \mathbf{w}_2 - \left(\frac{\langle\mathbf{w}_2, \mathbf{v}_1\rangle}{\langle\mathbf{v}_1, \mathbf{v}_1\rangle}\right)\mathbf{v}_1 = t - \left(\frac{\langle t, 1\rangle}{\langle 1, 1\rangle}\right)1.$$

Now, $\langle t, 1\rangle = \int_{-1}^{1} (t)\,(1)\, dt = \left.(t^2/2)\right|_{-1}^{1} = 0$. Hence, $\mathbf{v}_2 = t$. Next,

$$\mathbf{v}_3 = \mathbf{w}_3 - \left(\frac{\langle\mathbf{w}_3, \mathbf{v}_1\rangle}{\langle\mathbf{v}_1, \mathbf{v}_1\rangle}\right)\mathbf{v}_1 - \left(\frac{\langle\mathbf{w}_3, \mathbf{v}_2\rangle}{\langle\mathbf{v}_2, \mathbf{v}_2\rangle}\right)\mathbf{v}_2 = t^2 - \left(\frac{\langle t^2, 1\rangle}{\langle 1, 1\rangle}\right)1 - \left(\frac{\langle t^2, t\rangle}{\langle t, t\rangle}\right)t.$$

After a little calculation, we obtain $\langle t^2, 1 \rangle = \frac{2}{3}$, $\langle 1, 1 \rangle = 2$, and $\langle t^2, t \rangle = 0$. Hence, $\mathbf{v}_3 = t^2 - \left(\left(\frac{2}{3} \right) / 2 \right) 1 = t^2 - \frac{1}{3}$. Finally,

$$
\begin{aligned}
\mathbf{v}_4 \;&=\; \mathbf{w}_4 - \left(\frac{\langle \mathbf{w}_4, \mathbf{v}_1 \rangle}{\langle \mathbf{v}_1, \mathbf{v}_1 \rangle} \right) \mathbf{v}_1 - \left(\frac{\langle \mathbf{w}_4, \mathbf{v}_2 \rangle}{\langle \mathbf{v}_2, \mathbf{v}_2 \rangle} \right) \mathbf{v}_2 - \left(\frac{\langle \mathbf{w}_4, \mathbf{v}_3 \rangle}{\langle \mathbf{v}_3, \mathbf{v}_3 \rangle} \right) \mathbf{v}_3 \\
&=\; t^3 - \left(\frac{\langle t^3, 1 \rangle}{\langle 1, 1 \rangle} \right) 1 - \left(\frac{\langle t^3, t \rangle}{\langle t, t \rangle} \right) t - \left(\frac{\langle t^3, t^2 \rangle}{\langle t^2, t^2 \rangle} \right) t^2.
\end{aligned}
$$

Now, $\langle t^3, 1 \rangle = 0$, $\langle t^3, t \rangle = \frac{2}{5}$, $\langle t, t \rangle = \frac{2}{3}$, and $\langle t^3, t^2 \rangle = 0$. Hence, $\mathbf{v}_4 = t^3 - \left(\left(\frac{2}{5} \right) / \left(\frac{2}{3} \right) \right) t = t^3 - \frac{3}{5} t$.

Thus, the set $\{ \mathbf{v}_1, \mathbf{v}_2, \mathbf{v}_3, \mathbf{v}_4 \} = \left\{ 1, t, t^2 - \frac{1}{3}, t^3 - \frac{3}{5} t \right\}$ is an orthogonal set of vectors in this inner product space.[3]   ■

We saw in Theorem 6.7 that the transition matrix between orthonormal bases of $\mathbb{R}^n$ is an orthogonal matrix. This result generalizes to inner product spaces as follows:

THEOREM 7.17

Let $\mathcal{V}$ be a finite dimensional real [complex] inner product space, and let B and C be ordered orthonormal bases for $\mathcal{V}$. Then the transition matrix from B to C is an orthogonal [unitary] matrix.

Orthogonal Complements and Orthogonal Projections in Inner Product Spaces

We can generalize the notion of an orthogonal complement of a subspace to inner product spaces as follows:

DEFINITION

Let $\mathcal{W}$ be a subspace of a real (or complex) inner product space $\mathcal{V}$. Then the **orthogonal complement** $\mathcal{W}^\perp$ of $\mathcal{W}$ in $\mathcal{V}$ is the set of all vectors $\mathbf{x} \in \mathcal{V}$ with the property that $\langle \mathbf{x}, \mathbf{w} \rangle = 0$, for all $\mathbf{w} \in \mathcal{W}$.

---

[3] The polynomials $1, t, t^2 - \frac{1}{3}$, and $t^3 - \frac{3}{5} t$ from Example 11 are multiples of the first four **Legendre polynomials**: $1, t, \frac{3}{2} t^2 - \frac{1}{2}, \frac{5}{2} t^3 - \frac{3}{2} t$. All Legendre polynomials equal 1 when $t = 1$. To find the complete set of Legendre polynomials, we can continue the Generalized Gram-Schmidt Process with t^4, t^5, t^6, and so on, and take appropriate multiples so that the resulting polynomials equal 1 when $t = 1$. These polynomials form an (infinite) orthogonal set for the inner product space of Example 11.

EXAMPLE 12

Consider again the real vector space $\mathcal{P}_n$, with the inner product of Example 4 — for $\mathbf{p}_1 = a_n x^n + \cdots + a_1 x + a_0$ and $\mathbf{p}_2 = b_n x^n + \cdots + b_1 x + b_0$, $\langle \mathbf{p}_1, \mathbf{p}_2 \rangle = a_n b_n + \cdots + a_1 b_1 + a_0 b_0$. Example 9 showed that $\{x^n, x^{n-1}, \ldots, x, 1\}$ is an orthogonal basis for $\mathcal{P}_n$ under this inner product. Now, consider the subspace $\mathcal{W}$ spanned by $\{x, 1\}$. A little thought will convince you that $\mathcal{W}^{\perp} = \text{span}\{x^n, x^{n-1}, \ldots, x^2\}$ and so, $\dim(\mathcal{W}) + \dim(\mathcal{W}^{\perp}) = 2 + (n-1) = n + 1 = \dim(\mathcal{P}_n)$. ∎

The following properties of orthogonal complements are the analogs to Theorems 6.10 and 6.11 and Corollaries 6.12 and 6.13 and are proved in a similar manner (see Exercise 22):

THEOREM 7.18

Let $\mathcal{W}$ be a subspace of a real (or complex) inner product space $\mathcal{V}$. Then
    (1) $\mathcal{W}^{\perp}$ is a subspace of $\mathcal{V}$.
    (2) $\mathcal{W} \cap \mathcal{W}^{\perp} = \{\mathbf{0}\}$.
    (3) $\mathcal{W} \subseteq (\mathcal{W}^{\perp})^{\perp}$.
Furthermore, if $\mathcal{V}$ is finite dimensional, then
    (4) If $\{\mathbf{v}_1, \ldots, \mathbf{v}_k\}$ is an orthogonal basis for $\mathcal{W}$ contained in an orthogonal basis $\{\mathbf{v}_1, \ldots, \mathbf{v}_k, \mathbf{v}_{k+1}, \ldots, \mathbf{v}_n\}$ for $\mathcal{V}$, then $\{\mathbf{v}_{k+1}, \ldots, \mathbf{v}_n\}$ is an orthogonal basis for $\mathcal{W}^{\perp}$.
    (5) $\dim(\mathcal{W}) + \dim(\mathcal{W}^{\perp}) = \dim(\mathcal{V})$.
    (6) $(\mathcal{W}^{\perp})^{\perp} = \mathcal{W}$.

Note that if $\mathcal{V}$ is not finite dimensional, $(\mathcal{W}^{\perp})^{\perp}$ is not necessarily equal to $\mathcal{W}$, although it is always true that $\mathcal{W} \subseteq (\mathcal{W}^{\perp})^{\perp}$.[4]

The next theorem is the analog of Theorem 6.14. It holds for any inner product space $\mathcal{V}$ where the subspace $\mathcal{W}$ is finite dimensional. The proof is left for you to do in Exercise 25.

THEOREM 7.19 (Projection Theorem)

Let $\mathcal{W}$ be a finite dimensional subspace of an inner product space $\mathcal{V}$. Then every vector $\mathbf{v} \in \mathcal{V}$ can be expressed in a unique way as $\mathbf{w}_1 + \mathbf{w}_2$, where $\mathbf{w}_1 \in \mathcal{W}$ and $\mathbf{w}_2 \in \mathcal{W}^{\perp}$.

---

[4] The following is an example of a subspace $\mathcal{W}$ of of an infinite dimensional inner product space such that $\mathcal{W} \neq (\mathcal{W}^{\perp})^{\perp}$. Let $\mathcal{V}$ be the inner product space of Example 5 with $a = 0, b = 1$, and let

$$\mathbf{f}_n(x) = \begin{cases} 1, & \text{if } x > \frac{1}{n} \\ nx, & \text{if } 0 \leq x \leq \frac{1}{n} \end{cases}$$. Let $\mathcal{W}$ be the subspace of $\mathcal{V}$ spanned by $\{\mathbf{f}_1, \mathbf{f}_2, \mathbf{f}_3, \ldots\}$. It can be shown that $\mathbf{f}(x) = 1$ is not in $\mathcal{W}$, but $\mathbf{f}(x) \in (\mathcal{W}^{\perp})^{\perp}$. Hence, $\mathcal{W} \neq (\mathcal{W}^{\perp})^{\perp}$.

As before, we define the **orthogonal projection** of a vector $\mathbf{v}$ onto a subspace W as follows:

DEFINITION

If $\{\mathbf{v}_1, \ldots, \mathbf{v}_k\}$ is an orthonormal basis for W, a subspace of an inner product space V, then the vector $\mathbf{proj}_W \mathbf{v} = \langle \mathbf{v}, \mathbf{v}_1 \rangle \mathbf{v}_1 + \cdots + \langle \mathbf{v}, \mathbf{v}_k \rangle \mathbf{v}_k$ is called the **orthogonal projection of v onto** W.

It can be shown that the formula for $\mathbf{proj}_W \mathbf{v}$ yields the unique vector $\mathbf{w}_1$ in the Projection Theorem. Therefore, the choice of orthonormal basis in the definition does not matter because any choice leads to the same vector for $\mathbf{proj}_W \mathbf{v}$. Hence, the Projection Theorem can be restated as follows:

> If W is a finite dimensional subspace of an inner product space V, and if $\mathbf{v} \in V$, then $\mathbf{v}$ can be expressed as $\mathbf{w}_1 + \mathbf{w}_2$, where $\mathbf{w}_1 = \mathbf{proj}_W \mathbf{v} \in W$ and $\mathbf{w}_2 = \mathbf{v} - \mathbf{proj}_W \mathbf{v} \in W^\perp$.

EXAMPLE 13

Consider again the real vector space V of real continuous functions in Example 8, where $\langle \mathbf{f}, \mathbf{g} \rangle = \int_{-\pi}^{\pi} \mathbf{f}(t)\mathbf{g}(t)\, dt$. Notice from that example that the set $\left\{ 1/\sqrt{2\pi},\ (\sin t)/\sqrt{\pi} \right\}$ is an orthonormal (and hence, linearly independent) set of vectors in V. Let $W = \mathrm{span}\left(\left\{ 1/\sqrt{2\pi},\ (\sin t)/\sqrt{\pi} \right\} \right)$ in V. Then any continuous function $\mathbf{f}$ in V can be expressed uniquely as $\mathbf{f}_1 + \mathbf{f}_2$, where $\mathbf{f}_1 \in W$ and $\mathbf{f}_2 \in W^\perp$.

We illustrate this decomposition for the function $\mathbf{f} = t + 1$. Now,

$$\mathbf{f}_1 = \mathbf{proj}_W \mathbf{f} = c_1 \left(\frac{1}{\sqrt{2\pi}} \right) + c_2 \left(\frac{\sin t}{\sqrt{\pi}} \right),$$

where $c_1 = \left\langle (t + 1), 1/\sqrt{2\pi} \right\rangle$ and $c_2 = \left\langle (t + 1), (\sin t)/\sqrt{\pi} \right\rangle$. Then

$$
\begin{aligned}
c_1 &= \int_{-\pi}^{\pi} (t + 1) \left(\frac{1}{\sqrt{2\pi}} \right) dt = \frac{1}{\sqrt{2\pi}} \int_{-\pi}^{\pi} (t + 1)\, dt \\
&= \frac{1}{\sqrt{2\pi}} \left(\frac{t^2}{2} + t \right) \Bigg|_{-\pi}^{\pi} = \frac{2\pi}{\sqrt{2\pi}} = \sqrt{2\pi}.
\end{aligned}
$$

Also,

$$
\begin{aligned}
c_2 &= \int_{-\pi}^{\pi} (t + 1) \left(\frac{\sin t}{\sqrt{\pi}} \right) dt = \frac{1}{\sqrt{\pi}} \int_{-\pi}^{\pi} (t + 1) \sin t\, dt \\
&= \frac{1}{\sqrt{\pi}} \left(\int_{-\pi}^{\pi} t \sin t\, dt + \int_{-\pi}^{\pi} \sin t\, dt \right).
\end{aligned}
$$

The very last integral equals zero. Using integration by parts on the other integral, we obtain

$$c_2 = \frac{1}{\sqrt{\pi}} \left((-t \cos t)|_{-\pi}^{\pi} + \int_{-\pi}^{\pi} \cos t \, dt \right) = \left(\frac{1}{\sqrt{\pi}} \right) 2\pi = 2\sqrt{\pi}.$$

Hence,

$$\mathbf{f}_1 = c_1 \left(\frac{1}{\sqrt{2\pi}} \right) + c_2 \left(\frac{\sin t}{\sqrt{\pi}} \right) = \sqrt{2\pi} \left(\frac{1}{\sqrt{2\pi}} \right) + 2\sqrt{\pi} \left(\frac{\sin t}{\sqrt{\pi}} \right) = 1 + 2 \sin t.$$

Then by the Projection Theorem, $\mathbf{f}_2 = \mathbf{f} - \mathbf{f}_1 = (t+1) - (1 + 2 \sin t) = t - 2 \sin t$ is orthogonal to $\mathcal{W}$. We check that $\mathbf{f}_2 \in \mathcal{W}^\perp$ by showing that $\mathbf{f}_2$ is orthogonal to both $1/\sqrt{2\pi}$ and $(\sin t)/\sqrt{\pi}$.

$$\left\langle \mathbf{f}_2, \frac{1}{\sqrt{2\pi}} \right\rangle = \int_{-\pi}^{\pi} (t - 2 \sin t) \left(\frac{1}{\sqrt{2\pi}} \right) dt = \left(\frac{1}{\sqrt{2\pi}} \right) \left(\frac{t^2}{2} + 2 \cos t \right) \Bigg|_{-\pi}^{\pi} = 0.$$

Also,

$$\left\langle \mathbf{f}_2, \frac{\sin t}{\sqrt{\pi}} \right\rangle = \int_{-\pi}^{\pi} (t - 2 \sin t) \left(\frac{\sin t}{\sqrt{\pi}} \right) dt = \frac{1}{\sqrt{\pi}} \int_{-\pi}^{\pi} t \sin t \, dt - \frac{2}{\sqrt{\pi}} \int_{-\pi}^{\pi} \sin^2 t \, dt,$$

which equals $2\sqrt{\pi} - 2\sqrt{\pi} = 0$. ∎

Exercises for Section 7.5

1. (a) Let $\mathbf{A}$ be a nonsingular $n \times n$ real matrix. For $\mathbf{x}, \mathbf{y} \in \mathbb{R}^n$, define an operation $\langle \mathbf{x}, \mathbf{y} \rangle = (\mathbf{A}\mathbf{x}) \cdot (\mathbf{A}\mathbf{y})$ (dot product). Prove that this operation is a real inner product on $\mathbb{R}^n$.

   ★(b) For the inner product in part (a) with $\mathbf{A} = \begin{bmatrix} 5 & 4 & 2 \\ -2 & 3 & 1 \\ 1 & -1 & 0 \end{bmatrix}$, find $\langle \mathbf{x}, \mathbf{y} \rangle$ and $\|\mathbf{x}\|$, for $\mathbf{x} = [3, -2, 4]$ and $\mathbf{y} = [-2, 1, -1]$.

2. Define an operation $\langle \, , \, \rangle$ on $\mathcal{P}_n$ as follows: if $\mathbf{p}_1 = a_n x^n + \cdots + a_1 x + a_0$ and $\mathbf{p}_2 = b_n x^n + \cdots + b_1 x + b_0$, let $\langle \mathbf{p}_1, \mathbf{p}_2 \rangle = a_n b_n + \cdots + a_1 b_1 + a_0 b_0$. Prove that this operation is a real inner product on $\mathcal{P}_n$.

3. (a) Let a and b be fixed real numbers with $a < b$, and let $\mathcal{V}$ be the set of all real continuous functions on $[a, b]$. Define $\langle \, , \, \rangle$ on $\mathcal{V}$ by $\langle \mathbf{f}, \mathbf{g} \rangle = \int_a^b \mathbf{f}(t)\mathbf{g}(t) \, dt$. Prove that this operation is a real inner product on $\mathcal{V}$.

   ★(b) For the inner product of part (a) with $a = 0$ and $b = \pi$, find $\langle \mathbf{f}, \mathbf{g} \rangle$ and $\|\mathbf{f}\|$, for $\mathbf{f} = e^t$ and $\mathbf{g} = \sin t$.

4. Define $\langle \, , \, \rangle$ on the real vector space $\mathcal{M}_{mn}$ by $\langle \mathbf{A}, \mathbf{B} \rangle = \text{trace}(\mathbf{A}^T \mathbf{B})$. Prove that this operation is a real inner product on $\mathcal{M}_{mn}$. (Hint: Refer to Exercise 14 in Section 1.4 and Exercise 22 in Section 1.5.)

5. (a) Prove part (1) of Theorem 7.11. (Hint: $\mathbf{0} = \mathbf{0} + \mathbf{0}$. Use property (4) in the definition of an inner product space.)

  (b) Prove part (3) of Theorem 7.11. (Be sure to give a proof for both real and complex inner product spaces.)

▶**6.** Prove Theorem 7.12.

7. Let $\mathbf{x}, \mathbf{y} \in V$, a real inner product space.

  (a) Prove that $\|\mathbf{x} + \mathbf{y}\|^2 = \|\mathbf{x}\|^2 + 2\langle \mathbf{x}, \mathbf{y} \rangle + \|\mathbf{y}\|^2$.

  (b) Show that $\mathbf{x}$ and $\mathbf{y}$ are orthogonal in V if and only if $\|\mathbf{x} + \mathbf{y}\|^2 = \|\mathbf{x}\|^2 + \|\mathbf{y}\|^2$.

  (c) Show that $\frac{1}{2}\left(\|\mathbf{x} + \mathbf{y}\|^2 + \|\mathbf{x} - \mathbf{y}\|^2\right) = \|\mathbf{x}\|^2 + \|\mathbf{y}\|^2$.

8. The following formulas show how the value of the inner product can be derived from the norm (length):

  (a) Let $\mathbf{x}, \mathbf{y} \in V$, a real inner product space. Prove the following (real) **Polarization Identity:** $\langle \mathbf{x}, \mathbf{y} \rangle = \frac{1}{4}\left(\|\mathbf{x} + \mathbf{y}\|^2 - \|\mathbf{x} - \mathbf{y}\|^2\right)$.

  (b) Let $\mathbf{x}, \mathbf{y} \in V$, a complex inner product space. Prove the following **Complex Polarization Identity:**

$$\langle \mathbf{x}, \mathbf{y} \rangle = \frac{1}{4}\left(\left(\|\mathbf{x} + \mathbf{y}\|^2 - \|\mathbf{x} - \mathbf{y}\|^2\right) + i\left(\|\mathbf{x} + i\mathbf{y}\|^2 - \|\mathbf{x} - i\mathbf{y}\|^2\right)\right).$$

9. Consider the inner product space V of Example 5, with $a = 0$ and $b = \pi$.

  ★(a) Find the distance between $\mathbf{f} = t$ and $\mathbf{g} = \sin t$ in V.

  (b) Find the angle between $\mathbf{f} = e^t$ and $\mathbf{g} = \sin t$ in V.

10. Consider the inner product space V of Example 3, using

$$\mathbf{A} = \begin{bmatrix} -2 & 0 & 1 \\ 1 & -1 & 2 \\ 3 & -1 & -1 \end{bmatrix}.$$

  (a) Find the distance between $\mathbf{x} = [2, -1, 3]$ and $\mathbf{y} = [5, -2, 2]$ in V.

  ★(b) Find the angle between $\mathbf{x} = [2, -1, 3]$ and $\mathbf{y} = [5, -2, 2]$ in V.

11. Let V be an inner product space.

  (a) Prove part (1) of Theorem 7.13. (Hint: Modify the proof of Theorem 1.6.)

  (b) Prove part (2) of Theorem 7.13. (Hint: Modify the proof of Theorem 1.7.)

12. Let f and g be continuous real-valued functions defined on a closed interval $[a, b]$. Show that

$$\left(\int_a^b f(t)g(t)\,dt\right)^2 \le \int_a^b (f(t))^2\,dt \int_a^b (g(t))^2\,dt.$$

(Hint: Use the Cauchy-Schwarz Inequality in an appropriate inner product space.)

13. A **metric space** is a set in which every pair of elements x, y has been assigned a real number distance d with the following properties:

  (i) $d(x, y) = d(y, x)$.

(ii) $d(x, y) \geq 0$, with $d(x, y) = 0$ if and only if $x = y$.

(iii) $d(x, y) \leq d(x, z) + d(z, y)$, for all z in the set.

Prove that every inner product space is a metric space with $d(\mathbf{x}, \mathbf{y})$ taken to be $\|\mathbf{x} - \mathbf{y}\|$ for all vectors $\mathbf{x}$ and $\mathbf{y}$ in the space.

14. Determine whether the following sets of vectors are orthogonal:

   $\star$(a) $\{t^2, \ t+1, \ t-1\}$ in $\mathcal{P}_3$, under the inner product of Example 4

   (b) $\{[15, 9, 19], [-2, -1, -2], [-12, -9, -14]\}$ in $\mathbb{R}^3$, under the inner product of Example 3, with

$$\mathbf{A} = \begin{bmatrix} -3 & 1 & 2 \\ 0 & -2 & 1 \\ 2 & -1 & -1 \end{bmatrix}$$

   $\star$(c) $\{[5, -2], [3, 4]\}$ in $\mathbb{R}^2$, under the inner product of Example 2

   (d) $\{3t^2 - 1, \ 4t, \ 5t^3 - 3t\}$ in $\mathcal{P}_3$, under the inner product of Example 11

15. Prove Theorem 7.14. (Hint: Modify the proof of Result 7 in Section 1.3.)

16. (a) Show that $\int_{-\pi}^{\pi} \cos mt \, dt = 0$ and $\int_{-\pi}^{\pi} \sin nt \, dt = 0$, for all integers $m, n \geq 1$.

   (b) Show that $\int_{-\pi}^{\pi} \cos mt \cos nt \, dt = 0$ and $\int_{-\pi}^{\pi} \sin mt \sin nt \, dt = 0$, for any *distinct* integers $m, n \geq 1$. (Hint: Use trigonometric identities.)

   (c) Show that $\int_{-\pi}^{\pi} \cos mt \sin nt \, dt = 0$, for any integers $m, n \geq 1$.

   (d) Conclude from parts (a), (b), and (c) that $\{1, \cos t, \sin t, \cos 2t, \sin 2t, \cos 3t, \sin 3t, \ldots\}$ is an orthogonal set of real continuous functions on $[-\pi, \pi]$, as claimed after Example 8.

17. Prove Theorem 7.15. (Hint: Modify the proof of Theorem 6.3.)

18. Let $\{\mathbf{v}_1, \ldots, \mathbf{v}_k\}$ be an orthonormal basis for a complex inner product space $\mathcal{V}$. Prove that for all $\mathbf{v}, \mathbf{w} \in \mathcal{V}$,

$$\langle \mathbf{v}, \mathbf{w} \rangle = \langle \mathbf{v}, \mathbf{v}_1 \rangle \overline{\langle \mathbf{w}, \mathbf{v}_1 \rangle} + \langle \mathbf{v}, \mathbf{v}_2 \rangle \overline{\langle \mathbf{w}, \mathbf{v}_2 \rangle} + \cdots + \langle \mathbf{v}, \mathbf{v}_k \rangle \overline{\langle \mathbf{w}, \mathbf{v}_k \rangle}.$$

(Compare this with Exercise 9(a) in Section 6.1.)

$\star$**19.** Use the Generalized Gram-Schmidt Process to find an orthogonal basis for $\mathcal{P}_2$ containing $t^2 - t + 1$ under the inner product of Example 11.

20. Use the Generalized Gram-Schmidt Process to find an orthogonal basis for $\mathbb{R}^3$ containing $[-9, -4, 8]$ under the inner product of Example 3 with the matrix

$$\mathbf{A} = \begin{bmatrix} 2 & 1 & 3 \\ 3 & -1 & 3 \\ 2 & -1 & 2 \end{bmatrix}.$$

21. Prove Theorem 7.16. (Hint: Modify the proof of Theorem 6.4.)

22. (a) Prove parts (1) and (2) of Theorem 7.18. (Hint: Modify the proof of Theorem 6.10.)

   ▶(b) Prove parts (4) and (5) of Theorem 7.18. (Hint: Modify the proofs of Theorem 6.11 and Corollary 6.12.)

   (c) Prove part (3) of Theorem 7.18.

▶(d) Prove part (6) of Theorem 7.18. (Hint: Use part (5) of Theorem 7.18 to show that $\dim(\mathcal{W}) = \dim\left(\left(\mathcal{W}^\perp\right)^\perp\right)$. Then use part (c) and apply Theorem 4.17, or its complex analog.)

⋆23. Find $\mathcal{W}^\perp$ if $\mathcal{W} = \mathrm{span}\left(\{t^3 + t^2,\ t-1\}\right)$ in $\mathcal{P}_3$ with the inner product of Example 4.

24. Find an orthogonal basis for $\mathcal{W}^\perp$ if $\mathcal{W} = \mathrm{span}(\{(t-1)^2\})$ in $\mathcal{P}_2$, with the inner product $\langle \mathbf{f}, \mathbf{g} \rangle = \int_0^1 \mathbf{f}(t)\mathbf{g}(t)\,dt$, for all $\mathbf{f}, \mathbf{g} \in \mathcal{P}_2$.

▶**25.** Prove Theorem 7.19. (Hint: Choose an orthonormal basis $\{\mathbf{v}_1, \ldots, \mathbf{v}_k\}$ for $\mathcal{W}$. Then define $\mathbf{w}_1 = \mathbf{proj}_{\mathcal{W}}\mathbf{v} = \langle \mathbf{v}, \mathbf{v}_1 \rangle \mathbf{v}_1 + \cdots + \langle \mathbf{v}, \mathbf{v}_k \rangle \mathbf{v}_k$. Let $\mathbf{w}_2 = \mathbf{v} - \mathbf{w}_1$, and prove $\mathbf{w}_2 \in \mathcal{W}^\perp$. Finally, see the proof of Theorem 6.14 for uniqueness.)

⋆26. In the inner product space of Example 8, decompose $\mathbf{f} = \frac{1}{k}e^t$, where $k = e^\pi - e^{-\pi}$, as $\mathbf{w}_1 + \mathbf{w}_2$, where $\mathbf{w}_1 \in \mathcal{W} = \mathrm{span}(\{\cos t, \sin t\})$ and $\mathbf{w}_2 \in \mathcal{W}^\perp$. Check that $\langle \mathbf{w}_1, \mathbf{w}_2 \rangle = 0$. (Hint: First find an orthonormal basis for $\mathcal{W}$.)

27. Decompose $\mathbf{v} = 4t^2 - t + 3$ in $\mathcal{P}_2$ as $\mathbf{w}_1 + \mathbf{w}_2$, where $\mathbf{w}_1 \in \mathcal{W} = \mathrm{span}(\{2t^2 - 1,\ t+1\})$ and $\mathbf{w}_2 \in \mathcal{W}^\perp$, under the inner product of Example 11. Check that $\langle \mathbf{w}_1, \mathbf{w}_2 \rangle = 0$. (Hint: First find an orthonormal basis for $\mathcal{W}$.)

28. Bessel's Inequality: Let $\mathcal{V}$ be a real inner product space, and let $\{\mathbf{v}_1, \ldots, \mathbf{v}_k\}$ be an orthonormal set in $\mathcal{V}$. Prove that for any vector $\mathbf{v} \in \mathcal{V}$, $\sum_{i=1}^{k} \langle \mathbf{v}, \mathbf{v}_i \rangle^2 \le \|\mathbf{v}\|^2$. (Hint: Let $\mathcal{W} = \mathrm{span}(\{\mathbf{v}_1, \ldots, \mathbf{v}_k\})$. Now, $\mathbf{v} = \mathbf{w}_1 + \mathbf{w}_2$, where $\mathbf{w}_1 = \mathbf{proj}_{\mathcal{W}}\mathbf{v} \in \mathcal{W}$ and $\mathbf{w}_2 \in \mathcal{W}^\perp$. Expand $\langle \mathbf{v}, \mathbf{v} \rangle = \langle \mathbf{w}_1 + \mathbf{w}_2, \mathbf{w}_1 + \mathbf{w}_2 \rangle$. Show that $\|\mathbf{v}\|^2 \ge \|\mathbf{w}_1\|^2$, and use the definition of $\mathbf{proj}_{\mathcal{W}}\mathbf{v}$.)

29. Let $\mathcal{W}$ be a finite dimensional subspace of an inner product space $\mathcal{V}$. Consider the mapping $L\colon \mathcal{V} \to \mathcal{W}$ given by $L(\mathbf{v}) = \mathbf{proj}_{\mathcal{W}}\mathbf{v}$.

(a) Prove that L is a linear transformation.

⋆(b) What are the kernel and range of L?

(c) Show that $L \circ L = L$.

⋆30. True or False:

(a) If $\mathcal{V}$ is a complex inner product space, then for all $\mathbf{x} \in \mathcal{V}$ and all $k \in \mathbb{C}$, $\|k\mathbf{x}\| = \overline{k}\|\mathbf{x}\|$.

(b) In a complex inner product space, the distance between two distinct vectors can be a pure imaginary number.

(c) Every linearly independent set of unit vectors in an inner product space is an orthonormal set.

(d) It is possible to define more than one inner product on the same vector space.

(e) The uniqueness proof of the Projection Theorem shows that if $\mathcal{W}$ is a subspace of $\mathbb{R}^n$, then $\mathbf{proj}_W\mathbf{v}$ is independent of the particular inner product used on $\mathbb{R}^n$.

Chapter 8

Additional Applications

Mathematicians: Apply Within

Mathematics is everywhere. It is the language used to describe almost every aspect of our physical world and our society. It is a tool used to analyze and solve problems regarding almost everything we do. In particular, linear algebra is one of the most useful devices on the mathematician's toolbelt. There are important applications of linear algebra in almost every discipline.

In this chapter, we explore important uses of linear algebra in fields ranging from electronics to psychology. We show how linear algebra can be used to encode messages, fit functions to raw data, and rotate graphics on a computer screen. And, we illustrate how linear algebra and calculus can be used in tandem to change variables in multiple integrals, solve certain differential equations, and simplify the equation of a conic section. The applications given in this chapter are just a small sample of the myriad of problems in which linear algebra is used on a daily basis.

In this chapter, we present several additional practical applications of linear algebra in mathematics and the sciences.

8.1 Graph Theory

Prerequisite: Section 1.5, Matrix Multiplication

Multiplication of matrices is widely used in graph theory, a branch of mathematics that has come into prominence for modeling many situations in computer science, business, and the social sciences. We begin by introducing graphs and digraphs and then examine their relationship with matrices. Our main goal is to show how matrices are used to calculate the number of paths of a certain length between vertices of a graph or digraph.

Graphs and Digraphs

DEFINITION

A **graph** is a finite collection of **vertices** (points) together with a finite collection of **edges** (curves), each of which has two (not necessarily distinct) vertices as endpoints.

For example, Figure 8.1 depicts two graphs. Note that a graph may have an edge connecting some vertex to itself. Such edges are called **loops**. A graph with no loops, such as G_1 in Figure 8.1, is said to be **loop-free**.

Figure 8.1

Two examples of graphs

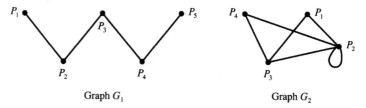

Graph G_1 Graph G_2

A **digraph**, or **directed graph**, is a special type of graph in which each edge is assigned a "direction." Some examples of digraphs appear in Figure 8.2.

Figure 8.2

Two examples of digraphs

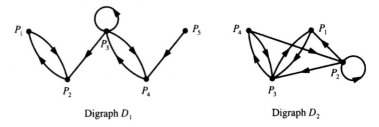

Digraph D_1 Digraph D_2

Although the edges in a digraph may resemble vectors, they are not necessarily vectors since there is usually no coordinate system present. One interpretation for graphs and digraphs is to consider the vertices as towns and the edges as roads connecting them. In the case of a digraph, we can think of the roads as one-way streets. Notice that some pairs of towns may not be connected by roads. Another interpretation for graphs and digraphs is to consider the vertices as relay stations and the edges as communication channels (for example, phone lines) between the stations. The stations could be individual people, homes, radio/TV installations, or even computer terminals hooked into a network. There are additional interpretations for graphs and digraphs in the exercises.

In this section, we consider only "simple" graphs and digraphs. A **simple graph** is one having at most one edge between each pair of vertices. Similarly, a **simple digraph** is one having at most one edge in each direction between each pair of vertices. All of the graphs and digraphs pictured in this section are simple.

The Adjacency Matrix

The pattern of edges between the vertices in a graph or digraph can be summarized in an algebraic way using matrices.

DEFINITION

The **adjacency matrix of a graph** having vertices $P_1, P_2, \ldots, P_n$ is the $n \times n$ matrix whose (i, j) entry is 1 if there is an edge between P_i and P_j and 0 otherwise.

The **adjacency matrix of a digraph** having vertices $P_1, P_2, \ldots, P_n$ is the $n \times n$ matrix whose (i, j) entry is 1 if there is an edge directed from P_i to P_j and 0 otherwise.

EXAMPLE 1   The adjacency matrices for the two graphs in Figure 8.1 and the two digraphs in Figure 8.2 are as follows:

$$
\begin{array}{c|ccccc}
 & P_1 & P_2 & P_3 & P_4 & P_5 \\
\hline
P_1 & 0 & 1 & 0 & 0 & 0 \\
P_2 & 1 & 0 & 1 & 0 & 0 \\
P_3 & 0 & 1 & 0 & 1 & 0 \\
P_4 & 0 & 0 & 1 & 0 & 1 \\
P_5 & 0 & 0 & 0 & 1 & 0
\end{array}
$$

Adjacency Matrix for G_1

$$
\begin{array}{c|cccc}
 & P_1 & P_2 & P_3 & P_4 \\
\hline
P_1 & 0 & 1 & 1 & 0 \\
P_2 & 1 & 1 & 1 & 1 \\
P_3 & 1 & 1 & 0 & 1 \\
P_4 & 0 & 1 & 1 & 0
\end{array}
$$

Adjacency Matrix for G_2

$$
\begin{array}{c|ccccc}
 & P_1 & P_2 & P_3 & P_4 & P_5 \\
\hline
P_1 & 0 & 1 & 0 & 0 & 0 \\
P_2 & 1 & 0 & 0 & 0 & 0 \\
P_3 & 0 & 1 & 1 & 1 & 0 \\
P_4 & 0 & 0 & 1 & 0 & 0 \\
P_5 & 0 & 0 & 0 & 1 & 0
\end{array}
$$

Adjacency Matrix for D_1

$$
\begin{array}{c|cccc}
 & P_1 & P_2 & P_3 & P_4 \\
\hline
P_1 & 0 & 0 & 1 & 0 \\
P_2 & 1 & 1 & 1 & 0 \\
P_3 & 1 & 0 & 0 & 1 \\
P_4 & 0 & 1 & 1 & 0
\end{array}
$$

Adjacency Matrix for D_2

■

The adjacency matrix of any graph is symmetric, for the obvious reason that there is an edge between P_i and P_j if and only if there is an edge (the same one) between P_j and P_i. However, the adjacency matrix for a digraph is usually not symmetric, since the existence of an edge from P_i to P_j does not necessarily imply the existence of an edge in the reverse direction.

Paths in a Graph or Digraph

We often want to know how many different routes exist between two given vertices in a graph or digraph.

DEFINITION

A **path** (or **chain**) between two vertices P_i and P_j in a graph or digraph is a finite sequence of edges with the following properties:
- (1) The first edge "begins" at P_i.
- (2) The last edge "ends" at P_j.
- (3) Each edge after the first one in the sequence "begins" at the vertex where the previous edge "ended."

The **length** of a path is the number of edges in the path.

EXAMPLE 2   Consider the digraph pictured in Figure 8.3. There are many different types of paths from P_1 to P_5. For example,

- (1) $P_1 \to P_2 \to P_5$
- (2) $P_1 \to P_2 \to P_3 \to P_5$
- (3) $P_1 \to P_4 \to P_3 \to P_5$
- (4) $P_1 \to P_4 \to P_4 \to P_3 \to P_5$
- (5) $P_1 \to P_2 \to P_5 \to P_4 \to P_3 \to P_5$.

(Can you find other paths from P_1 to P_5?) Path (1) is a path of length 2 (or a 2-chain); paths (2), (3), (4), and (5) are paths of lengths 3, 3, 4, and 5, respectively. ∎

Figure 8.3

Digraph for Examples 2, 3, and 4

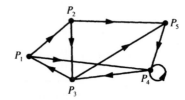

Counting Paths

Our goal is to calculate exactly how many paths of a given length exist between two vertices in a graph or digraph. For example, suppose we want to know precisely how many paths of length 4 from vertex P_2 to vertex P_4 exist in the digraph of Figure 8.3. We could attempt to list them, but the chance of making a mistake in counting them all can cast doubt on our final total. However, the next theorem, which you are asked to prove in Exercise 11, gives an algebraic method to get the exact count using the adjacency matrix.

THEOREM 8.1

Let $\mathbf{A}$ be the adjacency matrix for a graph or digraph having vertices $P_1, P_2, \ldots, P_n$. Then the total number of paths from P_i to P_j of length k is given by the (i, j) entry in the matrix $\mathbf{A}^k$.

EXAMPLE 3

Consider again the digraph in Figure 8.3. The adjacency matrix for this digraph is

$$\mathbf{A} = \begin{array}{c@{}c} & \begin{array}{ccccc} P_1 & P_2 & P_3 & P_4 & P_5 \end{array} \\ \begin{array}{c} P_1 \\ P_2 \\ P_3 \\ P_4 \\ P_5 \end{array} & \left[\begin{array}{ccccc} 0 & 1 & 0 & 1 & 0 \\ 0 & 0 & 1 & 0 & 1 \\ 1 & 0 & 0 & 0 & 1 \\ 0 & 0 & 1 & 1 & 0 \\ 0 & 0 & 0 & 1 & 0 \end{array} \right] \end{array}.$$

To find the number of paths of length 4 from P_1 to P_4, we need to calculate the $(1, 4)$ entry of $\mathbf{A}^4$. Now,

$$\mathbf{A}^4 = \left(\mathbf{A}^2\right)^2 = \left(\left[\begin{array}{ccccc} 0 & 0 & 2 & 1 & 1 \\ 1 & 0 & 0 & 1 & 1 \\ 0 & 1 & 0 & 2 & 0 \\ 1 & 0 & 1 & 1 & 1 \\ 0 & 0 & 1 & 1 & 0 \end{array} \right] \right)^2 = \begin{array}{c@{}c} & \begin{array}{ccccc} P_1 & P_2 & P_3 & P_4 & P_5 \end{array} \\ \begin{array}{c} P_1 \\ P_2 \\ P_3 \\ P_4 \\ P_5 \end{array} & \left[\begin{array}{ccccc} 1 & 2 & 2 & 6 & 1 \\ 1 & 0 & 4 & 3 & 2 \\ 3 & 0 & 2 & 3 & 3 \\ 1 & 1 & 4 & 5 & 2 \\ 1 & 1 & 1 & 3 & 1 \end{array} \right] \end{array}.$$

Since the $(1, 4)$ entry is 6, there are exactly six paths of length 4 from P_1 to P_4. Looking at the digraph, we can see that these paths are

$$P_1 \rightarrow P_2 \rightarrow P_3 \rightarrow P_1 \rightarrow P_4$$
$$P_1 \rightarrow P_2 \rightarrow P_3 \rightarrow P_5 \rightarrow P_4$$
$$P_1 \rightarrow P_2 \rightarrow P_5 \rightarrow P_4 \rightarrow P_4$$
$$P_1 \rightarrow P_4 \rightarrow P_3 \rightarrow P_1 \rightarrow P_4$$
$$P_1 \rightarrow P_4 \rightarrow P_3 \rightarrow P_5 \rightarrow P_4$$
$$P_1 \rightarrow P_4 \rightarrow P_4 \rightarrow P_4 \rightarrow P_4.$$

∎

Of course, we can generalize the result in Theorem 8.1. A little thought will convince you of the following:

The total number of paths of length $\leq k$ from a vertex P_i to a vertex P_j in a graph or digraph is the sum of the (i, j) entries of the matrices $\mathbf{A}, \mathbf{A}^2, \mathbf{A}^3, \ldots, \mathbf{A}^k$.

EXAMPLE 4   For the digraph in Figure 8.3, we will calculate the total number of paths of length ≤ 4 from P_2 to P_3. We listed the adjacency matrix $\mathbf{A}$ for this digraph in Example 3, as well as the products $\mathbf{A}^2$ and $\mathbf{A}^4$. You can verify that $\mathbf{A}^3$ is given by

$$
\mathbf{A}^3 = \begin{array}{c} \\ P_1 \\ P_2 \\ P_3 \\ P_4 \\ P_5 \end{array} \begin{array}{ccccc} P_1 & P_2 & P_3 & P_4 & P_5 \\ \left[\begin{array}{ccccc} 2 & 0 & 1 & 2 & 2 \\ 0 & 1 & 1 & 3 & 0 \\ 0 & 0 & 3 & 2 & 1 \\ 1 & 1 & 1 & 3 & 1 \\ 1 & 0 & 1 & 1 & 1 \end{array}\right]. \end{array}
$$

Then, a quick calculation gives

$$
\mathbf{A} + \mathbf{A}^2 + \mathbf{A}^3 + \mathbf{A}^4 = \begin{array}{c} \\ P_1 \\ P_2 \\ P_3 \\ P_4 \\ P_5 \end{array} \begin{array}{ccccc} P_1 & P_2 & P_3 & P_4 & P_5 \\ \left[\begin{array}{ccccc} 3 & 3 & 5 & 10 & 4 \\ 2 & 1 & 6 & 7 & 4 \\ 4 & 1 & 5 & 7 & 5 \\ 3 & 2 & 7 & 10 & 4 \\ 2 & 1 & 3 & 6 & 2 \end{array}\right]. \end{array}
$$

Hence, the number of paths of length ≤ 4 from P_2 to P_3 is the $(2, 3)$ entry of this matrix, which is 6. A list of these paths is as follows:

$$
\begin{aligned}
&P_2 \rightarrow P_3 \\
&P_2 \rightarrow P_5 \rightarrow P_4 \rightarrow P_3 \\
&P_2 \rightarrow P_3 \rightarrow P_1 \rightarrow P_2 \rightarrow P_3 \\
&P_2 \rightarrow P_3 \rightarrow P_1 \rightarrow P_4 \rightarrow P_3 \\
&P_2 \rightarrow P_3 \rightarrow P_5 \rightarrow P_4 \rightarrow P_3 \\
&P_2 \rightarrow P_5 \rightarrow P_4 \rightarrow P_4 \rightarrow P_3
\end{aligned}
$$

In fact, since we calculated all of the entries of the matrix $\mathbf{A} + \mathbf{A}^2 + \mathbf{A}^3 + \mathbf{A}^4$, we can now find the total number of paths of length ≤ 4 between any pair of given vertices. For example, the total number of paths of length ≤ 4 between P_3 and P_5 is 5 because that is the $(3, 5)$ entry of the sum.   Of course, if we only want to know the number of paths of length ≤ 4 from just one vertex to one other vertex, we would only need a single entry of $\mathbf{A} + \mathbf{A}^2 + \mathbf{A}^3 + \mathbf{A}^4$ and it would not be necessary to compute all of the entries of the sum.   ∎

Exercises for Section 8.1

Note: You may want to use a computer or calculator to perform the matrix computations in these exercises.

    *1. For each of the graphs and digraphs in Figure 8.4, give the corresponding adjacency matrix. Which of these matrices are symmetric?

Figure 8.4

Graphs and digraphs for Exercise 1

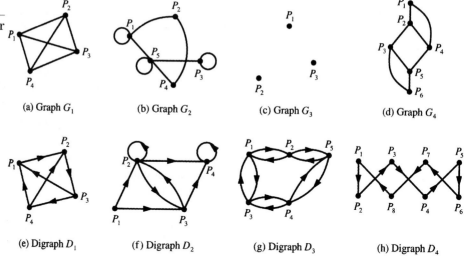

(a) Graph G_1 (b) Graph G_2 (c) Graph G_3 (d) Graph G_4

(e) Digraph D_1 (f) Digraph D_2 (g) Digraph D_3 (h) Digraph D_4

*2. Which of the given matrices could be the adjacency matrix for a simple graph or digraph? Draw the corresponding graph and/or digraph when appropriate.

$$\mathbf{A} = \begin{bmatrix} -1 & 4 \\ 0 & 1 \\ 6 & 0 \end{bmatrix} \quad \mathbf{B} = \begin{bmatrix} 2 & 0 \\ 0 & -1 \end{bmatrix} \quad \mathbf{C} = \begin{bmatrix} 6 & 0 & 0 & 0 \\ 0 & 6 & 0 & 0 \\ 0 & 0 & 6 & 0 \\ 0 & 0 & 0 & 6 \end{bmatrix} \quad \mathbf{D} = \begin{bmatrix} -1 \\ 4 \\ 2 \end{bmatrix}$$

$$\mathbf{E} = \begin{bmatrix} 0 & 0 & 0 & 6 \\ 0 & 0 & 6 & 0 \\ 0 & -6 & 0 & 0 \\ -6 & 0 & 0 & 0 \end{bmatrix} \quad \mathbf{F} = \begin{bmatrix} 1 & 0 & 1 & 0 & 1 \\ 0 & 1 & 0 & 0 & 1 \\ 1 & 0 & 0 & 1 & 1 \\ 0 & 0 & 1 & 0 & 0 \\ 1 & 1 & 1 & 0 & 1 \end{bmatrix} \quad \mathbf{G} = \begin{bmatrix} 1 & 1 & 1 \\ 0 & 1 & 1 \\ 0 & 0 & 1 \end{bmatrix}$$

$$\mathbf{H} = \begin{bmatrix} 0 & 0 & 0 \\ 1 & 0 & 0 \\ 1 & 1 & 0 \end{bmatrix} \quad \mathbf{I} = \begin{bmatrix} 1 & 0 & 0 \\ 0 & 1 & 0 \\ 0 & 0 & 1 \end{bmatrix} \quad \mathbf{J} = \begin{bmatrix} 1 & 2 & 3 & 4 \\ -2 & 1 & 5 & 6 \\ -3 & -5 & 1 & 7 \\ -4 & -6 & -7 & 1 \end{bmatrix}$$

$$\mathbf{K} = \begin{bmatrix} 0 & 1 \\ 1 & 0 \end{bmatrix} \quad \mathbf{L} = \begin{bmatrix} 0 & 1 & 0 & 0 \\ 1 & 0 & 1 & 1 \\ 0 & 1 & 1 & 1 \\ 0 & 1 & 1 & 0 \end{bmatrix} \quad \mathbf{M} = \begin{bmatrix} -2 & 0 & 0 \\ 4 & 0 & 0 \\ -1 & 2 & 3 \end{bmatrix}$$

*3. Suppose the writings of six authors — labeled A, B, C, D, E, and F — have been influenced by one another in the following ways:

A has been influenced by D and E.
B has been influenced by C and E.

C has been influenced by A.

D has been influenced by B, E, and F.

E has been influenced by B and C.

F has been influenced by D.

Draw the digraph that represents these relationships. What is its adjacency matrix? What would the transpose of this adjacency matrix represent?

4. Using the adjacency matrix for the digraph in Figure 8.5, find the following:

⋆(a) The number of paths of length 3 from P_2 to P_4

(b) The number of paths of length 4 from P_1 to P_5

⋆(c) The number of paths of length ≤ 3 from P_3 to P_2

(d) The number of paths of length ≤ 4 from P_3 to P_1

⋆(e) The length of the shortest path from P_4 to P_5

(f) The length of the shortest path from P_4 to P_1

Figure 8.5

Digraph for Exercises 4, 6, and 9

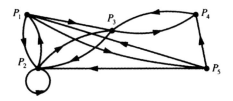

5. Repeat parts (a) through (f) of Exercise (4) for the digraph in Figure 8.6.

Figure 8.6

Digraph for Exercises 5, 6, and 9

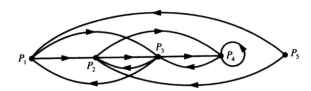

6. A **cycle** in a graph or digraph is a path connecting a vertex to itself. For the digraphs in each of Figures 8.5 and 8.6, find the following:

⋆(a) The number of cycles of length 3 connecting P_2 to itself

(b) The number of cycles of length 4 connecting P_1 to itself

⋆(c) The number of cycles of length ≤ 4 connecting P_4 to itself

⋆**7.** (a) Suppose that there is one vertex that is not connected to any other in a graph. How will this situation be reflected in the adjacency matrix for the graph?

(b) Suppose that there is one vertex that is not directed to any other in a digraph. How will this situation be reflected in the adjacency matrix for the digraph?

8. ⋆(a) Recall the definition of the trace of a matrix (Exercise 14 of Section 1.4). What information does the trace of the adjacency matrix of a graph or digraph give?

(b) Suppose **A** is the adjacency matrix of a graph or digraph, and $k > 0$. What information does the trace of $\mathbf{A}^k$ give? (Hint: See Exercise 6.)

9. ⋆(a) A **strongly connected digraph** is a digraph in which, given any pair of distinct vertices, there is a directed path (of some length) from each of these two vertices to the other. Determine whether the digraphs in Figures 8.5 and 8.6 are strongly connected.

(b) Prove that a digraph with n vertices having adjacency matrix **A** is strongly connected if and only if $\mathbf{A} + \mathbf{A}^2 + \mathbf{A}^3 + \cdots + \mathbf{A}^{n-1}$ has the property that all entries not on the main diagonal are nonzero.

10. (a) A **dominance digraph** is one with no loops in which, for any two distinct vertices P_i and P_j, there is either an edge from P_i to P_j, or an edge from P_j to P_i, but *not both*. (Dominance digraphs are useful in psychology, sociology, and communications.) Show that the following matrix is the adjacency matrix for a dominance digraph:

$$
\begin{array}{c@{\;}c}
 & \begin{array}{cccc} P_1 & P_2 & P_3 & P_4 \end{array} \\
\begin{array}{c} P_1 \\ P_2 \\ P_3 \\ P_4 \end{array} &
\left[\begin{array}{cccc}
0 & 1 & 0 & 1 \\
0 & 0 & 1 & 0 \\
1 & 0 & 0 & 1 \\
0 & 1 & 0 & 0
\end{array} \right].
\end{array}
$$

⋆(b) Suppose six teams in a league play a tournament in which each team plays every other team exactly once (with no tie games possible). Consider a digraph representing the outcomes of such a tournament in which an edge is drawn from the vertex for Team A to the vertex for Team B if Team A defeats Team B. Is this a dominance digraph? Why or why not?

(c) Suppose that **A** is a square matrix with each entry equal to 0 or to 1. Show that **A** is the adjacency matrix for a dominance digraph if and only if $\mathbf{A} + \mathbf{A}^T$ has all main diagonal entries equal to 0, and all other entries equal to 1.

▶11. Prove Theorem 8.1. (Hint: Use a proof by induction on the length of the path between vertices P_i and P_j. In the Inductive Step, use the fact that the total number of paths from P_i to P_j of length $t + 1$ is the sum of n products, where each product is the number of paths of length t from P_i to some vertex P_q $(1 \le q \le n)$ times the number of paths of length 1 from P_q to P_j.)

⋆12. True or False:

(a) The adjacency matrix of a simple graph must be symmetric.

(b) The adjacency matrix for a simple digraph may contain negative numbers.

(c) If **A** is the adjacency matrix for a simple digraph and the $(1, 2)$ entry of $\mathbf{A}^n$ is zero for all $n \ge 1$, then there is no path from vertex P_1 to P_2.

(d) The number of edges in any simple graph equals the number of 1's in its adjacency matrix.

(e) The number of edges in any simple digraph equals the number of 1's in its adjacency matrix.

(f) If a simple graph has a path of length k from P_1 to P_2 and a path of length j from P_2 to P_3, then it has a path of length $k + j$ from P_1 to P_3.

(g) The sum of the numbers in the ith column of the adjacency matrix for a simple graph gives the number of edges connected to P_i.

8.2 Ohm's Law

Prerequisite: Section 2.2, Gauss-Jordan Row Reduction and Reduced Row Echelon Form

In this section, we examine an important application of systems of linear equations to circuit theory in physics.

Circuit Fundamentals and Ohm's Law

In a simple electrical circuit, such as the one in Figure 8.7, **voltage sources** (for example, batteries) stimulate electric current to flow through the circuit. **Voltage** (V) is measured in **volts**, and **current** (I) is measured in **amperes**. The circuit in Figure 8.7 has two voltage sources: $48V$ and $9V$. Current flows from the positive $(+)$ end of the voltage source to the negative $(-)$ end.

Figure 8.7

Electrical circuit

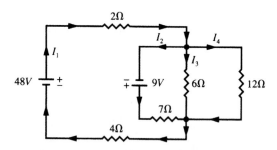

In contrast to voltage sources, there are **voltage drops**, or **sinks**, when **resistors** are present, because resistors impede the flow of current. In particular, the following principle holds:

OHM'S LAW

At any resistor, the amount of voltage V dropped is proportional to the amount of current I flowing through the resistor. That is, $V = IR$, where the proportionality constant R is a measure of the resistance to the current.

Resistance (R) is measured in **ohms**, or volts/ampere. The Greek letter Ω is used to denote ohms.

Any point in the circuit where current-carrying branches meet is called a **junction**. Any path along the branches of a circuit that begins and ends at the same location is called a **loop**. The following two principles involving junctions and loops are very important:

KIRCHHOFF'S LAWS

First Law: The sum of the currents flowing into a junction must equal the sum of the currents leaving a junction.
Second Law: The sum of the voltage sources and drops around any loop of a circuit is zero.

EXAMPLE 1

Consider the electrical circuit in Figure 8.7. We will use Ohm's Law to find the amount of current flowing through each branch of the circuit. The circuit has the following two junctions: the first where current I_1 branches into the three currents I_2, I_3, and I_4 and the second where these last three currents merge again into I_1. By Kirchhoff's First Law, both junctions produce the same equation: $I_1 = I_2 + I_3 + I_4$.

Also, three different loops start and end at the voltage source $48V$.

$$(1) \quad I_1 \to I_2 \to I_1$$
$$(2) \quad I_1 \to I_3 \to I_1$$
$$(3) \quad I_1 \to I_4 \to I_1$$

Kirchhoff's Second Law gives an Ohm's Law equation for each of these loops.

$$48V + 9V - I_1(2\Omega) - I_2(7\Omega) - I_1(4\Omega) = 0 \quad \text{(loop 1)}$$
$$48V - I_1(2\Omega) - I_3(6\Omega) - I_1(4\Omega) = 0 \quad \text{(loop 2)}$$
$$48V - I_1(2\Omega) - I_4(12\Omega) - I_1(4\Omega) = 0 \quad \text{(loop 3)}.$$

These equations lead to the following system of four equations and four variables.

$$\begin{cases} -I_1 & + & I_2 & + & I_3 & + & I_4 & = & 0 \\ 6I_1 & + & 7I_2 & & & & & = & 57 \\ 6I_1 & & & + & 6I_3 & & & = & 48 \\ 6I_1 & & & & & + & 12I_4 & = & 48 \end{cases}$$

After applying the Gauss-Jordan Method to the augmented matrix for this system, we obtain

$$\begin{bmatrix} 1 & 0 & 0 & 0 & | & 6 \\ 0 & 1 & 0 & 0 & | & 3 \\ 0 & 0 & 1 & 0 & | & 2 \\ 0 & 0 & 0 & 1 & | & 1 \end{bmatrix}.$$

Hence, $I_1 = 6$ amperes, $I_2 = 3$ amperes, $I_3 = 2$ amperes, and $I_4 = 1$ ampere. ∎

Exercises for Section 8.2

1. Use Ohm's Law to find the current in each branch of the electrical circuits in Figure 8.8, with the indicated voltage sources and resistances.

Figure 8.8

Electrical circuits for Exercise 1

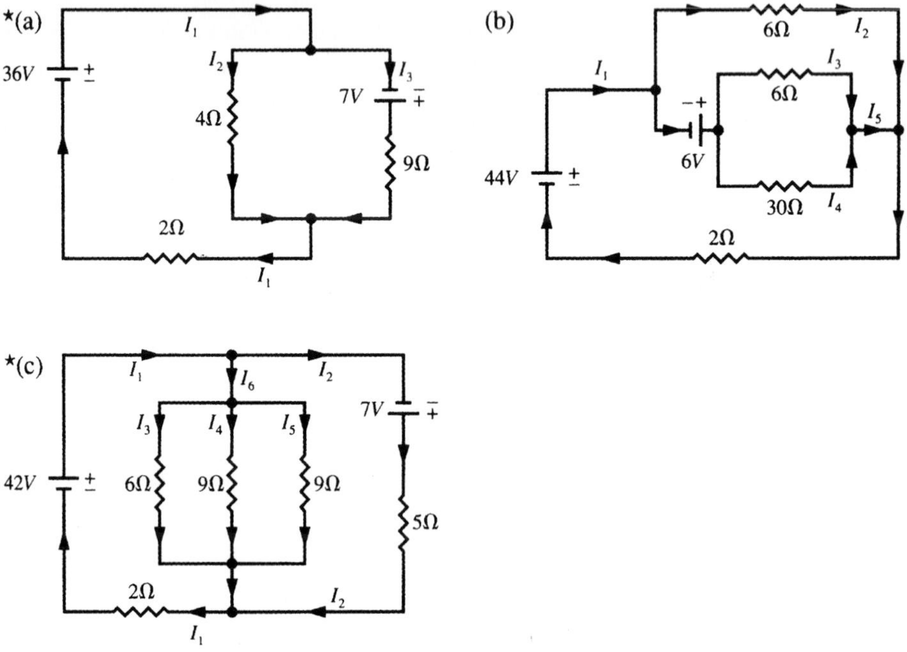

★2. True or False:

    (a) Kirchhoff's Laws produce one equation for each junction and one equation for each loop.

    (b) The resistance R is the constant of proportionality in Ohm's Law relating the current I and the voltage V.

8.3 Least-Squares Polynomials

Prerequisite: Section 2.2, Gauss-Jordan Row Reduction and Reduced Row Echelon Form

    In this section, we present the least-squares method for finding a polynomial "closest" to a given set of data points. You should have a calculator or computer handy as you work through some of the examples and exercises.

Least-Squares Polynomials

In science and business, we often need to predict the relationship between two given variables. In many cases, we begin by performing an appropriate laboratory experiment or statistical analysis to obtain the necessary data. However, even if a simple law governs the behavior of the variables, this law may not be easy to find because of errors introduced in measuring or sampling. In practice, therefore, we are often content with a polynomial equation that provides a close approximation to the data.

Suppose we have a set of points $(a_1, b_1), (a_2, b_2), (a_3, b_3), \ldots, (a_n, b_n)$. We want a method for finding polynomial equations $y = f(x)$ to fit these points as "closely" as possible. One approach would be to minimize the sum of the vertical distances $|f(a_1) - b_1|, |f(a_2) - b_2|, \ldots, |f(a_n) - b_n|$ between the graph of $y = f(x)$ and the data points. These distances are the lengths of the line segments in Figure 8.9. However, this is not the approach typically used. Instead, we will minimize the distance between the vectors $\mathbf{y} = [f(a_1), \ldots, f(a_n)]$ and $\mathbf{b} = [b_1, \ldots, b_n]$, which equals $\|\mathbf{y} - \mathbf{b}\|$. This is equivalent to minimizing the sum of the *squares* of the vertical distances shown in Figure 8.9.

Figure 8.9

Vertical distances from data points (a_k, b_k) to $y = f(x)$, for $1 \leq k \leq n$

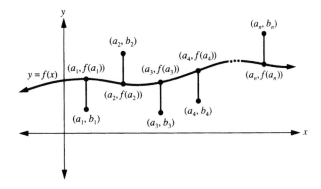

DEFINITION

A **degree t least-squares polynomial** for the points $(a_1, b_1), (a_2, b_2), \ldots, (a_n, b_n)$ is a polynomial $y = f(x) = c_t x^t + \cdots + c_2 x^2 + c_1 x + c_0$ for which the sum

$$S_f = (f(a_1) - b_1)^2 + (f(a_2) - b_2)^2 + (f(a_3) - b_3)^2 + \cdots + (f(a_n) - b_n)^2$$

of the squares of the vertical distances from each of the given points to the polynomial is less than or equal to the corresponding sum, S_g, for any other polynomial g of degree $\leq t$.

Note that it is possible for a "degree t least-squares polynomial" to actually have a degree less than t because there is no guarantee that its leading coefficient will be nonzero.

A method for calculating least-squares polynomials is stated in Theorem 8.2. This method is usually used to find a least-squares polynomial whose degree is less than the given number n of data points.

THEOREM 8.2

Let $(a_1, b_1), (a_2, b_2), \ldots, (a_n, b_n)$ be n points, and let $\mathbf{A}$ be the $n \times (t + 1)$ matrix

$$\begin{bmatrix} 1 & a_1 & a_1^2 & \cdots & a_1^t \\ 1 & a_2 & a_2^2 & \cdots & a_2^t \\ \vdots & \vdots & \vdots & \ddots & \vdots \\ 1 & a_n & a_n^2 & \cdots & a_n^t \end{bmatrix}.$$

A polynomial

$$c_t x^t + \cdots + c_2 x^2 + c_1 x + c_0$$

whose coefficients $c_0, c_1, \ldots, c_t$ satisfy the linear system

$$\left(\mathbf{A}^T \mathbf{A}\right) \begin{bmatrix} c_0 \\ c_1 \\ \vdots \\ c_t \end{bmatrix} = \mathbf{A}^T \begin{bmatrix} b_1 \\ b_2 \\ \vdots \\ b_n \end{bmatrix}$$

is a degree t least-squares polynomial for the given points. In addition, if $\mathbf{A}^T \mathbf{A}$ row reduces to $\mathbf{I}_{t+1}$, there is a unique degree t least-squares polynomial for the given points.

Notice in Theorem 8.2 that $\mathbf{A}^T$ is a $(t + 1) \times n$ matrix. Thus, $\mathbf{A}^T \mathbf{A}$ is a $(t + 1) \times (t + 1)$ matrix, and so the matrix products in Theorem 8.2 make sense.

We do not prove Theorem 8.2 here. However, the theorem follows in a straightforward manner from Theorem 8.8 in Section 8.10. You may want to prove Theorem 8.2 later if you study Section 8.10.

The following two examples show how this theorem is used to calculate least-squares polynomials. The first example calculates a least-squares line for a given set of points. Such a line is often called a **line of best fit**, or a **linear regression**.

EXAMPLE 1

We will find the least-squares line $y = c_1 x + c_0$ through the points $(a_1, b_1) = (-4, 6)$, $(a_2, b_2) = (-2, 4)$, $(a_3, b_3) = (1, 1)$, $(a_4, b_4) = (2, -1)$, and $(a_5, b_5) = (4, -3)$.

Since the degree of the desired polynomial is 1 in this case, the matrix $\mathbf{A}$

has only two columns, as follows:

$$\mathbf{A} = \begin{bmatrix} 1 & a_1 \\ 1 & a_2 \\ 1 & a_3 \\ 1 & a_4 \\ 1 & a_5 \end{bmatrix} = \begin{bmatrix} 1 & -4 \\ 1 & -2 \\ 1 & 1 \\ 1 & 2 \\ 1 & 4 \end{bmatrix}.$$

Then $\mathbf{A}^T = \begin{bmatrix} 1 & 1 & 1 & 1 & 1 \\ -4 & -2 & 1 & 2 & 4 \end{bmatrix}$ and $\mathbf{A}^T\mathbf{A} = \begin{bmatrix} 5 & 1 \\ 1 & 41 \end{bmatrix}$. Since

$$\begin{bmatrix} b_1 \\ b_2 \\ b_3 \\ b_4 \\ b_5 \end{bmatrix} = \begin{bmatrix} 6 \\ 4 \\ 1 \\ -1 \\ -3 \end{bmatrix}, \quad \text{we have} \quad \mathbf{A}^T \begin{bmatrix} b_1 \\ b_2 \\ b_3 \\ b_4 \\ b_5 \end{bmatrix} = \begin{bmatrix} 7 \\ -45 \end{bmatrix}.$$

Hence, the equation

$$\mathbf{A}^T\mathbf{A}\begin{bmatrix} c_0 \\ c_1 \end{bmatrix} = \mathbf{A}^T \begin{bmatrix} b_1 \\ b_2 \\ b_3 \\ b_4 \\ b_5 \end{bmatrix} \quad \text{becomes} \quad \begin{bmatrix} 5 & 1 \\ 1 & 41 \end{bmatrix}\begin{bmatrix} c_0 \\ c_1 \end{bmatrix} = \begin{bmatrix} 7 \\ -45 \end{bmatrix}.$$

Row reducing the augmented matrix

$$\left[\begin{array}{cc|c} 5 & 1 & 7 \\ 1 & 41 & -45 \end{array}\right] \quad \text{gives} \quad \left[\begin{array}{cc|c} 1 & 0 & 1.63 \\ 0 & 1 & -1.14 \end{array}\right],$$

and so the least-squares line is $y = -1.14x + 1.63$ (see Figure 8.10).

Notice that, in this example, for each given a_i value, this line produces a value "close" to the given b_i value. For example, when $x = a_1 = -4$, $y = -1.14(-4) + 1.63 = 6.19$, which is close to $b_1 = 6$. ∎

Once we have calculated the least-squares line, we can use it to find the values of other potential data points. This technique is called **extrapolation**. Returning to Example 1, if $x = 7$, the value of y is $-1.14(7) + 1.63 = -6.35$. Thus, we would expect the experiment that produced the original data to give a y-value close to -6.35 if an x-value of 7 were encountered.

In the next example, we use data that suggest a parabolic rather than a linear shape and find a second-degree least-squares polynomial to fit the data.

EXAMPLE 2 We will find the quadratic least-squares polynomial for the points $(-3, 7)$, $(-1, 4)$, $(2, 0)$, $(3, 1)$, and $(5, 6)$. We label these points (a_1, b_1) through (a_5, b_5), respectively. Since we want a quadratic polynomial, $t = 2$. Then, the matrix $\mathbf{A}$

Figure 8.10

Least-squares line for
the data points in
Example 1

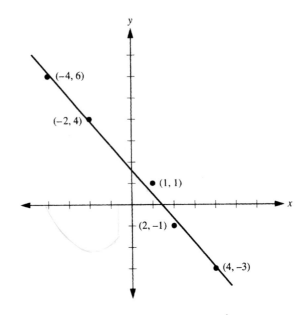

has three columns as follows:

$$\mathbf{A} = \begin{bmatrix} 1 & -3 & 9 \\ 1 & -1 & 1 \\ 1 & 2 & 4 \\ 1 & 3 & 9 \\ 1 & 5 & 25 \end{bmatrix}.$$

Hence,

$$\mathbf{A}^T = \begin{bmatrix} 1 & 1 & 1 & 1 & 1 \\ -3 & -1 & 2 & 3 & 5 \\ 9 & 1 & 4 & 9 & 25 \end{bmatrix}, \quad \text{and so} \quad \mathbf{A}^T\mathbf{A} = \begin{bmatrix} 5 & 6 & 48 \\ 6 & 48 & 132 \\ 48 & 132 & 804 \end{bmatrix}.$$

Also, the matrix

$$\begin{bmatrix} b_1 \\ b_2 \\ b_3 \\ b_4 \\ b_5 \end{bmatrix} = \begin{bmatrix} 7 \\ 4 \\ 0 \\ 1 \\ 6 \end{bmatrix}, \quad \text{and so} \quad \mathbf{A}^T \begin{bmatrix} b_1 \\ b_2 \\ b_3 \\ b_4 \\ b_5 \end{bmatrix} = \begin{bmatrix} 18 \\ 8 \\ 226 \end{bmatrix}.$$

Therefore, the coefficients of the least-squares polynomial $c_2 x^2 + c_1 x + c_0$ are
the solutions of the system

$$\mathbf{A}^T\mathbf{A} \begin{bmatrix} c_0 \\ c_1 \\ c_2 \end{bmatrix} = \mathbf{A}^T \begin{bmatrix} b_1 \\ b_2 \\ b_3 \\ b_4 \\ b_5 \end{bmatrix}, \quad \text{or} \quad \begin{bmatrix} 5 & 6 & 48 \\ 6 & 48 & 132 \\ 48 & 132 & 804 \end{bmatrix} \begin{bmatrix} c_0 \\ c_1 \\ c_2 \end{bmatrix} = \begin{bmatrix} 18 \\ 8 \\ 226 \end{bmatrix}.$$

Solving, we find $c_0 = 1.21$, $c_1 = -1.02$, and $c_2 = 0.38$. Hence, the least-squares quadratic polynomial is $y = 0.38x^2 - 1.02x + 1.21$ (see Figure 8.11). ∎

Figure 8.11

Least-squares quadratic polynomial for the data points in Example 2

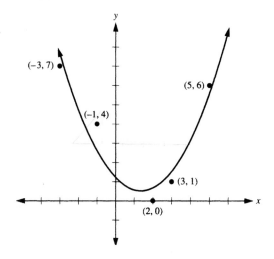

Exercises for Section 8.3

Note: You should have a calculator or computer handy for the computations in many of these exercises.

1. For each of the following sets of points, find the line of best fit (that is, the least-squares line). In each case, extrapolate to find the approximate y-value when $x = 5$.

   ⋆(a) $(3, -8)$, $(1, -5)$, $(0, -4)$, $(2, -1)$
   (b) $(-6, -6)$, $(-4, -3)$, $(-1, 0)$, $(1, 2)$
   ⋆(c) $(-4, 10)$, $(-3, 8)$, $(-2, 7)$, $(-1, 5)$, $(0, 4)$

2. For each of the following sets of points, find the least-squares quadratic polynomial:

   ⋆(a) $(-4, 8)$, $(-2, 5)$, $(0, 3)$, $(2, 6)$
   (b) $(-1, -4)$, $(0, -2)$, $(2, -2)$, $(3, -5)$
   ⋆(c) $(-4, -3)$, $(-3, -2)$, $(-2, -1)$, $(0, 0)$, $(1, 1)$

3. For each of the following sets of points, find the least-squares cubic (degree 3) polynomial:

   ⋆(a) $(-3, -3)$, $(-2, -1)$, $(-1, 0)$, $(0, 1)$, $(1, 4)$
   (b) $(-2, 5)$, $(-1, 4)$, $(0, 3)$, $(1, 3)$, $(2, 1)$

4. Use the points given for each function to find the desired approximation.

   ⋆(a) Least-squares quadratic polynomial for $y = x^4$, using $x = -2, -1, 0, 1, 2$
   (b) Least-squares quadratic polynomial for $y = e^x$, using $x = -2, -1, 0, 1, 2$
   ⋆(c) Least-squares quadratic polynomial for $y = \ln x$, using $x = 1, 2, 3, 4$

(d) Least-squares cubic polynomial for $y = \sin x$, using $x = -\frac{\pi}{2}, -\frac{\pi}{4}, 0,$ $\frac{\pi}{4}, \frac{\pi}{2}$

★(e) Least-squares cubic polynomial for $y = \cos x$, using $x = -\frac{\pi}{2}, -\frac{\pi}{4}, 0,$ $\frac{\pi}{4}, \frac{\pi}{2}$

5. An engineer is monitoring a leaning tower whose angle from the vertical over a period of months is given below.

Month	1	2	3	4	5
Angle from vertical	3°	3.3°	3.7°	4.1°	4.6°

★(a) Find the line of best fit for the data, and extrapolate to predict the month in which the angle will be 20° from the vertical.

★(b) Find a least-squares quadratic approximation for the data, and extrapolate to predict the month in which the angle will be 20° from the vertical.

(c) Compare your answers to parts (a) and (b). Which approximation do you think is more accurate? Why?

6. The population of the U.S. (in millions), according to the Census Bureau, is given below.

Year	1940	1950	1960	1970	1980	1990
Population	131.7	150.7	179.3	203.3	226.5	249.6

(a) Find the line of best fit for the data, and extrapolate to predict the population in 2010. (Hint: Renumber the years as 1 through 6 to simplify the computation.)

(b) Find a least-squares quadratic approximation for the data, and extrapolate to predict the population in 2010.

(c) Compare your answers to parts (a) and (b). Which approximation do you think is more accurate? Why?

★**7.** Show that the method of least-squares gives the *exact* quadratic polynomial that goes through the points $(-2, 6)$, $(0, 2)$, and $(3, 8)$.

8. Show that the following system has the same solutions for c_0 and c_1 as the system in Theorem 8.2 when $t = 1$:

$$\begin{cases} nc_0 + \left(\sum_{i=1}^{n} a_i\right) c_1 = \sum_{i=1}^{n} b_i \\ \left(\sum_{i=1}^{n} a_i\right) c_0 + \left(\sum_{i=1}^{n} a_i^2\right) c_1 = \sum_{i=1}^{n} a_i b_i \end{cases}.$$

9. Although an inconsistent system $\mathbf{AX} = \mathbf{B}$ has no solutions, the least-squares method is sometimes used to find values that come "close" to satisfying all the equations in the system. Solutions to the related system $\mathbf{A}^T \mathbf{AX} = \mathbf{A}^T \mathbf{B}$ (obtained by multiplying on the left by $\mathbf{A}^T$) are called **least-squares solutions** for the inconsistent system $\mathbf{AX} = \mathbf{B}$. For each inconsistent system, find a least-squares solution, and check that it comes close to satisfying each equation in the system.

★(a) $\begin{cases} 4x_1 - 3x_2 = 12 \\ 2x_1 + 5x_2 = 32 \\ 3x_1 + x_2 = 21 \end{cases}$

(b) $\begin{cases} 2x_1 & - & x_2 & + & x_3 & = & 11 \\ -x_1 & + & 3x_2 & - & x_3 & = & -9 \\ x_1 & - & 2x_2 & + & 3x_3 & = & 12 \\ 3x_1 & - & 4x_2 & + & 2x_3 & = & 21 \end{cases}$

*10. True or False:

(a) If a set of data points all lie on the same line, then that line will be the line of best fit for the data.

(b) The degree 3 least-squares polynomial for a set of points must have degree 3.

(c) The line of best fit for a set of points must pass through at least one of the points.

(d) The matrix $\mathbf{A}^T\mathbf{A}$ in the least-squares method of Theorem 8.2 is a $t \times t$ matrix.

8.4 Markov Chains

Prerequisite: Section 2.2, Gauss-Jordan Row Reduction and Reduced Row Echelon Form

In this section, we introduce Markov chains and demonstrate how they are used to predict the future states of an interdependent system. You should have a calculator or computer handy as you work through the examples and exercises.

An Introductory Example

The following example will introduce many of the ideas associated with Markov chains:

EXAMPLE 1 Suppose that three banks in a certain town are competing for investors. Currently, Bank A has 40% of the investors, Bank B has 10%, and Bank C has the remaining 50%. We can set up the following **probability** (or **state**) vector **p** to represent this distribution:

$$\mathbf{p} = \begin{bmatrix} .4 \\ .1 \\ .5 \end{bmatrix}.$$

Suppose the townsfolk are tempted by various promotional campaigns to switch banks. Records show that each year Bank A keeps half of its investors, with the remainder switching equally to Banks B and C. However, Bank B keeps two-thirds of its investors, with the remainder switching equally to Banks A and C. Finally, Bank C keeps half of its investors, with the remainder switching equally to Banks A and B. The following **transition matrix M** (rounded to three decimal places) keeps track of the changing investment patterns:

$$\mathbf{M} = \quad \text{Next Year} \begin{matrix} \\ A \\ B \\ C \end{matrix} \overset{\text{Current Year}}{\begin{matrix} \quad A \quad\; B \quad\; C \\ \begin{bmatrix} .500 & .167 & .250 \\ .250 & .667 & .250 \\ .250 & .167 & .500 \end{bmatrix} \end{matrix}}.$$

The (i, j) entry of $\mathbf{M}$ represents the fraction of current investors going *from* Bank j to Bank i next year.[1]

To find the distribution of investors after one year, consider

$$\mathbf{p}_1 = \mathbf{Mp} = \quad \text{Next Year} \begin{matrix} \\ A \\ B \\ C \end{matrix} \overset{\text{Current Year}}{\begin{matrix} \quad A \quad\; B \quad\; C \\ \begin{bmatrix} .500 & .167 & .250 \\ .250 & .667 & .250 \\ .250 & .167 & .500 \end{bmatrix} \end{matrix}} \begin{bmatrix} .4 \\ .1 \\ .5 \end{bmatrix} = \begin{bmatrix} .342 \\ .292 \\ .367 \end{bmatrix}.$$

The entries of $\mathbf{p}_1$ give the distribution of investors after one year. For example, the first entry of this product, .342, is obtained by taking the dot product of the first row of $\mathbf{M}$ with $\mathbf{p}$ as follows:

$$\underbrace{(.500)}_{\substack{\text{at Bank A,}\\ \text{fraction of}\\ \text{investors}\\ \text{who stay at}\\ \text{Bank A}}} \underbrace{(.4)}_{\substack{\text{fraction of}\\ \text{investors}\\ \text{currently at}\\ \text{Bank A}}} + \underbrace{(.167)}_{\substack{\text{at Bank B,}\\ \text{fraction of}\\ \text{investors}\\ \text{who switch}\\ \text{to Bank A}}} \underbrace{(.1)}_{\substack{\text{fraction of}\\ \text{investors}\\ \text{currently at}\\ \text{Bank B}}} + \underbrace{(.250)}_{\substack{\text{at Bank C,}\\ \text{fraction of}\\ \text{investors}\\ \text{who switch}\\ \text{to Bank A}}} \underbrace{(.5)}_{\substack{\text{fraction of}\\ \text{investors}\\ \text{currently at}\\ \text{Bank C}}},$$

which gives .342, the total fraction of investors at Bank A after one year. We can continue this process for another year, as follows:

$$\mathbf{p}_2 = \mathbf{Mp}_1 = \begin{matrix} \\ A \\ B \\ C \end{matrix} \begin{matrix} \quad A \quad\; B \quad\; C \\ \begin{bmatrix} .500 & .167 & .250 \\ .250 & .667 & .250 \\ .250 & .167 & .500 \end{bmatrix} \end{matrix} \begin{bmatrix} .342 \\ .292 \\ .367 \end{bmatrix} = \begin{bmatrix} .312 \\ .372 \\ .318 \end{bmatrix}.$$

Since multiplication by $\mathbf{M}$ gives the yearly change and the entries of $\mathbf{p}_1$ represent the distribution of investors at the end of the first year, we see that the entries of $\mathbf{p}_2$ represent the correct distribution of investors at the end of the second year. That is, after two years, 31.2% of the investors are at Bank A, 37.2% are at Bank B, and 31.8% are at Bank C. Notice that

$$\mathbf{p}_2 = \mathbf{Mp}_1 = \mathbf{M}(\mathbf{Mp}) = \mathbf{M}^2\mathbf{p}.$$

In other words, the matrix $\mathbf{M}^2$ takes us directly from $\mathbf{p}$ to $\mathbf{p}_2$. Similarly, if $\mathbf{p}_3$ is the distribution after three years, then

$$\mathbf{p}_3 = \mathbf{Mp}_2 = \mathbf{M}(\mathbf{M}^2\mathbf{p}) = \mathbf{M}^3\mathbf{p}.$$

---

[1] It may seem more natural to let the (i, j) entry of $\mathbf{M}$ represent the fraction going *from* Bank i *to* Bank j. However, we arrange the matrix entries this way to facilitate matrix multiplication.

A simple induction proof shows that, in general, if $\mathbf{p}_n$ represents the distribution after n years, then $\mathbf{p}_n = \mathbf{M}^n\mathbf{p}$. We can use this formula to find the distribution of investors after 6 years. After tedious calculation (rounding to three decimal places at each step), we find

$$\mathbf{M}^6 = \begin{bmatrix} .288 & .285 & .288 \\ .427 & .432 & .427 \\ .288 & .285 & .288 \end{bmatrix}.$$

Then

$$\mathbf{p}_6 = \mathbf{M}^6\mathbf{p} = \begin{bmatrix} .288 & .285 & .288 \\ .427 & .432 & .427 \\ .288 & .285 & .288 \end{bmatrix} \begin{bmatrix} .4 \\ .1 \\ .5 \end{bmatrix} = \begin{bmatrix} .288 \\ .428 \\ .288 \end{bmatrix}. \qquad \blacksquare$$

Formal Definitions

We now recap many of the ideas presented in Example 1 and give them a more formal treatment.

The notion of probability is important when discussing Markov chains. Probabilities of events are always given as values between $0 = 0\%$ and $1 = 100\%$, where a probability of 0 indicates no possibility, and a probability of 1 indicates certainty. For example, if we draw a random card from a standard deck of fifty-two playing cards, the probability that the card is an ace is $\frac{4}{52} = \frac{1}{13}$, because exactly four of the fifty-two cards are aces. The probability that the card is a red card is $\frac{26}{52} = \frac{1}{2}$, since there are twenty-six red cards in the deck. The probability that the card is both red and black (at the same time) is $\frac{0}{52} = 0$, since this event is impossible. Finally, the probability that the card is red or black is $\frac{52}{52} = 1$, since this event is certain.

Now consider a set of events that are completely "distinct" and "exhaustive" (that is, one and only one of them must occur at any time). The sum of all of their probabilities must total $100\% = 1$. For example, if we select a card at random, we have a $\frac{13}{52} = \frac{1}{4}$ chance each of choosing a club, diamond, heart, or spade. These represent the only distinct suit possibilities, and the sum of these four probabilities is 1.

Now recall that each column of the matrix $\mathbf{M}$ in Example 1 represents the probabilities that an investor switches assets to Bank A, B, or C. Since these are the only banks in town, the sum of the probabilities in each column of $\mathbf{M}$ must total 1, or Example 1 would not make sense as stated. Hence, $\mathbf{M}$ is a matrix of the following type:

DEFINITION

A **stochastic matrix** is a square matrix in which all entries are nonnegative and the entries of each column add up to 1.

A single-column stochastic matrix is often called a **stochastic vector**. The next theorem can be proven in a straightforward manner by induction (see Exercise 9).

THEOREM 8.3
The product of any finite number of stochastic matrices is a stochastic matrix.

Now we are ready to formally define a Markov chain.

DEFINITION
A **Markov chain** (or **Markov process**) is a system containing a finite number of distinct states $S_1, S_2, \ldots, S_n$ on which steps are performed such that
  (1) At any time, each element of the system resides in exactly one of the states.
  (2) At each step in the process, elements in the system can move from one state to another.
  (3) The probabilities of moving from state to state are fixed — that is, they are the same at each step in the process.

In Example 1, the distinct states of the Markov chain are the three banks, A, B, and C, and the elements of the system are the investors, each one keeping money in only one of the three banks at any given time. Each new year represents another step in the process, during which time investors could switch banks or remain with their current bank. Finally, we have assumed that the probabilities of switching banks do not change from year to year.

DEFINITION
A **probability** (or **state**) **vector p** for a Markov chain is a stochastic vector whose ith entry is the probability that an element in the system is currently in state S_i. A **transition matrix M** for a Markov chain is a stochastic matrix whose (i, j) entry is the probability that an element in state S_j will move to state S_i during the next step of the process.

The next theorem can be proven in a straightforward manner by induction (see Exercise 10).

THEOREM 8.4
Let **p** be the (current) probability vector and **M** be the transition matrix for a Markov chain. After n steps in the process, where $n \geq 1$, the (new) probability vector is given by $\mathbf{p}_n = \mathbf{M}^n \mathbf{p}$.

Theorem 8.4 asserts that once the initial probability vector **p** and the transition matrix **M** for a Markov chain are known, all future steps of the Markov chain are determined.

Limit Vectors and Fixed Points

A natural question to ask about a given Markov chain is whether we can discern any long-term trend.

> **EXAMPLE 2**   Consider the Markov chain from Example 1, with transition matrix
>
> $$\mathbf{M} = \begin{bmatrix} .500 & .167 & .250 \\ .250 & .667 & .250 \\ .250 & .167 & .500 \end{bmatrix}.$$
>
> What happens in the long run? To discern this, we calculate $\mathbf{p}_k$ for large values of k. Starting with $\mathbf{p} = [.4, .1, .5]$ and computing $\mathbf{p}_k = \mathbf{M}^k \mathbf{p}$ for increasing values of k (a calculator or computer is extremely useful here), we find that $\mathbf{p}_k$ approaches[2] the vector
>
> $$\mathbf{p}_f = [.286, .429, .286],$$
>
> where we are again rounding to three decimal places.[3]
>
> Alternatively, to calculate $\mathbf{p}_f$, we could have first shown that as k gets larger, $\mathbf{M}^k$ approaches the matrix
>
> $$\mathbf{M}_f = \begin{bmatrix} .286 & .286 & .286 \\ .429 & .429 & .429 \\ .286 & .286 & .286 \end{bmatrix},$$
>
> by multiplying out higher powers of $\mathbf{M}$ until successive powers agree to the desired number of decimal places. The probability vector $\mathbf{p}_f$ could then be found by
>
> $$\mathbf{p}_f = \mathbf{M}_f \mathbf{p} = \begin{bmatrix} .286 & .286 & .286 \\ .429 & .429 & .429 \\ .286 & .286 & .286 \end{bmatrix} \begin{bmatrix} .4 \\ .1 \\ .5 \end{bmatrix} = \begin{bmatrix} .286 \\ .429 \\ .286 \end{bmatrix}.$$
>
> Both techniques yield the same answer for $\mathbf{p}_f$. Ultimately, Banks A and C each capture 28.6%, or $\frac{2}{7}$, of the investors, and Bank B captures 42.9%, or $\frac{3}{7}$, of the investors. The vector $\mathbf{p}_f$ is called a **limit vector** of the Markov chain.   ∎

We now give a formal definition for a limit vector of a Markov chain.

---

[2] The intuitive concept of a sequence of vectors approaching a vector can be defined precisely using limits. We say that $\lim_{k \to \infty} \mathbf{p}_k = \mathbf{p}_f$ if and only if $\lim_{k \to \infty} \| \mathbf{p}_k - \mathbf{p}_f \| = 0$. It can be shown that this is equivalent to having the differences between the corresponding entries of $\mathbf{p}_k$ and $\mathbf{p}_f$ approach 0 as k grows larger. A similar approach can be used with matrices, where we say that $\lim_{k \to \infty} \mathbf{M}^k = \mathbf{M}_f$ if the differences between corresponding entries of $\mathbf{M}^k$ and $\mathbf{M}_f$ approach 0 as k grows larger.

[3] When raising matrices, such as $\mathbf{M}$, to high powers, roundoff error can quickly compound. Although we have printed $\mathbf{M}$ rounded to 3 significant digits, we actually performed the computations using $\mathbf{M}$ rounded to 12 digits of accuracy. In general, minimize your roundoff error by using as many digits as your calculator or software will provide.

DEFINITION

Let **M** be the transition matrix, and let **p** be the current probability vector for a Markov chain. Let $\mathbf{p}_k$ represent the probability vector after k steps of the Markov chain. If the sequence $\mathbf{p}, \mathbf{p}_1, \mathbf{p}_2, \ldots$ of vectors approaches some vector $\mathbf{p}_f$, then $\mathbf{p}_f$ is called a **limit vector** for the Markov chain.

The computation of $\mathbf{p}_k$ for large k, or equivalently, the computation of large powers of the transition matrix **M**, is not always an easy task, even with the use of a computer. We now show a quicker method to obtain the limit vector $\mathbf{p}_f$ for the Markov chain of Example 2. Notice that this vector $\mathbf{p}_f$ has the property that

$$\mathbf{M}\mathbf{p}_f = \begin{bmatrix} .500 & .167 & .250 \\ .250 & .667 & .250 \\ .250 & .167 & .500 \end{bmatrix} \begin{bmatrix} .286 \\ .429 \\ .286 \end{bmatrix} = \begin{bmatrix} .286 \\ .429 \\ .286 \end{bmatrix} = \mathbf{p}_f.$$

This remarkable property says that $\mathbf{p}_f$ is a vector that satisfies the equation $\mathbf{M}\mathbf{x} = \mathbf{x}$. Such a vector is called a **fixed point** for the Markov chain. Now, if we did not know $\mathbf{p}_f$, we could solve the equation

$$\mathbf{M} \begin{bmatrix} x_1 \\ x_2 \\ x_3 \end{bmatrix} = \begin{bmatrix} x_1 \\ x_2 \\ x_3 \end{bmatrix}$$

to find it. We can rewrite this as

$$\mathbf{M} \begin{bmatrix} x_1 \\ x_2 \\ x_3 \end{bmatrix} - \begin{bmatrix} x_1 \\ x_2 \\ x_3 \end{bmatrix} = \begin{bmatrix} 0 \\ 0 \\ 0 \end{bmatrix}, \quad \text{or} \quad (\mathbf{M} - \mathbf{I}_3) \begin{bmatrix} x_1 \\ x_2 \\ x_3 \end{bmatrix} = \begin{bmatrix} 0 \\ 0 \\ 0 \end{bmatrix}.$$

The augmented matrix for this system is

$$\begin{bmatrix} .500-1 & .167 & .250 & | & 0 \\ .250 & .667-1 & .250 & | & 0 \\ .250 & .167 & .500-1 & | & 0 \end{bmatrix} = \begin{bmatrix} -.500 & .167 & .250 & | & 0 \\ .250 & -.333 & .250 & | & 0 \\ .250 & .167 & -.500 & | & 0 \end{bmatrix}.$$

We can also add another condition, since we know that $x_1 + x_2 + x_3 = 1$. Thus, the augmented matrix gets a fourth row as follows:

$$\begin{bmatrix} -.500 & .167 & .250 & | & 0 \\ .250 & -.333 & .250 & | & 0 \\ .250 & .167 & -.500 & | & 0 \\ 1.000 & 1.000 & 1.000 & | & 1 \end{bmatrix}.$$

After row reduction, we find that the solution set is $x_1 = .286$, $x_2 = .429$, and $x_3 = .286$, as expected. Thus, the fixed point solution to $\mathbf{M}\mathbf{x} = \mathbf{x}$ equals the limit vector $\mathbf{p}_f$ we computed previously.

In general, if a limit vector $\mathbf{p}_f$ exists, it is a fixed point, and so this technique for finding the limit vector is especially useful where there is a unique fixed point. However, we must be careful because a given state vector for a Markov chain does not necessarily converge to a limit vector, as the next example shows.

EXAMPLE 3 Suppose that W, X, Y, and Z represent four train stations linked as shown in Figure 8.12. Suppose that twelve trains shuttle between these stations. Currently, there are six trains at station W, three trains at station X, two trains at station Y, and one train at station Z. The probability that a randomly chosen train is at each station is given by the probability vector

$$\mathbf{p} = \begin{array}{c} W \\ X \\ Y \\ Z \end{array} \left[\begin{array}{c} .500 \\ .250 \\ .167 \\ .083 \end{array} \right].$$

Figure 8.12

Four train stations

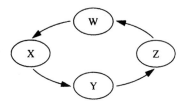

Suppose that during every hour, each train moves to the next station in Figure 8.12. Then we have a Markov chain whose transition matrix is

$$
\begin{array}{cc}
 & \text{Current State} \\
 & \begin{array}{cccc} W & X & Y & Z \end{array} \\
\mathbf{M} = \quad \text{Next State} \quad
\begin{array}{c} W \\ X \\ Y \\ Z \end{array}
& \left[\begin{array}{cccc}
0 & 0 & 0 & 1 \\
1 & 0 & 0 & 0 \\
0 & 1 & 0 & 0 \\
0 & 0 & 1 & 0
\end{array} \right].
\end{array}
$$

Intuitively, we can see there is no limit vector for this system, since the number of trains in each station never settles down to a fixed number but keeps rising and falling as the trains go around the "loop." This notion is borne out when we consider that the first few powers of the transition matrix are

$$
\mathbf{M}^2 = \left[\begin{array}{cccc}
0 & 0 & 1 & 0 \\
0 & 0 & 0 & 1 \\
1 & 0 & 0 & 0 \\
0 & 1 & 0 & 0
\end{array} \right], \quad
\mathbf{M}^3 = \left[\begin{array}{cccc}
0 & 1 & 0 & 0 \\
0 & 0 & 1 & 0 \\
0 & 0 & 0 & 1 \\
1 & 0 & 0 & 0
\end{array} \right], \quad
\text{and} \quad
\mathbf{M}^4 = \mathbf{I}_4 = \left[\begin{array}{cccc}
1 & 0 & 0 & 0 \\
0 & 1 & 0 & 0 \\
0 & 0 & 1 & 0 \\
0 & 0 & 0 & 1
\end{array} \right].
$$

Since $\mathbf{M}^4 = \mathbf{I}_4$, all higher powers of $\mathbf{M}$ are equal to $\mathbf{M}$, $\mathbf{M}^2$, $\mathbf{M}^3$, or $\mathbf{I}_4$. (Why?) Therefore, the only probability vectors produced by this Markov chain are $\mathbf{p}$,

$$
\mathbf{p}_1 = \mathbf{Mp} = \left[\begin{array}{c} .083 \\ .500 \\ .250 \\ .167 \end{array} \right], \quad
\mathbf{p}_2 = \mathbf{M}^2\mathbf{p} = \left[\begin{array}{c} .167 \\ .083 \\ .500 \\ .250 \end{array} \right], \quad
\text{and} \quad
\mathbf{p}_3 = \mathbf{M}^3\mathbf{p} = \left[\begin{array}{c} .250 \\ .167 \\ .083 \\ .500 \end{array} \right]
$$

because $\mathbf{p}_4 = \mathbf{M}^4\mathbf{p} = \mathbf{I}_4\mathbf{p} = \mathbf{p}$ again. Since $\mathbf{p}_k$ keeps changing to one of four distinct vectors, the initial state vector $\mathbf{p}$ does not converge to a limit vector.   ■

Regular Transition Matrices

DEFINITION

A square matrix $\mathbf{R}$ is **regular** if and only if $\mathbf{R}$ is a stochastic matrix and some power $\mathbf{R}^k$, for $k \geq 1$, has all entries nonzero.

EXAMPLE 4   The transition matrix $\mathbf{M}$ in Example 1 is a regular matrix, since $\mathbf{M}^1 = \mathbf{M}$ is a stochastic matrix with all entries nonzero. However, the transition matrix $\mathbf{M}$ in Example 3 is not regular because, as we saw in that example, all positive powers of $\mathbf{M}$ are equal to one of four matrices, each containing zero entries. Finally,

$$\mathbf{R} = \begin{bmatrix} 0 & \frac{1}{2} & 0 \\ \frac{1}{2} & 0 & 1 \\ \frac{1}{2} & \frac{1}{2} & 0 \end{bmatrix}$$

is regular since it is stochastic and

$$\mathbf{R}^4 = \left(\mathbf{R}^2\right)^2 = \left(\begin{bmatrix} \frac{1}{4} & 0 & \frac{1}{2} \\ \frac{1}{2} & \frac{3}{4} & 0 \\ \frac{1}{4} & \frac{1}{4} & \frac{1}{2} \end{bmatrix}\right)^2 = \begin{bmatrix} \frac{3}{16} & \frac{1}{8} & \frac{3}{8} \\ \frac{1}{2} & \frac{9}{16} & \frac{1}{4} \\ \frac{5}{16} & \frac{5}{16} & \frac{3}{8} \end{bmatrix},$$

which has all entries nonzero.   ■

The next theorem, stated without proof, shows that Markov chains with regular transition matrices always have a limit vector $\mathbf{p}_f$ for *every* choice of an initial probability vector $\mathbf{p}$.

THEOREM 8.5

If $\mathbf{R}$ is a regular $n \times n$ transition matrix for a Markov chain, then

(1) $\mathbf{R}_f = \lim_{k \to \infty} \mathbf{R}^k$ exists.

(2) $\mathbf{R}_f$ has all entries positive, and every column of $\mathbf{R}_f$ is identical.

(3) For all initial probability vectors $\mathbf{p}$, the Markov chain has a limit vector $\mathbf{p}_f$. Also, the limit vector $\mathbf{p}_f$ is the same for all $\mathbf{p}$.

(4) $\mathbf{p}_f$ is equal to any of the identical columns of $\mathbf{R}_f$.

(5) $\mathbf{p}_f$ is the unique stochastic n-vector such that $\mathbf{R}\mathbf{p}_f = \mathbf{p}_f$. That is, $\mathbf{p}_f$ is also the unique fixed point of the Markov chain.

When the matrix for a Markov chain is regular, Theorem 8.5 shows that the Markov chain has a unique fixed point, and that it agrees with the limit vector $\mathbf{p}_f$ for any initial state. When the transition matrix is regular, this unique vector $\mathbf{p}_f$ is called the **steady-state vector** for the Markov chain.

EXAMPLE 5

Consider a school of fish hunting for food in three adjoining lakes L_1, L_2, and L_3. Each day, the fish select a different lake to hunt in than the previous day, with probabilities given in the transition matrix below.

$$
\mathbf{M} = \begin{array}{c} \\ \\ L_1 \\ \text{Next Day } L_2 \\ L_3 \end{array}
\begin{array}{c} \text{Current Day} \\ \begin{array}{ccc} L_1 & L_2 & L_3 \end{array} \\ \left[\begin{array}{ccc} 0 & .5 & 0 \\ .5 & 0 & 1 \\ .5 & .5 & 0 \end{array}\right] \end{array}.
$$

Can we determine what percentage of time the fish will spend in each lake in the long run? Notice that $\mathbf{M}$ is equal to the matrix $\mathbf{R}$ in Example 4, and so $\mathbf{M}$ is regular. Theorem 8.5 asserts that the associated Markov chain has a steady-state vector. To find this vector, we solve the system

$$
(\mathbf{M} - \mathbf{I}_3) \begin{bmatrix} x_1 \\ x_2 \\ x_3 \end{bmatrix} = \begin{bmatrix} -1 & .5 & 0 \\ .5 & -1 & 1 \\ .5 & .5 & -1 \end{bmatrix} \begin{bmatrix} x_1 \\ x_2 \\ x_3 \end{bmatrix} = \begin{bmatrix} 0 \\ 0 \\ 0 \end{bmatrix},
$$

to find a fixed point for the Markov chain, under the extra condition that $x_1 + x_2 + x_3 = 1$. The solution is $x_1 = .222$, $x_2 = .444$, and $x_3 = .333$; that is, $\mathbf{p}_f = [.222, .444, .333]$. Therefore, in the long run, the fish will hunt $22.2\% = \frac{2}{9}$ of the time in lake L_1, $44.4\% = \frac{4}{9}$ of the time in lake L_2, and $33.3\% = \frac{1}{3}$ of the time in lake L_3. ∎

Notice in Example 5 that the initial probability state vector $\mathbf{p}$ was unneeded to find $\mathbf{p}_f$. The steady-state vector could also have been found by calculating larger and larger powers of $\mathbf{M}$ to see that they converge to the matrix

$$
\mathbf{M}_f = \begin{bmatrix} .222 & .222 & .222 \\ .444 & .444 & .444 \\ .333 & .333 & .333 \end{bmatrix}.
$$

Each of the identical columns of $\mathbf{M}_f$ is the steady-state vector for this Markov chain.

Exercises for Section 8.4

Note: You should have a calculator or computer handy for many of these exercises.

⋆1. Which of the following matrices are stochastic? Which are regular? Why?

$$A = \begin{bmatrix} \frac{1}{4} \\ \frac{1}{2} \\ \frac{1}{4} \end{bmatrix} \quad B = \begin{bmatrix} .2 & .4 & .5 \\ .5 & .1 & .4 \\ .3 & .4 & .1 \end{bmatrix} \quad C = \begin{bmatrix} \frac{1}{5} & \frac{2}{3} \\ \frac{4}{5} & \frac{1}{3} \end{bmatrix} \quad D = \begin{bmatrix} 0 & 1 & 0 \\ 0 & 0 & 1 \\ 1 & 0 & 0 \end{bmatrix}$$

$$E = \begin{bmatrix} \frac{1}{3} & \frac{2}{3} \\ \frac{1}{4} & \frac{3}{4} \end{bmatrix} \quad F = \begin{bmatrix} \frac{1}{3} & \frac{1}{3} & 1 \\ 0 & 0 & 0 \\ \frac{2}{3} & \frac{2}{3} & 0 \end{bmatrix} \quad G = \begin{bmatrix} \frac{1}{3} & 0 \\ 0 & \frac{2}{3} \\ \frac{2}{3} & 1 \\ & \frac{1}{3} \end{bmatrix} \quad H = \begin{bmatrix} \frac{1}{2} & 0 & \frac{1}{2} \\ 0 & \frac{1}{2} & \frac{1}{2} \\ \frac{1}{2} & \frac{1}{2} & 0 \end{bmatrix}$$

2. Suppose that each of the following represents the transition matrix **M** and the initial probability vector **p** for a Markov chain. Find the probability vectors $\mathbf{p}_1$ (after one step of the process) and $\mathbf{p}_2$ (after two steps).

⋆(a) $\mathbf{M} = \begin{bmatrix} \frac{1}{4} & \frac{1}{3} \\ \frac{3}{4} & \frac{2}{3} \end{bmatrix}$, $\mathbf{p} = \begin{bmatrix} \frac{2}{3} \\ \frac{1}{3} \end{bmatrix}$

(b) $\mathbf{M} = \begin{bmatrix} \frac{1}{2} & \frac{1}{3} & 0 \\ 0 & \frac{2}{3} & \frac{1}{2} \\ \frac{1}{2} & 0 & \frac{1}{2} \end{bmatrix}$, $\mathbf{p} = \begin{bmatrix} \frac{1}{3} \\ \frac{1}{6} \\ \frac{1}{2} \end{bmatrix}$

⋆(c) $\mathbf{M} = \begin{bmatrix} \frac{1}{4} & \frac{1}{3} & \frac{1}{2} \\ \frac{1}{2} & \frac{1}{3} & \frac{1}{6} \\ \frac{1}{4} & \frac{1}{3} & \frac{1}{3} \end{bmatrix}$, $\mathbf{p} = \begin{bmatrix} \frac{1}{4} \\ \frac{1}{2} \\ \frac{1}{4} \end{bmatrix}$

3. Suppose that each of the following regular matrices represents the transition matrix **M** for a Markov chain. Find the steady-state vector for the Markov chain by solving an appropriate system of linear equations.

⋆(a) $\begin{bmatrix} \frac{1}{2} & \frac{1}{3} \\ \frac{1}{2} & \frac{2}{3} \end{bmatrix}$ (b) $\begin{bmatrix} \frac{1}{3} & \frac{1}{4} & \frac{1}{3} \\ \frac{1}{6} & \frac{1}{2} & \frac{1}{3} \\ \frac{1}{2} & \frac{1}{4} & \frac{1}{3} \end{bmatrix}$

(c) $\begin{bmatrix} \frac{1}{5} & \frac{1}{2} & 0 & \frac{1}{3} \\ \frac{3}{5} & 0 & \frac{1}{2} & 0 \\ 0 & \frac{1}{2} & \frac{1}{2} & 0 \\ \frac{1}{5} & 0 & 0 & \frac{2}{3} \end{bmatrix}$

4. Find the steady-state vector for the Markov chains in parts (a) and (b) of Exercise 3 by calculating large powers of the transition matrix (using a computer or calculator).

*5. Suppose that the citizens in a certain community tend to switch their votes among political parties, as shown in the following transition matrix:

Current Election

		Party A	Party B	Party C	Nonvoting
	Party A	.7	.2	.2	.1
Next Election	Party B	.1	.6	.1	.1
	Party C	.1	.2	.6	.1
	Nonvoting	.1	0	.1	.7

(a) Suppose that in the last election 30% of the citizens voted for Party A, 15% voted for Party B, and 45% voted for Party C. What is the likely outcome of the next election? What is the likely outcome of the election after that?

(b) If current trends continue, what percentage of the citizens will vote for Party A one century from now? Party C?

*6. In a psychology experiment, a rat wanders in the maze in Figure 8.13. During each time interval, the rat is allowed to pass through exactly one doorway. Assume there is a 50% probability that the rat will switch rooms during each interval. If it does switch rooms, assume that it has an equally likely chance of using any doorway out of its current room.

(a) What is the transition matrix for the associated Markov chain?

(b) Show that the transition matrix from part (a) is regular.

(c) If the rat is known to be in room C, what is the probability it will be in room D after two time intervals have passed?

(d) What is the steady-state vector for this Markov chain? Over time, which room does the rat frequent the least? Which room does the rat frequent the most?

Figure 8.13

Maze with five rooms

7. Show that the converse to part (3) of Theorem 8.5 is not true by demonstrating that the transition matrix

$$\mathbf{M} = \begin{bmatrix} 1 & \frac{1}{2} & \frac{1}{4} \\ 0 & \frac{1}{2} & \frac{1}{4} \\ 0 & 0 & \frac{1}{2} \end{bmatrix}$$

has the same limit vector for any initial input but is not regular. Does this Markov chain have a unique fixed point?

8. (a) Show that the transition matrix $\begin{bmatrix} 1-a & b \\ a & 1-b \end{bmatrix}$ has $\left(\frac{1}{a+b}\right)\begin{bmatrix} b \\ a \end{bmatrix}$ as a steady-state vector if a and b are not both 0.

   (b) Use the result in part (a) to check that your answer for Exercise 3(a) is correct.

▶**9.** Prove Theorem 8.3.

▶**10.** Prove Theorem 8.4.

⋆**11.** True or False:

   (a) The transpose of a stochastic matrix is stochastic.

   (b) For $n > 1$, no upper triangular $n \times n$ matrix is regular.

   (c) If $\mathbf{M}$ is a regular $n \times n$ stochastic matrix, then there is a probability vector $\mathbf{p}$ such that $(\mathbf{M} - \mathbf{I}_n)\mathbf{p} = \mathbf{0}$.

   (d) If $\mathbf{M}$ is a stochastic matrix and $\mathbf{p}$ and $\mathbf{q}$ are distinct probability vectors such that $\mathbf{Mp} = \mathbf{q}$ and $\mathbf{Mq} = \mathbf{p}$, then $\mathbf{M}$ is not regular.

   (e) The entries of a transition matrix $\mathbf{M}$ give the probabilities of a Markov process being in each of its states.

8.5 Hill Substitution: An Introduction to Coding Theory

Prerequisite: Section 2.4, Inverses of Matrices

In this section, we show how matrix inverses can be used in a simple manner to encode and decode textual information.

Substitution Ciphers

The coding and decoding of secret messages has been important in times of warfare, of course, but it is also quite valuable in peacetime for keeping government and business secrets under tight security. Throughout history, many ingenious coding mechanisms have been proposed. One of the simplest is the **substitution cipher**, in which an array of symbols is used to assign each character of a given text (**plaintext**) to a corresponding character in coded text (**ciphertext**). For example, consider the **cipher array** in Figure 8.14. A message can be encoded by replacing every instance of the kth letter of the alphabet with the kth character in the cipher array. For example, the message

LINEAR ALGEBRA IS EXCITING

is encoded as

FXUSRI RFTSWIR XG SNEXVXUT.

This type of substitution can be extended to other characters, such as punctuation symbols and blanks.

Messages can be decoded by reversing the process. In fact, we can create an "inverse" array, or **decipher array**, as in Figure 8.15, to restore the symbols of FXUSRI RFTSWIR XG SNEXVXUT back to LINEAR ALGEBRA IS EXCITING.

Figure 8.14

A cipher array

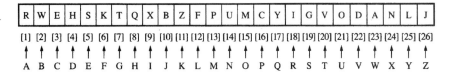

Figure 8.15

A decipher array

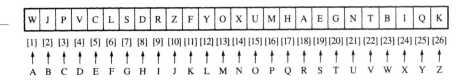

Cryptograms, a standard feature in newspapers and puzzle magazines, are substitution ciphers. However, these ciphers are relatively easy to "crack" because the relative frequencies (occurrences per length of text) of the letters of the English alphabet have been studied extensively.[4]

Hill Substitution

We now illustrate a method that uses matrices to create codes that are harder to break. This technique is known as **Hill substitution** after the mathematician Lester Hill, who developed it between the world wars. To begin, we choose any nonsingular $n \times n$ matrix **A**. (Usually **A** is chosen with integer entries.) We split the message into blocks of n symbols each and replace each symbol with an integer value. To simplify things, we replace each letter by its position in the alphabet. The last block may have to be "padded" with random values to ensure that each block contains exactly n integers. In effect, we are creating a set of n-vectors that we can label as $\mathbf{x}_1$, $\mathbf{x}_2$, and so on. We then multiply the matrix **A** by each of these vectors in turn to produce the following new set of n-vectors: $\mathbf{Ax}_1$, $\mathbf{Ax}_2$, and so on. When these vectors are concatenated together, they form the coded message. The matrix **A** used in the process is often called the **key matrix**, or **encoding matrix**.

EXAMPLE 1

Suppose we wish to encode the message LINEAR ALGEBRA IS EXCITING using the key matrix

$$\mathbf{A} = \begin{bmatrix} -7 & 5 & 3 \\ 3 & -2 & -2 \\ 3 & -2 & -1 \end{bmatrix}.$$

---

[4] The longer the enciphered text is, the easier it is to decode by comparing the number of times each letter appears. The actual frequency of the letters depends on the type of text, but the letters E, T, A, O, I, N, S, H, and R typically appear most often (about 70% of the time), with E usually the most common (about 12–13% of the time). Once a few letters have been deciphered, the rest of the text is usually easy to determine. Sample frequency tables can be found on p. 219 of *Cryptanalysis* by Gaines (published by Dover, 1956) and on p. 16 of *Cryptography: A Primer* by Konheim (published by Wiley, 1981).

Since we are using a 3×3 matrix, we break the characters of the message into blocks of length 3 and replace each character by its position in the alphabet. This procedure gives

$$
\begin{array}{c} L \\ I \\ N \end{array} \begin{bmatrix} 12 \\ 9 \\ 14 \end{bmatrix}, \qquad
\begin{array}{c} E \\ A \\ R \end{array} \begin{bmatrix} 5 \\ 1 \\ 18 \end{bmatrix}, \qquad
\begin{array}{c} A \\ L \\ G \end{array} \begin{bmatrix} 1 \\ 12 \\ 7 \end{bmatrix}, \qquad
\begin{array}{c} E \\ B \\ R \end{array} \begin{bmatrix} 5 \\ 2 \\ 18 \end{bmatrix},
$$

$$
\underbrace{}_{\mathbf{x}_1} \qquad \underbrace{}_{\mathbf{x}_2} \qquad \underbrace{}_{\mathbf{x}_3} \qquad \underbrace{}_{\mathbf{x}_4}
$$

$$
\begin{array}{c} A \\ I \\ S \end{array} \begin{bmatrix} 1 \\ 9 \\ 19 \end{bmatrix}, \qquad
\begin{array}{c} E \\ X \\ C \end{array} \begin{bmatrix} 5 \\ 24 \\ 3 \end{bmatrix}, \qquad
\begin{array}{c} I \\ T \\ I \end{array} \begin{bmatrix} 9 \\ 20 \\ 9 \end{bmatrix}, \qquad
\begin{array}{c} N \\ G \\ - \end{array} \begin{bmatrix} 14 \\ 7 \\ 27 \end{bmatrix},
$$

$$
\underbrace{}_{\mathbf{x}_5} \qquad \underbrace{}_{\mathbf{x}_6} \qquad \underbrace{}_{\mathbf{x}_7} \qquad \underbrace{}_{\mathbf{x}_8}
$$

where the last entry of the last vector was chosen outside the range from 1 to 26. Now, forming the products with $\mathbf{A}$, we have

$$
\mathbf{A}\mathbf{x}_1 = \begin{bmatrix} -7 & 5 & 3 \\ 3 & -2 & -2 \\ 3 & -2 & -1 \end{bmatrix} \begin{bmatrix} 12 \\ 9 \\ 14 \end{bmatrix} = \begin{bmatrix} 3 \\ -10 \\ 4 \end{bmatrix},
$$

$$
\mathbf{A}\mathbf{x}_2 = \begin{bmatrix} -7 & 5 & 3 \\ 3 & -2 & -2 \\ 3 & -2 & -1 \end{bmatrix} \begin{bmatrix} 5 \\ 1 \\ 18 \end{bmatrix} = \begin{bmatrix} 24 \\ -23 \\ -5 \end{bmatrix}, \qquad \text{and so on.}
$$

The final encoded text is

$$
\begin{array}{rrrrrrrrrrrr}
3 & -10 & 4 & 24 & -23 & -5 & 74 & -35 & -28 & 29 & -25 & -7 \\
95 & -53 & -34 & 94 & -39 & -36 & 64 & -31 & -22 & 18 & -26 & 1.
\end{array}
$$

  ■

  The code produced by a Hill substitution is much harder to break than a simple substitution cipher, since the coding of a given letter depends not only on the way the text is broken into blocks, but also on the letters adjacent to it. (Nevertheless, there are techniques to decode Hill substitutions using high-speed computers.) However, a Hill substitution is easy to decode if you know the inverse of the key matrix. In Example 6 of Section 2.4, we noted that

$$
\mathbf{A}^{-1} = \begin{bmatrix} 2 & 1 & 4 \\ 3 & 2 & 5 \\ 0 & -1 & 1 \end{bmatrix}.
$$

Breaking the encoded text back into 3-vectors and multiplying $\mathbf{A}^{-1}$ by each of these vectors in turn restores the original message. For example,

$$
\mathbf{A}^{-1}(\mathbf{A}\mathbf{x}_1) = \begin{bmatrix} 2 & 1 & 4 \\ 3 & 2 & 5 \\ 0 & -1 & 1 \end{bmatrix} \begin{bmatrix} 3 \\ -10 \\ 4 \end{bmatrix} = \begin{bmatrix} 12 \\ 9 \\ 14 \end{bmatrix} = \mathbf{x}_1,
$$

which represents the first three letters LIN.

Exercises for Section 8.5

1. Encode each message with the given key matrix.

   ★(a) PROOF BY INDUCTION   using the matrix $\begin{bmatrix} 3 & -4 \\ 5 & -7 \end{bmatrix}$

   (b) CONTACT HEADQUARTERS   using the matrix $\begin{bmatrix} 4 & 1 & 5 \\ 7 & 2 & 9 \\ 6 & 2 & 7 \end{bmatrix}$

2. Each of the following coded messages was produced with the key matrix shown. In each case, find the inverse of the key matrix, and use it to decode the message.

   ★(a) $\begin{matrix} -62 & 116 & 107 & -32 & 59 & 67 & -142 & 266 & 223 & -160 & 301 & 251 \\ -122 & 229 & 188 & -122 & 229 & 202 & -78 & 148 & 129 & -111 & 207 & 183 \end{matrix}$

   with key matrix $\begin{bmatrix} -8 & 1 & -1 \\ 15 & -2 & 2 \\ 12 & -1 & 2 \end{bmatrix}$

   (b) $\begin{matrix} 162 & 108 & 23 & 303 & 206 & 33 & 276 & 186 & 33 & 170 & 116 & 21 \\ 281 & 191 & 36 & 576 & 395 & 67 & 430 & 292 & 51 & 340 & 232 & 45 \end{matrix}$

   with key matrix $\begin{bmatrix} -10 & 19 & 16 \\ -7 & 13 & 11 \\ -1 & 2 & 2 \end{bmatrix}$

   (c) $\begin{matrix} 69 & 44 & -28 & -43 & 104 & 53 & -38 & -25 \\ 71 & 38 & -3 & -7 & 58 & 32 & -11 & -14 \end{matrix}$

   with key matrix $\begin{bmatrix} 1 & 2 & 5 & 1 \\ 0 & 1 & 3 & 1 \\ -2 & 0 & 0 & -1 \\ 0 & 0 & -1 & -2 \end{bmatrix}$

★**3.** True or False:
   (a) Text encoded with a Hill substitution is more difficult to decipher than text encoded with a substitution cipher.
   (b) The encoding matrix for a Hill substitution should not be singular.
   (c) To encode a message using Hill substitution that is n characters long, an $n \times n$ matrix is always used.

8.6 Change of Variables and the Jacobian

Prerequisite: Section 3.1, Introduction to Determinants

In this section, we show how the determinant of a matrix is used to perform a change of variables in a double or triple integral. This technique generalizes to a change of variables in higher dimensions as well. Although the prerequisite for this section is listed as Section 3.1, we will also need the fact that $|\mathbf{A}| = |\mathbf{A}^T|$ from Section 3.3.

Substitution in One Variable

The following example serves to recall the method of integration by substitution from calculus:

EXAMPLE 1 To compute $\int_1^5 \sqrt{3x+1}\, dx$, we first make the substitution $u = 3x + 1$. Then $du = 3\, dx$, and so

$$\int_1^5 \sqrt{3x+1}\, dx = \frac{1}{3} \int_1^5 \sqrt{3x+1}\,(3\, dx) = \frac{1}{3} \int_4^{16} \sqrt{u}\, du$$

$$= \frac{1}{3} \cdot \frac{2}{3} u^{\frac{3}{2}} \Big|_4^{16} = \frac{2}{9}(16^{\frac{3}{2}} - 4^{\frac{3}{2}}) = \frac{2}{9}(64 - 8) = \frac{112}{9}.$$

Note the factor of 3 in $du = 3\, dx$. This indicates that the variable u covers 3 units of distance for each single unit of x. (It is as if u is measured in feet, while x is measured in yards.) Note that the length of the x-interval is only 4 units (from 1 to 5), while the length of the u-interval is 12 units (from 4 to 16). The factor of 3 in the du term compensates for this change. ∎

In Example 1, the substitution variable u is a linear function of x, and so the change in units is constant throughout the given interval. In the next example, however, the substitution is non-linear.

EXAMPLE 2 Consider $\int_1^2 \frac{2x}{(x^2+1)^2}\, dx$. Let $u = x^2 + 1$. Then $du = 2x\, dx$. The integral is then calculated as

$$\int_1^2 \frac{2x}{(x^2+1)^2}\, dx = \int_2^5 \frac{du}{u^2} = -\frac{1}{u}\Big|_2^5 = -\frac{1}{5} - \left(-\frac{1}{2}\right) = \frac{3}{10}.$$

The factor $2x$ in $du = 2x\, dx$ indicates that the unit conversion from x to u is not constant. As the x-interval $[1, 2]$ is stretched into the u-interval $[2, 5]$, the stretching is done unevenly. For example, at $x = 1$, the scaling factor $2x = 2(1) = 2$, and so at this point, the length of a u unit is 2 times smaller than the length of an x unit. However, at $x = 1.5$, the scaling factor $2x = 2(1.5) = 3$, and so at this point, a u unit is 3 times smaller than an x unit.

In particular, the x-interval $[1.5, 1.51]$ (of length 0.01) is mapped to the u-interval $[3.25, 3.2801]$ (having length 0.0301). That is, the u-interval is approximately 3 times as long, because the scaling factor is 3 at $x = 1.5$. The error in using 3 as the scaling factor in this case is 0.0001, or $0.3\overline{3}\%$. As the length of the x-interval approaches 0, as it would in computing Riemann sums for integrals, the percent error in the scaling factor also approaches 0. ∎

In general, since $\frac{du}{dx}$ is the rate of change of u with respect to x, its presence in the formula $du = \frac{du}{dx}\, dx$ keeps track of the amount of stretching involved in converting from x-coordinates to u-coordinates. Thus, $\frac{du}{dx}$ is the desired scaling factor for a change of variable in single-variable integration.

Double Integrals

We now consider the analogous situation using two variables.

EXAMPLE 3 The area of the parallelogram P indicated in Figure 8.16 is given by the following double integral:

$$\text{Area} = \iint\limits_{P} 1 \, dx \, dy.$$

Figure 8.16

The parallelogram in the (x, y) system with vertices $(1, 1)$, $(0, 2)$, $(3, 2)$, $(2, 3)$

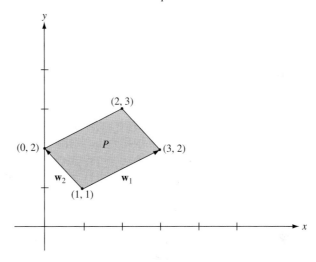

Converting this double integral into an iterated integral would be tedious. However, we can compute the area of P using Theorem 3.1. The vectors $\mathbf{w}_1 = [2, 1]$ and $\mathbf{w}_2 = [-1, 1]$ correspond to the sides of P, and so

$$\text{area of } P = \text{absolute value of } \begin{vmatrix} 2 & 1 \\ -1 & 1 \end{vmatrix} = |2 - (-1)| = 3.$$

Let us now examine the effect of a change of variables on the area. Since the sides of P are the vectors $\mathbf{w}_1$ and $\mathbf{w}_2$, we first create new variables u and v to satisfy the equation

$$[x, y] = u\mathbf{w}_1 + v\mathbf{w}_2 + [1, 1] = u[2, 1] + v[-1, 1] + [1, 1];$$

that is, $x = 2u - v + 1$, $y = u + v + 1$. Then, (x, y) vertices correspond to (u, v) vertices as follows:

(x, y)	(u, v)
$(1, 1)$	$(0, 0)$
$(0, 2)$	$(0, 1)$
$(3, 2)$	$(1, 0)$
$(2, 3)$	$(1, 1)$

Thus, in converting to the (u, v) coordinate system, the parallelogram P is mapped to the unit square S shown in Figure 8.17. Therefore, it follows that

$$\iint_S 1 \, du \, dv = \text{area of } S = 1.$$

Figure 8.17

The square in the (u, v) system with vertices $(0, 0), (0, 1), (1, 0), (1, 1)$

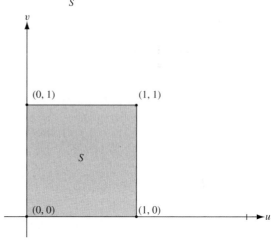

Since the parallelogram P does not have area 1, we must be missing a scaling factor of the type seen in the single variable case. Note that the scaling factor must be constant in this case, as in Example 1, because the change of coordinates involves only linear functions. Since the area of $P = 3(\text{area of } S)$, the scaling factor must be precisely 3.   ■

Note in Example 3 that we can work backwards to compute the vectors $\mathbf{w}_1$ and $\mathbf{w}_2$ from the formulas for x and y as $\mathbf{w}_1 = \left[\frac{\partial x}{\partial u}, \frac{\partial y}{\partial u} \right]$, and $\mathbf{w}_2 = \left[\frac{\partial x}{\partial v}, \frac{\partial y}{\partial v} \right]$. This will work in general for all change of variable transformations. The idea behind this is that a unit rectangle in (u, v) coordinates is mapped to a region in (x, y) coordinates that is approximated by a parallelogram whose sides are $\mathbf{w}_1$ and $\mathbf{w}_2$, as in Figure 8.18. The vectors $\mathbf{w}_1$ and $\mathbf{w}_2$ are tangent to the curved boundary of the actual image of the rectangle under the transformation. But differentiation, along with finding the tangent direction, also measures the rate of change, and so the lengths of $\mathbf{w}_1$ and $\mathbf{w}_2$ also represent the amount of stretching taking place in each of these directions. Hence, the scaling factor needed for the change of variable is the area of this approximating parallelogram, which, by Theorem 3.1, is the absolute value of $\begin{vmatrix} \frac{\partial x}{\partial u} & \frac{\partial y}{\partial u} \\ \frac{\partial x}{\partial v} & \frac{\partial y}{\partial v} \end{vmatrix}$.

In Section 3.3, it is proved that for any square matrix $\mathbf{A}$, $|\mathbf{A}| = |\mathbf{A}^T|$. Hence we could have also found the scaling factor as the absolute value of $\begin{vmatrix} \frac{\partial x}{\partial u} & \frac{\partial x}{\partial v} \\ \frac{\partial y}{\partial u} & \frac{\partial y}{\partial v} \end{vmatrix}$ instead. The matrix

$$\mathbf{J} = \begin{bmatrix} \dfrac{\partial x}{\partial u} & \dfrac{\partial x}{\partial v} \\ \dfrac{\partial y}{\partial u} & \dfrac{\partial y}{\partial v} \end{bmatrix}$$

Figure 8.18

Converting a rectangle in (u, v) coordinates to an approximate parallelogram in (x, y) coordinates

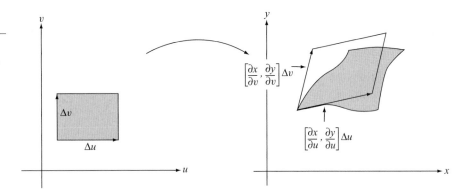

is called the **Jacobian matrix** of the change of coordinates function $\begin{cases} x = x(u, v) \\ y = y(u, v) \end{cases}$. We will refer to $|\mathbf{J}|$ as the **Jacobian determinant**. In general, the correct scaling factor to change an integral $\iint\limits_R f(x, y)\,dx\,dy$ over a region R into (u, v) coordinates is the absolute value of the Jacobian determinant, that is, $\left| |\mathbf{J}| \right|$.

Just as in the one-variable case, the scaling factor can vary if the change of coordinates is nonlinear, as we will see shortly.

Polar Coordinates

The polar coordinate system is frequently used to represent points in a two-dimensional space. In polar coordinates, each point $P = (x, y)$ in the plane is assigned a pair[5] of coordinates (r, θ), where r is the distance from the origin to P, and θ is the angle between the positive x-axis and the vector having initial point at the origin and terminal point P (see Figure 8.19). In all quadrants, the transformation from polar coordinates to standard (rectangular) coordinates is given by $\begin{cases} x &= r\cos\theta \\ y &= r\sin\theta \end{cases}$. We can also convert from rectangular coordinates to polar coordinates using $\begin{cases} r^2 &= x^2 + y^2 \\ \tan\theta &= \frac{y}{x} \ (\text{when } x \neq 0) \end{cases}$.

It is useful to express certain double integrals in polar coordinates if the region of integration (and/or the function involved) has radial or angular symmetry. In these instances, we need to compute the determinant of the Jacobian

---

[5] The assignment of polar coordinates to a given point (x, y) is not unique. For example, $(x, y) = \left(\sqrt{3}, 1\right)$ in rectangular coordinates can be represented as (r, θ) in polar coordinates as $(2, \frac{\pi}{6})$, $(2, \frac{13\pi}{6})$, or $(-2, \frac{7\pi}{6})$. In general, $\left(\sqrt{3}, 1\right)$ can be expressed in polar coordinates as (r, θ), where $r = \pm\sqrt{(\sqrt{3})^2 + 1^2} = \pm 2$, and $\theta = \frac{\pi}{6} + k\pi$, where k is an even integer when r is positive, and k is an odd integer when r is negative.

Figure 8.19

Relationship between
standard coordinates
and polar coordinates
in Quadrants I and II

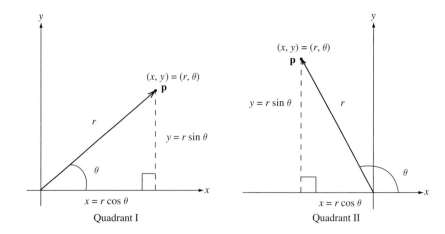

Quadrant I

Quadrant II

matrix in order to include the proper scaling factor when we change coordinates.

$$|\mathbf{J}| = \begin{vmatrix} \frac{\partial x}{\partial u} & \frac{\partial x}{\partial v} \\ \frac{\partial y}{\partial u} & \frac{\partial y}{\partial v} \end{vmatrix} = \begin{vmatrix} \cos\theta & -r\sin\theta \\ \sin\theta & r\cos\theta \end{vmatrix} = r\cos^2\theta + r\sin^2\theta = r.$$

If we are careful to ensure that $r \geq 0$, the absolute value of $|\mathbf{J}|$ is also r, and so this is our scaling factor. Hence,

$$\iint_R f(x, y)\, dx\, dy = \iint_{R^*} f(x(r, \theta),\, y(r, \theta))\, r\, dr\, d\theta,$$

where R^* is the region in the polar coordinate system corresponding to R. The next example illustrates this geometrically.

EXAMPLE 4   Consider the square S in the (r, θ) (polar) coordinate system with left bottom corner at $(2, \frac{\pi}{6})$, width $\Delta r = 0.1$, and height $\Delta\theta = 0.1$. The image R of this square in the (x, y) system under the polar coordinate map $\begin{cases} x &= r\cos\theta \\ y &= r\sin\theta \end{cases}$ is shown in Figure 8.20.

Now, the square S has area $\Delta r\, \Delta\theta = (0.1)(0.1) = 0.01$, and thus the area of R is approximately equal to the product of the Jacobian determinant, $r = 2$, with the area of S. Hence, the area of $R \approx 2(0.01) = 0.02$.

To understand this approximation, recall that the columns of the Jacobian matrix represent vectors tangent at the corner point to the curved edges of R. When these vectors are scaled properly by multiplying by Δr and $\Delta\theta$, respectively, they represent the sides of a parallelogram (shown in Figure 8.21) whose area approximates the area of R. (In this particular case, the dot product of the columns is zero, and so the parallelogram is a rectangle.)

Finally, we compute the actual area of R for comparison purposes. The actual area of R is $\frac{\Delta\theta}{2\pi}$ (the portion of the circle involved) times the differences of the areas of the circles of radii 2.1 and 2.0. Hence,

$$\text{area of } R = \frac{\Delta\theta}{2\pi}(\pi(2.1^2) - \pi(2^2)) = \frac{0.1}{2\pi}(\pi(0.41)) = \frac{0.041}{2} = 0.0205.$$

Figure 8.20

Image R in rectangular coordinates of polar coordinate system square S

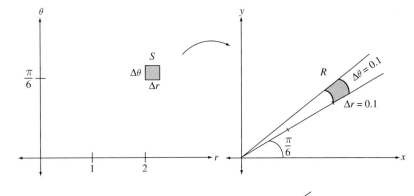

Figure 8.21

The parallelogram formed by the columns of the Jacobian at the point $(2, \frac{\pi}{6})$

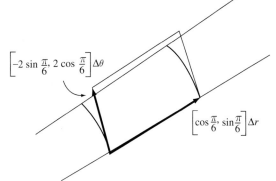

Thus, in this case, the scale factor obtained from the Jacobian induces an error of only 0.0005, or, 2.5%. Of course, in the actual integration, both $\Delta r \to 0$ and $\Delta \theta \to 0$, which makes the percent error approach 0 as well (although we do not prove this here). ∎

EXAMPLE 5

Consider $\iint_R \sqrt{x^2 + y^2}\, dx\, dy$ over the region R given by $0 \le r \le 1 + \cos\theta$ in polar coordinates (see Figure 8.22). Now, $\sqrt{x^2 + y^2} = r$, and so

$$\iint_R \sqrt{x^2 + y^2}\, dx\, dy = \iint_R r \cdot r\, dr\, d\theta = \int_0^{2\pi} \int_0^{1+\cos\theta} r^2\, dr\, d\theta$$

$$= \int_0^{2\pi} \left(\frac{r^3}{3}\right)\Bigg|_0^{1+\cos\theta} d\theta = \frac{1}{3}\int_0^{2\pi} (1 + \cos\theta)^3\, d\theta$$

$$= \frac{1}{3}\int_0^{2\pi} (1 + 3\cos\theta + 3\cos^2\theta + \cos^3\theta)\, d\theta$$

$$= \frac{1}{3}\int_0^{2\pi} (3\cos\theta + \cos^3\theta)\, d\theta + \frac{1}{3}\int_0^{2\pi} (1 + 3\cos^2\theta)\, d\theta.$$

Figure 8.22

The region R in polar coordinates given by $0 \le r \le 1 + \cos\theta$

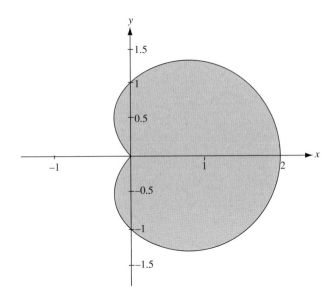

An appeal to symmetry considerations (or a tedious computation) shows the first integral equals 0. Using a double-angle formula on the second integral, we obtain

$$\frac{1}{3} \int_0^{2\pi} \left(1 + 3 \left(\frac{1}{2} + \frac{1}{2} \cos 2\theta \right) \right) d\theta = \left(\frac{5}{6} \theta + \frac{1}{4} \sin 2\theta \right) \Big|_0^{2\pi} = \frac{5\pi}{3}. \qquad \blacksquare$$

Triple Integrals

The situation for change of variables in three dimensions is similar. When converting an integral in (x, y, z) coordinates to an integral in (u, v, w) coordinates, any rectangular solid based at the point (x, y, z) and having sides Δx, Δy, and Δz is mapped to a region approximated by a parallelepiped. The sides of this parallelepiped are the columns of the Jacobian matrix evaluated at (x, y, z) multiplied by Δx, Δy, and Δz, respectively. Thus, by Theorem 3.1, the absolute value of the Jacobian determinant

$$|\mathbf{J}| = \begin{vmatrix} \frac{\partial x}{\partial u} & \frac{\partial x}{\partial v} & \frac{\partial x}{\partial w} \\ \frac{\partial y}{\partial u} & \frac{\partial y}{\partial v} & \frac{\partial y}{\partial w} \\ \frac{\partial z}{\partial u} & \frac{\partial z}{\partial v} & \frac{\partial z}{\partial w} \end{vmatrix}$$

provides the correct scaling factor for converting from xyz-space to uvw-space. That is, $dx\,dy\,dz = \Big| |\mathbf{J}| \Big|\, du\,dv\,dw$.

Spherical Coordinates

One coordinate system frequently used in three dimensions is spherical coordinates. If $P = (x, y, z)$ is a point in the rectangular coordinate system and $\mathbf{v}$ is a

Figure 8.23

Spherical coordinates for $P = (x, y, z)$

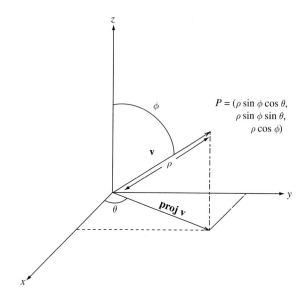

$P = (\rho \sin \phi \cos \theta,$
$\rho \sin \phi \sin \theta,$
$\rho \cos \phi)$

vector from the origin to P, then P is assigned coordinates (ρ, ϕ, θ) in spherical coordinates, where $\rho = ||\mathbf{v}||$, ϕ is the angle between the vector $[0, 0, 1]$ and $\mathbf{v}$, and θ is the angle between the vector $[1, 0, 0]$ and the projection of $\mathbf{v}$ onto the xy-plane (see Figure 8.23). From elementary trigonometry, we find that

$$x = \rho \sin \phi \cos \theta \qquad\qquad \rho^2 = x^2 + y^2 + z^2$$
$$y = \rho \sin \phi \sin \theta \qquad\qquad \tan \theta = \tfrac{y}{x}, \text{ when } x \neq 0$$
$$z = \rho \cos \phi \qquad\qquad\qquad \cos \phi = \frac{z}{\sqrt{x^2+y^2+z^2}}.$$

Hence,

$$
|\mathbf{J}| =
\begin{vmatrix}
\frac{\partial x}{\partial \rho} & \frac{\partial x}{\partial \phi} & \frac{\partial x}{\partial \theta} \\
\frac{\partial y}{\partial \rho} & \frac{\partial y}{\partial \phi} & \frac{\partial y}{\partial \theta} \\
\frac{\partial z}{\partial \rho} & \frac{\partial z}{\partial \phi} & \frac{\partial z}{\partial \theta}
\end{vmatrix}
=
\begin{vmatrix}
\sin \phi \cos \theta & \rho \cos \phi \cos \theta & -\rho \sin \phi \sin \theta \\
\sin \phi \sin \theta & \rho \cos \phi \sin \theta & \rho \sin \phi \cos \theta \\
\cos \phi & -\rho \sin \phi & 0
\end{vmatrix}
$$

$$
= \cos \phi
\begin{vmatrix}
\rho \cos \phi \cos \theta & -\rho \sin \phi \sin \theta \\
\rho \cos \phi \sin \theta & \rho \sin \phi \cos \theta
\end{vmatrix}
$$

$$
\quad -(-\rho \sin \phi)
\begin{vmatrix}
\sin \phi \cos \theta & -\rho \sin \phi \sin \theta \\
\sin \phi \sin \theta & \rho \sin \phi \cos \theta
\end{vmatrix}
$$

$$
= \cos \phi (\rho^2 \cos \phi \sin \phi \cos^2 \theta + \rho^2 \cos \phi \sin \phi \sin^2 \theta)
$$
$$
\quad + \rho \sin \phi (\rho \sin^2 \phi \cos^2 \theta + \rho \sin^2 \phi \sin^2 \theta)
$$
$$
= \rho^2 \cos^2 \phi \sin \phi (\cos^2 \theta + \sin^2 \theta) + \rho^2 \sin^3 \phi (\cos^2 \theta + \sin^2 \theta)
$$
$$
= \rho^2 \cos^2 \phi \sin \phi + \rho^2 \sin^3 \phi
$$
$$
= \rho^2 \sin \phi (\cos^2 \phi + \sin^2 \phi)
$$
$$
= \rho^2 \sin \phi.
$$

Since $0 \le \phi \le \pi$ in spherical coordinates, the quantity $\rho^2 \sin \phi$ is always nonnegative. Hence, when converting an integral from xyz-coordinates to $\rho\phi\theta$-coordinates, we have

$$dx \, dy \, dz = \rho^2 \sin \phi \, d\rho \, d\phi \, d\theta.$$

| **EXAMPLE 6** | We find the volume of the region R bounded below by the upper half of the cone $z^2 = x^2 + y^2$ and bounded above by the sphere $x^2 + y^2 + z^2 = 8$ (see Figure 8.24). Now,

$$\text{volume of } R = \iiint_R 1 \, dx \, dy \, dz.$$

Converting to spherical coordinates, we have

$$\text{volume of } R = \iiint_R \rho^2 \sin \phi \, d\rho \, d\phi \, d\theta.$$

Figure 8.24

Region R bounded below by the upper half of the cone $z^2 = x^2 + y^2$ and bounded above by the sphere $x^2 + y^2 + z^2 = 8$

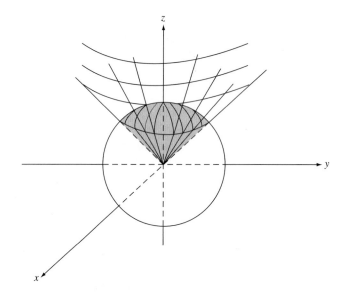

Since the radius of the sphere is $\sqrt{8}$, ρ ranges from 0 to $\sqrt{8}$. The sides of the cone are at a 45° angle from the z-axis, and so ϕ ranges from 0 to $\frac{\pi}{4}$. Hence, changing to an iterated integral, we obtain

$$\text{volume of } R = \int_0^{2\pi} \int_0^{\frac{\pi}{4}} \int_0^{\sqrt{8}} \rho^2 \sin \phi \, d\rho \, d\phi \, d\theta$$

$$= \int_0^{2\pi} \int_0^{\frac{\pi}{4}} \left(\frac{\rho^3}{3} \sin \phi \right) \Big|_0^{\sqrt{8}} d\phi \, d\theta$$

$$= \int_0^{2\pi} \int_0^{\frac{\pi}{4}} \frac{8\sqrt{8}}{3} \sin\phi \, d\phi \, d\theta$$

$$= -\frac{8\sqrt{8}}{3} \int_0^{2\pi} (\cos\phi) \Big|_0^{\frac{\pi}{4}} d\theta$$

$$= -\frac{8\sqrt{8}}{3} \int_0^{2\pi} \left(\frac{\sqrt{2}}{2} - 1 \right) d\theta$$

$$= -\frac{8\sqrt{8}}{3} \left(\frac{\sqrt{2}}{2} - 1 \right) (2\pi)$$

$$= \frac{32\pi}{3} (\sqrt{2} - 1).$$

■

Cylindrical Coordinates

Another frequently used three-dimensional coordinate system is cylindrical coordinates, (r, θ, z), in which the r and θ variables provide a polar coordinate system in the xy-plane, and z is unchanged from rectangular coordinates. Thus,

$$x = r\cos\theta$$
$$y = r\sin\theta \quad .$$
$$z = z$$

In Exercise 3, you are asked to show that the Jacobian determinant for a transformation from rectangular to cylindrical coordinates is r, and hence

$$dx \, dy \, dz = r \, dr \, d\theta \, dz.$$

Higher Dimensions

The method we have shown for changing variables in double and triple integrals also works in general for multiple integrals in $\mathbb{R}^n$. In particular, to change from $x_1 x_2 \ldots x_n$ coordinates to $u_1 u_2 \ldots u_n$ coordinates, we must calculate the absolute value of the determinant of the Jacobian matrix,

$$|\mathbf{J}| = \begin{vmatrix} \frac{\partial x_1}{\partial u_1} & \frac{\partial x_1}{\partial u_2} & \cdots & \frac{\partial x_1}{\partial u_n} \\ \frac{\partial x_2}{\partial u_1} & \frac{\partial x_2}{\partial u_2} & \cdots & \frac{\partial x_2}{\partial u_n} \\ \vdots & \vdots & \ddots & \cdots \\ \frac{\partial x_n}{\partial u_1} & \frac{\partial x_n}{\partial u_2} & \cdots & \frac{\partial x_n}{\partial u_n} \end{vmatrix},$$

and then we have,

$$dx_1 \, dx_2 \ldots dx_n = \Big| |\mathbf{J}| \Big| \, du_1 \, du_2 \ldots du_n.$$

Exercises for Section 8.6

1. For each change of variable formula, compute $dx\,dy$ in terms of $du\,dv$.

   $\star$(a) $x = u + v$, $y = u - v$
   (b) $x = u^2 + v^2$, $y = u^2 - v^2$
   $\star$(c) $x = u^2 - v^2$, $y = 2uv$
   (d) $x = \frac{u}{u^2+v^2}$, $y = \frac{-v}{u^2+v^2}$
   $\star$(e) $x = \frac{2u}{(u+1)^2+v^2}$, $y = \frac{1-(u^2+v^2)}{(u+1)^2+v^2}$

2. For each change of variable formula, compute $dx\,dy\,dz$ in terms of $du\,dv\,dw$.

   $\star$(a) $x = u + v$, $y = v + w$, $z = w + u$
   (b) $x = 3u + v + w$, $y = 3v + w$, $z = w$
   $\star$(c) $x = \frac{u}{u^2+v^2+w^2}$, $y = \frac{v}{u^2+v^2+w^2}$, $z = \frac{w}{u^2+v^2+w^2}$
   (d) $x = \frac{w}{u}$, $y = u$, $z = u\cos v$ (for $u > 0$)

3. Show that $|\mathbf{J}| = r$ for the change of variables from rectangular coordinates to cylindrical coordinates.

4. Compute each of the following integrals by changing to the indicated coordinate system:

   $\star$(a) $\iint\limits_{R}(x+y)\,dx\,dy$, where R is the region in the first quadrant between the circles $x^2 + y^2 = 1$ and $x^2 + y^2 = 9$; polar coordinates
   (b) $\iint\limits_{R} 1\,dx\,dy$, where R is the region inside the innermost ring of the spiral $r = \theta$ in the first quadrant (see Figure 8.25); polar coordinates
   $\star$(c) $\iiint\limits_{R} z\,dx\,dy\,dz$, where R is the half of the sphere of radius 1 centered at the origin which is above the xy-plane; spherical coordinates
   (d) $\iiint\limits_{R} \frac{1}{x^2+y^2+z^2}\,dx\,dy\,dz$, where R is the shell between the spheres of radii 2 and 3 centered at the origin; spherical coordinates
   $\star$(e) $\iiint\limits_{R}(x^2+y^2+z^2)\,dx\,dy\,dz$, where R is the region defined by $x^2+y^2 \le 4$ and $-3 \le z \le 5$; cylindrical coordinates

$\star$**5.** True or False:

   (a) A linear change of coordinates for an integration results in a constant scaling factor with respect to the associated integrals.
   (b) For the change of variables $u = y$, $v = x$, we have $du\,dv = 1\,dx\,dy$.
   (c) A rectangle in uv-coordinates with sides Δu and Δv is mapped by a change of coordinates to a region whose area is approximated by the area of the parallelogram with sides $\left[\frac{\partial x}{\partial u}, \frac{\partial y}{\partial u}\right]\Delta u$ and $\left[\frac{\partial x}{\partial v}, \frac{\partial y}{\partial v}\right]\Delta v$.
   (d) The scaling factor for a change of variables in integrals is always the determinant of the Jacobian matrix.

Figure 8.25

The spiral $r = \theta$

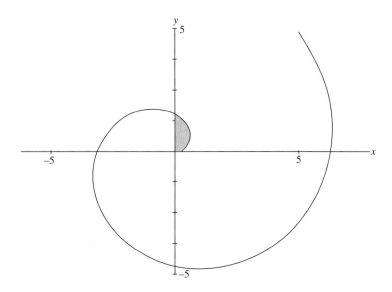

8.7 Rotation of Axes

Prerequisite: Section 4.7, Coordinatization

In this section, we show how to use rotation of the coordinate axes to simplify the equations of conic sections.

Simplifying the Equation of a Conic Section

The general form for the equation of a conic section in the xy-plane is

$$ax^2 + by^2 + cxy + dx + ey + f = 0.$$

If $c \neq 0$, the term cxy in this equation imposes a rotation on the graph of the conic section, putting its axes of symmetry on a slant. Our goal is to change to a different set of coordinates in $\mathbb{R}^2$ so that the equation of the conic expressed in these coordinates will no longer have an xy term. This will make the axes of symmetry horizontal and/or vertical with respect to the new coordinate axes.[6]

Let θ be the angle between the positive x-axis and the axis of symmetry of the conic section. We will construct a new coordinate grid by rotating the old coordinate grid counterclockwise through the angle θ. This is illustrated in Figure 8.26 for the hyperbola $xy = 1$.

Notice that, in converting from one set of coordinates to another, the standard basis $S = (\mathbf{i}, \mathbf{j})$ is replaced by a new basis $B = ([\cos \theta, \sin \theta], [-\sin \theta, \cos \theta])$ for $\mathbb{R}^2$ (see Figure 8.27). That is, a counterclockwise rotation of

---

[6] If the conic section is a circle, there is an axis of symmetry in every direction. However, the equation of a circle never contains an xy term.

Figure 8.26

Rotation of the
hyperbola $xy = 1$

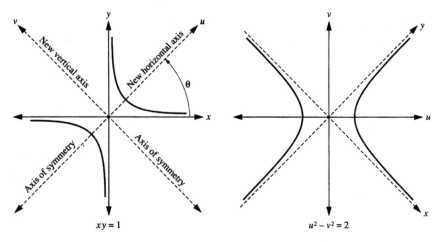

Figure 8.27

Vectors that replace the
standard basis vectors
in $\mathbb{R}^2$ after a
counterclockwise
rotation through the
angle θ

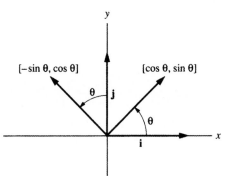

the axes through the angle θ is equivalent to introducing a new coordinate system with basis B whose vectors are obtained by rotating $\mathbf{i}, \mathbf{j}$ counterclockwise through θ.

We will use the variables u and v to describe the new coordinate system arising from the basis B. Our goal is to find an angle θ for which a given conic section has no uv term. Now, the transition matrix from B to S is

$$\mathbf{P} = \begin{bmatrix} \cos\theta & -\sin\theta \\ \sin\theta & \cos\theta \end{bmatrix}.$$

Thus, we convert points in B-coordinates (uv-coordinates) to points in S-coordinates (xy-coordinates) with the equation

$$\begin{bmatrix} x \\ y \end{bmatrix} = \begin{bmatrix} \cos\theta & -\sin\theta \\ \sin\theta & \cos\theta \end{bmatrix} \begin{bmatrix} u \\ v \end{bmatrix}, \quad \text{or,} \quad \begin{cases} x = & u\cos\theta - v\sin\theta \\ y = & u\sin\theta + v\cos\theta \end{cases}.$$

Substituting these equations into the general equation of a conic section, we obtain

$$a\,(u\cos\theta - v\sin\theta)^2 + b\,(u\sin\theta + v\cos\theta)^2$$
$$+ c\,(u\cos\theta - v\sin\theta)\,(u\sin\theta + v\cos\theta)$$
$$+ d(u\cos\theta - v\sin\theta) + e(u\sin\theta + v\cos\theta) + f = 0.$$

After expanding, we find that the uv term is

$$\left(2\sin\theta\cos\theta(b-a) + (\cos^2\theta - \sin^2\theta)c\right)uv$$

$$= \left((\sin 2\theta)(b-a) + (\cos 2\theta)c\right)uv.$$

One way of setting the coefficient of uv equal to zero in this expression is choosing the angle of rotation to be

$$\theta = \begin{cases} \frac{\pi}{4} & \text{if } a = b \\ \frac{1}{2}\arctan\left(\frac{c}{a-b}\right) & \text{otherwise} \end{cases}.$$

(Adding multiples of $\pi/2$ to this solution yields other solutions.)

EXAMPLE 1 Consider the conic section whose equation is

$$5x^2 + 7y^2 - 10xy - 3x + 2y - 8 = 0.$$

We want to find a simpler expression for this conic by changing to a different coordinate system. From the preceding formula, the appropriate angle of rotation to eliminate the xy term is $\theta = \frac{1}{2}\arctan(\frac{-10}{-2}) \approx 0.6867$ radians,[7] or about $39°\ 21'$. Now, $\cos\theta \approx 0.7733$ and $\sin\theta \approx 0.6340$. Hence, the relation between uv-coordinates and xy-coordinates is given by

$$\begin{cases} x = & 0.7733u - 0.6340v \\ y = & 0.6340u + 0.7733v \end{cases}.$$

Substituting the formulas for x and y in terms of u and v into the equation for the conic section, and simplifying, yields

$$0.9010u^2 + 11.10v^2 - 1.052u + 3.449v - 8 = 0.$$

Completing the squares gives

$$0.9010(u - 0.5838)^2 + 11.10(v + 0.1554)^2 = 8.575,$$

or

$$\frac{(u - 0.5838)^2}{(3.085)^2} + \frac{(v + 0.1554)^2}{(0.8790)^2} = 1.$$

The graph of this equation in the uv-plane, an ellipse centered at $(0.5838, -0.1554)$, is depicted in Figure 8.28.

The original conic section can be obtained by rotating the coordinate system in Figure 8.28 *counterclockwise* through the angle $\theta \approx 0.6867$ radians (see Figure 8.29). Hence, the major axis of the original ellipse has an angle of

---

[7] All computations in this example were done on a calculator rounding to twelve significant digits. However, we have printed only four significant digits in the text.

Figure 8.28

The ellipse
$\dfrac{(u-0.5838)^2}{(3.085)^2} +$
$\dfrac{(v+0.1554)^2}{(0.8790)^2} = 1$

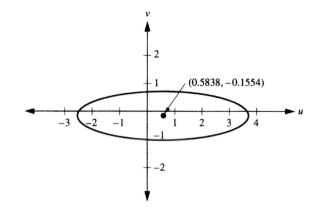

Figure 8.29

The ellipse $5x^2 + 7y^2 - 10xy - 3x + 2y - 8 = 0$

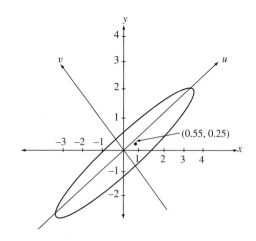

inclination with the x-axis of 0.6867 radians. Also, the center of the original conic section can be found by converting $(0.5838, -0.1554)$, the center of the uv-coordinate system ellipse, into xy-coordinates via the transition matrix

$$\mathbf{P} = \begin{bmatrix} \cos\theta & -\sin\theta \\ \sin\theta & \cos\theta \end{bmatrix} \approx \begin{bmatrix} 0.7733 & -0.6340 \\ 0.6340 & 0.7733 \end{bmatrix}.$$

That is, the center of

$$5x^2 + 7y^2 - 10xy - 3x + 2y - 8 = 0$$

is

$$\begin{bmatrix} x_0 \\ y_0 \end{bmatrix} = \begin{bmatrix} 0.7733 & -0.6340 \\ 0.6340 & 0.7733 \end{bmatrix} \begin{bmatrix} 0.5838 \\ -0.1554 \end{bmatrix} \approx \begin{bmatrix} 0.5500 \\ 0.2500 \end{bmatrix}.$$

∎

Notice in Example 1 that the transition matrix

$$\mathbf{P} = \begin{bmatrix} \cos\theta & -\sin\theta \\ \sin\theta & \cos\theta \end{bmatrix}$$

allows us to convert directly from uv-coordinates to xy-coordinates. Hence,

$$\mathbf{P}^{-1} = \begin{bmatrix} \cos\theta & \sin\theta \\ -\sin\theta & \cos\theta \end{bmatrix}$$

provides the means for converting from xy-coordinates to uv-coordinates. For example, with the angle $\theta \approx 0.6867$ radians in Example 1, the point $(-1, 0)$ on the ellipse in xy-coordinates corresponds to the point

$$\begin{bmatrix} u_0 \\ v_0 \end{bmatrix} = \begin{bmatrix} \cos\theta & \sin\theta \\ -\sin\theta & \cos\theta \end{bmatrix} \begin{bmatrix} -1 \\ 0 \end{bmatrix} \approx \begin{bmatrix} 0.7733 & 0.6340 \\ -0.6340 & 0.7733 \end{bmatrix} \begin{bmatrix} -1 \\ 0 \end{bmatrix}$$

$$\approx \begin{bmatrix} -0.7733 \\ 0.6340 \end{bmatrix}$$

in uv-coordinates.

The material in this section is revisited in a more general, abstract manner in Section 10.3, "Quadratic Forms."

Exercises for Section 8.7

1. For each of the conic sections described below, perform the following steps:

 (i) Find an appropriate angle θ through which to rotate the coordinate system counterclockwise from xy-coordinates into uv-coordinates so that the conic has no uv term.

 (ii) Calculate the transition matrix $\mathbf{P}$ from uv-coordinates to xy-coordinates.

 (iii) Solve for the equation of the conic in uv-coordinates.

 (iv) Determine the center of the conic in uv-coordinates if it is an ellipse or hyperbola, or the vertex in uv-coordinates if it is a parabola. Graph the conic in uv-coordinates.

 (v) Use the transition matrix $\mathbf{P}$ to solve for the center or vertex of the conic in xy-coordinates. Draw the graph of the conic in xy-coordinates.

 (a) $3x^2 - 3y^2 - 2\sqrt{3}(xy) - 4\sqrt{3} = 0$ (hyperbola)
 (b) $13x^2 + 13y^2 - 10xy - 8\sqrt{2}x - 8\sqrt{2}y - 64 = 0$ (ellipse)
 ★(c) $3x^2 + y^2 - 2\sqrt{3}xy - (1 + 12\sqrt{3})x + (12 - \sqrt{3})y + 26 = 0$ (parabola)
 ★(d) $29x^2 + 36y^2 - 24xy - 118x + 24y - 55 = 0$ (ellipse)
 (e) $-16x^2 - 9y^2 + 24xy - 60x + 420y = 0$ (parabola)
 ★(f) $8x^2 - 5y^2 + 16xy - 37 = 0$ (hyperbola)

★**2.** True or False:

 (a) The conic section $x^2 + xy + y^2 = 12$ has an axis of symmetry that makes a $45°$ angle with the positive x-axis.

 (b) The coordinates of the center of a hyperbola always stay fixed when changing from xy-coordinates to uv-coordinates.

   (c) If **P** is the transition matrix that converts from uv-coordinates to xy-coordinates, then $\mathbf{P}^{-1}$ is the matrix that will convert from xy-coordinates to uv-coordinates.

   (d) The equation of a conic section with no xy term has a graph in xy-coordinates that is symmetric with respect to the x-axis.

8.8 Computer Graphics

Prerequisite: Section 5.2, The Matrix of a Linear Transformation

   In this section, we give some insight into how linear algebra is used to manipulate objects on a computer screen. We will see that, in many cases, shifting the position or size of objects can be accomplished using matrix multiplication. However, to represent all possible movements by matrix multiplication, we will find it necessary to work in higher dimensions and use a somewhat different method of coordinatizing vectors, known as "homogeneous coordinates."

Introduction to Computer Graphics

   Computer screens consist of **pixels**, tiny areas of the screen arranged in rows and columns. Pixels are turned "off" and "on" to create patterns on the screen.[8] A typical 1024×768 screen, for example, would have 1024 pixels in each row (labeled "0" through "1023") and 768 pixels in each column (labeled "0" through "767") (see Figure 8.30). We can think of the screen pixels as forming a lattice (grid), with a single pixel at the intersection of each row and column.

Figure 8.30

A typical 1024×768 computer screen, with labeled pixels

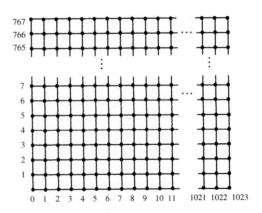

---

[8] When a pixel is "on," commands can be given that adjust its brightness and color to produce a desired effect. However, to avoid complications, we will ignore brightness and color in what follows, and simply consider a pixel to be "off" or "on."

Today, the most common computer graphics technique is **raster graphics**, in which the current screen content (text, figures, icons, etc.) is stored in the memory of the computer and updated and displayed whenever a change of screen contents is necessary. In this system, algorithms have been created to draw fundamental geometric figures at specified areas on the screen. For example, given two different points (pixels), we can display the line connecting them by calling an algorithm to turn on the appropriate pixels. Similarly, given the points that represent the vertices of a triangle (or any polygon), we can have the computer connect them to form the appropriate screen figure.

In this system, we can represent a polygon algebraically by storing its n vertices as columns in a $2 \times n$ matrix, as in the next example.

EXAMPLE 1   The polygon in Figure 8.31 (a "Knee") can be associated with the 2×6 matrix

$$\begin{bmatrix} 6 & 8 & 8 & 8 & 10 & 10 \\ 10 & 6 & 8 & 12 & 6 & 10 \end{bmatrix}.$$

Each column lists a pair of x- and y-coordinates representing a different vertex of the figure. ■

Figure 8.31

Graphic with 6 vertices and 6 edges

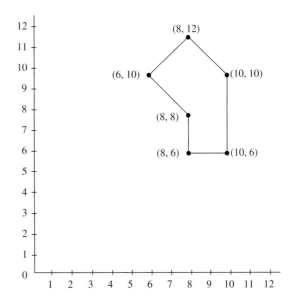

The "edges" of a polygonal figure could also be represented in computer memory. For example, we could represent the "edges" with a 6×6 matrix, with (i, j) entry equal to 1 if the ith and jth vertices are connected by an edge, and 0 otherwise. However, we will focus on the vertices only in this section.

Whenever we rotate a given figure on the screen, each computed vertex for the new figure may not land "exactly" on a single pixel, since the new x- and y-coordinates may not be integers. For simplicity, we assume that whenever a figure is manipulated, we round off each computation of a pixel coordinate to the nearest integer. Also, a figure must be "clipped" whenever portions of the figure extend beyond the current screen window. Powerful algorithms have been developed to address such problems, but these and many similar issues are beyond the scope of this text.

In this section, we will illustrate how to manipulate two-dimensional figures on the screen. Similar methods are used to manipulate three-dimensional figures, although we will not consider them here. For further details, consult Chapter 5 of *Computer Graphics: Principles and Practice in C*, 2nd edition, by Foley, vanDam, Feiner, and Hughes, published by Addison-Wesley, 1996.

Fundamental Movements in the Plane

Geometric arguments can be given to show that any **similarity** (a movement that takes any given figure to a figure similar in shape) in the plane can be accomplished by composing one or more of the following mappings[9]:

(1) **Translation:** shifting all points of a figure along a fixed vector.

(2) **Rotation:** rotating all points of a figure about a given center point, through a given angle θ. We will assume that all rotations are in a *counterclockwise* direction in the plane unless otherwise specified.

(3) **Reflection:** reflecting all points of a figure about a given line.

Finally, we also consider a fourth type of movement, which can change the size of a figure.

(4) **Scaling:** dilating/contracting the distance of all points in the figure from a given center point.

Each of these first three fundamental movements is actually an **isometry**; it maps a given figure to a *congruent* figure.

We consider each movement briefly in turn. As we will see, all translations are straightforward, but we begin with only the simplest possible type of rotation (about the origin), reflection (about a line through the origin), and scaling (with the origin as center point).

(1) **Translation:** To perform a translation of a vertex along a vector $\begin{bmatrix} a \\ b \end{bmatrix}$, we simply add $\begin{bmatrix} a \\ b \end{bmatrix}$ to the vertex.

(2) **Rotation about the origin:** In Section 5.1, we saw that multiplying on the left by the matrix

$$\begin{bmatrix} \cos\theta & -\sin\theta \\ \sin\theta & \cos\theta \end{bmatrix}$$

rotates a vertex through an angle θ about the origin.

---

[9] In fact, it can be shown that any translation or rotation can be expressed as the composition of appropriate reflections. However, translations and rotations are used so often in computer graphics that it is useful to consider these mappings separately.

(3) **Reflection about a line through the origin:** In Exercise 22 of Section 5.2, we found that multiplying on the left by the matrix

$$\frac{1}{1+m^2} \begin{bmatrix} 1-m^2 & 2m \\ 2m & m^2-1 \end{bmatrix}$$

reflects a vertex about the line $y = mx$. In the special case where the line of reflection is the y-axis, the reflection matrix is simply

$$\begin{bmatrix} -1 & 0 \\ 0 & 1 \end{bmatrix}. \quad \text{(Why?)}$$

(4) **Scaling from the origin:** In what follows, we will allow different scale factors in the x- and y-directions. We multiply distances from the center point by c in the x-direction and d in the y-direction. With the origin as center point, we can achieve the desired scaling of a vertex simply by multiplying the vertex by the matrix

$$\begin{bmatrix} c & 0 \\ 0 & d \end{bmatrix}.$$

We have seen that the last three types of mappings (rotation about the origin, reflection about a line through the origin, scaling with the origin as center) can all be performed using matrix multiplication. Of course, by Example 10 in Section 5.1, these are linear transformations. However, (nontrivial) translations are not linear transformations, and neither are rotations, reflections, or scaling when they are not centered at the origin. Nevertheless, there is a way to represent all of these movements using matrix multiplication in a different type of coordinate system taken from projective geometry, "homogeneous coordinates."

Homogeneous Coordinates

Our goal is to create a useful coordinate representation for the points in two-dimensional space by "going up" one dimension. We define any three-dimensional "point" of the form $(tx, ty, t) = t(x, y, 1)$, where $t \neq 0$, to be **equivalent** to the ordinary two-dimensional point (x, y). That is, as far as we are concerned, $(3, 4, 1)$, $(6, 8, 2) = 2(3, 4, 1)$ and $(9, 12, 3) = 3(3, 4, 1)$ are all equivalent to $(3, 4)$. Similarly, the point $(2, -5)$ has three-dimensional representations, such as $(2, -5, 1)$, $(4, -10, 2)$, and $(-8, 20, -4)$. This three-dimensional coordinate system, taken from projective geometry, gives each two-dimensional point a corresponding set of **homogeneous coordinates**. Notice that there is an infinite set of homogeneous coordinates for each two-dimensional point. However, by dividing all three coordinates of a triple by its last coordinate, any point in homogeneous coordinates can be **normalized** so that its last coordinate equals 1. Each two-dimensional point has a unique corresponding normalized point in homogeneous coordinates, which is said to be its **standard form**. Thus, $(5/2, -3/2, 1)$ is the standard form for the equivalent triples $(15, -9, 6)$ and $(10, -6, 4)$.

Representing Movements with Matrix Multiplication in Homogeneous Coordinates

Translation: To translate vertex (x, y) along a given vector $[a, b]$, we first convert (x, y) to homogeneous coordinates. The simplest way to do this is to replace (x, y) with the equivalent vector $[x, y, 1]$. Then, multiplication on the left by the matrix

$$\begin{bmatrix} 1 & 0 & a \\ 0 & 1 & b \\ 0 & 0 & 1 \end{bmatrix} \quad \text{gives} \quad \begin{bmatrix} 1 & 0 & a \\ 0 & 1 & b \\ 0 & 0 & 1 \end{bmatrix} \begin{bmatrix} x \\ y \\ 1 \end{bmatrix} = \begin{bmatrix} x + a \\ y + b \\ 1 \end{bmatrix},$$

which is equivalent to the two-dimensional point $(x + a, y + b)$, the desired result.

Rotation, Reflection, Scaling: You can verify that multiplying $[x, y, 1]$ on the left by the following matrices performs, respectively, a rotation of (x, y) about the origin through angle θ, a reflection of (x, y) about the line $y = mx$, and a scaling of (x, y) about the origin by a factor of c in the x-direction and d in the y-direction.

$$\begin{bmatrix} \cos\theta & -\sin\theta & 0 \\ \sin\theta & \cos\theta & 0 \\ 0 & 0 & 1 \end{bmatrix}, \quad \left(\tfrac{1}{1+m^2}\right) \begin{bmatrix} 1 - m^2 & 2m & 0 \\ 2m & m^2 - 1 & 0 \\ 0 & 0 & 1 + m^2 \end{bmatrix},$$

$$\begin{bmatrix} c & 0 & 0 \\ 0 & d & 0 \\ 0 & 0 & 1 \end{bmatrix}$$

Finally, the special case of a reflection about the y-axis can be accomplished by multiplying on the left by the matrix

$$\begin{bmatrix} -1 & 0 & 0 \\ 0 & 1 & 0 \\ 0 & 0 & 1 \end{bmatrix}.$$

Recall that for any matrix $\mathbf{A}$ and vector $\mathbf{v}$ (of compatible sizes) and any scalar t, we have $\mathbf{A}(t\mathbf{v}) = t(\mathbf{A}\mathbf{v})$. Hence, multiplying a 3×3 matrix $\mathbf{A}$ by any two vectors of the form $t[x, y, 1] = [tx, ty, t]$ equivalent to (x, y) always produces two results that are equivalent in homogeneous coordinates.

Movements Not Centered at the Origin

Our next goal is to determine the matrices for rotations, reflections, and scaling that are not centered about the origin. This can be done by combining appropriate translation matrices with the matrices for origin-centered rotations, reflections, and scaling.

> **SIMILARITY METHOD**
>
> **Step 1:** Use a translation to move the figure so that the rotation, reflection, or scaling to be performed is "about the origin." (This means moving the figure so that the center of rotation/scaling is the origin, or so that the line of reflection goes through the origin.)
> **Step 2:** Perform the desired rotation, reflection, or scaling "about the origin."
> **Step 3:** Translate the altered figure back to the position of the original figure by reversing the translation in Step 1.

The Similarity Method requires the composition of three movements. Theorem 5.8 shows that the matrix for a composition is the product of the corresponding matrices for the individual mappings in *reverse* order, as we will illustrate in Examples 2, 3, and 4. A little thought will convince you that the Similarity Method also has the overall effect of multiplying a vertex in homogeneous coordinates by a matrix *similar* (see Section 5.2) to the matrix for the movement in Step 2 (see Exercise 10).

We will demonstrate the Similarity Method for each type of movement in turn.

EXAMPLE 2

Rotation: Suppose we rotate the vertices of the "Knee" from Example 1 through an angle of $\theta = 90°$ about the point $(r, s) = (12, 6)$. We first replace each (x, y) with its vector $[x, y, 1]$ in homogeneous coordinates and follow the Similarity Method. In Step 1, we translate from $(12, 6)$ to $(0, 0)$ in order to establish the origin as center. In Step 2, we perform a rotation through angle $\theta = 90°$ about the origin. Finally, in Step 3, we translate from $(0, 0)$ back to $(12, 6)$. The net effect of these three operations is to rotate each vertex about $(12, 6)$. (Why?) The combined result of these operations is

$$
\begin{bmatrix} 1 & 0 & 12 \\ 0 & 1 & 6 \\ 0 & 0 & 1 \end{bmatrix}
\begin{bmatrix} \cos 90° & -\sin 90° & 0 \\ \sin 90° & \cos 90° & 0 \\ 0 & 0 & 1 \end{bmatrix}
\begin{bmatrix} 1 & 0 & -12 \\ 0 & 1 & -6 \\ 0 & 0 & 1 \end{bmatrix}
\begin{bmatrix} x \\ y \\ 1 \end{bmatrix}.
$$

$$\underbrace{\text{translate from}}_{(0, 0) \text{ back to } (12, 6)} \qquad \underbrace{\text{rotate about } (0, 0)}_{\text{through angle } 90°} \qquad \underbrace{\text{translate from}}_{(12, 6) \text{ to } (0, 0)}$$

This reduces to

$$
\begin{bmatrix} 0 & -1 & 18 \\ 1 & 0 & -6 \\ 0 & 0 & 1 \end{bmatrix}
\begin{bmatrix} x \\ y \\ 1 \end{bmatrix}. \text{ (Verify!)}
$$

Therefore, performing the rotation on all vertices of the figure simultaneously, we obtain

$$
\begin{bmatrix} 0 & -1 & 18 \\ 1 & 0 & -6 \\ 0 & 0 & 1 \end{bmatrix}
\begin{bmatrix} 6 & 8 & 8 & 8 & 10 & 10 \\ 10 & 6 & 8 & 12 & 6 & 10 \\ 1 & 1 & 1 & 1 & 1 & 1 \end{bmatrix}
=
\begin{bmatrix} 8 & 12 & 10 & 6 & 12 & 8 \\ 0 & 2 & 2 & 2 & 4 & 4 \\ 1 & 1 & 1 & 1 & 1 & 1 \end{bmatrix}.
$$

The columns of the final matrix (ignoring the last row entries) give the vertices of the rotated figure, as illustrated in Figure 8.32(a). ∎

Figure 8.32

Movements of "Knee":
(a) rotation through 90°
about (12, 6);
(b) reflection about line
$y = -3x + 30$;
(c) scaling with $c = 1/2$,
$d = 4$ about (6, 10)

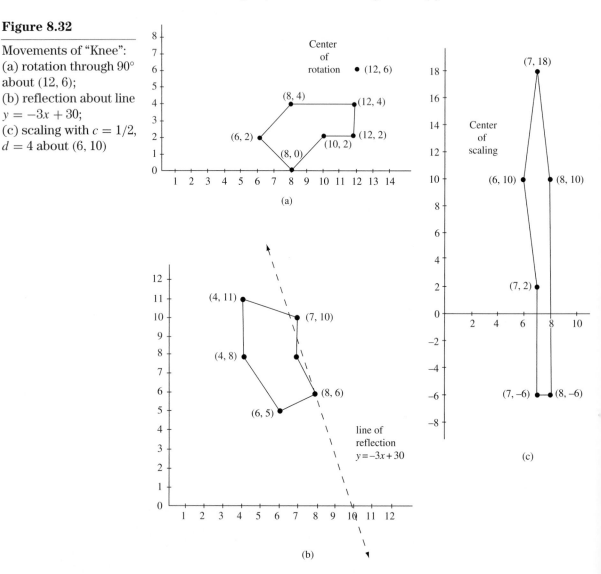

EXAMPLE 3

Reflection: Suppose we reflect the vertices of the "Knee" in Example 1 about the line $y = -3x + 30$. In this case, $m = -3$ and $b = 30$. As before, we replace (x, y) with its equivalent vector $[x, y, 1]$, and follow the Similarity Method. In Step 1, we translate from $(0, 30)$ to $(0, 0)$ in order to "drop" the line 30 units vertically so that it passes through the origin. In Step 2, we perform a reflection about the corresponding line $y = -3x$. Finally, in Step 3, we translate from $(0, 0)$ back to $(0, 30)$. The net effect of these three operations is to reflect each vertex about the line $y = -3x + 30$. (Why?) The combined result of these operations is

$$
\begin{bmatrix} 1 & 0 & 0 \\ 0 & 1 & 30 \\ 0 & 0 & 1 \end{bmatrix} \left(\frac{1}{1+(-3)^2} \right) \begin{bmatrix} 1-(-3)^2 & 2(-3) & 0 \\ 2(-3) & (-3)^2-1 & 0 \\ 0 & 0 & 1+(-3)^2 \end{bmatrix}
$$

translate from
(0, 0) back
to (0, 30)

reflect about
the line $y = -3x$

$$
\begin{bmatrix} 1 & 0 & 0 \\ 0 & 1 & -30 \\ 0 & 0 & 1 \end{bmatrix} \begin{bmatrix} x \\ y \\ 1 \end{bmatrix}.
$$

translate from
(0, 30) to (0, 0)

This reduces to

$$
\left(\frac{1}{10} \right) \begin{bmatrix} -8 & -6 & 180 \\ -6 & 8 & 60 \\ 0 & 0 & 10 \end{bmatrix} \begin{bmatrix} x \\ y \\ 1 \end{bmatrix}.
$$

Performing the reflection on all vertices of the figure simultaneously, we obtain

$$
\frac{1}{10} \begin{bmatrix} -8 & -6 & 180 \\ -6 & 8 & 60 \\ 0 & 0 & 10 \end{bmatrix} \begin{bmatrix} 6 & 8 & 8 & 8 & 10 & 10 \\ 10 & 6 & 8 & 12 & 6 & 10 \\ 1 & 1 & 1 & 1 & 1 & 1 \end{bmatrix} \approx \begin{bmatrix} 7 & 8 & 7 & 4 & 6 & 4 \\ 10 & 6 & 8 & 11 & 5 & 8 \\ 1 & 1 & 1 & 1 & 1 & 1 \end{bmatrix},
$$

after rounding the results for each vertex to the nearest integer. The columns of the final matrix (ignoring the last row entries) give the vertices of the reflected figure, as illustrated in Figure 8.32(b). Notice that the reflected figure is slightly distorted because of the rounding involved. For simplicity in this example, small pixel values were used, but a larger figure on the screen would probably undergo less distortion after such a reflection. ∎

The special case of a reflection about a line parallel to the y-axis is treated in Exercise 8.

EXAMPLE 4

Scaling: Suppose we scale the vertices of the "Knee" in Example 1 about the point $(r, s) = (6, 10)$ with a factor of $c = 1/2$ in the x-direction and $d = 4$ in the y-direction. In a manner similar to Examples 2 and 3 we obtain

$$
\begin{bmatrix} 1 & 0 & 6 \\ 0 & 1 & 10 \\ 0 & 0 & 1 \end{bmatrix} \begin{bmatrix} \frac{1}{2} & 0 & 0 \\ 0 & 4 & 0 \\ 0 & 0 & 1 \end{bmatrix} \begin{bmatrix} 1 & 0 & -6 \\ 0 & 1 & -10 \\ 0 & 0 & 1 \end{bmatrix} \begin{bmatrix} x \\ y \\ 1 \end{bmatrix}
$$

translate from
(0, 0) back
to (6, 10)

scale about (0, 0)
using scale factors $\frac{1}{2}$
and 4, respectively

translate from
(6, 10) to (0, 0)

$$
= \begin{bmatrix} \frac{1}{2} & 0 & 3 \\ 0 & 4 & -30 \\ 0 & 0 & 1 \end{bmatrix} \begin{bmatrix} x \\ y \\ 1 \end{bmatrix}.
$$

Therefore, scaling all vertices of the figure simultaneously, we obtain

$$
\begin{bmatrix} \frac{1}{2} & 0 & 3 \\ 0 & 4 & -30 \\ 0 & 0 & 1 \end{bmatrix}
\begin{bmatrix} 6 & 8 & 8 & 8 & 10 & 10 \\ 10 & 6 & 8 & 12 & 6 & 10 \\ 1 & 1 & 1 & 1 & 1 & 1 \end{bmatrix}
=
\begin{bmatrix} 6 & 7 & 7 & 7 & 8 & 8 \\ 10 & -6 & 2 & 18 & -6 & 10 \\ 1 & 1 & 1 & 1 & 1 & 1 \end{bmatrix},
$$

as illustrated in Figure 8.32(c). Two of the scaled vertices have negative y-values, and so would not be displayed on the computer screen. ■

Composition of Movements

Now that we have established that all translations, rotations, reflections, and scaling operations can be performed by appropriate matrix multiplications in homogeneous coordinates, we can find the matrix for a composition of such movements.

EXAMPLE 5

Suppose we rotate the "Knee" in Example 1 through an angle of $300°$ about the point $(8, 10)$, and then reflect the resulting figure about the line $y = -(1/2)x + 20$. With $\theta = 300°$, $m = -1/2$, and $b = 20$, the matrix for this composition is the product of the following six matrices:

$$
\underbrace{\begin{bmatrix} 1 & 0 & 0 \\ 0 & 1 & 20 \\ 0 & 0 & 1 \end{bmatrix}}_{\substack{\text{translate from} \\ (0,0)\text{ back} \\ \text{to }(0,20)}}
\underbrace{\left(\frac{1}{1+(-\frac{1}{2})^2}\right)\begin{bmatrix} 1-(-\frac{1}{2})^2 & 2(-\frac{1}{2}) & 0 \\ 2(-\frac{1}{2}) & (-\frac{1}{2})^2-1 & 0 \\ 0 & 0 & 1+(-\frac{1}{2})^2 \end{bmatrix}}_{\substack{\text{reflect about} \\ \text{the line }y=-\frac{1}{2}x}}
$$

$$
\underbrace{\begin{bmatrix} 1 & 0 & 0 \\ 0 & 1 & -20 \\ 0 & 0 & 1 \end{bmatrix}}_{\substack{\text{translate from} \\ (0,20)\text{ to }(0,0)}}
\underbrace{\begin{bmatrix} 1 & 0 & 8 \\ 0 & 1 & 10 \\ 0 & 0 & 1 \end{bmatrix}}_{\substack{\text{translate from} \\ (0,0)\text{ back to }(8,10)}}
\underbrace{\begin{bmatrix} \cos 300° & -\sin 300° & 0 \\ \sin 300° & \cos 300° & 0 \\ 0 & 0 & 1 \end{bmatrix}}_{\substack{\text{rotate about }(0,0) \\ \text{through angle }300°}}
$$

$$
\underbrace{\begin{bmatrix} 1 & 0 & -8 \\ 0 & 1 & -10 \\ 0 & 0 & 1 \end{bmatrix}}_{\substack{\text{translate from} \\ (8,10)\text{ to }(0,0)}}.
$$

This reduces to (approximately)

$$
\begin{bmatrix} 0.9928 & 0.1196 & 3.6613 \\ 0.1196 & -0.9928 & 28.5713 \\ 0 & 0 & 1 \end{bmatrix}.
$$

Multiplying this matrix by all vertices of the figure simultaneously and rounding the results for each vertex to the nearest integer, we have

$$
\begin{bmatrix}
0.9928 & 0.1196 & 3.6613 \\
0.1196 & -0.9928 & 28.5713 \\
0 & 0 & 1
\end{bmatrix}
\begin{bmatrix}
6 & 8 & 8 & 8 & 10 & 10 \\
10 & 6 & 8 & 12 & 6 & 10 \\
1 & 1 & 1 & 1 & 1 & 1
\end{bmatrix}
$$

$$
\approx
\begin{bmatrix}
11 & 12 & 13 & 13 & 14 & 15 \\
19 & 24 & 22 & 18 & 24 & 20 \\
1 & 1 & 1 & 1 & 1 & 1
\end{bmatrix}.
$$

The columns of the final matrix (ignoring the last row entries) give the vertices of the final figure after the indicated rotation and reflection. These are illustrated in Figure 8.33. ■

Figure 8.33

Movement of "Knee" after rotation through an angle of $300°$ about the point $(8, 10)$, followed by reflection about the line $y = -(1/2)x + 20$.

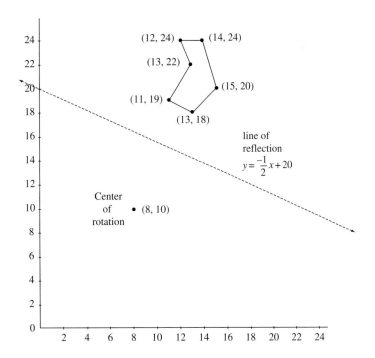

Exercises for Section 8.8

Round all calculations of pixel coordinates to the nearest integer. Some of the resulting coordinate values may be "outside" a typical pixel configuration.

1. For the graphic in Figure 8.34(a), use ordinary coordinates in $\mathbb{R}^2$ to find the new vertices after performing each indicated operation.

   ★(a) translation along the vector $[4, -2]$

   (b) rotation about the origin through $\theta = 30°$

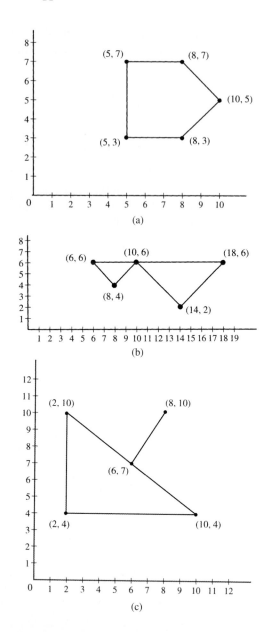

(a)

(b)

(c)

★(c) reflection about the line $y = 3x$

(d) scaling about the origin with scale factors of 4 in the x-direction
and 2 in the y-direction

2. For the graphic in Figure 8.34(b), use ordinary coordinates in $\mathbb{R}^2$ to find
the new vertices after performing each indicated operation. Then sketch
the figure that would result from this movement.

(a) translation along the vector $[-3, 5]$

★(b) rotation about the origin through $\theta = 120°$

(c) reflection about the line $y = \frac{1}{2}x$

★(d) scaling about the origin with scale factors of $\frac{1}{2}$ in the x-direction and 3 in the y-direction

3. For the graphic in Figure 8.34(c), use homogeneous coordinates to find the new vertices after performing each indicated sequence of operations.

   ★(a) rotation about the origin through $\theta = 45°$, followed by a reflection about the line $y = \frac{1}{2}x$

   (b) reflection about the line $y = \frac{1}{2}x$, followed by a rotation about the origin through $\theta = 45°$

   ★(c) scaling about the origin with scale factors of 3 in the x-direction and $\frac{1}{2}$ in the y-direction, followed by a reflection about the line $y = 2x$

   (d) translation along the vector $[-2, 3]$, followed by a rotation about the origin through $\theta = 300°$

4. For the graphic in Figure 8.35(a), use homogeneous coordinates to find the new vertices after performing each indicated operation.

   ★(a) rotation about $(8, 9)$ through $\theta = 120°$

   (b) reflection about the line $y = 2 - x$

   ★(c) scaling about $(8, 4)$ with scale factors of 2 in the x-direction and $\frac{1}{3}$ in the y-direction

5. For the graphic in Figure 8.35(b), use homogeneous coordinates to find the new vertices after performing each indicated operation.

   (a) rotation about $(10, 8)$ through $\theta = 315°$

   ★(b) reflection about the line $y = 4x - 10$

   (c) scaling about $(7, 3)$ with scale factors of $\frac{1}{2}$ in the x-direction and 3 in the y-direction

6. For the graphic in Figure 8.35(c), use homogeneous coordinates to find the new vertices after performing each indicated sequence of operations. Then sketch the final figure that would result from these movements.

   ★(a) rotation about $(12, 8)$ through $\theta = 60°$, followed by a reflection about the line $y = \frac{1}{2}x + 6$

   (b) reflection about the line $y = 2x - 1$, followed by a rotation about $(10, 10)$ through $\theta = 210°$

   ★(c) scaling about $(9, 4)$ with scale factors $\frac{1}{3}$ in the x-direction and 2 in the y-direction, followed by a rotation about $(2, 9)$ through $\theta = 150°$

   (d) reflection about the line $y = 3x - 2$, followed by scaling about $(8, 6)$ using scale factors of 3 in the x-direction and $\frac{1}{2}$ in the y-direction

7. Use the Similarity Method to verify each of the following assertions:

   (a) A rotation about (r, s) through angle θ is represented by the matrix

$$\begin{bmatrix} \cos\theta & -\sin\theta & r(1 - \cos\theta) + s(\sin\theta) \\ \sin\theta & \cos\theta & s(1 - \cos\theta) - r(\sin\theta) \\ 0 & 0 & 1 \end{bmatrix}.$$

Figure 8.35

(a) Figure for
Exercise 4;
(b) figure for
Exercise 5;
(c) Figure for
Exercise 6

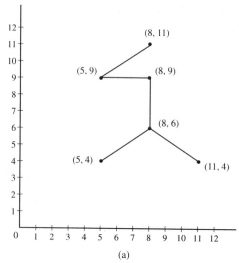

(a)

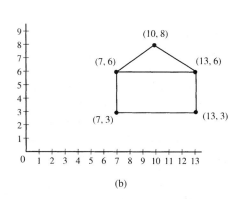

(b)

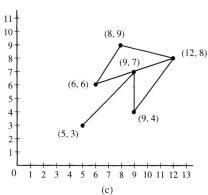

(c)

(b) A reflection about the line $y = mx + b$ is represented by the matrix

$$\left(\frac{1}{1+m^2}\right)\begin{bmatrix} 1-m^2 & 2m & -2mb \\ 2m & m^2-1 & 2b \\ 0 & 0 & 1+m^2 \end{bmatrix}.$$

(c) A scaling about (r, s) with scale factors c in the x-direction and d in
the y-direction is represented by the matrix

$$\begin{bmatrix} c & 0 & r(1-c) \\ 0 & d & s(1-d) \\ 0 & 0 & 1 \end{bmatrix}.$$

8. Show that a reflection about the line $x = k$ is represented by the matrix

$$\begin{bmatrix} -1 & 0 & 2k \\ 0 & 1 & 0 \\ 0 & 0 & 1 \end{bmatrix}.$$

(Hint: First, translate from $(k, 0)$ to $(0, 0)$, then, reflect about the y-axis, and finally, translate from $(0, 0)$ back to $(k, 0)$.)

9. Redo each part of Exercise 5 with a single matrix multiplication by using an appropriate matrix from Exercise 7 in each case.

10. (a) Verify computationally that the translation matrices

$$\begin{bmatrix} 1 & 0 & a \\ 0 & 1 & b \\ 0 & 0 & 1 \end{bmatrix} \quad \text{and} \quad \begin{bmatrix} 1 & 0 & -a \\ 0 & 1 & -b \\ 0 & 0 & 1 \end{bmatrix}$$

are inverses of each other.

(b) Explain geometrically why it makes sense that the translation matrices from part (a) are inverses.

(c) Explain why the matrices for a rotation about the origin through a given angle θ and a rotation about any other point (r, s) through the same angle θ must be similar. (Hint: Use part (a).)

11. (a) Let L_1 be a scaling about the point (r, s) with equal scale factors in the x- and y-directions, and let L_2 be a rotation about the point (r, s) through angle θ. Show that L_1 and L_2 commute. (That is, show $L_1 \circ L_2 = L_2 \circ L_1$.)

★(b) Give a counterexample to show that, in general, a reflection and a rotation do not commute.

(c) Give a counterexample to show that, in general, a scaling and a reflection do not commute.

12. An $n \times n$ matrix $\mathbf{A}$ is an **orthogonal matrix** if and only if $\mathbf{A}\mathbf{A}^T = \mathbf{I}_n$.

(a) Show that the 2×2 matrix for rotation about the origin through an angle θ, and its 3×3 counterpart in homogeneous coordinates (as given in this section), are both orthogonal matrices.

(b) Show that the single matrix for the rotation of the plane through an angle of $90°$ about the point $(12, 6)$ given in Example 2 is *not* an orthogonal matrix.

(c) Is either the 2×2 matrix for a reflection about a line through the origin, or its 3×3 counterpart in homogeneous coordinates (as given in this section), an orthogonal matrix? Why? (Hint: Let $\mathbf{A}$ be either matrix. Note that $\mathbf{A}^2 = \mathbf{I}$.)

★13. True or False?

(a) We may use vectors in homogeneous coordinates having third coordinate 0 to represent pixels on the screen.

(b) Every pixel on the screen has a unique representation in homogeneous coordinates.

(c) Every rotation has a unique 3×3 matrix representing it in homogeneous coordinates.

(d) Every isometry in the plane can be expressed using the basic motions of rotation, reflection, and translation.

(e) Non-identity translations are not linear transformations.

(f) All rotations and reflections in the plane are linear transformations.

Prerequisite: Section 5.5, Diagonalization of Linear Operators

In this section, we use the diagonalization process to solve certain first-order linear homogeneous systems of differential equations. We then adjust this technique to solve higher-order homogeneous differential equations as well.

First-Order Linear Homogeneous Systems

DEFINITION

Let

$$\mathbf{F}(t) = \begin{bmatrix} f_1(t) \\ \vdots \\ f_n(t) \end{bmatrix}$$

represent an $n \times 1$ matrix whose entries are real-valued functions, and let **A** be an $n \times n$ matrix of real numbers. Then the equation $\mathbf{F}'(t) - \mathbf{A}\mathbf{F}(t) = \mathbf{0}$, or $\mathbf{F}'(t) = \mathbf{A}\mathbf{F}(t)$, is called a **first-order linear homogeneous system of differential equations**. A **solution** for such a system is a particular function $\mathbf{F}(t)$ that satisfies the equation for all values of t.

For brevity, in the remainder of this section we will refer to an equation of the form $\mathbf{F}'(t) = \mathbf{A}\mathbf{F}(t)$ as a **first-order system**.

EXAMPLE 1 Let $\mathbf{F} = \begin{bmatrix} f_1(t) \\ f_2(t) \end{bmatrix}$ and $\mathbf{A} = \begin{bmatrix} 13 & -45 \\ 6 & -20 \end{bmatrix}$, and consider the first-order system $\mathbf{F}'(t) = \mathbf{A}\mathbf{F}(t)$, or

$$\begin{bmatrix} f_1'(t) \\ f_2'(t) \end{bmatrix} = \begin{bmatrix} 13 & -45 \\ 6 & -20 \end{bmatrix} \begin{bmatrix} f_1(t) \\ f_2(t) \end{bmatrix}.$$

Multiplying yields

$$\begin{cases} f_1'(t) &= 13 f_1(t) - 45 f_2(t) \\ f_2'(t) &= 6 f_1(t) - 20 f_2(t) \end{cases}.$$

A solution for this system consists of a pair of functions, $f_1(t)$ and $f_2(t)$, that satisfy both of these differential equations. One such solution is

$$\mathbf{F}(t) = \begin{bmatrix} f_1(t) \\ f_2(t) \end{bmatrix} = \begin{bmatrix} 5e^{-5t} \\ 2e^{-5t} \end{bmatrix}.$$

(Verify.) We will see how to obtain such solutions later in this section. ■

In what follows, we concern ourselves only with solutions that are *continuously differentiable* (that is, solutions having continuous derivatives). First, we state, without proof, a well-known result from the theory of differential equations about solutions of a single first-order equation.

LEMMA 8.6

A real-valued continuously differentiable function $f(t)$ is a solution to the differential equation $f'(t) = af(t)$ if and only if $f(t) = be^{at}$ for some real number b.

A first-order system of the form $\mathbf{F}'(t) = \mathbf{A}\mathbf{F}(t)$ is more complicated than the differential equation in Lemma 8.6 since it involves a matrix $\mathbf{A}$ instead of a real number a. However, in the special case when $\mathbf{A}$ is a diagonal matrix, the system $\mathbf{F}'(t) = \mathbf{A}\mathbf{F}(t)$ can be written as

$$\begin{cases} f_1'(t) = a_{11} f_1(t) \\ f_2'(t) = a_{22} f_2(t) \\ \quad\vdots \\ f_n'(t) = a_{nn} f_n(t) \end{cases}.$$

Each of the differential equations in this system can be solved separately using Lemma 8.6. Hence, when $\mathbf{A}$ is diagonal, the general solution has the form

$$\mathbf{F}(t) = [b_1 e^{a_{11}t}, b_2 e^{a_{22}t}, \dots, b_n e^{a_{nn}t}],$$

for some $b_1, \dots, b_n \in \mathbb{R}$.

EXAMPLE 2 Consider the first-order system $\mathbf{F}'(t) = \begin{bmatrix} 3 & 0 \\ 0 & -2 \end{bmatrix} \mathbf{F}(t)$, whose matrix is diagonal. This system is equivalent to

$$\begin{cases} f_1'(t) = \quad 3 f_1(t) \\ f_2'(t) = \quad -2 f_2(t) \end{cases}.$$

Using Lemma 8.6, we see that the solutions are all functions of the form

$$\mathbf{F}(t) = [f_1(t), f_2(t)] = [b_1 e^{3t}, b_2 e^{-2t}].$$

■

Since first-order systems $\mathbf{F}'(t) = \mathbf{A}\mathbf{F}(t)$ are easily solved when the matrix $\mathbf{A}$ is diagonal, it is natural to consider the case when $\mathbf{A}$ is diagonalizable. Thus, suppose $\mathbf{A}$ is a diagonalizable $n \times n$ matrix with (not necessarily distinct) eigenvalues $\lambda_1, \dots, \lambda_n$ corresponding to the eigenvectors in the ordered basis $B = (\mathbf{v}_1, \dots, \mathbf{v}_n)$ for $\mathbb{R}^n$. The matrix $\mathbf{P}$ having columns $\mathbf{v}_1, \dots, \mathbf{v}_n$ is the transition matrix from B to standard coordinates, and $\mathbf{P}^{-1}\mathbf{A}\mathbf{P} = \mathbf{D}$, the diagonal matrix having eigenvalues $\lambda_1, \lambda_2, \ldots, \lambda_n$ along its main diagonal. Hence,

$$\mathbf{F}'(t) = \mathbf{AF}(t) \Longleftrightarrow \mathbf{F}'(t) = (\mathbf{PP}^{-1}\mathbf{APP}^{-1})\mathbf{F}(t)$$

$$\Longleftrightarrow \mathbf{F}'(t) = \mathbf{PDP}^{-1}\mathbf{F}(t)$$

$$\Longleftrightarrow \mathbf{P}^{-1}\mathbf{F}'(t) = \mathbf{DP}^{-1}\mathbf{F}(t).$$

Letting $\mathbf{G}(t) = \mathbf{P}^{-1}\mathbf{F}(t)$, we see that the original system $\mathbf{F}'(t) = \mathbf{AF}(t)$ is equivalent to the system $\mathbf{G}'(t) = \mathbf{DG}(t)$. Since $\mathbf{D}$ is diagonal, with diagonal entries $\lambda_1, \ldots, \lambda_n$, the latter system is solved as follows:

$$\mathbf{G}(t) = \left[b_1 e^{\lambda_1 t}, \ b_2 e^{\lambda_2 t}, \ldots, \ b_n e^{\lambda_n t}\right].$$

But, $\mathbf{F}(t) = \mathbf{PG}(t)$. Since the columns of $\mathbf{P}$ are the eigenvectors $\mathbf{v}_1, \mathbf{v}_2, \ldots, \mathbf{v}_n$, we obtain

$$\mathbf{F}(t) = b_1 e^{\lambda_1 t}\mathbf{v}_1 + b_2 e^{\lambda_2 t}\mathbf{v}_2 + \cdots + b_n e^{\lambda_n t}\mathbf{v}_n.$$

Thus, we have proved the following:

THEOREM 8.7

Let $\mathbf{A}$ be a diagonalizable $n \times n$ matrix and let $(\mathbf{v}_1, \ldots, \mathbf{v}_n)$ be an ordered basis for $\mathbb{R}^n$ consisting of eigenvectors for $\mathbf{A}$ corresponding to the (not necessarily distinct) eigenvalues $\lambda_1, \ldots, \lambda_n$. Then the continuously differentiable solutions for the first-order system $\mathbf{F}'(t) = \mathbf{AF}(t)$ are all functions of the form

$$\mathbf{F}(t) = b_1 e^{\lambda_1 t}\mathbf{v}_1 + b_2 e^{\lambda_2 t}\mathbf{v}_2 + \cdots + b_n e^{\lambda_n t}\mathbf{v}_n,$$

where $b_1, \ldots, b_n \in \mathbb{R}$.

EXAMPLE 3   We will solve the first-order system $\mathbf{F}'(t) = \mathbf{AF}(t)$, where

$$\mathbf{A} = \begin{bmatrix} 1 & 0 & -2 & 6 \\ 4 & -1 & -4 & 12 \\ -32 & 9 & 40 & -114 \\ -11 & 3 & 14 & -40 \end{bmatrix}.$$

Following Steps 3 through 6 of the method in Section 5.5 for diagonalizing a linear operator, we find that $\mathbf{A}$ has the following eigenvectors and corresponding eigenvalues:

$$\begin{array}{ll} \mathbf{v}_1 = [-2, -4, 5, 2] & \text{corresponding to } \lambda_1 = 0 \\ \mathbf{v}_2 = [-3, 2, 0, 1] & \text{corresponding to } \lambda_2 = -1 \\ \mathbf{v}_3 = [1, -1, 1, 0] & \text{corresponding to } \lambda_3 = -1 \\ \mathbf{v}_4 = [0, 0, 3, 1] & \text{corresponding to } \lambda_4 = 2. \end{array}$$

(Notice that $\mathbf{v}_2$ and $\mathbf{v}_3$ were chosen to be linearly independent eigenvectors for the eigenvalue -1, so that $\{\mathbf{v}_2, \mathbf{v}_3\}$ forms a basis for E_{-1}.) Therefore,

Theorem 8.7 tells us that the continuously differentiable solutions to the first-order system $\mathbf{F}'(t) = \mathbf{A}\mathbf{F}(t)$ consist precisely of all functions of the form

$$
\begin{aligned}
\mathbf{F}(t) &= [f_1(t),\ f_2(t),\ f_3(t),\ f_4(t)] \\
&= b_1[-2, -4, 5, 2] + b_2 e^{-t}[-3, 2, 0, 1] + b_3 e^{-t}[1, -1, 1, 0] \\
&\quad + b_4 e^{2t}[0, 0, 3, 1] \\
&= [-2b_1 - 3b_2 e^{-t} + b_3 e^{-t},\ -4b_1 + 2b_2 e^{-t} - b_3 e^{-t},\ 5b_1 + b_3 e^{-t} \\
&\quad + 3b_4 e^{2t},\ 2b_1 + b_2 e^{-t} + b_4 e^{2t}].
\end{aligned}
$$

∎

Notice that in order to use Theorem 8.7 to solve a first-order system $\mathbf{F}'(t) = \mathbf{A}\mathbf{F}(t)$, $\mathbf{A}$ must be a diagonalizable matrix. If it is not, you can still find some of the solutions to the system using an analogous process. If $\{\mathbf{v}_1, \ldots, \mathbf{v}_k\}$ is a linearly independent set of eigenvectors for $\mathbf{A}$ corresponding to the eigenvalues $\lambda_1, \ldots, \lambda_k$, then functions of the form

$$\mathbf{F}(t) = b_1 e^{\lambda_1 t}\mathbf{v}_1 + b_2 e^{\lambda_2 t}\mathbf{v}_2 + \cdots + b_k e^{\lambda_k t}\mathbf{v}_k$$

are solutions (see Exercise 3). However, these are not all the possible solutions for the system. To find all the solutions, you must use complex eigenvalues and eigenvectors, as well as *generalized eigenvectors*. Complex eigenvalues are studied in Section 7.2; generalized eigenvectors are not covered in this book.

Higher-Order Homogeneous Differential Equations

Our next goal is to solve higher-order homogeneous differential equations of the form

$$y^{(n)} + a_{n-1}y^{(n-1)} + \cdots + a_2 y'' + a_1 y' + a_0 y = 0.$$

EXAMPLE 4   Consider the differential equation $y''' - 6y'' + 3y' + 10y = 0$. To find solutions for this equation, we define the functions $f_1(t)$, $f_2(t)$, and $f_3(t)$ as follows: $f_1 = y$, $f_2 = y'$, and $f_3 = y''$. We then have the system

$$
\begin{cases}
f_1' = & f_2 \\
f_2' = & f_3 \\
f_3' = -10f_1 - 3f_2 + 6f_3
\end{cases}.
$$

The first two equations in this system come directly from the definitions of f_1, f_2, and f_3. The third equation is obtained from the original differential equation by moving all terms except y''' to the right side. But this system can be expressed as

$$
\begin{bmatrix}
f_1'(t) \\
f_2'(t) \\
f_3'(t)
\end{bmatrix}
=
\begin{bmatrix}
0 & 1 & 0 \\
0 & 0 & 1 \\
-10 & -3 & 6
\end{bmatrix}
\begin{bmatrix}
f_1(t) \\
f_2(t) \\
f_3(t)
\end{bmatrix},
$$

that is, as $\mathbf{F}'(t) = \mathbf{A}\mathbf{F}(t)$, with

$$\mathbf{F}(t) = \begin{bmatrix} f_1(t) \\ f_2(t) \\ f_3(t) \end{bmatrix} \quad \text{and} \quad \mathbf{A} = \begin{bmatrix} 0 & 1 & 0 \\ 0 & 0 & 1 \\ -10 & -3 & 6 \end{bmatrix}.$$

We now use the method of Theorem 8.7 to solve this first-order system.

A quick calculation yields $p_{\mathbf{A}}(x) = x^3 - 6x^2 + 3x + 10 = (x+1)(x-2)(x-5)$, giving the eigenvalues $\lambda_1 = -1$, $\lambda_2 = 2$, and $\lambda_3 = 5$. Solving for the corresponding eigenvectors for $\mathbf{A}$ produces

$$\begin{aligned} \mathbf{v}_1 &= [1, -1, 1] & \text{corresponding to } \lambda_1 = -1 \\ \mathbf{v}_2 &= [1, 2, 4] & \text{corresponding to } \lambda_2 = 2 \\ \mathbf{v}_3 &= [1, 5, 25]. & \text{corresponding to } \lambda_3 = 5 \end{aligned}$$

Hence, Theorem 8.7 gives us the general solution

$$\begin{aligned} \mathbf{F}(t) &= b_1 e^{-t}[1, -1, 1] + b_2 e^{2t}[1, 2, 4] + b_3 e^{5t}[1, 5, 25] \\ &= [b_1 e^{-t} + b_2 e^{2t} + b_3 e^{5t}, \ -b_1 e^{-t} + 2b_2 e^{2t} + 5b_3 e^{5t}, \\ & \quad b_1 e^{-t} + 4b_2 e^{2t} + 25b_3 e^{5t}]. \end{aligned}$$

Since the first entry of this result equals $f_1(t) = y$, the general continuously differentiable solution to the original third-order differential equation is

$$y = b_1 e^{-t} + b_2 e^{2t} + b_3 e^{5t}. \qquad \blacksquare$$

The method of Example 4 can be generalized to many homogeneous higher-order differential equations $y^{(n)} + a_{n-1} y^{(n-1)} + \cdots + a_1 y' + a_0 y = 0$. In Exercise 5(a), you are asked to show that this equation can be represented as a linear system $\mathbf{F}'(t) = \mathbf{A}\mathbf{F}(t)$, where $\mathbf{F}(t) = [f_1(t), f_2(t), \ldots, f_n(t)]$, with $f_1(t) = y$, $f_2(t) = y', \ldots, f_n(t) = y^{(n-1)}$ and where

$$\mathbf{A} = \begin{bmatrix} 0 & 1 & 0 & 0 & \cdots & 0 \\ 0 & 0 & 1 & 0 & \cdots & 0 \\ \vdots & \vdots & \vdots & \vdots & \ddots & \vdots \\ 0 & 0 & 0 & 0 & \cdots & 1 \\ -a_0 & -a_1 & -a_2 & -a_3 & \cdots & -a_{n-1} \end{bmatrix}.$$

The corresponding linear system can then be solved using the method of Theorem 8.7, as in Example 4.

Several startling patterns were revealed in Example 4. First, notice the similarity between the original differential equation $y''' - 6y'' + 3y' + 10y = 0$ and $p_{\mathbf{A}}(x) = x^3 - 6x^2 + 3x + 10$. This observation leads to the following general principle, which you are asked to prove in Exercise 5(b):

If $y^{(n)} + a_{n-1} y^{(n-1)} + \cdots + a_1 y' + a_0 y$ is represented as a linear system $\mathbf{F}'(t) = \mathbf{A}\mathbf{F}(t)$, where $\mathbf{F}(t)$ and $\mathbf{A}$ are as described above, then

$$p_{\mathbf{A}}(x) = x^n + a_{n-1} x^{n-1} + \cdots + a_1 x + a_0.$$

Hence, from now on, we can avoid the long calculations necessary to determine $p_A(x)$. When solving differential equations, $p_A(x)$ is always derived from this shortcut. The equation $p_A(x) = 0$ is called the **characteristic equation** of the original differential equation. The roots of this equation, the eigen-values of A, are frequently called the **characteristic values** of the differential equation.

Also, notice in Example 4 that the eigenspace E_λ for each eigenvalue λ is one-dimensional and is spanned by the vector $[1, \lambda, \lambda^2]$. More generally, you are asked to prove the following in Exercise 6:

If $y^{(n)} + a_{n-1}y^{(n-1)} + \cdots + a_1y' + a_0y$ is represented as a linear system $F'(t) = AF(t)$, where $F(t)$ and A are as described above, and if λ is any eigenvalue for A, then the eigenspace E_λ is one-dimensional and is spanned by the vector $[1, \lambda, \lambda^2, \ldots, \lambda^{n-1}]$.

Combining the preceding facts, we can state the solution set for many higher-order homogeneous differential equations directly (and avoid linear algebra techniques altogether), as follows:

Consider the differential equation

$$y^{(n)} + a_{n-1}y^{(n-1)} + \cdots + a_2y'' + a_1y' + a_0y = 0.$$

Suppose that $\lambda_1, \ldots, \lambda_n$ are n distinct solutions to the characteristic equation

$$x^n + a_{n-1}x^{n-1} + \cdots + a_2x^2 + a_1x + a_0 = 0.$$

Then all continuously differentiable solutions of the differential equation have the form

$$y = b_1e^{\lambda_1 t} + b_2e^{\lambda_2 t} + \cdots + b_ne^{\lambda_n t}.$$

EXAMPLE 5 To solve the homogeneous differential equation

$$y'''' + 2y''' - 28y'' - 50y' + 75y = 0,$$

we first find its characteristic values by solving the characteristic equation

$$x^4 + 2x^3 - 28x^2 - 50x + 75 = 0.$$

By factoring, or using an appropriate numerical technique, we obtain four distinct characteristic values. These are $\lambda_1 = -5$, $\lambda_2 = -3$, $\lambda_3 = 1$, and $\lambda_4 = 5$. Thus, the continuously differentiable solutions for the original differential equation are precisely those functions of the form

$$y = b_1e^{-5t} + b_2e^{-3t} + b_3e^t + b_4e^{5t}.$$

∎

Notice that the method in Example 5 cannot be used if the differential equation has fewer than n distinct characteristic values. If you can find only k distinct characteristic values for an nth-order equation, with $k < n$, then the method yields only a k-dimensional subspace of the full n-dimensional solution space. As with first-order systems, finding the complete solution set in such a case requires the use of complex eigenvalues, complex eigenvectors, and generalized eigenvectors.

Exercises for Section 8.9

1. In each part of this exercise, the given matrix represents $\mathbf{A}$ in a first-order system of the form $\mathbf{F}'(t) = \mathbf{A}\mathbf{F}(t)$. Use Theorem 8.7 to find the general form of a solution to each system.

   ⋆(a) $\begin{bmatrix} 13 & -28 \\ 6 & -13 \end{bmatrix}$

   (b) $\begin{bmatrix} 18 & -15 \\ 20 & -17 \end{bmatrix}$

   ⋆(c) $\begin{bmatrix} 1 & 4 & 4 \\ -1 & 2 & 2 \\ 1 & 1 & 1 \end{bmatrix}$

   ⋆(d) $\begin{bmatrix} -5 & -6 & 15 \\ -6 & -5 & 15 \\ -6 & -6 & 16 \end{bmatrix}$

   (e) $\begin{bmatrix} -1 & 0 & -2 & 2 \\ -3 & 5 & 1 & -9 \\ 0 & 4 & 5 & -12 \\ -1 & 4 & 3 & -10 \end{bmatrix}$

2. Find the solution set for each given homogeneous differential equation.

   ⋆(a) $y'' + y' - 6y = 0$

   (b) $y''' - 5y'' - y' + 5y = 0$

   ⋆(c) $y'''' - 6y'' + 8y = 0$

3. Let $\mathbf{A}$ be an $n \times n$ matrix with linearly independent eigenvectors $\mathbf{v}_1, \ldots, \mathbf{v}_k$ corresponding, respectively, to the eigenvalues $\lambda_1, \ldots, \lambda_k$. Prove that

$$\mathbf{F}(t) = b_1 e^{\lambda_1 t} \mathbf{v}_1 + b_2 e^{\lambda_2 t} \mathbf{v}_2 + \cdots + b_k e^{\lambda_k t} \mathbf{v}_k$$

   is a solution for the first-order system $\mathbf{F}'(t) = \mathbf{A}\mathbf{F}(t)$, for every choice of $b_1, \ldots, b_k \in \mathbb{R}$.

4. (a) Let $\mathbf{A}$ be a diagonalizable $n \times n$ matrix, and let $\mathbf{v}$ be a fixed vector in $\mathbb{R}^n$. Show there is a unique function $\mathbf{F}(t)$ that satisfies the first-order system $\mathbf{F}'(t) = \mathbf{A}\mathbf{F}(t)$ such that $\mathbf{F}(0) = \mathbf{v}$. (The vector $\mathbf{v}$ is called an **initial condition** for the system.)

   ⋆(b) Find the unique solution to $\mathbf{F}'(t) = \mathbf{A}\mathbf{F}(t)$ with initial condition $\mathbf{F}(0) = \mathbf{v}$, where

$$\mathbf{A} = \begin{bmatrix} -11 & -6 & 16 \\ -4 & -1 & 4 \\ -12 & -6 & 17 \end{bmatrix} \quad \text{and} \quad \mathbf{v} = [1, -4, 0].$$

5. (a) Verify that the homogeneous differential equation

$$y^{(n)} + a_{n-1} y^{(n-1)} + \cdots + a_1 y' + a_0 y = 0$$

can be represented as $\mathbf{F}'(t) = \mathbf{A}\mathbf{F}(t)$, where $\mathbf{F}(t) = [f_1(t),\ f_2(t),\dots,$
$f_n(t)]$, with $f_1(t) = y$, $f_2(t) = y',\dots,$ $f_n(t) = y^{(n-1)}$, and where

$$\mathbf{A} = \begin{bmatrix} 0 & 1 & 0 & 0 & \cdots & 0 \\ 0 & 0 & 1 & 0 & \cdots & 0 \\ \vdots & \vdots & \vdots & \vdots & \ddots & \vdots \\ 0 & 0 & 0 & 0 & \cdots & 1 \\ -a_0 & -a_1 & -a_2 & -a_3 & \cdots & -a_{n-1} \end{bmatrix}.$$

▶(b) If $\mathbf{A}$ is the matrix given in part (a), prove that

$$p_{\mathbf{A}}(x) = x^n + a_{n-1}x^{n-1} + \cdots + a_1 x + a_0.$$

(Hint: Use induction on n and a cofactor expansion on the first
column of $(x\mathbf{I}_n - \mathbf{A})$.)

6. Let $\mathbf{A}$ be the $n \times n$ matrix from Exercise 5, for some $a_0, a_1, \dots, a_{n-1} \in \mathbb{R}$.

(a) Calculate $\mathbf{A} \begin{bmatrix} b_1 \\ b_2 \\ \vdots \\ b_n \end{bmatrix}$, for a general n-vector $\begin{bmatrix} b_1 \\ b_2 \\ \vdots \\ b_n \end{bmatrix}$.

(b) Let λ be an eigenvalue for $\mathbf{A}$. Show that $[1, \lambda, \lambda^2, \dots, \lambda^{n-1}]$ is an
eigenvector corresponding to λ. (Hint: Use part (b) of Exercise 5.)

(c) Show that if $\mathbf{v}$ is a vector with first coordinate c such that $\mathbf{A}\mathbf{v} = \lambda\mathbf{v}$,
for some $\lambda \in \mathbb{R}$, then $\mathbf{v} = c[1, \lambda, \lambda^2, \dots, \lambda^{n-1}]$.

(d) Conclude that the eigenspace E_λ for an eigenvalue λ of $\mathbf{A}$ is always
one-dimensional.

★**7.** True or False:

(a) $\mathbf{F}(t) = \mathbf{0}$ is always a solution of $\mathbf{F}'(t) = \mathbf{A}\mathbf{F}(t)$.

(b) The set of all continuously differentiable solutions of $\mathbf{F}'(t) = \mathbf{A}\mathbf{F}(t)$
is a vector space.

(c) $\mathbf{F}'(t) = \begin{bmatrix} 1 & 2 \\ 0 & 3 \end{bmatrix} \mathbf{F}(t)$ has solution set $\left\{ \begin{bmatrix} b_1 e^t + b_2 e^{3t} \\ b_2 e^{3t} \end{bmatrix} \middle| b_1, b_2 \in \mathbb{R} \right\}$.

(d) $\mathbf{F}'(t) = \begin{bmatrix} 0 & 1 \\ -1 & 0 \end{bmatrix} \mathbf{F}(t)$ has no nontrivial solutions because $\begin{bmatrix} 0 & 1 \\ -1 & 0 \end{bmatrix}$
is not diagonalizable.

8.10 Least-Squares Solutions for Inconsistent Systems

Prerequisite: Section 6.2, Orthogonal Complements

When attempting to solve a system of linear equations $\mathbf{A}\mathbf{x} = \mathbf{b}$, there is
always the possibility that the system is inconsistent. However, in practical
situations, even if no solutions to $\mathbf{A}\mathbf{x} = \mathbf{b}$ exist, it is usually helpful to find an
approximate solution; that is, a vector $\mathbf{v}$ such that $\mathbf{A}\mathbf{v}$ is as close as possible to $\mathbf{b}$.

Finding Approximate Solutions

If $\mathbf{A}$ is an $m \times n$ matrix, consider the linear transformation $L: \mathbb{R}^n \longrightarrow \mathbb{R}^m$ given by $L(\mathbf{x}) = \mathbf{A}\mathbf{x}$. If $\mathbf{b} \in \mathbb{R}^m$, then any solution to the linear system $\mathbf{A}\mathbf{x} = \mathbf{b}$ is a pre-image for $\mathbf{b}$ under L. However, if $\mathbf{b} \notin \text{range}(L)$, the system is inconsistent, but we can calculate an approximate solution to the system $\mathbf{A}\mathbf{x} = \mathbf{b}$ by finding a pre-image under L of a vector in the subspace $\mathcal{W} = \text{range}(L)$ that is as close as possible to $\mathbf{b}$. Theorem 6.16 implies that, among the vectors in $\mathcal{W}$, $\mathbf{proj}_{\mathcal{W}}\mathbf{b}$ has minimal distance to $\mathbf{b}$. The following theorem shows that $\mathbf{proj}_{\mathcal{W}}\mathbf{b}$ is the *unique* closest vector in $\mathcal{W}$ to $\mathbf{b}$ and that the set of pre-images $L^{-1}(\{\mathbf{proj}_{\mathcal{W}}\mathbf{b}\})$ can be found by solving the linear system $(\mathbf{A}^T\mathbf{A})\mathbf{x} = \mathbf{A}^T\mathbf{b}$.

THEOREM 8.8

Let $\mathbf{A}$ be an $m \times n$ matrix, let $\mathbf{b} \in \mathbb{R}^m$, and let $\mathcal{W}$ be the subspace $\{\mathbf{A}\mathbf{x} \mid \mathbf{x} \in \mathbb{R}^n\}$. Then the following three conditions on a vector $\mathbf{v} \in \mathbb{R}^n$ are equivalent:
  (1) $\mathbf{A}\mathbf{v} = \mathbf{proj}_{\mathcal{W}}\mathbf{b}$
  (2) $\|\mathbf{A}\mathbf{v} - \mathbf{b}\| \leq \|\mathbf{A}\mathbf{z} - \mathbf{b}\|$ for all $\mathbf{z} \in \mathbb{R}^n$
  (3) $(\mathbf{A}^T\mathbf{A})\mathbf{v} = \mathbf{A}^T\mathbf{b}$.
Such a vector $\mathbf{v}$ is called a **least-squares solution** to the linear system $\mathbf{A}\mathbf{x} = \mathbf{b}$.

The inequality $\|\mathbf{A}\mathbf{v} - \mathbf{b}\| \leq \|\mathbf{A}\mathbf{z} - \mathbf{b}\|$ in Theorem 8.8 implies that there is no better approximation than $\mathbf{v}$ for a solution to $\mathbf{A}\mathbf{x} = \mathbf{b}$ because the distance from $\mathbf{A}\mathbf{v}$ to $\mathbf{b}$ is never larger than the distance from $\mathbf{A}\mathbf{z}$ to $\mathbf{b}$ for any other vector $\mathbf{z}$. Of course, if $\mathbf{A}\mathbf{x} = \mathbf{b}$ is consistent, then $\mathbf{v}$ is an actual solution to $\mathbf{A}\mathbf{x} = \mathbf{b}$ (see Exercise 4).

The inequality $\|\mathbf{A}\mathbf{v} - \mathbf{b}\| \leq \|\mathbf{A}\mathbf{z} - \mathbf{b}\|$ also shows why $\mathbf{v}$ is called a least-squares solution. Since calculating a norm involves finding a sum of squares, this inequality implies that the solution $\mathbf{v}$ produces the least possible value for the sum of the squares of the differences in each coordinate between $\mathbf{A}\mathbf{z}$ and $\mathbf{b}$ over all possible vectors $\mathbf{z}$.

PROOF OF THEOREM 8.8

Let $\mathbf{A}$ and $\mathbf{b}$ be as given in the statement of the theorem, and let $L: \mathbb{R}^n \longrightarrow \mathbb{R}^m$ be the linear transformation given by $L(\mathbf{x}) = \mathbf{A}\mathbf{x}$. Then $\mathcal{W} = \{\mathbf{A}\mathbf{x} \mid \mathbf{x} \in \mathbb{R}^n\} = \text{range}(L)$.

Our first goal is to prove (1) if and only if (2). Now, let $\mathbf{A}\mathbf{v} = \mathbf{proj}_{\mathcal{W}}\mathbf{b}$. Since $\mathbf{A}\mathbf{z} \in \mathcal{W}$, Theorem 6.16 shows that $\|\mathbf{A}\mathbf{v} - \mathbf{b}\| \leq \|\mathbf{A}\mathbf{z} - \mathbf{b}\|$ for all $\mathbf{z} \in \mathbb{R}^n$.

Conversely, suppose $\|\mathbf{A}\mathbf{v} - \mathbf{b}\| \leq \|\mathbf{A}\mathbf{z} - \mathbf{b}\|$ for all $\mathbf{z} \in \mathbb{R}^n$. Let $\mathbf{p} = \mathbf{proj}_{\mathcal{W}}\mathbf{b}$. We need to show that $\mathbf{A}\mathbf{v} = \mathbf{p}$. Now $\mathbf{p} \in \mathcal{W}$, so $\mathbf{p}$ is a vector of the form $\mathbf{A}\mathbf{z}$ for some $\mathbf{z} \in \mathbb{R}^n$. Hence, $\|\mathbf{A}\mathbf{v} - \mathbf{b}\| \leq \|\mathbf{p} - \mathbf{b}\|$ by assumption. But $\|\mathbf{p} - \mathbf{b}\| \leq \|\mathbf{A}\mathbf{v} - \mathbf{b}\|$ by Theorem 6.16. Therefore, $\|\mathbf{A}\mathbf{v} - \mathbf{b}\| = \|\mathbf{p} - \mathbf{b}\|$.

Now $\mathbf{A}\mathbf{v}, \mathbf{p} \in \mathcal{W}$, so $\mathbf{A}\mathbf{v} - \mathbf{p} \in \mathcal{W}$. Also, $\mathbf{p} - \mathbf{b} = -(\mathbf{b} - \mathbf{p}) = -\mathbf{proj}_{\mathcal{W}^\perp}\mathbf{b} \in \mathcal{W}^\perp$, from the remark just before Example 7 in Section 6.2. Thus, $(\mathbf{A}\mathbf{v} - \mathbf{p}) \cdot (\mathbf{p} - \mathbf{b}) = 0$.

Therefore,

$$\begin{aligned}
\|\mathbf{Av} - \mathbf{b}\|^2 &= \|(\mathbf{Av} - \mathbf{p}) + (\mathbf{p} - \mathbf{b})\|^2 \\
&= ((\mathbf{Av} - \mathbf{p}) + (\mathbf{p} - \mathbf{b})) \cdot ((\mathbf{Av} - \mathbf{p}) + (\mathbf{p} - \mathbf{b})) \\
&= \|\mathbf{Av} - \mathbf{p}\|^2 + 2(\mathbf{Av} - \mathbf{p}) \cdot (\mathbf{p} - \mathbf{b}) + \|\mathbf{p} - \mathbf{b}\|^2 \\
&= \|\mathbf{Av} - \mathbf{p}\|^2 + \|\mathbf{p} - \mathbf{b}\|^2 .
\end{aligned}$$

But $\|\mathbf{Av} - \mathbf{b}\| = \|\mathbf{p} - \mathbf{b}\|$, implying $\|\mathbf{Av} - \mathbf{p}\|^2 = 0$. Hence, $\mathbf{Av} - \mathbf{p} = \mathbf{0}$, or $\mathbf{Av} = \mathbf{p}$. This completes our first goal.

To finish the proof, we will prove (1) if and only if (3). First, suppose $\mathbf{A}^T \mathbf{Av} = \mathbf{A}^T \mathbf{b}$. We will prove that $\mathbf{Av} = \mathbf{proj}_{\mathcal{W}} \mathbf{b}$. Let $\mathbf{u} = \mathbf{b} - \mathbf{Av}$, and hence $\mathbf{b} = \mathbf{Av} + \mathbf{u}$. If we can show that $\mathbf{Av} \in \mathcal{W}$ and $\mathbf{u} \in \mathcal{W}^{\perp}$, then we will have $\mathbf{Av} = \mathbf{proj}_{\mathcal{W}} \mathbf{b}$ by the uniqueness assertion in the Projection Theorem (Theorem 6.14). But $\mathbf{Av} \in \mathcal{W}$, since $\mathcal{W}$ consists precisely of vectors of this form. Also, $\mathbf{u} = \mathbf{b} - \mathbf{Av}$, and so $\mathbf{A}^T \mathbf{u} = \mathbf{A}^T \mathbf{b} - \mathbf{A}^T \mathbf{Av} = \mathbf{0}$ since $\mathbf{A}^T \mathbf{Av} = \mathbf{A}^T \mathbf{b}$. Now $\mathbf{A}^T \mathbf{u} = \mathbf{0}$ implies that $\mathbf{u}$ is orthogonal to every row of $\mathbf{A}^T$, and hence $\mathbf{u}$ is orthogonal to every column of $\mathbf{A}$. But recall from Section 5.3 that the columns of $\mathbf{A}$ span $\mathcal{W} = \text{range}(L)$. Hence, $\mathbf{u} \in \mathcal{W}^{\perp}$ by Theorem 6.9, completing this half of the proof.

Conversely, suppose $\mathbf{Av} = \mathbf{proj}_{\mathcal{W}} \mathbf{b}$. Then $\mathbf{b} = \mathbf{Av} + \mathbf{u}$, where $\mathbf{u} \in \mathcal{W}^{\perp}$. Hence, $\mathbf{A}^T \mathbf{u} = \mathbf{0}$ since $\mathbf{u}$ must be orthogonal to the rows of $\mathbf{A}^T$, which form a spanning set for $\mathcal{W}$. Therefore,

$$\mathbf{b} = \mathbf{Av} + \mathbf{u} \implies \mathbf{A}^T \mathbf{b} = \mathbf{A}^T \mathbf{Av} + \mathbf{A}^T \mathbf{u} \implies \mathbf{A}^T \mathbf{b} = \mathbf{A}^T \mathbf{Av}.$$

$\blacksquare$

EXAMPLE 1   Consider the inconsistent linear system

$$\begin{cases}
7x &+& 7y &+& 5z &=& 15 \\
4x & & &+& z &=& 1 \\
2x &+& y &+& z &=& 4 \\
5x &+& 8y &+& 5z &=& 16
\end{cases} .$$

Letting $\mathbf{A} = \begin{bmatrix} 7 & 7 & 5 \\ 4 & 0 & 1 \\ 2 & 1 & 1 \\ 5 & 8 & 5 \end{bmatrix}$ and $\mathbf{b} = \begin{bmatrix} 15 \\ 1 \\ 4 \\ 16 \end{bmatrix}$, we will find a least-squares solution to $\mathbf{Ax} = \mathbf{b}$. By part (3) of Theorem 8.8, we need to solve the linear system $\mathbf{A}^T \mathbf{Ax} = \mathbf{A}^T \mathbf{b}$. Now,

$$\mathbf{A}^T \mathbf{A} = \begin{bmatrix} 94 & 91 & 66 \\ 91 & 114 & 76 \\ 66 & 76 & 52 \end{bmatrix} \text{ and } \mathbf{A}^T \mathbf{b} = \begin{bmatrix} 197 \\ 237 \\ 160 \end{bmatrix} .$$

Row reducing $\begin{bmatrix} 94 & 91 & 66 & 197 \\ 91 & 114 & 76 & 237 \\ 66 & 76 & 52 & 160 \end{bmatrix}$ to obtain $\begin{bmatrix} 1 & 0 & 0 & -7 \\ 0 & 1 & 0 & -12 \\ 0 & 0 & 1 & 29.5 \end{bmatrix}$

shows that $\mathbf{v} = [-7, -12, 29.5]$ is the desired solution. Notice that

$$\mathbf{Av} = \begin{bmatrix} 7 & 7 & 5 \\ 4 & 0 & 1 \\ 2 & 1 & 1 \\ 5 & 8 & 5 \end{bmatrix} \begin{bmatrix} -7 \\ -12 \\ 29.5 \end{bmatrix} = \begin{bmatrix} 14.5 \\ 1.5 \\ 3.5 \\ 16.5 \end{bmatrix},$$

and so $\mathbf{Av}$ comes close to producing the vector $\mathbf{b}$.

In fact, for any $\mathbf{z} \in \mathbb{R}^3$, $\|\mathbf{Av} - \mathbf{b}\| \leq \|\mathbf{Az} - \mathbf{b}\|$. For example, if $\mathbf{z} = [-11, -19, 45]$, which is the unique solution to the first three equations in the system, then $\|\mathbf{Az} - \mathbf{b}\| = \|[15, 1, 4, 18] - [15, 1, 4, 16]\| = \|[0, 0, 0, 2]\| = 2$. However, $\|\mathbf{Av} - \mathbf{b}\| = \|[14.5, 1.5, 3.5, 16.5] - [15, 1, 4, 16]\| = \|[-0.5, 0.5, -0.5, 0.5]\| = 1$, which is less than $\|\mathbf{Az} - \mathbf{b}\|$. ■

Non-unique Least-Squares Solutions

Theorem 8.8 shows that if $\mathbf{v}$ is a least-squares solution for a linear system $\mathbf{Ax} = \mathbf{b}$, then $\mathbf{Av} = \mathbf{proj}_{\mathcal{W}}\mathbf{b}$, where $\mathcal{W} = \{\mathbf{Ax} \mid \mathbf{x} \in \mathbb{R}^n\}$. Now, even though $\mathbf{proj}_{\mathcal{W}}\mathbf{b}$ is uniquely determined, there may be more than one vector $\mathbf{v}$ with $\mathbf{Av} = \mathbf{proj}_{\mathcal{W}}\mathbf{b}$. In such a case, there are infinitely many least-squares solutions for $\mathbf{Ax} = \mathbf{b}$, all of which produce the same value for $\mathbf{Ax}$.

 EXAMPLE 2   Consider the system $\mathbf{Ax} = \mathbf{b}$, where

$$\mathbf{A} = \begin{bmatrix} 2 & 3 & -1 \\ 4 & 1 & 3 \\ 2 & -7 & 9 \end{bmatrix} \quad \text{and} \quad \mathbf{b} = \begin{bmatrix} 9 \\ 8 \\ -1 \end{bmatrix}.$$

We find a least-squares solution to $\mathbf{Ax} = \mathbf{b}$ by solving the linear system $\mathbf{A}^T\mathbf{Ax} = \mathbf{A}^T\mathbf{b}$. Now,

$$\mathbf{A}^T\mathbf{A} = \begin{bmatrix} 24 & -4 & 28 \\ -4 & 59 & -63 \\ 28 & -63 & 91 \end{bmatrix} \quad \text{and} \quad \mathbf{A}^T\mathbf{b} = \begin{bmatrix} 48 \\ 42 \\ 6 \end{bmatrix}.$$

Row reducing $\begin{bmatrix} 24 & -4 & 28 & 48 \\ -4 & 59 & -63 & 42 \\ 28 & -63 & 91 & 6 \end{bmatrix}$ to obtain $\begin{bmatrix} 1 & 0 & 1 & \frac{15}{7} \\ 0 & 1 & -1 & \frac{6}{7} \\ 0 & 0 & 0 & 0 \end{bmatrix}$

shows that this system has infinitely many solutions. The solution set is $S = \{[\frac{15}{7} - c, \frac{6}{7} + c, c] \mid c \in \mathbb{R}\}$. Two particular solutions are $\mathbf{v}_1 = [\frac{15}{7}, \frac{6}{7}, 0]$, and $\mathbf{v}_2 = [3, 0, -\frac{6}{7}]$. You can verify that $\mathbf{Av}_1 = \mathbf{Av}_2 = [\frac{48}{7}, \frac{66}{7}, -\frac{12}{7}]$. In general, multiplying $\mathbf{A}$ by any vector in S produces the result $[\frac{48}{7}, \frac{66}{7}, -\frac{12}{7}]$. Every vector in S is a least-squares solution for $\mathbf{Ax} = \mathbf{b}$. They all produce the same result for $\mathbf{Ax}$, which is as close as possible to $\mathbf{b}$. ■

Approximate Eigenvalues and Eigenvectors

When solving for eigenvalues and eigenvectors for a square matrix $\mathbf{C}$, a problem can arise if the exact value of an eigenvalue λ is not known, but only a close approximation λ' instead. Then, since λ' is not the precise eigenvalue, the matrix $\lambda'\mathbf{I} - \mathbf{C}$ is nonsingular. This makes it impossible to solve $(\lambda'\mathbf{I} - \mathbf{C})\mathbf{x} = \mathbf{0}$ directly for an eigenvector because only the trivial solution exists. One of several possible approaches to this problem[10] is to use the method of least-squares to find an approximate eigenvector associated with the approximate eigenvalue λ'. To do this, first add an extra equation to the system $(\lambda'\mathbf{I}-\mathbf{C})\mathbf{x} = \mathbf{0}$ to force the solution to be nontrivial. One possibility is to require that the sum of the coordinates of the solution equals 1. Even though this new nonhomogeneous system formed is inconsistent, a least-squares solution for this expanded system frequently serves as the desired approximate eigenvector. We illustrate this technique in the following example:

EXAMPLE 3 Consider the matrix
$$\mathbf{C} = \begin{bmatrix} 2 & -3 & -1 \\ 7 & -6 & -1 \\ -16 & 14 & 3 \end{bmatrix},$$
which has eigenvalues $\sqrt{5}$, $-\sqrt{5}$, and -1.

Suppose the best estimate we have for the eigenvalue $\lambda = \sqrt{5} \approx 2.23606$ is $\lambda' = \frac{9}{4} = 2.25$. Then

$$\lambda'\mathbf{I}_3 - \mathbf{C} = \begin{bmatrix} \frac{1}{4} & 3 & 1 \\ -7 & \frac{33}{4} & 1 \\ 16 & -14 & -\frac{3}{4} \end{bmatrix},$$

which is nonsingular. (Its determinant is $\frac{13}{64}$.) Hence, the system $(\lambda'\mathbf{I}_3 - \mathbf{C})\mathbf{x} = \mathbf{0}$ has only the trivial solution. We now force a nontrivial solution $\mathbf{x}$ by adding the condition that the sum of the coordinates of $\mathbf{x}$ equals 1. This produces the system

$$\begin{cases} \frac{1}{4}x_1 & + & 3x_2 & + & x_3 & = & 0 \\ -7x_1 & + & \frac{33}{4}x_2 & + & x_3 & = & 0 \\ 16x_1 & - & 14x_2 & - & \frac{3}{4}x_3 & = & 0 \\ x_1 & + & x_2 & + & x_3 & = & 1 \end{cases}.$$

However, this new system is inconsistent since the first three equations together have only the trivial solution, which does not satisfy the last equation.

---

[10]Numerical techniques exist for finding approximate eigenvectors that produce more accurate results than the method of least-squares. The major problem with the least-squares technique is that the accuracy of the approximate eigenvector is limited by the accuracy of the approximate eigenvalue used. Other numerical methods, such as an adaptation of the inverse power method, are iterative and adjust the approximation for the eigenvalue while solving for the eigenvector. For more information on the inverse power method and other numerical techniques for solving for eigenvalues and eigenvectors, consult a text on numerical methods in your library. One classic text is *Numerical Analysis*, 7th ed., by Burden and Faires (published by Brooks/Cole, 2001).

We will find a least-squares solution to this system.

$$\text{Let } \mathbf{A} = \begin{bmatrix} \frac{1}{4} & 3 & 1 \\ -7 & \frac{33}{4} & 1 \\ 16 & -14 & -\frac{3}{4} \\ 1 & 1 & 1 \end{bmatrix} \text{ and } \mathbf{b} = \begin{bmatrix} 0 \\ 0 \\ 0 \\ 1 \end{bmatrix}. \text{ Then}$$

$$\mathbf{A}^T\mathbf{A} = \begin{bmatrix} \frac{4897}{16} & -280 & -\frac{71}{4} \\ -280 & \frac{4385}{16} & \frac{91}{4} \\ -\frac{71}{4} & \frac{91}{4} & \frac{57}{16} \end{bmatrix} \text{ and } \mathbf{A}^T\mathbf{b} = \begin{bmatrix} 1 \\ 1 \\ 1 \end{bmatrix}.$$

$$\text{Row reducing } \begin{bmatrix} \frac{4897}{16} & -280 & -\frac{71}{4} & 1 \\ -280 & \frac{4385}{16} & \frac{91}{4} & 1 \\ -\frac{71}{4} & \frac{91}{4} & \frac{57}{16} & 1 \end{bmatrix} \text{ produces } \begin{bmatrix} 1 & 0 & 0 & -0.50 \\ 0 & 1 & 0 & -0.69 \\ 0 & 0 & 1 & 2.19 \end{bmatrix},$$

where we have rounded the results to two places after the decimal point. Hence, $\mathbf{v} = [-0.50, -0.69, 2.19]$ is an approximate eigenvector for $\mathbf{C}$ corresponding to the approximate eigenvalue $\lambda' = \frac{9}{4}$. In fact, $(\lambda'\mathbf{I}_3 - \mathbf{C})\mathbf{v} = [-0.005, -0.0025, 0.0175]$, which is close to the zero vector. This implies that $\mathbf{Cv}$ is very close to $\lambda'\mathbf{v}$. In fact, the maximum difference among the three coordinates (≈ 0.0175) is about the same magnitude as the error in the estimation of the eigenvalue (≈ 0.01394). Also, a lengthy computation would show that the unit vector $\mathbf{v}/\|\mathbf{v}\| \approx [-0.21, -0.29, 0.93]$ agrees with an actual unit eigenvector for $\mathbf{C}$ corresponding to $\lambda = \sqrt{5}$ in every coordinate, after rounding to the first two places after the decimal point. ∎

There may be a problem with the technique described in Example 3 if the actual eigenspace for λ is orthogonal to the vector $\mathbf{t} = [1, 1, \ldots, 1]$ since our added requirement implies that the dot product of the approximate eigenvector with $\mathbf{t}$ equals 1. If this problem arises, simply change the requirement to specify that the dot product with any nonzero vector of your choice (other than $\mathbf{t}$) equals 1 and try again.

Least-Squares Polynomials

In Theorem 8.2 of Section 8.3, we presented a method for finding a polynomial function $\mathbf{p}$ in $\mathcal{P}_k$ that comes closest to passing through a given set of data points $(a_1, b_1), (a_2, b_2), \ldots, (a_n, b_n)$. This method sets up a linear system whose intended solution is a polynomial that passes through all n data points. However, if the desired degree k of the polynomial is less than $n + 1$, then the linear system is inconsistent (in most cases). Thus, we find that a least-squares solution to the system produces a **least-squares polynomial** that approximates the given data.

Theorem 8.2 is a corollary of Theorem 8.8 in this section. We ask you to prove Theorem 8.2 using Theorem 8.8 in Exercise 6. See Section 8.3 if you have further interest in least-squares polynomials.

Exercises for Section 8.10

We strongly recommend that you use a computer or calculator to help you perform the required computations in these exercises.

1. In each part, find the set of all least-squares solutions for the linear system $\mathbf{Ax} = \mathbf{b}$ for the given matrix $\mathbf{A}$ and vector $\mathbf{b}$. If there is more than one least-squares solution, find at least two particular least-squares solutions. Finally, illustrate the inequality $\|\mathbf{Av} - \mathbf{b}\| \le \|\mathbf{Az} - \mathbf{b}\|$ by computing $\|\mathbf{Av} - \mathbf{b}\|$ for a particular least-squares solution $\mathbf{v}$ and $\|\mathbf{Az} - \mathbf{b}\|$ for the given vector $\mathbf{z}$.

   ★(a) $\mathbf{A} = \begin{bmatrix} 2 & 3 \\ 1 & -1 \\ 4 & 1 \end{bmatrix}$, $\mathbf{b} = \begin{bmatrix} 5 \\ 0 \\ 4 \end{bmatrix}$, $\mathbf{z} = \begin{bmatrix} 1 \\ 1 \end{bmatrix}$

   (b) $\mathbf{A} = \begin{bmatrix} 5 & 2 \\ 3 & 1 \\ 4 & 3 \end{bmatrix}$, $\mathbf{b} = \begin{bmatrix} 12 \\ 15 \\ 14 \end{bmatrix}$, $\mathbf{z} = \begin{bmatrix} 3 \\ 0 \end{bmatrix}$

   ★(c) $\mathbf{A} = \begin{bmatrix} 2 & 1 & -1 \\ 3 & 2 & 5 \\ 1 & 0 & -7 \end{bmatrix}$, $\mathbf{b} = \begin{bmatrix} 3 \\ 2 \\ 6 \end{bmatrix}$, $\mathbf{z} = \begin{bmatrix} 1 \\ 1 \\ -1 \end{bmatrix}$

   (d) $\mathbf{A} = \begin{bmatrix} 3 & 1 & 0 & -1 \\ 5 & 3 & 2 & 2 \\ 2 & 2 & 2 & 3 \\ 7 & 5 & 4 & 5 \end{bmatrix}$, $\mathbf{b} = \begin{bmatrix} 3 \\ 14 \\ 10 \\ 25 \end{bmatrix}$, $\mathbf{z} = \begin{bmatrix} -1 \\ 6 \\ 0 \\ 0 \end{bmatrix}$

2. In practical applications, we are frequently interested in only those solutions having nonnegative entries in every coordinate. In each part, find the set of all such least-squares solutions to the linear system $\mathbf{Ax} = \mathbf{b}$ for the given matrix $\mathbf{A}$ and vector $\mathbf{b}$.

   ★(a) $\mathbf{A} = \begin{bmatrix} 2 & 1 & 0 \\ 3 & -2 & 4 \\ 7 & 0 & 4 \end{bmatrix}$, $\mathbf{b} = \begin{bmatrix} 1 \\ 2 \\ 3 \end{bmatrix}$

   (b) $\mathbf{A} = \begin{bmatrix} 2 & 3 & 3 \\ 1 & -1 & 1 \\ 1 & 9 & 3 \end{bmatrix}$, $\mathbf{b} = \begin{bmatrix} 5 \\ 2 \\ 7 \end{bmatrix}$

3. In each part, find an approximate eigenvector $\mathbf{v}$ for the given matrix $\mathbf{C}$ corresponding to the given approximate eigenvalue λ' using the method of Example 3. Round the entries of $\mathbf{v}$ to two places after the decimal point. Then compute $(\lambda'\mathbf{I} - \mathbf{C})\mathbf{v}$ to estimate the error in your answer.

   (a) $\mathbf{C} = \begin{bmatrix} 3 & -2 \\ -1 & 1 \end{bmatrix}$, $\lambda' = \frac{15}{4}$

$$\star\text{(b)} \quad \mathbf{C} = \begin{bmatrix} 3 & -3 & -2 \\ -5 & 5 & 4 \\ 11 & -12 & -9 \end{bmatrix}, \quad \lambda' = \frac{3}{2}$$

$$\text{(c)} \quad \mathbf{C} = \begin{bmatrix} 1 & 18 & -7 \\ -1 & 12 & -5 \\ -3 & 32 & -13 \end{bmatrix}, \quad \lambda' = \frac{9}{4}$$

4. Prove that if a linear system $\mathbf{Ax} = \mathbf{b}$ is consistent, then the set of least-squares solutions for the system equals the set of actual solutions.

5. Let $\mathbf{A}$ be an $m \times n$ matrix, and let $\mathbf{v}_1, \mathbf{v}_2 \in \mathbb{R}^n$. Prove that if $\mathbf{A}^T \mathbf{Av}_1 = \mathbf{A}^T \mathbf{Av}_2$, then $\mathbf{Av}_1 = \mathbf{Av}_2$.

▶**6.** Use Theorem 8.8 to prove Theorem 8.2 in Section 8.3.

★**7.** True or False:

(a) A least-squares solution to an inconsistent system is a vector $\mathbf{v}$ that satisfies as many equations in the system as possibly can be satisfied.

(b) For any matrix $\mathbf{A}$, the matrix $\mathbf{A}^T \mathbf{A}$ is square and symmetric.

(c) Every system $\mathbf{Ax} = \mathbf{b}$ must have at least one least-squares solution.

(d) If $\mathbf{v}_1$ and $\mathbf{v}_2$ are both least-squares solutions to $\mathbf{Ax} = \mathbf{b}$, then $\mathbf{Av}_1 = \mathbf{Av}_2$.

(e) In this section, the least-squares method is applied to solving for eigenvectors in cases in which only an estimate of the eigenvalue is known.

8.11 Max-Min Problems in $\mathbb{R}^n$ and the Hessian Matrix

Prerequisite: Section 6.3, Orthogonal Diagonalization

In this section, we study the problem of finding local maxima and minima for real-valued functions on $\mathbb{R}^n$. The method we describe is the higher-dimensional analogue to finding critical points and applying the second derivative test to functions on $\mathbb{R}$ studied in first-semester calculus.

Taylor's Theorem in $\mathbb{R}^n$

Let $f \in C^2(\mathbb{R}^n)$, where $C^2(\mathbb{R}^n)$ is the set of real-valued functions defined on $\mathbb{R}^n$ having continuous second partial derivatives. The method for solving for local extreme points of f relies upon Taylor's Theorem with second degree remainder terms, which we state here without proof. (In the following theorem, an **open hypersphere** centered at $\mathbf{x}_0$ is a set of the form $\{\mathbf{x} \in \mathbb{R}^n \mid \|\mathbf{x} - \mathbf{x}_0\| < r\}$ for some positive real number r.)

THEOREM 8.9 (Taylor's Theorem in $\mathbb{R}^n$)

Let A be an open hypersphere centered at $\mathbf{x}_0 \in \mathbb{R}^n$, let $\mathbf{u}$ be a unit vector in $\mathbb{R}^n$, and let $t \in \mathbb{R}$ such that $\mathbf{x}_0 + t\mathbf{u} \in A$. Suppose $f : A \to \mathbb{R}$ has continuous second partial derivatives throughout A; that is, $f \in C^2(A)$. Then there is a c with $0 \le c \le t$ such that

$$
f(\mathbf{x}_0 + t\mathbf{u}) \;=\; f(\mathbf{x}_0) + \sum_{i=1}^{n} \left.\frac{\partial f}{\partial x_i}\right|_{\mathbf{x}_0} (tu_i) + \frac{1}{2}\sum_{i=1}^{n} \left.\frac{\partial^2 f}{\partial x_i^2}\right|_{\mathbf{x}_0 + c\mathbf{u}} (t^2 u_i^2)
$$

$$
+ \sum_{i=1}^{n}\sum_{j=i+1}^{n} \left.\frac{\partial^2 f}{\partial x_i \partial x_j}\right|_{\mathbf{x}_0 + c\mathbf{u}} (t^2 u_i u_j).
$$

Taylor's Theorem in $\mathbb{R}^n$ is derived from the familiar Taylor's Theorem in $\mathbb{R}$ by applying it to the function $g(t) = f(\mathbf{x}_0 + t\mathbf{u})$. In $\mathbb{R}^2$, the formula in Taylor's Theorem is

$$
f(\mathbf{x}_0 + t\mathbf{u}) \;=\; f(\mathbf{x}_0) + \left.\frac{\partial f}{\partial x}\right|_{\mathbf{x}_0} (tu_1) + \left.\frac{\partial f}{\partial y}\right|_{\mathbf{x}_0} (tu_2)
$$

$$
+ \frac{1}{2}\left.\frac{\partial^2 f}{\partial x^2}\right|_{\mathbf{x}_0 + c\mathbf{u}} (t^2 u_1^2) + \frac{1}{2}\left.\frac{\partial^2 f}{\partial y^2}\right|_{\mathbf{x}_0 + c\mathbf{u}} (t^2 u_2^2)
$$

$$
+ \left.\frac{\partial^2 f}{\partial x \partial y}\right|_{\mathbf{x}_0 + c\mathbf{u}} (t^2 u_1 u_2).
$$

Recall that the **gradient** of f is defined by $\nabla f = \left[\frac{\partial f}{\partial x_1}, \frac{\partial f}{\partial x_2}, \ldots, \frac{\partial f}{\partial x_n} \right]$. If we let $\mathbf{v} = t\mathbf{u}$, then, in $\mathbb{R}^2$, $\mathbf{v} = [v_1, v_2] = [tu_1, tu_2]$, and so the sum $\left.\frac{\partial f}{\partial x}\right|_{\mathbf{x}_0} (tu_1) + \left.\frac{\partial f}{\partial y}\right|_{\mathbf{x}_0} (tu_2)$ simplifies to $\left(\left.\nabla f\right|_{\mathbf{x}_0} \right) \cdot \mathbf{v}$. Also, since f has continuous second partial derivatives, we have $\frac{\partial^2 f}{\partial x \partial y} = \frac{\partial^2 f}{\partial y \partial x}$. Therefore,

$$
\frac{1}{2}\frac{\partial^2 f}{\partial x^2}(t^2 u_1^2) + \frac{1}{2}\frac{\partial^2 f}{\partial y^2}(t^2 u_2^2) + \frac{\partial^2 f}{\partial x \partial y}(t^2 u_1 u_2)
$$

$$
= \frac{1}{2}v_1\left(\frac{\partial^2 f}{\partial x^2}v_1 + \frac{\partial^2 f}{\partial x \partial y}v_2 \right) + \frac{1}{2}v_2\left(\frac{\partial^2 f}{\partial y \partial x}v_1 + \frac{\partial^2 f}{\partial y^2}v_2 \right)
$$

$$
= \frac{1}{2}\mathbf{v}^T \begin{bmatrix} \frac{\partial^2 f}{\partial x^2} & \frac{\partial^2 f}{\partial x \partial y} \\[2mm] \frac{\partial^2 f}{\partial y \partial x} & \frac{\partial^2 f}{\partial y^2} \end{bmatrix} \mathbf{v},
$$

where $\mathbf{v}$ is considered to be a column vector. The matrix

$$
\mathbf{H} = \begin{bmatrix} \frac{\partial^2 f}{\partial x^2} & \frac{\partial^2 f}{\partial x \partial y} \\[2mm] \frac{\partial^2 f}{\partial y \partial x} & \frac{\partial^2 f}{\partial y^2} \end{bmatrix}
$$

in this expression is called the **Hessian matrix** for f. Thus, in the $\mathbb{R}^2$ case, with $\mathbf{v} = t\mathbf{u}$, the formula in Taylor's Theorem can be written as

$$f(\mathbf{x}_0 + \mathbf{v}) = f(\mathbf{x}_0) + \left(\nabla f\Big|_{\mathbf{x}_0}\right) \cdot \mathbf{v} + \frac{1}{2}\mathbf{v}^T \left(\mathbf{H}\Big|_{\mathbf{x}_0 + k\mathbf{v}}\right)\mathbf{v},$$

for some k with $0 \le k \le 1$ (where $k = \frac{c}{t}$). While we have derived this result in $\mathbb{R}^2$, the same formula holds in $\mathbb{R}^n$, where the Hessian $\mathbf{H}$ is the matrix whose (i, j) entry is $\frac{\partial^2 f}{\partial x_i \partial x_j}$.

Critical Points

If A is a subset of $\mathbb{R}^n$, then we say that $f : A \rightarrow \mathbb{R}$ has a **local maximum** at a point $\mathbf{x}_0 \in A$ if and only if there is an open neighborhood $\mathcal{U}$ of $\mathbf{x}_0$ such that $f(\mathbf{x}_0) \ge f(\mathbf{x})$ for all $\mathbf{x} \in \mathcal{U}$. A **local minimum** for a function f is defined analogously.

> **THEOREM 8.10**
>
> Let A be an open hypersphere centered at $\mathbf{x}_0 \in \mathbb{R}^n$, and let $f : A \rightarrow \mathbb{R}$ have continuous first partial derivatives on A. If f has a local maximum or a local minimum at $\mathbf{x}_0$, then $\nabla f(\mathbf{x}_0) = \mathbf{0}$.

PROOF OF THEOREM 8.10

If $\mathbf{x}_0$ is a local maximum, then $f(\mathbf{x}_0 + h\mathbf{e}_i) - f(\mathbf{x}_0) \le 0$ for small h. Then, $\lim_{h \to 0^+} \frac{f(\mathbf{x}_0 + h\mathbf{e}_i) - f(\mathbf{x}_0)}{h} \le 0$. Similarly, $\lim_{h \to 0^-} \frac{f(\mathbf{x}_0 + h\mathbf{e}_i) - f(\mathbf{x}_0)}{h} \ge 0$. Hence, for the limit to exist, we must have $\frac{\partial f}{\partial x_i}\Big|_{\mathbf{x}_0} = 0$. Since this is true for each i, $\nabla f\Big|_{\mathbf{x}_0} = \mathbf{0}$. A similar proof works for local minimums. ∎

Points $\mathbf{x}_0$ at which $\nabla f(\mathbf{x}_0) = \mathbf{0}$ are called **critical points**.

EXAMPLE 1   Let $f : \mathbb{R}^2 \rightarrow \mathbb{R}$ be given by

$$f(x, y) = 7x^2 + 6xy + 2x + 7y^2 - 22y + 23.$$

Then $\nabla f = [14x + 6y + 2, \; 6x + 14y - 22]$. We find critical points for f by solving $\nabla f = \mathbf{0}$. This is the linear system

$$\begin{cases} 14x & + & 6y & + & 2 & = & 0 \\ 6x & + & 14y & - & 22 & = & 0 \end{cases},$$

which has the unique solution $\mathbf{x}_0 = [-1, 2]$. Hence, by Theorem 8.10, $(-1, 2)$ is the only possible extreme point for f. (We will see later that $(-1, 2)$ is a local minimum.) ∎

Sufficient Conditions for Local Extreme Points

If $\mathbf{x}_0$ is a critical point for a function f, how can we determine whether $\mathbf{x}_0$ is a local maximum or a local minimum? For functions on $\mathbb{R}$, we have the second derivative test from calculus, which says that if $f''(\mathbf{x}_0) < 0$, then $\mathbf{x}_0$ is a local maximum, but if $f''(\mathbf{x}_0) > 0$, then $\mathbf{x}_0$ is a local minimum. We now derive a similar test in $\mathbb{R}^n$.

Consider the following formula from Taylor's Theorem:

$$f(\mathbf{x}_0 + \mathbf{v}) = f(\mathbf{x}_0) + \nabla f(\mathbf{x}_0) \cdot \mathbf{v} + \frac{1}{2}\mathbf{v}^T \left(\mathbf{H}\Big|_{\mathbf{x}_0 + k\mathbf{v}} \right) \mathbf{v}.$$

At a critical point, $\nabla f(\mathbf{x}_0) = \mathbf{0}$, and so

$$f(\mathbf{x}_0 + \mathbf{v}) = f(\mathbf{x}_0) + \frac{1}{2}\mathbf{v}^T \left(\mathbf{H}\Big|_{\mathbf{x}_0 + k\mathbf{v}} \right) \mathbf{v}.$$

Hence, if $\mathbf{v}^T \left(\mathbf{H}\Big|_{\mathbf{x}_0 + k\mathbf{v}} \right) \mathbf{v}$ is positive for all small nonzero vectors $\mathbf{v}$, then f will have a local minimum at $\mathbf{x}_0$. (Similarly, if $\mathbf{v}^T \left(\mathbf{H}\Big|_{\mathbf{x}_0 + k\mathbf{v}} \right) \mathbf{v}$ is negative, f will have a local maximum.) But since we assume that f has continuous second partial derivatives, $\mathbf{v}^T \left(\mathbf{H}\Big|_{\mathbf{x}_0 + k\mathbf{v}} \right) \mathbf{v}$ is continuous in $\mathbf{v}$ and k, and will be positive for small $\mathbf{v}$ if $\mathbf{v}^T \left(\mathbf{H}\Big|_{\mathbf{x}_0} \right) \mathbf{v}$ is positive for all nonzero $\mathbf{v}$. Hence,

---

THEOREM 8.11

Given the conditions of Taylor's Theorem for a set A and a function $f : A \rightarrow \mathbb{R}$, f has a local minimum at a critical point $\mathbf{x}_0$ if $\mathbf{v}^T \left(\mathbf{H}\Big|_{\mathbf{x}_0} \right) \mathbf{v} > 0$ for all nonzero vectors $\mathbf{v}$. Similarly, f has a local maximum at a critical point $\mathbf{x}_0$ if $\mathbf{v}^T \left(\mathbf{H}\Big|_{\mathbf{x}_0} \right) \mathbf{v} < 0$ for all nonzero vectors $\mathbf{v}$.

Positive Definite Quadratic Forms

The expression $\mathbf{v}^T \left(\mathbf{H}\Big|_{\mathbf{x}_0} \right) \mathbf{v}$ in Theorem 8.11 is called a **quadratic form**. (For more details on the general theory of quadratic forms, see Section 10.3.) A quadratic form such that $\mathbf{v}^T \left(\mathbf{H}\Big|_{\mathbf{x}_0} \right) \mathbf{v} > 0$ for all nonzero vectors $\mathbf{v}$ is said to be **positive definite**. Theorem 8.11 says that if $\mathbf{v}^T \left(\mathbf{H}\Big|_{\mathbf{x}_0} \right) \mathbf{v}$ is a positive definite quadratic form at a critical point $\mathbf{x}_0$, then f has a local minimum at $\mathbf{x}_0$. So, how can we determine whether a quadratic form is positive definite?

Now, the Hessian matrix $\left(\mathbf{H}\Big|_{\mathbf{x}_0} \right)$, which we will abbreviate as $\mathbf{H}$, is symmetric because $\frac{\partial^2 f}{\partial x_i \partial x_j} = \frac{\partial^2 f}{\partial x_j \partial x_i}$ (since $f \in C^2(A)$). Hence, by Theorem 6.19, $\mathbf{H}$ can be orthogonally diagonalized. That is, there is an orthogonal matrix

P such that $\mathbf{PHP}^T = \mathbf{D}$, a diagonal matrix, and so, $\mathbf{H} = \mathbf{P}^T\mathbf{DP}$. Hence, $\mathbf{v}^T\mathbf{Hv} = \mathbf{v}^T\mathbf{P}^T\mathbf{DPv} = (\mathbf{Pv})^T\mathbf{D}(\mathbf{Pv})$. Letting $\mathbf{w} = \mathbf{Pv}$, we get $\mathbf{v}^T\mathbf{Hv} = \mathbf{w}^T\mathbf{Dw}$. But **P** is nonsingular, so as **v** ranges over all of $\mathbb{R}^n$, so does **w**, and vice-versa. Thus, $\mathbf{v}^T\mathbf{Hv} > 0$ for all nonzero **v** if and only if $\mathbf{w}^T\mathbf{Dw} > 0$ for all nonzero **w**. Now, **D** is diagonal, and so $\mathbf{w}^T\mathbf{Dw} = d_{11}w_1^2 + d_{22}w_2^2 + \cdots + d_{nn}w_n^2$. But the d_{ii}'s are the eigenvalues of **H**. Thus, it follows that $\mathbf{w}^T\mathbf{Dw} > 0$ for *all* nonzero **w** if and only if all of these eigenvalues are positive. (Set $\mathbf{w} = \mathbf{e}_i$ for each i to prove the "only if" part of this statement.) Hence,

---

THEOREM 8.12

A symmetric matrix **A** defines a positive definite quadratic form $\mathbf{v}^T\mathbf{Av}$ if and only if all of the eigenvalues of **A** are positive.

Hence, Theorem 8.11 can be restated as follows: Given the conditions of Taylor's Theorem for a set A and a function $f : A \to \mathbb{R}$, if all of the eigenvalues of **H** are positive at a critical point $\mathbf{x}_0$, then f has a local minimum at $\mathbf{x}_0$.

EXAMPLE 2   Consider the function

$$f(x, y) = 7x^2 + 6xy + 2x + 7y^2 - 22y + 23.$$

In Example 1, we found that f has a critical point at $\mathbf{x}_0 = [-1, 2]$. Now, the Hessian matrix for f at $\mathbf{x}_0$ is

$$\mathbf{H} = \left. \begin{bmatrix} \frac{\partial^2 f}{\partial x^2} & \frac{\partial^2 f}{\partial x \partial y} \\ \frac{\partial^2 f}{\partial y \partial x} & \frac{\partial^2 f}{\partial y^2} \end{bmatrix} \right|_{\mathbf{x}_0} = \begin{bmatrix} 14 & 6 \\ 6 & 14 \end{bmatrix}.$$

But $p_{\mathbf{H}}(x) = x^2 - 28x + 160$, which has roots $x = 8$ and $x = 20$. Thus, **H** has all eigenvalues positive, and hence, $\mathbf{v}^T\mathbf{Hv}$ is positive definite. Theorem 8.12 then tells us that $\mathbf{x}_0 = [-1, 2]$ is a local minimum for f.   ∎

Local Maxima and Minima in $\mathbb{R}^2$

It can be shown (see Exercise 3) that a 2×2 symmetric matrix **A** defines a positive definite quadratic form if and only if $a_{11} > 0$ and $|\mathbf{A}| > 0$. Similarly, a 2×2 symmetric matrix defines a **negative definite quadratic form** ($\mathbf{v}^T\mathbf{Av} < 0$ for all nonzero **v**) if and only if $a_{11} < 0$ and $|\mathbf{A}| > 0$.

EXAMPLE 3   Suppose $f(x, y) = 2x^2 - 2x^2y^2 + 2y^2 + 24y - x^4 - y^4$. First, we look for critical points by solving the system

$$\begin{cases} \frac{\partial f}{\partial x} = 4x - 4xy^2 - 4x^3 = 4x(1 - (y^2 + x^2)) = 0 \\\\ \frac{\partial f}{\partial y} = -4x^2y + 4y + 24 - 4y^3 = -4y(x^2 + y^2) + 4y + 24 = 0 \end{cases}$$

Now $\frac{\partial f}{\partial x} = 0$ yields $x = 0$ or $y^2 + x^2 = 1$. If $x = 0$, then $\frac{\partial f}{\partial y} = 0$ gives $4y + 24 - 4y^3 = 0$. The unique real solution to this equation is $y = 2$. Thus, $[0, 2]$ is a critical point.

If $x \neq 0$, then $y^2 + x^2 = 1$. From $\frac{\partial f}{\partial y} = 0$, we have $0 = -4y(1) + 4y + 24 = 24$, a contradiction, so there is no critical point when $x \neq 0$.

Next, we compute the Hessian matrix at the critical point $[0, 2]$.

$$
\mathbf{H} \;=\; \left[\begin{array}{cc} \frac{\partial^2 f}{\partial x^2} & \frac{\partial^2 f}{\partial x \partial y} \\[2mm] \frac{\partial^2 f}{\partial y \partial x} & \frac{\partial^2 f}{\partial y^2} \end{array} \right]\Bigg|_{[0,2]}
$$

$$
= \left[\begin{array}{cc} 4 - 4y^2 - 12x^2 & -8xy \\[2mm] -8xy & -4x^2 + 4 - 12y^2 \end{array} \right]\Bigg|_{[0,2]} = \left[\begin{array}{cc} -12 & 0 \\ 0 & -44 \end{array} \right].
$$

Since the $(1, 1)$ entry is negative and $|\mathbf{H}| > 0$, $\mathbf{H}$ defines a negative definite quadratic form and so f has a local maximum at $[0, 2]$. ∎

An Example in $\mathbb{R}^3$

EXAMPLE 4   Consider the function

$$
g(x, y, z) = 5x^2 + 2xz + 4xy + 10x + 3z^2 - 6yz - 6z + 5y^2 + 12y + 21.
$$

We find the critical points by solving the system

$$
\begin{cases}
\frac{\partial g}{\partial x} = 10x + 2z + 4y + 10 = 0 \\[2mm]
\frac{\partial g}{\partial y} = 4x - 6z + 10y + 12 = 0 \\[2mm]
\frac{\partial g}{\partial z} = 2x + 6z - 6y - 6 = 0
\end{cases}.
$$

Using row reduction to solve this linear system yields the unique critical point $[-9, 12, 16]$. The Hessian matrix at $[-9, 12, 16]$ is

$$
\mathbf{H} = \left[\begin{array}{ccc} \frac{\partial^2 g}{\partial x^2} & \frac{\partial^2 g}{\partial x \partial y} & \frac{\partial^2 g}{\partial x \partial z} \\[2mm] \frac{\partial^2 g}{\partial y \partial x} & \frac{\partial^2 g}{\partial y^2} & \frac{\partial^2 g}{\partial y \partial z} \\[2mm] \frac{\partial^2 g}{\partial z \partial x} & \frac{\partial^2 g}{\partial z \partial y} & \frac{\partial^2 g}{\partial z^2} \end{array} \right]\Bigg|_{[-9,12,16]} = \left[\begin{array}{ccc} 10 & 4 & 2 \\ 4 & 10 & -6 \\ 2 & -6 & 6 \end{array} \right].
$$

A lengthy computation produces $p_{\mathbf{H}}(x) = x^3 - 26x^2 + 164x - 8$. The roots of $p_{\mathbf{H}}(x)$ are approximately 0.04916, 10.6011, and 15.3497. Since all of these eigenvalues for $\mathbf{H}$ are positive, $[-9, 12, 16]$ is a local minimum for g. ∎

Failure of the Hessian Matrix Test

In calculus, we discovered that the second derivative test fails when the second derivative is zero at a critical point. A similar situation is true in $\mathbb{R}^n$. If the Hessian matrix at a critical point has 0 as an eigenvalue, and all other eigenvalues have the same sign, then the function f could have a local maximum, a local minimum, or neither at this critical point. Of course, if the Hessian matrix at a critical point has two eigenvalues with opposite signs, the critical point is not a local extreme point (why?). Exercise 2 illustrates these   concepts.

Exercises for Section 8.11

1. In each part, solve for all critical points for the given function. Then, for each critical point, use the Hessian matrix to determine whether the critical point is a local maximum, a local minimum, or neither.
   ⋆(a) $f(x, y) = x^3 + x^2 + 2xy - 3x + y^2$
    (b) $f(x, y) = 6x^2 + 4xy + 3y^2 + 8x - 9y$
   ⋆(c) $f(x, y) = 2x^2 + 2xy + 2x + y^2 - 2y + 5$
    (d) $f(x, y) = x^3 + 3x^2y - x^2 + 3xy^2 + 2xy - 3x + y^3 - y^2 - 3y$ (Hint: To solve for critical points, first set $\frac{\partial f}{\partial x} - \frac{\partial f}{\partial y} = 0$.)
   ⋆(e) $f(x, y, z) = 2x^2 + 2xy + 2xz + y^4 + 4y^3z + 6y^2z^2 - y^2 + 4yz^3 - 4yz + z^4 - z^2$

2. (a) Show that $f(x, y) = (x - 2)^4 + (y - 3)^2$ has a local minimum at $[2, 3]$, but its Hessian matrix at $[2, 3]$ has 0 as an eigenvalue.
   (b) Show that $f(x, y) = -(x - 2)^4 + (y - 3)^2$ has a critical point at $[2, 3]$, its Hessian matrix at $[2, 3]$ has all nonnegative eigenvalues, but $[2, 3]$ is not a local extreme point for f.
   (c) Show that $f(x, y) = -(x + 1)^4 - (y + 2)^4$ has a local maximum at $[-1, -2]$, but its Hessian matrix at $[-1, -2]$ is $\mathbf{O}$ and thus has all of its eigenvalues equal to zero.
   (d) Show that $f(x, y, z) = (x - 1)^2 - (y - 2)^2 + (z - 3)^4$ does not have any local extreme points. Then verify that its Hessian matrix has eigenvalues of opposite sign at the function's only critical point.

3. (a) Prove that a symmetric 2×2 matrix $\mathbf{A} = \begin{bmatrix} a & b \\ b & c \end{bmatrix}$ defines a positive definite quadratic form if and only if $a > 0$ and $|\mathbf{A}| > 0$. (Hint: Compute $p_{\mathbf{A}}(x)$ and show that both roots are positive if and only if $a > 0$ and $|\mathbf{A}| > 0$.)
   (b) Prove that a symmetric 2×2 matrix $\mathbf{A}$ defines a negative definite quadratic form if and only if $a_{11} < 0$ and $|\mathbf{A}| > 0$.

⋆4. True or False:
   (a) If $f : \mathbb{R}^n \to \mathbb{R}$ has continuous second partial derivatives, then the Hessian matrix is symmetric.
   (b) Every symmetric matrix $\mathbf{A}$ defines either a positive definite or a negative definite quadratic form.

(c) A Hessian matrix for a function with continuous second partial derivatives evaluated at any point is diagonalizable.

(d) $\mathbf{v}^T \begin{bmatrix} 5 & 3 \\ 3 & 2 \end{bmatrix} \mathbf{v}$ is a positive definite quadratic form.

(e) $\mathbf{v}^T \begin{bmatrix} 3 & 0 & 0 \\ 0 & -9 & 0 \\ 0 & 0 & 4 \end{bmatrix} \mathbf{v}$ is a positive definite quadratic form.

Chapter 9

Numerical Methods

A Calculating Mindset

Although we have focused on many theoretical results in this book, computation is also an extremely important part of mathematics. Some mathematical problems that can not be solved with perfect precision can be solved numerically to within a certain specified margin of error. As an example, recall that the familiar Quadratic Formula allows us to find the roots of every possible second-degree polynomial. However, because of the work of mathematicians such as Niels Abel and Evariste Galois in the first half of the nineteenth century, we now know that no formula analogous to the Quadratic Formula exists that computes the exact roots for every given fifth-degree polynomial. Nevertheless, there are numerous computational techniques that can be used to approximate the roots of any polynomial to any desired degree of accuracy.

In this chapter, we present several computational techniques for solving linear systems and for approximating eigenvalues. While these methods do not always give the exact solution, they usually provide adequate enough answers for most practical purposes.

T hroughout the book, we have urged you to use a calculator or computer with appropriate software to perform tedious calculations after you have mastered a computational technique. A calculator or computer is especially useful when solving a linear system or when finding eigenvalues and eigenvectors for a linear operator. In this chapter, we discuss additional numerical methods for solving systems and finding eigenvalues that are best suited for the calculator or computer. If you have some programming experience, you should find it a straightforward task to write your own programs to implement these algorithms.

9.1 Numerical Methods for Solving Systems

Prerequisite: Section 2.3, Equivalent Systems, Rank, and Row Space

In this section, we discuss some considerations for solving linear systems by calculator or computer and investigate some alternate methods for solving systems, including partial pivoting, the Jacobi Method, and the Gauss-Seidel Method.

Computational Accuracy

One basic problem in using a computational device in linear algebra is that real numbers cannot always be represented exactly in its memory. Because the physical storage space of any device is limited, a predetermined amount of space is assigned in the memory for the storage of any real number. Thus, only the most significant digits of any real number can be stored.[1] Nonterminating decimals, such as $\frac{1}{3} = 0.333333\ldots$ or $e = 2.718281828459045\ldots$, can never be represented fully. Using the first few decimal places of such numbers may be enough for most practical purposes, but it is not completely accurate.

As calculations are performed, all computational results are truncated and rounded to fit within the limited storage space allotted. Numerical errors caused by this process are called **roundoff errors**. Unfortunately, if many operations are performed, roundoff errors can compound, thus producing a significant error in the final result. This is one reason that Gaussian elimination is computationally more accurate than the Gauss-Jordan Method. Since fewer

---

[1] The first n significant digits of a decimal number are its leftmost n digits, beginning with the first nonzero digit. For example, consider the real numbers $r_1 = 47.26835$, $r_2 = 9.00473$, and $r_3 = 0.000456$. Approximating these by stopping after the first three significant digits and rounding to the nearest digit, we get $r_1 \approx 47.3$, $r_2 \approx 9.00$, and $r_3 \approx 0.000456$ (since the first nonzero digit in r_3 is 4).

arithmetic operations generally need to be performed, Gaussian elimination allows less chance for roundoff errors to compound.

Ill-Conditioned Systems

Sometimes the number of significant digits used in computations has a great effect on the answers. For example, consider the similar systems

$$(A) \begin{cases} 2x_1 + x_2 = 2 \\ 2.005x_1 + x_2 = 7 \end{cases} \quad \text{and} \quad (B) \begin{cases} 2x_1 + x_2 = 2 \\ 2.01x_1 + x_2 = 7 \end{cases}.$$

The linear equations of these systems are graphed in Figure 9.1.

Even though the coefficients of systems (A) and (B) are almost identical, the solutions to the systems are very different.

Solution to (A) $= (1000, -1998)$ and solution to (B) $= (500, -998)$.

Systems like these, in which a very small change in a coefficient leads to a very large change in the solution set, are called **ill-conditioned systems**. In this case, there is a geometric way to see that these systems are ill-conditioned; the pair of lines in each system are almost parallel. Therefore, a small change in one line can move the point of intersection very far along the other line, as in Figure 9.1.

Figure 9.1

(a) Lines of system (A);
(b) lines of system (B)

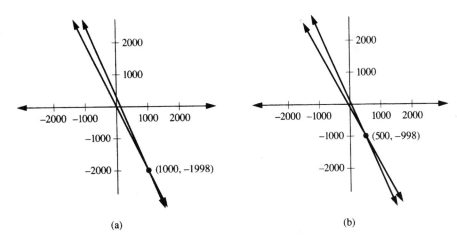

(a)　　　　　(b)

Suppose the coefficients in system (A) had been obtained after a series of long calculations. A slight difference in the roundoff error of those calculations could have led to a very different final solution set. Thus, we need to be very careful when working with ill-conditioned systems. Special methods have been developed for recognizing ill-conditioned systems, and a technique known as

iterative refinement is used when the coefficients are known only to a certain degree of accuracy. These methods are beyond the scope of this book, but further details can be found in *Numerical Analysis*, 7th ed., by Burden and Faires (published by Brooks/Cole, 2001).

Partial Pivoting

A common problem in numerical linear algebra occurs when dividing by real numbers that are very close to zero — for example, during the row reduction process when a pivot element is extremely small. This small number might be inaccurate itself because of a previous roundoff error. Performing a type (I) row operation with this number might result in additional roundoff error.

Even when dealing with accurate small numbers, we can still have problems. When we divide every entry of a row by a very small pivot value, the remaining row entries could become much larger (in absolute value) than the other matrix entries. Then, when these larger row entries are added to the (smaller) entries of another row in a type (II) operation, the most significant digits of the larger row entries may not be affected at all. That is, the data stored in smaller row entries may not be playing their proper role in determining the final solution set. As more computations are performed, these roundoff errors can accumulate, making the final result inaccurate.

EXAMPLE 1

Consider the linear system

$$\begin{cases} 0.0006x_1 & - & x_2 & + & x_3 & = & 10 \\ 0.03x_1 & + & 30x_2 & - & 5x_3 & = & 15 \\ 0.04x_1 & + & 40x_2 & - & 7x_3 & = & 19 \end{cases}.$$

The unique solution is $(x_1, x_2, x_3) = (5000, -4, 3)$. But if we attempt to solve the system by row reduction and round all computations to four significant figures, we get an inaccurate result. For example, using Gaussian elimination, the augmented matrices are

$$\begin{bmatrix} 0.0006 & -1 & 1 & | & 10 \\ 0.03 & 30 & -5 & | & 15 \\ 0.04 & 40 & -7 & | & 19 \end{bmatrix}$$

(I): $\langle 1 \rangle \leftarrow (1/0.0006)\,\langle 1 \rangle$
$$\begin{bmatrix} 1 & -1667 & 1667 & | & 16670 \\ 0.03 & 30 & -5 & | & 15 \\ 0.04 & 40 & -7 & | & 19 \end{bmatrix}$$

(II): $\langle 2 \rangle \leftarrow -0.03\,\langle 1 \rangle + \langle 2 \rangle$
(II): $\langle 3 \rangle \leftarrow -0.04\,\langle 1 \rangle + \langle 3 \rangle$
$$\begin{bmatrix} 1 & -1667 & 1667 & | & 16670 \\ 0 & 80.01 & -55.01 & | & -485.1 \\ 0 & 106.7 & -73.68 & | & -647.8 \end{bmatrix}$$

$$(I): \langle 2 \rangle \leftarrow (1/80.01) \langle 2 \rangle \qquad \left[\begin{array}{ccc|c} 1 & -1667 & 1667 & 16670 \\ 0 & 1 & -0.6876 & -6.064 \\ 0 & 106.7 & -73.68 & -647.8 \end{array} \right]$$

$$(II): \langle 3 \rangle \leftarrow -106.7 \langle 2 \rangle + \langle 3 \rangle \qquad \left[\begin{array}{ccc|c} 1 & -1667 & 1667 & 16670 \\ 0 & 1 & -0.6876 & -6.064 \\ 0 & 0 & -0.3131 & -0.7712 \end{array} \right]$$

$$(I): \langle 3 \rangle \leftarrow (-1/0.3131) + \langle 3 \rangle \qquad \left[\begin{array}{ccc|c} 1 & -1667 & 1667 & 16670 \\ 0 & 1 & -0.6876 & -6.064 \\ 0 & 0 & 1 & 2.463 \end{array} \right].$$

Back substitution produces the solution $(x_1, x_2, x_3) = (5279, -4.370, 2.463)$. This inaccurate answer is largely the result of dividing row 1 through by 0.0006, a number much smaller than the other entries of the matrix, in the first step of the row reduction. ■

A method known as **partial pivoting** is employed to avoid roundoff errors like those encountered in Example 1. In this method, when choosing a pivot element, we first determine whether there are any entries below the next pivot candidate that have a greater absolute value. If so, we switch rows to move the entry with the highest absolute value into the pivot position.

EXAMPLE 2   We use partial pivoting on the system in Example 1. The initial augmented matrix is

$$\left[\begin{array}{ccc|c} 0.0006 & -1 & 1 & 10 \\ 0.03 & 30 & -5 & 15 \\ 0.04 & 40 & -7 & 19 \end{array} \right].$$

The entry in the first column with the largest absolute value is in the third row, so we interchange the first and third rows to obtain

$$(III): \langle 1 \rangle \leftrightarrow \langle 3 \rangle \qquad \left[\begin{array}{ccc|c} 0.04 & 40 & -7 & 19 \\ 0.03 & 30 & -5 & 15 \\ 0.0006 & -1 & 1 & 10 \end{array} \right].$$

Continuing the row reduction, we obtain

$$(I): \langle 1 \rangle \leftarrow (1/0.04) \langle 1 \rangle \qquad \left[\begin{array}{ccc|c} 1 & 1000 & -175.0 & 475.0 \\ 0.03 & 30 & -5 & 15 \\ 0.0006 & -1 & 1 & 10 \end{array} \right]$$

$$\begin{array}{l} (II): \langle 2 \rangle \leftarrow -0.03 \langle 1 \rangle + \langle 2 \rangle \\ (II): \langle 3 \rangle \leftarrow -0.0006 \langle 1 \rangle + \langle 3 \rangle \end{array} \qquad \left[\begin{array}{ccc|c} 1 & 1000 & -175.0 & 475.0 \\ 0 & 0.000 & 0.2500 & 0.7500 \\ 0 & -1.600 & 1.105 & 9.715 \end{array} \right]$$

$$(\text{III}): \langle 2 \rangle \longleftrightarrow \langle 3 \rangle \qquad \begin{bmatrix} 1 & 1000 & -175.0 & 475.0 \\ 0 & -1.600 & 1.105 & 9.715 \\ 0 & 0 & 0.2500 & 0.7500 \end{bmatrix}$$

$$(\text{I}): \langle 2 \rangle \leftarrow (-1/1.600) \langle 2 \rangle \qquad \begin{bmatrix} 1 & 1000 & -175.0 & 475.0 \\ 0 & 1 & -0.6906 & -6.072 \\ 0 & 0 & 0.2500 & 0.7500 \end{bmatrix}$$

$$(\text{I}): \langle 3 \rangle \leftarrow (1/0.2500) \langle 3 \rangle \qquad \begin{bmatrix} 1 & 1000 & -175.0 & 475.0 \\ 0 & 1 & -0.6906 & -6.072 \\ 0 & 0 & 1 & 3.000 \end{bmatrix}.$$

Back substitution produces the solution $(x_1, x_2, x_3) = (5000, -4.000, 3.000)$. Therefore, by partial pivoting, we have obtained the correct solution, a big improvement over the answer obtained in Example 1 without partial pivoting. ∎

For many systems, the method of partial pivoting is powerful enough to provide reasonably accurate answers. However, in more difficult cases, partial pivoting is not enough. An even more useful technique is **total pivoting** (also called **full pivoting** or **complete pivoting**), in which columns as well as rows are interchanged. The strategy in total pivoting is to select the entry with the largest absolute value from all the remaining rows and columns to be the next pivot.

Iterative Techniques: Jacobi and Gauss-Seidel Methods

When we have a rough approximation of the unique solution to a certain $n \times n$ linear system, an **iterative method** may be the fastest way to obtain the actual solution. We use the initial approximation to generate a second (preferably better) approximation. We then use the second approximation to generate a third, and so on. The process stops if the approximations "stabilize" — that is, if the difference between successive approximations becomes negligible. In this section, we illustrate the following two iterative methods: the **Jacobi Method** and the **Gauss-Seidel Method**.

For these iterative methods, it is convenient to express linear systems in a slightly different form. Suppose we are given the following system of n equations in n unknowns:

$$\begin{cases} a_{11}x_1 + a_{12}x_2 + a_{13}x_3 + \cdots + a_{1n}x_n = b_1 \\ a_{21}x_1 + a_{22}x_2 + a_{23}x_3 + \cdots + a_{2n}x_n = b_2 \\ \vdots \qquad \vdots \qquad \vdots \qquad \vdots \qquad \vdots \qquad \vdots \\ a_{n1}x_1 + a_{n2}x_2 + a_{n3}x_3 + \cdots + a_{nn}x_n = b_n \end{cases}.$$

If the coefficient matrix has rank n, every row and column of the reduced row echelon form of the coefficient matrix contains a (nonzero) pivot. In this case, it is always possible to rearrange the equations so that the coefficient of x_i is nonzero in the ith equation, for $1 \le i \le n$. Let us assume that the equations have already been rearranged in this way.[2] Solving for x_i in the ith equation in terms of the remaining unknowns, we obtain

$$
\begin{cases}
x_1 = & & c_{12}x_2 & + & c_{13}x_3 & + \cdots + & c_{1n}x_n & + & d_1 \\
x_2 = & c_{21}x_1 & & + & c_{23}x_3 & + \cdots + & c_{2n}x_n & + & d_2 \\
x_3 = & c_{31}x_1 & + & c_{32}x_2 & & + \cdots + & c_{3n}x_n & + & d_3 \\
\vdots & \vdots & & \vdots & & \vdots \quad \vdots & & & \vdots \\
x_n = & c_{n1}x_1 & + & c_{n2}x_2 & + & c_{n3}x_3 & + \cdots & & + & d_n
\end{cases},
$$

where each c_{ij} and d_i represents a new coefficient obtained after we reorder the equations and solve for each x_i.

For example, suppose we are given the system

$$
\begin{cases}
3x_1 & - & 2x_2 & + & x_3 & = & 11 \\
2x_1 & + & 7x_2 & - & 3x_3 & = & -14 \\
9x_1 & - & x_2 & - & 4x_3 & = & 17
\end{cases}.
$$

Solving for x_1 in the first equation, x_2 in the second equation, and x_3 in the third equation, we obtain

$$
\begin{cases}
x_1 & = & & \frac{2}{3}x_2 & - & \frac{1}{3}x_3 & + & \frac{11}{3} \\
x_2 & = & -\frac{2}{7}x_1 & & + & \frac{3}{7}x_3 & - & 2 \\
x_3 & = & \frac{9}{4}x_1 & - & \frac{1}{4}x_2 & & - & \frac{17}{4}
\end{cases}.
$$

For the **Jacobi Method**, we solve a system in the form

$$
\begin{cases}
x_1 = & & c_{12}x_2 & + & c_{13}x_3 & + \cdots + & c_{1n}x_n & + & d_1 \\
x_2 = & c_{21}x_1 & & + & c_{23}x_3 & + \cdots + & c_{2n}x_n & + & d_2 \\
x_3 = & c_{31}x_1 & + & c_{32}x_2 & & + \cdots + & c_{3n}x_n & + & d_3 \\
\vdots & \vdots & & \vdots & & \vdots \quad \vdots & & & \vdots \\
x_n = & c_{n1}x_1 & + & c_{n2}x_2 & + & c_{n3}x_3 & + \cdots & & + & d_n
\end{cases}
$$

by substituting an initial approximation for $x_1, x_2, \ldots, x_n$ into the right-hand side to obtain new values for $x_1, x_2, \ldots, x_n$ on the left-hand side. These new values are then substituted into the right-hand side to obtain another set of values for $x_1, x_2, \ldots, x_n$ on the left-hand side. This process is repeated as many times as necessary. If the values on the left-hand side "stabilize," they are a good approximation for a solution.

EXAMPLE 3 We solve

$$
\begin{cases}
8x_1 & + & x_2 & - & 2x_3 & = & -11 \\
2x_1 & + & 9x_2 & + & x_3 & = & 22 \\
-x_1 & - & 2x_2 & + & 11x_3 & = & -15
\end{cases}
$$

---

[2] In fact, the Jacobi and Gauss-Seidel Methods often require fewer steps if the equations are rearranged so that the coefficient of x_i in the ith row is as large as possible.

with the Jacobi Method. The true solution is $(x_1, x_2, x_3) = (-2, 3, -1)$. Let us use $x_1 = -1.5$, $x_2 = 2.5$, and $x_3 = -0.5$ as an initial approximation (or guess) of the solution.

First, we rewrite the system in the form

$$\begin{cases} x_1 = -\frac{1}{8}x_2 + \frac{1}{4}x_3 - \frac{11}{8} \\ x_2 = -\frac{2}{9}x_1 - \frac{1}{9}x_3 + \frac{22}{9} \\ x_3 = \frac{1}{11}x_1 + \frac{2}{11}x_2 - \frac{15}{11} \end{cases}.$$

In the following calculations, we round all results to three decimal places. Plugging the initial guess into the right-hand side of each equation, we get

$$\begin{cases} x_1 = -\frac{1}{8}(2.5) + \frac{1}{4}(-0.5) - \frac{11}{8} \\ x_2 = -\frac{2}{9}(-1.5) - \frac{1}{9}(-0.5) + \frac{22}{9} \\ x_3 = \frac{1}{11}(-1.5) + \frac{2}{11}(2.5) - \frac{15}{11} \end{cases},$$

yielding the new values $x_1 = -1.813$, $x_2 = 2.833$, $x_3 = -1.045$. We then plug these values into the right-hand side of each equation to obtain

$$\begin{cases} x_1 = -\frac{1}{8}(2.833) + \frac{1}{4}(-1.045) - \frac{11}{8} \\ x_2 = -\frac{2}{9}(-1.813) - \frac{1}{9}(-1.045) + \frac{22}{9} \\ x_3 = \frac{1}{11}(-1.813) + \frac{2}{11}(2.833) - \frac{15}{11} \end{cases},$$

yielding the values $x_1 = -1.990$, $x_2 = 2.963$, $x_3 = -1.013$. Repeating this process, we get the values in the following chart:

	x_1	x_2	x_3
Initial values	−1.500	2.500	−0.500
After 1 step	−1.813	2.833	−1.045
After 2 steps	−1.990	2.963	−1.013
After 3 steps	−1.999	2.999	−1.006
After 4 steps	−2.001	3.000	−1.000
After 5 steps	−2.000	3.000	−1.000
After 6 steps	−2.000	3.000	−1.000

After six steps, the values for x_1, x_2, and x_3 have stabilized at the true solution. ■

In Example 3, we could have used any starting values for x_1, x_2, and x_3 as the initial approximation. In the absence of any information about the solution, we can begin with $x_1 = x_2 = x_3 = 0$. If we use the Jacobi Method on the system in

Example 3 with $x_1 = x_2 = x_3 = 0$ as the initial values, we obtain the following chart (again, rounding each result to three decimal places):

	x_1	x_2	x_3
Initial values	0.000	0.000	0.000
After 1 step	−1.375	2.444	−1.364
After 2 steps	−2.022	2.902	−1.044
After 3 steps	−1.999	3.010	−1.020
After 4 steps	−2.006	3.002	−0.998
After 5 steps	−2.000	3.001	−1.000
After 6 steps	−2.000	3.000	−1.000
After 7 steps	−2.000	3.000	−1.000

In this case, the Jacobi Method still produces the correct solution, although an extra step is required.

The **Gauss-Seidel Method** is similar to the Jacobi Method except that as each new value x_i is obtained, it is used immediately in place of the previous value for x_i when plugging values into the right-hand side of the equations.

EXAMPLE 4 Consider the system

$$\begin{cases} 8x_1 + x_2 - 2x_3 = -11 \\ 2x_1 + 9x_2 + x_3 = 22 \\ -x_1 - 2x_2 + 11x_3 = -15 \end{cases}$$

of Example 3. We solve this system with the Gauss-Seidel Method, using the initial approximation $x_1 = x_2 = x_3 = 0$. Again, we begin by rewriting the system in the form

$$\begin{cases} x_1 = -\frac{1}{8}x_2 + \frac{1}{4}x_3 - \frac{11}{8} \\ x_2 = -\frac{2}{9}x_1 - \frac{1}{9}x_3 + \frac{22}{9} \\ x_3 = \frac{1}{11}x_1 + \frac{2}{11}x_2 - \frac{15}{11} \end{cases}.$$

Plugging the initial approximation into the right-hand side of the first equation, we get

$$x_1 = -\tfrac{1}{8}(0) + \tfrac{1}{4}(0) - \tfrac{11}{8} = -1.375.$$

We now plug this new value for x_1 and the current values for x_2 and x_3 into the right-hand side of the second equation to get

$$x_2 = -\tfrac{2}{9}(-1.375) - \tfrac{1}{9}(0) + \tfrac{22}{9} = 2.750.$$

We then plug the new values for x_1 and x_2 and the current value for x_3 into the right-hand side of the third equation to get

$$x_3 = \tfrac{1}{11}(-1.375) + \tfrac{2}{11}(2.750) - \tfrac{15}{11} = -0.989.$$

The process is then repeated as many times as necessary with the newest values of x_1, x_2, and x_3 used in each case. The results are given in the following chart (rounding all results to three decimal places):

	x_1	x_2	x_3
Initial values	0.000	0.000	0.000
After 1 step	−1.375	2.750	−0.989
After 2 steps	−1.966	2.991	−0.999
After 3 steps	−1.999	3.000	−1.000
After 4 steps	−2.000	3.000	−1.000
After 5 steps	−2.000	3.000	−1.000

After five steps, we see that the values for x_1, x_2, and x_3 have stabilized to the correct solution.   ■

For certain classes of linear systems, the Jacobi and Gauss-Seidel Methods will always stabilize to the correct solution for any given initial approximation (see Exercise 7). In most ordinary applications, the Gauss-Seidel Method takes fewer steps than the Jacobi Method, but for some systems, the Jacobi Method is superior to the Gauss-Seidel Method. However, for other systems, neither method produces the correct answer (see Exercise 8).[3]

Comparing Iterative and Row Reduction Methods

When are iterative methods useful? A major advantage of iterative methods is that roundoff errors are not given a chance to "accumulate," as they are in Gaussian elimination and the Gauss-Jordan Method, because each iteration essentially creates a new approximation to the solution. The only roundoff error that we need to consider with an iterative method is the error involved in the most recent step.

Also, in many applications, the coefficient matrix for a given system contains a large number of zeroes. Such matrices are said to be **sparse**. When a linear system has a sparse matrix, each equation in the system may involve very few variables. If so, each step of the iterative process is relatively easy. However, neither the Gauss-Jordan Method nor Gaussian elimination would be very attractive in such a case because the cumulative effect of many row operations would tend to replace the zero coefficients with nonzero numbers. But even if the coefficient matrix is not sparse, iterative methods often give more accurate answers when large matrices are involved because fewer arithmetic operations are performed overall.

On the other hand, when iterative methods take an extremely large number of steps to stabilize or do not stabilize at all, it is much better to use the Gauss-Jordan Method or Gaussian elimination.

---

[3] In cases where the Jacobi and Gauss-Seidel Methods do not stabilize, related iterative methods (known as **relaxation methods**) may still work. For further details, see *Numerical Analysis*, 7th ed., by Burden and Faires (published by Brooks/Cole, 2001).

Exercises for Section 9.1

Note: You should use a calculator or appropriate computer software to solve these problems.

1. In each part of this exercise, find the exact solution sets for the two given systems. Are the systems ill-conditioned? Why or why not?

   ⋆(a) $\begin{cases} 5x - 2y = 10 \\ 5x - 1.995y = 17.5 \end{cases}$, $\begin{cases} 5x - 2y = 10 \\ 5x - 1.99y = 17.5 \end{cases}$

   (b) $\begin{cases} 6x - z = 400 \\ 3y - z = 400 \\ 25x + 12y - 8z = 3600 \end{cases}$, $\begin{cases} 6x - 1.01z = 400 \\ 3y - z = 400 \\ 25x + 12y - 8z = 3600 \end{cases}$

2. First, use Gaussian elimination *without* partial pivoting to solve each of the following systems. Then, solve each system using Gaussian elimination *with* partial pivoting. Which solution is more accurate? In each case, round all numbers in the problem to three significant digits before beginning, and round the results after each row operation is performed to three significant digits.

   ⋆(a) $\begin{cases} 0.00072x - 4.312y = -0.9846 \\ 2.31x - 9876.0y = -130.8 \end{cases}$

   (b) $\begin{cases} 0.0004x_1 - 0.6234x_2 - 2.123x_3 = 5.581 \\ 0.0832x_1 - 26.17x_2 - 1.759x_3 = -3.305 \\ 0.09512x_1 + 0.1458x_2 + 55.13x_3 = 11.168 \end{cases}$

   ⋆(c) $\begin{cases} 0.00032x_1 + 0.2314x_2 + 0.127x_3 = -0.03456 \\ -241x_1 - 217x_2 - 8x_3 = -576 \\ 49x_1 + 45x_2 + 2.4x_3 = 283.2 \end{cases}$

3. Repeat Exercise 2, but round all computations to four significant digits.

4. Solve each of the following systems using the Jacobi Method. Round all results to three decimal places, and stop when successive values of the variables agree to three decimal places. Let the initial values of all variables be zero. List the values of the variables after each step of the iteration.

   ⋆(a) $\begin{cases} 5x_1 + x_2 = 26 \\ 3x_1 + 7x_2 = -42 \end{cases}$

   (b) $\begin{cases} 9x_1 - x_2 - x_3 = -7 \\ 2x_1 - 8x_2 - x_3 = 35 \\ x_1 + 2x_2 + 11x_3 = 22 \end{cases}$

   ⋆(c) $\begin{cases} 7x_1 + x_2 - 2x_3 = -62 \\ -x_1 + 6x_2 + x_3 = 27 \\ 2x_1 - x_2 - 6x_3 = 26 \end{cases}$

   (d) $\begin{cases} 10x_1 + x_2 - 2x_3 + x_4 = 9 \\ -x_1 - 9x_2 + x_3 - 2x_4 = 15 \\ -2x_1 + x_2 + 7x_3 + x_4 = 21 \\ x_1 - x_2 - x_3 + 13x_4 = -27 \end{cases}$

5. Repeat Exercise 4 using the Gauss-Seidel Method instead of the Jacobi Method.

★6. A square matrix is **strictly diagonally dominant** if the absolute value of each diagonal entry is larger than the sum of the absolute values of the remaining entries in its row. That is, if $\mathbf{A}$ is an $n \times n$ matrix, then $\mathbf{A}$ is strictly diagonally dominant if, for $1 \le i \le n$, $|a_{ii}| > \sum_{\substack{1 \le j \le n \\ j \ne i}} |a_{ij}|$.

Which of the following matrices are strictly diagonally dominant?

(a) $\begin{bmatrix} -3 & 1 \\ -2 & 4 \end{bmatrix}$
(b) $\begin{bmatrix} 2 & 2 \\ 4 & 3 \end{bmatrix}$

(c) $\begin{bmatrix} -6 & 2 & 1 \\ 2 & 5 & -2 \\ -1 & 4 & 7 \end{bmatrix}$
(d) $\begin{bmatrix} 15 & 9 & -3 \\ 3 & 6 & 4 \\ 7 & -2 & 11 \end{bmatrix}$

(e) $\begin{bmatrix} 6 & 2 & 3 \\ 4 & 5 & 1 \\ 7 & 1 & 9 \end{bmatrix}$

7. The Jacobi and Gauss-Seidel Methods stabilize to the correct solution (for any choice of initial values) if the equations can be rearranged to make the coefficient matrix for the system strictly diagonally dominant (see Exercise 6). For the following systems, rearrange the equations accordingly, and then perform the Gauss-Seidel Method. Use initial values of zero for all variables. Round all results to three decimal places. List the values of the variables after each step of the iteration, and give the final solution set in each case.

★(a) $\begin{cases} 2x_1 + 13x_2 + x_3 = 0 \\ x_1 - 2x_2 + 15x_3 = 26 \\ 8x_1 - x_2 + 3x_3 = 25 \end{cases}$

(b) $\begin{cases} -3x_1 - x_2 - 7x_3 = -39 \\ 10x_1 + x_2 + x_3 = 37 \\ x_1 + 9x_2 + 2x_3 = -58 \end{cases}$

★(c) $\begin{cases} x_1 + x_2 + 13x_3 + 2x_4 = 120 \\ 9x_1 + 2x_2 - x_3 + x_4 = 49 \\ -2x_1 + 3x_2 - x_3 - 14x_4 = 110 \\ -x_1 - 17x_2 - 3x_3 + 2x_4 = 86 \end{cases}$

★8. Show that neither the Jacobi Method nor the Gauss-Seidel Method seems to stabilize when applied to the following system by observing what happens during the first six steps of the Jacobi Method and the first four steps of the Gauss-Seidel Method. Let the initial value of all variables be zero, and round all results to three decimal places. Then find the solution using Gaussian elimination.

$$\begin{cases} x_1 - 5x_2 - x_3 = 16 \\ 6x_1 - x_2 - 2x_3 = 13 \\ 7x_1 + x_2 + x_3 = 12 \end{cases}$$

9. (a) For the following system, show that with initial values of zero for each variable, the Gauss-Seidel Method stabilizes to the correct solution. Round all results to three decimal places, and give the values of the variables after each step of the iteration.

$$\begin{cases} 2x_1 + x_2 + x_3 = 7 \\ x_1 + 2x_2 + x_3 = 8 \\ x_1 + x_2 + 2x_3 = 9 \end{cases}$$

(b) Work out the first eight steps of the Jacobi Method for the system in part (a) (again using initial values of zero for each variable), and observe that this method does not stabilize. On alternate passes, the results oscillate between values near $x_1 = 3$, $x_2 = 4$, $x_3 = 5$ and $x_1 = -1$, $x_2 = 0$, $x_3 = 1$.

★10. True or False:

(a) Roundoff error occurs when fewer digits are used to represent a number than are actually required.

(b) An ill-conditioned system of linear equations is a system in which some of the coefficients are unknown.

(c) In partial pivoting, we use row swaps to ensure that each pivot element is as small as possible in absolute value.

(d) Iterative methods generally tend to introduce less roundoff error than Gauss-Jordan row reduction.

(e) In the Jacobi Method, the new value of x_i is immediately used to compute x_{i+1} (for $i < n$) on the same iteration.

(f) The first approximate solution obtained using initial values of 0 for all variables in the system $\begin{cases} x - 2y = 6 \\ 2x + 3y = 15 \end{cases}$ using the Gauss-Seidel Method is $x = 6$, $y = 5$.

9.2 LDU Decomposition

Prerequisite: Section 2.4, Inverses of Matrices

In this section, we show that many nonsingular matrices can be written as the product of a lower triangular matrix **L**, a diagonal matrix **D**, and an upper triangular matrix **U**. As you will see, this **LDU** decomposition is useful in solving certain types of linear systems. Although **LDU** decomposition is used here only to solve systems having square coefficient matrices, the method can be generalized to solve systems with nonsquare coefficient matrices as well.

Calculating the LDU Decomposition

For a given matrix **A**, we can find matrices **L**, **D**, and **U** such that $\mathbf{A} = \mathbf{LDU}$ by using row reduction. It is not necessary to bring **A** completely to reduced row echelon form. Instead, we put **A** into row echelon form.

In our discussion, we need to give a name to a row operation of type (II) in which the pivot row is used to zero out an entry *below* it. Let us call this a

lower type (II) row operation. Notice that a matrix can be put in row echelon form using only type (I) and lower type (II) operations if you do not need to interchange any rows.

Throughout this section, we assume that row reduction into row echelon form is performed exactly as described in Section 2.1 for Gaussian elimination. Beware! If you try to be "creative" in your choice of row operations and stray from this standard method of row reduction, you may obtain incorrect answers.

We can now state the **LDU** decomposition theorem, as follows:

THEOREM 9.1

Let $\mathbf{A}$ be a nonsingular $n \times n$ matrix. If $\mathbf{A}$ can be row reduced to row echelon form using only type (I) and lower type (II) operations, then $\mathbf{A} = \mathbf{LDU}$ where $\mathbf{L}$ is an $n \times n$ lower triangular matrix, $\mathbf{D}$ is an $n \times n$ diagonal matrix, and $\mathbf{U}$ is an $n \times n$ upper triangular matrix and where all main diagonal entries of $\mathbf{L}$ and $\mathbf{U}$ equal 1.

Furthermore, this decomposition of $\mathbf{A}$ is unique; that is, if $\mathbf{A} = \mathbf{L'D'U'}$, where $\mathbf{L'}$ is $n \times n$ lower triangular, $\mathbf{D'}$ is $n \times n$ diagonal, and $\mathbf{U'}$ is $n \times n$ upper triangular with all main diagonal entries of $\mathbf{L'}$ and $\mathbf{U'}$ equal to 1, then $\mathbf{L'} = \mathbf{L}$, $\mathbf{D'} = \mathbf{D}$, and $\mathbf{U'} = \mathbf{U}$.

We outline the proof of this theorem below, which illustrates how to calculate the **LDU** decomposition for a matrix $\mathbf{A}$ when it exists. We omit the proof of uniqueness since that property is not needed for the applications.

PROOF OF THEOREM 9.1

(outline) Suppose that $\mathbf{A}$ is a nonsingular $n \times n$ matrix and we can reduce $\mathbf{A}$ to row echelon form using only type (I) and lower type (II) row operations. Let $\mathbf{U}$ be the row echelon form matrix obtained from this process. Then $\mathbf{U}$ is an upper triangular matrix (why?). Since $\mathbf{A}$ is nonsingular, all of the main diagonal entries of $\mathbf{U}$ must equal 1 (why?). Now, $\mathbf{U} = R_t(R_{t-1}(\cdots(R_2(R_1(\mathbf{A})))\cdots))$ where $R_1, \ldots, R_t$ are the type (I) and lower type (II) row operations used to obtain $\mathbf{U}$ from $\mathbf{A}$. Hence,

$$\mathbf{A} = R_1^{-1}(R_2^{-1}(\cdots(R_{t-1}^{-1}(R_t^{-1}(\mathbf{U})))\cdots))$$
$$= R_1^{-1}(R_2^{-1}(\cdots(R_{t-1}^{-1}(R_t^{-1}(\mathbf{I}_n\mathbf{U})))\cdots))$$
$$= R_1^{-1}(R_2^{-1}(\cdots(R_{t-1}^{-1}(R_t^{-1}(\mathbf{I}_n)))\cdots))\mathbf{U},$$

by Theorem 2.1. Let $\mathbf{K} = R_1^{-1}(R_2^{-1}(\cdots(R_{t-1}^{-1}(R_t^{-1}(\mathbf{I}_n)))\cdots))$. Then $\mathbf{A} = \mathbf{KU}$.

Consulting Table 2.1 in Section 2.3, we see that each of $R_1^{-1}, R_2^{-1}, \ldots, R_t^{-1}$ is also either type (I) or lower type (II). Now, since $\mathbf{I}_n$ is lower triangular and applying type (I) and lower type (II) row operations to a lower triangular matrix always produces a lower triangular matrix (why?), it follows that $\mathbf{K}$ is a lower triangular matrix. Thus, $\mathbf{K}$ has the general form

$$
\begin{bmatrix}
k_{11} & 0 & 0 & \cdots & 0 \\
k_{21} & k_{22} & 0 & \cdots & 0 \\
k_{31} & k_{32} & k_{33} & \cdots & 0 \\
\vdots & \vdots & \vdots & \ddots & \vdots \\
k_{n1} & k_{n2} & k_{n3} & \cdots & k_{nn}
\end{bmatrix}.
$$

In fact, if we are careful to follow the standard method of row reduction, we get the following values for the entries of $\mathbf{K}$:

$$
\begin{cases}
k_{ii} = \frac{1}{c} & \text{if we performed } \langle i \rangle \leftarrow c \langle i \rangle \text{ to convert} \\
& \text{the pivot to 1 in column } i \\
k_{ij} = -c & \text{if we performed } \langle j \rangle \leftarrow c \langle i \rangle + \langle j \rangle \\
& \text{to zero out the } (i,\, j) \text{ entry (where } i > j)
\end{cases}
$$

Thus, the main diagonal entries of $\mathbf{K}$ are the reciprocals of the constants used in the type (I) operations, and the entries of $\mathbf{K}$ below the main diagonal are the additive inverses of the constants used in the lower type (II) operations (verify!). In particular, all of the main diagonal entries of $\mathbf{K}$ are nonzero.

Finally, $\mathbf{K}$ can be expressed as $\mathbf{LD}$, where

$$
\mathbf{L} =
\begin{bmatrix}
1 & 0 & 0 & \cdots & 0 \\
\frac{k_{21}}{k_{11}} & 1 & 0 & \cdots & 0 \\
\frac{k_{31}}{k_{11}} & \frac{k_{32}}{k_{22}} & 1 & \cdots & 0 \\
\vdots & \vdots & \vdots & \ddots & \vdots \\
\frac{k_{n1}}{k_{11}} & \frac{k_{n2}}{k_{22}} & \frac{k_{n3}}{k_{33}} & \cdots & 1
\end{bmatrix}
\quad \text{and} \quad
\mathbf{D} =
\begin{bmatrix}
k_{11} & 0 & 0 & \cdots & 0 \\
0 & k_{22} & 0 & \cdots & 0 \\
0 & 0 & k_{33} & \cdots & 0 \\
\vdots & \vdots & \vdots & \ddots & \vdots \\
0 & 0 & 0 & \cdots & k_{nn}
\end{bmatrix}.
$$

Therefore, we have $\mathbf{A} = \mathbf{KU} = \mathbf{LDU}$, with $\mathbf{L}$ lower triangular, $\mathbf{D}$ diagonal, $\mathbf{U}$ upper triangular, and all main diagonal entries of $\mathbf{L}$ and $\mathbf{U}$ equal to 1. ■

In the next example, we decompose a nonsingular matrix $\mathbf{A}$ into $\mathbf{LDU}$ form. As in the proof of Theorem 9.1, we first decompose $\mathbf{A}$ into $\mathbf{KU}$ form, with $\mathbf{K} = \mathbf{LD}$. We then find the matrices $\mathbf{L}$ and $\mathbf{D}$ using $\mathbf{K}$.

EXAMPLE 1 Let us express

$$
\mathbf{A} =
\begin{bmatrix}
2 & 1 & 4 \\
3 & 2 & 5 \\
4 & 1 & 9
\end{bmatrix}
$$

in $\mathbf{LDU}$ form. To do this, we convert $\mathbf{A}$ into row echelon form $\mathbf{U}$. Notice that only type (I) and lower type (II) row operations are used.

$$\mathbf{A} = \begin{bmatrix} 2 & 1 & 4 \\ 3 & 2 & 5 \\ 4 & 1 & 9 \end{bmatrix}$$

(I): $\langle 1 \rangle \leftarrow \frac{1}{2} \langle 1 \rangle$

$$\begin{bmatrix} 1 & \frac{1}{2} & 2 \\ 3 & 2 & 5 \\ 4 & 1 & 9 \end{bmatrix}$$

(II): $\langle 2 \rangle \leftarrow -3 \langle 1 \rangle + \langle 2 \rangle$

$$\begin{bmatrix} 1 & \frac{1}{2} & 2 \\ 0 & \frac{1}{2} & -1 \\ 4 & 1 & 9 \end{bmatrix}$$

(II): $\langle 3 \rangle \leftarrow -4 \langle 1 \rangle + \langle 3 \rangle$

$$\begin{bmatrix} 1 & \frac{1}{2} & 2 \\ 0 & \frac{1}{2} & -1 \\ 0 & -1 & 1 \end{bmatrix}$$

(I): $\langle 2 \rangle \leftarrow 2 \langle 2 \rangle$

$$\begin{bmatrix} 1 & \frac{1}{2} & 2 \\ 0 & 1 & -2 \\ 0 & -1 & 1 \end{bmatrix}$$

(II): $\langle 3 \rangle \leftarrow 1 \langle 2 \rangle + \langle 3 \rangle$

$$\begin{bmatrix} 1 & \frac{1}{2} & 2 \\ 0 & 1 & -2 \\ 0 & 0 & -1 \end{bmatrix}$$

(I): $\langle 3 \rangle \leftarrow -1 \langle 3 \rangle$

$$\begin{bmatrix} 1 & \frac{1}{2} & 2 \\ 0 & 1 & -2 \\ 0 & 0 & 1 \end{bmatrix} = \mathbf{U}.$$

Using the formulas in the proof of Theorem 9.1 for k_{ii} and k_{ij}, we have

$$\mathbf{K} = \begin{bmatrix} 2 & 0 & 0 \\ 3 & \frac{1}{2} & 0 \\ 4 & -1 & -1 \end{bmatrix}.$$

For example, $k_{22} = \frac{1}{2}$ because it is the reciprocal of the constant $c = 2$ used in the row operation $\langle 2 \rangle \longleftarrow 2\langle 2 \rangle$ to make the pivot equal 1 in the $(2, 2)$ position. Similarly, $k_{31} = 4$ because it is the additive inverse of the constant $c = -4$ used in the row operation $\langle 3 \rangle \longleftarrow -4\langle 1 \rangle + \langle 3 \rangle$ to zero out the $(3, 1)$ entry of $\mathbf{A}$.

Finally, $\mathbf{K}$ can be broken into a product $\mathbf{LD}$ as follows: take the main diagonal entries of $\mathbf{D}$ to be those of $\mathbf{K}$ and create $\mathbf{L}$ by dividing each column of $\mathbf{K}$ by the main diagonal entry in that column. Performing these steps yields

$$L = \begin{bmatrix} 1 & 0 & 0 \\ \frac{3}{2} & 1 & 0 \\ 2 & -2 & 1 \end{bmatrix} \quad \text{and} \quad D = \begin{bmatrix} 2 & 0 & 0 \\ 0 & \frac{1}{2} & 0 \\ 0 & 0 & -1 \end{bmatrix}.$$

You should verify that $A = LDU$. ∎

Solving a System Using LDU Decomposition

When solving a system of linear equations with coefficient matrix A, it is often useful to leave the **LDU** decomposition of A in **KU** form. We can then find the solution of the system using substitution techniques, as in the next example.

EXAMPLE 2 We solve

$$\begin{cases} -4x_1 & + & 5x_2 & - & 2x_3 & = & 5 \\ -3x_1 & + & 2x_2 & - & x_3 & = & 4 \\ x_1 & + & x_2 & & & = & -1 \end{cases}$$

by decomposing the coefficient matrix into **KU** form. Let A be the coefficient matrix. First, putting A into row echelon form U, we have

$$A = \begin{bmatrix} -4 & 5 & -2 \\ -3 & 2 & -1 \\ 1 & 1 & 0 \end{bmatrix}$$

$$(\text{I}): \langle 1 \rangle \leftarrow -\tfrac{1}{4}\langle 1 \rangle \qquad \begin{bmatrix} 1 & -\frac{5}{4} & \frac{1}{2} \\ -3 & 2 & -1 \\ 1 & 1 & 0 \end{bmatrix}$$

$$\begin{aligned} (\text{II}): \langle 2 \rangle \leftarrow 3\langle 1 \rangle + \langle 2 \rangle \\ (\text{II}): \langle 3 \rangle \leftarrow -1\langle 1 \rangle + \langle 3 \rangle \end{aligned} \qquad \begin{bmatrix} 1 & -\frac{5}{4} & \frac{1}{2} \\ 0 & -\frac{7}{4} & \frac{1}{2} \\ 0 & \frac{9}{4} & -\frac{1}{2} \end{bmatrix}$$

$$(\text{I}): \langle 2 \rangle \leftarrow -\tfrac{4}{7}\langle 2 \rangle \qquad \begin{bmatrix} 1 & -\frac{5}{4} & \frac{1}{2} \\ 0 & 1 & -\frac{2}{7} \\ 0 & \frac{9}{4} & -\frac{1}{2} \end{bmatrix}$$

$$(\text{II}): \langle 3 \rangle \leftarrow -\tfrac{9}{4}\langle 2 \rangle + \langle 3 \rangle \qquad \begin{bmatrix} 1 & -\frac{5}{4} & \frac{1}{2} \\ 0 & 1 & -\frac{2}{7} \\ 0 & 0 & \frac{1}{7} \end{bmatrix}$$

$$(\text{I}): \langle 3 \rangle \leftarrow 7\langle 3 \rangle \qquad \begin{bmatrix} 1 & -\frac{5}{4} & \frac{1}{2} \\ 0 & 1 & -\frac{2}{7} \\ 0 & 0 & 1 \end{bmatrix} = U.$$

Then

$$\mathbf{K} = \begin{bmatrix} -4 & 0 & 0 \\ -3 & -\frac{7}{4} & 0 \\ 1 & \frac{9}{4} & \frac{1}{7} \end{bmatrix}$$

because the main diagonal entries of $\mathbf{K}$ are the reciprocals of the constants used in the type (I) operations and the entries of $\mathbf{K}$ below the main diagonal are the additive inverses of the constants used in the lower type (II) operations.

Now the original system can be written as

$$\mathbf{A} \begin{bmatrix} x_1 \\ x_2 \\ x_3 \end{bmatrix} = \begin{bmatrix} 5 \\ 4 \\ -1 \end{bmatrix}, \quad \text{or} \quad \mathbf{KU} \begin{bmatrix} x_1 \\ x_2 \\ x_3 \end{bmatrix} = \begin{bmatrix} 5 \\ 4 \\ -1 \end{bmatrix}.$$

If we let

$$\begin{bmatrix} y_1 \\ y_2 \\ y_3 \end{bmatrix} = \mathbf{U} \begin{bmatrix} x_1 \\ x_2 \\ x_3 \end{bmatrix}, \quad \text{then we have} \quad \mathbf{K} \begin{bmatrix} y_1 \\ y_2 \\ y_3 \end{bmatrix} = \begin{bmatrix} 5 \\ 4 \\ -1 \end{bmatrix}.$$

Both of the last two systems can be solved using substitution. We solve the second system for the y-values, and once they are known, we solve the first system for the x-values.

The second system,

$$\mathbf{K} \begin{bmatrix} y_1 \\ y_2 \\ y_3 \end{bmatrix} = \begin{bmatrix} 5 \\ 4 \\ -1 \end{bmatrix},$$

is equivalent to

$$\begin{cases} -4y_1 & & & = & 5 \\ -3y_1 & - & \frac{7}{4}y_2 & & = & 4 \\ y_1 & + & \frac{9}{4}y_2 & + & \frac{1}{7}y_3 & = & -1 \end{cases}.$$

The first equation gives $y_1 = -\frac{5}{4}$. Substituting this solution into the second equation and solving for y_2, we get $-3\left(-\frac{5}{4}\right) - \frac{7}{4}y_2 = 4$, or $y_2 = -\frac{1}{7}$. Finally, substituting for y_1 and y_2 in the third equation, we get $-\frac{5}{4} + \frac{9}{4}\left(-\frac{1}{7}\right) + \frac{1}{7}y_3 = -1$, or $y_3 = 4$. But then the first system,

$$\mathbf{U} \begin{bmatrix} x_1 \\ x_2 \\ x_3 \end{bmatrix} = \begin{bmatrix} y_1 \\ y_2 \\ y_3 \end{bmatrix},$$

is equivalent to

$$\begin{cases} x_1 & - & \frac{5}{4}x_2 & + & \frac{1}{2}x_3 & = & -\frac{5}{4} \\ & & x_2 & - & \frac{2}{7}x_3 & = & -\frac{1}{7} \\ & & & & x_3 & = & 4 \end{cases}.$$

This time, we solve the equations in *reverse* order. The last equation gives $x_3 = 4$. Then $x_2 - \frac{2}{7}(4) = -\frac{1}{7}$, or $x_2 = 1$. Finally, $x_1 - \frac{5}{4}(1) + \frac{1}{2}(4) = -\frac{5}{4}$, or $x_1 = -2$. Therefore, $(x_1, x_2, x_3) = (-2, 1, 4)$. ∎

Solving a system of linear equations using $(\mathbf{KU} =) \mathbf{LDU}$ decomposition has an advantage over Gaussian elimination when there are many systems to be solved with the same coefficient matrix $\mathbf{A}$. In that case, $\mathbf{K}$ and $\mathbf{U}$ need to be calculated just once, and the solutions to each system can be obtained relatively efficiently using substitution. We saw a similar philosophy in Section 2.4 when we discussed the practicality of solving several systems that had the same coefficient matrix by using the inverse of that matrix.

In our discussion of $\mathbf{LDU}$ decomposition, we have not encountered type (III) row operations. If we need to use type (III) row operations to reduce a non-singular matrix $\mathbf{A}$ to row echelon form, it turns out that $\mathbf{A} = \mathbf{PLDU}$, for some matrix $\mathbf{P}$ formed by rearranging the rows of the $n \times n$ identity matrix, and with $\mathbf{L}, \mathbf{D}$, and $\mathbf{U}$ as before. (Rearranging the rows of $\mathbf{P}$ essentially corresponds to putting the equations of the system in the correct order first so that no type (III) row operations are needed thereafter.) However, the $\mathbf{PLDU}$ decomposition thus obtained is not necessarily unique.

Exercises for Section 9.2

1. Find the **LDU** decomposition for each of the following matrices:

$\star$(a) $\begin{bmatrix} 2 & -4 \\ -6 & 17 \end{bmatrix}$

(b) $\begin{bmatrix} 3 & 1 \\ \frac{3}{2} & -\frac{3}{2} \end{bmatrix}$

$\star$(c) $\begin{bmatrix} -1 & 4 & -2 \\ 2 & -6 & -4 \\ 2 & 0 & -25 \end{bmatrix}$

(d) $\begin{bmatrix} 2 & 6 & -4 \\ 5 & 11 & 10 \\ 1 & 9 & -29 \end{bmatrix}$

$\star$(e) $\begin{bmatrix} -3 & -12 & 6 & 9 \\ -6 & -26 & 12 & 20 \\ 9 & 42 & -17 & -28 \\ 3 & 8 & -8 & -18 \end{bmatrix}$

(f) $\begin{bmatrix} 2 & 6 & -4 \\ 5 & 11 & 10 \\ 1 & 9 & -29 \end{bmatrix}$

2. (a) Show that the matrix $\begin{bmatrix} 0 & 1 \\ 1 & 0 \end{bmatrix}$ has no **LDU** decomposition by showing that there are no values w, x, y, and z such that

$$\begin{bmatrix} 0 & 1 \\ 1 & 0 \end{bmatrix} = \underbrace{\begin{bmatrix} 1 & 0 \\ w & 1 \end{bmatrix}}_{\mathbf{L}} \underbrace{\begin{bmatrix} x & 0 \\ 0 & y \end{bmatrix}}_{\mathbf{D}} \underbrace{\begin{bmatrix} 1 & z \\ 0 & 1 \end{bmatrix}}_{\mathbf{U}}.$$

(b) The result of part (a) does not contradict Theorem 9.1. Why not?

3. For each system, find the **KU** decomposition (where **K** = **LD**) for the coefficient matrix, and use it to solve the system by substitution, as in Example 2.

⋆(a) $\begin{cases} -x_1 + 5x_2 = -9 \\ 2x_1 - 13x_2 = 21 \end{cases}$

(b) $\begin{cases} 2x_1 - 4x_2 + 10x_3 = 34 \\ 2x_1 - 5x_2 + 7x_3 = 29 \\ x_1 - 5x_2 - x_3 = 8 \end{cases}$

⋆(a) $\begin{cases} -x_1 + 3x_2 - 2x_3 = -13 \\ 4x_1 - 9x_2 - 7x_3 = 28 \\ -2x_1 + 11x_2 - 31x_3 = -68 \end{cases}$

(d) $\begin{cases} 3x_1 - 15x_2 + 6x_3 + 6x_4 = 60 \\ x_1 - 7x_2 + 8x_3 + 2x_4 = 30 \\ -5x_1 + 24x_2 - 3x_3 - 18x_4 = -115 \\ x_1 - 2x_2 - 7x_3 - x_4 = -4 \end{cases}$

⋆**4.** True or False:

(a) Every nonsingular matrix has a unique **LDU** decomposition.

(b) The entries of the matrix **K** (as defined in this section) can be obtained just by examining the row operations that were used to reduce **A** to upper triangular form.

(c) The operation R given by $\langle 2 \rangle \leftarrow -2\langle 3 \rangle + \langle 2 \rangle$ is a lower type (II) row operation.

(d) If **A** = **KU** (as described in this section), then **AX** = **B** is solved by first solving for **Y** in **UY** = **B** and then solving for **X** in **KX** = **Y**.

9.3 The Power Method for Finding Eigenvalues

Prerequisite: Section 5.5, Diagonalization of Linear Operators

The only method   given in Section 5.5 for finding the eigenvalues of an $n \times n$ matrix **A** is to calculate the characteristic polynomial of **A** and find its roots. However, if n is large, $p_{\mathbf{A}}(x)$ is often difficult to calculate. Also, numerical techniques may be required to find its roots. Finally, if an eigenvalue λ is not known to a high enough degree of accuracy, we may have difficulty finding a corresponding eigenvector **v**, because the matrix $\lambda \mathbf{I} - \mathbf{A}$ in the equation $(\lambda \mathbf{I} - \mathbf{A})\mathbf{v} = \mathbf{0}$ may not be singular for the given value of λ.

Therefore, in this section we present a numerical technique known as the Power Method for finding the largest eigenvalue (in absolute value) of a matrix and a corresponding eigenvector. Such an eigenvalue is called a **dominant eigenvalue**.

All calculations for the examples and exercises in this section were performed on a calculator that stores numbers with twelve-digit accuracy, but only the first four significant digits are printed here. Your own computations may differ slightly if you are using a different number of significant digits. If you do not have a calculator with the ability to perform matrix calculations, use an appropriate linear algebra software package. You might also consider writing your own Power Method program, since the algorithm involved is not difficult.

The Power Method

Suppose $\mathbf{A}$ is a diagonalizable $n \times n$ matrix having (not necessarily distinct) eigenvalues $\lambda_1, \lambda_2, \ldots, \lambda_n$, with λ_1 being the dominant eigenvalue. The **Power Method** can be used to find λ_1 and an associated eigenvector. In fact, it often works in cases where $\mathbf{A}$ is not diagonalizable, but it is not guaranteed to work in such a case.

The idea behind the Power Method is as follows: choose any unit n-vector $\mathbf{v}$ and calculate $(\mathbf{A}^k \mathbf{v}) / \left\| \mathbf{A}^k \mathbf{v} \right\|$ for some large positive integer k. The result should be a good approximation for a unit eigenvector corresponding to λ_1.

To see why, first express $\mathbf{v}$ in the form $\mathbf{v} = a_1 \mathbf{v}_1 + a_2 \mathbf{v}_2 + \cdots + a_n \mathbf{v}_n$, where $\{\mathbf{v}_1, \ldots, \mathbf{v}_n\}$ is a basis of eigenvectors for $\mathbf{A}$ corresponding to the eigenvalues $\lambda_1, \ldots, \lambda_n$. Then

$$\mathbf{A}^k \mathbf{v} = a_1 \mathbf{A}^k \mathbf{v}_1 + a_2 \mathbf{A}^k \mathbf{v}_2 + \cdots + a_n \mathbf{A}^k \mathbf{v}_n$$

$$= a_1 \lambda_1^k \mathbf{v}_1 + a_2 \lambda_2^k \mathbf{v}_2 + \cdots + a_n \lambda_n^k \mathbf{v}_n.$$

Because $|\lambda_1| > |\lambda_i|$ for $2 \leq i \leq n$, we see that for large k, $\left| \lambda_1^k \right|$ is significantly larger than $\left| \lambda_i^k \right|$, since the ratio $|\lambda_i|^k / |\lambda_1|^k$ approaches 0 as $k \to \infty$. Thus, the term $a_1 \lambda_1^k \mathbf{v}_1$ dominates the expression for $\mathbf{A}^k \mathbf{v}$ for large enough values of k.[4] If we normalize $\mathbf{A}^k \mathbf{v}$, we have $\mathbf{u} = \left(\mathbf{A}^k \mathbf{v} \right) / \left\| \mathbf{A}^k \mathbf{v} \right\| \approx \left(a_1 \lambda_1^k \mathbf{v}_1 \right) / \left\| a_1 \lambda_1^k \mathbf{v}_1 \right\|$, which is a scalar multiple of $\mathbf{v}_1$, and thus, $\mathbf{u}$ is a unit eigenvector corresponding to λ_1.

Finally, $\mathbf{A}\mathbf{u} \approx \lambda_1 \mathbf{u}$, and so $\|\mathbf{A}\mathbf{u}\|$ approximates $|\lambda_1|$. The sign of λ_1 is determined by checking whether $\mathbf{A}\mathbf{u}$ is in the same direction as $\mathbf{u}$ or in the opposite direction. We now outline the Power Method in detail.

---

[4]Theoretically, a problem may arise if $a_1 = 0$. However, in most practical situations, this will not happen. If the method does not work and you suspect it is because $a_1 = 0$, try using instead some $\mathbf{v}$ that is linearly independent from those you have already tried.

> ## POWER METHOD FOR FINDING THE DOMINANT EIGENVALUE OF A SQUARE MATRIX (POWER METHOD)
>
> Let A be an $n \times n$ matrix.
>
> **Step 1:** Choose an arbitrary unit n-vector $\mathbf{u}_0$.
>
> **Step 2:** Create a sequence of unit n-vectors $\mathbf{u}_1, \mathbf{u}_2, \mathbf{u}_3, \ldots$ by repeating Steps 2(a) through 2(d) until one of the terminal conditions in Steps 2(c) or 2(d) is reached or until it becomes clear that the method is not converging to an answer.
>
> (a) Given $\mathbf{u}_{k-1}$, calculate $\mathbf{w}_k = \mathbf{A}\mathbf{u}_{k-1}$.
>
> (b) Calculate $\mathbf{u}_k = \mathbf{w}_k / \|\mathbf{w}_k\|$.
>
> (c) If $\mathbf{u}_{k-1}$ equals $\mathbf{u}_k$ to the desired degree of accuracy, let $\lambda = \|\mathbf{w}_k\|$ and go to Step 3.
>
> (d) If $\mathbf{u}_{k-1}$ equals $-\mathbf{u}_k$ to the desired degree of accuracy, let $\lambda = -\|\mathbf{w}_k\|$ and go to Step 3.
>
> **Step 3:** The last $\mathbf{u}_k$ vector calculated in Step 2 is an approximate eigenvector of A corresponding to the (approximate) eigenvalue λ.

Notice that in the Power Method, we normalize each new vector *after* multiplying by A, while in our prior discussion we normalized the final vector $\mathbf{A}^k\mathbf{v}$. However, the fact that matrix and scalar multiplication commute and that both methods result in a unit vector should convince you that the two techniques are equivalent.

It is possible (but unlikely) to get $\mathbf{w}_k = \mathbf{0}$ in Step 2(a) of the Power Method, which makes Step 2(b) impossible to perform. In this case, $\mathbf{u}_{k-1}$ is an eigenvector for A corresponding to $\lambda = 0$. You can then return to Step 1, choosing a different $\mathbf{u}_0$, in hope of finding another eigenvalue for A.

EXAMPLE 1   Let

$$\mathbf{A} = \begin{bmatrix} -16 & 6 & 30 \\ 4 & 1 & -8 \\ -9 & 3 & 17 \end{bmatrix}.$$

We use the Power Method to find the dominant eigenvalue for A and a corresponding eigenvector correct to four decimal places.

Step 1: We choose $\mathbf{u}_0 = [1, 0, 0]$.

Step 2: A first pass through this step gives the following:

(a) $\mathbf{w}_1 = \mathbf{A}\mathbf{u}_0 \approx [-16, 4, -9]$.

(b) $\|\mathbf{w}_1\| = \sqrt{(-16)^2 + 4^2 + (-9)^2} \approx 18.79$.

So $\mathbf{u}_1 = \mathbf{w}_1 / \|\mathbf{w}_1\| \approx [-0.8516, 0.2129, -0.4790]$.

Because $\mathbf{u}_0$ and $\pm\mathbf{u}_1$ do not agree to four decimal places, we return to Step 2(a). Subsequent iterations of Step 2 lead to the results in the following table:

k	$\mathbf{w}_k = \mathbf{A}\mathbf{u}_{k-1}$	$\|\mathbf{w}_k\|$	$\mathbf{u}_k = \dfrac{\mathbf{w}_k}{\|\mathbf{w}_k\|}$
1	$[-16, 4, -9]$	18.79	$[-0.8516, 0.2129, -0.4790]$
2	$[0.5322, 0.6387, 0.1597]$	0.8466	$[0.6287, 0.7544, 0.1886]$
3	$[0.1257, 1.760, -0.1886]$	1.775	$[0.0708, 0.9918, -0.1063]$
4	$[1.629, 2.125, 0.5313]$	2.730	$[0.5968, 0.7784, 0.1946]$
5	$[0.9601, 1.609, 0.2725]$	1.893	$[0.5071, 0.8498, 0.1439]$
6	$[1.302, 1.727, 0.4317]$	2.205	$[0.5904, 0.7830, 0.1958]$
7	$[1.125, 1.578, 0.3635]$	1.972	$[0.5704, 0.8004, 0.1843]$
8	$[1.207, 1.607, 0.4018]$	2.050	$[0.5889, 0.7841, 0.1960]$
9	$[1.164, 1.571, 0.3851]$	1.993	$[0.5840, 0.7884, 0.1932]$
10	$[1.184, 1.578, 0.3946]$	2.012	$[0.5885, 0.7844, 0.1961]$
11	$[1.173, 1.570, 0.3905]$	1.998	$[0.5873, 0.7855, 0.1954]$
12	$[1.179, 1.571, 0.3928]$	2.003	$[0.5884, 0.7844, 0.1961]$
13	$[1.176, 1.569, 0.3918]$	2.000	$[0.5881, 0.7847, 0.1959]$
14	$[1.177, 1.570, 0.3924]$	2.001	$[0.5884, 0.7845, 0.1961]$
15	$[1.176, 1.569, 0.3921]$	2.000	$[0.5883, 0.7845, 0.1961]$
16	$[1.177, 1.569, 0.3923]$	2.000	$[0.5884, 0.7845, 0.1961]$
17	$[1.177, 1.569, 0.3922]$	2.000	$[0.5883, 0.7845, 0.1961]$
18	$[1.177, 1.569, 0.3922]$	2.000	$[0.5883, 0.7845, 0.1961]$

After eighteen iterations, we find that $\mathbf{u}_{17}$ and $\mathbf{u}_{18}$ agree to four decimal places. Therefore, Step 2 terminates with $\lambda = 2.000$.

Step 3: Thus, $\lambda = 2.000$ is the dominant eigenvalue for $\mathbf{A}$ with corresponding unit eigenvector $\mathbf{u}_{18} = [0.5883, 0.7845, 0.1961]$. ∎

We can check that the Power Method gives the correct result in this particular case. A quick calculation shows that for the given matrix $\mathbf{A}$, $p_{\mathbf{A}}(x) = x^3 - 2x^2 - x + 2 = (x-2)(x-1)(x+1)$. Thus, $\lambda_1 = 2$ is the dominant eigenvalue for $\mathbf{A}$.

Solving the system $(2\mathbf{I}_3 - \mathbf{A})\mathbf{v} = \mathbf{0}$ produces an eigenvector $\mathbf{v} = [3, 4, 1]$ corresponding to $\lambda_1 = 2$. Normalizing $\mathbf{v}$ yields a unit eigenvector $\mathbf{v}/\|\mathbf{v}\| \approx [0.5883, 0.7845, 0.1961]$.

Problems with the Power Method

Unfortunately, the Power Method does not always work. Note that it depends on the fact that multiplying by $\mathbf{A}$ magnifies the size of an eigenvector for the dominant eigenvalue more than for any other vector in $\mathbb{R}^n$. For example, if $\mathbf{A}$ is a diagonalizable matrix, the Power Method fails if both $\pm\lambda$ are eigenvalues of $\mathbf{A}$ with the largest absolute value. In particular, suppose $\mathbf{A}$ is a 3×3 matrix with eigenvalues $\lambda_1 = 2$, $\lambda_2 = -2$, and $\lambda_3 = 1$ and corresponding eigenvectors $\mathbf{v}_1$, $\mathbf{v}_2$, and $\mathbf{v}_3$. Multiplying $\mathbf{A}$ by any vector $\mathbf{v} = a_1\mathbf{v}_1 + a_2\mathbf{v}_2 + a_3\mathbf{v}_3$ produces

$\mathbf{Av} = 2a_1\mathbf{v}_1 - 2a_2\mathbf{v}_2 + a_3\mathbf{v}_3$. The contribution of neither eigenvector $\mathbf{v}_1$ nor $\mathbf{v}_2$ dominates over the other, since both terms are doubled simultaneously.

The next example illustrates that the Power Method is not guaranteed to work for a nondiagonalizable matrix.

EXAMPLE 2   Consider the matrix

$$\mathbf{A} = \begin{bmatrix} 7 & -15 & -24 \\ -12 & 25 & 42 \\ 6 & -15 & -23 \end{bmatrix}.$$

This matrix has only one eigenvalue, $\lambda = 1$, with a corresponding one-dimensional eigenspace spanned by $\mathbf{v}_1 = [3, -2, 2]$. The Power Method cannot be used to find this eigenvalue, since some vectors in $\mathbb{R}^3$ that are not eigenvectors have their magnitudes increased when multiplied by $\mathbf{A}$ while the magnitude of $\mathbf{v}_1$ is fixed by $\mathbf{A}$. If we attempt the Power Method anyway, starting with $\mathbf{u}_0 = [1, 0, 0]$, the following results are produced:

k	$\mathbf{w}_k = \mathbf{Au}_{k-1}$	$\|\mathbf{w}_k\|$	$\mathbf{u}_k = \frac{\mathbf{w}_k}{\|\mathbf{w}_k\|}$
1	$[7, -12, 6]$	15.13	$[0.4626, -0.7930, 0.3965]$
2	$[5.617, -8.723, 5.551]$	11.77	$[0.4774, -0.7413, 0.4718]$
3	$[3.139, -4.448, 3.134]$	6.282	$[0.4998, -0.7081, 0.4989]$
$\vdots$	$\vdots$	$\vdots$	$\vdots$
25	$[0.3434, 0.3341, 0.3434]$	0.5894	$[0.5825, 0.5668, 0.5825]$
26	$[-18.41, 31.65, -18.41]$	40.98	$[-0.4492, 0.7723, -0.4492]$
27	$[-3.949, 5.833, -3.949]$	8.075	$[-0.4890, 0.7223, -0.4890]$
$\vdots$	$\vdots$	$\vdots$	$\vdots$
50	$[2.589, -5.325, 2.589]$	6.462	$[0.4006, -0.8240, 0.4006]$
51	$[5.551, -8.583, 5.551]$	11.63	$[0.4772, -0.7379, 0.4772]$
52	$[2.957, -4.132, 2.957]$	5.879	$[0.5029, -0.7029, 0.5029]$
$\vdots$	$\vdots$	$\vdots$	$\vdots$

As you can see, there is no evidence of any convergence at all in either the $\|\mathbf{w}_k\|$ or $\mathbf{u}_k$ columns. If the Power Method were successful, these would be converging to, respectively, the absolute value of the dominant eigenvalue and a corresponding unit eigenvector. ∎

One disadvantage of the Power Method is that it can only be used to find the dominant eigenvalue for a matrix. There are additional numerical techniques for calculating other eigenvalues. One such technique is the **Inverse Power Method,** which finds the *smallest* eigenvalue of a matrix essentially by using the Power Method on the inverse of the matrix. If you are interested in learning more about this technique and other more sophisticated methods for finding eigenvalues, check the numerical analysis books in your library. One classic reference is *Numerical Analysis*, 7th ed., by Burden and Faires (published by Brooks/Cole, 2001).

Exercises for Section 9.3

1. Use the Power Method on each of the given matrices, starting with the given vector,[5] to find the dominant eigenvalue and a corresponding unit eigenvector for each matrix. Perform as many iterations as needed until two successive vectors agree in every entry in the first m digits after the decimal point for the given value of m. Carry out all calculations using as many significant digits as are feasible with your calculator or computer software.

   ★(a) $\begin{bmatrix} 2 & 36 \\ 36 & 23 \end{bmatrix}, \begin{bmatrix} 1 \\ 0 \end{bmatrix}, \quad m = 2$

   (b) $\begin{bmatrix} 3 & 5 \\ 2 & 1 \end{bmatrix}, \begin{bmatrix} 0 \\ 1 \end{bmatrix}, \quad m = 2$

   ★(c) $\begin{bmatrix} 2 & 3 & -1 \\ 1 & 0 & 1 \\ 3 & 1 & 1 \end{bmatrix}, \begin{bmatrix} 0 \\ 1 \\ 0 \end{bmatrix}, \quad m = 2$

   (d) $\begin{bmatrix} 3 & 1 & 1 & 2 \\ 1 & 1 & 0 & 4 \\ 0 & 1 & 0 & -1 \\ 2 & 3 & 2 & 1 \end{bmatrix}, \begin{bmatrix} 1 \\ 0 \\ 0 \\ 0 \end{bmatrix}, \quad m = 2$

   ★(e) $\begin{bmatrix} -10 & 2 & -1 & 11 \\ 4 & 2 & -3 & 6 \\ -44 & 7 & 3 & 28 \\ -17 & 4 & 1 & 12 \end{bmatrix}, \begin{bmatrix} 3 \\ 8 \\ 2 \\ 3 \end{bmatrix}, \quad m = 3$

   (f) $\begin{bmatrix} 5 & 3 & -4 & 6 \\ -2 & -1 & 6 & -10 \\ -6 & -6 & 8 & -7 \\ -2 & -2 & 1 & 2 \end{bmatrix}, \begin{bmatrix} 4 \\ -5 \\ -6 \\ -1 \end{bmatrix}, \quad m = 4$

2. In each part of this exercise, show that the Power Method does not work on the given matrix using $[1, 0, 0]$ as an initial vector. Explain why the method fails in each case.

   (a) $\begin{bmatrix} -21 & 10 & -74 \\ 25 & -9 & 80 \\ 10 & -4 & 33 \end{bmatrix}$ (b) $\begin{bmatrix} 13 & -10 & 8 \\ -8 & 11 & -4 \\ -40 & 40 & -23 \end{bmatrix}$

3. (a) Suppose that $\mathbf{A}$ is a diagonalizable 2×2 matrix with eigenvalues λ_1 and λ_2 such that $|\lambda_1| > |\lambda_2| \neq 0$. Let $\{\mathbf{v}_1, \mathbf{v}_2\}$ be a basis of unit eigenvectors for $\mathbb{R}^2$ corresponding to λ_1 and λ_2, respectively. Then

---

[5] In parts (e) and (f), the initial vector $\mathbf{u}_0$ is not a unit vector. This does not affect the outcome of the Power Method since all subsequent vectors $\mathbf{u}_1, \mathbf{u}_2, \ldots$ will be unit vectors.

each vector $\mathbf{x} \in \mathbb{R}^2$ can be expressed uniquely in the form $\mathbf{x} = a\mathbf{v}_1 + b\mathbf{v}_2$. Finally, suppose $\mathbf{u}_0$ is the initial vector used in the Power Method for finding the dominant eigenvalue of $\mathbf{A}$. Expressing $\mathbf{u}_i$ in that method as $a_i\mathbf{v}_1 + b_i\mathbf{v}_2$, prove that for all $i \geq 0$,

$$\frac{|a_i|}{|b_i|} = \left|\frac{\lambda_1}{\lambda_2}\right|^i \cdot \frac{|a_0|}{|b_0|},$$

assuming that $b_i \neq 0$. Explain what this result implies about the rate of convergence of the Power Method in this case.

$\star$(a) State and prove a similar formula for an $n \times n$ diagonalizable matrix $\mathbf{A}$.

$\star$**4.** True or False:

(a) If the Power Method succeeds in finding a dominant eigenvalue λ for a matrix $\mathbf{A}$, then we must have $\lambda = \|\mathbf{A}\mathbf{u}_{k-1}\|$, where $\mathbf{u}_k$ is the final vector found in the process.

(b) The Power Method does not find the dominant eigenvalue of a matrix $\mathbf{A}$ if the initial vector used is an eigenvector for a different eigenvalue for $\mathbf{A}$.

(c) Starting with the vector $[1, 0, 0, 0]$, the Power Method produces the eigenvalue 4 for the 4×4 matrix $\mathbf{A}$ having all entries equal to 1.

(d) If 2 and -3 are eigenvalues for a 2×2 matrix $\mathbf{A}$, then the Power Method produces an eigenvector corresponding to the eigenvalue 2 because $2 > -3$.

Chapter 10

Further Horizons

Converging Perspectives

Mathematicians are sometimes fortunate enough to discover important connections by looking at things in just the right way. Revisiting old ideas from a new perspective can bring a broader understanding of familiar concepts and may ultimately lead to further useful applications. In this chapter, we revisit some former ideas with new insight. The section on elementary matrices provides a way of understanding row operations from a matrix multiplication viewpoint. The section on function spaces illustrates techniques for determining linear independence and span that are unique to those spaces. The final section on quadratic forms generalizes the usefulness of orthogonal diagonalization to a non-linear setting and thereby illustrates that linear algebra can be used even in certain non-linear situations.

10.1 Elementary Matrices

Prerequisite: Section 2.4, Inverses of Matrices

In this section, we introduce elementary matrices and show that performing a row operation on a matrix is equivalent to multiplying it by an elementary matrix. We conclude with some useful properties of elementary matrices.

Elementary Matrices

DEFINITION

An $n \times n$ matrix is an **elementary matrix of type (I), (II)**, or **(III)** if and only if it is obtained by performing a single row operation of type (I), (II), or (III), respectively, on the identity matrix $\mathbf{I}_n$.

That is, an elementary matrix is a matrix that is one step away from an identity matrix in terms of row operations.

EXAMPLE 1   The type (I) row operation $\langle 2 \rangle \leftarrow -3\langle 2 \rangle$ converts the identity matrix

$$\mathbf{I}_3 = \begin{bmatrix} 1 & 0 & 0 \\ 0 & 1 & 0 \\ 0 & 0 & 1 \end{bmatrix} \quad \text{into} \quad \mathbf{A} = \begin{bmatrix} 1 & 0 & 0 \\ 0 & -3 & 0 \\ 0 & 0 & 1 \end{bmatrix}.$$

Hence, $\mathbf{A}$ is an elementary matrix of type (I) because it is the result of a single row operation of that type on $\mathbf{I}_3$. Next, consider

$$\mathbf{B} = \begin{bmatrix} 1 & 0 & -2 \\ 0 & 1 & 0 \\ 0 & 0 & 1 \end{bmatrix}.$$

Since $\mathbf{B}$ is obtained from $\mathbf{I}_3$ by performing the single type (II) row operation $\langle 1 \rangle \leftarrow -2\langle 3 \rangle + \langle 1 \rangle$, $\mathbf{B}$ is an elementary matrix of type (II). Finally,

$$\mathbf{C} = \begin{bmatrix} 0 & 1 \\ 1 & 0 \end{bmatrix}$$

is an elementary matrix of type (III) because it is obtained by performing the single type (III) row operation $\langle 1 \rangle \leftrightarrow \langle 2 \rangle$ on $\mathbf{I}_2$. ∎

Representing a Row Operation as Multiplication by an Elementary Matrix

The next theorem shows that there is a connection between row operations and matrix multiplication.

> **THEOREM 10.1**
> Let **A** and **B** be $m \times n$ matrices. If **B** is obtained from **A** by performing a single row operation and if **E** is the $m \times m$ elementary matrix obtained by performing that same row operation on $\mathbf{I}_m$, then $\mathbf{B} = \mathbf{EA}$.

In other words, the effect of a single row operation on **A** can be obtained by multiplying **A** on the left by the appropriate elementary matrix.

PROOF OF THEOREM 10.1

Suppose **B** is obtained from **A** by performing the row operation R. Then $\mathbf{E} = R(\mathbf{I}_m)$. Hence, by Theorem 2.1, $\mathbf{B} = R(\mathbf{A}) = R(\mathbf{I}_m \mathbf{A}) = (R(\mathbf{I}_m))\mathbf{A} = \mathbf{EA}$. ∎

EXAMPLE 2 Consider the matrices

$$\mathbf{A} = \begin{bmatrix} 2 & -3 & 0 & 1 \\ 1 & 6 & -2 & -2 \\ 0 & 5 & 3 & 4 \end{bmatrix} \quad \text{and} \quad \mathbf{B} = \begin{bmatrix} 2 & -3 & 0 & 1 \\ 1 & 6 & -2 & -2 \\ -3 & -13 & 9 & 10 \end{bmatrix}.$$

Notice that **B** is obtained from **A** by performing the operation (II): $\langle 3 \rangle \leftarrow -3\langle 2 \rangle + \langle 3 \rangle$. The elementary matrix

$$\mathbf{E} = \begin{bmatrix} 1 & 0 & 0 \\ 0 & 1 & 0 \\ 0 & -3 & 1 \end{bmatrix}$$

is obtained by performing this same row operation on $\mathbf{I}_3$. Notice that

$$\mathbf{EA} = \begin{bmatrix} 1 & 0 & 0 \\ 0 & 1 & 0 \\ 0 & -3 & 1 \end{bmatrix} \begin{bmatrix} 2 & -3 & 0 & 1 \\ 1 & 6 & -2 & -2 \\ 0 & 5 & 3 & 4 \end{bmatrix} = \begin{bmatrix} 2 & -3 & 0 & 1 \\ 1 & 6 & -2 & -2 \\ -3 & -13 & 9 & 10 \end{bmatrix} = \mathbf{B}.$$

That is, **B** can also be obtained from **A** by multiplying **A** on the left by the appropriate elementary matrix. ∎

Inverses of Elementary Matrices

Recall that every row operation has a corresponding inverse row operation. The exact form for the inverse of a row operation of each type was given in Table 2.1 in Section 2.3. These inverse row operations can be used to find inverses of elementary matrices, as we see in the next theorem.

THEOREM 10.2
Every elementary matrix $\mathbf{E}$ is nonsingular, and its inverse $\mathbf{E}^{-1}$ is an elementary matrix of the same type ((I), (II), or (III)).

PROOF OF THEOREM 10.2
Any $n \times n$ elementary matrix $\mathbf{E}$ is formed by performing a single row operation (of type (I), (II), or (III)) on $\mathbf{I}_n$. If we then perform its inverse operation on $\mathbf{E}$, the result is $\mathbf{I}_n$ again. But the inverse row operation has the same type as the original row operation, and so its corresponding $n \times n$ elementary matrix $\mathbf{F}$ has the same type as $\mathbf{E}$. Now by Theorem 10.1, the product $\mathbf{FE}$ must equal $\mathbf{I}_n$. Hence $\mathbf{F}$ and $\mathbf{E}$ are inverses and have the same type. ∎

EXAMPLE 3   Suppose we want the inverse of the elementary matrix

$$\mathbf{B} = \begin{bmatrix} 1 & 0 & -2 \\ 0 & 1 & 0 \\ 0 & 0 & 1 \end{bmatrix}.$$

The row operation corresponding to $\mathbf{B}$ is (II): $\langle 1 \rangle \leftarrow -2\langle 3 \rangle + \langle 1 \rangle$. Hence, the inverse operation is (II): $\langle 1 \rangle \leftarrow 2\langle 3 \rangle + \langle 1 \rangle$, whose elementary matrix is

$$\mathbf{B}^{-1} = \begin{bmatrix} 1 & 0 & 2 \\ 0 & 1 & 0 \\ 0 & 0 & 1 \end{bmatrix}.$$

∎

Using Elementary Matrices to Show Row Equivalence

If two matrices $\mathbf{A}$ and $\mathbf{B}$ are row equivalent, there is some finite sequence of, say, k row operations that converts $\mathbf{A}$ into $\mathbf{B}$. But according to Theorem 10.1, performing each of these row operations is equivalent to multiplying (on the left) by an appropriate elementary matrix. Hence, there must be a sequence of k elementary matrices $\mathbf{E}_1, \mathbf{E}_2, \ldots, \mathbf{E}_k$, such that $\mathbf{B} = \mathbf{E}_k(\cdots(\mathbf{E}_3(\mathbf{E}_2(\mathbf{E}_1\mathbf{A})))\cdots)$. In fact, the converse is true as well since if $\mathbf{B} = \mathbf{E}_k(\cdots(\mathbf{E}_3(\mathbf{E}_2(\mathbf{E}_1\mathbf{A})))\cdots)$ for some collection of elementary matrices $\mathbf{E}_1, \mathbf{E}_2, \ldots, \mathbf{E}_k$, then $\mathbf{B}$ can be obtained from $\mathbf{A}$ through a sequence of k row operations. Hence we have the following result:

THEOREM 10.3
Two $m \times n$ matrices $\mathbf{A}$ and $\mathbf{B}$ are row equivalent if and only if there is a (finite) sequence $\mathbf{E}_1, \mathbf{E}_2, \cdots, \mathbf{E}_k$ of elementary matrices such that $\mathbf{B} = \mathbf{E}_k \cdots \mathbf{E}_2 \mathbf{E}_1 \mathbf{A}$.

EXAMPLE 4 Consider the matrix $\mathbf{A} = \begin{bmatrix} 0 & 1 & -4 \\ 2 & 5 & 9 \end{bmatrix}$. We perform a series of row operations to obtain a row equivalent matrix $\mathbf{B}$. Next to each operation we give its corresponding elementary matrix.

$$\mathbf{A} = \begin{bmatrix} 0 & 1 & -4 \\ 2 & 5 & 9 \end{bmatrix}$$

(III): $\langle 1 \rangle \leftrightarrow \langle 2 \rangle$ $\qquad \begin{bmatrix} 2 & 5 & 9 \\ 0 & 1 & -4 \end{bmatrix} \qquad \mathbf{E}_1 = \begin{bmatrix} 0 & 1 \\ 1 & 0 \end{bmatrix}$

(I): $\langle 1 \rangle \leftarrow \frac{1}{2}\langle 1 \rangle$ $\qquad \begin{bmatrix} 1 & \frac{5}{2} & \frac{9}{2} \\ 0 & 1 & -4 \end{bmatrix} \qquad \mathbf{E}_2 = \begin{bmatrix} \frac{1}{2} & 0 \\ 0 & 1 \end{bmatrix}$

(II): $\langle 1 \rangle \leftarrow -\frac{5}{2}\langle 2 \rangle + \langle 1 \rangle$ $\qquad \begin{bmatrix} 1 & 0 & \frac{29}{2} \\ 0 & 1 & -4 \end{bmatrix} = \mathbf{B}. \qquad \mathbf{E}_3 = \begin{bmatrix} 1 & -\frac{5}{2} \\ 0 & 1 \end{bmatrix}$

Alternatively, the same result $\mathbf{B}$ is obtained if we multiply $\mathbf{A}$ on the left by the product of the elementary matrices $\mathbf{E}_3\mathbf{E}_2\mathbf{E}_1$.

$$\mathbf{B} = \begin{bmatrix} 1 & 0 & \frac{29}{2} \\ 0 & 1 & -4 \end{bmatrix} = \underbrace{\begin{bmatrix} 1 & -\frac{5}{2} \\ 0 & 1 \end{bmatrix}}_{\mathbf{E}_3} \underbrace{\begin{bmatrix} \frac{1}{2} & 0 \\ 0 & 1 \end{bmatrix}}_{\mathbf{E}_2} \underbrace{\begin{bmatrix} 0 & 1 \\ 1 & 0 \end{bmatrix}}_{\mathbf{E}_1} \underbrace{\begin{bmatrix} 0 & 1 & -4 \\ 2 & 5 & 9 \end{bmatrix}}_{\mathbf{A}}.$$

(Verify that the final product really does equal $\mathbf{B}$.) Note that the product is written in the *reverse* of the order in which the row operations were performed. ∎

Nonsingular Matrices Expressed as a Product of Elementary Matrices

Suppose that we can convert a matrix $\mathbf{A}$ to a matrix $\mathbf{B}$ using row operations. Then, by Theorem 10.3, $\mathbf{B} = \mathbf{E}_k \cdots \mathbf{E}_2\mathbf{E}_1\mathbf{A}$, for some elementary matrices $\mathbf{E}_1, \mathbf{E}_2, \ldots, \mathbf{E}_k$. But we can multiply both sides by $\mathbf{E}_k^{-1}, \ldots, \mathbf{E}_2^{-1}, \mathbf{E}_1^{-1}$ (in that order) to obtain $\mathbf{E}_1^{-1}\mathbf{E}_2^{-1} \cdots \mathbf{E}_k^{-1}\mathbf{B} = \mathbf{A}$. Now, by Theorem 10.2, each of the inverses $\mathbf{E}_1^{-1}, \mathbf{E}_2^{-1}, \ldots, \mathbf{E}_k^{-1}$ is also an elementary matrix. Therefore, we have found a product of elementary matrices that converts $\mathbf{B}$ back into the original matrix $\mathbf{A}$. We can use this fact to express a nonsingular matrix as a product of elementary matrices, as in the next example.

EXAMPLE 5 Suppose that we want to express the nonsingular matrix $\mathbf{A} = \begin{bmatrix} -5 & -2 \\ 7 & 3 \end{bmatrix}$ as a product of elementary matrices. We begin by row reducing $\mathbf{A}$, keeping track of the row operations used.

$$\mathbf{A} = \begin{bmatrix} -5 & -2 \\ 7 & 3 \end{bmatrix}$$

(I): $\langle 1 \rangle \leftarrow -\frac{1}{5}\langle 1 \rangle$ $\begin{bmatrix} 1 & \frac{2}{5} \\ 7 & 3 \end{bmatrix}$

(II): $\langle 2 \rangle \leftarrow -7\langle 1 \rangle + \langle 2 \rangle$ $\begin{bmatrix} 1 & \frac{2}{5} \\ 0 & \frac{1}{5} \end{bmatrix}$

(I): $\langle 2 \rangle \leftarrow 5\langle 2 \rangle$ $\begin{bmatrix} 1 & \frac{2}{5} \\ 0 & 1 \end{bmatrix}$

(II): $\langle 1 \rangle \leftarrow -\frac{2}{5}\langle 2 \rangle + \langle 1 \rangle$ $\begin{bmatrix} 1 & 0 \\ 0 & 1 \end{bmatrix} = \mathbf{I}_2.$

Reversing this process, we get a series of row operations that start with $\mathbf{I}_2$ and end with $\mathbf{A}$. The inverse of each row operation above, in reverse order, is listed below together with its corresponding elementary matrix.

(II): $\langle 1 \rangle \leftarrow \frac{2}{5}\langle 2 \rangle + \langle 1 \rangle$ $\mathbf{F}_1 = \begin{bmatrix} 1 & \frac{2}{5} \\ 0 & 1 \end{bmatrix}$

(I): $\langle 2 \rangle \leftarrow \frac{1}{5}\langle 2 \rangle$ $\mathbf{F}_2 = \begin{bmatrix} 1 & 0 \\ 0 & \frac{1}{5} \end{bmatrix}$

(II): $\langle 2 \rangle \leftarrow 7\langle 1 \rangle + \langle 2 \rangle$ $\mathbf{F}_3 = \begin{bmatrix} 1 & 0 \\ 7 & 1 \end{bmatrix}$

(I): $\langle 1 \rangle \leftarrow -5\langle 1 \rangle$ $\mathbf{F}_4 = \begin{bmatrix} -5 & 0 \\ 0 & 1 \end{bmatrix}$

Therefore, we can express $\mathbf{A}$ as the product

$$\mathbf{A} = \underbrace{\begin{bmatrix} -5 & 0 \\ 0 & 1 \end{bmatrix}}_{\mathbf{F}_4} \underbrace{\begin{bmatrix} 1 & 0 \\ 7 & 1 \end{bmatrix}}_{\mathbf{F}_3} \underbrace{\begin{bmatrix} 1 & 0 \\ 0 & \frac{1}{5} \end{bmatrix}}_{\mathbf{F}_2} \underbrace{\begin{bmatrix} 1 & \frac{2}{5} \\ 0 & 1 \end{bmatrix}}_{\mathbf{F}_1} \underbrace{\begin{bmatrix} 1 & 0 \\ 0 & 1 \end{bmatrix}}_{\mathbf{I}_2}.$$

You should verify that this product is really equal to $\mathbf{A}$.   ∎

Example 5 motivates the following corollary of Theorem 10.3. We leave the proof for you to do in Exercise 7.

> **COROLLARY 10.4**
> An $n \times n$ matrix $\mathbf{A}$ is nonsingular if and only if $\mathbf{A}$ is the product of a finite collection of $n \times n$ elementary matrices.

Exercises for Section 10.1

1. For each elementary matrix below, determine its corresponding row operation. Also, use the inverse operation to find the inverse of the given matrix.

   $\star$(a) $\begin{bmatrix} 1 & 0 & 0 \\ 0 & 0 & 1 \\ 0 & 1 & 0 \end{bmatrix}$

   $\star$(b) $\begin{bmatrix} 1 & 0 & 0 \\ 0 & -2 & 0 \\ 0 & 0 & 1 \end{bmatrix}$

   (c) $\begin{bmatrix} 1 & 0 & 0 \\ 0 & 1 & 0 \\ -4 & 0 & 1 \end{bmatrix}$

   (d) $\begin{bmatrix} 1 & 0 & 0 & 0 \\ 0 & 6 & 0 & 0 \\ 0 & 0 & 1 & 0 \\ 0 & 0 & 0 & 1 \end{bmatrix}$

   $\star$(e) $\begin{bmatrix} 1 & 0 & 0 & 0 \\ 0 & 1 & 0 & 0 \\ 0 & 0 & 1 & -2 \\ 0 & 0 & 0 & 1 \end{bmatrix}$

   (f) $\begin{bmatrix} 0 & 0 & 0 & 1 \\ 0 & 1 & 0 & 0 \\ 0 & 0 & 1 & 0 \\ 1 & 0 & 0 & 0 \end{bmatrix}$

2. Express each of the following as a product of elementary matrices (if possible), in the manner of Example 5:

   $\star$(a) $\begin{bmatrix} 4 & 9 \\ 3 & 7 \end{bmatrix}$

   (b) $\begin{bmatrix} -3 & 2 & 1 \\ 13 & -8 & -9 \\ 1 & -1 & 2 \end{bmatrix}$

   $\star$(c) $\begin{bmatrix} 0 & 0 & 5 & 0 \\ -3 & 0 & 0 & -2 \\ 0 & 6 & -10 & -1 \\ 3 & 0 & 0 & 3 \end{bmatrix}$

3. Let $\mathbf{A}$ and $\mathbf{B}$ be $m \times n$ matrices. Prove that $\mathbf{A}$ and $\mathbf{B}$ are row equivalent if and only if $\mathbf{B} = \mathbf{PA}$, for some nonsingular $m \times m$ matrix $\mathbf{P}$.

4. Prove that if $\mathbf{U}$ is an upper triangular matrix with all main diagonal entries nonzero, then $\mathbf{U}^{-1}$ exists and is upper triangular. (Hint: Show that the method for calculating the inverse of a matrix does not produce a row of zeroes on the left side of the augmented matrix. Also, show that for each row reduction step, the corresponding elementary matrix is upper triangular. Conclude that $\mathbf{U}^{-1}$ is the product of upper triangular matrices, and is therefore upper triangular (see Exercise 14(b) in Section 1.5).)

5. If $\mathbf{E}$ is an elementary matrix, show that $\mathbf{E}^T$ is also an elementary matrix. What is the relationship between the row operation corresponding to $\mathbf{E}$ and the row operation corresponding to $\mathbf{E}^T$?

6. Let $\mathbf{F}$ be an elementary $n \times n$ matrix. Show that the product $\mathbf{AF}^T$ is the matrix obtained by performing a "column" operation on $\mathbf{A}$ analogous to one of the three types of row operations. (Hint: What is $(\mathbf{AF}^T)^T$?)

►**7.** Prove Corollary 10.4.

8. Consider the homogeneous system $\mathbf{AX} = \mathbf{O}$, where $\mathbf{A}$ is an $n \times n$ matrix. Show that this system has a nontrivial solution if and only if $\mathbf{A}$ cannot be expressed as the product of elementary $n \times n$ matrices.

9. Let $\mathbf{A}$ and $\mathbf{B}$ be $m \times n$ and $n \times p$ matrices, respectively, and let $\mathbf{E}$ be an $m \times m$ elementary matrix.

(a) Prove that $\text{rank}(\mathbf{EA}) = \text{rank}(\mathbf{A})$.

(b) Show that if $\mathbf{A}$ has k rows of all zeroes, then $\text{rank}(\mathbf{A}) \leq m - k$.

(c) Show that if $\mathbf{A}$ is in reduced row echelon form, then $\text{rank}(\mathbf{AB}) \leq \text{rank}(\mathbf{A})$. (Use part (b).)

(d) Use parts (a) and (c) to prove that for a general matrix $\mathbf{A}$, $\text{rank}(\mathbf{AB}) \leq \text{rank}(\mathbf{A})$.

(e) Compare this exercise with Exercise 18 in Section 2.3.

$\star$**10.** True or False:

(a) Every elementary matrix is square.

(b) If $\mathbf{A}$ and $\mathbf{B}$ are row equivalent matrices, then there must be an elementary matrix $\mathbf{E}$ such that $\mathbf{B} = \mathbf{EA}$.

(c) If $\mathbf{E}_1, \ldots, \mathbf{E}_k$ are $n \times n$ elementary matrices, then the inverse of $\mathbf{E}_1\mathbf{E}_2 \cdots \mathbf{E}_k$ is $\mathbf{E}_k \cdots \mathbf{E}_2\mathbf{E}_1$.

(d) If $\mathbf{A}$ is a nonsingular matrix, then $\mathbf{A}^{-1}$ can be expressed as a product of elementary matrices.

(e) If R is a row operation, $\mathbf{E}$ is its corresponding $m \times m$ matrix, and $\mathbf{A}$ is any $m \times n$ matrix, then the reverse row operation R^{-1} has the property $R^{-1}(\mathbf{A}) = \mathbf{E}^{-1}\mathbf{A}$.

10.2 Function Spaces

Prerequisite: Section 4.7, Coordinatization

In this section, we apply the techniques of Chapter 4 to vector spaces whose elements are functions. The vector spaces $\mathcal{P}_n$ and $\mathcal{P}$ are familiar examples of such spaces. Other important examples are $C^0(\mathbb{R}) = \{$all continuous real-valued functions on $\mathbb{R}\}$ and $C^1(\mathbb{R}) = \{$all continuously differentiable real-valued functions on $\mathbb{R}\}$.

Linear Independence in Function Spaces

Proving that a finite subset S of a function space is linearly independent usually requires a modification of the strategy used in $\mathbb{R}^n$.

EXAMPLE 1   Consider the subset $S = \left\{x^3 - x,\ xe^{-x^2},\ \sin\left(\frac{\pi}{2}x\right)\right\}$ of $C^1(\mathbb{R})$. We will show that S is linearly independent using Theorem 4.7. Let a, b, and c be real numbers such that

$$a\left(x^3 - x\right) + b\left(xe^{-x^2}\right) + c\left(\sin\left(\frac{\pi}{2}x\right)\right) = 0$$

for every value of x. We must show that $a = b = c = 0$.

The above equation must be satisfied for every value of x. In particular, it is true for $x = 1$, $x = 2$, and $x = 3$. This yields the following system:

$$\begin{cases} \text{(Letting } x = 1 \Longrightarrow) & a(0) + b\left(\dfrac{1}{e}\right) + c(1) = 0 \\[2mm] \text{(Letting } x = 2 \Longrightarrow) & a(6) + b\left(\dfrac{2}{e^4}\right) + c(0) = 0 \\[2mm] \text{(Letting } x = 3 \Longrightarrow) & a(24) + b\left(\dfrac{3}{e^9}\right) + c(-1) = 0 \end{cases} \quad .$$

Row reducing the matrix

$$\begin{array}{ccc} a & b & c \\ \end{array}$$
$$\left[\begin{array}{ccc|c} 0 & \frac{1}{e} & 1 & 0 \\ 6 & \frac{2}{e^4} & 0 & 0 \\ 24 & \frac{3}{e^9} & -1 & 0 \end{array}\right] \quad \text{to} \quad \begin{array}{ccc} a & b & c \\ \end{array} \left[\begin{array}{ccc|c} 1 & 0 & 0 & 0 \\ 0 & 1 & 0 & 0 \\ 0 & 0 & 1 & 0 \end{array}\right]$$

shows that the trivial solution $a = b = c = 0$ is the only solution to this homogeneous system. Hence, the set S is linearly independent by Theorem 4.7. ∎

When proving linear independence using the technique of Example 1, we try to choose "nice" values of x to make computations easier. Even so, the use of a calculator or computer is often desirable when working with function spaces.

Other problems may occur because of the choice of x-values. Returning to Example 1, if instead we had plugged in $x = -1$, $x = 0$, and $x = 1$, we would have obtained the system

$$\begin{cases} (x = -1 \Longrightarrow) & a(0) + b\left(-\dfrac{1}{e}\right) + c(-1) = 0 \\[2mm] (x = 0 \Longrightarrow) & a(0) + b(0) + c(0) = 0 \\[2mm] (x = 1 \Longrightarrow) & a(0) + b\left(\dfrac{1}{e}\right) + c(1) = 0 \end{cases} \quad ,$$

which has infinitely many nontrivial solutions. To prove linear independence, we must examine further values of x, generating more equations for the system, until the new system we obtain has only the trivial solution, as in Example 1.

Suppose, however, that after substituting many values for x and creating a huge homogeneous system, we still have nontrivial solutions. We cannot conclude that the set of functions is linearly dependent, although we may suspect that it is. In general, to *prove* that a set of functions $\{\mathbf{f}_1, \ldots, \mathbf{f}_n\}$ is linearly dependent, we must find real numbers $a_1, \ldots, a_n$, not all zero, such that

$$a_1\mathbf{f}_1(x) + a_2\mathbf{f}_2(x) + \cdots + a_n\mathbf{f}_n(x) = 0$$

is a functional identity for every value of x, not just those we have tried.

EXAMPLE 2   Let $S = \{\sin 2x,\ \cos 2x,\ \sin^2 x,\ \cos^2 x\}$, a subset of $C^1(\mathbb{R})$. Suppose we attempt to show that S is linearly independent using Theorem 4.7. Let a, b, c, and d represent real numbers such that

$$a(\sin 2x) + b(\cos 2x) + c(\sin^2 x) + d(\cos^2 x) = 0.$$

Since we have four vectors in S, we substitute four different values for x into this equation to obtain the following system:

$$\begin{cases} (x = 0 \implies) & a(0) + b(1) + c(0) + d(1) = 0 \\ (x = \frac{\pi}{4} \implies) & a(1) + b(0) + c\left(\frac{1}{2}\right) + d\left(\frac{1}{2}\right) = 0 \\ (x = \frac{\pi}{2} \implies) & a(0) + b(-1) + c(1) + d(0) = 0 \\ (x = \frac{3\pi}{4} \implies) & a(-1) + b(0) + c\left(\frac{1}{2}\right) + d\left(\frac{1}{2}\right) = 0 \end{cases}.$$

Since the coefficient matrix for this homogeneous system row reduces to

$$\begin{array}{cccc} a & b & c & d \end{array}$$
$$\begin{bmatrix} 1 & 0 & 0 & 0 \\ 0 & 1 & 0 & 1 \\ 0 & 0 & 1 & 1 \\ 0 & 0 & 0 & 0 \end{bmatrix},$$

there are nontrivial solutions to the system, such as $a = 0$, $b = -1$, $c = -1$, $d = 1$.

At this point, we cannot infer that S is linearly independent because we have nontrivial solutions. We also cannot conclude that S is linearly dependent because we have tested only a few values for x. We could try more values, such as $x = \frac{\pi}{6}$ and $x = \pi$, but we would still find that $a = 0$, $b = -1$, $c = -1$, $d = 1$ satisfies each equation we generate. This situation leads us to believe that the set S is linearly dependent. To be certain, we must check that the values $a = 0$, $b = -1$, $c = -1$, and $d = 1$ yield a functional identity when plugged into the original functional equation. Substituting these values yields

$$0(\sin 2x) + (-1)(\cos 2x) + (-1)(\sin^2 x) + (1)(\cos^2 x) = 0,$$

or $\cos 2x = \cos^2 x - \sin^2 x$, a well-known trigonometric identity. Thus, one vector in S can be expressed as a linear combination of the other vectors in S, and S is linearly dependent.   ■

Exercises for Section 10.2

1. In each part of this exercise, determine whether the given subset S of $C^1(\mathbb{R})$ is linearly independent. If S is linearly independent, prove that it is. If S is linearly dependent, solve for a functional identity that expresses one function in S as a linear combination of the others.
   ★(a) $S = \{e^x, e^{2x}, e^{3x}\}$
   (b) $S = \{\sin x, \sin 2x, \sin 3x, \sin 4x\}$

⋆(c) $S = \{(5x - 1)/(1 + x^2), (3x + 1)/(2 + x^2),$
$(7x^3 - 3x^2 + 17x - 5)/(x^4 + 3x^2 + 2)\}$

(d) $S = \{\sin x, \sin(x + 1), \sin(x + 2), \sin(x + 3)\}$

2. Recall that a function $\mathbf{f}(x) \in C^0(\mathbb{R})$ is **even** if $\mathbf{f}(x) = \mathbf{f}(-x)$ for all $x \in \mathbb{R}$ and is **odd** if $\mathbf{f}(x) = -\mathbf{f}(-x)$ for all $x \in \mathbb{R}$. Suppose we want to prove that a finite subset S of $C^0(\mathbb{R})$ is linearly independent by the method of Example 1.

   (a) Suppose that every element of S is an odd function of x (as in Example 1). Explain why we would not want to substitute both 1 and -1 for x into the appropriate functional equation. Also explain why $x = 0$ would be a poor choice.

   (b) Suppose that every element of S is an even function. Would we want to substitute both 1 and -1 for x into the appropriate functional equation? Why? How about $x = 0$?

3. Let S be the subset $\{\cos(x + 1), \cos(x + 2), \cos(x + 3)\}$ of $C^1(\mathbb{R})$.

   (a) Show that span(S) has $\{\cos x, \sin x\}$ for a basis. (Hint: The identity $\cos(\alpha + \beta) = \cos\alpha\cos\beta - \sin\alpha\sin\beta$ is useful.)

   (b) Use part (a) to prove that S is linearly dependent.

4. For each given subset S of $C^1(\mathbb{R})$, find a subset B of S that is a basis for $V = \text{span}(S)$.

   ⋆(a) $S = \{\sin 2x, \cos 2x, \sin^2 x, \cos^2 x, \sin x \cos x, 1\}$

   (b) $S = \{e^x, 1, e^{-x}\}$

   ⋆(c) $S = \{\sin(x + 1), \cos(x + 1), \sin(x + 2), \cos(x + 2)\}$

5. In each part of this exercise, let B represent an ordered basis for a subspace V of $C^1(\mathbb{R})$ and find $[\mathbf{v}]_B$ for the given $\mathbf{v} \in V$.

   ⋆(a) $B = (e^x, e^{2x}, e^{3x})$, $\mathbf{v} = 5e^x - 7e^{3x}$

   (b) $B = (\sin 2x, \cos 2x, \sin^2 x)$, $\mathbf{v} = 1$

   ⋆(c) $B = (\sin(x + 1), \sin(x + 2))$, $\mathbf{v} = \cos x$

⋆6. True or False:

   (a) A subset $\{\mathbf{f}_1, \mathbf{f}_2\}$ of nonzero functions in $C^0(\mathbb{R})$ is linearly dependent if and only if $\mathbf{f}_1$ is a nonzero constant multiple of $\mathbf{f}_2$.

   (b) The set $\{x^2, x^3, x^4, x^5\}$ is a linearly independent subset of $C^1(\mathbb{R})$.

   (c) Let $\mathbf{f}_1, \mathbf{f}_2, \mathbf{f}_3 \in C^0(\mathbb{R})$. If plugging values for x into $a\mathbf{f}_1(x) + b\mathbf{f}_2(x) + c\mathbf{f}_3(x) = 0$ leads to $a = b = c = 0$, then $\mathbf{f}_1, \mathbf{f}_2$, and $\mathbf{f}_3$ are linearly dependent.

   (d) Let $\mathbf{f}_1, \mathbf{f}_2, \mathbf{f}_3 \in C^0(\mathbb{R})$. If plugging 3 different values for x into $a\mathbf{f}_1(x) + b\mathbf{f}_2(x) + c\mathbf{f}_3(x) = 0$ does not allow us to conclude that $a = b = c = 0$, then $\mathbf{f}_1, \mathbf{f}_2$, and $\mathbf{f}_3$ are linearly dependent.

10.3 Quadratic Forms

Prerequisite: Section 6.3, Orthogonal Diagonalization

In Section 8.7, we used a change of basis to simplify a general second-degree equation (conic section) in two variables x and y. In this section, we generalize this process to any finite number of variables, using orthogonal diagonalization.

Quadratic Forms

DEFINITION

A **quadratic form** on $\mathbb{R}^n$ is a function $Q: \mathbb{R}^n \to \mathbb{R}$ of the form

$$Q([x_1, \ldots, x_n]) = \sum_{1 \le i \le j \le n} c_{ij} x_i x_j,$$

for some real numbers c_{ij}, $1 \le i \le j \le n$.

Thus, a quadratic form on $\mathbb{R}^n$ is a polynomial in n variables in which each term has degree 2.

EXAMPLE 1

The function $Q_1([x_1, x_2, x_3]) = 7x_1^2 + 5x_1x_2 - 6x_2^2 + 9x_2x_3 + 14x_3^2$ is a quadratic form on $\mathbb{R}^3$. Q_1 is a polynomial in three variables in which each term has degree 2. Note that the coefficient c_{13} of the x_1x_3 term is zero.

The function $Q_2([x, y]) = 8x^2 - 3y^2 + 12xy$ is a quadratic form on $\mathbb{R}^2$ with coefficients $c_{11} = 8$, $c_{22} = -3$, and $c_{12} = 12$. On $\mathbb{R}^2$, a quadratic form consists of the x^2, y^2, and xy terms from the general form for the equation of a conic section. ∎

In general, a quadratic form Q on $\mathbb{R}^n$ can be expressed as $Q(\mathbf{x}) = \mathbf{x}^T \mathbf{C} \mathbf{x}$, where $\mathbf{x}$ is a column matrix and $\mathbf{C}$ is the upper triangular matrix whose entries on and above the main diagonal are given by the coefficients c_{ij} in the definition of a quadratic form above. For example, the quadratic forms Q_1 and Q_2 in Example 1 can be expressed as

$$Q_1\left(\begin{bmatrix} x_1 \\ x_2 \\ x_3 \end{bmatrix}\right) = [x_1, x_2, x_3] \begin{bmatrix} 7 & 5 & 0 \\ 0 & -6 & 9 \\ 0 & 0 & 14 \end{bmatrix} \begin{bmatrix} x_1 \\ x_2 \\ x_3 \end{bmatrix} \quad \text{and}$$

$$Q_2\left(\begin{bmatrix} x \\ y \end{bmatrix}\right) = [x, y] \begin{bmatrix} 8 & 12 \\ 0 & -3 \end{bmatrix} \begin{bmatrix} x \\ y \end{bmatrix}.$$

However, this representation for a quadratic form is not the most useful one for our purposes. Instead, we will replace the upper triangular matrix $\mathbf{C}$ with a symmetric matrix $\mathbf{A}$.

THEOREM 10.5

Let $Q: \mathbb{R}^n \to \mathbb{R}$ be a quadratic form. Then there is a unique symmetric $n \times n$ matrix $\mathbf{A}$ such that $Q(\mathbf{x}) = \mathbf{x}^T \mathbf{A} \mathbf{x}$.

PROOF OF THEOREM 10.5

(abridged) The uniqueness of the matrix $\mathbf{A}$ in the theorem is unimportant in what follows. Its proof is left for you to provide in Exercise 3.

To prove the existence of $\mathbf{A}$, let $Q([x_1, \ldots, x_n]) = \sum_{1 \le i \le j \le n} c_{ij} x_i x_j$. If $\mathbf{C} = [c_{ij}]$ is the upper triangular matrix of coefficients for Q, then define $\mathbf{A} = \frac{1}{2}(\mathbf{C} + \mathbf{C}^T)$. Notice that $\mathbf{A}$ is symmetric (verify!). A straightforward calculation of $\mathbf{x}^T \mathbf{A} \mathbf{x}$ shows that the coefficient of its $x_i x_j$ term is c_{ij}. (Verify.) Hence, $\mathbf{x}^T \mathbf{A} \mathbf{x} = Q(\mathbf{x})$. ∎

EXAMPLE 2 Let $Q_3\left(\begin{bmatrix} x_1 \\ x_2 \end{bmatrix}\right) = 17x_1^2 + 8x_1 x_2 - 9x_2^2$. Then the corresponding symmetric

matrix $\mathbf{A}$ for Q_3 is $\begin{bmatrix} c_{11} & \frac{1}{2}c_{12} \\ \frac{1}{2}c_{12} & c_{22} \end{bmatrix} = \begin{bmatrix} 17 & 4 \\ 4 & -9 \end{bmatrix}$. You can verify that

$$Q_3\left(\begin{bmatrix} x_1 \\ x_2 \end{bmatrix}\right) = [x_1, x_2]\begin{bmatrix} 17 & 4 \\ 4 & -9 \end{bmatrix}\begin{bmatrix} x_1 \\ x_2 \end{bmatrix}.$$ ∎

Orthogonal Change of Basis

The next theorem indicates how the symmetric matrix for a quadratic form is altered when we perform an orthogonal change of coordinates.

THEOREM 10.6

Let $Q: \mathbb{R}^n \rightarrow \mathbb{R}$ be a quadratic form given by $Q(x) = \mathbf{x}^T \mathbf{A} \mathbf{x}$, for some symmetric matrix $\mathbf{A}$. Let B be an orthonormal basis for $\mathbb{R}^n$. Let $\mathbf{P}$ be the transition matrix from B-coordinates to standard coordinates, and let $\mathbf{K} = \mathbf{P}^{-1}\mathbf{A}\mathbf{P}$. Then $\mathbf{K}$ is symmetric and $Q(\mathbf{x}) = [\mathbf{x}]_B^T \mathbf{K}[\mathbf{x}]_B$.

PROOF OF THEOREM 10.6
Since B is an orthonormal basis, $\mathbf{P}$ is an orthogonal matrix by Theorem 6.7. Hence, $\mathbf{P}^{-1} = \mathbf{P}^T$. Now, $[\mathbf{x}]_B = \mathbf{P}^{-1}\mathbf{x} = \mathbf{P}^T\mathbf{x}$, and thus, $[\mathbf{x}]_B^T = (\mathbf{P}^T\mathbf{x})^T = \mathbf{x}^T\mathbf{P}$. Therefore,

$$Q(\mathbf{x}) = \mathbf{x}^T \mathbf{A} \mathbf{x} = \mathbf{x}^T \mathbf{P}\mathbf{P}^{-1}\mathbf{A}\mathbf{P}\mathbf{P}^{-1}\mathbf{x} = [\mathbf{x}]_B^T\mathbf{P}^{-1}\mathbf{A}\mathbf{P}[\mathbf{x}]_B.$$

Letting $\mathbf{K} = \mathbf{P}^{-1}\mathbf{A}\mathbf{P}$, we have $Q(\mathbf{x}) = [\mathbf{x}]_B^T\mathbf{K}[\mathbf{x}]_B$. Finally, notice that $\mathbf{K}$ is symmetric, since

$$\mathbf{K}^T = \left(\mathbf{P}^{-1}\mathbf{A}\mathbf{P}\right)^T = \left(\mathbf{P}^T\mathbf{A}\mathbf{P}\right)^T = \mathbf{P}^T\mathbf{A}^T\left(\mathbf{P}^T\right)^T = \mathbf{P}^{-1}\mathbf{A}\mathbf{P} = \mathbf{K}.$$ ∎

EXAMPLE 3 Consider the quadratic form $Q([x, y, z]) = 2xy + 4xz + 2yz - y^2 + 3z^2$. Then

$$Q\left(\begin{bmatrix} x \\ y \\ z \end{bmatrix}\right) = [x, y, z]\mathbf{A}\begin{bmatrix} x \\ y \\ z \end{bmatrix}, \quad \text{where} \quad \mathbf{A} = \begin{bmatrix} 0 & 1 & 2 \\ 1 & -1 & 1 \\ 2 & 1 & 3 \end{bmatrix}.$$

Consider the orthonormal basis $B = \left(\frac{1}{3}[2, 1, 2], \frac{1}{3}[2, -2, -1], \frac{1}{3}[1, 2, -2]\right)$ for $\mathbb{R}^3$. We will find the symmetric matrix for Q with respect to this new basis B.

The transition matrix from B-coordinates to standard coordinates is the orthogonal matrix

$$\mathbf{P} = \frac{1}{3}\begin{bmatrix} 2 & 2 & 1 \\ 1 & -2 & 2 \\ 2 & -1 & -2 \end{bmatrix} \quad \text{and so} \quad \mathbf{P}^{-1} = \mathbf{P}^T = \frac{1}{3}\begin{bmatrix} 2 & 1 & 2 \\ 2 & -2 & -1 \\ 1 & 2 & -2 \end{bmatrix}.$$

Then,

$$\mathbf{K} = \mathbf{P}^{-1}\mathbf{AP} = \frac{1}{9}\begin{bmatrix} 35 & -7 & -11 \\ -7 & -13 & 4 \\ -11 & 4 & -4 \end{bmatrix}.$$

Let $[u, v, w]$ be the representation of the vector $[x, y, z]$ in B-coordinates; that is, $[x, y, z]_B = [u, v, w]$. Then, by Theorem 10.6, Q can be expressed as

$$Q\left(\begin{bmatrix} u \\ v \\ w \end{bmatrix}\right) = [u, v, w]\left(\frac{1}{9}\begin{bmatrix} 35 & -7 & -11 \\ -7 & -13 & 4 \\ -11 & 4 & -4 \end{bmatrix}\right)\begin{bmatrix} u \\ v \\ w \end{bmatrix}$$

$$= \frac{35}{9}u^2 - \frac{13}{9}v^2 - \frac{4}{9}w^2 - \frac{14}{9}uv - \frac{22}{9}uw + \frac{8}{9}vw.$$

Let us check this formula for Q in a particular case. If $[x, y, z] = [9, 2, -1]$, then the original formula for Q yields

$$Q([9, 2, -1]) = (2)(9)(2) + (4)(9)(-1) + (2)(2)(-1) - (2)^2$$
$$+ (3)(-1)^2 = -5.$$

On the other hand,

$$\begin{bmatrix} u \\ v \\ w \end{bmatrix} = \begin{bmatrix} 9 \\ 2 \\ -1 \end{bmatrix}_B = \mathbf{P}^{-1}\begin{bmatrix} 9 \\ 2 \\ -1 \end{bmatrix}$$

$$= \frac{1}{3}\begin{bmatrix} 2 & 1 & 2 \\ 2 & -2 & -1 \\ 1 & 2 & -2 \end{bmatrix}\begin{bmatrix} 9 \\ 2 \\ -1 \end{bmatrix} = \begin{bmatrix} 6 \\ 5 \\ 5 \end{bmatrix}.$$

Calculating Q using the formula for B-coordinates, we get

$$Q([u, v, w]) = \frac{35}{9}(6)^2 - \frac{13}{9}(5)^2 - \frac{4}{9}(5)^2 - \frac{14}{9}(6)(5) - \frac{22}{9}(6)(5) + \frac{8}{9}(5)(5) = -5,$$

which agrees with our previous calculation for Q.   ∎

The Principal Axes Theorem

We are now ready to prove the main result of this section — given any quadratic form Q on $\mathbb{R}^n$, an orthonormal basis B for $\mathbb{R}^n$ can be chosen so that the expression for Q in B-coordinates contains no "mixed-product" terms (that is, Q contains only "square" terms).

THEOREM 10.7 (Principal Axes Theorem)

Let $Q: \mathbb{R}^n \to \mathbb{R}$ be a quadratic form. Then there is an orthonormal basis B for $\mathbb{R}^n$ such that $Q(\mathbf{x}) = [\mathbf{x}]_B^T \mathbf{D} [\mathbf{x}]_B$ for some diagonal matrix $\mathbf{D}$. That is, if $[\mathbf{x}]_B = \mathbf{y} = [y_1, y_2, \ldots, y_n]$, then

$$Q(\mathbf{x}) = d_{11} y_1^2 + d_{22} y_2^2 + \cdots + d_{nn} y_n^2.$$

PROOF OF THEOREM 10.7

Let Q be a quadratic form on $\mathbb{R}^n$. Then by Theorem 10.5, there is a symmetric $n \times n$ matrix $\mathbf{A}$ such that $Q(\mathbf{x}) = \mathbf{x}^T \mathbf{A} \mathbf{x}$. Now, by Theorems 6.17 and 6.19, $\mathbf{A}$ can be orthogonally diagonalized; that is, there is an orthogonal matrix $\mathbf{P}$ such that $\mathbf{P}^{-1} \mathbf{A} \mathbf{P} = \mathbf{D}$ is diagonal. Let B be the orthonormal basis for $\mathbb{R}^n$ given by the columns of $\mathbf{P}$. Then Theorem 10.6 implies that $Q(\mathbf{x}) = [\mathbf{x}]_B^T \mathbf{D} [\mathbf{x}]_B$. ■

The process of finding a diagonal matrix for a given quadratic form Q is referred to as **diagonalizing** Q. We now outline the method for diagonalizing a quadratic form, as presented in the proof of Theorem 10.7.

METHOD FOR DIAGONALIZING A QUADRATIC FORM (QUADRATIC FORM METHOD)

Given a quadratic form $Q: \mathbb{R}^n \to \mathbb{R}$,

Step 1: Find a symmetric $n \times n$ matrix $\mathbf{A}$ such that $Q(\mathbf{x}) = \mathbf{x}^T \mathbf{A} \mathbf{x}$.

Step 2: Apply Steps 3 through 8 of the method in Section 6.3 for orthogonally diagonalizing a symmetric operator, using the matrix $\mathbf{A}$. This process yields an orthonormal basis B, an orthogonal matrix $\mathbf{P}$ whose columns are the vectors in B, and a diagonal matrix $\mathbf{D}$ with $\mathbf{D} = \mathbf{P}^{-1} \mathbf{A} \mathbf{P}$.

Step 3: Then $Q(\mathbf{x}) = [\mathbf{x}]_B^T \mathbf{D} [\mathbf{x}]_B$, with $[\mathbf{x}]_B = \mathbf{P}^{-1} \mathbf{x} = \mathbf{P}^T \mathbf{x}$. If $[\mathbf{x}]_B = [y_1, y_2, \ldots, y_n]$, then $Q(\mathbf{x}) = d_{11} y_1^2 + d_{22} y_2^2 + \cdots + d_{nn} y_n^2$.

EXAMPLE 4　　Let $Q([x, y, z]) = \frac{1}{121}(183x^2 + 266y^2 + 35z^2 + 12xy + 408xz + 180yz)$. We will diagonalize Q.

　　Step 1: Note that $Q(\mathbf{x}) = \mathbf{x}^T \mathbf{A} \mathbf{x}$, where $\mathbf{A}$ is the symmetric matrix

$$\frac{1}{121} \begin{bmatrix} 183 & 6 & 204 \\ 6 & 266 & 90 \\ 204 & 90 & 35 \end{bmatrix}.$$

Step 2: We apply Steps 3 through 8 of the method for orthogonally diagonalizing **A**. We list the results here but leave the details of the calculations for you to check.

(3) A quick computation gives

$$p_{\mathbf{A}}(x) = x^3 - 4x^2 + x + 6 = (x - 3)(x - 2)(x + 1).$$

Therefore, the eigenvalues of **A** are $\lambda_1 = 3$, $\lambda_2 = 2$, and $\lambda_3 = -1$.

(4) Next, we find a basis for each eigenspace for **A**. To find a basis for E_{λ_1}, we solve the system $(3\mathbf{I}_3 - \mathbf{A})\mathbf{x} = \mathbf{0}$, which yields the basis $\{[7, 6, 6]\}$. Similarly, we solve appropriate systems to find

$$\text{Basis for } E_{\lambda_2} = \{[6, -9, 2]\}$$
$$\text{Basis for } E_{\lambda_3} = \{[6, 2, -9]\}.$$

(5) Since each eigenspace from (4) is one-dimensional, we need only normalize each basis vector to find orthonormal bases for E_{λ_1}, E_{λ_2}, and E_{λ_3}.

$$\text{Orthonormal basis for } E_{\lambda_1} = \left\{ \tfrac{1}{11}[7, 6, 6] \right\}$$
$$\text{Orthonormal basis for } E_{\lambda_2} = \left\{ \tfrac{1}{11}[6, -9, 2] \right\}$$
$$\text{Orthonormal basis for } E_{\lambda_3} = \left\{ \tfrac{1}{11}[6, 2, -9] \right\}.$$

(6) Let B be the ordered orthonormal basis $\left(\tfrac{1}{11}[7, 6, 6], \tfrac{1}{11}[6, -9, 2], \tfrac{1}{11}[6, 2, -9] \right)$.

(7) The desired diagonal matrix for Q with respect to the basis B is

$$\mathbf{D} = \begin{bmatrix} 3 & 0 & 0 \\ 0 & 2 & 0 \\ 0 & 0 & -1 \end{bmatrix},$$

which has eigenvalues $\lambda_1 = 3$, $\lambda_2 = 2$, and $\lambda_3 = -1$ along the main diagonal.

(8) The transition matrix **P** from B-coordinates to standard coordinates is the matrix whose columns are the vectors in B, namely,

$$\mathbf{P} = \frac{1}{11} \begin{bmatrix} 7 & 6 & 6 \\ 6 & -9 & 2 \\ 6 & 2 & -9 \end{bmatrix}.$$

Of course, $\mathbf{D} = \mathbf{P}^{-1}\mathbf{A}\mathbf{P}$. In this case, **P** is not only orthogonal but is symmetric as well, so $\mathbf{P}^{-1} = \mathbf{P}^T = \mathbf{P}$. (Be careful! **P** will not always be symmetric.) This concludes Step 2.

Step 3: Let $[x, y, z]_B = [u, v, w]$. Then using **D**, we have $Q = 3u^2 + 2v^2 - w^2$. Notice that Q has only "square" terms, since **D** is diagonal.

For a particular example, let $[x, y, z] = [2, 6, -1]$. Then

$$\begin{bmatrix} u \\ v \\ w \end{bmatrix} = \mathbf{P}^{-1} \begin{bmatrix} 2 \\ 6 \\ -1 \end{bmatrix} = \frac{1}{11} \begin{bmatrix} 7 & 6 & 6 \\ 6 & -9 & 2 \\ 6 & 2 & -9 \end{bmatrix} \begin{bmatrix} 2 \\ 6 \\ -1 \end{bmatrix} = \begin{bmatrix} 4 \\ -4 \\ 3 \end{bmatrix}.$$

Hence, $Q([2, 6, -1]) = 3(4)^2 + 2(-4)^2 - (3)^2 = 71$. As an independent check, notice that plugging $[2, 6, -1]$ into the original equation for Q produces the same result.   ■

Exercises for Section 10.3

1. In each part of this exercise, a quadratic form $Q: \mathbb{R}^n \to \mathbb{R}$ is given. Find an upper triangular matrix $\mathbf{C}$ and a symmetric matrix $\mathbf{A}$ such that, for every $\mathbf{x} \in \mathbb{R}^n$, $Q(\mathbf{x}) = \mathbf{x}^T \mathbf{C} \mathbf{x} = \mathbf{x}^T \mathbf{A} \mathbf{x}$.
   ⋆(a) $Q([x, y]) = 8x^2 - 9y^2 + 12xy$
   ⋆(b) $Q([x, y]) = 7x^2 + 11y^2 - 17xy$
   ⋆(c) $Q([x_1, x_2, x_3]) = 5x_1^2 - 2x_2^2 + 4x_1x_2 - 3x_1x_3 + 5x_2x_3$

2. In each part of this exercise, diagonalize the given quadratic form $Q: \mathbb{R}^n \to \mathbb{R}$ by following the three-step method described in the text. Your answers should include the matrices $\mathbf{A}$, $\mathbf{P}$, and $\mathbf{D}$ defined in that method, as well as the orthonormal basis B. Finally, calculate $Q(\mathbf{x})$ for the given vector $\mathbf{x}$ in the following two different ways: first, using the given formula for Q, and second, calculating $Q = [\mathbf{x}]_B^T \mathbf{D} [\mathbf{x}]_B$ where $[\mathbf{x}]_B = \mathbf{P}^{-1}\mathbf{x}$ and $\mathbf{D} = \mathbf{P}^{-1}\mathbf{A}\mathbf{P}$.
   ⋆(a) $Q([x, y]) = 43x^2 + 57y^2 - 48xy$; $\mathbf{x} = [1, -8]$
   (b) $Q([x_1, x_2, x_3]) = -5x_1^2 + 37x_2^2 + 49x_3^2 + 32x_1x_2 + 80x_1x_3 + 32x_2x_3$; $\mathbf{x} = [7, -2, 1]$
   ⋆(c) $Q([x_1, x_2, x_3]) = 18x_1^2 - 68x_2^2 + x_3^2 + 96x_1x_2 - 60x_1x_3 + 36x_2x_3$; $\mathbf{x} = [4, -3, 6]$
   (d) $Q([x_1, x_2, x_3, x_4]) = x_1^2 + 5x_2^2 + 864x_3^2 + 864x_4^2 - 24x_1x_3 + 24x_1x_4 + 120x_2x_3 + 120x_2x_4 + 1152x_3x_4$; $\mathbf{x} = [5, 9, -3, -2]$

3. Let $Q: \mathbb{R}^n \to \mathbb{R}$ be a quadratic form, and let $\mathbf{A}$ and $\mathbf{B}$ be symmetric matrices such that $Q(\mathbf{x}) = \mathbf{x}^T \mathbf{A} \mathbf{x} = \mathbf{x}^T \mathbf{B} \mathbf{x}$. Prove that $\mathbf{A} = \mathbf{B}$ (the uniqueness assertion from Theorem 10.5). (Hint: Use $\mathbf{x} = \mathbf{e}_i$ to show that $a_{ii} = b_{ii}$. Then use $\mathbf{x} = \mathbf{e}_i + \mathbf{e}_j$ to prove that $a_{ij} = b_{ij}$ when $i \neq j$.)

⋆4. Let $Q: \mathbb{R}^n \to \mathbb{R}$ be a quadratic form. Is the upper triangular representation for Q necessarily unique? That is, if $\mathbf{C}_1$ and $\mathbf{C}_2$ are upper triangular $n \times n$ matrices with $Q(\mathbf{x}) = \mathbf{x}^T \mathbf{C}_1 \mathbf{x} = \mathbf{x}^T \mathbf{C}_2 \mathbf{x}$, for all $\mathbf{x} \in \mathbb{R}^n$, must $\mathbf{C}_1 = \mathbf{C}_2$? Prove your answer.

5. A quadratic form $Q(\mathbf{x})$ on $\mathbb{R}^n$ is **positive definite** if and only if both of the following conditions hold:
   (i) $Q(\mathbf{x}) \geq 0$, for all $\mathbf{x} \in \mathbb{R}^n$.
   (ii) $Q(\mathbf{x}) = 0$ if and only if $\mathbf{x} = \mathbf{0}$.

   A quadratic form having only property (i) is said to be **positive semidefinite**.

   Let Q be a quadratic form on $\mathbb{R}^n$, and let $\mathbf{A}$ be the symmetric matrix such that $Q(\mathbf{x}) = \mathbf{x}^T \mathbf{A} \mathbf{x}$.
   (a) Prove that Q is positive definite if and only if every eigenvalue of $\mathbf{A}$ is positive.

   (b) Prove that Q is positive semidefinite if and only if every eigenvalue of $\mathbf{A}$ is nonnegative.

$\star$**6.** True or False:

   (a) If $Q(\mathbf{x}) = \mathbf{x}^T \mathbf{C} \mathbf{x}$ is a quadratic form, and $\mathbf{A} = \frac{1}{2}(\mathbf{C} + \mathbf{C}^T)$, then $Q(\mathbf{x}) = \mathbf{x}^T \mathbf{A} \mathbf{x}$.

   (b) $Q(x, y) = xy$ is not a quadratic form because it has no x^2 or y^2 terms.

   (c) If $\mathbf{x}^T \mathbf{A} \mathbf{x} = \mathbf{x}^T \mathbf{B} \mathbf{x}$ for every $\mathbf{x} \in \mathbb{R}^n$, then $\mathbf{A} = \mathbf{B}$.

   (d) Every quadratic form can be diagonalized.

   (e) If $\mathbf{A}$ is a symmetric matrix and $Q(\mathbf{x}) = \mathbf{x}^T \mathbf{A} \mathbf{x}$ is a quadratic form that diagonalizes to $Q(\mathbf{x}) = [\mathbf{x}]_B^T \mathbf{D} [\mathbf{x}]_B$, then the main diagonal entries of $\mathbf{D}$ are the eigenvalues of $\mathbf{A}$.

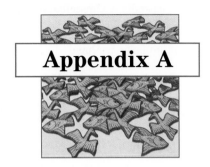

Appendix A

Miscellaneous Proofs

$\mathbf{I}$n this appendix, we present some proofs of theorems that were omitted from the text.

Proof of Theorem 1.14, Part (1)

Part (1) of Theorem 1.14 can be restated as follows:

THEOREM 1.14, Part (1)

If $\mathbf{A}$ is an $m \times n$ matrix, $\mathbf{B}$ is an $n \times p$ matrix, and $\mathbf{C}$ is a $p \times r$ matrix, then $\mathbf{A}(\mathbf{BC}) = (\mathbf{AB})\mathbf{C}$.

PROOF OF THEOREM 1.14, Part (1)

We must show that the (i, j) entry of $\mathbf{A}(\mathbf{BC})$ is the same as the (i, j) entry of $(\mathbf{AB})\mathbf{C}$. Now,

$$
\begin{aligned}
(i, j) \text{ entry of } \mathbf{A}(\mathbf{BC}) &= [i\text{th row of } \mathbf{A}] \cdot [j\text{th column of } \mathbf{BC}] \\
&= [i\text{th row of } \mathbf{A}] \quad \cdot \\
&\qquad \left[\sum_{k=1}^{p} b_{1k}c_{kj}, \sum_{k=1}^{p} b_{2k}c_{kj}, \ldots, \sum_{k=1}^{p} b_{nk}c_{kj} \right] \\
&= a_{i1}\left(\sum_{k=1}^{p} b_{1k}c_{kj} \right) + a_{i2}\left(\sum_{k=1}^{p} b_{2k}c_{kj} \right) + \cdots \\
&\qquad + a_{in}\left(\sum_{k=1}^{p} b_{nk}c_{kj} \right) \\
&= \sum_{k=1}^{p} \left(a_{i1}b_{1k}c_{kj} + a_{i2}b_{2k}c_{kj} + \cdots + a_{in}b_{nk}c_{kj} \right).
\end{aligned}
$$

Similarly, we have

$$(i, j) \text{ entry of } (\mathbf{AB})\mathbf{C} = [i\text{th row of } \mathbf{AB}] \cdot [j\text{th column of } \mathbf{C}]$$

$$= \left[\sum_{k=1}^{n} a_{ik}b_{k1}, \sum_{k=1}^{n} a_{ik}b_{k2}, \ldots, \sum_{k=1}^{n} a_{ik}b_{kp} \right] \cdot$$

$$[j\text{th column of } \mathbf{C}]$$

$$= \left(\sum_{k=1}^{n} a_{ik}b_{k1} \right) c_{1j} + \left(\sum_{k=1}^{n} a_{ik}b_{k2} \right) c_{2j} + \cdots$$

$$+ \left(\sum_{k=1}^{n} a_{ik}b_{kp} \right) c_{pj}$$

$$= \sum_{k=1}^{n} \left(a_{ik}b_{k1}c_{1j} + a_{ik}b_{k2}c_{2j} + \cdots + a_{ik}b_{kp}c_{pj} \right).$$

It then follows that the final sums for the (i, j) entries of $\mathbf{A}(\mathbf{BC})$ and $(\mathbf{AB})\mathbf{C}$ are equal because both are equal to the giant sum of terms

$$\begin{Bmatrix} a_{i1}b_{11}c_{1j} + a_{i1}b_{12}c_{2j} + a_{i1}b_{13}c_{3j} + \cdots + a_{i1}b_{1p}c_{pj} \\ a_{i2}b_{21}c_{1j} + a_{i2}b_{22}c_{2j} + a_{i2}b_{23}c_{3j} + \cdots + a_{i2}b_{2p}c_{pj} \\ \vdots \\ a_{in}b_{n1}c_{1j} + a_{in}b_{n2}c_{2j} + a_{in}b_{n3}c_{3j} + \cdots + a_{in}b_{np}c_{pj} \end{Bmatrix}.$$

Notice that the ith term in the sum for $\mathbf{A}(\mathbf{BC})$ represents the ith column of terms in the giant sum, whereas the ith term in the sum for $(\mathbf{AB})\mathbf{C}$ represents the ith row of terms in the giant sum. Hence, the (i, j) entries of $\mathbf{A}(\mathbf{BC})$ and $(\mathbf{AB})\mathbf{C}$ agree. ∎

Proof of Theorem 2.9

> **THEOREM 2.9**
> Let $\mathbf{A}$ and $\mathbf{B}$ be $n \times n$ matrices. If either product $\mathbf{AB}$ or $\mathbf{BA}$ equals $\mathbf{I}_n$, then the other product also equals $\mathbf{I}_n$ and $\mathbf{A}$ and $\mathbf{B}$ are inverses of each other.

We say that $\mathbf{B}$ is a **left inverse** of $\mathbf{A}$ and $\mathbf{A}$ is a **right inverse** of $\mathbf{B}$ whenever $\mathbf{BA} = \mathbf{I}_n$.

PROOF OF THEOREM 2.9
We need to show that any left inverse of a matrix is also a right inverse and vice versa.

First, suppose that $\mathbf{B}$ is a left inverse of $\mathbf{A}$; that is, $\mathbf{BA} = \mathbf{I}_n$. We will show that $\mathbf{AB} = \mathbf{I}_n$. To do this, we show that $\text{rank}(\mathbf{A}) = n$, then use this to find a right inverse $\mathbf{C}$ of $\mathbf{A}$, and finally show $\mathbf{B} = \mathbf{C}$.

Consider the homogeneous system $\mathbf{AX} = \mathbf{O}$ of n equations and n unknowns. This system has only the trivial solution, because multiplying both sides of $\mathbf{AX} = \mathbf{O}$ by $\mathbf{B}$ on the left, we obtain

$$\mathbf{B}(\mathbf{AX}) = \mathbf{BO} \implies (\mathbf{BA})\mathbf{X} = \mathbf{O} \implies \mathbf{I}_n\mathbf{X} = \mathbf{O} \implies \mathbf{X} = \mathbf{O}.$$

By Theorem 2.5, rank$(\mathbf{A}) = n$, and every column of $\mathbf{A}$ becomes a pivot column during the Gauss-Jordan method. Therefore, each of the augmented matrices

$$\left[\begin{array}{c|c} \mathbf{A} & \begin{array}{c} 1\text{st} \\ \text{column} \\ \text{of } \mathbf{I}_n \end{array} \end{array}\right], \left[\begin{array}{c|c} \mathbf{A} & \begin{array}{c} 2\text{nd} \\ \text{column} \\ \text{of } \mathbf{I}_n \end{array} \end{array}\right], \ldots, \left[\begin{array}{c|c} \mathbf{A} & \begin{array}{c} n\text{th} \\ \text{column} \\ \text{of } \mathbf{I}_n \end{array} \end{array}\right]$$

represents a system with a unique solution. Consider the matrix $\mathbf{C}$, whose ith column is the solution to the ith of these systems. Then $\mathbf{C}$ is a right inverse for $\mathbf{A}$ because the product $\mathbf{AC} = \mathbf{I}_n$. But then

$$\mathbf{B} = \mathbf{B}(\mathbf{I}_n) = \mathbf{B}(\mathbf{AC}) = (\mathbf{BA})\mathbf{C} = \mathbf{I}_n\mathbf{C} = \mathbf{C}.$$

Hence, $\mathbf{B}$ is also a right inverse for $\mathbf{A}$.

Conversely, suppose that $\mathbf{B}$ is a right inverse for $\mathbf{A}$; that is, $\mathbf{AB} = \mathbf{I}_n$. We must show that $\mathbf{B}$ is also a left inverse for $\mathbf{A}$. By assumption, $\mathbf{A}$ is a left inverse for $\mathbf{B}$. However, we have already shown that any left inverse is also a right inverse. Therefore, $\mathbf{A}$ must be a (full) inverse for $\mathbf{B}$, and $\mathbf{AB} = \mathbf{BA} = \mathbf{I}_n$. Hence, $\mathbf{B}$ is a left (and a full) inverse for $\mathbf{A}$.   ∎

Proof of Theorem 3.3, Part (3), Case (2)

THEOREM 3.3, Part (3), Case (2)
Let $\mathbf{A}$ be an $n \times n$ matrix with $n > 2$. If R is the row operation $\langle n - 1 \rangle \leftrightarrow \langle n \rangle$, then $|R(\mathbf{A})| = -|\mathbf{A}|$.

PROOF OF THEOREM 3.3, Part(3), Case (2)
Suppose R is the row operation $\langle n - 1 \rangle \leftrightarrow \langle n \rangle$, switching the last two rows of $\mathbf{A}$. Let $\mathbf{B} = R(\mathbf{A})$. Define the notation $\mathbf{A}^{i,j}$ to represent the $(n - 2) \times (n - 2)$ submatrix formed by deleting rows $n - 1$ and n, as well as deleting columns i and j from $\mathbf{A}$. Define $\mathbf{B}^{i,j}$ similarly. Notice that because the first $n - 2$ rows of $\mathbf{A}$ and $\mathbf{B}$ are identical, $\mathbf{A}^{i,j} = \mathbf{B}^{i,j}$, for $1 \leq i, j \leq n$.

The following observation is useful throughout the rest of the proof: Since the ith column of $\mathbf{B}$ is removed from the submatrix $\mathbf{B}_{ni}$, any element of the form b_{kj} is in the jth column of $\mathbf{B}_{ni}$ if $j < i$, but b_{kj} is in the $(j - 1)$st column of $\mathbf{B}_{ni}$ if $j > i$. Similarly, since the jth column of $\mathbf{A}$ is removed from $\mathbf{A}_{nj}$, any element of the form a_{ki} is in the ith column of $\mathbf{A}_{nj}$ if $i < j$, but a_{ki} is in the $(i - 1)$st column of $\mathbf{A}_{nj}$ if $i > j$.

Now,

$$
\begin{aligned}
|\mathbf{B}| &= \sum_{i=1}^{n} b_{ni}\,\mathcal{B}_{ni} = \sum_{i=1}^{n}(-1)^{n+i} b_{ni}\,|\mathbf{B}_{ni}| \\[2mm]
&= \sum_{i=1}^{n}(-1)^{n+i} b_{ni}\left(\sum_{j=1}^{i-1}(-1)^{(n-1)+j} b_{(n-1)j}\,|\mathbf{B}^{i,j}|\right. \\[2mm]
&\quad \left. + \sum_{j=i+1}^{n}(-1)^{(n-1)+(j-1)} b_{(n-1)j}\,|\mathbf{B}^{i,j}|\right) \\[2mm]
&= \sum_{i=1}^{n}\sum_{j=1}^{i-1}(-1)^{2n+i+j-1} b_{ni}\,b_{(n-1)j}\,|\mathbf{B}^{i,j}| \\[2mm]
&\quad + \sum_{i=1}^{n}\sum_{j=i+1}^{n}(-1)^{2n+i+j-2} b_{ni}\,b_{(n-1)j}\,|\mathbf{B}^{i,j}| \\[2mm]
&= \sum_{\substack{i,j \\ j<i}}(-1)^{2n+i+j-1} b_{ni}\,b_{(n-1)j}\,|\mathbf{B}^{i,j}| + \sum_{\substack{i,j \\ j>i}}(-1)^{2n+i+j-2} b_{ni}\,b_{(n-1)j}\,|\mathbf{B}^{i,j}|.
\end{aligned}
$$

But $b_{ni} = a_{(n-1)i}$, and $b_{(n-1)j} = a_{nj}$ because we are switching rows n and $n-1$. Also, recall that $\mathbf{A}^{i,j} = \mathbf{B}^{i,j}$. Making these substitutions and then reversing the previous steps, we have

$$
\begin{aligned}
|\mathbf{B}| &= \sum_{\substack{i,j \\ j<i}}(-1)^{2n+i+j-1} a_{(n-1)i}\,a_{nj}\,|\mathbf{A}^{i,j}| + \sum_{\substack{i,j \\ j>i}}(-1)^{2n+i+j-2} a_{(n-1)i}\,a_{nj}\,|\mathbf{A}^{i,j}| \\[2mm]
&= (-1)\sum_{\substack{i,j \\ j<i}}(-1)^{2n+i+j-2} a_{nj}\,a_{(n-1)i}\,|\mathbf{A}^{i,j}| \\[2mm]
&\quad + (-1)\sum_{\substack{i,j \\ j>i}}(-1)^{2n+i+j-1} a_{nj}\,a_{(n-1)i}\,|\mathbf{A}^{i,j}| \\[2mm]
&= -\left(\sum_{j=1}^{n}\sum_{i=j+1}^{n}(-1)^{2n+i+j-2} a_{nj}\,a_{(n-1)i}\,|\mathbf{A}^{i,j}|\right. \\[2mm]
&\quad \left. + \sum_{j=1}^{n}\sum_{i=1}^{j-1}(-1)^{2n+i+j-1} a_{nj}\,a_{(n-1)i}\,|\mathbf{A}^{i,j}|\right) \\[2mm]
&= -\left(\sum_{j=1}^{n}(-1)^{n+j} a_{nj}\left(\sum_{i=j+1}^{n}(-1)^{n+i-2} a_{(n-1)i}\,|\mathbf{A}^{i,j}|\right.\right. \\[2mm]
&\quad \left.\left. + \sum_{i=1}^{j-1}(-1)^{n+i-1} a_{(n-1)i}\,|\mathbf{A}^{i,j}|\right)\right)
\end{aligned}
$$

$$= -\sum_{j=1}^{n}(-1)^{n+j}a_{nj}\left(\sum_{i=j+1}^{n}(-1)^{(n-1)+(i-1)}a_{(n-1)i}|\mathbf{A}^{i,j}|\right.$$

$$\left.+\sum_{i=1}^{j-1}(-1)^{(n-1)+i}a_{(n-1)i}|\mathbf{A}^{i,j}|\right)$$

$$= -\sum_{j=1}^{n}(-1)^{n+j}a_{nj}|\mathbf{A}_{nj}| = -\sum_{j=1}^{n}a_{nj}\mathcal{A}_{nj} = -|\mathbf{A}|.$$

This completes Case 2. ∎

Proof of Theorem 5.28

THEOREM 5.28 (Cayley-Hamilton Theorem)
Let $\mathbf{A}$ be an $n \times n$ matrix, and let $p_{\mathbf{A}}(x)$ be its characteristic polynomial. Then $p_{\mathbf{A}}(\mathbf{A}) = \mathbf{O}_n$.

PROOF OF THEOREM 5.28
Let $\mathbf{A}$ be an $n \times n$ matrix with characteristic polynomial $p_{\mathbf{A}}(x) = |x\mathbf{I}_n - \mathbf{A}| = x^n + a_{n-1}x^{n-1} + a_{n-2}x^{n-2} + \cdots + a_1 x + a_0$, for some real numbers $a_0, \ldots, a_{n-1}$. Consider the (classical) adjoint $\mathbf{B}(x)$ of $x\mathbf{I}_n - \mathbf{A}$ (see Section 3.3). By Theorem 3.11,

$$(x\mathbf{I}_n - \mathbf{A})\,\mathbf{B}(x) = p_{\mathbf{A}}(x)\mathbf{I}_n,$$

for every $x \in \mathbb{R}$. We will find an expanded form for $\mathbf{B}(x)$ and then use the preceding equation to show that $p_{\mathbf{A}}(\mathbf{A})$ reduces to $\mathbf{O}_n$.

Now, each entry of $\mathbf{B}(x)$ is defined as $\pm$ the determinant of an $(n-1) \times (n-1)$ minor of $x\mathbf{I}_n - \mathbf{A}$ and hence is a polynomial in x of degree $\leq n-1$ (see Exercise 22 in Section 3.4). For each k, $0 \leq k \leq n-1$, create the matrix $\mathbf{B}_k$ whose (i, j) entry is the coefficient of x^k in the (i, j) entry of $\mathbf{B}(x)$. Thus,

$$\mathbf{B}(x) = x^{n-1}\mathbf{B}_{n-1} + x^{n-2}\mathbf{B}_{n-2} + \cdots + x\mathbf{B}_1 + \mathbf{B}_0.$$

Therefore,

$$(x\mathbf{I}_n - \mathbf{A})\mathbf{B}(x) = (x^n\mathbf{B}_{n-1} - x^{n-1}\mathbf{A}\mathbf{B}_{n-1}) + (x^{n-1}\mathbf{B}_{n-2} - x^{n-2}\mathbf{A}\mathbf{B}_{n-2})$$
$$+ \cdots + (x^2\mathbf{B}_1 - x\mathbf{A}\mathbf{B}_1) + (x\mathbf{B}_0 - \mathbf{A}\mathbf{B}_0)$$
$$= x^n\mathbf{B}_{n-1} + x^{n-1}(-\mathbf{A}\mathbf{B}_{n-1} + \mathbf{B}_{n-2}) + x^{n-2}(-\mathbf{A}\mathbf{B}_{n-2} + \mathbf{B}_{n-3})$$
$$+ \cdots + x(-\mathbf{A}\mathbf{B}_1 + \mathbf{B}_0) + (-\mathbf{A}\mathbf{B}_0).$$

Setting the coefficient of x^k in this expression equal to the coefficient of x^k in $p_{\mathbf{A}}(x)\mathbf{I}_n$ yields

$$\begin{cases} \mathbf{B}_{n-1} &= \mathbf{I}_n \\ -\mathbf{A}\mathbf{B}_k + \mathbf{B}_{k-1} &= a_k\mathbf{I}_n, \quad \text{for } 1 \leq k \leq n-1 \\ -\mathbf{A}\mathbf{B}_0 &= a_0\mathbf{I}_n \end{cases}$$

Hence,

$$
\begin{aligned}
p_{\mathbf{A}}(\mathbf{A}) &= \mathbf{A}^n + a_{n-1}\mathbf{A}^{n-1} + a_{n-2}\mathbf{A}^{n-2} + \cdots + a_1\mathbf{A} + a_0\mathbf{I}_n \\
&= \mathbf{A}^n\mathbf{I}_n + \mathbf{A}^{n-1}(a_{n-1}\mathbf{I}_n) + \mathbf{A}^{n-2}(a_{n-2}\mathbf{I}_n) + \cdots + \mathbf{A}(a_1\mathbf{I}_n) + a_0\mathbf{I}_n \\
&= \mathbf{A}^n(\mathbf{B}_{n-1}) + \mathbf{A}^{n-1}(-\mathbf{A}\mathbf{B}_{n-1} + \mathbf{B}_{n-2}) + \mathbf{A}^{n-2}(-\mathbf{A}\mathbf{B}_{n-2} + \mathbf{B}_{n-3}) \\
&\qquad + \cdots + \mathbf{A}(-\mathbf{A}\mathbf{B}_1 + \mathbf{B}_0) + (-\mathbf{A}\mathbf{B}_0) \\
&= \mathbf{A}^n\mathbf{B}_{n-1} + (-\mathbf{A}^n\mathbf{B}_{n-1} + \mathbf{A}^{n-1}\mathbf{B}_{n-2}) + (-\mathbf{A}^{n-1}\mathbf{B}_{n-2} + \mathbf{A}^{n-2}\mathbf{B}_{n-3}) \\
&\qquad + \cdots + (-\mathbf{A}^2\mathbf{B}_1 + \mathbf{A}\mathbf{B}_0) + (-\mathbf{A}\mathbf{B}_0) \\
&= \mathbf{A}^n(\mathbf{B}_{n-1} - \mathbf{B}_{n-1}) + \mathbf{A}^{n-1}(\mathbf{B}_{n-2} - \mathbf{B}_{n-2}) + \mathbf{A}^{n-2}(\mathbf{B}_{n-3} - \mathbf{B}_{n-3}) \\
&\qquad + \cdots + \mathbf{A}^2(\mathbf{B}_1 - \mathbf{B}_1) + \mathbf{A}(\mathbf{B}_0 - \mathbf{B}_0) \\
&= \mathbf{O}_n.
\end{aligned}
$$

∎

Proof of Theorem 6.17

> **THEOREM 6.17**
> Let $\mathcal{V}$ be a nontrivial subspace of $\mathbb{R}^n$, and let L be a linear operator on $\mathcal{V}$. Let B be an ordered orthonormal basis for $\mathcal{V}$, and let $\mathbf{A}$ be the matrix for L with respect to B. Then L is a symmetric operator if and only if $\mathbf{A}$ is a symmetric matrix.

PROOF OF THEOREM 6.17

Let $\mathcal{V}$, L, B, and $\mathbf{A}$ be given in the statement of the theorem, and let $k = \dim(\mathcal{V})$. Also, suppose that $B = (\mathbf{v}_1, \ldots, \mathbf{v}_k)$.

First, we claim that for all $\mathbf{w}_1, \mathbf{w}_2 \in \mathcal{V}$, $[\mathbf{w}_1]_B \cdot [\mathbf{w}_2]_B = \mathbf{w}_1 \cdot \mathbf{w}_2$, where the first dot product is in $\mathbb{R}^k$ and the second is in $\mathbb{R}^n$. To prove this statement, suppose that $[\mathbf{w}_1]_B = [a_1, \ldots, a_k]$ and $[\mathbf{w}_2]_B = [b_1, \ldots, b_k]$. Then,

$$
\begin{aligned}
\mathbf{w}_1 \cdot \mathbf{w}_2 &= (a_1\mathbf{v}_1 + \cdots + a_k\mathbf{v}_k) \cdot (b_1\mathbf{v}_1 + \cdots + b_k\mathbf{v}_k) \\
&= \sum_{i=1}^{k}\sum_{j=1}^{k}\left(a_ib_j\right)\mathbf{v}_i \cdot \mathbf{v}_j = \sum_{i=1}^{k}(a_ib_i)\mathbf{v}_i \cdot \mathbf{v}_i \quad (\text{since } \mathbf{v}_i \cdot \mathbf{v}_j = 0 \text{ if } i \neq j) \\
&= \sum_{i=1}^{k} a_ib_i \quad (\text{since } \mathbf{v}_i \cdot \mathbf{v}_i = 1) \\
&= [\mathbf{w}_1]_B \cdot [\mathbf{w}_2]_B.
\end{aligned}
$$

Now suppose that L is a symmetric operator on $\mathcal{V}$. We will prove that $\mathbf{A}$ is symmetric by showing that its (i, j) entry equals its (j, i) entry. We have

$$
\begin{aligned}
(i, j) \text{ entry of } \mathbf{A} &= \mathbf{e}_i \cdot (\mathbf{A}\mathbf{e}_j) = [\mathbf{v}_i]_B \cdot \left(\mathbf{A}[\mathbf{v}_j]_B\right) \\
&= [\mathbf{v}_i]_B \cdot [L(\mathbf{v}_j)]_B \\
&= \mathbf{v}_i \cdot L(\mathbf{v}_j) \qquad \text{by the claim verified} \\
&\qquad\qquad\qquad\qquad\qquad \text{earlier in this proof}
\end{aligned}
$$

$$= L(\mathbf{v}_i) \cdot \mathbf{v}_j \qquad \text{since } L \text{ is symmetric}$$
$$= [L(\mathbf{v}_i)]_B \cdot [\mathbf{v}_j]_B \qquad \text{by the claim}$$
$$= (\mathbf{A}[\mathbf{v}_i]_B) \cdot [\mathbf{v}_j]_B$$
$$= (\mathbf{A}\mathbf{e}_i) \cdot \mathbf{e}_j = (j, i) \text{ entry of } \mathbf{A}.$$

Conversely, if $\mathbf{A}$ is a symmetric matrix and $\mathbf{w}_1, \mathbf{w}_2 \in \mathcal{V}$, we have

$$
\begin{aligned}
L(\mathbf{w}_1) \cdot \mathbf{w}_2 \quad &= [L(\mathbf{w}_1)]_B \cdot [\mathbf{w}_2]_B \qquad && \text{by the claim} \\
&= (\mathbf{A}[\mathbf{w}_1]_B) \cdot [\mathbf{w}_2]_B \\
&= (\mathbf{A}[\mathbf{w}_1]_B)^T [\mathbf{w}_2]_B \qquad && \text{changing vector dot product} \\
& && \qquad \text{to matrix multiplication} \\
&= [\mathbf{w}_1]_B^T \mathbf{A}^T [\mathbf{w}_2]_B \\
&= [\mathbf{w}_1]_B^T \mathbf{A}[\mathbf{w}_2]_B \qquad && \text{since } \mathbf{A} \text{ is symmetric} \\
&= [\mathbf{w}_1]_B \cdot (\mathbf{A}[\mathbf{w}_2]_B) \qquad && \text{changing matrix multiplication} \\
& && \qquad \text{to vector dot product} \\
&= [\mathbf{w}_1]_B \cdot [L(\mathbf{w}_2)]_B \\
&= \mathbf{w}_1 \cdot L(\mathbf{w}_2) \qquad && \text{by the claim}
\end{aligned}
$$

Thus, L is a symmetric operator on $\mathcal{V}$, and the proof is complete. ∎

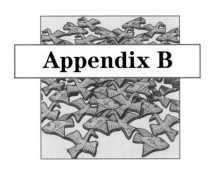

Appendix B

Functions

In this appendix, we define some basic terms associated with functions — domain, codomain, range, image, pre-image, one-to-one, onto, composition, and inverses. It is a good idea to review this material thoroughly before beginning Chapter 5.

Functions: Domain, Codomain, and Range

A **function** f from a set X to a set Y, expressed as $f : X \rightarrow Y$, is a mapping (assignment) of elements of X (called the **domain**) to elements of Y (called the **codomain**) in such a way that each element of X is assigned to some (single) chosen element of Y. That is, every element of X must be assigned to *some* element of Y and *only one* element of Y. For example, $f : \mathbb{Z} \rightarrow \mathbb{R}$ (where $\mathbb{Z}$ represents the set $\{\ldots, -3, -2, -1, 0, 1, 2, 3, \ldots\}$ of all integers) given by $f(x) = x^2$ is a function since each integer in $\mathbb{Z}$ is assigned by f to one and only one element of $\mathbb{R}$.

Notice that the definition of a function allows two different elements of X to map (be assigned) to the same element of Y, as in the function $f : \mathbb{Z} \rightarrow \mathbb{R}$ given by $f(x) = x^2$, where $f(3) = f(-3) = 9$. However, no function allows any member of the domain to map to more than one element of the codomain. Hence, the rule $x \rightarrow \pm\sqrt{x}$, for $x \in \mathbb{R}^+$ (positive real numbers) is not a function, since, for example, 4 would have to map to both 2 and -2.

The **image** of a domain element is the unique codomain element to which it is mapped, and the **pre-images** of a codomain element are the domain elements that map to it. With the function $f : \mathbb{Z} \rightarrow \mathbb{R}$ given by $f(x) = x^2$, 4 is the image of 2, and both 2 and -2 are pre-images of 4 since $2^2 = (-2)^2 = 4$.

If $f : X \rightarrow Y$ is a function, not every element of Y necessarily has a pre-image. For the function $f : \mathbb{Z} \rightarrow \mathbb{R}$ given by $f(x) = x^2$ given above, the element 5 in the codomain has no pre-image because no integer squared equals 5.

The **image of a subset** S of the domain under a function f, written as $f(S)$, is the set of all values in the codomain that are mapped to by elements of S. The **pre-image of a subset** T of the codomain under f, or $f^{-1}(T)$, is the set of *all* values in the domain that map to elements of T under f. For example, for the function $f\colon \mathbb{Z} \to \mathbb{R}$ given by $f(x) = x^2$, the image of the subset $\{-5, -3, 3, 5\}$ of the domain is $\{9, 25\}$, and the pre-image of $\{15, 16, 17\}$ is $\{4, -4\}$.

The image of the entire domain is called the **range** of the function. For the function $f\colon \mathbb{Z} \to \mathbb{R}$ given by $f(x) = x^2$, the range is the set of all squares of integers. In this case, the range is a proper subset of the codomain. This situation is depicted in Figure B.1. For some functions, however, the range is the whole codomain, as we will see shortly.

Figure B.1

The domain X, codomain Y, and range of a function $f\colon X \to Y$

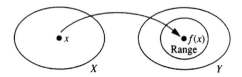

One-to-One and Onto Functions

We now consider two very important types of functions — one-to-one and onto functions. We say that a function $f\colon X \to Y$ is **one-to-one** if and only if distinct elements of X map to distinct elements of Y. That is, f is one-to-one if and only if no two different elements of X map to the same element of Y. For example, $f\colon \mathbb{R} \to \mathbb{R}$ given by $f(x) = x^3$ is one-to-one since no two distinct real numbers have the same cube.

A standard method of proving that a function f is one-to-one is the following:

To show that $f\colon X \to Y$ is one-to-one: Prove that for arbitrary elements $x_1, x_2 \in X$, if $f(x_1) = f(x_2)$, then $x_1 = x_2$.

In other words, the only way x_1 and x_2 can have the same image is if they are not really distinct. We will use this technique to show that $f\colon \mathbb{R} \to \mathbb{R}$ given by $f(x) = 3x - 7$ is one-to-one. Suppose that $f(x_1) = f(x_2)$, for some $x_1, x_2 \in \mathbb{R}$. Then $3x_1 - 7 = 3x_2 - 7$. Hence, $3x_1 = 3x_2$, which implies $x_1 = x_2$. Thus, f is one-to-one.

On the other hand, we sometimes need to show that a function is *not* one-to-one. The usual method for doing this is the following:

To show that $f\colon X \to Y$ is not one-to-one: Find two different elements x_1 and x_2 in the domain X such that $f(x_1) = f(x_2)$.

For example, $g\colon \mathbb{R} \to \mathbb{R}$ given by $g(x) = x^2$ is not one-to-one because $g(3) = g(-3) = 9$. That is, both elements 3 and -3 in the domain $\mathbb{R}$ of g have the same image 9, so g is not one-to-one.

We say that a function $f\colon X \to Y$ is **onto** if and only if every element of Y is an image of some element in X. That is, f is onto if and only if the range of f equals the codomain of f. For example, the function $f\colon \mathbb{R} \to \mathbb{R}$ given by $f(x) = 2x$ is onto since every real number y_1 in the codomain $\mathbb{R}$ is the image of the real number $x_1 = \frac{1}{2}y_1$; that is, $f(x_1) = f\left(\frac{1}{2}y_1\right) = y_1$. Here we are using the standard method of proving that a given function is onto, as follows:

To show that $f: X \rightarrow Y$ **is onto**: Choose an arbitrary element $y_1 \in Y$, and show that there is some $x_1 \in X$ such that $y_1 = f(x_1)$.

On the other hand, we sometimes need to show that a function is *not* onto. The usual method for doing this is the following:

To show that $f: X \rightarrow Y$ **is not onto**: Find an element y_1 in the codomain Y that is not the image of any element x_1 in the domain X.

For example, $f: \mathbb{R} \rightarrow \mathbb{R}$ given by $f(x) = x^2$ is not onto since the real number -4 in the codomain $\mathbb{R}$ is never the image of any real number in the domain; that is, for all $x \in \mathbb{R}$, $f(x) \neq -4$.

Composition and Inverses of Functions

If $f: X \rightarrow Y$ and $g: Y \rightarrow Z$ are functions, we define the **composition** of f and g to be the function $g \circ f: X \rightarrow Z$ given by $(g \circ f)(x) = g(f(x))$. This composition mapping is pictured in Figure B.2. For example, if $f: \mathbb{R} \rightarrow \mathbb{R}$ is given by $f(x) = 1 - x^2$ and $g: \mathbb{R} \rightarrow \mathbb{R}$ is given by $g(x) = 5\cos x$, then $(g \circ f)(x) = g(f(x)) = g(1 - x^2) = 5\cos(1 - x^2)$. In particular, $(g \circ f)(2) = g(f(2)) = g(1 - 2^2) = g(-3) = 5\cos(-3) \approx -4.95$.

Figure B.2

Composition $g \circ f$
of $f : X \rightarrow Y$ and
$g : Y \rightarrow Z$

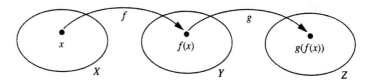

THEOREM B.1
(1) If $f: X \rightarrow Y$ and $g: Y \rightarrow Z$ are both one-to-one, then $g \circ f: X \rightarrow Z$ is one-to-one.
(2) If $f: X \rightarrow Y$ and $g: Y \rightarrow Z$ are both onto, then $g \circ f: X \rightarrow Z$ is onto.

PROOF OF THEOREM B.1
Part (1): Assume that f and g are both one-to-one. To prove $g \circ f$ is one-to-one, we assume that $(g \circ f)(x_1) = (g \circ f)(x_2)$ for two elements $x_1, x_2 \in X$, and prove that $x_1 = x_2$. However, $(g \circ f)(x_1) = (g \circ f)(x_2)$ implies that $g(f(x_1)) = g(f(x_2))$. Hence, $f(x_1)$ and $f(x_2)$ have the same image under g. Since g is one-to-one, we must have $f(x_1) = f(x_2)$. Then x_1 and x_2 have the same image under f. Since f is one-to-one, $x_1 = x_2$. Hence, $g \circ f$ is one-to-one.

Part (2): Assume that f and g are both onto. To prove that $g \circ f : X \to Z$ is onto, we choose an arbitrary element $z_1 \in Z$ and try to find some element in X that $g \circ f$ maps to z_1. Now, since g is onto, there is some $y_1 \in Y$ for which $g(y_1) = z_1$. Also, since f is onto, there is some $x_1 \in X$ for which $f(x_1) = y_1$. Therefore, $(g \circ f)(x_1) = g(f(x_1)) = g(y_1) = z_1$, and so $g \circ f$ maps x_1 to z_1. Hence, $g \circ f$ is onto. ∎

Two functions $f : X \to Y$ and $g : Y \to X$ are **inverses** of each other if $(g \circ f)(x) = x$ and $(f \circ g)(y) = y$, for every $x \in X$ and $y \in Y$. For example, $f : \mathbb{R} \to \mathbb{R}$ given by $f(x) = x^3$ and $g : \mathbb{R} \to \mathbb{R}$ given by $g(x) = \sqrt[3]{x}$ are inverses of each other because $(g \circ f)(x) = g(f(x)) = g(x^3) = \sqrt[3]{x^3} = x$, and $(f \circ g)(x) = f(g(x)) = f\left(\sqrt[3]{x}\right) = (\sqrt[3]{x})^3 = x$.

Not every function can be paired with an inverse function. The next theorem characterizes those functions that do have an inverse.

THEOREM B.2

The function $f : X \to Y$ has an inverse $g : Y \to X$ if and only if f is both one-to-one and onto.

Notice that the inverse functions $f : \mathbb{R} \to \mathbb{R}$ given by $f(x) = x^3$ and $g : \mathbb{R} \to \mathbb{R}$ given by $g(x) = \sqrt[3]{x}$ are both one-to-one and onto. However, a function such as $f : \mathbb{R} \to \mathbb{R}$ given by $f(x) = x^2$ has no inverse, since it is not one-to-one. In this case, we could also have shown that f has no inverse since it is not onto.

PROOF OF THEOREM B.2

First, suppose that $f : X \to Y$ has an inverse $g : Y \to X$. We show that f is one-to-one and onto. To prove f is one-to-one, we assume that $f(x_1) = f(x_2)$, for some $x_1, x_2 \in X$, and try to prove $x_1 = x_2$. Since $f(x_1) = f(x_2)$, we have $g(f(x_1)) = g(f(x_2))$. However, since g is an inverse for f, $x_1 = (g \circ f)(x_1) = g(f(x_1)) = g(f(x_2)) = x_2$, and so $x_1 = x_2$. Hence, f is one-to-one. To prove f is onto, we choose an arbitrary $y_1 \in Y$. We must show that y_1 is the image of some $x_1 \in X$. Now, g maps y_1 to an element x_1 of X; that is, $g(y_1) = x_1$. However, $f(x_1) = f(g(y_1)) = (f \circ g)(y_1) = y_1$, since f and g are inverses. Hence, f maps x_1 to y_1, and f is onto.

Conversely, we assume that $f : X \to Y$ is one-to-one and onto and show that f has an inverse $g : Y \to X$. Let y_1 be an arbitrary element of Y. Since f is onto, the element y_1 in Y is the image of some element in X. Since f is one-to-one, y_1 is the image of precisely one element, say x_1, in X. Hence, y_1 has a *unique* pre-image under f. Now consider the mapping $g : Y \to X$, which maps each element y_1 in Y to its unique pre-image x_1 in X under f. Then $(f \circ g)(y_1) = f(g(y_1)) = f(x_1) = y_1$.

To finish the proof, we must show that $(g \circ f)(x_1) = x_1$ for any $x_1 \in X$. But $(g \circ f)(x_1) = g(f(x_1))$ is defined to be the unique pre-image of $f(x_1)$ under f. Since x_1 is this pre-image, we have $(g \circ f)(x_1) = x_1$. Thus, g and f are inverses. ∎

The next result assures us that when inverses exist, they are unique.

THEOREM B.3

If $f: X \to Y$ has an inverse $g: Y \to X$, then g is the only inverse of f.

PROOF OF THEOREM B.3

Suppose that $g_1: Y \to X$ and $g_2: Y \to X$ are both inverse functions for f. Our goal is to show that $g_1(y) = g_2(y)$ for all $y \in Y$ because then g_1 and g_2 are identical functions, and the inverse of f is unique.

Now, $(g_2 \circ f)(x) = x$, for every $x \in X$, since f and g_2 are inverses. Thus, since $g_1(y) \in X$, $g_1(y) = (g_2 \circ f)(g_1(y)) = g_2(f(g_1(y))) = g_2((f \circ g_1)(y)) = g_2(y)$, since f and g_1 are inverses. ∎

Whenever a function $f: X \to Y$ has an inverse, we denote this unique inverse by $f^{-1}: Y \to X$.

THEOREM B.4

If $f: X \to Y$ and $g: Y \to Z$ both have inverses, then $g \circ f: X \to Z$ has an inverse, and $(g \circ f)^{-1} = f^{-1} \circ g^{-1}$.

PROOF OF THEOREM B.4

Because $g^{-1}: Z \to Y$ and $f^{-1}: Y \to X$, it follows that $f^{-1} \circ g^{-1}$ is a well-defined function from Z to X. We need to show that the inverse of $g \circ f$ is $f^{-1} \circ g^{-1}$. If we can show that both

$$\left((g \circ f) \circ \left(f^{-1} \circ g^{-1} \right) \right)(z) \quad = \quad z, \quad \text{for all } z \in Z,$$

and

$$\left(\left(f^{-1} \circ g^{-1} \right) \circ (g \circ f) \right)(x) \quad = \quad x, \quad \text{for all } x \in X,$$

then by definition, $g \circ f$ and $f^{-1} \circ g^{-1}$ are inverses. Now,

$$\left((g \circ f) \circ \left(f^{-1} \circ g^{-1} \right) \right)(z) = g\left(f\left(f^{-1}\left(g^{-1}(z) \right) \right) \right)$$
$$= g\left(g^{-1}(z) \right) \qquad \text{since } f \text{ and } f^{-1} \text{ are inverses}$$
$$= z. \qquad \text{since } g \text{ and } g^{-1} \text{ are inverses}$$

A similar argument establishes the other statement. ∎

As an example of Theorem B.4, consider $f: \mathbb{R} \to \mathbb{R}$ given by $f(x) = x^3$ and $g: \mathbb{R} \to \mathbb{R}^+$ given by $g(x) = e^x$. Then, $g \circ f: \mathbb{R} \to \mathbb{R}^+$ is $(g \circ f)(x) = e^{x^3}$. However, since $f^{-1}(x) = \sqrt[3]{x}$ and $g^{-1}(x) = \ln x$, $(g \circ f)^{-1}: \mathbb{R}^+ \to \mathbb{R}$ is given by

$$(g \circ f)^{-1}(x) = (f^{-1} \circ g^{-1})(x) = f^{-1}(g^{-1}(x)) = \sqrt[3]{\ln x}.$$

Exercises for Appendix B

1. Which of the following are functions? For those that are functions, determine the range, as well as the image and all pre-images of the value 2.

For those that are not functions, explain why with a precise reason. (Note: $\mathbb{N}$ represents the set $\{0, 1, 2, 3, \ldots\}$ of natural numbers, and $\mathbb{Z}$ represents the set $\{\ldots, -2, -1, 0, 1, 2, \ldots\}$ of integers.)

★(a) $f\colon \mathbb{R} \to \mathbb{R}$, given by $f(x) = \sqrt{x-1}$

(b) $g\colon \mathbb{R} \to \mathbb{R}$, given by $g(x) = \sqrt{|x-1|}$

★(c) $h\colon \mathbb{R} \to \mathbb{R}$, given by $h(x) = \pm\sqrt{|x-1|}$

(d) $j\colon \mathbb{N} \to \mathbb{Z}$, given by $j(a) = \begin{cases} a-5, & \text{if } a \text{ is odd} \\ a-4, & \text{if } a \text{ is even} \end{cases}$

★(e) $k\colon \mathbb{R} \to \mathbb{R}$, given by $k(\theta) = \tan\theta$ (where θ is in radians)

★(f) $l\colon \mathbb{N} \to \mathbb{N}$, where $l(t)$ is the smallest prime number $\geq t$

(g) $m\colon \mathbb{R} \to \mathbb{R}$, given by $m(x) = \begin{cases} x-3 & \text{if } x \leq 2 \\ x+4 & \text{if } x \geq 2 \end{cases}$

2. Let $f\colon \mathbb{Z} \to \mathbb{N}$ (with $\mathbb{Z}$ and $\mathbb{N}$ as in Exercise 1) be given by $f(x) = 2|x|$.

★(a) Find the pre-image of the set $\{10, 20, 30\}$.

(b) Find the pre-image of the set $\{10, 11, 12, \ldots, 19\}$.

★(c) Find the pre-image of the multiples of 4 in $\mathbb{N}$.

★3. Let $f, g\colon \mathbb{R} \to \mathbb{R}$ be given by $f(x) = (5x-1)/4$ and $g(x) = \sqrt{3x^2+2}$. Find $g \circ f$ and $f \circ g$.

★4. Let $f\colon \mathbb{R}^2 \to \mathbb{R}^2$ be given by $f\left(\begin{bmatrix} x \\ y \end{bmatrix}\right) = \begin{bmatrix} 3 & -2 \\ 1 & 4 \end{bmatrix}\begin{bmatrix} x \\ y \end{bmatrix}$. Let $g\colon \mathbb{R}^2 \to \mathbb{R}^2$ be given by $g\left(\begin{bmatrix} x \\ y \end{bmatrix}\right) = \begin{bmatrix} -4 & 4 \\ 0 & 2 \end{bmatrix}\begin{bmatrix} x \\ y \end{bmatrix}$. Describe $g \circ f$ and $f \circ g$.

5. Let $A = \{1, 2, 3\}$, $B = \{4, 5, 6, 7\}$, and $C = \{8, 9, 10\}$.

(a) Give an example of functions $f\colon A \to B$ and $g\colon B \to C$ such that $g \circ f$ is onto but f is not onto.

(b) Give an example of functions $f\colon A \to B$ and $g\colon B \to C$ such that $g \circ f$ is one-to-one but g is not one-to-one.

6. For $n \geq 2$, show that $f\colon \mathcal{M}_{nn} \to \mathbb{R}$ given by $f(\mathbf{A}) = |\mathbf{A}|$ is onto but not one-to-one.

7. Show that $f\colon \mathcal{M}_{33} \to \mathcal{M}_{33}$ given by $f(\mathbf{A}) = \mathbf{A} + \mathbf{A}^T$ is neither one-to-one nor onto.

★8. For $n \geq 1$, show that the function $f\colon \mathcal{P}_n \to \mathcal{P}_n$ given by $f(\mathbf{p}) = \mathbf{p}'$ is neither one-to-one nor onto. When $n \geq 3$, what is the pre-image of the subset $\mathcal{P}_2$ of the codomain?

9. Prove that $f\colon \mathbb{R} \to \mathbb{R}$ given by $f(x) = 3x^3 - 5$ has an inverse by showing that it is both one-to-one and onto. Give a formula for $f^{-1}\colon \mathbb{R} \to \mathbb{R}$.

★10. Let $\mathbf{B}$ be a fixed nonsingular matrix in $\mathcal{M}_{nn}$. Show that the map $f\colon \mathcal{M}_{nn} \to \mathcal{M}_{nn}$ given by $f(\mathbf{A}) = \mathbf{B}^{-1}\mathbf{A}\mathbf{B}$ is both one-to-one and onto. What is the inverse of f?

11. Let $f\colon A \to B$ and $g\colon B \to C$ be functions.

(a) Prove that if $g \circ f$ is onto, then g is onto. (Compare this exercise with Exercise 5(a).)

(b) Prove that if $g \circ f$ is one-to-one, then f is one-to-one. (Compare this exercise with Exercise 5(b).)

★**12.** True or False:

(a) If f assigns elements of X to elements of Y, and two different elements of X are assigned by f to the same element of Y, then f is not a function.

(b) If f assigns elements of X to elements of Y and each element of X is assigned to exactly one element of Y but not every element of Y corresponds to an element of X, then f is a function.

(c) If $f: \mathbb{R} \to \mathbb{R}$ is a function and $f(5) = f(6)$, then $f^{-1}(5) = 6$.

(d) If $f: X \to Y$ and the domain of f equals the codomain of f, then f must be onto.

(e) If $f: X \to Y$ then f is one-to-one if $x_1 = x_2$ implies $f(x_1) = f(x_2)$.

(f) If $f: X \to Y$ and $g: Y \to Z$ are functions and $g \circ f: X \to Z$ is one-to-one, then both f and g are one-to-one.

(g) If $f: X \to Y$ is a function, then f has an inverse if f is either one-to-one or onto.

(h) If $f: X \to Y$ and $g: Y \to Z$ both have inverses and $g \circ f: X \to Z$ has an inverse, then $(g \circ f)^{-1} = g^{-1} \circ f^{-1}$.

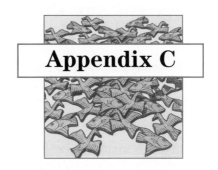

Appendix C

Complex Numbers

$\mathbf{I}$n this appendix, we define complex numbers and list their most important operations and properties. Complex numbers employ the use of the number i, which is outside the real number system and has the property that $i^2 = -1$.

DEFINITION

The set of **complex numbers** is the set of all numbers of the form $a + bi$, where $i^2 = -1$ and where a and b are real numbers. The **real part** of $a + bi$ is a, and the **imaginary part** of $a + bi$ is b.

Some examples of complex numbers are $2 + 3i$, $-\frac{1}{2} + \frac{1}{4}i$, and $\sqrt{3} - i$. Any real number a can be expressed as $a + 0i$, so the real numbers are a subset of the complex numbers; that is, $\mathbb{R} \subset \mathbb{C}$. A complex number of the form $0 + bi = bi$ is called a **pure imaginary** complex number.

Two complex numbers $a + bi$ and $c + di$ are **equal** if and only if $a = c$ and $b = d$. For example, if $3 + bi = c - 4i$, then $b = -4$ and $c = 3$.

The **magnitude**, or **absolute value**, of $a + bi$ is defined to be $|a + bi| = \sqrt{a^2 + b^2}$, a nonnegative real number. For example, the magnitude of $3 - 2i$ is $|3 - 2i| = \sqrt{3^2 + (-2)^2} = \sqrt{13}$.

We define **addition** of complex numbers by

$$(a + bi) + (c + di) = (a + c) + (b + d)i,$$

where $a, b, c, d \in \mathbb{R}$. Complex number **multiplication** is defined by

$$(a + bi)(c + di) = (ac - bd) + (ad + bc)i.$$

For example,

$$
\begin{aligned}
(3 - 2i)[(2 - i) + (-3 + 5i)] &= (3 - 2i)(-1 + 4i) \\
&= [(3)(-1) - (-2)(4)] + [(3)(4) + (-2)(-1)]i \\
&= 5 + 14i.
\end{aligned}
$$

If $z = a + bi$, we let $-z$ denote the special product $-1z = -a - bi$. The **complex conjugate** of a complex number $a + bi$ is defined as

$$\overline{a + bi} = a - bi.$$

For example, $\overline{-4 - 3i} = -4 + 3i$. Notice that if $z = a + bi$, then $\overline{z} = a - bi$, and so $z\overline{z} = (a + bi)(a - bi) = a^2 + b^2 = |a + bi|^2 = |z|^2$, a real number. We can use this property to calculate the **multiplicative inverse,** or **reciprocal,** of a complex number, as follows:

If $z = a + bi \neq 0$, then

$$\frac{1}{z} = \frac{1}{a + bi} = \frac{1}{a + bi} \cdot \frac{a - bi}{a - bi} = \frac{a - bi}{a^2 + b^2} = \frac{\overline{z}}{|z|^2}.$$

It is a straightforward matter to show that the operations of complex addition and multiplication satisfy the commutative, associative, and distributive laws. Some other useful properties are listed in the next theorem, whose proof is left as Exercise 3. You are asked to prove further properties in Exercise 4.

THEOREM C.1

Let $z_1, z_2, z_3 \in \mathbb{C}$. Then

(1)	$\overline{z_1 + z_2} = \overline{z_1} + \overline{z_2}$	Additive Conjugate Law
(2)	$\overline{(z_1 z_2)} = \overline{z_1}\, \overline{z_2}$	Multiplicative Conjugate Law
(3)	If $z_1 z_2 = 0$, then either $z_1 = 0$ or $z_2 = 0$	Zero Product Property
(4)	$z_1 = \overline{z_1}$ if and only if z_1 is real	Condition for complex number to be real
(5)	$z_1 = -\overline{z_1}$ if and only if z_1 is pure imaginary	Condition for complex number to be pure imaginary

---

Exercises for Appendix C

1. Perform the following computations involving complex numbers:
   - ⋆(a) $(6 - 3i) + (5 + 2i)$
   - (b) $8(3 - 4i)$
   - ⋆(c) $4((8 - 2i) - (3 + i))$
   - (d) $-3((-2 + i) - (4 - 2i))$
   - ⋆(e) $(5 + 3i)(3 + 2i)$
   - (f) $(-6 + 4i)(3 - 5i)$
   - ⋆(g) $(7 - i)(-2 - 3i)$
   - (h) $\overline{5 + 4i}$
   - ⋆(i) $\overline{9 - 2i}$
   - (j) $\overline{-6}$
   - ⋆(k) $\overline{(6 + i)(2 - 4i)}$
   - (l) $|8 - 3i|$
   - ⋆(m) $|-2 + 7i|$
   - (n) $\left|\overline{3 + 4i}\right|$

2. Find the multiplicative inverse (reciprocal) of each of the following:

⋆(a) $6 - 2i$ (b) $3 + 4i$

⋆(c) $-4 + i$ (d) $-5 - 3i$

▶**3.** (a) Prove parts (1) and (2) of Theorem C.1.

    (b) Prove part (3) of Theorem C.1.

    (c) Prove parts (4) and (5) of Theorem C.1.

4. Let z_1 and z_2 be complex numbers.

    (a) Prove that $|z_1 z_2| = |z_1||z_2|$.

    (b) If $z_1 \neq 0$, prove that $\left|\dfrac{1}{z_1}\right| = \dfrac{1}{|z_1|}$.

    (c) If $z_2 \neq 0$, prove that $\overline{\left(\dfrac{z_1}{z_2}\right)} = \dfrac{\overline{z_1}}{\overline{z_2}}$.

⋆**5.** True or False:

    (a) The magnitude (absolute value) of a complex number is the product of the number and its conjugate.

    (b) A complex number equals its conjugate if and only if it is zero.

    (c) The conjugate of a pure imaginary number is equal to its negative.

    (d) Every complex number has an additive inverse.

    (e) Every complex number has a multiplicative inverse.

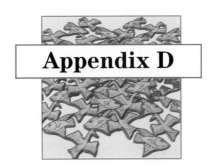

Appendix D

Computers and Calculators

Most students taking a first course in linear algebra have access to a computer algebra system or a graphing calculator. In this appendix, we explore the fundamental linear algebra operations within the following four popular computer algebra systems: *Maple 8 (Version 8.00)*, *Derive 5 (Version 5.04)*, *Mathematica 4.2*, and *MATLAB 6.5*. We also examine several widely used graphing calculators from Texas Instruments — the *TI-86*, *TI-89*, *TI-92*, *TI-92 Plus*, and the *Voyage 200*. The material related to each computer package or calculator can be read independently of the others, except the *TI-89*, *TI-92*, *TI-92 Plus*, and *Voyage 200* are treated together. Although we do not discuss the *TI-85* graphing calculator in this text, it is similar to the *TI-86*. The few differences between the two are explained at our web site.

This Appendix is not designed to be an encyclopedic reference for each of these computer programs or calculators but is intended merely to provide a general introduction. We focus on a small subset of commands that, when taken together, will perform the nontrivial calculations for almost all the examples and exercises in this text. After illustrating the input of vectors and matrices and performing several fundamental operations, we work through the same three examples in each case: row reducing a matrix (for a system with an infinite solution set), finding the eigenvalues for a matrix (and related eigenvectors), and replacing a given basis with an orthogonal basis (Gram-Schmidt Process).

Throughout this appendix, we use "typewriter" font (such as `vector` or `matrix`) to indicate the actual commands or keys used as input to a computer package or calculator. To conserve space, input and output will be displayed here in tables, even though most of the systems discussed do not display input and output in this way. Also, if you are using a different calculator or a different version of one of the computer packages discussed in this Appendix, you may not be able to execute commands exactly as they are listed here. In addition, if the commands do execute, you may get very different results. The remarks in this section are intended to apply only to the specific calculators and software versions mentioned. Each of the computer algebra systems was tested in the Windows environment. You may find differences in the interface in these programs if you run the software under a different operating system.

With most computer algebra systems and graphing calculators, there are many equivalent ways of performing the same operations. For brevity, we usually just indicate one of those ways. All of these computer algebra systems and calculators are capable of doing much more than we describe here. We encourage you to explore the available manuals and help menus for further information.

Maple 8

You are prompted for each line of input in *Maple 8* with the ">" symbol, and each input line must be followed by a semicolon. Users of *Maple 8* who want to perform linear algebra computations should begin by loading a utility named "linalg." This is done simply by typing

`with(linalg);`

We assume this has been done for all that follows.

Input of Vectors and Matrices; Fundamental Operations
To enter a vector, use the `vector` command. Type the word `vector`, and then, within parentheses, type a pair of brackets containing the vector entries (separated by commas). To enter a matrix, use the `matrix` command. First indicate the number of rows and the number of columns, followed by a pair of brackets "[ ]" containing the entries (row by row). It is often useful to assign names to vectors and matrices as they are entered. Use a colon followed by an equal sign (":=") to give a symbolic name to a vector or matrix.

Figures D.1 and D.2 illustrate the input of vectors/matrices as well as the fundamental operations of dot product, scalar multiplication, matrix multiplication, transpose, and inverse. The `evalm` function evaluates the result of vector/matrix calculations. The `dotprod` function calculates the complex dot product of two vectors. Of course, if the vectors are both real n-vectors, the result is the same as the real dot product. Addition and subtraction are performed using "+" and "−", respectively. Scalar multiplication can never be implied simply by putting a scalar next to a vector; instead, you must use "*". Matrix multiplication is indicated by "&*." The "^" symbol is used to find integer powers of a square matrix. In particular, "^(−1)" is used to calculate the inverse of a square matrix. The `transpose` function calculates the transpose of a matrix.

The number of significant digits displayed by *Maple 8* is controlled by the value of the variable `Digits`. *Maple 8* attempts to express all calculations in

Figure D.1

Maple 8 session: vectors; fundamental vector operations

Input	Output
>`with(linalg);`	(opens linear algebra utility)
>`v1:=vector([5,7,-4]);`	$v1 := [5, 7, -4]$
>`v2:=vector([x,y,z]);`	$v2 := [x, y, z]$
>`v3:= evalm(2*v1+3*v2);`	$v3 := [10 + 3x, 14 + 3y, -8 + 3z]$
>`dotprod(v1,v2);`	$5\overline{x} + 7\overline{y} - 4\overline{z}$

Figure D.2

Maple 8 session:
matrices; fundamental
matrix operations

Input	Output
`>M1:=matrix(3,4,[4,-1,6,-2,` `    -3,2,-3,2,-6,8,1,3]);`	$M1 := \begin{bmatrix} 4 & -1 & 6 & -2 \\ -3 & 2 & -3 & 2 \\ -6 & 8 & 1 & 3 \end{bmatrix}$
`>M2:=matrix(4,3,[2,-3,0,6,8,-1,` `    3,1,-2,2,-4,-2]);`	$M2 := \begin{bmatrix} 2 & -3 & 0 \\ 6 & 8 & -1 \\ 3 & 1 & -2 \\ 2 & -4 & -2 \end{bmatrix}$
`>M3:=evalm(M1-2*transpose(M2));`	$M3 := \begin{bmatrix} 0 & -13 & 0 & -6 \\ 3 & -14 & -5 & 10 \\ -6 & 10 & 5 & 7 \end{bmatrix}$
`>M4:=evalm(M1&*M2);`	$M4 := \begin{bmatrix} 16 & -6 & -7 \\ 1 & 14 & 0 \\ 45 & 71 & -16 \end{bmatrix}$
`>M5:=evalm(M4^(-1));`	$M5 := \begin{bmatrix} \frac{-224}{233} & \frac{-593}{233} & \frac{98}{233} \\ \frac{16}{233} & \frac{59}{233} & \frac{-7}{233} \\ \frac{-559}{233} & \frac{-1406}{233} & \frac{230}{233} \end{bmatrix}$
`>Digits:=4;`	$Digits := 4;$
`>M5:=evalm(1.0*M4^(-1));`	$M5 := \begin{bmatrix} -.9614 & -2.545 & .4206 \\ .06867 & .2532 & -.03004 \\ -2.399 & -6.034 & .9871 \end{bmatrix}$

fractional form if possible. To force an expression into decimal form, simply include a decimal number somewhere in the expression, as in the second expression for matrix M5 in Figure D.2.

Solving a Linear System; Gauss-Jordan Row Reduction Method
You can solve a linear system using the `solve` function by simply listing a set (in braces such as "{ }") containing each equation in the system in turn. When printing out the solution set of such a system, *Maple 8* expresses each dependent variable in terms of the independent variables. If a system has no solution, *Maple 8* returns no output at all.

You can also solve a linear system using the `rref` function. This function calculates the reduced row echelon form of a (possibly augmented) matrix. The matrix M6 in Figure D.3 is the augmented matrix for a linear system with an infinite solution set.

These functions are illustrated in Figure D.3, in which the same linear system is solved using both `solve` and `rref`. Using either method, the general solution set is $\{(-3c + 4,\ 2c + 5,\ c,\ -2)\}$.

Determinants; Eigenvalues/Eigenvectors; Characteristic Polynomial
The `det` function calculates the determinant of a square matrix. The characteristic polynomial for a square matrix M can be found by using the function `charpoly(M,x)`. Eigenvalues for a square matrix can be found by factoring the characteristic polynomial using the function `factor`. The percent mark in

Figure D.3

Maple 8 session: solution of a linear system; row reduction

Input	Output

>solve({3*x+y+7*z+2*w=13,
  2*x-4*y+14*z-w=-10,
  5*x+11*y-7*z+8*w=59,
  2*x+5*y-4*z-3*w=39});
	$\{y = 5 + 2z,\ x = 4 - 3z,\ z = z,\ w = -2\}$

>M6:=matrix(4,5,[3,1,7,2,13,
  2,-4,14,-1,-10,5,11,-7,8,59,
  2,5,-4,-3,39]);
	$M6 := \begin{bmatrix} 3 & 1 & 7 & 2 & 13 \\ 2 & -4 & 14 & -1 & -10 \\ 5 & 11 & -7 & 8 & 59 \\ 2 & 5 & -4 & -3 & 39 \end{bmatrix}$

>M7:=rref(M6);	$M7 := \begin{bmatrix} 1 & 0 & 3 & 0 & 4 \\ 0 & 1 & -2 & 0 & 5 \\ 0 & 0 & 0 & 1 & -2 \\ 0 & 0 & 0 & 0 & 0 \end{bmatrix}$

Maple 8 represents the most recent output, and so the command factor(%) indicates that the previous result is to be factored.

A more direct way to calculate the eigenvalues is to use the eigenvals function. The eigenvects function is even more powerful; it lists each eigenvalue, its algebraic multiplicity, and a basis for its eigenspace, respectively. Most of these functions are illustrated in Figure D.4, where the matrix M8 has two eigenvalues, $\lambda_1 = -5$, having algebraic and geometric multiplicity 1, and $\lambda_2 = 3$, having algebraic multiplicity 3 and geometric multiplicity 2.

The characteristic polynomial of a matrix M can also be computed directly by calculating the determinant of $x\mathbf{I}_n - $ M. The identity matrix is represented by "&*( )", where *Maple 8* determines its correct size from context.

Of course, a basis of eigenvectors for each eigenvalue λ of a square matrix M can also be calculated by row reducing the matrix $\lambda\mathbf{I}_n - $ M, setting each independent variable in turn equal to 1 with all others equal to 0, and then solving for the dependent variables.

These operations are illustrated in Figure D.5. Linearly independent eigenvectors for the earlier matrix M8, for eigenvalue $\lambda_2 = 3$, are found from the reduced row echelon form matrix M10 for M9 $= 3\mathbf{I}_4 - $ M8. First, by letting the third column variable of M10 equal 1 and its fourth column variable equal 0, we obtain $[\frac{1}{2}, -\frac{1}{2}, 1, 0]$, and then by letting its third column variable equal 0

Figure D.4

Maple 8 session: characteristic polynomial; eigenvalues; eigenvectors

Input	Output
>M8:=matrix(4,4,[5,2,0,1,-2,1,0,-1, 4,4,3,2,16,0,-8,-5]);	$M8 := \begin{bmatrix} 5 & 2 & 0 & 1 \\ -2 & 1 & 0 & -1 \\ 4 & 4 & 3 & 2 \\ 16 & 0 & -8 & -5 \end{bmatrix}$
>charpoly(M8,x);	$x^4 - 4x^3 - 18x^2 + 108x - 135$
>factor(%);	$(x+5)(x-3)^3$
>eigenvals(M8);	$-5, 3, 3, 3$
>eigenvects(M8);	$[3, 3, \{[0, 1, 2, -2], [1, 0, 4, -2]\}],$ $[-5, 1, \{[-1, 1, -2, 8]\}]$

and fourth column variable equal 1, we obtain $[\frac{1}{2}, -1, 0, 1]$. (You can verify that $\{[\frac{1}{2}, -\frac{1}{2}, 1, 0], [\frac{1}{2}, -1, 0, 1]\}$ spans the same two-dimensional subspace of $\mathbb{R}^4$ as the set $\{[1, 0, 4, -2], [0, 1, 2, -2]\}$ of eigenvectors obtained earlier from the eigenvects function.)

Figure D.5

Maple 8 session: characteristic polynomial via determinant; direct calculation of eigenspace

Input	Output
`>det(x*&*( )-M8);`	$x^4 - 4x^3 - 18x^2 + 108x - 135$
`>factor(%);`	$(x+5)(x-3)^3$
`>M9:=evalm(3*&*()-M8);` $M9 :=$	$\begin{bmatrix} -2 & -2 & 0 & -1 \\ 2 & 2 & 0 & 1 \\ -4 & -4 & 0 & -2 \\ -16 & 0 & 8 & 8 \end{bmatrix}$
`>M10:=rref(M9);` $M10 :=$	$\begin{bmatrix} 1 & 0 & -\frac{1}{2} & -\frac{1}{2} \\ 0 & 1 & \frac{1}{2} & 1 \\ 0 & 0 & 0 & 0 \\ 0 & 0 & 0 & 0 \end{bmatrix}$

Gram-Schmidt Process

You can perform the Gram-Schmidt Process in *Maple 8* for a given set of vectors simply by using the fundamental operations for vector addition and scalar multiplication. However, there is a built-in function, GramSchmidt (no hyphen), that replaces a given set of linearly independent vectors with an orthogonal basis for the span of those vectors. Note that *Maple 8* does not eliminate the fractions in the vectors it computes as we did in Section 6.1, and thus the orthogonal basis it generates may be different from what you might obtain by hand. An orthonormal basis for the span can then be produced by dividing each orthogonal basis vector by its length. The function norm(v,2) calculates the length of the vector v. These operations are illustrated in Figure D.6, where

Figure D.6

Maple 8 session: Gram-Schmidt Process; orthogonal and orthonormal bases

Input	Output
`>v5:=vector([2,1,0,-1]);`	$v5 := [2, 1, 0, -1]$
`>v6:=vector([1,0,2,-1]);`	$v6 := [1, 0, 2, -1]$
`>v7:=vector([0,-2,1,0]);`	$v7 := [0, -2, 1, 0]$
`>GramSchmidt({v5,v6,v7});`	$\{[2, 1, 0, -1], [0, \frac{-1}{2}, 2, \frac{-1}{2}], [\frac{2}{3}, \frac{-4}{3}, \frac{-1}{3}, 0]\}$
`>v8:=evalm(v5/norm(v5,2));`	$v8 := \left[\frac{\sqrt{6}}{3}, \frac{\sqrt{6}}{6}, 0, -\frac{\sqrt{6}}{6}\right]$
`>v9:=vector([0,-1/2,2,-1/2]);`	$v9 := \left[0, \frac{-1}{2}, 2, \frac{-1}{2}\right]$
`>v10:=evalm(v9/norm(v9,2));`	$v10 := \left[0, -\frac{\sqrt{2}}{6}, \frac{2\sqrt{2}}{3}, -\frac{\sqrt{2}}{6}\right]$
`>v11:=vector([2/3,-4/3,-1/3,0]);`	$v11 := \left[\frac{2}{3}, \frac{-4}{3}, \frac{-1}{3}, 0\right]$
`>v12:=evalm(v11/norm(v11,2));`	$v12 := \left[\frac{2\sqrt{21}}{21}, -\frac{4\sqrt{21}}{21}, -\frac{\sqrt{21}}{21}, 0\right]$

we find an orthogonal, and then an orthonormal, basis for the subspace of $\mathbb{R}^4$ spanned by the vectors labeled $v5$, $v6$, and $v7$.

You can verify that $\left\{ [2, 1, 0, -1], \left[0, -\frac{1}{2}, 2, -\frac{1}{2}\right], \left[\frac{2}{3}, -\frac{4}{3}, -\frac{1}{3}, 0\right] \right\}$ (vectors $v5$, $v9$, $v11$) is an orthogonal set of vectors spanning the same subspace of $\mathbb{R}^4$ as $\{[2, 1, 0, -1], [1, 0, 2, -1], [0, -2, 1, 0]\}$. An orthonormal basis for the same subspace is given by the vectors $v8$, $v10$, and $v12$. *Maple 8* can also generate an orthonormal basis by using the command $GramSchmidt(\{v5,v6,v7\},$ $normalized);$. When we tried this, the orthonormal basis obtained was numerically the same as $\{v8, \ v10, \ v12\}$; however, the form of the output was substantially different (try it!).

Derive 5

Algebraic expressions in *Derive 5* must be "authored" before operators can be applied to them. This is convenient to do because Derive 5 maintains an "expression entry bar" at the bottom of the active worksheet. Expressions entered into the expression entry bar are "authored" into the active worksheet. Note that the previous entry in the expression entry bar must be deleted before entering the next expression. You can also author expressions by clicking on the "author expression" icon on the toolbar at the top of the screen. The toolbar also contains shortcuts for many of the most frequently used commands.

Derive 5 numbers each line of input and output as it is created on the screen. The first line is referred to as "#1," the second as "#2," and so on. The line numbering system in *Derive 5* makes it handy to refer to any previously created expression, as you will observe in the sessions below.

At any given time in *Derive 5*, at least one line or part of a line is highlighted. You can change the highlighted area by using the arrow keys to scroll up or down. Some commands, such as $Simplify$, are designed to apply to the currently highlighted expression. In *Derive 5*, the most recent output line is highlighted by default, which makes it convenient to simplify the most recently computed expression. For example, if line #6 contains a dot product such as $[2, -3, 4] \cdot [-1, 0, -2]$, and if this dot product is highlighted on the screen, then clicking on $Simplify$ (and then on $Basic$, if using the menu) produces a new line (#7) containing the dot product -10. ("$Simplify, \ Basic$" is "=" on the toolbar.)

Input of Vectors and Matrices; Fundamental Operations

To enter a vector or matrix, simply click on the vector or the matrix icon on the toolbar. After indicating the dimension(s) of the vector or matrix, a template of the correct size will form, which you fill in with the desired vector or matrix entries. Use the Tab key to move from entry to entry. Alternatively, you could enter the vector $[1, 2, 3]$, by merely entering $[1,2,3]$ in the expression entry bar at the bottom of the screen. The matrix $\begin{bmatrix} 1 & 2 & 3 \\ 4 & 5 & 6 \end{bmatrix}$ would be entered as $[[1,2,3],[4,5,6]]$.

Use the "+" and "−" keys between vectors or matrices for addition and subtraction, respectively. Place a scalar before a vector or matrix to perform scalar multiplication. Use a " . " (period) between vectors to perform dot product and between matrices to perform matrix multiplication. Use the " ' " (left apostrophe) symbol after a matrix to calculate its transpose. The "^" symbol is used to find powers of a square matrix. In particular, use "^(−1)" to find the inverse of a square matrix. To convert fractions to decimals, use the Approximate command ("≈" on the toolbar). To specify the desired number of significant digits displayed, you must use Approximate in the Simplify menu at the top of the screen. These operations are illustrated in Figures D.7 and D.8. Lines 1, 2, 3, and 5 in Figure D.7 are entered in the expression entry bar, while lines 4 and 6 are entered by clicking on the appropriate items in the menus or the toolbar.

Figure D.7

Derive 5 session: vectors; fundamental vector operations

Input	Output
[5,7,-4]	#1: $[5, 7, -4]$
[x,y,z]	#2: $[x, y, z]$
2#1+3#2	#3: $2 \cdot [5, 7, -4] + 3 \cdot [x, y, z]$
Simplify, Basic (=)	#4: $[3x + 10,\ 3y + 14,\ 3z - 8]$
#1.#2	#5: $[5, 7, -4] \cdot [x, y, z]$
=	#6: $5x + 7y - 4z$

To save space in Figure D.8, we combined entries #13 and #14 (an expression entry and a Simplify command), only displaying the output on line #14. We will skip over similar lines of output in later figures as well.

Solving a Linear System; Gauss-Jordan Row Reduction Method

You can solve a linear system directly using the Solve command. Click on Solve and then System, and then indicate the number of equations to be solved. A template will appear, into which you type each equation in turn. Under the template, indicate the variables for which you wish to solve. If the system has no solution, the result "[ ]" will appear. If the system has more than one solution, a new system is returned corresponding to the rows of the reduced row echelon form of the augmented matrix for the original system. Equations in this new system are separated by a ^ sign.

You can also solve a system using the row_reduce function. This function computes the reduced row echelon form of a (possibly augmented) matrix.

The matrix on output line #18 below is the augmented matrix for a linear system with an infinite solution set. In Figure D.9, this linear system is solved using both Solve and row_reduce. You should verify that the general solution set obtained from either method is equivalent to $\{(-3c + 4,\ 2c + 5,\ c,\ -2)\}$.

Determinants; Eigenvalues/Eigenvectors; Characteristic Polynomial

The det function calculates the determinant of a square matrix. If M is an $n \times n$ matrix where n is even, the charpoly(M,x) function computes the characteristic polynomial for M, using x as the variable. In *Derive 5*, the charpoly(M,x) function actually returns the determinant $|M - xI_n|$, which is $(-1)^n |xI_n - M|$. Thus, for n odd, *Derive 5* computes the negative of the

Figure D.8

Derive 5 session: matrices; fundamental matrix operations

Input		Output
`[[4,-1,6,-2],` `[-3,2,-3,2],` `[-6,8,1,3]]`	#7:	$\begin{bmatrix} 4 & -1 & 6 & -2 \\ -3 & 2 & -3 & 2 \\ -6 & 8 & 1 & 3 \end{bmatrix}$
`[[2,-3,0],` `[6,8,-1],` `[3,1,-2],` `[2,-4,-2]]`	#8:	$\begin{bmatrix} 2 & -3 & 0 \\ 6 & 8 & -1 \\ 3 & 1 & -2 \\ 2 & -4 & -2 \end{bmatrix}$
`#7-2(#8')`	#9:	$\begin{bmatrix} 4 & -1 & 6 & -2 \\ -3 & 2 & -3 & 2 \\ -6 & 8 & 1 & 3 \end{bmatrix} - 2 \cdot \begin{bmatrix} 2 & -3 & 0 \\ 6 & 8 & -1 \\ 3 & 1 & -2 \\ 2 & -4 & -2 \end{bmatrix}$
=	#10:	$\begin{bmatrix} 0 & -13 & 0 & -6 \\ 3 & -14 & -5 & 10 \\ -6 & 10 & 5 & 7 \end{bmatrix}$
`#7.#8`	#11:	$\begin{bmatrix} 4 & -1 & 6 & -2 \\ -3 & 2 & -3 & 2 \\ -6 & 8 & 1 & 3 \end{bmatrix} \cdot \begin{bmatrix} 2 & -3 & 0 \\ 6 & 8 & -1 \\ 3 & 1 & -2 \\ 2 & -4 & -2 \end{bmatrix}$
=	#12:	$\begin{bmatrix} 16 & -6 & -7 \\ 1 & 14 & 0 \\ 45 & 71 & -16 \end{bmatrix}$
`#12^(-1), =`	#14:	$\begin{bmatrix} -\frac{224}{233} & -\frac{593}{233} & \frac{98}{233} \\ \frac{16}{233} & \frac{59}{233} & -\frac{7}{233} \\ -\frac{559}{233} & -\frac{1406}{233} & \frac{230}{233} \end{bmatrix}$
`Simplify,` `Approximate,` `4,Approximate`	#15:	$\begin{bmatrix} -0.9613 & -2.545 & 0.4206 \\ 0.06866 & 0.2532 & -0.03004 \\ -2.399 & -6.034 & -0.9871 \end{bmatrix}$

Figure D.9

Derive 5 session: solution of a linear system; row reduction

Input	Output
`Solve,System,4,OK,3x+y+7z+2w` `=13,Tab,2x-4y+14z-w=-10,` `Tab,5x+11y-7z+8w=59,Tab,` `2x+5y-4z-3w=39,Tab` `(x, y, z, w are highlighted),Solve`	#17: $[x + 3z = 4 \;\hat{}\; y - 2z = 5 \;\hat{}\; w = -2]$
`[[3,1,7,2,13],` `[2,-4,14,-1,-10],` `[5,11,-7,8,59],` `[2,5,-4,-3,39]]`	#18: $\begin{bmatrix} 3 & 1 & 7 & 2 & 13 \\ 2 & -4 & 14 & -1 & -10 \\ 5 & 11 & -7 & 8 & 59 \\ 2 & 5 & -4 & -3 & 39 \end{bmatrix}$
`row_reduce(#18), =`	#20: $\begin{bmatrix} 1 & 0 & 3 & 0 & 4 \\ 0 & 1 & -2 & 0 & 5 \\ 0 & 0 & 0 & 1 & -2 \\ 0 & 0 & 0 & 0 & 0 \end{bmatrix}$

characteristic polynomial that we have studied in this text.[1] The function `eigenvalues(M,x)` calculates the eigenvalues for M. Finally, the functions `exact_eigenvector(M,λ)` and `approx_eigenvector(M,λ)` give a general eigenvector of M for eigenvalue λ. Use the former function when the exact value of λ is known, but use the latter only when a close approximation of λ is known. (Note: The eigenvector functions are in a utility file known as "vector.mth". This file can be loaded by clicking on `File`, then `Load`, then `Utility File`. We found the file "vector.mth" in the directory "C:\DfW5\Math.")

Most of these functions are illustrated in Figure D.10, where the matrix in output line #21 has two eigenvalues, $\lambda_1 = -5$, having algebraic and geometric multiplicity 1, and $\lambda_2 = 3$, having algebraic multiplicity 3 and geometric multiplicity 2. Output line #29 displays a general eigenvector for the (exact) eigenvalue 3 of the matrix in line #21. This is written in a form using the symbols "@1" and "@2" to denote independent variables. Specific eigenvectors can be found by substituting values for @1 and @2. Letting "@1" equal 1 and "@2" equal 0, we obtain the eigenvector $[1, 0, 4, -2]$, while letting "@1" equal 0, and "@2" equal 1, we obtain the eigenvector $[0, 1, 2, -2]$. These two eigenvectors form a basis in $\mathbb{R}^4$ for the eigenspace E_3.

Figure D.10

Derive 5 session: characteristic polynomial; eigenvalues; eigenvectors

Input	Output	

`[[5,2,0,1],`
`[-2,1,0,-1],`
`[4,4,3,2],`
`[16,0,-8,-5]]`
	#21:	$\begin{bmatrix} 5 & 2 & 0 & 1 \\ -2 & 1 & 0 & -1 \\ 4 & 4 & 3 & 2 \\ 16 & 0 & -8 & -5 \end{bmatrix}$

`charpoly(#21,x), =`	#23: $x^4 - 4x^3 - 18x^2$ $+108x - 135$	
`Simplify,Factor` (with default `Rational`),`Factor`	#24: $(x+5)(x-3)^3$	
`eigenvalues(#21,x), =`	#26: $[x = 3, x = -5]$	
`File,Load,Utility File...,` (select: `vector.mth`),`Open`	#27: LOAD(C:\DfW5\Math\Vector.mth)	
`exact_eigenvector(#21,3), =`	#29: $[[@1,\ @2,\ 2 \cdot (2 \cdot @1 + @2),$ $-2 \cdot (@1 + @2)]]$	

The characteristic polynomial can also be computed directly using determinants. The $n \times n$ identity matrix is created using the function `identity_matrix(n)`.

If output line #k consists of a square matrix, a basis of eigenvectors for any eigenvalue λ of the matrix can also be calculated by row reducing the matrix "$\lambda \mathbf{I}_n - $#$k$", setting each independent variable in turn equal to 1 with all others equal to 0, and then solving for the dependent variables.

---

[1] Some linear algebra texts use $|\mathbf{M} - x\mathbf{I}_n|$ to define the characteristic polynomial of $\mathbf{M}$ rather than $|x\mathbf{I}_n - \mathbf{M}|$.

These operations are illustrated in Figure D.11 for the matrix in output line #21. Linearly independent eigenvectors for the eigenvector $\lambda_2 = 3$ are found from the reduced row echelon form matrix for $3\mathbf{I}_4 - \#21$, given in output line #35. Letting the third column variable equal 1 for the matrix in output line #35 and its fourth column variable equal 0, we obtain $[\frac{1}{2}, -\frac{1}{2}, 1, 0]$, and letting its third column variable equal 0 and its fourth column variable equal 1, we obtain $[\frac{1}{2}, -1, 0, 1]$. (You can verify that $\{[\frac{1}{2}, -\frac{1}{2}, 1, 0], [\frac{1}{2}, -1, 0, 1]\}$ spans the same two-dimensional subspace of $\mathbb{R}^4$ as the set $\{[1, 0, 4, -2], [0, 1, 2, -2]\}$ of eigenvectors obtained earlier from the exact_eigenvector function.)

Figure D.11

Derive 5 session: characteristic polynomial via determinant; direct calculation of eigenspace

Input	Output
det(x*identity_matrix(4)-#21), =	#31: $x^4 - 4x^3 - 18x^2 + 108x - 135$
3*identity_matrix(4)-#21, =	#33: $\begin{bmatrix} -2 & -2 & 0 & -1 \\ 2 & 2 & 0 & 1 \\ -4 & -4 & 0 & -2 \\ -16 & 0 & 8 & 8 \end{bmatrix}$
row_reduce(#33), =	#35: $\begin{bmatrix} 1 & 0 & -\frac{1}{2} & -\frac{1}{2} \\ 0 & 1 & \frac{1}{2} & 1 \\ 0 & 0 & 0 & 0 \\ 0 & 0 & 0 & 0 \end{bmatrix}$

Gram-Schmidt Process

The necessary calculations for the Gram-Schmidt Process can be performed in *Derive 5* in the manner illustrated in Figure D.12. We begin with a given set of three linearly independent vectors $\{[2, 1, 0, -1], [1, 0, 2, -1], [0, -2, 1, 0]\}$ in $\mathbb{R}^4$ (output lines #36 through #38) and construct an orthogonal basis for the span of those vectors. We then produce an orthonormal basis for the span by dividing each vector in the orthogonal basis by its length. The abs function calculates the length of a given vector.

You can verify that $\{[2, 1, 0, -1], [0, -\frac{1}{2}, 2, -\frac{1}{2}], [\frac{2}{3}, -\frac{4}{3}, -\frac{1}{3}, 0]\}$ (output lines #36, #40, and #42) is an orthogonal set of vectors spanning the same subspace

Figure D.12

Derive 5 session: Gram-Schmidt Process

Input	Output
[2,1,0,-1]	#36: $[2, 1, 0, -1]$
[1,0,2,-1]	#37: $[1, 0, 2, -1]$
[0,-2,1,0]	#38: $[0, -2, 1, 0]$
#37-(#37.#36)/(#36.#36)#36, =	#40: $[0, -\frac{1}{2}, 2, -\frac{1}{2}]$
#38-(#38.#36)/(#36.#36)#36 -(#38.#40)/(#40.#40)#40, =	#42: $[\frac{2}{3}, -\frac{4}{3}, -\frac{1}{3}, 0]$
#36/abs(#36), =	#44: $\frac{\sqrt{6}}{3}, \frac{\sqrt{6}}{6}, 0, -\frac{\sqrt{6}}{6}$
#40/abs(#40), =	#46: $0, -\frac{\sqrt{2}}{6}, \frac{2\sqrt{2}}{3}, -\frac{\sqrt{2}}{6}$
#42/abs(#42), =	#48: $\frac{2\sqrt{21}}{21}, -\frac{4\sqrt{21}}{21}, -\frac{\sqrt{21}}{21}, 0$

of $\mathbb{R}^4$ as {[2, 1, 0, −1], [1, 0, 2, −1], [0, −2, 1, 0]}. An orthonormal basis for the same subspace is given by the vectors in output lines #44, #46, and #48.

Mathematica 4.2

Mathematica 4.2 allows you to type in several lines of input before executing them. Each line of input is ordinarily followed by Enter, but when you are ready to execute the input lines, type Shift and Enter together after the last input line in the series. For example, in Figure D.13, you could type in all four lines of input and then execute them as a group. In Figures D.13 through D.18, we list the output from each input line next to that line, even though several input lines may, in fact, appear consecutively on the computer screen. Input into a *Mathematica 4.2* notebook is case sensitive; that is, *Mathematica 4.2* distinguishes between upper and lower case letters of the alphabet. This is especially important to keep in mind when entering commands.

Figure D.13

Mathematica 4.2 session: vectors; fundamental vector operations

Input	Output
`In[1]:=v1 = {5,7,-4}`	$\text{Out}[1] = \{5, 7, -4\}$
`In[2]:=v2= {x,y,z}`	$\text{Out}[2] = \{x, y, z\}$
`In[3]:=2*v1+3*v2`	$\text{Out}[3] = \{10 + 3x, 14 + 3y, -8 + 3z\}$
`In[4]:=v1.v2`	$\text{Out}[4] = 5x + 7y - 4z$
`In[5]:= In[1].Out[2]`	$\text{Out}[5] = 5x + 7y - 4z$

Mathematica 4.2 assigns a label of the form `In[k]:=` (for some positive integer k) to each input line and, similarly, an `Out[k]=` label to each output line. The `In[k]` label will not print on the screen until you have used Shift+Enter instead of just Enter, and only the first `In[k]` label will appear in each group. In Figure D.13, do not type in the `In[k]:=` labels. A line label often is a handy way to refer to the contents of a previous line in a *Mathematica 4.2* expression, as shown in line `In[5]` in Figure D.13. Even if a line number is not displayed, it can often be inferred for use in subsequent calculations. To save space, we illustrate the `In[ ]` and `Out[ ]` labels in Figure D.13 only.

In *Mathematica 4.2*, you can refer to the most recently computed expression using the symbol "%".

Input of Vectors and Matrices; Fundamental Operations

Vectors and matrices are entered with set notation (using braces, such as "{ }"). A matrix is expressed as a set of vectors representing successive rows of the matrix. For this reason, in *Mathematica 4.2* matrices are not typically displayed with their columns aligned vertically but as sets of vectors. However, the command `MatrixForm[M]` will cause *Mathematica 4.2* to display the matrix M in the customary matrix format. It is often useful to assign names to vectors and matrices as they are entered. Use an equal sign ("=") to give a symbolic name to a vector or matrix. For matrices, use the "=" sign *inside* the `MatrixForm` command, as illustrated in Figure D.14.

Figure D.14

Mathematica 4.2 session: matrices; fundamental matrix operations

Input	Output
`M1 = {{4,-1,6,-2},{-3,2,-3,2},` `{-6,8,1,3}}`	$\{\{4,-1,6,-2\},\{-3,2,-3,2\},$ $\{-6,8,1,3\}\}$
`MatrixForm[M1]`	$\begin{pmatrix} 4 & -1 & 6 & -2 \\ -3 & 2 & -3 & 2 \\ -6 & 8 & 1 & 3 \end{pmatrix}$
`MatrixForm[M2 = {{2,-3,0},{6,8,-1},` `{3,1,-2},{2,-4,-2}}]`	$\begin{pmatrix} 2 & -3 & 0 \\ 6 & 8 & -1 \\ 3 & 1 & -2 \\ 2 & -4 & -2 \end{pmatrix}$
`MatrixForm[M3 = M1-2*Transpose[M2]]`	$\begin{pmatrix} 0 & -13 & 0 & -6 \\ 3 & -14 & -5 & 10 \\ -6 & 10 & 5 & 7 \end{pmatrix}$
`MatrixForm[M4 = M1. M2]`	$\begin{pmatrix} 16 & -6 & -7 \\ 1 & 14 & 0 \\ 45 & 71 & -16 \end{pmatrix}$
`MatrixForm[M5=MatrixPower[M4, -1]]`	$\begin{pmatrix} -\frac{224}{233} & -\frac{593}{233} & \frac{98}{233} \\ \frac{16}{233} & \frac{59}{233} & -\frac{7}{233} \\ -\frac{559}{233} & -\frac{1406}{233} & \frac{230}{233} \end{pmatrix}$
`MatrixForm[M5 = Inverse[M4]]`	$\begin{pmatrix} -\frac{224}{233} & -\frac{593}{233} & \frac{98}{233} \\ \frac{16}{233} & \frac{59}{233} & -\frac{7}{233} \\ -\frac{559}{233} & -\frac{1406}{233} & \frac{230}{233} \end{pmatrix}$
`MatrixForm[M5] // N`	$\begin{pmatrix} -0.961373 & -2.54506 & 0.420601 \\ 0.0686695 & 0.253219 & -0.0300429 \\ -2.39914 & -6.03433 & 0.987124 \end{pmatrix}$

Matrices can also be entered using the `Create/Table/Matrix/Palette` subheading under the `Input` menu. A shortcut to this is `Shift+Ctrl+C` (hold down the `Shift` and `Ctrl` keys while typing `C`).[2] You choose the size of the matrix and are then presented with a template to fill in the entries. Use the `Tab` key to move from entry to entry within the matrix. Use the arrow key to place the cursor outside the matrix before hitting `Enter` or `Shift+Enter` to finish entering the matrix. You can edit any previous input or output to form a new input by merely moving the cursor to the object to be edited.

Figures D.13 and D.14 illustrate the input of vectors/matrices as well as the fundamental operations of dot product, scalar multiplication, matrix multiplication, transpose, and inverse. Addition and subtraction are performed using "+" and "−", respectively. You may use "*" for scalar multiplication or merely type the scalar in front of the vector or matrix. A dot product of two vectors

---

[2] There are similar alternatives, shortcuts, and/or toolbar icons for many commands, but we will not discuss all the possibilities for every command we illustrate.

or a multiplication of two matrices is performed by placing a "." (period) between them. The function `MatrixPower[M, n]` function calculates the nth power of the matrix M. The function `Inverse[M]` calculates the inverse of the square matrix M. The `Transpose[M]` function calculates the transpose of the matrix M. Note the use of brackets (" [ ] ") in these last three functions. Placing "//N" after a calculation causes a decimal approximation of the result to be printed out.

Notice that both ways of calculating matrix M5 in Figure D.14 (using `MatrixPower` and `Inverse`) produce the same result.

Solving a Linear System; Gauss-Jordan Row Reduction Method

You can solve a linear system using the `Solve[{eqns},{vars}]` function. The set {eqns} contains each equation of the system in turn but with "==" (double equal sign) used in place of each equal sign. The set {vars} contains all the variables of the system. When printing out the solution set of such a system, *Mathematica 4.2* expresses each dependent variable in terms of the independent variables. If there is no solution to the system, the output is "{ }".

You can also solve a linear system using the `RowReduce[M]` function. This function calculates the reduced row echelon form of the (possibly augmented) matrix M. The matrix M6 in Figure D.15 is the augmented matrix for a linear system with an infinite solution set.

In Figure D.15, the same linear system is solved using both `Solve` and `RowReduce`. Using either method, the general solution set is $\{(-3c+4, 2c+5, c, -2)\}$.

Figure D.15

Mathematica 4.2 session: solution of a linear system; row reduction

Input	Output

`Solve[{3x+y+7z+2w==13,`
`2x-4y+14z-w==-10,`
`5x+11y-7z+8w==59,`
`2x+5y-4z-3w==39},{x,y,z,w}]`
	$\{\{x \longrightarrow 4 - 3z,$
	$y \longrightarrow 5 + 2z,$
	$w \longrightarrow -2\}\}$

`MatrixForm[M6={{3,1,7,2,13},`
`{2,-4,14,-1,-10},{5,11,-7,8,59},`
`{2,5,-4,-3,39}}]`
	$\begin{pmatrix} 3 & 1 & 7 & 2 & 13 \\ 2 & -4 & 14 & -1 & -10 \\ 5 & 11 & -7 & 8 & 59 \\ 2 & 5 & -4 & -3 & 39 \end{pmatrix}$

`MatrixForm[M7 = RowReduce[M6]]`	$\begin{pmatrix} 1 & 0 & 3 & 0 & 4 \\ 0 & 1 & -2 & 0 & 5 \\ 0 & 0 & 0 & 1 & -2 \\ 0 & 0 & 0 & 0 & 0 \end{pmatrix}$

Determinants; Eigenvalues/Eigenvectors; Characteristic Polynomial

The `Det[M]` function calculates the determinant of the square matrix M. The function `Eigenvalues[M]` calculates the eigenvalues of a square matrix M. The `Eigenvectors[M]` function returns a set of n eigenvectors for the $n \times n$ matrix M. However, if M is not diagonalizable, then `Eigenvectors` returns as many linearly independent eigenvectors for M as possible but includes the zero vector as often as necessary to ensure that a total of n vectors is output.

The matrix M8 in Figure D.16 has two eigenvalues, $\lambda_1 = -5$, with algebraic and geometric multiplicity 1, and $\lambda_2 = 3$, with algebraic multiplicity 3 and geometric multiplicity 2. Eigenvectors corresponding to these eigenvalues appear as *rows* of the output matrix from the command MatrixForm [Eigenvectors[M8]].

Figure D.16

Mathematica 4.2 session: eigenvalues; eigenvectors

Input	Output
MatrixForm[M8={{5,2,0,1},{-2,1,0,-1}, {4,4,3,2},{16,0,-8,-5}}]	$\begin{pmatrix} 5 & 2 & 0 & 1 \\ -2 & 1 & 0 & -1 \\ 4 & 4 & 3 & 2 \\ 16 & 0 & -8 & -5 \end{pmatrix}$
Eigenvalues[M8]	$\{-5, 3, 3, 3\,\}$
MatrixForm[Eigenvectors[M8]]	$\begin{pmatrix} -1 & 1 & -2 & 8 \\ 1 & -2 & 0 & 2 \\ 1 & -1 & 2 & 0 \\ 0 & 0 & 0 & 0 \end{pmatrix}$

Notice in the list of eigenvectors for M8 that $[-1, 1, -2, 8]$ is an eigenvector for $\lambda_1 = -5$, while the other two nonzero vectors are eigenvectors for $\lambda_2 = 3$. Since the algebraic multiplicity of $\lambda_2 = 3$ is exactly 1 greater than its geometric multiplicity, a single zero vector is included in the list.

The characteristic polynomial of a square matrix can be computed directly in *Mathematica 4.2* using determinants. The $k \times k$ identity matrix is created using the function IdentityMatrix[k]. If the most recent output line consists of a polynomial, you can factor it using the command Factor[%].

Of course, a basis of the eigenspace for any eigenvalue λ of a square matrix M can also be calculated by row reducing the matrix $\lambda \mathbf{I}_n - M$, setting each independent variable in turn equal to 1 with all others equal to 0, and then solving for the dependent variables. These operations are illustrated in Figure D.17.

Linearly independent eigenvectors for the earlier matrix M8, for the eigenvalue $\lambda_2 = 3$, are found from the reduced row echelon form matrix M10 for $M9 = 3\mathbf{I}_4 - M8$ in Figure D.17. Letting the third column variable in M10 equal 1 and fourth column variable equal 0, we obtain $[\frac{1}{2}, -\frac{1}{2}, 1, 0]$, and then by letting the third column variable equal 0 and fourth column variable equal 1, we obtain $[\frac{1}{2}, -1, 0, 1]$. Since these are scalar multiples of $[1, -1, 2, 0]$ and $[1, -2, 0, 2]$, respectively, it follows that $\{[\frac{1}{2}, -\frac{1}{2}, 1, 0], [\frac{1}{2}, -1, 0, 1]\}$ spans the same two-dimensional subspace of $\mathbb{R}^4$ as the set $\{[1, -2, 0, 2], [1, -1, 2, 0]\}$ of eigenvectors for $\lambda_2 = 3$, obtained earlier from the Eigenvectors function.

Gram-Schmidt Process

There is a function to perform the Gram-Schmidt Process in *Mathematica 4.2*, but to use it, the utility LinearAlgebra'Orthogonalization' must be

Input	Output
`Det[x*IdentityMatrix[4]-M8]`	$-135 + 108x - 18x^2 - 4x^3 + x^4$
`Factor[%]`	$(-3+x)^3(5+x)$
`MatrixForm[M9 = 3*IdentityMatrix[4]-M8]`	$\begin{pmatrix} -2 & -2 & 0 & -1 \\ 2 & 2 & 0 & 1 \\ -4 & -4 & 0 & -2 \\ -16 & 0 & 8 & 8 \end{pmatrix}$
`MatrixForm[M10 = RowReduce[M9]]`	$\begin{pmatrix} 1 & 0 & -\frac{1}{2} & -\frac{1}{2} \\ 0 & 1 & \frac{1}{2} & 1 \\ 0 & 0 & 0 & 0 \\ 0 & 0 & 0 & 0 \end{pmatrix}$

loaded. (This is done by typing "`<<`" before the name of the utility, as in Figure
D.18. Note that a single left quotation mark is used before and after the word
`Orthogonalization`.)[3]

Input	Output
`<<LinearAlgebra'Orthogonalization'`	(loads necessary utility)
`v5 = {2,1,0,-1}`	$\{2,1,0,-1\}$
`v6 = {1,0,2,-1}`	$\{1,0,2,-1\}$
`v7 = {0,-2,1,0}`	$\{0,-2,1,0\}$
`GramSchmidt[{v5,v6,v7},` `        Normalized->False]`	$\left\{\{2,1,0,-1\},\left\{0,-\frac{1}{2},2,-\frac{1}{2}\right\},\right.$ $\left.\left\{\frac{2}{3},-\frac{4}{3},-\frac{1}{3},0\right\}\right\}$
`Normalize[v5]`	$\left\{\sqrt{\frac{2}{3}},\frac{1}{\sqrt{6}},0,-\frac{1}{\sqrt{6}}\right\}$
`GramSchmidt[{v5,v6,v7}]`	$\left\{\left\{\sqrt{\frac{2}{3}},\frac{1}{\sqrt{6}},0,-\frac{1}{\sqrt{6}}\right\},\right.$ $\left\{0,-\frac{1}{3\sqrt{2}},\frac{2\sqrt{2}}{3},-\frac{1}{3\sqrt{2}}\right\},$ $\left.\left\{\frac{2}{\sqrt{21}},-\frac{4}{\sqrt{21}},-\frac{1}{\sqrt{21}},0\right\}\right\}$

The `GramSchmidt[vecs]` (no hyphen) function replaces a given set `vecs`
of linearly independent vectors with an orthonormal basis for the span of
those vectors. To produce an orthogonal basis instead, include `Normalized`
`-> False` after the set `vecs`. The function `Normalize[v]` normalizes a
given vector v. (The `LinearAlgebra'Orthogonalization'` utility is also
needed for the `Normalize` function.)

These operations are illustrated in Figure D.18. We first find an orthogonal basis for the subspace of $\mathbb{R}^4$ spanned by the vectors v5, v6, and v7 using the `GramSchmidt` function. Each vector in the orthogonal basis can be

---

[3] The minimal installation of *Mathematica 4.2* does not include this utility. Thus, we recommend
the full installation. But even without this utility, you can still perform the Gram-Schmidt
process just by calculating dot products and linear combinations of the vectors involved.

normalized individually to produce an orthonormal basis, and we illustrate this for the vector v5. Finally, we display the entire orthonormal basis using the GramSchmidt function.

You should verify that $\{[2, 1, 0, -1], [0, -\frac{1}{2}, 2, -\frac{1}{2}], [\frac{2}{3}, -\frac{4}{3}, -\frac{1}{3}, 0]\}$ is an orthogonal set of vectors spanning the same subspace of $\mathbb{R}^4$ as $\{[2, 1, 0, -1],$ $[1, 0, 2, -1], [0, -2, 1, 0]\}$, and that $\left\{\left[\sqrt{\frac{2}{3}}, \frac{1}{\sqrt{6}}, 0, -\frac{1}{\sqrt{6}}\right], \left[0, -\frac{1}{3\sqrt{2}}, \frac{2\sqrt{2}}{3}, -\frac{1}{3\sqrt{2}}\right],\right.$ $\left.\left[\frac{2}{\sqrt{21}}, -\frac{4}{\sqrt{21}}, -\frac{1}{\sqrt{21}}, 0\right]\right\}$ is an orthonormal set of vectors spanning this same subspace.

MATLAB 6.5

MATLAB 6.5 divides the screen into the following three windows: the Workspace, the Command Window, and the Command History. New commands are entered in the Command Window. The prompt symbol ">>" will appear there. Once entered, the command will also appear in the Command History window, and the result will appear in the Workspace window. You may click on objects in the Workspace window, edit them and use them. You may also use the variable "ans" to refer to the answer generated from the last command. Note that *MATLAB 6.5* distinguishes between uppercase and lowercase letters. For example, the variable M9 is not the same in *MATLAB 6.5* as the variable m9.

Input of Vectors and Matrices; Fundamental Operations

To enter a vector, simply type the list of entries (separated by commas) within a pair of brackets (such as "[ ]"). Matrix input is done similarly, with a semicolon at the end of each row. It is often useful to assign names to vectors and matrices as they are entered. Use an equal sign ("=") to give a symbolic name to a vector or matrix. Notice that usually no brackets/braces are displayed when vectors and matrices are printed out.

Figures D.19 and D.20 illustrate the input of vectors/matrices as well as the fundamental operations of dot product, scalar multiplication, matrix multiplication, transpose, and inverse. The dot function calculates the dot product of two vectors. Addition and subtraction are performed using "+" and "−", respectively. Scalar multiplication can never be implied simply by putting a scalar next to a vector; instead, you must use "*". Matrix multiplication is also indicated by "*". The "^" symbol is used to find integer powers of a square matrix. In particular, "^(−1)" can be used to calculate the inverse of a square matrix, although the answer may be given as a decimal approximation. The inv function can also be used to find the inverse of a square matrix. The "'" symbol denotes the transpose of a matrix.

The sym function is used to create a symbolic matrix or vector. This is used for vectors or matrices that contain variables or fractions, as illustrated with the vector v2 in Figure D.19. When typing in such a vector or matrix, it must be put inside single quotes, as shown for the vector v2 in Figure D.19. Using sym on an existing vector or matrix will turn it into a symbolic vector or matrix.

Figure D.19

MATLAB 6.5 session:
vectors; fundamental
vector operations

Input	Output
>> v1=[5,7,-4]	v1 = 5 7 −4
>> v2=sym('[x,y,z]')	$v2 = [x, y, z]$
>> v3=2*v1+3*v2	$v3 = [10 + 3 * x, \ 14 + 3 * y, \ −8 + 3 * z]$
>> dot(v1,v2)	$\text{ans} = 5 * x + 7 * y − 3 * z$

Figure D.20

MATLAB 6.5 session:
matrices; fundamental
matrix operations

Input	Output
>> M1=[4,-1,6,-2; -3,2,-3,2;-6,8,1,3]	$M1 = \begin{matrix} 4 & −1 & 6 & −2 \\ −3 & 2 & −3 & 2 \\ −6 & 8 & 1 & 3 \end{matrix}$
>> M2=[2,-3,0; 6,8,-1; 3,1,-2; 2,-4,-2]	$M2 = \begin{matrix} 2 & −3 & 0 \\ 6 & 8 & −1 \\ 3 & 1 & −2 \\ 2 & −4 & −2 \end{matrix}$
>> M3=M1-2*M2'	$M3 = \begin{matrix} 0 & −13 & 0 & −6 \\ 3 & −14 & −5 & 10 \\ −6 & 10 & 5 & 7 \end{matrix}$
>> M4=M1*M2	$M4 = \begin{matrix} 16 & −6 & −7 \\ 1 & 14 & 0 \\ 45 & 71 & −16 \end{matrix}$
>> M5=sym(inv(M4))	$M5 = \begin{bmatrix} −224/233, & −593/233, & 98/233 \\ 16/233, & 59/233, & −7/233 \\ −559/233, & −1406/233 & 230/233 \end{bmatrix}$
>> M5=sym(M4^(-1))	$M5 = \begin{bmatrix} −224/233, & −593/233, & 98/233 \\ 16/233, & 59/233, & −7/233 \\ −559/233, & −1406/233 & 230/233 \end{bmatrix}$
>> M5=M4^(-1)	$M5 = \begin{matrix} −0.9614 & −2.5451 & 0.4206 \\ 0.0687 & 0.2532 & −0.0300 \\ −2.3991 & −6.0343 & 0.9871 \end{matrix}$

Symbolic vectors and matrices are displayed with brackets around each row,
while nonsymbolic vectors and matrices have no such brackets displayed.
Note that the combination of sym and inv used on matrix M4 in Figure D.20
displays the exact inverse of that matrix.

Note the difference in output in Figure D.20 between the symbolic version
of M5 and the nonsymbolic (final) version. Although we do not illustrate it
here, the numeric function gives approximate values for a symbolic object
containing only numbers.

Solving a Linear System; Gauss-Jordan Row Reduction Method
You can solve a linear system with the solve function. Type a pair of brackets
containing the variables of the system in lexicographic order, then an equal sign

followed by solve, and finally a pair of parentheses containing each equation in the system in turn (separated by commas). Note that each equation in the system should be enclosed within a pair of right apostrophes ("'"). When printing out the solution set of such a system, *MATLAB 6.5* expresses each dependent variable in terms of the independent variables. If there is no solution to the system, the output is "x = [empty sym]", where "x" represents the first variable in the system.

You can also solve a system using the rref function. This function calculates the reduced row echelon form of a (possibly augmented) matrix. The matrix M6 in Figure D.21 is the augmented matrix for a linear system with an infinite solution set.

Figure D.21

MATLAB 6.5 session: solution of a linear system; row reduction

Input	Output

>> [w,x,y,z] = solve
    ('3*x+y+7*z+2*w=13',
     '2*x-4*y+14*z-w=-10',
     '5*x+11*y-7*z+8*w=59',
     '2*x+5*y-4*z-3*w=39')
	$w = -2$
	$x = -3*z + 4$
	$y = 2*z + 5$
	$z = z$

>> M6=[3,1,7,2,13;
    2,-4,14,-1,-10;
    5,11,-7,8,59; 2,5,-4,-3,39]
	$M6 = \begin{matrix} 3 & 1 & 7 & 2 & 13 \\ 2 & -4 & 14 & -1 & -10 \\ 5 & 11 & -7 & 8 & 59 \\ 2 & 5 & -4 & -3 & 39 \end{matrix}$

>> M7=rref(M6)	$M7 = \begin{matrix} 1 & 0 & 3 & 0 & 4 \\ 0 & 1 & -2 & 0 & 5 \\ 0 & 0 & 0 & 1 & -2 \\ 0 & 0 & 0 & 0 & 0 \end{matrix}$

These functions are illustrated in Figure D.21, in which the same linear system is solved using both solve and rref. Using either method, the general solution set is $\{(-3c + 4, 2c + 5, c, -2)\}$.

Determinants; Eigenvalues/Eigenvectors; Characteristic Polynomial: The det function calculates the determinant of a square matrix. The coefficients (alone) of the characteristic polynomial (from highest to lowest degree) are computed using the poly function. Using poly on a symbolic matrix causes the characteristic polynomial to be displayed in the customary polynomial form.

Eigenvalues for a given square matrix can be found by calculating the roots of its characteristic polynomial. One way to do this is to use the roots function on the coefficients found by poly. To find the roots of a symbolic polynomial, use the solve command instead.

A more direct way to calculate eigenvalues (and eigenvectors as well) is to create an appropriate equation involving the eigensys function. Place [V, E] on the left side of the equation and eigensys(M) on the right side, where M is the given matrix. This will create a matrix E containing all the eigenvalues of M on its main diagonal and a matrix V whose *columns* are a set of linearly independent eigenvectors for these eigenvalues. Regardless of the method you use to find eigenvalues and eigenvectors, the complex number "i" (see Appendix C) may appear in the result.

These functions are illustrated in Figure D.22, where the matrix M8 has two eigenvalues, $\lambda_1 = -5$, with algebraic and geometric multiplicity 1, and $\lambda_2 = 3$, with algebraic multiplicity 3 and geometric multiplicity 2. The *columns* of V give a linearly independent set of eigenvectors for M8: $[1, 0, 4, -2]$, $[0, 1, 2, -2]$, and $[-1, 1, -2, 8]$.

Figure D.22

MATLAB 6.5 session: characteristic polynomial; eigenvalues; eigenvectors

Input	Output
`>> M8=[5,2,0,1;-2,1,0,-1;` `4,4,3,2;16,0,-8,-5]`	$\text{M8} = \begin{array}{rrrr} 5 & 2 & 0 & 1 \\ -2 & 1 & 0 & -1 \\ 4 & 4 & 3 & 2 \\ 16 & 0 & -8 & -5 \end{array}$
`>> det(M8)`	ans $= -135$
`>> poly(M8)`	ans $= 1.0000 \quad -4.0000 \quad -18.0000$ $108.0000 \quad -135.0000$
`>> roots(poly(M8))`	ans $= -5.0000$ $3.0000 + 0.0000i$ $3.0000 - 0.0000i$ 3.0000
`>> poly(sym(M8))`	ans $= x\hat{}4 - 4*x\hat{}3 - 18*x\hat{}2 + 108*x - 135$
`>> solve(poly(sym(M8)))`	ans $= \begin{array}{r}[-5]\\ [\ 3]\\ [\ 3]\\ [\ 3]\end{array}$
`>> [V,E] = eigensys(M8)`	$V = \begin{array}{rrr}[1, & 0, -1]\\ [0, & 1, \ 1]\\ [4, & 2, -2]\\ [-2, & -2, 8]\end{array} \quad E = \begin{array}{l}[3,0,0,0]\\ [0,3,0,0]\\ [0,0,3,0]\\ [0,0,0,-5]\end{array}$

Of course, a basis of eigenvectors for any eigenvalue λ of a (square) matrix M can also be calculated by row reducing the matrix $\lambda \mathbf{I}_n - \text{M}$, setting each independent variable in turn equal to 1 with all others equal to 0 and then solving for the dependent variables. The $k \times k$ identity matrix is represented by `eye(k)`.

In Figure D.23, linearly independent eigenvectors for the earlier matrix M8, for eigenvalue $\lambda_2 = 3$, are found from the reduced row echelon form matrix M10 of M9 $= 3\mathbf{I}_4 - \text{M8}$. Letting the third column variable of M10 equal 1 and its fourth column variable equal 0, we obtain $[\frac{1}{2}, -\frac{1}{2}, 1, 0]$, and letting its third column variable equal 0 and fourth column variable equal 1, we obtain $[\frac{1}{2}, -1, 0, 1]$. (You can verify that $\{[\frac{1}{2}, -\frac{1}{2}, 1, 0], [\frac{1}{2}, -1, 0, 1]\}$ spans the same

Figure D.23

MATLAB 6.5 session: direct calculation of eigenspace

Input	Output
>> M9 = 3*eye(4)-M8 M9 =	$\begin{array}{rrrr} -2 & -2 & 0 & -1 \\ 2 & 2 & 0 & 1 \\ -4 & -4 & 0 & -2 \\ -16 & 0 & 8 & 8 \end{array}$
>> M10 = rref(M9) M10 =	$\begin{array}{rrrr} 1.0000 & 0 & -0.5000 & -0.5000 \\ 0 & 1.0000 & 0.5000 & 1.0000 \\ 0 & 0 & 0 & 0 \\ 0 & 0 & 0 & 0 \end{array}$

two-dimensional subspace of $\mathbb{R}^4$ as the set $\{[1, 0, 4, -2], [0, 1, 2, -2]\}$ of eigenvectors obtained earlier from the `eigensys` function.)

Gram-Schmidt Process

The necessary calculations for the Gram-Schmidt Process can be performed in *MATLAB 6.5* in the manner illustrated in Figure D.24. We begin with a given set of 3 linearly independent vectors $\{[2, 1, 0, -1], [1, 0, 2, -1], [0, -2, 1, 0]\}$ in $\mathbb{R}^4$ (vectors v5, v6, and v7), and construct an orthogonal basis (v5, v8, v9) for the span of those vectors. We then produce an orthonormal basis (v10, v11, v12) for the span by dividing each orthogonal basis vector by its length. The `norm` function calculates the length of a given vector. The results of all of these computations can be converted to symbolic vectors using the `sym` function.

Figure D.24

MATLAB 6.5 session: Gram-Schmidt Process; orthogonal and orthonormal bases

Input	Output
>> v5=[2,1,0,-1]	v5 = 2 1 0 -1
>> v6=[1,0,2,-1]	v6 = 1 0 2 -1
>> v7=[0,-2,1,0]	v7 = 0 -2 1 0
>> v8=v6-(dot(v6,v5)/dot(v5,v5))*v5	v8 = 0 -0.5000 2.0000 -0.5000
>> sym(v8)	ans = $[0, -1/2, 2, -1/2]$
>> v9=v7-(dot(v7,v5)/dot(v5,v5))*v5 -(dot(v7,v8)/dot(v8,v8))*v8	v9 = 0.6667 -1.3333 -0.3333 0
>> sym(v9)	ans = $[2/3, -4/3, -1/3, 0]$
>> v10=v5 / norm(v5)	v10 = 0.8165 0.4082 0 -0.4082
>> sym(v10)	ans = $[\mathrm{sqrt}(2/3), \mathrm{sqrt}(1/6),$ $0, -\mathrm{sqrt}(1/6)]$
>> v11=v8 / norm(v8)	v11 = 0 -0.2357 0.9428 -0.2357
>> sym(v11)	ans = $[0, -\mathrm{sqrt}(1/18),$ $\mathrm{sqrt}(8/9), -\mathrm{sqrt}(1/18)]$
>> v12=v9 / norm(v9)	v12 = 0.4364 -0.8729 -0.2182 0
>> sym(v12)	ans = $[\mathrm{sqrt}(4/21), -\mathrm{sqrt}(16/21),$ $-\mathrm{sqrt}(1/21), 0]$

Analogous calculations can be made if you begin with symbolic vectors. However, the `norm` function is not defined for symbolic vectors. Instead, the norm of a symbolic vector can be calculated by taking the square root of the dot product of the vector with itself. (Another approach is to first apply the `numeric` function to the symbolic vector, take the `norm`, and then convert that result to a symbolic value.)

You can verify that $\{[2, 1, 0, -1], [0, -\frac{1}{2}, 2, -\frac{1}{2}], [\frac{2}{3}, -\frac{4}{3}, -\frac{1}{3}, 0]\}$ (vectors v5, v8, and v9) is an orthogonal set of vectors spanning the same subspace of $\mathbb{R}^4$ as $\{[2, 1, 0, -1], [1, 0, 2, -1], [0, -2, 1, 0]\}$. An orthonormal basis for the same subspace is given by the set $\left\{\left[\sqrt{\frac{2}{3}}, \sqrt{\frac{1}{6}}, 0, -\sqrt{\frac{1}{6}}\right], \left[0, -\sqrt{\frac{1}{18}}, \sqrt{\frac{8}{9}}, -\sqrt{\frac{1}{18}}\right], \left[\sqrt{\frac{4}{21}}, -\sqrt{\frac{16}{21}}, -\sqrt{\frac{1}{21}}, 0\right]\right\}$ (vectors `sym(v10)`, `sym(v11)`, and `sym(v12)`).

TI-86 Graphing Calculator

Most keys on the *TI-86* calculator have more than one function. An orange command/symbol above a key can be activated by first pressing (where necessary) the orange `2nd` key. Similarly, a blue command/symbol above a key can be activated by first pressing (where necessary) the blue `ALPHA` key. Occasionally, a menu will appear on the bottom of the calculator screen. The various menu items are accessed using the `F1` through `F5` keys. Most functions on the *TI-86* can be accessed, if desired, via the orange `CATLG-VARS` (`2nd+CUSTOM`) command.

To adjust the number of decimal places displayed, type `2nd`, then `More`, use the gray arrow keys to move down to the `Float` line and over to the desired number of decimal places, and then hit `ENTER`. Finally, hit `EXIT` to return to the home screen. In all of Figures D.25 through D.30, we assume that exactly three decimal places are chosen.

There are two types of "−" keys on the calculator: the gray "−" key is used to negate a quantity (a unary operation), while the black "−" key is used for subtracting one quantity from another (a binary operation). To repeat the last input, type `2nd`, then `ENTER`.

In Figures D.25 to D.29, the "Keystrokes" column indicates the exact sequence of keys pressed on the *TI-86* to create each line of input, while the "Input" column illustrates how each line of input actually appears on the calculator screen.

Figure D.25

TI-86 session: vectors; fundamental vector operations

Keystrokes	Input	Output
2,ALPHA,2,+,3,ALPHA, 3,STO→,+,ENTER	2V+3W→X	[1.000 20.000 −26.000]
ALPHA,+,ENTER	X	[1.000 20.000 −26.000]
2nd,8,F3,F4,ALPHA,2, comma,ALPHA,3,),ENTER	dot(V,W)	23.000

Input of Vectors and Matrices; Fundamental Operations

To enter a vector, hit the `2nd` key and then the "8" key. (This opens the VECTR menu.) Next, hit `F2` (`EDIT`) and the calculator enters alphanumeric (blue) mode so that you can type in a name using the blue letters above the keys for the vector (names should be no more than 8 characters long), and then hit ENTER. Then type in the size of the vector, and a template will appear into which you type the vector entries. (Hit ENTER or the "down" arrow key after each entry.) When finished, hit EXIT to return to the home screen. A similar process is used to enter a matrix. First, hit the 2nd key and then the "7" key. (This opens the MATRX menu.) Then hit F2, type a name for the matrix, and then hit ENTER. Then type in the dimensions of the matrix (hitting ENTER after each dimension), and a template will appear into which you type the matrix entries. (Hit ENTER after each entry.) When finished, hit EXIT.

Figures D.25 and D.26 illustrate the fundamental operations of dot product, scalar multiplication, matrix multiplication, transpose, and inverse. Assume that the vectors $V = [5, 7, -4]$ and $W = [-3, 2, -6]$ have been entered, as well as the matrices

$$M = \begin{bmatrix} 4 & -1 & 6 & -2 \\ -3 & 2 & -3 & 2 \\ -6 & 8 & 1 & 3 \end{bmatrix} \quad \text{and} \quad N = \begin{bmatrix} 2 & -3 & 0 \\ 6 & 8 & -1 \\ 3 & 1 & -2 \\ 2 & -4 & -2 \end{bmatrix}.$$

Figure D.26

TI-86 session: matrices; fundamental matrix operations

Keystrokes	Input	Output
ALPHA,8,-,2,ALPHA, 9,2nd,7,F3,F2, ENTER	$M-2N^T$	[[0.000 −13.000 0.000 −6.000] [3.000 −14.000 −5.000 10.000] [−6.000 10.000 5.000 7.000]]
ALPHA,8,×,ALPHA,9, STO→,comma, ENTER	$M*N{\to}P$	[[16.000 −6.000 −7.000] [1.000 14.000 0.000] [45.000 71.000 −16.000]]
ALPHA,comma,2nd, EE,2nd,×, F5, MORE,F1,ENTER	$P^{-1}\blacktriangleright$Frac	[[−224/233 − 593/233 98/233] [16/233 59/233 −7/233] [−559/233 −1406/233 230/233]]
ALPHA,comma,2nd, EE,ENTER	P^{-1}	[[−.961 −2.545 .421] [.069 .253 −.030] [−2.399 −6.034 .987]]

There are various vector and matrix functions/operations listed under the VECTR and MATRX menus. Figures D.25 and D.26 give the correct sequence of keystrokes for several of these. The dot function (in the MATH submenu of VECTR) calculates the dot product of two vectors. Addition and subtraction are performed using the "+" and (the black) "−" keys, respectively. Scalar multiplication can be implied simply by putting a scalar in front of a vector or matrix. Matrix multiplication is performed using the "×" key, and displayed on the screen as "*". The "^" key is used to find positive integer powers of a square matrix. The orange "x^{-1}" key is used to calculate the inverse of a square matrix. The "T" function (in the MATH submenu of MATRX) calculates the transpose of a matrix.

To assign a name to a vector or matrix result, type STO→ and the desired name before hitting the ENTER key. (The STO→ key places the calculator in ALPHA (alphanumeric) mode; that is, you do not need to hit the ALPHA key before entering a name.) For example, in Figure D.25, the result of the first calculation is stored as the vector X, and in Figure D.26, the result of the second calculation is stored as the matrix P.

To print out a vector or matrix in fractional form (where possible), use the ▶Frac command (in the MISC submenu of the MATH menu (above the "×" key)), as shown in Figure D.26 for the matrix P.

To delete a vector or matrix from the calculator memory, type 2nd, 3 (MEM), F2 (DELETE). Then type F5 to delete a vector or type MORE, F1 to delete a matrix. Finally, use the "down" arrow key to move the cursor to the desired vector or matrix to be deleted, and then hit ENTER followed by EXIT.

Solving a Linear System; Gauss-Jordan Row Reduction Method

The rref function (in the OPS submenu of the MATRX menu) calculates the reduced row echelon form of a (possibly augmented) matrix. Assume for Figure D.27 that the matrix

$$R = \begin{bmatrix} 3 & 1 & 7 & 2 & 13 \\ 2 & -4 & 14 & -1 & -10 \\ 5 & 11 & -7 & 8 & 59 \\ 2 & 5 & -4 & -3 & 39 \end{bmatrix}$$

has been entered into the calculator. R is the augmented matrix for a linear system with an infinite solution set. From the result in Figure D.27, you can see that the general solution set of this system is $\{(-3c + 4,\ 2c + 5,\ c,\ -2)\}$.

Figure D.27

TI-86 session: solution of a linear system; row reduction

Keystrokes	Input	Output
2nd,7,F4,F5,	rref R	[[1.000 0.000 3.000 0.000 4.000]
ALPHA,5,		[0.000 1.000 −2.000 0.000 5.000]
ENTER		[0.000 0.000 0.000 1.000 −2.000]
		[0.000 0.000 0.000 0.000 0.000]]

If a linear system has a nonsingular coefficient matrix and hence has a unique solution, you can solve the system using the orange SIMULT function. SIMULT works for any linear system having no more than 30 equations or 30 unknowns. This function asks you to first type in the number of linear equations in the system, and a template appears in which you can enter the coefficients of each row in turn.

Determinants; Eigenvalues/Eigenvectors

The det function (under the MATH submenu of the MATRX menu) calculates the determinant of a square matrix. Eigenvalues for a given square matrix can be found using the eigVl function. On the *TI-86*, the eigVc function returns a matrix whose columns are normalized eigenvectors for a given square matrix.

These functions are illustrated in Figure D.28, for the matrix

$$T = \begin{bmatrix} 5 & 2 & 0 & 1 \\ -2 & 1 & 0 & -1 \\ 4 & 4 & 3 & 2 \\ 16 & 0 & -8 & -5 \end{bmatrix},$$

which has two eigenvalues, $\lambda_1 = -5$, having algebraic and geometric multiplicity 1, and $\lambda_2 = 3$, having algebraic multiplicity 3 and geometric multiplicity 2. Assume that matrix T has already been entered into the calculator.

Figure D.28

TI-86 session: eigenvalues and eigenvectors

Keystrokes	Input	Output
2nd,7,F3,F1, ALPHA,-,ENTER	det T	−135.000
2nd,7,F3,F4, ALPHA,-,ENTER	eigVl T	{(3.000, 1.462E−6), (3.000, −1.462E−6), (−5.000, 0.000), (3.000, 0.000)}
2nd,7,F3,F5, ALPHA,-,ENTER	eigVc T	[[(.408, 1.492E−7) (.408, −1.492E−7) [(−.408, 5.000E−19) (−.408, −5.000E−19) [(.816, 0.000) (.816, 0.000) [(−2.377E−13, 2.983E−7) (−2.377E−13, −2.983E−7) (−.120, 0.000) (.348, 0.000)] (.120, 0.000) (−.255, 0.000)] (−.239, 0.000) (.883, 0.000)] (.956, 0.000) (−.186, 0.000)]]

The eigenvalues and the eigenvector entries are expressed as ordered pairs because they are written as complex numbers, where the first entry of each ordered pair is the real part, and the second entry is the imaginary part (see Appendix C). The notation "E−k" indicates that the immediately preceding number should be multiplied by 10^{-k}. If k is large, these numbers are extremely small. Therefore, all of the entries in Figure D.28 containing "E" are zero, for all practical purposes. It follows that the imaginary part of each complex number in the output of Figure D.28 is zero. Hence, these complex numbers are actually real numbers, and the second entry of each ordered pair can be ignored. This also implies that the first two columns of eigenvectors produced by eigVc are, in this case, equal.

Thus, we can use the results of the eigVc function on the *TI-86* to get a set of eigenvectors for T: {[.408, −.408, .816, 0], [−.120, .120, −.239, .956], [.348, −.255, .883, −.186]}. The first eigenvector corresponds to the eigenvalue $\lambda_2 = 3$ and is derived from the first two columns of the output in Figure D.28. The second eigenvector corresponds to $\lambda_1 = -5$ and is derived from the third column. The third eigenvector corresponds to $\lambda_2 = 3$ and comes from the fourth column of this output. You should verify that these are indeed eigenvectors for T.

Of course, a basis of eigenvectors for each eigenvalue λ of a (square) matrix A can also be calculated by row reducing the matrix $\lambda\mathbf{I}_n-$ A, setting each independent variable in turn equal to 1 with all others equal to 0, and then solving for the dependent variables. The function ident k (ident is in the OPS submenu of the MATRX menu) creates a $k \times k$ identity matrix.

These operations are illustrated in Figure D.29. Linearly independent eigenvectors for the earlier matrix T, for eigenvalue $\lambda_2 = 3$, are found in Figure D.29 from the reduced row echelon form matrix R for S = $3\mathbf{I}_4-$T. First, by letting the third column variable of R equal 1 and its fourth column variable equal 0, we obtain $[\frac{1}{2}, -\frac{1}{2}, 1, 0]$, and then by letting its third column variable equal 0 and fourth column variable equal 1, we obtain $[\frac{1}{2}, -1, 0, 1]$. Although it is not readily apparent, it can be shown that the set $\{[\frac{1}{2}, -\frac{1}{2}, 1, 0], [\frac{1}{2}, -1, 0, 1]\}$ spans the same two-dimensional subspace of $\mathbb{R}^4$ as the set $\{[.408, -.408, .816, 0],$ $[.348, -.255, .883, -.186]\}$ of eigenvectors for $\lambda_2 = 3$ obtained earlier from eigVc (ignoring error due to roundoff). For example, normalizing $[\frac{1}{2}, -\frac{1}{2}, 1, 0]$ and rounding to three significant digits produces $[.408, -.408, .816, 0]$.

Keystrokes	Input	Output
2nd,7,F4,F3, 4,STO→,), ENTER	ident 4 → I	$\begin{bmatrix} 1.000 & 0.000 & 0.000 & 0.000 \\ 0.000 & 1.000 & 0.000 & 0.000 \\ 0.000 & 0.000 & 1.000 & 0.000 \\ 0.000 & 0.000 & 0.000 & 1.000 \end{bmatrix}$
3,×,ALPHA,),-, ALPHA,-,STO→, 6,ENTER	3*I-T→S	$\begin{bmatrix} -2.000 & -2.000 & 0.000 & -1.000 \\ 2.000 & 2.000 & 0.000 & 1.000 \\ -4.000 & -4.000 & 0.000 & -2.000 \\ -16.000 & 0.000 & 8.000 & 8.000 \end{bmatrix}$
2nd,7,F4,F5, ALPHA,6,STO→, 5,ENTER	rref S → R	$\begin{bmatrix} 1.000 & 0.000 & -.500 & -.500 \\ 0.000 & 1.000 & .500 & 1.000 \\ 0.000 & 0.000 & 0.000 & 0.000 \\ 0.000 & 0.000 & 0.000 & 0.000 \end{bmatrix}$

Gram-Schmidt Process

Assume the vectors C = [2, 1, 0, -1], D = [1, 0, 2, -1], and E =[0, -2, 1, 0] have already been stored in the calculator. In Figure D.30, we perform a Gram-Schmidt Process to create an orthogonal basis for the span of these vectors in $\mathbb{R}^4$. An orthonormal basis for the span can be produced by dividing each orthogonal basis vector by its length. The norm function (under the MATH submenu of the VECTR menu) calculates the length of a given vector.

You can verify that $\{[2, 1, 0, -1], [0, -\frac{1}{2}, 2, -\frac{1}{2}], [\frac{2}{3}, -\frac{4}{3}, -\frac{1}{3}, 0]\}$ (vectors C, F, G) is an orthogonal set of vectors spanning the same subspace of $\mathbb{R}^4$ as $\{[2, 1, 0, -1], [1, 0, 2, -1], [0, -2, 1, 0]\}$. An orthonormal basis for the same subspace is given by the set of vectors $\{$J, K, L$\}$.

Figure D.30

TI-86 session:
Gram-Schmidt Process;
orthogonal and
orthonormal bases

Input	Output
`D - (dot(D,C) / dot(C,C))*C → F`	$[0.000 \quad -.500 \quad 2.000 \quad -.500]$
`E - (dot(E,C) / dot(C,C))*C` `   - (dot(E,F) / dot(F,F))*F → G`	$[.667 \quad -1.333 \quad -.333 \quad 1.000E-14]$
`C / norm(C) → J`	$[.816 \quad .408 \quad 0.000 \quad -.408]$
`F / norm(F) → K`	$[0.000 \quad -.236 \quad .943 \quad -.236]$
`G / norm(G) → L`	$[.436 \quad -.873 \quad -.218 \quad 6.547E-15]$

TI-92, TI-92 Plus, TI-89, and Voyage 200 Graphing Calculators

In what follows, we will generally focus on the *TI-92* graphing calculator. The *TI-92 Plus* graphing calculator is more powerful than the *TI-92*, but all of our examples apply to both the *TI-92* and the *TI-92 Plus*, except where specifically noted. The *Voyage 200* graphing calculator is almost identical to the *TI-92 Plus* with respect to all of the functions we consider here. The *TI-89* graphing calculator is also similar in function (although not in appearance) to the *TI-92 Plus*. Hence, comments regarding the *TI-92 Plus* also apply to the *Voyage 200* and the *TI-89*, unless we state otherwise.[4]

Most keys on the *TI-92* calculator have more than one function. A yellow command/symbol above a key can be activated by pressing (where necessary) either of the two yellow 2nd keys. (There is only one yellow 2nd key on the *TI-89*, and blue is used instead of yellow on the *Voyage 200*.) Similarly, a green command/symbol above a key can be activated by pressing (where necessary) the green "diamond" (♦) key. The "cursor pad" (the large blue cross-shaped button in the upper right corner of the *TI-92*) is used to scroll vertically or horizontally on the display screen, and is useful for moving up and down through screen menus. (You must use the blue arrow keys on the *TI-89* and the light gray arrow keys on the *Voyage 200*.) The "entry line" (along the bottom of the display) contains the current input text. Hitting the CLEAR key erases the entry line contents.

To adjust the number of decimal places displayed, type Mode, then use the cursor pad to move down to Display Digits, and then over to the right and down to the desired number of decimal places, and then hit ENTER twice. In Figures D.31 through D.36, we assume exactly four decimal places have been chosen. (This choice is denoted FLOAT 4.)

There are two types of "-" keys on the calculator: the gray "-" key (black on the *Voyage 200*) is used to negate a quantity (a unary operation), while the black "−" key (light gray on the *Voyage 200*) is used for subtracting one quantity from another (a binary operation).

---

[4]The *Voyage 200* displays a number of icons on the screen when you first turn it on. Use the light gray arrow keys to highlight the HOME icon (with a picture of a calculator) and press ENTER. To return to the icon display, press the light gray APPS key. The *Voyage 200* might not display the icons when you turn it on again if you turned it off without them displayed.

Variable names may consist of 1 to 8 characters, not beginning with a digit, and not containing any spaces. There is no distinction between uppercase and lowercase; that is, AAA and aaa both refer to the same variable.

One nice feature of the *TI-92* is that the input expressions remain on the screen after they have been executed. In this way, any previous expression can easily be modified to create a similar expression. Use the "cursor pad" to scroll upward on the display to choose any previous input or output line, and then hit ENTER to copy the highlighted contents to the entry line. (Another way to retrieve a previous input expression is to type 2nd, then ENTER.)

Most functions on the *TI-92* can be accessed, if desired, via the orange CATALOG (2nd+2) command. (The *TI-89* has its own separate CATALOG button, while CATALOG is in blue on the *Voyage 200*.)

Input of Vectors and Matrices; Fundamental Operations

The *TI-92* provides two ways to enter vectors and matrices. The simplest method to enter a vector is to type a pair of brackets ("[ ]") containing the elements (separated by commas). Matrix input is handled similarly, with the entries entered row by row, except that the last element of every row (except the last row) is followed by a semicolon instead of a comma. (The yellow (blue on *Voyage 200*) semicolon key is above the M key on the *TI-92*, and above the 9 key on the *TI-89*.) This method of input is illustrated in Figure D.31 and Figure D.32.

The *TI-92* will also produce a template of the right size into which you can type the entries of a vector or matrix. This is done with the rectangular APPS key, choosing menu item 6:Data/Matrix Editor, (Data/Matrix Editor icon on the *Voyage 200*) moving to the right and choosing 3:New, then ENTER. Under Type choose 2:Matrix, and hit ENTER. Then scroll down to Variable and type in a name for the vector or matrix. Finally, scroll down in order to enter the row and column dimensions, and hit ENTER twice. A template will appear of the appropriate size, and you can simply enter the entries row by row, hitting ENTER after each entry. When finished, return to the home screen by using the Home key (♦Q).

Figures D.31 and D.32 illustrate the fundamental operations of dot product, scalar multiplication, matrix multiplication, transpose, and inverse. Various matrix functions/operations are listed under the MATH, 4:Matrix menu, and the F:Vector ops (on the *TI-92 Plus*, *Voyage 200*, and *TI-89*, L:Vector ops) submenu of this menu contains various vector functions/operations. Figure D.31 gives the correct sequence of keystrokes for several of these.[5] The dotP function (under F:Vector ops on the *TI-92*, and under L:Vector ops on the *TI-92 Plus*, *Voyage 200*, and *TI-89*) calculates the dot product of two vectors. Addition and subtraction are performed using the "+" and (the black) "−" keys (light gray on the *Voyage 200*), respectively. Scalar multiplication can be implied simply by putting a scalar in front of a vector or matrix. Matrix multiplication is performed using the "×" key and is displayed on the screen using "*". The "^" key is used to find positive integer powers of a square matrix.

---

[5] Note that most TI-92 commands can also be typed in directly without using the menus.

The yellow (or blue) "x^{-1}" key (equivalent to "$\char94 -1$") is used to calculate the inverse of a square matrix. (The *TI-89* has no such key. Just type "$X\char94 -1$" to find the inverse of matrix X.) The "T" function (item #1 in the MATH, 4:Matrix menu) calculates the transpose of a matrix.

To assign a name to a vector or matrix result, type STO→ and the desired name, before hitting the ENTER key. For example, in Figure D.31, a linear combination of vectors v1 and v2 is stored as the vector v3, and in Figure D.32, the calculation m1−2(m2)^T is stored as the matrix m3. The *TI-92* has individual keys for each letter of the alphabet. However, the *TI-89* has individual keys for only the letters X, Y, and Z. Other letters are entered by pressing the purple alpha key prior to entering the letter. Thus, to enter the letter V in Figure D.31 on the *TI-89*, the alpha key must be pressed first.

Figure D.31

TI-92/TI-92 Plus/ Voyage 200/TI-89 session: vectors; fundamental vector operations

Keystrokes	Input	Output
2nd,comma,5,comma,7,comma,-, 4,2nd, ÷ ,STO → ,V,1,ENTER	$[5,7,-4] \to$ v1	$[5 \quad 7 \quad -4]$
2nd,comma,x,comma,y,comma, z,2nd, ÷ ,STO → ,V,2,ENTER	$[x,y,z] \to$ v2	$[x \quad y \quad z]$
2,V,1,+,3,V,2,STO→,V,3,ENTER	2v1+3v2 → v3	$[3x + 10 \quad 3y + 14 \quad 3z - 8]$
2nd,5,4,F (or L on *TI-92 Plus* and *TI-89*),3, V,1,comma, V,2,),ENTER	dotP(v1,v2)	$5x + 7y - 4z$

Figure D.32

TI-92/TI-92 Plus/ Voyage 200/TI-89 session: matrices; fundamental matrix operations

Input	Output
[4,-1,6,-2;-3,2,-3,2;-6,8,1,3] → m1	$\begin{bmatrix} 4 & -1 & 6 & -2 \\ -3 & 2 & -3 & 2 \\ -6 & 8 & 1 & 3 \end{bmatrix}$
[2,-3,0;6,8,-1;3,1,-2;2,-4,-2] → m2	$\begin{bmatrix} 2 & -3 & 0 \\ 6 & 8 & -1 \\ 3 & 1 & -2 \\ 2 & -4 & -2 \end{bmatrix}$
m1-2(m2)^T → m3	$\begin{bmatrix} 0 & -13 & 0 & -6 \\ 3 & -14 & -5 & 10 \\ -6 & 10 & 5 & 7 \end{bmatrix}$
m1*m2 → m4	$\begin{bmatrix} 16 & -6 & -7 \\ 1 & 14 & 0 \\ 45 & 71 & -16 \end{bmatrix}$
(m4)^-1 → m5	$\begin{bmatrix} \frac{-224}{233} & \frac{-593}{233} & \frac{98}{233} \\ \frac{16}{233} & \frac{59}{233} & \frac{-7}{233} \\ \frac{-559}{233} & \frac{-1406}{233} & \frac{230}{233} \end{bmatrix}$
1.0((m4)^-1) → m5	$\begin{bmatrix} -.9614 & -2.545 & .4206 \\ .0687 & .2532 & -.03 \\ -2.399 & -6.034 & .9871 \end{bmatrix}$

The *TI-92* attempts to express output in fractional form, where feasible. To force an expression into decimal form, simply include a decimal number somewhere in the expression, as in the second expression for matrix m5 in Figure D.32. In all figures, we assume that the calculator is in "Auto" mode, in contrast to "Exact" or "Approximate" modes. Different outputs than shown here can result when the calculator is not in "Auto" mode.

To delete a vector or matrix from the calculator memory, type F4,4:DelVar, then the name of the vector or matrix to be deleted, and hit ENTER.

Solving a Linear System; Gauss-Jordan Row Reduction Method

The rref function (in the MATH,4:Matrix menu) calculates the reduced row echelon form of a (possibly augmented) matrix. In Figure D.33, m6 is the augmented matrix for a linear system with an infinite solution set, and m7 is the reduced row echelon form of m6. Figure D.33 shows that the general solution set of this system is $\{(-3c+4, 2c+5, c, -2)\}$.

Figure D.33

TI-92/TI-92 Plus/ Voyage 200/TI-89 session: solution of a linear system; row reduction

Input	Output
[3,1,7,2,13;2,-4,14,-1,-10; 5,11,-7,8,59;2,5,-4,-3,39] → m6	$\begin{bmatrix} 3 & 1 & 7 & 2 & 13 \\ 2 & -4 & 14 & -1 & -10 \\ 5 & 11 & -7 & 8 & 59 \\ 2 & 5 & -4 & -3 & 39 \end{bmatrix}$
rref(m6) → m7	$\begin{bmatrix} 1 & 0 & 3 & 0 & 4 \\ 0 & 1 & -2 & 0 & 5 \\ 0 & 0 & 0 & 1 & -2 \\ 0 & 0 & 0 & 0 & 0 \end{bmatrix}$

If AX=B represents a linear system having a nonsingular coefficient matrix A, you can solve the system by typing simult(A,B). Note that B must be entered as a matrix having a single column. The function simult is found in the MATH, 4:Matrix menu.

Determinants; Eigenvalues/Eigenvectors; Characteristic Polynomial

The det function (in the MATH,4:Matrix menu) calculates the determinant of a square matrix. The characteristic polynomial of a square matrix M can be calculated directly, as in Figure D.34. The $k \times k$ identity matrix is created with the expression identity(k). (The identity function is found in the MATH, 4:Matrix menu.) To factor the characteristic polynomial (or any algebraic expression), use the factor command in the F2 (Algebra) menu. After factoring, we see in Figure D.34 that matrix m8 has two eigenvalues, $\lambda_1 = -5$ and $\lambda_2 = 3$, with algebraic multiplicities 1 and 3, respectively.

On the *TI-92 Plus*, *Voyage 200*, and *TI-89*, you can type eigVl(M) (in the MATH, 4:Matrix menu) to obtain the eigenvalues of square matrix M. Also, on the *TI-92 Plus*, *Voyage 200*, and *TI-89*, the function eigVc(M) (also in the MATH, 4:Matrix menu) returns a matrix whose columns are normalized eigenvectors for the square matrix M.

The notation "E$-k$" indicates that the immediately preceding number should be multiplied by 10^{-k}. If k is large, these numbers are extremely small. Hence,

Figure D.34

TI-92/TI-92 Plus/
Voyage 200/TI-89
session: determinant;
characteristic
polynomial; eigenvalues
and eigenvectors

Input	Output
`[5,2,0,1;-2,1,0,-1;` `4,4,3,2;16,0,-8,-5] → m8`	$\begin{bmatrix} 5 & 2 & 0 & 1 \\ -2 & 1 & 0 & -1 \\ 4 & 4 & 3 & 2 \\ 16 & 0 & -8 & -5 \end{bmatrix}$
`det(m8)`	-135
`det(x*identity(4)-m8)`	$(x^2 - 6x + 9)(x^2 + 2x - 15)$
`factor(det(x*identity(4)-m8))`	$(x-3)^3(x+5)$
For the *TI-92 Plus*, **Voyage 200, and *TI-89*:**	
`eigVl(m8)`	$\{-5. \quad 3. \quad 3. \quad 3.\}$
`eigVc(m8)`	$\begin{bmatrix} -.1195 & .4082 & -.4082 & -.2642 \\ .1195 & -.4082 & .4082 & .6212 \\ -.239 & .8165 & -.8165 & .1856 \\ .9562 & -2.154E-7 & -2.154E-7 & -.714 \end{bmatrix}$

all of the entries in Figure D.34 containing "E" are zero, for all practical purposes. Thus, the third column of the matrix output for `eigVc(m8)` is a scalar
multiple of the second column.

Thus, we can use the results of the `EigVc` function to get a set of eigenvectors for m8: $\{[-.1195, .1195, -.239, .9562], [.4082, -.4082, .8165, 0],$
$[-.2642, .6212, .1856, -.714]\}$. The first eigenvector corresponds to the eigenvalue $\lambda_1 = -5$ and is derived from the first column of output. The second
eigenvector corresponds to $\lambda_2 = 3$ and is derived from the second and third
columns. The third eigenvector corresponds to $\lambda_2 = 3$ and comes from the
fourth column of output. You should verify that these are indeed eigenvectors
for m8.

Of course, a basis of eigenvectors for each eigenvalue λ of a (square) matrix
A can be calculated by row reducing the matrix $\lambda \mathbf{I}_n - A$, setting each independent variable in turn equal to 1 with all others equal to 0, and then solving for
the dependent variables. This technique is illustrated in Figure D.35. Linearly
independent eigenvectors for the earlier matrix m8, for eigenvalue $\lambda_2 = 3$,
are found in Figure D.35 from the reduced row echelon form matrix m10 of
m9 $= 3\mathbf{I}_4 - $m8. First, by letting the third column variable of m10 equal 1
and its fourth column variable equal 0, we obtain $[\frac{1}{2}, -\frac{1}{2}, 1, 0]$, and then by
letting its third column variable equal 0 and fourth column variable equal 1,
we obtain $[\frac{1}{2}, -1, 0, 1]$. Thus, $\{[\frac{1}{2}, -\frac{1}{2}, 1, 0], [\frac{1}{2}, -1, 0, 1]\}$ spans the eigenspace
for $\lambda_2 = 3$, and so the eigenvalue $\lambda_2 = 3$ for matrix m8 has geometric mul-
tiplicity 2. Although it is not readily apparent, it can be shown that the set
$\{[\frac{1}{2}, -\frac{1}{2}, 1, 0], [\frac{1}{2}, -1, 0, 1]\}$ spans the same two-dimensional subspace of $\mathbb{R}^4$
as the set $\{[.4082, -.4082, .8165, 0], [-.2642, .6212, .1856, -.714]\}$ of eigen-
vectors for $\lambda_2 = 3$ obtained earlier from `eigVc` (ignoring error due to roundoff). For example, normalizing $[\frac{1}{2}, -\frac{1}{2}, 1, 0]$ and rounding to four significant
digits produces $[.4082, -.4082, .8165, 0]$.

Figure D.35

TI-92/TI-92 Plus/
Voyage 200/TI-89
session: direct
calculation of
eigenspace

Input	Output
`(3*identity(4)-m8) → m9`	$\begin{bmatrix} -2 & -2 & 0 & -1 \\ 2 & 2 & 0 & 1 \\ -4 & -4 & 0 & -2 \\ -16 & 0 & 8 & 8 \end{bmatrix}$
`rref(m9) ⟶ m10`	$\begin{bmatrix} 1 & 0 & -\frac{1}{2} & -\frac{1}{2} \\ 0 & 1 & \frac{1}{2} & 1 \\ 0 & 0 & 0 & 0 \\ 0 & 0 & 0 & 0 \end{bmatrix}$

Gram-Schmidt Process

We can carry out the necessary calculations for the Gram-Schmidt Process on the *TI-92* using the `dotP` function. In Figure D.36, we create an orthogonal basis using the Gram-Schmidt Process for the span in $\mathbb{R}^4$ of the vectors v5, v6, and v7. An orthonormal basis for the span can then be produced by dividing each orthogonal basis vector by its length. The `norm` function (found in the MATH, 4:Matrix, B:Norms menu on the *TI-92*, and in the MATH, 4:Matrix, H:Norms menu on the *TI-92 Plus*, *Voyage 200*, and *TI-89*) calculates the length of a given vector.

Figure D.36

TI-92/TI-92 Plus/
Voyage 200/TI-89
session: Gram-Schmidt
Process; orthogonal and
orthonormal bases

Input	Output
`[2,1,0,-1]→ v5`	$[2 \quad 1 \quad 0 \quad -1]$
`[1,0,2,-1]→ v6`	$[1 \quad 0 \quad 2 \quad -1]$
`[0,-2,1,0]→ v7`	$[0 \quad -2 \quad 1 \quad 0]$
`v6-(dotP(v6,v5)/dotP(v5,v5))v5 → v8`	$[0 \quad -\frac{1}{2} \quad 2 \quad -\frac{1}{2}]$
`v7-(dotP(v7,v5)/dotP(v5,v5))v5` `  -(dotP(v7,v8)/dotP(v8,v8))v8 → v9`	$[\frac{2}{3} \quad -\frac{4}{3} \quad -\frac{1}{3} \quad 0]$
`v5 / norm(v5) → v10`	$[\frac{\sqrt{6}}{3} \quad \frac{\sqrt{6}}{6} \quad 0 \quad -\frac{\sqrt{6}}{6}]$
`v8 / norm(v8) → v11`	$[0 \quad -\frac{\sqrt{2}}{6} \quad \frac{2\sqrt{2}}{3} \quad -\frac{\sqrt{2}}{6}]$
`v9 / norm(v9) → v12`	$[\frac{2\sqrt{21}}{21} \quad -\frac{4\sqrt{21}}{21} \quad -\frac{\sqrt{21}}{21} \quad 0]$

You can verify that $\{[2, 1, 0, -1], [0, -\frac{1}{2}, 2, -\frac{1}{2}], [\frac{2}{3}, -\frac{4}{3}, -\frac{1}{3}, 0]\}$ (vectors v5, v8, v9) is an orthogonal set of vectors spanning the same subspace of $\mathbb{R}^4$ as $\{[2, 1, 0, -1], [1, 0, 2, -1], [0, -2, 1, 0]\}$. An orthonormal basis for the same subspace is given by vectors v10, v11, and v12.

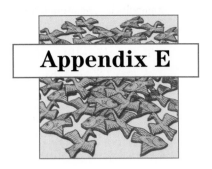

Appendix E

Answers to Selected Exercises

Section 1.1 (p.14–17)

1. (a) $[9, -4]$; distance $= \sqrt{97}$

(c) $[-1, -1, 2, -3, -4]$; distance $= \sqrt{31}$

2. (a) $(3, 4, 2)$ (see accompanying figure)

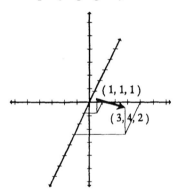

(c) $(1, -2, 0)$ (see accompanying figure)

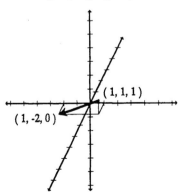

3. (a) $(7, -13)$ (c) $(-1, 3, -1, 4, 6)$

4. (a) $(\frac{16}{3}, -\frac{13}{3}, 8)$

5. (a) $\left[\frac{3}{\sqrt{70}}, -\frac{5}{\sqrt{70}}, \frac{6}{\sqrt{70}}\right]$; shorter, since length of original vector is > 1
   (c) $[0.6, -0.8]$; neither, since given vector is a unit vector

6. (a) Parallel   (c) Not parallel

7. (a) $[-6, 12, 15]$ (c) $[-3, 4, 8]$

   (e) $[6, -20, -13]$

8. (a) $\mathbf{x} + \mathbf{y} = [1, 1]$; $\mathbf{x} - \mathbf{y} = [-3, 9]$; $\mathbf{y} - \mathbf{x} = [3, -9]$ (see accompanying figure)

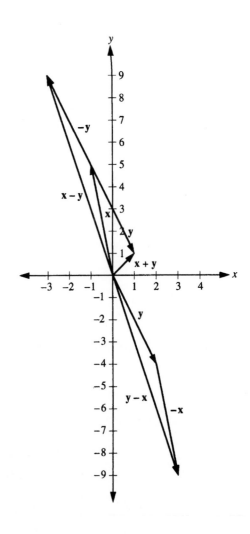

   (c) $\mathbf{x} + \mathbf{y} = [1, 8, -5]$; $\mathbf{x} - \mathbf{y} = [3, 2, -1]$; $\mathbf{y} - \mathbf{x} = [-3, -2, 1]$ (see accompanying figure)

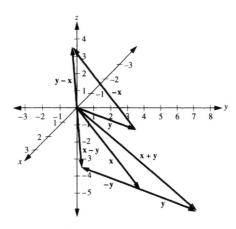

10. (a) $[10, -10]$ (b) $[-5\sqrt{3}, -15]$

13. $[0.5 - 0.6\sqrt{2}, -0.4\sqrt{2}] \approx [-0.3485, -0.5657]$

15. Net velocity = $[-2\sqrt{2}, -3 + 2\sqrt{2}]$; resultant speed ≈ 2.83 km/hr

17. $[-8 - \sqrt{2}, -\sqrt{2}]$

18. Acceleration = $\frac{1}{20}\left[\frac{12}{13}, -\frac{344}{65}, \frac{392}{65}\right] \approx [0.0462, -0.2646, 0.3015]$

20. $\mathbf{a} = [\frac{-mg}{1+\sqrt{3}}, \frac{mg}{1+\sqrt{3}}]$; $\mathbf{b} = [\frac{mg}{1+\sqrt{3}}, \frac{mg\sqrt{3}}{1+\sqrt{3}}]$

26. (a) F   (b) T
        (c) T   (d) F
        (e) T   (f) F
        (g) F

Section 1.2 (p.25–27)

1. (a) $\arccos(-\frac{27}{5\sqrt{37}}) \approx 152.6°$, or 2.66 radians

   (c) $\arccos(0) = 90°$, or $\frac{\pi}{2}$ radians

4. (b) $\frac{1040\sqrt{5}}{9} \approx 258.4$ joules

7. No; consider $\mathbf{x} = [1, 0]$, $\mathbf{y} = [0, 1]$, and $\mathbf{z} = [1, 1]$.

13. $\cos\theta_1 = \frac{a}{\sqrt{a^2+b^2+c^2}}$, $\cos\theta_2 = \frac{b}{\sqrt{a^2+b^2+c^2}}$, and $\cos\theta_3 = \frac{c}{\sqrt{a^2+b^2+c^2}}$

14. (a) Length of diagonal = $\sqrt{3}s$

   (b) Angle = $\arccos(\frac{\sqrt{3}}{3}) \approx 54.7°$, or 0.955 radians

15. (a) $\left[-\frac{3}{5}, -\frac{3}{10}, -\frac{3}{2}\right]$

   (c) $\left[\frac{1}{6}, 0, -\frac{1}{6}, \frac{1}{3}\right]$

17. $a\mathbf{i}, b\mathbf{j}, c\mathbf{k}$

18. (a) Parallel: $\left[\frac{20}{29}, -\frac{30}{29}, \frac{40}{29}\right]$; orthogonal: $\left[-\frac{194}{29}, \frac{88}{29}, \frac{163}{29}\right]$

      (c) Parallel: $\left[\frac{60}{49}, -\frac{40}{49}, \frac{120}{49}\right]$; orthogonal: $\left[-\frac{354}{49}, \frac{138}{49}, \frac{223}{49}\right]$

23. (a) T   (b) T
      (c) F   (d) F
      (e) T   (f) F

Section 1.3 (p.40–43)

1. (b) Let $m = \max\{|c|, |d|\}$. Then $\|c\mathbf{x} \pm d\mathbf{y}\| \leq m(\|\mathbf{x}\| + \|\mathbf{y}\|)$.

2. (b) Consider the number 4.

5. (a) Consider $\mathbf{x} = [1, 0, 0]$ and $\mathbf{y} = [1, 1, 0]$.

      (b) If $\mathbf{x} \neq \mathbf{y}$, then $\mathbf{x} \cdot \mathbf{y} \neq \|\mathbf{x}\|^2$.

      (c) Yes

8. (a) Contrapositive: If $\mathbf{x} = \mathbf{0}$, then $\mathbf{x}$ is not a unit vector.
         Converse: If $\mathbf{x}$ is nonzero, then $\mathbf{x}$ is a unit vector.
         Inverse: If $\mathbf{x}$ is not a unit vector, then $\mathbf{x} = \mathbf{0}$.

      (c) (Let $\mathbf{x}, \mathbf{y}$ be nonzero vectors.)
         Contrapositive: If $\mathbf{proj_y x} \neq \mathbf{0}$, then $\mathbf{proj_x y} \neq \mathbf{0}$.
         Converse: If $\mathbf{proj_y x} = \mathbf{0}$, then $\mathbf{proj_x y} = \mathbf{0}$.
         Inverse: If $\mathbf{proj_x y} \neq \mathbf{0}$, then $\mathbf{proj_y x} \neq \mathbf{0}$.

10. (b) Converse: Let $\mathbf{x}$ and $\mathbf{y}$ be vectors in $\mathbb{R}^n$. If $\|\mathbf{x} + \mathbf{y}\| \geq \|\mathbf{y}\|$, then $\mathbf{x} \cdot \mathbf{y} = 0$. The original statement is true, but the converse is false in general. Proof of the original statement follows from

$$\begin{aligned} \|\mathbf{x} + \mathbf{y}\|^2 &= (\mathbf{x} + \mathbf{y}) \cdot (\mathbf{x} + \mathbf{y}) \\ &= \|\mathbf{x}\|^2 + 2(\mathbf{x} \cdot \mathbf{y}) + \|\mathbf{y}\|^2 \\ &= \|\mathbf{x}\|^2 + \|\mathbf{y}\|^2 \geq \|\mathbf{y}\|^2. \end{aligned}$$

         Counterexample to converse: let $\mathbf{x} = [1, 0]$, $\mathbf{y} = [1, 1]$

18. Step 1 cannot be reversed because y could equal $\pm(x^2 + 2)$.
Step 2 cannot be reversed because y^2 could equal $x^4 + 4x^2 + c$.
Step 4 cannot be reversed because in general y does not have to equal $x^2 + 2$.
Step 6 cannot be reversed since $\frac{dy}{dx}$ could equal $2x + c$.
All other steps remain true when reversed.

19. (a) For every unit vector $\mathbf{x}$ in $\mathbb{R}^3$, $\mathbf{x} \cdot [1, -2, 3] \neq 0$.

      (c) $\mathbf{x} = \mathbf{0}$ or $\|\mathbf{x} + \mathbf{y}\| \neq \|\mathbf{y}\|$, for all $\mathbf{x}, \mathbf{y} \in \mathbb{R}^n$.

      (e) There is an $\mathbf{x} \in \mathbb{R}^3$ such that for every nonzero $\mathbf{y} \in \mathbb{R}^3$, $\mathbf{x} \cdot \mathbf{y} \neq 0$.

20. (a) Contrapositive: If $\mathbf{x} \neq \mathbf{0}$ and $\|\mathbf{x} - \mathbf{y}\| \leq \|\mathbf{y}\|$, then $\mathbf{x} \cdot \mathbf{y} \neq 0$.
Converse: If $\mathbf{x} = \mathbf{0}$ or $\|\mathbf{x} - \mathbf{y}\| > \|\mathbf{y}\|$, then $\mathbf{x} \cdot \mathbf{y} = 0$.
Inverse: If $\mathbf{x} \cdot \mathbf{y} \neq 0$, then $\mathbf{x} \neq \mathbf{0}$ and $\|\mathbf{x} - \mathbf{y}\| \leq \|\mathbf{y}\|$.

25. (a) F   (b) T
(c) T   (d) F
(e) F   (f) F
(g) F   (h) T
(i) F

Section 1.4 (p.50–53)

1. (a) $\begin{bmatrix} 2 & 1 & 3 \\ 2 & 7 & -5 \\ 9 & 0 & -1 \end{bmatrix}$ (c) $\begin{bmatrix} -16 & 8 & 12 \\ 0 & 20 & -4 \\ 24 & 4 & -8 \end{bmatrix}$

(e) Impossible   (g) $\begin{bmatrix} -23 & 14 & -9 \\ -5 & 8 & 8 \\ -9 & -18 & 1 \end{bmatrix}$

(i) $\begin{bmatrix} -1 & 1 & 12 \\ -1 & 5 & 8 \\ 8 & -3 & -4 \end{bmatrix}$ (l) Impossible

(n) $\begin{bmatrix} 13 & -6 & 2 \\ 3 & -3 & -5 \\ 3 & 5 & 1 \end{bmatrix}$

2. Square: $\mathbf{B}, \mathbf{C}, \mathbf{E}, \mathbf{F}, \mathbf{G}, \mathbf{H}, \mathbf{J}, \mathbf{K}, \mathbf{L}, \mathbf{M}, \mathbf{N}, \mathbf{P}, \mathbf{Q}$
Diagonal: $\mathbf{B}, \mathbf{G}, \mathbf{N}$
Upper triangular: $\mathbf{B}, \mathbf{G}, \mathbf{L}, \mathbf{N}$
Lower triangular: $\mathbf{B}, \mathbf{G}, \mathbf{M}, \mathbf{N}, \mathbf{Q}$
Symmetric: $\mathbf{B}, \mathbf{F}, \mathbf{G}, \mathbf{J}, \mathbf{N}, \mathbf{P}$
Skew-symmetric: $\mathbf{H}$ (but not $\mathbf{E}, \mathbf{C}, \mathbf{K}$)
Transposes: $\mathbf{A}^T = \begin{bmatrix} -1 & 0 & 6 \\ 4 & 1 & 0 \end{bmatrix}$, $\mathbf{B}^T = \mathbf{B}$, $\mathbf{C}^T = \begin{bmatrix} -1 & -1 \\ 1 & 1 \end{bmatrix}$, and so on

3. (a) $\begin{bmatrix} 3 & -\frac{1}{2} & \frac{5}{2} \\ -\frac{1}{2} & 2 & 1 \\ \frac{5}{2} & 1 & 2 \end{bmatrix} + \begin{bmatrix} 0 & -\frac{1}{2} & \frac{3}{2} \\ \frac{1}{2} & 0 & 4 \\ -\frac{3}{2} & -4 & 0 \end{bmatrix}$

5. (d) The matrix must be a square zero matrix.

14. (a) Trace $(\mathbf{B}) = 1$; trace $(\mathbf{C}) = 0$; trace $(\mathbf{E}) = -6$; trace $(\mathbf{F}) = 2$; trace $(\mathbf{G}) = 18$; trace $(\mathbf{H}) = 0$; trace $(\mathbf{J}) = 1$; trace $(\mathbf{K}) = 4$; trace $(\mathbf{L}) = 3$; trace $(\mathbf{M}) = 0$; trace $(\mathbf{N}) = 3$; trace $(\mathbf{P}) = 0$; trace $(\mathbf{Q}) = 1$

(c) No; consider matrices $\mathbf{L}$ and $\mathbf{N}$ in Exercise 2.

15. (a) F   (b) T
(c) F   (d) T
(e) T

Section 1.5 (p. 60–63)

1. (b) $\begin{bmatrix} 34 & -24 \\ 42 & 49 \\ 8 & -22 \end{bmatrix}$ (c) Impossible

(e) $[-38]$ (f) $\begin{bmatrix} -24 & 48 & -16 \\ 3 & -6 & 2 \\ -12 & 24 & -8 \end{bmatrix}$

(g) Impossible   (j) Impossible

(l) $\begin{bmatrix} 5 & 3 & 2 & 5 \\ 4 & 1 & 3 & 1 \\ 1 & 1 & 0 & 2 \\ 4 & 1 & 3 & 1 \end{bmatrix}$ (n) $\begin{bmatrix} 146 & 5 & -603 \\ 154 & 27 & -560 \\ 38 & -9 & -193 \end{bmatrix}$

2. (a) No   (c) No
(d) Yes

3. (a) $[15, -13, -8]$ (c) $[4]$

4. (a) Valid, by Theorem 1.14, part (1)

(b) Invalid

(c) Valid, by Theorem 1.14, part (1)

(d) Valid, by Theorem 1.14, part (2)

(e) Valid, by Theorem 1.16

(f) Invalid

(g) Valid, by Theorem 1.14, part (3)

(h) Invalid

(i) Valid, by Theorem 1.14, part (2)

(j) Invalid

(k) Valid, by Theorem 1.14, part (3), and Theorem 1.16

5.
$$\begin{array}{c} \\ \text{Outlet 1} \\ \text{Outlet 2} \\ \text{Outlet 3} \\ \text{Outlet 4} \end{array} \begin{array}{cc} \text{Salary} & \text{Fringe Benefits} \\ \begin{bmatrix} \$367,500 & \$78,000 \\ \$225,000 & \$48,000 \\ \$765,000 & \$162,000 \\ \$360,000 & \$76,500 \end{bmatrix} \end{array}$$

6.
$$\begin{array}{c} \\ \text{Nitrogen} \\ \text{Phosphate} \\ \text{Potash} \end{array} \begin{array}{ccc} \text{Field 1} & \text{Field 2} & \text{Field 3} \\ \begin{bmatrix} 1.00 & 0.45 & 0.65 \\ 0.90 & 0.35 & 0.75 \\ 0.95 & 0.35 & 0.85 \end{bmatrix} \end{array} \text{(in tons)}$$

7. (a) One example: $\begin{bmatrix} 1 & 1 \\ 0 & -1 \end{bmatrix}$

  (b) One example: $\begin{bmatrix} 1 & 1 & 0 \\ 0 & -1 & 0 \\ 0 & 0 & 1 \end{bmatrix}$

  (c) Consider $\begin{bmatrix} 0 & 0 & 1 \\ 1 & 0 & 0 \\ 0 & 1 & 0 \end{bmatrix}$.

8. (a) Third row, fourth column entry of **AB**

  (c) Third row, second column entry of **BA**

9. (a) $\sum_{k=1}^{n} a_{3k}b_{k2}$

23. (a) Consider any matrix of the form $\begin{bmatrix} 1 & 0 \\ x & 0 \end{bmatrix}$.

24. (b) Consider $\mathbf{A} = \begin{bmatrix} 1 & 2 & -1 \\ 2 & 4 & -2 \end{bmatrix}$ and $\mathbf{B} = \begin{bmatrix} 1 & -2 \\ 0 & 1 \\ 1 & 0 \end{bmatrix}$.

25. See Exercise 26(c).

27. (a) T   (b) T
  (c) T   (d) F
  (e) F   (f) F

Section 2.1 (p.81–83)

1. (a) Consistent; solution set = $\{(-2, 3, 5)\}$

  (c) Inconsistent; solution set = $\{\,\}$

  (e) Consistent; solution set = $\{(2b - d - 4,\, b,\, 2d + 5,\, d,\, 2) \,|\, b, d \in \mathbb{R}\}$; three particular solutions are: $(-4, 0, 5, 0, 2)$ (with $b = d = 0$), $(-2, 1, 5, 0, 2)$ (with $b = 1,\, d = 0$), and $(-5, 0, 7, 1, 2)$ (with $b = 0,\, d = 1$)

  (g) Consistent; solution set = $\{(6, -1, 3)\}$

2. (a) Solution set = $\{(3c + 11e + 46,\, c + e + 13,\, c,\, -2e + 5,\, e) \,|\, c, e \in \mathbb{R}\}$

  (c) Solution set = $\{(-20c + 9d - 153f - 68,\, 7c - 2d + 37f + 15,\, c,\, d,\, 4f + 2,\, f) \,|\, c, d, f \in \mathbb{R}\}$

3. 51 nickels, 62 dimes, 31 quarters

4. $y = 2x^2 - x + 3$

6. $x^2 + y^2 - 6x - 8y = 0$, or $(x - 3)^2 + (y - 4)^2 = 25$

7. (a) $R(\mathbf{AB}) = (R(\mathbf{A}))\mathbf{B} = \begin{bmatrix} 26 & 15 & -6 \\ 6 & 4 & 1 \\ 0 & -6 & 12 \\ 10 & 4 & -14 \end{bmatrix}$.

11. (a) T   (b) F
     (c) F   (d) F
     (e) T   (f) T

Section 2.2 (p.90–93)

1. Matrices in (a), (b), (c), (d), and (f) are not in reduced row echelon form.
Matrix in (a) fails condition 2 of the definition.
Matrix in (b) fails condition 4 of the definition.
Matrix in (c) fails condition 1 of the definition.
Matrix in (d) fails conditions 1, 2, and 3 of the definition.
Matrix in (f) fails condition 3 of the definition.

2. (a) $\left[\begin{array}{ccc|c} 1 & 4 & 0 & -13 \\ 0 & 0 & 1 & -3 \\ 0 & 0 & 0 & 0 \end{array}\right]$

    (b) $\begin{bmatrix} 1 & 0 & 0 & 0 \\ 0 & 1 & 0 & 0 \\ 0 & 0 & 1 & 0 \\ 0 & 0 & 0 & 1 \end{bmatrix} = \mathbf{I}_4$

    (c) $\left[\begin{array}{ccccc|c} 1 & -2 & 0 & 11 & -23 \\ 0 & 0 & 1 & -2 & 5 \\ 0 & 0 & 0 & 0 & 0 \\ 0 & 0 & 0 & 0 & 0 \end{array}\right]$

    (e) $\left[\begin{array}{ccccc|c} 1 & -2 & 0 & 2 & -1 & 1 \\ 0 & 0 & 1 & -1 & 3 & 2 \end{array}\right]$

3. (a) Solution set $= \{(-2, 3, 5)\}$

    (e) Solution set $= \{(2b - d - 4, b, 2d + 5, d, 2) \mid b, d \in \mathbb{R}\}$; three particular solutions are: $(-4, 0, 5, 0, 2)$ (with $b = d = 0$), $(-2, 1, 5, 0, 2)$ (with $b = 1, d = 0$), and $(-5, 0, 7, 1, 2)$ (with $b = 0, d = 1$)

    (g) Solution set $= \{(6, -1, 3)\}$

4. (a) Solution set $= \{(c - 2d, -3d, c, d) \mid c, d \in \mathbb{R}\}$; one particular solution $= (-3, -6, 1, 2)$

    (c) Solution set $= \{(-4b + 2d - f, b, -3d + 2f, d, -2f, f) \mid b, d, f \in \mathbb{R}\}$; one particular solution $= (-3, 1, 0, 2, -6, 3)$

5. (a) Solution set $= \{(2c, -4c, c) \mid c \in \mathbb{R}\} = \{c(2, -4, 1) \mid c \in \mathbb{R}\}$

    (c) Solution set $= \{(0, 0, 0, 0)\}$

6. (a) $a = 2$, $b = 15$, $c = 12$, $d = 6$

   (c) $a = 4$, $b = 2$, $c = 4$, $d = 1$, $e = 4$

7. (a) $A = 3$, $B = 4$, $C = -2$

8. Solution for system $\mathbf{AX} = \mathbf{B}_1$: $(6, -51, 21)$;
   Solution for system $\mathbf{AX} = \mathbf{B}_2$: $(\frac{35}{3}, -98, \frac{79}{2})$

11. (b) Any nonhomogeneous system with two equations and two unknowns that has a unique solution will serve as a counterexample. For instance, consider
$$\begin{cases} x + y = 1 \\ x - y = 1 \end{cases}.$$
This system has $(1, 0)$ as its unique solution. Let (s_1, s_2) and (t_1, t_2) both equal $(1, 0)$. Then the sum of solutions is not a solution in this case. Also, let c be any real number other than 1. The scalar multiple of a solution by c is not a solution in this case.

14. (a) T (b) T
   (c) F (d) T
   (e) F (f) F

Section 2.3 (p.103–107)

1. (a) A row operation of type (I) converts $\mathbf{A}$ to $\mathbf{B}$: $\langle 2 \rangle \longleftarrow -5\langle 2 \rangle$.

   (c) A row operation of type (II) converts $\mathbf{A}$ to $\mathbf{B}$: $\langle 2 \rangle \longleftarrow \langle 3 \rangle + \langle 2 \rangle$.

2. (b) The sequence of row operations converting $\mathbf{B}$ to $\mathbf{A}$ is:
   (II): $\langle 1 \rangle \longleftarrow -5\langle 3 \rangle + \langle 1 \rangle$
   (III): $\langle 2 \rangle \longleftrightarrow \langle 3 \rangle$
   (II): $\langle 3 \rangle \longleftarrow 3\langle 1 \rangle + \langle 3 \rangle$
   (II): $\langle 2 \rangle \longleftarrow -2\langle 1 \rangle + \langle 2 \rangle$
   (I): $\langle 1 \rangle \longleftarrow 4\langle 1 \rangle$

3. (a) Common reduced row echelon form is $\mathbf{I}_3$.

   (b) The sequence of row operations is as follows:
   (II): $\langle 3 \rangle \longleftarrow 2\langle 2 \rangle + \langle 3 \rangle$
   (I): $\langle 3 \rangle \longleftarrow -1\langle 3 \rangle$
   (II): $\langle 1 \rangle \longleftarrow -9\langle 3 \rangle + \langle 1 \rangle$
   (II): $\langle 2 \rangle \longleftarrow 3\langle 3 \rangle + \langle 2 \rangle$
   (II): $\langle 3 \rangle \longleftarrow -\frac{9}{5}\langle 2 \rangle + \langle 3 \rangle$
   (II): $\langle 1 \rangle \longleftarrow -\frac{3}{5}\langle 2 \rangle + \langle 1 \rangle$
   (I): $\langle 2 \rangle \longleftarrow -\frac{1}{5}\langle 2 \rangle$
   (II): $\langle 3 \rangle \longleftarrow -3\langle 1 \rangle + \langle 3 \rangle$
   (II): $\langle 2 \rangle \longleftarrow -2\langle 1 \rangle + \langle 2 \rangle$
   (I): $\langle 1 \rangle \longleftarrow -5\langle 1 \rangle$

5. (a) 2   **(c)** 2
    (e) 3

6. (a) Corollary 2.6 does not apply here. Rank = 3. Thus, Theorem 2.5 predicts the system has only the trivial solution. In fact, solution set = $\{(0, 0, 0)\}$.

7. In the following answers, the asterisk represents any real entry:

(a) Smallest rank = 1: $\left[\begin{array}{ccc|c} 1 & * & * & * \\ 0 & 0 & 0 & 0 \\ 0 & 0 & 0 & 0 \\ 0 & 0 & 0 & 0 \end{array}\right]$;

   largest rank = 4: $\left[\begin{array}{ccc|c} 1 & 0 & 0 & 0 \\ 0 & 1 & 0 & 0 \\ 0 & 0 & 1 & 0 \\ 0 & 0 & 0 & 1 \end{array}\right]$

(c) Smallest rank = 2: $\left[\begin{array}{cccc|c} 1 & * & * & * & 0 \\ 0 & 0 & 0 & 0 & 1 \\ 0 & 0 & 0 & 0 & 0 \end{array}\right]$;

   largest rank = 3: $\left[\begin{array}{cccc|c} 1 & 0 & * & * & 0 \\ 0 & 1 & * & * & 0 \\ 0 & 0 & 0 & 0 & 1 \end{array}\right]$

8. (a) $\mathbf{x} = -\frac{21}{11}\mathbf{a}_1 + \frac{6}{11}\mathbf{a}_2$

   (c) Not possible

   (e) The answer is not unique; one possible answer is $\mathbf{x} = -3\mathbf{a}_1 + 2\mathbf{a}_2 + 0\mathbf{a}_3$.

   (g) $\mathbf{x} = 2\mathbf{a}_1 - \mathbf{a}_2 - \mathbf{a}_3$

9. (a) Yes: $5(\text{row } 1) - 3(\text{row } 2) - 1(\text{row } 3)$

   (c) Not in row space

   (e) Yes, but the linear combination of the rows is not unique; one possible expression for the given vector is $-3(\text{row } 1) + 1(\text{row } 2) + 0(\text{row } 3)$.

10. (a) $[13, -23, 60] = -2\mathbf{q}_1 + \mathbf{q}_2 + 3\mathbf{q}_3$

   (b) $\mathbf{q}_1 = 3\mathbf{r}_1 - \mathbf{r}_2 - 2\mathbf{r}_3$
   $\mathbf{q}_2 = 2\mathbf{r}_1 + 2\mathbf{r}_2 - 5\mathbf{r}_3$
   $\mathbf{q}_3 = \mathbf{r}_1 - 6\mathbf{r}_2 + 4\mathbf{r}_3$

   (c) $[13, -23, 60] = -\mathbf{r}_1 - 14\mathbf{r}_2 + 11\mathbf{r}_3$

11. (a) $\mathbf{B} = \left[\begin{array}{ccc} 1 & 0 & -1 & 2 \\ 0 & 1 & 3 & 2 \\ 0 & 0 & 0 & 0 \end{array}\right]$; $[1, 0, -1, 2] = -\frac{7}{8}[0, 4, 12, 8] + \frac{1}{2}[2, 7, 19, 18] +$

   $0[1, 2, 5, 6]$; $[0, 1, 3, 2] = \frac{1}{4}[0, 4, 12, 8] + 0[2, 7, 19, 18] + 0[1, 2, 5, 6]$ (other solutions are possible for $[1, 0, -1, 2]$ and $[0, 1, 3, 2]$);

$[0, 4, 12, 8] = 0[1, 0, -1, 2] + 4[0, 1, 3, 2]; [2, 7, 19, 18] = 2[1, 0, -1, 2] + 7[0, 1, 3, 2]; [1, 2, 5, 6] = 1[1, 0, -1, 2] + 2[0, 1, 3, 2].$

14. The zero vector is a solution to $\mathbf{AX} = \mathbf{O}$, but it is not a solution for $\mathbf{AX} = \mathbf{B}$.

15. Consider the systems

$$\begin{cases} x + y = 1 \\ x + y = 0 \end{cases} \text{ and } \begin{cases} x - y = 1 \\ x - y = 2 \end{cases}.$$

The reduced row echelon matrices for these inconsistent systems are, respectively,

$$\left[\begin{array}{cc|c} 1 & 1 & 0 \\ 0 & 0 & 1 \end{array}\right] \text{ and } \left[\begin{array}{cc|c} 1 & -1 & 0 \\ 0 & 0 & 1 \end{array}\right].$$

Thus, the original augmented matrices are not row equivalent, since their reduced row echelon forms are different.

22. (a) T (b) T
  (c) F (d) F
  (e) F (f) T

Section 2.4 (p.116–119)

2. (a) Rank = 2; nonsingular

  (c) Rank = 3; nonsingular

  (e) Rank = 3; singular

3. (a) $\begin{bmatrix} \frac{1}{10} & \frac{1}{15} \\ \frac{3}{10} & -\frac{2}{15} \end{bmatrix}$ (c) $\begin{bmatrix} -\frac{2}{21} & -\frac{5}{84} \\ \frac{1}{7} & -\frac{1}{28} \end{bmatrix}$

  (e) No inverse exists.

4. (a) $\begin{bmatrix} 1 & 3 & 2 \\ -1 & 0 & 2 \\ 2 & 2 & -1 \end{bmatrix}$ (c) $\begin{bmatrix} \frac{3}{2} & 0 & \frac{1}{2} \\ -3 & \frac{1}{2} & -\frac{1}{2} \\ -\frac{8}{3} & \frac{1}{3} & -\frac{2}{3} \end{bmatrix}$

  (e) No inverse exists.

5. (c) $\begin{bmatrix} \frac{1}{a_{11}} & 0 & \cdots & 0 \\ 0 & \frac{1}{a_{22}} & \cdots & 0 \\ \vdots & \vdots & \ddots & \vdots \\ 0 & 0 & \cdots & \frac{1}{a_{nn}} \end{bmatrix}$

6. (a) The general inverse is $\begin{bmatrix} \cos\theta & \sin\theta \\ -\sin\theta & \cos\theta \end{bmatrix}$.

  When $\theta = \frac{\pi}{6}$, matrix = $\begin{bmatrix} \frac{\sqrt{3}}{2} & -\frac{1}{2} \\ \frac{1}{2} & \frac{\sqrt{3}}{2} \end{bmatrix}$; inverse = $\begin{bmatrix} \frac{\sqrt{3}}{2} & \frac{1}{2} \\ -\frac{1}{2} & \frac{\sqrt{3}}{2} \end{bmatrix}$.

When $\theta = \frac{\pi}{4}$, matrix $= \begin{bmatrix} \frac{\sqrt{2}}{2} & -\frac{\sqrt{2}}{2} \\ \frac{\sqrt{2}}{2} & \frac{\sqrt{2}}{2} \end{bmatrix}$; inverse $= \begin{bmatrix} \frac{\sqrt{2}}{2} & \frac{\sqrt{2}}{2} \\ -\frac{\sqrt{2}}{2} & \frac{\sqrt{2}}{2} \end{bmatrix}$.

When $\theta = \frac{\pi}{2}$, matrix $= \begin{bmatrix} 0 & -1 \\ 1 & 0 \end{bmatrix}$; inverse $= \begin{bmatrix} 0 & 1 \\ -1 & 0 \end{bmatrix}$.

(b) The general inverse is $\begin{bmatrix} \cos\theta & \sin\theta & 0 \\ -\sin\theta & \cos\theta & 0 \\ 0 & 0 & 1 \end{bmatrix}$.

When $\theta = \frac{\pi}{6}$, matrix $= \begin{bmatrix} \frac{\sqrt{3}}{2} & -\frac{1}{2} & 0 \\ \frac{1}{2} & \frac{\sqrt{3}}{2} & 0 \\ 0 & 0 & 1 \end{bmatrix}$; inverse $= \begin{bmatrix} \frac{\sqrt{3}}{2} & \frac{1}{2} & 0 \\ -\frac{1}{2} & \frac{\sqrt{3}}{2} & 0 \\ 0 & 0 & 1 \end{bmatrix}$.

When $\theta = \frac{\pi}{4}$, matrix $= \begin{bmatrix} \frac{\sqrt{2}}{2} & -\frac{\sqrt{2}}{2} & 0 \\ \frac{\sqrt{2}}{2} & \frac{\sqrt{2}}{2} & 0 \\ 0 & 0 & 1 \end{bmatrix}$; inverse $= \begin{bmatrix} \frac{\sqrt{2}}{2} & \frac{\sqrt{2}}{2} & 0 \\ -\frac{\sqrt{2}}{2} & \frac{\sqrt{2}}{2} & 0 \\ 0 & 0 & 1 \end{bmatrix}$.

When $\theta = \frac{\pi}{2}$, matrix $= \begin{bmatrix} 0 & -1 & 0 \\ 1 & 0 & 0 \\ 0 & 0 & 1 \end{bmatrix}$; inverse $= \begin{bmatrix} 0 & 1 & 0 \\ -1 & 0 & 0 \\ 0 & 0 & 1 \end{bmatrix}$.

7. (a) Inverse $= \begin{bmatrix} \frac{2}{3} & \frac{1}{3} \\ \frac{7}{3} & \frac{5}{3} \end{bmatrix}$; solution set $= \{(3, -5)\}$

(c) Inverse $= \begin{bmatrix} 1 & -13 & -15 & 5 \\ -3 & 3 & 0 & -7 \\ -1 & 2 & 1 & -3 \\ 0 & -4 & -5 & 1 \end{bmatrix}$; solution set $= \{(5, -8, 2, -1)\}$

8. (a) Consider $\begin{bmatrix} 0 & 1 \\ 1 & 0 \end{bmatrix}$.

(b) Consider $\begin{bmatrix} 0 & 1 & 0 \\ 1 & 0 & 0 \\ 0 & 0 & 1 \end{bmatrix}$.

(c) $\mathbf{A} = \mathbf{A}^{-1}$ if $\mathbf{A}$ is involutory.

10. (a) $\mathbf{B}$ must be the zero matrix.

(b) No, since $\mathbf{A}^{-1} = \mathbf{B}$ exists, $\mathbf{AC} = \mathbf{O}_n \Longrightarrow \mathbf{A}^{-1}\mathbf{AC} = \mathbf{A}^{-1}\mathbf{O}_n \Longrightarrow \mathbf{C} = \mathbf{O}_n$.

11. $\ldots, \mathbf{A}^{-11}, \mathbf{A}^{-6}, \mathbf{A}^{-1}, \mathbf{A}^4, \mathbf{A}^9, \mathbf{A}^{14}, \ldots$

12. $\mathbf{B}^{-1}\mathbf{A}$ is the inverse of $\mathbf{A}^{-1}\mathbf{B}$.

14. (a) All steps in the row reduction process will not alter the column of zeroes, and so the matrix cannot be reduced to $\mathbf{I}_n$.

21. (a) F   (b) T
    (c) T   (d) F
    (e) F   (f) T

Section 3.1 (p.128–132)

1. (a) -17 (c) 0
    (e) -108 (g) -40
    (i) 0 (j) -3

2. (a) $\begin{vmatrix} 4 & 3 \\ -2 & 4 \end{vmatrix} = 22$ (c) $\begin{vmatrix} -3 & 0 & 5 \\ 2 & -1 & 4 \\ 6 & 4 & 0 \end{vmatrix} = 118$

3. (a) $(-1)^{2+2} \begin{vmatrix} 4 & -3 \\ 9 & -7 \end{vmatrix} = -1$

    (c) $(-1)^{4+3} \begin{vmatrix} -5 & 2 & 13 \\ -8 & 2 & 22 \\ -6 & -3 & -16 \end{vmatrix} = 222$

    (d) $(-1)^{1+2} \begin{vmatrix} x-4 & x-3 \\ x-1 & x+2 \end{vmatrix} = -2x + 11$

5. (a) 0 (d) 352

7. Let $\mathbf{A} = \begin{bmatrix} 1 & 1 \\ 1 & 1 \end{bmatrix}$, and let $\mathbf{B} = \begin{bmatrix} 1 & 0 \\ 0 & 1 \end{bmatrix}$.

9. (a) 7 (c) 12

11. (a) 18 (c) 63

15. (a) $x = -5$ or $x = 2$
    (c) $x = 3, x = 1,$ or $x = 2$

16. (b) 20

18. (a) F   (b) T
    (c) F   (d) F
    (e) T

Section 3.2 (p.138–141)

1. (a) (II): $\langle 1 \rangle \longleftarrow -3\langle 2 \rangle + \langle 1 \rangle$; determinant $= 1$
    (c) (I): $\langle 3 \rangle \longleftarrow -4\langle 3 \rangle$; determinant $= -4$
    (f) (III): $\langle 1 \rangle \longleftrightarrow \langle 2 \rangle$; determinant $= -1$

2. (a) 30   (c) −4
  (e) 35

3. (a) Determinant = −2; matrix is nonsingular because determinant is nonzero

  (c) Determinant = −79; matrix is nonsingular

4. (a) Determinant = −1; system has only the trivial solution

6. $-a_{16}a_{25}a_{34}a_{43}a_{52}a_{61}$

16. (a) F   (b) T
  (c) F   (d) F
  (e) F   (f) T

Section 3.3 (p.148–153)

1. (a) $a_{31}(-1)^{3+1}|\mathbf{A}_{31}| + a_{32}(-1)^{3+2}|\mathbf{A}_{32}| + a_{33}(-1)^{3+3}|\mathbf{A}_{33}|$
  $+a_{34}(-1)^{3+4}|\mathbf{A}_{34}|$

  (c) $a_{14}(-1)^{1+4}|\mathbf{A}_{14}| + a_{24}(-1)^{2+4}|\mathbf{A}_{24}| + a_{34}(-1)^{3+4}|\mathbf{A}_{34}|$
  $+a_{44}(-1)^{4+4}|\mathbf{A}_{44}|$

2. (a) −76   (c) 102

3. (a) Adjoint $= \begin{bmatrix} -6 & 9 & 3 \\ 6 & -42 & 0 \\ -4 & 8 & 2 \end{bmatrix}$; determinant $= -6$;

  inverse $= \begin{bmatrix} 1 & -\frac{3}{2} & -\frac{1}{2} \\ -1 & 7 & 0 \\ \frac{2}{3} & -\frac{4}{3} & -\frac{1}{3} \end{bmatrix}$

  (c) Adjoint $= \begin{bmatrix} -3 & 0 & 3 & -3 \\ 0 & 0 & 0 & 0 \\ -3 & 0 & 3 & -3 \\ 6 & 0 & -6 & 6 \end{bmatrix}$; determinant = 0; no inverse

  (e) Adjoint $= \begin{bmatrix} 3 & -1 & -2 \\ 0 & -3 & -6 \\ 0 & 0 & -9 \end{bmatrix}$; determinant = 9; inverse $= \begin{bmatrix} \frac{1}{3} & -\frac{1}{9} & -\frac{2}{9} \\ 0 & -\frac{1}{3} & -\frac{2}{3} \\ 0 & 0 & -1 \end{bmatrix}$

4. (a) $\{(-4, 3, -7)\}$ (d) $\{(4, -1, -3, 6)\}$

8. (b) Consider $\mathbf{A} = \begin{bmatrix} 0 & -1 \\ 1 & 0 \end{bmatrix}$.

9. (b) Consider $\begin{bmatrix} 0 & 1 & 0 \\ 1 & 0 & 0 \\ 0 & 0 & 1 \end{bmatrix}$.

13. (b) For example, consider $\mathbf{B} = \begin{bmatrix} 1 & -1 \\ -1 & 2 \end{bmatrix} \mathbf{A} \begin{bmatrix} 1 & -1 \\ -1 & 2 \end{bmatrix}^{-1} = \begin{bmatrix} -6 & -4 \\ 16 & 11 \end{bmatrix}$,

or $\mathbf{B} = \begin{bmatrix} 3 & 5 \\ 1 & 2 \end{bmatrix} \mathbf{A} \begin{bmatrix} 3 & 5 \\ 1 & 2 \end{bmatrix}^{-1} = \begin{bmatrix} 10 & -12 \\ 4 & -5 \end{bmatrix}$.

14. $(\mathcal{B}\mathcal{A})/(|\mathbf{AB}|)$

18. (b) Consider $\mathbf{A} = \begin{bmatrix} 0 & 1 & 1 \\ -1 & 0 & 1 \\ -1 & -1 & 0 \end{bmatrix}$. Then $\mathcal{A} = \begin{bmatrix} 1 & -1 & 1 \\ -1 & 1 & -1 \\ 1 & -1 & 1 \end{bmatrix}$, which is not skew-symmetric.

22. (a) T   (b) T
   (c) F   (d) T
   (e) F   (f) T

Section 3.4 (p.162–166)

1. (a) $x^2 - 7x + 14$

(c) $x^3 - 8x^2 + 21x - 18$

(e) $x^4 - 3x^3 - 4x^2 + 12x$

2. (a) $E_2 = \{a[1, 1] \mid a \in \mathbb{R}\}$

(c) $E_{-1} = \{a[1, 2, 0] + b[0, 0, 1] \mid a, b \in \mathbb{R}\}$

3. (a) $\lambda = 1$; $E_1 = \{a[1, 0] \mid a \in \mathbb{R}\}$; algebraic multiplicity of λ is 2

(c) $\lambda_1 = 1$; $E_1 = \{a[1, 0, 0] \mid a \in \mathbb{R}\}$; algebraic multiplicity of λ_1 is 1; $\lambda_2 = 2$; $E_2 = \{b[0, 1, 0] \mid b \in \mathbb{R}\}$; algebraic multiplicity of λ_2 is 1; $\lambda_3 = -5$; $E_{-5} = \{c[-\frac{1}{6}, \frac{3}{7}, 1] \mid c \in \mathbb{R}\}$; algebraic multiplicity of λ_3 is 1

(e) $\lambda_1 = 0$; $E_0 = \{a[1, 3, 2] \mid a \in \mathbb{R}\}$; algebraic multiplicity of λ_1 is 1; $\lambda_2 = 2$; $E_2 = \{c[1, 0, 1] + b[0, 1, 0] \mid c, b \in \mathbb{R}\}$; algebraic multiplicity of λ_2 is 2

(h) $\lambda_1 = 0$; $E_0 = \{c[-1, 1, 1, 0] + d[0, -1, 0, 1] \mid c, d \in \mathbb{R}\}$; algebraic multiplicity of λ_1 is 2; $\lambda_2 = -3$; $E_{-3} = \{d[-1, 0, 2, 2] \mid d \in \mathbb{R}\}$; algebraic multiplicity of λ_2 is 2

4. (a) $\mathbf{P} = \begin{bmatrix} 3 & 2 \\ 1 & 1 \end{bmatrix}$; $\mathbf{D} = \begin{bmatrix} 3 & 0 \\ 0 & -5 \end{bmatrix}$

(c) Not diagonalizable

(d) $\mathbf{P} = \begin{bmatrix} 6 & 1 & 1 \\ 2 & 2 & 1 \\ 5 & 1 & 1 \end{bmatrix}$; $\mathbf{D} = \begin{bmatrix} 1 & 0 & 0 \\ 0 & -1 & 0 \\ 0 & 0 & 2 \end{bmatrix}$

(f) Not diagonalizable

(g) $\mathbf{P} = \begin{bmatrix} 2 & 1 & 0 \\ 3 & 0 & -1 \\ 0 & 3 & 1 \end{bmatrix}; \mathbf{D} = \begin{bmatrix} 2 & 0 & 0 \\ 0 & 2 & 0 \\ 0 & 0 & 3 \end{bmatrix}$

(i) $\mathbf{P} = \begin{bmatrix} 2 & 1 & 1 & 1 \\ 2 & 0 & 2 & -1 \\ 1 & 0 & 1 & 0 \\ 0 & 1 & 0 & 1 \end{bmatrix}; \mathbf{D} = \begin{bmatrix} 1 & 0 & 0 & 0 \\ 0 & 1 & 0 & 0 \\ 0 & 0 & -1 & 0 \\ 0 & 0 & 0 & 0 \end{bmatrix}$

5. (a) $\begin{bmatrix} 32770 & -65538 \\ 32769 & -65537 \end{bmatrix}$

(c) $\mathbf{A}^{49} = \mathbf{A}$

(e) $\begin{bmatrix} 4188163 & 6282243 & -9421830 \\ 4192254 & 6288382 & -9432060 \\ 4190208 & 6285312 & -9426944 \end{bmatrix}$

7. (b) $\mathbf{A}$ has a square root if and only if $\mathbf{A}$ has all eigenvalues nonnegative.

8. One possible answer: $\begin{bmatrix} 3 & -2 & -2 \\ -7 & 10 & 11 \\ 8 & -10 & -11 \end{bmatrix}$

10. (b) Consider the matrix $\mathbf{A} = \begin{bmatrix} 0 & -1 \\ 1 & 0 \end{bmatrix}$, which represents a rotation about the origin in $\mathbb{R}^2$ through an angle of $\frac{\pi}{2}$ radians, or $90°$. Although $\mathbf{A}$ has no eigenvalues, $\mathbf{A}^4 = \mathbf{I}_2$ has 1 as an eigenvalue.

24. (a) T   (b) F
    (c) T   (d) T
    (e) F   (f) T
    (g) T   (h) F

Section 4.1 (p.178–180)

5. The set of singular 2×2 matrices is not closed under addition. For example, $\begin{bmatrix} 1 & 0 \\ 0 & 0 \end{bmatrix}$ and $\begin{bmatrix} 0 & 0 \\ 0 & 1 \end{bmatrix}$ are both singular, but their sum $\begin{bmatrix} 1 & 0 \\ 0 & 1 \end{bmatrix} = \mathbf{I}_2$ is nonsingular.

8. Properties (2), (3), and (6) are not satisfied, and property (4) makes no sense without property (3). The following is a counterexample for property (2): $3 \oplus (4 \oplus 5) = 3 \oplus 18 = 42$, but $(3 \oplus 4) \oplus 5 = 14 \oplus 5 = 38$.

20. (a) F   (b) F
    (c) T   (d) T
    (e) F   (f) T
    (g) T

Section 4.2 (p.187–190)

    1. (a) Not a subspace; no zero vector

        (c) Subspace

        (e) Not a subspace; no zero vector

        (g) Not a subspace; not closed under addition

        (j) Not a subspace; not closed under addition

        (l) Not a subspace; not closed under scalar multiplication

    2. Only starred parts are listed:
        Subspaces: (a), (c), (e), (g)
        Part (h) is not a subspace because it is not closed under addition.

    3. Only starred parts are listed:
        Subspaces: (a), (b), (g)
        Part (e) is not a subspace because it does not contain the zero polynomial.
        Also, it is not closed under addition.

   12. (e) No; if $|\mathbf{A}| \neq 0$ and $c = 0$, then $|c\mathbf{A}| = 0$.

   15. $S = \{\mathbf{0}\}$, the trivial subspace of $\mathbb{R}^n$.

   22. (a) F   (b) T
         (c) F   (d) T
         (e) T   (f) F
         (g) T   (h) T

Section 4.3 (p.198–201)

    1. (a) $\{[a, b, -a + b] \mid a, b \in \mathbb{R}\}$

        (c) $\{[a, b, -b] \mid a, b \in \mathbb{R}\}$

        (e) $\{[a, b, c, -2a + b + c] \mid a, b, c \in \mathbb{R}\}$

    2. (a) $\{ax^3 + bx^2 + cx - (a + b + c) \mid a, b, c \in \mathbb{R}\}$

        (c) $\{ax^3 - ax + b \mid a, b \in \mathbb{R}\}$

    3. (a) $\left\{ \begin{bmatrix} a & b \\ c & -a - b - c \end{bmatrix} \,\middle|\, a, b, c \in \mathbb{R} \right\}$

        (c) $\left\{ \begin{bmatrix} a & b \\ c & d \end{bmatrix} \,\middle|\, a, b, c, d \in \mathbb{R} \right\} = \mathcal{M}_{22}$

    4. (a) $[a + b, a + c, b + c, c] = a[1, 1, 0, 0] + b[1, 0, 1, 0] + c[0, 1, 1, 1]$.
        The set of vectors of this form is the row space of

$$\mathbf{A} = \begin{bmatrix} 1 & 1 & 0 & 0 \\ 1 & 0 & 1 & 0 \\ 0 & 1 & 1 & 1 \end{bmatrix}.$$

(b) $\mathbf{B} = \begin{bmatrix} 1 & 0 & 0 & -\frac{1}{2} \\ 0 & 1 & 0 & \frac{1}{2} \\ 0 & 0 & 1 & \frac{1}{2} \end{bmatrix}$.

(c) Row space of $\mathbf{B}$ = $\{a[1,0,0,-\frac{1}{2}] + b[0,1,0,\frac{1}{2}] + c[0,0,1,\frac{1}{2}] \mid a,b,c \in \mathbb{R}\} = \{[a,b,c,-\frac{1}{2}a+\frac{1}{2}b+\frac{1}{2}c] \mid a,b,c \in \mathbb{R}\}$

11. One answer is: $-1(x^3 - 2x^2 + x - 3) + 2(2x^3 - 3x^2 + 2x + 5) - 1(4x^2 + x - 3) + 0(4x^3 - 7x^2 + 4x - 1)$.

14. (a) Hint: Use Theorem 1.13.

16. (a) $S = \{[-3,2,0],[4,0,5]\}$

24. (b) $S_1 = \{[1,0,0],[0,1,0]\}$, $S_2 = \{[0,1,0],[0,0,1]\}$

(c) $S_1 = \{[1,0,0],[0,1,0]\}$, $S_2 = \{[1,0,0],[1,1,0]\}$

25. (c) $S_1 = \{x^5\}$, $S_2 = \{x^4\}$

28. (a) F   (b) T
(c) F   (d) F
(e) F   (f) T
(g) F

Section 4.4 (p.208–212)

1. Linearly independent: (a), (b)
Linearly dependent: (c), (d), (e)

2. Answers given for starred parts only:
Linearly independent: (b)
Linearly dependent: (a), (e)

3. Answers given for starred parts only:
Linearly independent: (a)
Linearly dependent: (c)

4. Answers given for starred parts only:
Linearly independent: (a), (e)
Linearly dependent: (c)

7. (b) $[0,1,0]$

(c) No; $[0,0,1]$ also works.

(d) Any linear combination of $[1,1,0]$ and $[-2,0,1]$ works, other than $[1,1,0]$ and $[-2,0,1]$ themselves.

11. (a) One answer is $\{[1,0,0,0],[0,1,0,0],[0,0,1,0],[0,0,0,1]\}$.

(c) One answer is $\{1,x,x^2,x^3\}$.

(e) One answer is $\left\{ \begin{bmatrix} 1 & 0 & 0 \\ 0 & 0 & 0 \\ 0 & 0 & 0 \end{bmatrix}, \begin{bmatrix} 0 & 1 & 0 \\ 1 & 0 & 0 \\ 0 & 0 & 0 \end{bmatrix}, \begin{bmatrix} 0 & 0 & 1 \\ 0 & 0 & 0 \\ 1 & 0 & 0 \end{bmatrix}, \begin{bmatrix} 0 & 0 & 0 \\ 0 & 1 & 0 \\ 0 & 0 & 0 \end{bmatrix} \right\}.$

(Notice that each matrix is symmetric.)

13. (b) $[0, 0, -6, 0]$ is redundant because $[0, 0, -6, 0] = 6[1, 1, 0, 0] - 6[1, 1, 1, 0]$. Hence, $a[1, 1, 0, 0] + b[1, 1, 1, 0] + c[0, 0, -6, 0] = (a + 6c)[1, 1, 0, 0] + (b - 6c)[1, 1, 1, 0]$.

19. (b) Let **A** be the zero matrix.

28. (a) F (b) T
(c) T (d) F
(e) T (f) T
(g) F (h) T
(i) T

Section 4.5 (p.221–224)

4. (a) Not a basis (linearly independent but does not span)

(c) Basis

(e) Not a basis (linearly dependent but spans)

5. (b) 2

(c) No; $\dim(\mathrm{span}(S)) = 2 \neq 4 = \dim(\mathbb{R}^4)$

11. (b) 5

(c) $\{(x - 2)(x - 3), x(x - 2)(x - 3), x^2(x - 2)(x - 3), x^3(x - 2)(x - 3)\}$

(d) 4

12. (a) Let $\mathcal{V} = \mathbb{R}^3$, and let $S = \{[1, 0, 0], [2, 0, 0], [3, 0, 0]\}$.

(b) Let $\mathcal{V} = \mathbb{R}^3$, and let $T = \{[1, 0, 0], [2, 0, 0], [3, 0, 0]\}$.

25. (a) T (b) F
(c) F (d) F
(e) F (f) T
(g) F (h) F
(i) F (j) T

Section 4.6 (p.231–234)

1. (a) $\{[1, 0, 0, 2, -2], [0, 1, 0, 0, 1], [0, 0, 1, -1, 0]\}$

(d) $\{[1, 0, 0, -2, -\frac{13}{4}], [0, 1, 0, 3, \frac{9}{2}], [0, 0, 1, 0, -\frac{1}{4}]\}$

2. $\{x^3 - 3x, x^2 - x, 1\}$

3. $\left\{\begin{bmatrix} 1 & 0 \\ \frac{4}{3} & \frac{1}{3} \\ 2 & 0 \end{bmatrix}, \begin{bmatrix} 0 & 1 \\ -\frac{1}{3} & -\frac{1}{3} \\ 0 & 0 \end{bmatrix}, \begin{bmatrix} 0 & 0 \\ 0 & 0 \\ 0 & 1 \end{bmatrix}\right\}$

4. (a) $\{[3, 1, -2], [6, 2, -3]\}$

(c) One answer is $\{[1, 3, -2], [2, 1, 4], [0, 1, -1]\}$.

(e) One answer is $\{[3, -2, 2], [1, 2, -1], [3, -2, 7]\}$.

(h) One answer is $\{[1, -3, 0], [0, 1, 1]\}$.

5. (a) One answer is $\{x^3 - 8x^2 + 1, 3x^3 - 2x^2 + x, 4x^3 + 2x - 10, x^3 - 20x^2 - x + 12\}$.

(c) One answer is $\{x^3, x^2, x\}$.

(e) One answer is $\{x^3 + x^2, x, 1\}$.

6. (a) One answer is $\{\Psi_{ij} \mid 1 \le i, j \le 3\}$, where Ψ_{ij} is the 3×3 matrix with (i, j) entry $= 1$ and all other entries 0.

(c) One answer is $\left\{\begin{bmatrix} 1 & 0 & 0 \\ 0 & 0 & 0 \\ 0 & 0 & 0 \end{bmatrix}, \begin{bmatrix} 0 & 1 & 0 \\ 1 & 0 & 0 \\ 0 & 0 & 0 \end{bmatrix}, \begin{bmatrix} 0 & 0 & 1 \\ 0 & 0 & 0 \\ 1 & 0 & 0 \end{bmatrix}, \begin{bmatrix} 0 & 0 & 0 \\ 0 & 1 & 0 \\ 0 & 0 & 0 \end{bmatrix},\right.$
$\left.\begin{bmatrix} 0 & 0 & 0 \\ 0 & 0 & 1 \\ 0 & 1 & 0 \end{bmatrix}, \begin{bmatrix} 0 & 0 & 0 \\ 0 & 0 & 0 \\ 0 & 0 & 1 \end{bmatrix}\right\}$.

7. (a) One answer is $\{[1, -3, 0, 1, 4], [2, 2, 1, -3, 1], [1, 0, 0, 0, 0], [0, 1, 0, 0, 0], [0, 0, 1, 0, 0]\}$.

(c) One answer is $\{[1, 0, -1, 0, 0], [0, 1, -1, 1, 0], [2, 3, -8, -1, 0], [1, 0, 0, 0, 0], [0, 0, 0, 0, 1]\}$.

8. (a) One answer is $\{x^3 - x^2, x^4 - 3x^3 + 5x^2 - x, x^4, x^3, 1\}$.

(c) One answer is $\{x^4 - x^3 + x^2 - x + 1, x^3 - x^2 + x - 1, x^2 - x + 1, x^2, x\}$.

9. (a) One answer is $\left\{\begin{bmatrix} 1 & -1 \\ -1 & 1 \\ 0 & 0 \end{bmatrix}, \begin{bmatrix} 0 & 0 \\ 1 & -1 \\ -1 & 1 \end{bmatrix}, \begin{bmatrix} 1 & 0 \\ 0 & 0 \\ 0 & 0 \end{bmatrix}, \begin{bmatrix} 0 & 1 \\ 0 & 0 \\ 0 & 0 \end{bmatrix},\right.$
$\left.\begin{bmatrix} 0 & 0 \\ 1 & 0 \\ 0 & 0 \end{bmatrix}, \begin{bmatrix} 0 & 0 \\ 0 & 0 \\ 1 & 0 \end{bmatrix}\right\}$.

(c) One answer is $\left\{\begin{bmatrix} 3 & 0 \\ -1 & 7 \\ 0 & 1 \end{bmatrix}, \begin{bmatrix} -1 & 0 \\ 1 & 3 \\ 0 & -2 \end{bmatrix}, \begin{bmatrix} 2 & 0 \\ 3 & 1 \\ 0 & -1 \end{bmatrix}, \begin{bmatrix} 6 & 0 \\ 0 & 1 \\ 0 & -1 \end{bmatrix},\right.$
$\left.\begin{bmatrix} 0 & 1 \\ 0 & 0 \\ 0 & 0 \end{bmatrix}, \begin{bmatrix} 0 & 0 \\ 0 & 0 \\ 1 & 0 \end{bmatrix}\right\}$.

10. $\{\Psi_{ij} \mid 1 \le i \le j \le 4\}$, where Ψ_{ij} is the 4×4 matrix with (i, j) entry $= 1$ and all other entries 0. Notice that the condition $1 \le i \le j \le 4$ assures that only upper triangular matrices are included.

11. (b) 8   (d) 3

12. (b) $(n^2 - n)/2$

15. (b) No; consider the subspace $\mathcal{W}$ of $\mathbb{R}^3$ given by $\mathcal{W} = \{[a, 0, 0] \mid a \in \mathbb{R}\}$. No subset of $B = \{[1, 1, 0], [1, -1, 0], [0, 0, 1]\}$ (a basis for $\mathbb{R}^3$) is a basis for $\mathcal{W}$.

   (c) Yes; consider $\mathcal{Y} = \operatorname{span}(B')$.

16. (b) In $\mathbb{R}^3$, consider $\mathcal{W} = \{[a, b, 0] \mid a, b \in \mathbb{R}\}$. We could let $\mathcal{W}' = \{[0, 0, c] \mid c \in \mathbb{R}\}$ or $\mathcal{W}' = \{[0, c, c] \mid c \in \mathbb{R}\}$.

20. (a) T   (b) T
    (c) F   (d) T
    (e) T   (f) F
    (g) F

Section 4.7 (p.246–249)

1. (a) $[\mathbf{v}]_B = [7, -1, -5]$ (c) $[\mathbf{v}]_B = [-2, 4, -5]$
   (e) $[\mathbf{v}]_B = [4, -5, 3]$ (g) $[\mathbf{v}]_B = [-1, 4, -2]$
   (h) $[\mathbf{v}]_B = [2, -3, 1]$ (j) $[\mathbf{v}]_B = [5, -2]$

2. (a) $\begin{bmatrix} -102 & 20 & 3 \\ 67 & -13 & -2 \\ 36 & -7 & -1 \end{bmatrix}$ (c) $\begin{bmatrix} 20 & -30 & -69 \\ 24 & -24 & -80 \\ -9 & 11 & 31 \end{bmatrix}$

   (d) $\begin{bmatrix} -1 & -4 & 2 & -9 \\ 4 & 5 & 1 & 3 \\ 0 & 2 & -3 & 1 \\ -4 & -13 & 13 & -15 \end{bmatrix}$ (f) $\begin{bmatrix} 6 & 1 & 2 \\ 1 & 1 & 2 \\ -1 & -1 & -3 \end{bmatrix}$

4. (a) $\mathbf{P} = \begin{bmatrix} 13 & 31 \\ -18 & -43 \end{bmatrix}$; $\mathbf{Q} = \begin{bmatrix} -11 & -8 \\ 29 & 21 \end{bmatrix}$; $\mathbf{T} = \begin{bmatrix} 1 & 3 \\ -1 & -4 \end{bmatrix}$

   (c) $\mathbf{P} = \begin{bmatrix} 2 & 8 & 13 \\ -6 & -25 & -43 \\ 11 & 45 & 76 \end{bmatrix}$; $\mathbf{Q} = \begin{bmatrix} -24 & -2 & 1 \\ 30 & 3 & -1 \\ 139 & 13 & -5 \end{bmatrix}$;

   $\mathbf{T} = \begin{bmatrix} -25 & -97 & -150 \\ 31 & 120 & 185 \\ 145 & 562 & 868 \end{bmatrix}$

5. (a) $C = ([1, -4, 0, -2, 0], [0, 0, 1, 4, 0], [0, 0, 0, 0, 1]);$　$\mathbf{P} = \begin{bmatrix} 1 & 6 & 3 \\ 1 & 5 & 3 \\ 1 & 3 & 2 \end{bmatrix};$

$$\mathbf{Q} = \mathbf{P}^{-1} = \begin{bmatrix} 1 & -3 & 3 \\ 1 & -1 & 0 \\ -2 & 3 & -1 \end{bmatrix}; [\mathbf{v}]_B = [17, 4, -13]; [\mathbf{v}]_C = [2, -2, 3]$$

(c) $C = ([1, 0, 0, 0], [0, 1, 0, 0], [0, 0, 1, 0], [0, 0, 0, 1]);$

$$\mathbf{P} = \begin{bmatrix} 3 & 6 & -4 & -2 \\ -1 & 7 & -3 & 0 \\ 4 & -3 & 3 & 1 \\ 6 & -2 & 4 & 2 \end{bmatrix}; \mathbf{Q} = \mathbf{P}^{-1} = \begin{bmatrix} 1 & -4 & -12 & 7 \\ -2 & 9 & 27 & -\frac{31}{2} \\ -5 & 22 & 67 & -\frac{77}{2} \\ 5 & -23 & -71 & 41 \end{bmatrix};$$

$$[\mathbf{v}]_B = [2, 1, -3, 7]; [\mathbf{v}]_C = [10, 14, 3, 12]$$

7. (a) Transition matrix to $C_1 = \begin{bmatrix} 0 & 1 & 0 \\ 0 & 0 & 1 \\ 1 & 0 & 0 \end{bmatrix}$

Transition matrix to $C_2 = \begin{bmatrix} 0 & 0 & 1 \\ 1 & 0 & 0 \\ 0 & 1 & 0 \end{bmatrix}$

Transition matrix to $C_3 = \begin{bmatrix} 1 & 0 & 0 \\ 0 & 0 & 1 \\ 0 & 1 & 0 \end{bmatrix}$

Transition matrix to $C_4 = \begin{bmatrix} 0 & 1 & 0 \\ 1 & 0 & 0 \\ 0 & 0 & 1 \end{bmatrix}$

Transition matrix to $C_5 = \begin{bmatrix} 0 & 0 & 1 \\ 0 & 1 & 0 \\ 1 & 0 & 0 \end{bmatrix}$

10. $C = ([-142, 64, 167], [-53, 24, 63], [-246, 111, 290])$

11. (b) $\mathbf{D}[\mathbf{v}]_B = [\mathbf{Av}]_B = [2, -2, 3].$

15. (a) F　　(b) T
　　(c) T　　(d) F
　　(e) F　　(f) T
　　(g) F

Section 5.1 (p.260–264)

1. Only starred parts are listed.
Linear transformations: (a), (d), (h)
Linear operators: (a), (d)

10. (c) $\begin{bmatrix} \cos\theta & 0 & -\sin\theta \\ 0 & 1 & 0 \\ \sin\theta & 0 & \cos\theta \end{bmatrix}$

26. $L(\mathbf{i}) = \frac{7}{5}\mathbf{i} - \frac{11}{5}\mathbf{j}; L(\mathbf{j}) = -\frac{2}{5}\mathbf{i} - \frac{4}{5}\mathbf{j}$

30. (b) Consider the zero linear transformation.

36. (a) F   (b) T
    (c) F   (d) F
    (e) T   (f) F
    (g) T (h) T

Section 5.2 (p.274–279)

2. (a) $\begin{bmatrix} -6 & 4 & -1 \\ -2 & 3 & -5 \\ 3 & -1 & 7 \end{bmatrix}$ (c) $\begin{bmatrix} 4 & -1 & 3 & 3 \\ 1 & 3 & -1 & 5 \\ -2 & -7 & 5 & -1 \end{bmatrix}$

3. (a) $\begin{bmatrix} -47 & 128 & -288 \\ -18 & 51 & -104 \end{bmatrix}$ (c) $\begin{bmatrix} 22 & 14 \\ 62 & 39 \\ 68 & 43 \end{bmatrix}$

(e) $\begin{bmatrix} 5 & 6 & 0 \\ -11 & -26 & -6 \\ -14 & -19 & -1 \\ 6 & 3 & -2 \\ -1 & 1 & 1 \\ 11 & 13 & 0 \end{bmatrix}$

4. (a) $\begin{bmatrix} -202 & -32 & -43 \\ -146 & -23 & -31 \\ 83 & 14 & 18 \end{bmatrix}$ (b) $\begin{bmatrix} 21 & 7 & 21 & 16 \\ -51 & -13 & -51 & -38 \end{bmatrix}$

6. (a) $\begin{bmatrix} 67 & -123 \\ 37 & -68 \end{bmatrix}$ (b) $\begin{bmatrix} -7 & 2 & 10 \\ 5 & -2 & -9 \\ -6 & 1 & 8 \end{bmatrix}$

7. (a) $\begin{bmatrix} 3 & 0 & 0 & 0 \\ 0 & 2 & 0 & 0 \\ 0 & 0 & 1 & 0 \end{bmatrix}$; $12x^2 - 10x + 6$

8. (a) $\begin{bmatrix} \frac{\sqrt{3}}{2} & -\frac{1}{2} \\ \frac{1}{2} & \frac{\sqrt{3}}{2} \end{bmatrix}$

9. (b) $\frac{1}{2}$
$$\begin{bmatrix} 1 & 0 & 0 & -1 & 0 & 0 \\ 1 & 0 & 0 & 1 & 0 & 0 \\ 0 & 1 & 1 & 0 & -1 & 0 \\ 0 & 1 & 1 & 0 & 1 & 0 \\ 0 & -1 & 0 & 0 & -1 & 1 \\ 0 & -1 & 0 & 0 & 1 & -1 \end{bmatrix}$$

10.
$$\begin{bmatrix} -12 & 12 & -2 \\ -4 & 6 & -2 \\ -10 & -3 & 7 \end{bmatrix}$$

13. (a) $\mathbf{I}_n$

  (c) $c\mathbf{I}_n$

  (e) The $n \times n$ matrix whose columns are $\mathbf{e}_n, \mathbf{e}_1, \mathbf{e}_2, \ldots, \mathbf{e}_{n-1}$, respectively

23. (a) $p_{\mathbf{A}_{BB}}(x) = x^3 - 2x^2 + x = x(x-1)^2$

  (b) Basis for $E_1 = ([2, 1, 0], [2, 0, 1])$; basis for $E_0 = ([-1, 2, 2])$

  (c) One answer is $\mathbf{P} = \begin{bmatrix} 2 & 2 & -1 \\ 1 & 0 & 2 \\ 0 & 1 & 2 \end{bmatrix}^{-1} = \frac{1}{9} \begin{bmatrix} 2 & 5 & -4 \\ 2 & -4 & 5 \\ -1 & 2 & 2 \end{bmatrix}.$

31. (a) T **(b)** T
  (c) F **(d)** F
  (e) T **(f)** F
  (g) T **(h)** T
  (i) T **(j)** F

Section 5.3 (p.287–290)

1. (a) Yes, because $L([1, -2, 3]) = [0, 0, 0]$.

  (c) No, because the system
$$\begin{cases} 5x_1 + x_2 - x_3 = 2 \\ -3x_1 + x_3 = -1 \\ x_1 - x_2 - x_3 = 4 \end{cases}$$

  has no solutions.

2. (a) No, since $L(x^3 - 5x^2 + 3x - 6) \neq 0$

  (c) Yes, because, for example, $L(x^3 + 4x + 3) = 8x^3 - x - 1$

3. (a) $\dim(\ker(L)) = 1$; basis for $\ker(L) = \{[-2, 3, 1]\}$; $\dim(\text{range}(L)) = 2$; basis for $\text{range}(L) = \{[1, -2, 3], [-1, 3, -3]\}$

  (d) $\dim(\ker(L)) = 2$; basis for $\ker(L) = \{[1, -3, 1, 0], [-1, 2, 0, 1]\}$; $\dim(\text{range}(L)) = 2$; basis for $\text{range}(L) = \{[-14, -4, -6, 3, 4], [-8, -1, 2, -7, 2]\}$

4. (a) $\dim(\ker(L)) = 2$; basis for $\ker(L) = \{[1, 0, 0], [0, 0, 1]\}$; $\dim(\text{range}(L))$ $= 1$; basis for range$(L) = \{[0, 1]\}$

(d) $\dim(\ker(L)) = 2$; basis for $\ker(L) = \{x^4, x^3\}$; $\dim(\text{range}(L)) = 3$; basis for range$(L) = \{x^2, x, 1\}$

(f) $\dim(\ker(L)) = 1$; basis for $\ker(L) = \{[0, 1, 1]\}$; $\dim(\text{range}(L)) = 2$; basis for range$(L) = \{[1, 0, 1], [0, 0, -1]\}$ (A simpler basis for range(L) $= \{[1, 0, 0], [0, 0, 1]\}$.)

(g) $\dim(\ker(L)) = 0$; basis for $\ker(L) = \{\ \}$ (empty set); $\dim(\text{range}(L))$ $= 4$; basis for range$(L) = $ standard basis for $\mathcal{M}_{22}$

(i) $\dim(\ker(L)) = 1$; basis for $\ker(L) = \{x^2 - 2x + 1\}$; $\dim(\text{range}(L)) = 2$; basis for range$(L) = \{[1, 2], [1, 1]\}$ (A simpler basis for range$(L) = $ standard basis for $\mathbb{R}^2$.)

6. $\ker(L) = \left\{ \begin{bmatrix} a & b & c \\ d & e & f \\ g & h & -a-e \end{bmatrix} \middle| a, b, c, d, e, f, g, h \in \mathbb{R} \right\}$; $\dim(\ker(L)) = 8$;

range$(L) = \mathbb{R}$; $\dim(\text{range}(L)) = 1$

8. $\ker(L) = \{\mathbf{0}\}$; range$(L) = \{ax^4 + bx^3 + cx^2\}$; $\dim(\ker(L)) = 0$; $\dim(\text{range}\ (L)) = 3$

10. When $k \leq n$, $\ker(L) = $ all polynomials of degree less than k, $\dim(\ker(L)) = k$, range$(L) = \mathcal{P}_{n-k}$, and $\dim(\text{range}(L)) = n - k + 1$. When $k > n$, $\ker(L) = \mathcal{P}_n$, $\dim(\ker(L)) = n + 1$, range$(L) = \{\mathbf{0}\}$, and $\dim(\text{range}\ (L)) = 0$.

12. $\ker(L) = \{[0, 0, \ldots, 0]\}$; range$(L) = \mathbb{R}^n$ (Note: Every vector $\mathbf{X}$ is in the range since $L(\mathbf{A}^{-1}\mathbf{X}) = \mathbf{A}(\mathbf{A}^{-1}\mathbf{X}) = \mathbf{X}$.)

16. Consider $L\left(\begin{bmatrix} x \\ y \end{bmatrix}\right) = \begin{bmatrix} 1 & -1 \\ 1 & -1 \end{bmatrix}\begin{bmatrix} x \\ y \end{bmatrix}$. Then, $\ker(L) = $ range$(L) = $ $\{[a, a] \mid a \in \mathbb{R}\}$.

18. (d) The statement to be proven is as follows: Let $L : \mathcal{V} \longrightarrow \mathcal{W}$ be a linear transformation between finite dimensional vector spaces. Suppose that $\mathbf{A}$ is the matrix for L with respect to some bases for $\mathcal{V}$ and $\mathcal{W}$. Then, $\dim(\text{range}(L)) = \text{rank}(\mathbf{A})$ and $\dim(\ker(L)) = n - \text{rank}(\mathbf{A})$.

20. (a) F   (b) F
(c) T   (d) F
(e) T   (f) F
(g) F   (h) F

Section 5.4 (p.299–304)

1. (a) Not one-to-one, because $L([1, 0, 0]) = L([0, 0, 0]) = [0, 0, 0, 0]$; not onto, because $[0, 0, 0, 1]$ is not in range(L)

(c) One-to-one, because $L([x, y, z]) = [0, 0, 0]$ implies that $[2x, x + y + z, -y] = [0, 0, 0]$, which gives $x = y = z = 0$; onto, because every vector $[a, b, c]$ can be expressed as $[2x, x + y + z, -y]$, where $x = \frac{a}{2}$, $y = -c$, and $z = b - \frac{a}{2} + c$

(e) One-to-one, because $L(ax^2 + bx + c) = 0$ implies that $a + b = b + c = a + c = 0$, which gives $b = c$ and hence $a = b = c = 0$; onto, because every polynomial $Ax^2 + Bx + C$ can be expressed as $(a+b)x^2 + (b+c)x + (a + c)$, where $a = (A - B + C)/2$, $b = (A + B - C)/2$, and $c = (-A + B + C)/2$

(g) Not one-to-one, because $L\left(\begin{bmatrix} 0 & 1 & 0 \\ 1 & 0 & -1 \end{bmatrix}\right) = L\left(\begin{bmatrix} 0 & 0 & 0 \\ 0 & 0 & 0 \end{bmatrix}\right) = \begin{bmatrix} 0 & 0 \\ 0 & 0 \end{bmatrix}$;

onto, because every 2×2 matrix $\begin{bmatrix} A & B \\ C & D \end{bmatrix}$ can be expressed as

$\begin{bmatrix} a & -c \\ 2e & d+f \end{bmatrix}$, where $a = A$, $c = -B$, $e = C/2$, $d = D$, and $f = 0$

(h) One-to-one, because $L(ax^2 + bx + c) = \begin{bmatrix} 0 & 0 \\ 0 & 0 \end{bmatrix}$ implies that $a + c = b - c = -3a = 0$, which gives $a = b = c = 0$; not onto, because $\begin{bmatrix} 0 & 1 \\ 0 & 0 \end{bmatrix}$ is not in range(L)

2. (a) One-to-one; onto; isomorphism; the matrix row reduces to $\mathbf{I}_2$, which means that $\dim(\ker(L)) = 0$ and $\dim(\mathrm{range}(L)) = 2$.

(b) One-to-one; not onto; not an isomorphism; the matrix row reduces to $\begin{bmatrix} 1 & 0 \\ 0 & 1 \\ 0 & 0 \end{bmatrix}$, which means that $\dim(\ker(L)) = 0$ and $\dim(\mathrm{range}\,(L)) = 2$.

(c) Not one-to-one; not onto; not an isomorphism; the matrix row reduces to $\begin{bmatrix} 1 & 0 & -\frac{2}{5} \\ 0 & 1 & -\frac{6}{5} \\ 0 & 0 & 0 \end{bmatrix}$, which means that $\dim(\ker(L)) = 1$ and $\dim(\mathrm{range}(L)) = 2$.

3. (a) One-to-one; onto; isomorphism; the matrix row reduces to $\mathbf{I}_3$, which means that $\dim(\ker(L)) = 0$ and $\dim(\mathrm{range}(L)) = 3$.

(c) Not one-to-one; not onto; not an isomorphism; the matrix row reduces to $\begin{bmatrix} 1 & 0 & -\frac{10}{11} & \frac{19}{11} \\ 0 & 1 & \frac{3}{11} & -\frac{9}{11} \\ 0 & 0 & 0 & 0 \\ 0 & 0 & 0 & 0 \end{bmatrix}$, which means that $\dim(\ker(L)) = 2$ and $\dim(\mathrm{range}(L)) = 2$.

4. (a) No, because $\dim(\mathbb{R}^6) = \dim(\mathcal{P}_5)$

(b) No, because $\dim(\mathcal{M}_{22}) = \dim(\mathcal{P}_3)$

8. (a) $L_1^{-1}\left(\begin{bmatrix} x_1 \\ x_2 \\ x_3 \end{bmatrix}\right) = \begin{bmatrix} 0 & 0 & 1 \\ 0 & -1 & 0 \\ 1 & -2 & 0 \end{bmatrix}\begin{bmatrix} x_1 \\ x_2 \\ x_3 \end{bmatrix}$,

$L_2^{-1}\left(\begin{bmatrix} x_1 \\ x_2 \\ x_3 \end{bmatrix}\right) = \begin{bmatrix} 1 & 0 & 0 \\ 0 & 0 & -\frac{1}{3} \\ 2 & 1 & 0 \end{bmatrix}\begin{bmatrix} x_1 \\ x_2 \\ x_3 \end{bmatrix}$,

$(L_2 \circ L_1)\left(\begin{bmatrix} x_1 \\ x_2 \\ x_3 \end{bmatrix}\right) = \begin{bmatrix} 0 & -2 & 1 \\ 1 & 4 & -2 \\ 0 & 3 & 0 \end{bmatrix}\begin{bmatrix} x_1 \\ x_2 \\ x_3 \end{bmatrix}$,

$(L_2 \circ L_1)^{-1}\left(\begin{bmatrix} x_1 \\ x_2 \\ x_3 \end{bmatrix}\right) = (L_1^{-1} \circ L_2^{-1})\left(\begin{bmatrix} x_1 \\ x_2 \\ x_3 \end{bmatrix}\right) = \begin{bmatrix} 2 & 1 & 0 \\ 0 & 0 & \frac{1}{3} \\ 1 & 0 & \frac{2}{3} \end{bmatrix}\begin{bmatrix} x_1 \\ x_2 \\ x_3 \end{bmatrix}$

(c) $L_1^{-1}\left(\begin{bmatrix} x_1 \\ x_2 \\ x_3 \end{bmatrix}\right) = \begin{bmatrix} 2 & -4 & -1 \\ 7 & -13 & -3 \\ 5 & -10 & -3 \end{bmatrix}\begin{bmatrix} x_1 \\ x_2 \\ x_3 \end{bmatrix}$,

$L_2^{-1}\left(\begin{bmatrix} x_1 \\ x_2 \\ x_3 \end{bmatrix}\right) = \begin{bmatrix} 1 & 0 & -1 \\ 3 & 1 & -3 \\ -1 & -2 & 2 \end{bmatrix}\begin{bmatrix} x_1 \\ x_2 \\ x_3 \end{bmatrix}$,

$(L_2 \circ L_1)\left(\begin{bmatrix} x_1 \\ x_2 \\ x_3 \end{bmatrix}\right) = \begin{bmatrix} 29 & -6 & -4 \\ 21 & -5 & -2 \\ 38 & -8 & -5 \end{bmatrix}\begin{bmatrix} x_1 \\ x_2 \\ x_3 \end{bmatrix}$,

$(L_2 \circ L_1)^{-1}\left(\begin{bmatrix} x_1 \\ x_2 \\ x_3 \end{bmatrix}\right) = (L_1^{-1} \circ L_2^{-1})\left(\begin{bmatrix} x_1 \\ x_2 \\ x_3 \end{bmatrix}\right)$

$= \begin{bmatrix} -9 & -2 & 8 \\ -29 & -7 & 26 \\ -22 & -4 & 19 \end{bmatrix}\begin{bmatrix} x_1 \\ x_2 \\ x_3 \end{bmatrix}$

12. (a) $\begin{bmatrix} 0 & 1 \\ 1 & 0 \end{bmatrix}$

26. (a) $T = F_2 \circ L \circ F_1^{-1}$

31. (a) F (b) F
(c) T (d) F
(e) T (f) T
(g) F (h) T

Section 5.5 (p.318–320)

1. (a) $\lambda_1 = 2$; basis for $E_2 = ([1, 0])$; algebraic multiplicity of λ_1 is 2; geometric multiplicity of λ_1 is 1

(c) $\lambda_1 = 1$; basis for $E_1 = ([2, 1, 1])$; $\lambda_2 = -1$; basis for $E_{-1} = ([-1, 0, 1])$; $\lambda_3 = 2$; basis for $E_2 = ([1, 1, 1])$; all three eigenvalues have algebraic multiplicity = geometric multiplicity = 1.

(d) $\lambda_1 = 2$; basis for $E_2 = ([5, 4, 0], [3, 0, 2])$; algebraic multiplicity of λ_1 is 2; geometric multiplicity of λ_1 is 2; $\lambda_2 = 3$; basis for $E_3 = ([0, -1, 1])$; algebraic multiplicity of λ_2 is 1; geometric multiplicity of λ_2 is 1

2. (b) $C = (x^2, x, 1)$; $\mathbf{A} = \begin{bmatrix} 2 & 0 & 0 \\ -2 & 1 & 0 \\ 0 & -1 & 0 \end{bmatrix}$; $B = (x^2 - 2x + 1, -x + 1, 1)$;

$$\mathbf{D} = \begin{bmatrix} 2 & 0 & 0 \\ 0 & 1 & 0 \\ 0 & 0 & 0 \end{bmatrix}; \mathbf{P} = \begin{bmatrix} 1 & 0 & 0 \\ -2 & -1 & 0 \\ 1 & 1 & 1 \end{bmatrix}$$

(d) $C = (x^2, x, 1)$; $\mathbf{A} = \begin{bmatrix} -1 & 0 & 0 \\ -12 & -4 & 0 \\ 18 & 0 & -5 \end{bmatrix}$; $B = (2x^2 - 8x + 9, x, 1)$;

$$\mathbf{D} = \begin{bmatrix} -1 & 0 & 0 \\ 0 & -4 & 0 \\ 0 & 0 & -5 \end{bmatrix}; \mathbf{P} = \begin{bmatrix} 2 & 0 & 0 \\ -8 & 1 & 0 \\ 9 & 0 & 1 \end{bmatrix}$$

(e) $C = (\mathbf{i}, \mathbf{j})$; $\mathbf{A} = \frac{1}{2}\begin{bmatrix} 1 & -\sqrt{3} \\ \sqrt{3} & 1 \end{bmatrix}$; no eigenvalues; not diagonalizable

(h) $C = \left(\begin{bmatrix} 1 & 0 \\ 0 & 0 \end{bmatrix}, \begin{bmatrix} 0 & 1 \\ 0 & 0 \end{bmatrix}, \begin{bmatrix} 0 & 0 \\ 1 & 0 \end{bmatrix}, \begin{bmatrix} 0 & 0 \\ 0 & 1 \end{bmatrix} \right)$; $\mathbf{A} = \begin{bmatrix} -4 & 0 & 3 & 0 \\ 0 & -4 & 0 & 3 \\ -10 & 0 & 7 & 0 \\ 0 & -10 & 0 & 7 \end{bmatrix}$;

$B = \left(\begin{bmatrix} 3 & 0 \\ 5 & 0 \end{bmatrix}, \begin{bmatrix} 0 & 3 \\ 0 & 5 \end{bmatrix}, \begin{bmatrix} 1 & 0 \\ 2 & 0 \end{bmatrix}, \begin{bmatrix} 0 & 1 \\ 0 & 2 \end{bmatrix} \right)$; $\mathbf{D} = \begin{bmatrix} 1 & 0 & 0 & 0 \\ 0 & 1 & 0 & 0 \\ 0 & 0 & 2 & 0 \\ 0 & 0 & 0 & 2 \end{bmatrix}$;

$$\mathbf{P} = \begin{bmatrix} 3 & 0 & 1 & 0 \\ 0 & 3 & 0 & 1 \\ 5 & 0 & 2 & 0 \\ 0 & 5 & 0 & 2 \end{bmatrix}$$

4. (a) The only eigenvalue is $\lambda = 1$; $E_1 = \{1\}$

7. (a) $\begin{bmatrix} 1 & 1 & -1 \\ 0 & 1 & 0 \\ 0 & 0 & 1 \end{bmatrix}$; eigenvalue $\lambda = 1$; basis for $E_1 = \{[0, 1, 1], [1, 0, 0]\}$; λ has algebraic multiplicity 3 and geometric multiplicity 2

(b) $\begin{bmatrix} 1 & 0 & 0 \\ 0 & 1 & 0 \\ 0 & 0 & 0 \end{bmatrix}$; eigenvalues $\lambda_1 = 1$, $\lambda_2 = 0$; basis for $E_1 = \{[1, 0, 0],$

$[0, 1, 0]\}$; λ_1 has algebraic and geometric multiplicity 2

17. (a) F   (b) T
     (c) T   (d) F
     (e) T   (f) T
     (g) T   (h) F
     (i) F   (j) T

Section 6.1 (p.331–334)

1. (a) Orthogonal, not orthonormal

   (c) Neither

   (f) Orthogonal, not orthonormal

2. (a) Orthogonal

   (c) Not orthogonal: columns not normalized

   (e) Orthogonal

3. (a) $[\mathbf{v}]_B = \left[\frac{2\sqrt{3}+3}{2}, \frac{3\sqrt{3}-2}{2}\right]$

   (c) $[\mathbf{v}]_B = [3, \frac{13\sqrt{3}}{3}, \frac{5\sqrt{6}}{3}, 4\sqrt{2}]$

4. (a) $\{[5, -1, 2], [5, -3, -14]\}$

   (c) $\{[2, 1, 0, -1], [-1, 1, 3, -1], [5, -7, 5, 3]\}$

5. (a) $\{[2, 2, -3], [13, -4, 6], [0, 3, 2]\}$

   (c) $\{[1, -3, 1], [2, 5, 13], [4, 1, -1]\}$

   (e) $\{[2, 1, -2, 1], [3, -1, 2, -1], [0, 5, 2, -1], [0, 0, 1, 2]\}$

7. (a) $[-1, 3, 3]$ (c) $[5, 1, 1]$

8. (b) No

16. (b) $\begin{bmatrix} \frac{\sqrt{6}}{6} & \frac{\sqrt{6}}{3} & \frac{\sqrt{6}}{6} \\ \frac{\sqrt{30}}{6} & -\frac{\sqrt{30}}{15} & -\frac{\sqrt{30}}{30} \\ 0 & \frac{\sqrt{5}}{5} & -\frac{2\sqrt{5}}{5} \end{bmatrix}$

21. (a) F   (b) T
      (c) T   (d) F
      (e) T   (f) T
      (g) T   (h) F
      (i) T   (j) T

Section 6.2 (p.345–348)

1. (a) $\mathcal{W}^{\perp} = \text{span}(\{[2, 3]\})$

   (c) $\mathcal{W}^{\perp} = \text{span}(\{[2, 3, 7]\})$

   (e) $\mathcal{W}^{\perp} = \text{span}(\{[-2, 5, -1]\})$

   (f) $\mathcal{W}^{\perp} = \text{span}(\{[7, 1, -2, -3], [0, 4, -1, 2]\})$

2. (a) $\mathbf{w}_1 = \mathbf{proj}_W\mathbf{v} = \left[-\frac{33}{35}, \frac{111}{35}, \frac{12}{7}\right]$; $\mathbf{w}_2 = \left[-\frac{2}{35}, -\frac{6}{35}, \frac{2}{7}\right]$

   (b) $\mathbf{w}_1 = \mathbf{proj}_W\mathbf{v} = \left[-\frac{17}{9}, -\frac{10}{9}, \frac{14}{9}\right]$; $\mathbf{w}_2 = \left[\frac{26}{9}, -\frac{26}{9}, \frac{13}{9}\right]$

4. (a) $\frac{3\sqrt{129}}{43}$ (d) $\frac{8\sqrt{17}}{17}$

5. (a) Orthogonal projection onto $3x + y + z = 0$

   (d) Neither

6. $\frac{1}{9}\begin{bmatrix} 5 & 2 & -4 \\ 2 & 8 & 2 \\ -4 & 2 & 5 \end{bmatrix}$

9. (a) $x^3 - 2x^2 + x$

   (c) $x^3 - x^2 - x + 1$

10. (a) $\frac{1}{59}\begin{bmatrix} 50 & -21 & -3 \\ -21 & 10 & -7 \\ -3 & -7 & 58 \end{bmatrix}$

   (c) $\frac{1}{9}\begin{bmatrix} 2 & 2 & 3 & -1 \\ 2 & 8 & 0 & 2 \\ 3 & 0 & 6 & -3 \\ -1 & 2 & -3 & 2 \end{bmatrix}$

25. (a) T   (b) T
   (c) F   (d) F
   (e) T   (f) F
   (g) T   (h) T
   (i) T   (j) F
   (k) T   (l) T

Section 6.3 (p.357–360)

1. (a) Symmetric, because the matrix for L with respect to the standard basis is symmetric

   (d) Symmetric, since L is orthogonally diagonalizable

   (e) Not symmetric, since L is not diagonalizable, and hence not orthogonally diagonalizable

   (g) Symmetric, because the matrix with respect to the standard basis is symmetric

2. (a) $\frac{1}{25}\begin{bmatrix} -7 & 24 \\ 24 & 7 \end{bmatrix}$

(d) $\frac{1}{169}\begin{bmatrix} -119 & -72 & -96 & 0 \\ -72 & 119 & 0 & 96 \\ -96 & 0 & 119 & -72 \\ 0 & 96 & -72 & -119 \end{bmatrix}$

3. (a) $B = (\frac{1}{13}[5, 12], \frac{1}{13}[-12, 5]); \mathbf{P} = \frac{1}{13}\begin{bmatrix} 5 & -12 \\ 12 & 5 \end{bmatrix}; \mathbf{D} = \begin{bmatrix} 0 & 0 \\ 0 & 169 \end{bmatrix}$

(c) $B = \left(\frac{1}{\sqrt{2}}[-1, 1, 0], \frac{1}{3\sqrt{2}}[1, 1, 4], \frac{1}{3}[-2, -2, 1]\right)$ (other bases are possible, since E_1 is two-dimensional), $\mathbf{P} = \begin{bmatrix} -\frac{1}{\sqrt{2}} & \frac{1}{3\sqrt{2}} & -\frac{2}{3} \\ \frac{1}{\sqrt{2}} & \frac{1}{3\sqrt{2}} & -\frac{2}{3} \\ 0 & \frac{4}{3\sqrt{2}} & \frac{1}{3} \end{bmatrix}$,

$\mathbf{D} = \begin{bmatrix} 1 & 0 & 0 \\ 0 & 1 & 0 \\ 0 & 0 & 3 \end{bmatrix}$

(e) $B = \left(\frac{1}{\sqrt{14}}[3, 2, 1, 0], \frac{1}{\sqrt{14}}[-2, 3, 0, 1], \frac{1}{\sqrt{14}}[1, 0, -3, 2],\right.$

$\left.\frac{1}{\sqrt{14}}[0, -1, 2, 3]\right); \mathbf{P} = \frac{1}{\sqrt{14}}\begin{bmatrix} 3 & -2 & 1 & 0 \\ 2 & 3 & 0 & -1 \\ 1 & 0 & -3 & 2 \\ 0 & 1 & 2 & 3 \end{bmatrix}; \mathbf{D} = \begin{bmatrix} 2 & 0 & 0 & 0 \\ 0 & 2 & 0 & 0 \\ 0 & 0 & -3 & 0 \\ 0 & 0 & 0 & 5 \end{bmatrix}$

(g) $B = \left(\frac{1}{\sqrt{5}}[1, 2, 0], \frac{1}{\sqrt{6}}[-2, 1, 1], \frac{1}{\sqrt{30}}[2, -1, 5]\right)$ (other bases are possible, since E_{15} is two-dimensional); $\mathbf{P} = \begin{bmatrix} \frac{1}{\sqrt{5}} & -\frac{2}{\sqrt{6}} & \frac{2}{\sqrt{30}} \\ \frac{2}{\sqrt{5}} & \frac{1}{\sqrt{6}} & -\frac{1}{\sqrt{30}} \\ 0 & \frac{1}{\sqrt{6}} & \frac{5}{\sqrt{30}} \end{bmatrix}$;

$\mathbf{D} = \begin{bmatrix} 15 & 0 & 0 \\ 0 & 15 & 0 \\ 0 & 0 & -15 \end{bmatrix}$

4. (a) $C = (\frac{1}{19}[-10, 15, 6], \frac{1}{19}[15, 6, 10]); \mathbf{A} = \begin{bmatrix} -2 & 2 \\ 2 & 1 \end{bmatrix}$;

$B = (\frac{1}{19\sqrt{5}}[20, 27, 26], \frac{1}{19\sqrt{5}}[35, -24, -2]); \mathbf{P} = \frac{1}{\sqrt{5}}\begin{bmatrix} 1 & -2 \\ 2 & 1 \end{bmatrix}$;

$\mathbf{D} = \begin{bmatrix} 2 & 0 \\ 0 & -3 \end{bmatrix}$

5. (a) $\frac{1}{25}\begin{bmatrix} 23 & -36 \\ -36 & 2 \end{bmatrix}$

(c) $\frac{1}{3}\begin{bmatrix} 11 & 4 & -4 \\ 4 & 17 & -8 \\ -4 & -8 & 17 \end{bmatrix}$

6. For example, the matrix **A** in Example 7 of Section 5.5 is diagonalizable, but not symmetric, and hence, not orthogonally diagonalizable.

7. $\frac{1}{2}\begin{bmatrix} a+c+\sqrt{(a-c)^2+4b^2} & 0 \\ 0 & a+c-\sqrt{(a-c)^2+4b^2} \end{bmatrix}$

8. (b) L must be the zero linear operator. Since L is diagonalizable, the eigenspace for 0 must be all of $\mathcal{V}$.

13. (a) T   (b) F
    (c) T   (d) T
    (e) T   (f) T

Section 7.1 (p.367–369)

1. (a) $[1+4i, 1+i, 6-i]$

   (b) $[-12-32i, -7+30i, 53-29i]$

   (d) $[-24-12i, -28-8i, -32i]$

   (e) $1+28i$

3. (a) $\begin{bmatrix} 11+4i & -4-2i \\ 2-4i & 12 \end{bmatrix}$

   (c) $\begin{bmatrix} 1-i & 0 & 10i \\ 2i & 3-i & 0 \\ 6-4i & 5 & 7+3i \end{bmatrix}$

   (d) $\begin{bmatrix} -3-15i & -3 & 9i \\ 9-6i & 0 & 3+12i \end{bmatrix}$

   (f) $\begin{bmatrix} 1+40i & -4-14i \\ 13-50i & 23+21i \end{bmatrix}$

   (i) $\begin{bmatrix} 4+36i & -5+39i \\ 1-7i & -6-4i \\ 5+40i & -7-5i \end{bmatrix}$

5. (a) Skew-Hermitian

   (b) Neither

   (c) Hermitian

   (d) Skew-Hermitian

   (e) Hermitian

11. (a) F   (b) F
     (c) T   (d) F
     (e) F   (f) T

Section 7.2 (p.373–374)

1. (a) $w = \frac{1}{5} + \frac{13}{5}i$, $z = \frac{28}{5} - \frac{3}{5}i$

   (c) $x = (2 + 5i) - (4 - 3i)c$, $y = (5 + 2i) + ic$, $z = c$

   (e) Solution set = { }

2. (b) $|\mathbf{A}| = -15 - 23i$; $\mathbf{A}$ is nonsingular; $|\mathbf{A}^*| = -15 + 23i = \overline{|\mathbf{A}|}$

3. (a) Eigenvalues: $\lambda_1 = i$; $\lambda_2 = -1$, with respective eigenvectors $[1+i, 2]$ and $[7 + 6i, 17]$. Hence, $E_i = \{c[1 + i, 2] \mid c \in \mathbb{C}\}$, and $E_{-1} = \{c[7 + 6i, 17] \mid c \in \mathbb{C}\}$.

   (c) Eigenvalues: $\lambda_1 = i$ and $\lambda_2 = -2$, with eigenvectors $[1, 1, 0]$ and $[(-3 - 2i), 0, 2]$ for λ_1, and eigenvector $[-1, i, 1]$ for λ_2. Hence, $E_i = \{c[1, 1, 0] + d[(-3-2i), 0, 2] \mid c, d \in \mathbb{C}\}$ and $E_{-2} = \{c[-1, i, 1] \mid c \in \mathbb{C}\}$.

4. (a) The 2×2 matrix $\mathbf{A}$ is diagonalizable since two eigenvectors were found in the diagonalization process; $\mathbf{P} = \begin{bmatrix} 1+i & 7+6i \\ 2 & 17 \end{bmatrix}$; $\mathbf{D} = \begin{bmatrix} i & 0 \\ 0 & -1 \end{bmatrix}$

6. (a) T   (b) F
   (c) F   (d) F

Section 7.3 (p.376–377)

2. (b) Not linearly independent, dim $= 1$

   (d) Not linearly independent, dim $= 2$

3. (b) Linearly independent, dim $= 2$

   (d) Linearly independent, dim $= 3$

4. (b) $[i, 1 + i, -1]$

5. Ordered basis $= ([1, 0], [i, 0], [0, 1], [0, i])$; matrix $= \begin{bmatrix} 0 & 0 & 1 & 0 \\ 0 & 0 & 0 & -1 \\ 1 & 0 & 0 & 0 \\ 0 & -1 & 0 & 0 \end{bmatrix}$.

8. $\begin{bmatrix} -3 + i & -\frac{2}{5} - \frac{11}{5}i \\ \frac{1}{2} - \frac{3}{2}i & -i \\ -\frac{1}{2} + \frac{7}{2}i & -\frac{8}{5} - \frac{4}{5}i \end{bmatrix}$

9. (a) T   (b) F
   (c) T   (d) F

Section 7.4 (p.381–383)

1. (a) Not orthogonal

   (c) Orthogonal

3. (a) $\{[1+i, i, 1], [2, -1-i, -1+i], [0, 1, i]\}$

   (b) $\begin{bmatrix} \frac{1+i}{2} & \frac{i}{2} & \frac{1}{2} \\ \frac{2}{\sqrt{8}} & \frac{-1-i}{\sqrt{8}} & \frac{-1+i}{\sqrt{8}} \\ 0 & \frac{1}{\sqrt{2}} & \frac{i}{\sqrt{2}} \end{bmatrix}$

10. (b) $\mathbf{P} = \begin{bmatrix} \frac{-1+i}{\sqrt{6}} & \frac{1-i}{\sqrt{3}} \\ \frac{2}{\sqrt{6}} & \frac{1}{\sqrt{3}} \end{bmatrix}$; the corresponding diagonal matrix is $\begin{bmatrix} 9+6i & 0 \\ 0 & -3-12i \end{bmatrix}$.

13. Eigenvalues are $-4, 2+\sqrt{6}$ and $2-\sqrt{6}$.

15. (a) F   (b) T
    (c) T   (d) T
    (e) F

Section 7.5 (p.393–396)

1. (b) $<\mathbf{x}, \mathbf{y}> = -183; \|\mathbf{x}\| = \sqrt{314}$

3. (b) $<\mathbf{f}, \mathbf{g}> = \frac{1}{2}(e^{\pi} + 1); \|\mathbf{f}\| = \sqrt{\frac{1}{2}\left(e^{2\pi} - 1\right)}$

9. (a) $\sqrt{\frac{\pi^3}{3} - \frac{3\pi}{2}}$

10. (b) 0.586 radians, or 33.6°

14. (a) Orthogonal

    (c) Not orthogonal

19. Using $\mathbf{w}_1 = t^2 - t + 1$, $\mathbf{w}_2 = 1$, and $\mathbf{w}_3 = t$ yields the orthogonal basis $\{\mathbf{v}_1, \mathbf{v}_2, \mathbf{v}_3\}$, with $\mathbf{v}_1 = t^2 - t + 1$, $\mathbf{v}_2 = -20t^2 + 20t + 13$, and $\mathbf{v}_3 = 15t^2 + 4t - 5$.

23. $\mathcal{W}^{\perp} = \text{span}(\{t^3 - t^2, t + 1\})$

26. $\mathbf{w}_1 = \frac{1}{2\pi}(\sin t - \cos t), \mathbf{w}_2 = \frac{1}{k}e^t - \frac{1}{2\pi}\sin t + \frac{1}{2\pi}\cos t$

29. (b) $\ker(L) = \mathcal{W}^{\perp}; \text{range}(L) = \mathcal{W}$

30. (a) F   (b) F
    (c) F   (d) T
    (e) F

1. Symmetric: (a), (b), (c), (d), (g)

(a) Matrix for $G_1 = \begin{bmatrix} 0 & 1 & 1 & 1 \\ 1 & 0 & 1 & 1 \\ 1 & 1 & 0 & 1 \\ 1 & 1 & 1 & 0 \end{bmatrix}$

(b) Matrix for $G_2 = \begin{bmatrix} 1 & 1 & 0 & 0 & 1 \\ 1 & 0 & 0 & 1 & 0 \\ 0 & 0 & 1 & 0 & 1 \\ 0 & 1 & 0 & 0 & 1 \\ 1 & 0 & 1 & 1 & 1 \end{bmatrix}$

(c) Matrix for $G_3 = \begin{bmatrix} 0 & 0 & 0 \\ 0 & 0 & 0 \\ 0 & 0 & 0 \end{bmatrix}$

(d) Matrix for $G_4 = \begin{bmatrix} 0 & 1 & 0 & 1 & 0 & 0 \\ 1 & 0 & 1 & 1 & 0 & 0 \\ 0 & 1 & 0 & 0 & 1 & 1 \\ 1 & 1 & 0 & 0 & 1 & 0 \\ 0 & 0 & 1 & 1 & 0 & 1 \\ 0 & 0 & 1 & 0 & 1 & 0 \end{bmatrix}$

(e) Matrix for $D_1 = \begin{bmatrix} 0 & 1 & 0 & 0 \\ 0 & 0 & 1 & 0 \\ 1 & 0 & 0 & 1 \\ 1 & 1 & 0 & 0 \end{bmatrix}$

(f) Matrix for $D_2 = \begin{bmatrix} 0 & 1 & 1 & 0 \\ 0 & 1 & 1 & 1 \\ 0 & 1 & 0 & 1 \\ 0 & 0 & 0 & 1 \end{bmatrix}$

(g) Matrix for $D_3 = \begin{bmatrix} 0 & 1 & 1 & 0 & 0 \\ 1 & 0 & 0 & 0 & 1 \\ 1 & 0 & 0 & 1 & 0 \\ 0 & 0 & 1 & 0 & 1 \\ 0 & 1 & 0 & 1 & 0 \end{bmatrix}$

(h) Matrix for $D_4 = \begin{bmatrix} 0 & 1 & 0 & 0 & 0 & 0 & 0 & 0 \\ 0 & 0 & 1 & 0 & 0 & 0 & 0 & 0 \\ 0 & 0 & 0 & 1 & 0 & 0 & 0 & 0 \\ 0 & 0 & 0 & 0 & 1 & 0 & 0 & 0 \\ 0 & 0 & 0 & 0 & 0 & 1 & 0 & 0 \\ 0 & 0 & 0 & 0 & 0 & 0 & 1 & 0 \\ 0 & 0 & 0 & 0 & 0 & 0 & 0 & 1 \\ 1 & 0 & 0 & 0 & 0 & 0 & 0 & 0 \end{bmatrix}$

2. F can be the adjacency matrix for either a graph or digraph.

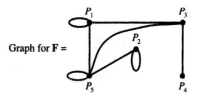

Graph for **F** =

G can be the adjacency matrix for a digraph (only).

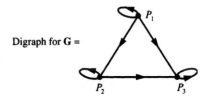

Digraph for **G** =

H can be the adjacency matrix for a digraph (only).

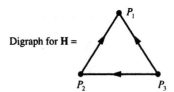

Digraph for **H** =

I can be the adjacency matrix for a graph or digraph.

Graph for **I** =

$$
\bigcirc \quad \bigcirc \quad \bigcirc
$$

$P_1 \qquad P_2 \qquad P_3$

K can be the adjacency matrix for a graph or digraph.

Graph for **K** =

$P_1 \qquad\qquad P_2$

L can be the adjacency matrix for a graph or digraph.

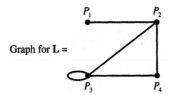

Graph for **L** =

3. The digraph is shown in the accompanying figure, and the adjacency matrix is

$$
\begin{array}{c c}
 & \begin{array}{c c c c c c} A & B & C & D & E & F \end{array} \\
\begin{array}{c} A \\ B \\ C \\ D \\ E \\ F \end{array} &
\left[\begin{array}{c c c c c c}
0 & 0 & 1 & 0 & 0 & 0 \\
0 & 0 & 0 & 1 & 1 & 0 \\
0 & 1 & 0 & 0 & 1 & 0 \\
1 & 0 & 0 & 0 & 0 & 1 \\
1 & 1 & 0 & 1 & 0 & 0 \\
0 & 0 & 0 & 1 & 0 & 0
\end{array}\right]
\end{array}.
$$

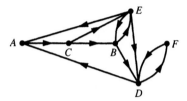

The transpose gives no new information, but it does suggest a different interpretation of the results; namely, the (i, j) entry of the transpose equals 1 if author j influences author i.

4. (a) 3

(c) $6 = 1 + 1 + 4$

(e) Length 4

5. (a) 4

(c) $5 = 1 + 1 + 3$

(e) No such path exists.

6. (a) Figure 8.5: 7; Figure 8.6: 2

(c) Figure 8.5: $3 = 0 + 1 + 0 + 2$; Figure 8.6: $17 = 1 + 2 + 4 + 10$

7. (a) If the vertex is the ith vertex, then the ith row and ith column entries of the adjacency matrix all equal 0, except possibly for the (i, i) entry.

(b) If the vertex is the ith vertex, then the ith row entries of the adjacency matrix all equal 0, except possibly for the (i, i) entry. (Note: The ith column entries may be nonzero.)

8. (a) The trace equals the number of loops in the graph or digraph.

9. (a) Figure 8.5: strongly connected; Figure 8.6: not strongly connected (since there is no path to P_5 from any other vertex)

10. (b) Yes, it is a dominance digraph because no tie games are possible and because each team plays every other team. Thus, if P_i and P_j are two given teams, either P_i defeats P_j or vice versa.

12. (a) T   (b) F
(c) T   (d) F
(e) T   (f)] T
(g) T

Section 8.2 (p.408)

1. (a) $I_1 = 8$, $I_2 = 5$, $I_3 = 3$

(c) $I_1 = 12$, $I_2 = 5$, $I_3 = 3$, $I_4 = 2$, $I_5 = 2$, $I_6 = 7$

2. (a) T   (b) T

Section 8.3 (p.413–415)

1. (a) $y = -0.8x - 3.3$, $y = -7.3$ when $x = 5$

(c) $y = -1.5x + 3.8$, $y = -3.7$ when $x = 5$

2. (a) $y = 0.375x^2 + 0.35x + 3.60$

(c) $y = -0.042x^2 + 0.633x + 0.266$

3. (a) $y = \frac{1}{4}x^3 + \frac{25}{28}x^2 + \frac{25}{14}x + \frac{37}{35}$

4. (a) $y = 4.4286x^2 - 2.0571$

(c) $y = -0.1014x^2 + 0.9633x - 0.8534$

(e) $y = 0x^3 - 0.3954x^2 + 0.9706$

5. (a) $y = 0.4x + 2.54$; the angle reaches 20° in the 44th month

(b) $y = 0.02857x^2 + 0.2286x + 2.74$; the angle reaches 20° in the 21st month

7. The least-squares polynomial is $y = \frac{4}{5}x^2 - \frac{2}{5}x + 2$, which is the *exact* quadratic through the three given points.

9. (a) $x_1 = \frac{230}{39}$, $x_2 = \frac{155}{39}$; $\begin{cases} 4x_1 - 3x_2 = 11\frac{2}{3}, & \text{which is almost 12} \\ 2x_1 + 5x_2 = 31\frac{2}{3}, & \text{which is almost 32} \\ 3x_1 + x_2 = 21\frac{2}{3}, & \text{which is close to 21} \end{cases}$

10. (a) T   (b) F
(c) F   (d) F

Section 8.4 (p.424–426)

1. **A** is not stochastic since **A** is not square; **A** is not regular since **A** is not stochastic.
   B is not stochastic since the entries of column 2 do not sum to 1; **B** is not regular since **B** is not stochastic.
   C is stochastic; **C** is regular since **C** is stochastic and has all nonzero entries.
   D is stochastic; **D** is not regular since every positive power of **D** is a matrix whose rows are the rows of **D** rearranged in some order, and hence, every such power contains zero entries.
   E is not stochastic since the entries of column 1 do not sum to 1; **E** is not regular since **E** is not stochastic.
   F is stochastic; **F** is not regular since every positive power of **F** has all second row entries zero.
   G is not stochastic since **G** is not square; **G** is not regular since **G** is not stochastic.

   H is stochastic; **H** is regular since **H** is stochastic and $\mathbf{H}^2 = \begin{bmatrix} \frac{1}{2} & \frac{1}{4} & \frac{1}{4} \\ \frac{1}{4} & \frac{1}{2} & \frac{1}{4} \\ \frac{1}{4} & \frac{1}{4} & \frac{1}{2} \end{bmatrix}$,

   which has all nonzero entries.

2. (a) $\mathbf{p}_1 = \left[\frac{5}{18}, \frac{13}{18} \right]$, $\mathbf{p}_2 = \left[\frac{67}{216}, \frac{149}{216} \right]$

   (c) $\mathbf{p}_1 = \left[\frac{17}{48}, \frac{1}{3}, \frac{5}{16} \right]$, $\mathbf{p}_2 = \left[\frac{205}{576}, \frac{49}{144}, \frac{175}{576} \right]$

3. (a) $\left[\frac{2}{5}, \frac{3}{5} \right]$

5. (a) $[0.34, 0.175, 0.34, 0.145]$ in the next election; $[0.3555, 0.1875, 0.2875, 0.1695]$ in the election after that

   (b) The steady-state vector is $[0.36, 0.20, 0.24, 0.20]$. After a century, the votes would be 36% for Party A and 24% for Party C.

6. (a) $\mathbf{M} = \begin{bmatrix} \frac{1}{2} & \frac{1}{6} & \frac{1}{6} & \frac{1}{5} & 0 \\ \frac{1}{8} & \frac{1}{2} & 0 & 0 & \frac{1}{5} \\ \frac{1}{8} & 0 & \frac{1}{2} & \frac{1}{10} & \frac{1}{10} \\ \frac{1}{4} & 0 & \frac{1}{6} & \frac{1}{2} & \frac{1}{5} \\ 0 & \frac{1}{3} & \frac{1}{6} & \frac{1}{5} & \frac{1}{2} \end{bmatrix}$

   (b) $\mathbf{M}^2$ has all nonzero entries

(c) $\frac{29}{120}$, since the probability vector after 2 time intervals is
$$\left[\frac{1}{5}, \frac{13}{240}, \frac{73}{240}, \frac{29}{120}, \frac{1}{5}\right]$$

(d) $\left[\frac{1}{5}, \frac{3}{20}, \frac{3}{20}, \frac{1}{4}, \frac{1}{4}\right]$; over time, the rat frequents rooms B and C the least, and rooms D and E the most.

11. (a) F   (b) T
(c) T   (d) T
(e) F

Section 8.5 (p.429)

1. (a) -24 -46 -15 -30 10 16 39 62 26 42 51 84 24 37 -11 -23

2. (a) HOMEWORK IS GOOD FOR THE SOUL

3. (a) T   (b) T
(c) F

Section 8.6 (p.440)

1. (a) $dx\,dy = 2\,du\,dv$

(c) $dx\,dy = 4(u^2 + v^2)\,du\,dv$

(e) $dx\,dy = \left(\frac{8|v|}{((u+1)^2+v^2)^3}\right)du\,dv$

2. (a) $dx\,dy\,dz = 2\,du\,dv\,dw$

(c) $dx\,dy\,dz = \left(\frac{1}{(u^2+v^2+w^2)^3}\right)du\,dv\,dw$

4. (a) $\frac{52}{3}$

(c) $\frac{\pi}{4}$

(e) $\frac{800}{3}\pi$

5. (a) T   (b) T
(c) T   (d) F

Section 8.7 (p.445–446)

1. (c) $\theta = \frac{1}{2}\arctan(-\sqrt{3}) = -\frac{\pi}{6}$; $\mathbf{P} = \begin{bmatrix} \frac{\sqrt{3}}{2} & \frac{1}{2} \\ -\frac{1}{2} & \frac{\sqrt{3}}{2} \end{bmatrix}$; equation in uv-coordinates: $v = 2u^2 - 12u + 13$, or, $(v+5) = 2(u-3)^2$; vertex in uv-coordinates: $(3, -5)$; vertex in xy-coordinates: $(0.0981, -5.830)$ (see accompanying figures)

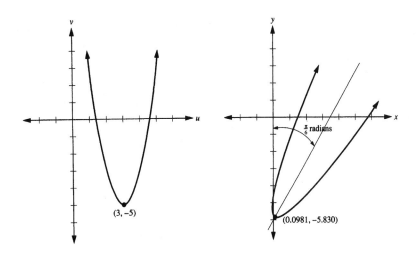

(d) $\theta \approx 0.6435$ radians (about $36°52'$); $\mathbf{P} = \frac{1}{5}\begin{bmatrix} 4 & -3 \\ 3 & 4 \end{bmatrix}$; equation in uv-coordinates: $\frac{(u-2)^2}{9} + \frac{(v+1)^2}{4} = 1$; center in uv-coordinates $= (2, -1)$; center in xy-coordinates $= (\frac{11}{5}, \frac{2}{5})$ (see accompanying figures)

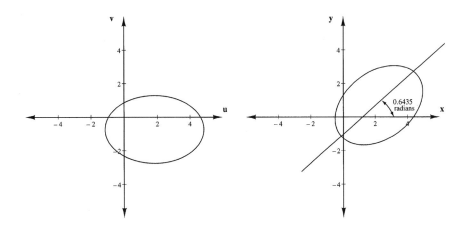

(f) All answers rounded to 4 significant digits: $\theta = 0.4442$ radians (about $25°27'$); $\mathbf{P} = \begin{bmatrix} 0.9029 & -0.4298 \\ 0.4298 & 0.9029 \end{bmatrix}$; equation in uv-coordinates: $\frac{u^2}{(1.770)^2} - \frac{v^2}{(2.050)^2} = 1$; center in uv-coordinates: $(0, 0)$; center in xy-coordinates $= (0, 0)$ (see accompanying figures)

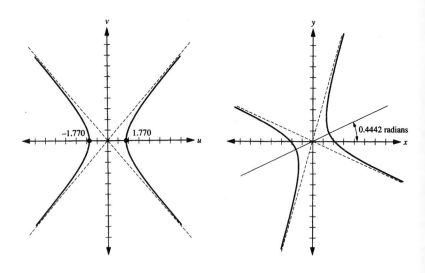

2. (a) T   (b) F
    (c) T   (d) F

Section 8.8 (p.455–459)

1. (a) $(9, 1)$, $(9, 5)$, $(12, 1)$, $(12, 5)$, $(14, 3)$

   (c) $(-2, 5)$, $(0, 9)$, $(-5, 7)$, $(-2, 10)$, $(-5, 10)$

2. (b) $(-8, 2)$, $(-7, 5)$, $(-10, 6)$, $(-9, 11)$, $(-14, 13)$ (see accompanying figure)

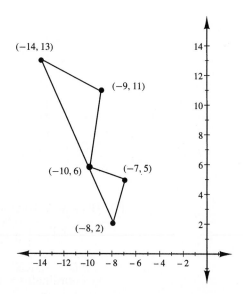

(d) $(3, 18)$, $(4, 12)$, $(5, 18)$, $(7, 6)$, $(9, 18)$ (see accompanying figure)

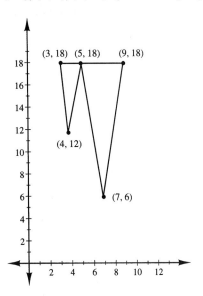

3. (a) $(3, -4)$, $(3, -10)$, $(7, -6)$, $(9, -9)$, $(10, -3)$

(c) $(-2, 6)$, $(0, 8)$, $(-8, 17)$, $(-10, 22)$, $(-16, 25)$

4. (a) $(14, 9)$, $(10, 6)$, $(11, 11)$, $(8, 9)$, $(6, 8)$, $(11, 14)$

(c) $(2, 4)$, $(2, 6)$, $(8, 5)$, $(8, 6)$, $(8, 6)$, $(14, 4)$

5. (b) $(0, 5)$, $(1, 7)$, $(0, 11)$, $(-5, 8)$, $(-4, 10)$

6. (a) $(2, 20)$, $(3, 17)$, $(5, 14)$, $(6, 19)$, $(6, 16)$, $(9, 14)$ (see accompanying figure)

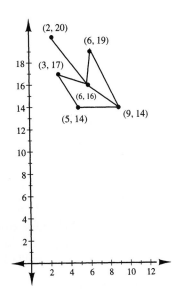

(c) (1, 18), (−3, 13), (−6, 8), (−2, 17), (−5, 12), (−6, 10) (see accompanying figure)

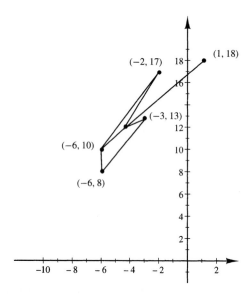

11. (b) Consider the reflection about the y-axis and a counterclockwise rotation of $90°$ about the origin. Starting from the point $(1, 0)$, performing the rotation and then the reflection yields $(0, 1)$. However, performing the reflection followed by the rotation produces $(0, -1)$. Hence, the two transformations do not commute.

13. (a) F   (b) F
(c) F   (d) T
(e) T   (f) F

Section 8.9 (p.466–467)

1. (a) $b_1 e^t \begin{bmatrix} 7 \\ 3 \end{bmatrix} + b_2 e^{-t} \begin{bmatrix} 2 \\ 1 \end{bmatrix}$

(c) $b_1 \begin{bmatrix} 0 \\ -1 \\ 1 \end{bmatrix} + b_2 e^t \begin{bmatrix} 1 \\ -1 \\ 1 \end{bmatrix} + b_3 e^{3t} \begin{bmatrix} 2 \\ 0 \\ 1 \end{bmatrix}$

(d) $b_1 e^t \begin{bmatrix} -1 \\ 1 \\ 0 \end{bmatrix} + b_2 e^t \begin{bmatrix} 5 \\ 0 \\ 2 \end{bmatrix} + b_3 e^{4t} \begin{bmatrix} 1 \\ 1 \\ 1 \end{bmatrix}$ (There are other possible

answers. For example, the first two vectors in the sum could be any basis for the two-dimensional eigenspace corresponding to the eigenvalue 1.)

2. (a) $y = b_1 e^{2t} + b_2 e^{-3t}$

(c) $y = b_1 e^{2t} + b_2 e^{-2t} + b_3 e^{\left(\sqrt{2}\right)t} + b_4 e^{-\left(\sqrt{2}\right)t}$

4. (b) $\mathbf{F}(t) = 2e^{5t} \begin{bmatrix} 1 \\ 0 \\ 1 \end{bmatrix} - e^t \begin{bmatrix} -1 \\ 2 \\ 0 \end{bmatrix} - 2e^{-t} \begin{bmatrix} 1 \\ 1 \\ 1 \end{bmatrix}$

7. (a) T   **(b)** T
   (c) T   **(d)** F

Section 8.10 (p.473–474)

1. (a) Unique least-squares solution: $\mathbf{v} = \begin{bmatrix} \frac{23}{30}, & \frac{11}{10} \end{bmatrix}$; $||\mathbf{Av} - \mathbf{b}|| = \frac{\sqrt{6}}{6} \approx$ 0.408; $||\mathbf{Az} - \mathbf{b}|| = 1$

   (c) Infinite number of least-squares solutions, all of the form $[7c + \frac{17}{3},$ $-13c - \frac{23}{3}, c]$. Two particular least-squares solutions are: $[\frac{17}{3}, -\frac{23}{3}, 0]$ and $[8, -12, \frac{1}{3}]$. Also, with $\mathbf{v}$ as either of these vectors, $||\mathbf{Av} - \mathbf{b}|| = \frac{\sqrt{6}}{3} \approx 0.816$; $||\mathbf{Az} - \mathbf{b}|| = 3$.

2. (a) Infinite number of least-squares solutions, all of the form $[-\frac{4}{7}c + \frac{19}{42}, \frac{8}{7}c - \frac{5}{21}, c]$, with $\frac{5}{24} \leq c \leq \frac{19}{24}$.

3. (b) $\mathbf{v} \approx [0.46, -0.36, 0.90]$; $(\lambda'\mathbf{I} - \mathbf{C})\mathbf{v} \approx [0.03, -0.04, 0.07]$.

7. (a) F   **(b)** T
   (c) T   **(d)** T
   (e) T

Section 8.11 (p.480–481)

1. (a) Critical points: $(1, -1)$, $(-1, 1)$; local minimum at $(1, -1)$

   (c) Critical point: $(-2, 3)$; local minimum at $(-2, 3)$

   (e) Critical points: $(0, 0, 0)$, $(\frac{1}{2}, -\frac{1}{2}, -\frac{1}{2})$, $(-\frac{1}{2}, \frac{1}{2}, \frac{1}{2})$; local minimums at $(\frac{1}{2}, -\frac{1}{2}, -\frac{1}{2})$, $(-\frac{1}{2}, \frac{1}{2}, \frac{1}{2})$

4. (a) T   **(b)** F
   (c) T   **(d)** T
   (e) F

Section 9.1 (p.493–495)

1. (a) Solution to first system: $(602, 1500)$; solution to second system: $(302, 750)$. The systems are ill-conditioned because a very small change in the coefficient of y leads to a very large change in the solution.

2. Answers to this problem may differ significantly from the following depending on where rounding is done in the algorithm:

   (a) Without partial pivoting: $(3210, 0.765)$; with partial pivoting: $(3230, 0.767)$. (Actual solution is $(3214, 0.765)$.)

   (c) Without partial pivoting: $(2.26, 1.01, -2.11)$; with partial pivoting: $(277, -327, 595)$. (Actual solution is $(267, -315, 573)$.)

3. Answers to this problem may differ significantly from the following depending on where rounding is done in the algorithm:

   (a) Without partial pivoting: $(3214, 0.7651)$; with partial pivoting: $(3213, 0.7648)$. (Actual solution is $(3214, 0.765)$.)

   (c) Without partial pivoting: $(-2.380, 8.801, -16.30)$; with partial pivoting: $(267.8, -315.9, 574.6)$. (Actual solution is $(267, -315, 573)$.)

4. (a)

	x_1	x_2
Initial values	0.000	0.000
After 1 step	5.200	-6.000
After 2 steps	6.400	-8.229
After 3 steps	6.846	-8.743
After 4 steps	6.949	-8.934
After 5 steps	6.987	-8.978
After 6 steps	6.996	-8.994
After 7 steps	6.999	-8.998
After 8 steps	7.000	-9.000
After 9 steps	7.000	-9.000

(c)

	x_1	x_2	x_3
Initial values	0.000	0.000	0.000
After 1 step	-8.857	4.500	-4.333
After 2 steps	-10.738	3.746	-8.036
After 3 steps	-11.688	4.050	-8.537
After 4 steps	-11.875	3.975	-8.904
After 5 steps	-11.969	4.005	-8.954
After 6 steps	-11.988	3.998	-8.991
After 7 steps	-11.997	4.001	-8.996
After 8 steps	-11.999	4.000	-8.999
After 9 steps	-12.000	4.000	-9.000
After 10 steps	-12.000	4.000	-9.000

5. (a)

	x_1	x_2
Initial values	0.000	0.000
After 1 step	5.200	−8.229
After 2 steps	6.846	−8.934
After 3 steps	6.987	−8.994
After 4 steps	6.999	−9.000
After 5 steps	7.000	−9.000
After 6 steps	7.000	−9.000

(c)

	x_1	x_2	x_3
Initial values	0.000	0.000	0.000
After 1 step	−8.857	3.024	−7.790
After 2 steps	−11.515	3.879	−8.818
After 3 steps	−11.931	3.981	−8.974
After 4 steps	−11.990	3.997	−8.996
After 5 steps	−11.998	4.000	−8.999
After 6 steps	−12.000	4.000	−9.000
After 7 steps	−12.000	4.000	−9.000

6. Strictly diagonally dominant: (a), (c)

7. (a) Put the third equation first, and move the other two down to get the following:

	x_1	x_2	x_3
Initial values	0.000	0.000	0.000
After 1 step	3.125	−0.481	1.461
After 2 steps	2.517	−0.500	1.499
After 3 steps	2.500	−0.500	1.500
After 4 steps	2.500	−0.500	1.500

(c) Put the second equation first, the fourth equation second, the first equation third, and the third equation fourth to get the following:

	x_1	x_2	x_3	x_4
Initial values	0.000	0.000	0.000	0.000
After 1 step	5.444	−5.379	9.226	−10.447
After 2 steps	8.826	−8.435	10.808	−11.698
After 3 steps	9.820	−8.920	10.961	−11.954
After 4 steps	9.973	−8.986	10.994	−11.993
After 5 steps	9.995	−8.998	10.999	−11.999
After 6 steps	9.999	−9.000	11.000	−12.000
After 7 steps	10.000	−9.000	11.000	−12.000
After 8 steps	10.000	−9.000	11.000	−12.000

8. Jacobi method yields the following:

	x_1	x_2	x_3
Initial values	0.0	0.0	0.0
After 1 step	16.0	−13.0	12.0
After 2 steps	−37.0	59.0	−87.0
After 3 steps	224.0	−61.0	212.0
After 4 steps	−77.0	907.0	−1495.0
After 5 steps	3056.0	2515.0	−356.0
After 6 steps	12235.0	19035.0	−23895.0

Gauss-Seidel method yields the following:

	x_1	x_2	x_3
Initial values	0.0	0.0	0.0
After 1 step	16.0	83.0	−183.0
After 2 steps	248.0	1841.0	−3565.0
After 3 steps	5656.0	41053.0	−80633.0
After 4 steps	124648.0	909141.0	−1781665.0

The actual solution is: $(2, -3, 1)$.

10. (a) T

    (b) F

    (c) F

    (d) T

    (e) F

    (f) F

Section 9.2 (p.501–502)

1. (a) $\mathbf{LDU} = \begin{bmatrix} 1 & 0 \\ -3 & 1 \end{bmatrix} \begin{bmatrix} 2 & 0 \\ 0 & 5 \end{bmatrix} \begin{bmatrix} 1 & -2 \\ 0 & 1 \end{bmatrix}$

    (c) $\mathbf{LDU} = \begin{bmatrix} 1 & 0 & 0 \\ -2 & 1 & 0 \\ -2 & 4 & 1 \end{bmatrix} \begin{bmatrix} -1 & 0 & 0 \\ 0 & 2 & 0 \\ 0 & 0 & 3 \end{bmatrix} \begin{bmatrix} 1 & -4 & 2 \\ 0 & 1 & -4 \\ 0 & 0 & 1 \end{bmatrix}$

    (e) $\mathbf{LDU} = \begin{bmatrix} 1 & 0 & 0 & 0 \\ -\frac{4}{3} & 1 & 0 & 0 \\ -2 & -\frac{3}{2} & 1 & 0 \\ \frac{2}{3} & -2 & 0 & 1 \end{bmatrix} \begin{bmatrix} -3 & 0 & 0 & 0 \\ 0 & -\frac{2}{3} & 0 & 0 \\ 0 & 0 & \frac{1}{2} & 0 \\ 0 & 0 & 0 & 1 \end{bmatrix} \begin{bmatrix} 1 & -\frac{1}{3} & -\frac{1}{3} & \frac{1}{3} \\ 0 & 1 & \frac{5}{2} & -\frac{11}{2} \\ 0 & 0 & 1 & 3 \\ 0 & 0 & 0 & 1 \end{bmatrix}$

3. (a) $\mathbf{KU} = \begin{bmatrix} -1 & 0 \\ 2 & -3 \end{bmatrix} \begin{bmatrix} 1 & -5 \\ 0 & 1 \end{bmatrix}$; the solution is $\{(4, -1)\}$.

(c) $\mathbf{KU} = \begin{bmatrix} -1 & 0 & 0 \\ 4 & 3 & 0 \\ -2 & 5 & -2 \end{bmatrix} \begin{bmatrix} 1 & -3 & 2 \\ 0 & 1 & -5 \\ 0 & 0 & 1 \end{bmatrix}$; the solution is $\{(2, -3, 1)\}$.

4. (a) F (b) T
(c) F (d) F

Section 9.3 (p.507–508)

1. (a) After 9 iterations, eigenvector $= [0.60, 0.80]$, eigenvalue $= 50$

(c) After 7 iterations, eigenvector $= [0.41, 0.41, 0.82]$, eigenvalue $= 3.0$

(e) After 15 iterations, eigenvector $= [0.346, 0.852, 0.185, 0.346]$, eigenvalue $= 5.405$

3. (b) Let $\lambda_1, \ldots \lambda_n$ be the eigenvalues of $\mathbf{A}$ with $|\lambda_1| > |\lambda_j|$ for $2 \le j \le n$. Let $\{\mathbf{v}_1, \ldots, \mathbf{v}_n\}$ be a basis of unit eigenvectors for $\mathbb{R}^n$ corresponding to $\lambda_1, \ldots, \lambda_n$ respectively. Suppose the initial vector in the Power Method is $\mathbf{u}_0 = a_{01}\mathbf{v}_1 + \cdots + a_{0n}\mathbf{v}_n$, and after the ith iteration, we have $\mathbf{u}_i = a_{i1}\mathbf{v}_1 + \cdots + a_{in}\mathbf{v}_n$. Then for $2 \le j \le n$, $a_{0j} \ne 0$, and $\lambda_j \ne 0$ we have $\frac{|a_{i1}|}{|a_{ij}|} = \left|\frac{\lambda_1}{\lambda_j}\right|^i \frac{|a_{01}|}{|a_{0j}|}$.

4. (a) F (b) T
(c) T (d) F

Section 10.1 (p.515–516)

1. (a) (III): $\langle 2 \rangle \leftrightarrow \langle 3 \rangle$; inverse operation is (III): $\langle 2 \rangle \leftrightarrow \langle 3 \rangle$. The matrix is its own inverse.

(b) (I): $\langle 2 \rangle \longleftarrow -2\langle 2 \rangle$; inverse operation is (I): $\langle 2 \rangle \longleftarrow -\frac{1}{2}\langle 2 \rangle$.

The inverse matrix is $\begin{bmatrix} 1 & 0 & 0 \\ 0 & -\frac{1}{2} & 0 \\ 0 & 0 & 1 \end{bmatrix}$.

(e) (II): $\langle 3 \rangle \longleftarrow -2\langle 4 \rangle + \langle 3 \rangle$; inverse operation is (II): $\langle 3 \rangle \longleftarrow 2\langle 4 \rangle + \langle 3 \rangle$.

The inverse matrix is $\begin{bmatrix} 1 & 0 & 0 & 0 \\ 0 & 1 & 0 & 0 \\ 0 & 0 & 1 & 2 \\ 0 & 0 & 0 & 1 \end{bmatrix}$.

2. (a) $\begin{bmatrix} 4 & 9 \\ 3 & 7 \end{bmatrix} = \begin{bmatrix} 4 & 0 \\ 0 & 1 \end{bmatrix} \begin{bmatrix} 1 & 0 \\ 3 & 1 \end{bmatrix} \begin{bmatrix} 1 & 0 \\ 0 & \frac{1}{4} \end{bmatrix} \begin{bmatrix} 1 & \frac{9}{4} \\ 0 & 1 \end{bmatrix} \begin{bmatrix} 1 & 0 \\ 0 & 1 \end{bmatrix}$

(c) The product of the following matrices in the order listed:

$$\begin{bmatrix} 0 & 1 & 0 & 0 \\ 1 & 0 & 0 & 0 \\ 0 & 0 & 1 & 0 \\ 0 & 0 & 0 & 1 \end{bmatrix}, \begin{bmatrix} -3 & 0 & 0 & 0 \\ 0 & 1 & 0 & 0 \\ 0 & 0 & 1 & 0 \\ 0 & 0 & 0 & 1 \end{bmatrix}, \begin{bmatrix} 1 & 0 & 0 & 0 \\ 0 & 1 & 0 & 0 \\ 0 & 0 & 1 & 0 \\ 3 & 0 & 0 & 1 \end{bmatrix},$$

$$\begin{bmatrix} 1 & 0 & 0 & 0 \\ 0 & 0 & 1 & 0 \\ 0 & 1 & 0 & 0 \\ 0 & 0 & 0 & 1 \end{bmatrix}, \begin{bmatrix} 1 & 0 & 0 & 0 \\ 0 & 6 & 0 & 0 \\ 0 & 0 & 1 & 0 \\ 0 & 0 & 0 & 1 \end{bmatrix}, \begin{bmatrix} 1 & 0 & 0 & 0 \\ 0 & 1 & 0 & 0 \\ 0 & 0 & 5 & 0 \\ 0 & 0 & 0 & 1 \end{bmatrix},$$

$$\begin{bmatrix} 1 & 0 & 0 & 0 \\ 0 & 1 & -\frac{5}{3} & 0 \\ 0 & 0 & 1 & 0 \\ 0 & 0 & 0 & 1 \end{bmatrix}, \begin{bmatrix} 1 & 0 & 0 & \frac{2}{3} \\ 0 & 1 & 0 & 0 \\ 0 & 0 & 1 & 0 \\ 0 & 0 & 0 & 1 \end{bmatrix}, \begin{bmatrix} 1 & 0 & 0 & 0 \\ 0 & 1 & 0 & -\frac{1}{6} \\ 0 & 0 & 1 & 0 \\ 0 & 0 & 0 & 1 \end{bmatrix}.$$

10. (a) T (b) F
(c) F (d) T
(e) T

Section 10.2 (p.518–519)

1. (a) Linearly independent; to prove that it is, substitute the values $x = 0$, $x = 1$, $x = 2$, and follow the method of Example 1

(c) Linearly dependent ($a = -2, b = 1, c = 1$)

4. (a) $B = \{\sin(2x), \cos(2x), \sin^2 x\}$

(c) $B = \{\sin(x + 1), \cos(x + 1)\}$

5. (a) $[\mathbf{v}]_B = [5, 0, -7]$

(c) $[\mathbf{v}]_B = [-\frac{\cos 2}{\sin 1}, \frac{\cos 1}{\sin 1}] \approx [0.4945, 0.6421]$. (If your answer is more complicated than this, compare numerical approximations.)

6. (a) T (b) T
(c) F (d) F

Section 10.3 (p.525–526)

1. (a) $\mathbf{C} = \begin{bmatrix} 8 & 12 \\ 0 & -9 \end{bmatrix}; \mathbf{A} = \begin{bmatrix} 8 & 6 \\ 6 & -9 \end{bmatrix}$

(c) $\mathbf{C} = \begin{bmatrix} 5 & 4 & -3 \\ 0 & -2 & 5 \\ 0 & 0 & 0 \end{bmatrix}; \mathbf{A} = \begin{bmatrix} 5 & 2 & -\frac{3}{2} \\ 2 & -2 & \frac{5}{2} \\ -\frac{3}{2} & \frac{5}{2} & 0 \end{bmatrix}$

2. (a) $\mathbf{A} = \begin{bmatrix} 43 & -24 \\ -24 & 57 \end{bmatrix}; \mathbf{P} = \frac{1}{5}\begin{bmatrix} -3 & 4 \\ 4 & 3 \end{bmatrix}; \mathbf{D} = \begin{bmatrix} 75 & 0 \\ 0 & 25 \end{bmatrix};$

$B = \left(\frac{1}{5}[-3, 4], \frac{1}{5}[4, 3]\right); [\mathbf{x}]_B = [-7, -4]; Q(\mathbf{x}) = 4075$

(c) $\mathbf{A} = \begin{bmatrix} 18 & 48 & -30 \\ 48 & -68 & 18 \\ -30 & 18 & 1 \end{bmatrix}; \mathbf{P} = \frac{1}{7}\begin{bmatrix} 2 & -6 & 3 \\ 3 & -2 & -6 \\ 6 & 3 & 2 \end{bmatrix}; \mathbf{D} = \begin{bmatrix} 0 & 0 & 0 \\ 0 & 49 & 0 \\ 0 & 0 & -98 \end{bmatrix};$

$B = \left(\frac{1}{7}[2, 3, 6], \frac{1}{7}[-6, -2, 3], \frac{1}{7}[3, -6, 2]\right); [\mathbf{x}]_B = [5, 0, 6];$
$Q(\mathbf{x}) = -3528$

4. Yes. If $Q(\mathbf{x}) = \sum\limits_{1 \le i \le j \le n} a_{ij}x_i x_j$, then $Q(\mathbf{x}) = \mathbf{x}^T \mathbf{C}_1 \mathbf{x}$ and $\mathbf{C}_1$ upper triangular imply that the (i, j) entry for $\mathbf{C}_1$ is 0 if $i > j$ and a_{ij} if $i \le j$.
A similar argument describes $\mathbf{C}_2$. Thus $\mathbf{C}_1 = \mathbf{C}_2$.

6. (a) T   **(b)** F
   (c) F   **(d)** T
   (e) T

Appendix B (p.539–541)

1. (a) Not a function; undefined for $x < 1$

(c) Not a function; two values assigned to each $x \ne 1$

(e) Not a function (k is undefined at $\theta = \frac{\pi}{2}$)

(f) Function; range $=$ all prime numbers; image of 2 is 2; pre-image of 2 $= \{0, 1, 2\}$

2. (a) $\{-15, -10, -5, 5, 10, 15\}$

(c) $\{\ldots, -8, -6, -4, -2, 0, 2, 4, 6, 8, \ldots\}$

3. $(g \circ f)(x) = \frac{1}{4}\sqrt{75x^2 - 30x + 35}; (f \circ g)(x) = \frac{1}{4}(5\sqrt{3x^2 + 2} - 1)$

4. $(g \circ f)\left(\begin{bmatrix} x \\ y \end{bmatrix}\right) = \begin{bmatrix} -8 & 24 \\ 2 & 8 \end{bmatrix}\begin{bmatrix} x \\ y \end{bmatrix}; (f \circ g)\left(\begin{bmatrix} x \\ y \end{bmatrix}\right) = \begin{bmatrix} -12 & 8 \\ -4 & 12 \end{bmatrix}\begin{bmatrix} x \\ y \end{bmatrix}$

8. f is not one-to-one because $f(x + 1) = f(x + 2) = 1$; f is not onto
because there is no pre-image for x^n. For $n \ge 3$, the pre-image of $\mathcal{P}_2$ is
$\mathcal{P}_3$.

10. f is one-to-one, because if $f(\mathbf{A}_1) = f(\mathbf{A}_2)$, then $\mathbf{B}^{-1}\mathbf{A}_1\mathbf{B} = \mathbf{B}^{-1}\mathbf{A}_2\mathbf{B} \Rightarrow \mathbf{B}(\mathbf{B}^{-1}\mathbf{A}_1\mathbf{B})\mathbf{B}^{-1} = \mathbf{B}(\mathbf{B}^{-1}\mathbf{A}_2\mathbf{B})\mathbf{B}^{-1} \Rightarrow (\mathbf{B}\mathbf{B}^{-1})\mathbf{A}_1(\mathbf{B}\mathbf{B}^{-1}) = (\mathbf{B}\mathbf{B}^{-1})\mathbf{A}_2(\mathbf{B}\mathbf{B}^{-1}) \Rightarrow \mathbf{I}_n\mathbf{A}_1\mathbf{I}_n = \mathbf{I}_n\mathbf{A}_2\mathbf{I}_n \Rightarrow \mathbf{A}_1 = \mathbf{A}_2$; f is onto, because
for any $\mathbf{C} \in \mathcal{M}_{nn}$, $f(\mathbf{B}\mathbf{C}\mathbf{B}^{-1}) = \mathbf{B}^{-1}(\mathbf{B}\mathbf{C}\mathbf{B}^{-1})\mathbf{B} = \mathbf{C}$. Also, $f^{-1}(\mathbf{A}) = \mathbf{B}\mathbf{A}\mathbf{B}^{-1}$.

12. (a) F   (b) T
    (c) F   (d) F
    (e) F   (f) F
    (g) F   (h) F

Appendix C (p.544–545)

1. (a) $11 - i$ (c) $20 - 12i$
    (e) $9 + 19i$ (g) $-17 - 19i$
    (i) $9 + 2i$ (k) $16 + 22i$
    (m) $\sqrt{53}$

2. (a) $\frac{3}{20} + \frac{1}{20}i$ (c) $-\frac{4}{17} - \frac{1}{17}i$

5. (a) F   (b) F
    (c) T   (d) T
    (e) F

Index

Equivalent Conditions on Square Matrices

Let $\mathbf{A}$ be an $n \times n$ matrix. Any pair of statements in the same column are equivalent.

$\mathbf{A}$ is singular ($\mathbf{A}^{-1}$ does not exist).	$\mathbf{A}$ is nonsingular ($\mathbf{A}^{-1}$ exists).
Rank($\mathbf{A}$) $\neq n$.	Rank($\mathbf{A}$) $= n$.
$|\mathbf{A}| = 0$.	$|\mathbf{A}| \neq 0$.
$\mathbf{A}$ is not row equivalent to $\mathbf{I}_n$.	$\mathbf{A}$ is row equivalent to $\mathbf{I}_n$.
$\mathbf{AX} = \mathbf{O}$ has a nontrivial solution for $\mathbf{X}$.	$\mathbf{AX} = \mathbf{O}$ has only the trivial solution for $\mathbf{X}$.
$\mathbf{AX} = \mathbf{B}$ does not have a unique solution (no solutions or infinitely many solutions).	$\mathbf{AX} = \mathbf{B}$ has a unique solution for $\mathbf{X}$ (namely, $\mathbf{X} = \mathbf{A}^{-1}\mathbf{B}$).

Tests for Linear Independence

Let S be a set of vectors in a vector space. Any pair of statements in the same column are equivalent.

Linear Independence of S	Linear Dependence of S
For every $\mathbf{v} \in S$, we have $\mathbf{v} \notin \text{span}(S - \{\mathbf{v}\})$.	There is a $\mathbf{v} \in S$ such that $\mathbf{v} \in \text{span}(S - \{\mathbf{v}\})$.
If $\{\mathbf{v}_1, \ldots, \mathbf{v}_n\} \subseteq S$ and $a_1\mathbf{v}_1 + \cdots + a_n\mathbf{v}_n = \mathbf{0}$, then $a_1 = a_2 = \cdots = a_n = 0$. (The zero vector requires zero coefficients.)	There is a subset $\{\mathbf{v}_1, \ldots, \mathbf{v}_n\}$ of S such that $a_1\mathbf{v}_1 + \cdots + a_n\mathbf{v}_n = \mathbf{0}$, for scalars $a_1, a_2, \ldots, a_n$, with some $a_i \neq 0$. (The zero vector does not require all coefficients to be zero.)
Every vector in span(S) can be uniquely expressed as a linear combination of the vectors in S.	*Some* vector in span(S) can be expressed in more than one way as a linear combination of the vectors in S.
If $S = \{\mathbf{v}_1, \ldots, \mathbf{v}_n\}$, then for each k, $\mathbf{v}_k \notin \text{span}(\{\mathbf{v}_1, \ldots, \mathbf{v}_{k-1}\})$. (Each $\mathbf{v}_k$ is not a linear combination of the previous vectors in S.)	If $S = \{\mathbf{v}_1, \ldots, \mathbf{v}_n\}$, some $\mathbf{v}_k$ can be expressed as $\mathbf{v}_k = a_1\mathbf{v}_1 + \cdots + a_{k-1}\mathbf{v}_{k-1}$. (Some $\mathbf{v}_k$ is a linear combination of the previous vectors in S.)
For every $\mathbf{v} \in S$, span($S - \{\mathbf{v}\}$) does not contain all the vectors of span(S).	There is some $\mathbf{v} \in S$ such that span($S - \{\mathbf{v}\}$) $=$ span(S).
Every finite subset of S is linearly independent.	*Some* finite subset of S is linearly dependent.